A Specialist Periodical Report

# Organic Compounds of Sulphur, Selenium, and Tellurium
Volume 1

A Review of the Literature Published between April 1969 and March 1970

Senior Reporter
**D. H. Reid,** Department of Chemistry, The University, St. Andrews

Reporters
**G. C. Barrett,** West Ham College of Technology
**R. J. S. Beer,** The University of Liverpool
**D. T. Clark,** University of Durham
**A. Wm. Johnson,** University of North Dakota
**D. N. Jones,** The University, Sheffield
**F. Kurzer,** Royal Free Hospital School of Medicine, University of London
**S. McKenzie,** B.P. Research Centre, Sunbury-on-Thames

SBN: 85186 259 4

© Copyright 1970

**The Chemical Society**
Burlington House, London W1V 0BN.

*Organic formulae composed by Wright's Symbolset method*

PRINTED IN GREAT BRITAIN BY JOHN WRIGHT AND SONS LTD., AT THE STONEBRIDGE PRESS, BRISTOL

# *Preface*

This is the first volume of a series which will appear biennially on 'Organic Compounds of Sulphur, Selenium, and Tellurium'. It covers the literature of the period April 1969 to March 1970. Material not available to the authors in time for inclusion in Volume 1 will be considered for inclusion in Volume 2. Although the main intention has been to review the literature of the twelvemonth period stated, it appeared desirable in this first volume of the series to look back more frequently and rather further than should be necessary in future Reports, in order to provide some continuity with existing reviews and with the literature of the recent past. This is especially true of Chapters 2, 5—8, and 12—16. Henceforth reviewing will be on a biennial basis.

Chemical aspects of the subject have been emphasised, and naturally occurring compounds have not generally been covered, nor have biochemical aspects such as biosynthesis, metabolism, enzymology, and physiological activity. Future inclusion of these topics is not ruled out but would be the result of demand by a substantial number of readers. To this extent therefore the coverage is selective. Within these limits the authors have attempted to be both comprehensive, in order that the Report be useful as a reference or source of information, and critical. While care has been taken to be comprehensive, notification of any important omissions would be welcomed so that the relevant topics may be included in future volumes. Regrettably, circumstances prevented the writer of the projected chapter on Thiophen and related compounds from completing his review in time for inclusion in Volume 1. This gap will be filled by an extended review of the subject in Volume 2.

Chapter 1 comprises a review of the whole subject from a theoretical standpoint. Organisation of the remainder of the Report is broadly according to chemical structure and functional group classification. The subjects of Chapters 5—8 have been given individual and more detailed treatment in view of their unusual importance or topicality. Readers may feel that certain other topics, which in this volume are covered in more than one chapter in relation to different classes of compounds, have sufficient general importance to justify their being reviewed in a unified manner in individual chapters. Constructive criticism, comments, and advice which will enable subsequent reports to be orientated more exactly with the requirements of research workers are welcomed with a view to our improving future volumes.

The following Introductory Review, based on notes compiled by each contributor to the Report, is intended to present to the reader a rapidly readable, broad survey of recent progress in the chemistry of Organic Compounds of Sulphur, Selenium, and Tellurium. Particularly active areas of research and topics of novelty, general interest, and importance are outlined.

Thanks are due to Mr. P. J. Robinson, B.P. Research Centre, Sunbury-on-Thames, for translations of Russian papers.

D. H. R.

# Introductory Review

Significant theoretical developments fall into two related but distinct groups. First, the development of optimised exponents for Gaussian basis sets for sulphur now offers the possibility of carrying out good non-empirical calculations on a wide variety of organosulphur compounds. Secondly, the results of these calculations have been applied to larger molecules, a notable example being thiophene for which the question of *d*-orbital participation in the ground state appears to have been settled (*1*, 22, 23).\* Detailed theoretical treatments of molecular structure are of limited value if they are not related in some way to experimental situations, and the *ab initio* calculations on potential energy surfaces for transformation involving organosulphur compounds published by Csizmadia (*1*, 19a—d) and Scrocco (*1*, 14) and their coworkers represent outstanding contributions to the chemical literature in this field.

Aliphatic organosulphur compounds have for some time provided suitable ground for mechanistic studies, and substitution reactions at sulphur are well represented in the recent literature. The work of Kice is notable (*2*, 218, 244, 525—531). Some very simple reactions are demonstrated for the first time. In particular, a Claisen-type condensation of a simple sulphonate ester yielded a sulphone (*2*, 211), and a haloform reaction has been carried out successfully with a methyl sulphone (*2*, 253). Insofar as these reactions are mechanistically analogous to the corresponding reactions of carbonyl compounds there is ground for treating at least some aspects of the chemistry of methyl(ene) sulphones and carbonyl compounds in a unified manner in future textbooks. Notable use has been made of bivalent sulphur groups in synthetic work, vinyl sulphides serving as masking groups for eventual transformations into carbonyl compounds (*2*, 134, 167), and phenyl sulphones finding application in the construction of cyclopropanes with stereochemical preference (*2*, 271).

In the area of small-ring organosulphur chemistry new light has been shed on the mechanism of the thermal extrusion of sulphur dioxide from thiiran 1,1-dioxides, a well-known synthetically useful reaction. The reaction proceeds stereospecifically, although the concerted elimination of sulphur dioxide is forbidden by the Woodward–Hoffmann rules. It is now suggested (*3*, 33) that ring-expansion to a 1,2-oxathietan and thence to a

---

\* Italicised numerals in brackets refer to chapter numbers, ordinary numerals to reference numbers.

1,3,2-dioxathiolan occurs, either of which could decompose to the observed products. Orbital correlation diagrams indicate that concerted loss of sulphur dioxide from the 1,3,2-dioxathiolan is an allowed process. Episulphonium salts are widely accepted as intermediates in the ionic addition of sulphenyl halides to olefins, but doubt is now cast on this assumption by the observation that the stable isolable *cis*-cyclo-octene-*S*-methylepisulphonium ion underwent predominant or exclusive attack (depending on the nucleophile) at sulphur to give mainly *cis*-cyclo-octene and methane–sulphenyl compounds, and not at carbon, as analogy with the accepted mechanism for olefin–sulphenyl halide additions would suggest (*3*, 50, 51). Mislow and his group have adduced evidence that the nucleophilic displacement with inversion of configuration at sulphur in thietanium salts probably occurs directly, without the intervention of a transient pentaco-ordinate intermediate of trigonal bipyramidal geometry (*3*, 83). Detailed physico-chemical studies have led to the elucidation of the configuration, conformation, and $^1$H n.m.r. spectral characteristics of 3-substituted thietan 1-oxides (*3*, 80—82).

Transannular elimination of sulphur dioxide from monocyclic halogeno-hydrocarbons spanned by a sulphoxide bridge leads to bicyclic hydrocarbons unsaturated at the bond common to the two rings, in a synthetically useful variant of the Ramberg–Backlund reaction (*4*, 129, 130). Significant results of physico-chemical studies of saturated cyclic compounds of sulphur include the determination of the relative rates of hydrogen–deuterium exchange of protons α to the sulphoxide group in thiane 1-oxides (*4*,1, 22); the elucidation of the conformations, barriers to ring-inversion, conformational preferences of alkyl substituents, and the chair–boat energy differences in 1,3-dithianes (*4*, 144—146); and the accumulation of data leading to a better understanding of the requirements for $R_2S$-4 participation in the heterolysis of alkyl toluene-*p*-sulphonates (*4*, 132).

Despite numerous investigations since about 1850 thioaldehydes have eluded isolation and characterisation until recently. The first authentic thiioaldehyde to be described was a complex dipyrrylmethane used as an intermediate in Woodward's synthesis of chlorophyll-a. A series of stable heterocyclic thioaldehydes of simple structure has now been obtained by a modification of the Vilsmeier reaction, and spectral properties are cited (*5*, 62, 63). Isolation of a stable heterocyclic telluroketone, probably the first authentic one, is reported by a Russian group (*5*, 412). The reaction of ketimine anions with carbon disulphide or *NN*-dimethylthioformamide represents a fresh approach to the synthesis of thioketones (*5*, 72). Reaction proceeds via four-membered ring intermediates analogous to those involved in the Wittig reaction. The mechanism of the photocycloaddition of thiones to olefins has been clarified to a great extent (*5*, 98). The course of the addition is shown to depend on the nature and size of the substituents in the olefin addend. Magnetic anisotropy studies of thioformamide and thioacetamide constitute significant advances in the chemistry of thioamides (*5*, 448). Evidence for the transient existence of sulphenes is

# Introductory Review

accumulating, based mainly on trapping experiments. Methylsulphonylsulphene can be stored for relatively long periods at low temperatures, and is trapped by trimethylamine at room temperature to form a stable 1 : 1 zwitterionic complex (6, 174, 175). The species detected in the gas phase, produced by flash thermolysis of thietan and thiete S-dioxides, is believed to be sulphene (5, 122). Although several aryl sulphines have been prepared, simple alkyl sulphines have hitherto eluded isolation. Evidence for the transient existence of dimethylsulphine (thioacetone S-oxide) in solution has now been adduced (6, 192).

Data bearing on our knowledge of the structure of sulphur ylides come from an X-ray crystallographic structure determination of dicyanomethylenedimethylsulphurane, the first sulphur ylid whose structure has been determined. The molecular geometry (planar carbanion centre, pyramidal sulphur) and important bond distances have been reported independently by two groups of workers (6, 23, 25). Geometrical isomerism, restricted rotation, inversion at sulphur, and magnetic non-equivalence of $\alpha$-methylene groups in $\beta$-carbonyl sulphonium ylides have been studied by n.m.r. spectroscopy (6, 22, 23, 28, 33—35). Speculation that 1,5-rearrangements of sulphonium ylides of polyenes might be part of a biosynthetic mechanism for natural polyene formation has led to demonstrations by two groups (6, 57, 65, 66) of the chemical feasibility of this proposal and to a synthesis (6, 66) of squalene in which the key step is the overall coupling of two triene units to form a symmetrical hexaene unit. Elaboration of the sulphide, sulphoxide, sulphidimine, sulphoxime cycle by the groups of Andersen, Cram, Johnson, and Kjaer constitutes an impressive stereochemical study of organosulphur compounds (6, 230—234).

Interest has grown recently in the chemistry of those sulphur-heteroaromatic systems for which the only possible neutral covalent formulations are ones containing tetracovalent sulphur. Early examples of one such system were derivatives of thiabenzene, and the researches of Price and his group have culminated in the recent synthesis and study of 1-phenylthiabenzene (7, 8). Systems of more recent vintage, studied principally by the groups of Cava and Schlessinger, include the 2-thiaphenalenes (7, 19, 20), thieno[3,4-c]thiophenes (7, 23), and thieno[3,4-c]-1,2,5-thiadiazoles (7, 24). Stable derivatives of all three systems have been synthesised during the period under review, but information which might shed light on the bonding in these heterocycles is still lacking. The recent past has seen the appearance of numerous publications on X-ray crystallographic structure determinations of 6a-thiathiophthenes and related compounds. Equal S—S distances for 2,5-dimethyl-6a-thiathiophthene has been reaffirmed (8, 5). Gleiter and Hoffmann (8, 21) have made a novel contribution to theories of bonding in 6a-thiathiophthenes. The essential feature of their theory is to regard the three sulphur atoms as a linear system of atoms which uses three orbitals occupied by four electrons for $\sigma$-bonding, with superimposed $\pi$-bonding. Attempts to ascertain whether a linear arrangement of more than three sulphur atoms in suitable heterocyclic structures might be a source of unusual stability has resulted in the synthesis of two compounds with a potentially linear sequence of five sulphur atoms (*8*, 33, 34). Results of X-ray crystallographic studies are eagerly awaited.

In an interesting paper Grigg rationalises a number of known reactions in organosulphur chemistry, involving the elimination of sulphur and sulphur dioxide from heterocyclic structures, in terms of a thermal electrocyclic process followed by cheletropic elimination of sulphur or sulphur dioxide (*11*, 16). The thermal ($4n+2$) disrotatory electrocyclic processes for 1,4-dithiins, thiepines, and thiepine 1,1-dioxides are discussed.

Although a great proportion of the work on thiazoles, thiadiazoles and related compounds is primarily concerned with synthesis, interest in the correlation between structure, physical properties, and reactivities is clearly on the increase. The contributions and discussions of Metzger and his associates emphasise this approach in the thiazole field (*13*, 7). Among more recent preparative methods, the use of aziridine derivatives as starting-materials in the synthesis of 2(and 4)-thiazolines (*13*, 72—74, 86), and of sulphur dioxide–ketene adducts in that of thiazolidines (*13*, 98), is noteworthy. Interest in mesoionic thiazoles (*13*, 63—65) and thiadiazoles (*16*, 37—45) continues actively. In the benzothiazole field current researches on metal complexes (*14*, 34—40), in spirane derivatives (*14*, 28—31), and on the possible role of benzothiazole residues in the structures of the reddish-brown pigments of bird feathers and red human hair (*14*, 41—43), have given results of interest.

A number of synthetically important and mechanistically interesting papers on the chemistry of thiazines and thiazepines have appeared, prompted by the occurrence of the 1,3-thiazine structure in the cephalosporins. Notable are the ring-enlargement reactions of penicillin *S*-oxides in acetic anhydride to give 1,3-thiazines (*17*, 7), and a novel base-catalysed rearrangement of a penicillanic acid derivative into a 1,4-thiazepine and thence into a 1,4-thiazine (*17*, 9, 38). The thermal rearrangement of a 2,2-dimethyl-1,4-thiazine into a 2,3-dihydro-1,4-thiazine (*17*, 21) is also noteworthy. Cadogan and his group have extended their interesting work on the decomposition of aryl 2-azidophenyl sulphides and on the phosphite deoxygenation of 2-nitrophenyl aryl sulphides, which give rearranged phenothiazines (probably *via* nitrenes), to the 'blocked' derivative 2-azidophenyl 2,6-dimethylphenyl sulphide. A 5,11-dihydrodibenzo[*be*]-1,4-thiazepine results (*17*, 42).

G. C. B.
R. J. S. B.
D. T. C.
A. W. J.
D. N. J.
F. K.
S. McK.
D. H. R.

# Contents

**Chapter 1 Theoretical Aspects of Organosulphur, Organo-
selenium, and Organotellurium Compounds**
*By D. T. Clark*

| | |
|---|---|
| **1 Introduction** | 1 |
| **2 $d$-Orbital Participation in the Ground States of Organo-sulphur Compounds** | 3 |
|     Calculations with Slater-type Basis Functions | 3 |
|     Calculations with Gaussian-type Basis Functions | 5 |
|     Organosulphur Compounds | 8 |
|         Carbon Monosulphide | 9 |
|         Carbonyl Sulphide and Carbon Disulphide | 9 |
|         Thiiran | 9 |
|         Thiophen | 9 |
|         Hydrogen Methyl Sulphoxide and its Anion | 12 |
|         Dimethyl Sulphoxide and its Anion | 12 |
|         Hydrogen Sulphonyl Carbanion | 12 |
|         Thiomethoxide Anion and Dimethyl Sulphide Dicarbanion | 12 |
| **3 Theoretical Treatments of Organosulphur Compounds** | 12 |
|     Non-empirical Calculations | 13 |
|         Carbon Monosulphide | 13 |
|         Carbonyl Sulphide and Carbon Disulphide | 13 |
|         Thiiren and Thiiran | 15 |
|         Hydrogen Methyl Sulphoxide and its Anion | 19 |
|         Dimethyl Sulphoxide and its Anion | 25 |
|         Hydrogen Sulphonyl Carbanion | 25 |
|         Methanethiol | 26 |
|         Dimethyl Sulphide Dicarbanion | 27 |
|         Thiophen | 29 |
|         Molecular Core Binding Energies | 31 |
|     Semi-empirical All-valence-electron Treatments | 33 |
|         ($a$) Introduction | 33 |
|         ($b$) CNDO SCF MO Calculations: Thiophen, 2,3-Thiophyne, and 3,4-Thiophyne | 34 |

|  |  |  |
|---|---|---|
| (c) Extended Hückel Calculations. | | 38 |
| (i) Thiathiophthene | | 38 |
| (ii) Episulphonium cation | | 41 |
| (d) Semi-empirical π-Electron Treatments | | 41 |

4 Semi-empirical PPP SCF MO Calculations on Organoselenium Compounds — 48

## Chapter 2 Aliphatic Organo-sulphur Compounds, Compounds with Exocyclic Sulphur Functional Groups, and their Selenium and Tellurium Analogues
### By G. C. Barrett

| | |
|---|---|
| 1 General | 49 |
| 2 Thiols | 50 |
| Preparation | 50 |
| Addition of Thiols to Olefins, Carbonyl Compounds, and Epoxides | 53 |
| Nucleophilic Substitution Reactions of Thiols | 56 |
| Reactivity of —SH Groups in Proteins | 57 |
| Estimation and Identification of Thiol Groups | 57 |
| Protection of Thiol Groups in Synthesis | 58 |
| Other Reactions of Thiols | 59 |
| 3 Sulphides | 60 |
| Preparation | 60 |
| Reactions | 62 |
| Rearrangements of Sulphides | 66 |
| Naturally-occurring Acyclic Sulphides (excluding Methionine Derivatives) | 68 |
| 4 Thioacetals, Dithioacetals, Orthothioformates, and Orthothiocarbonates | 69 |
| Preparation | 69 |
| Reactions | 69 |
| 5 Sulphoxides and Sulphones | 71 |
| Preparation | 71 |
| Physical Properties of Sulphoxides and Sulphones | 74 |
| Stereochemistry of Sulphinyl Compounds | 75 |
| Reactions of Sulphoxides and Sulphones | 78 |
| 6 Thiocyanates and Isothiocyanates; Sulphonyl Cyanides | 83 |
| Preparation | 83 |
| Reactions | 84 |
| Sulphonyl Cyanides | 85 |

*Contents* xi

| | |
|---|---|
| 7 **Sulphenic Acids and their Derivatives** | 86 |
| Sulphenyl Halides | 86 |
| Sulphenamides | 89 |
| 8 **Sulphinic Acids and their Derivatives** | 91 |
| 9 **Sulphonic Acids and their Derivatives** | 93 |
| Preparation | 93 |
| Reactions of Sulphonic Acids | 95 |
| Sulphonate Esters | 95 |
| Sulphonyl Halides | 97 |
| Sulphonamides | 98 |
| Sulphonyl Isocyanates | 101 |
| $N$-Sulphinyl Sulphonamides and $NN'$-Disulphonyl Sulphur Di-imides | 101 |
| 10 **Disulphides and Polysulphides; their Oxy-sulphur Analogues** | 102 |
| Preparation | 102 |
| Reactions of Di- and Poly-sulphides | 104 |
| Oxy-sulphur Analogues of Di- and Tri-sulphides | 106 |

**Chapter 3 Small-ring Compounds of Sulphur and Selenium**
*By D. N. Jones*

| | |
|---|---|
| 1 **Three-membered Rings** | 109 |
| Formation of Thiirans and Oxathiirans | 109 |
| Reactions of Thiirans | 113 |
| Thiiran Oxides | 116 |
| Thiiran-1,1-dioxides and Thiiren-1,1-dioxides | 118 |
| Thiiranium, Thiirenium, and Seleniranium Ions | 120 |
| 2 **Four-membered Rings** | 122 |
| Formation of Thietan Derivatives | 122 |
| Formation of 1,3-Dithietan and 1,3-Diselenetan Derivatives | 126 |
| Formation of 1,2-Oxathietan and 1,3-Thietidine Derivatives | 127 |
| Configurational and Conformational Aspects of Thietan and Selenetan Derivatives | 128 |
| Oxidation and Cleavage of Thietan Derivatives | 131 |
| Pyrolysis of Thiet-1,1-dioxides | 133 |

**Chapter 4 Saturated Cyclic Compounds of Sulphur, Selenium, and Tellurium**
*By D. N. Jones*

| | |
|---|---|
| 1 **Five-membered Rings** | 134 |
| Thiolan Derivatives | 134 |

| | |
|---|---:|
| 1,2-Dithiolan Derivatives | 139 |
| 1,3-Dithiolan Derivatives | 141 |
| 1,2,3- and 1,2,4-Trithiolan Derivatives | 144 |
| 1,2-Oxathiolan Derivatives | 145 |
| 1,3-Oxathiolan Derivatives | 146 |
| Derivatives of 1,3,2-Dioxathiolan, 1,2,4-Dioxathiolan, and 1,2,5-Oxadithiolan | 148 |
| 1,2- and 1,3-Thiazolidine Derivatives | 150 |
| Miscellaneous Five-membered Rings containing Sulphur | 155 |
| **2 Six-membered Rings** | **157** |
| Thian Derivatives | 157 |
| 1,2-Dithian Derivatives | 164 |
| 1,3-Dithian and 1,3-Diselenan Derivatives | 164 |
| 1,4-Dithian and 1,4-Diselenan Derivatives | 171 |
| 1,3,5-Trithian and 1,2,4,5-Tetrathian Derivatives | 172 |
| Six-membered Rings containing Sulphur and Other Heteroatoms | 173 |
| **3 Seven- and Higher-membered Rings** | **177** |

Chapter 5 Thiocarbonyls, Selenocarbonyls, and Tellurocarbonyls
By S. McKenzie

| | |
|---|---:|
| **1 Introduction** | **180** |
| Reviews | 182 |
| **2 Thioaldehydes** | **184** |
| Synthesis | 184 |
| Transient Species | 184 |
| **3 Thioketones** | **186** |
| Synthesis | 186 |
| Transient Species | 189 |
| Reactions | 191 |
| **4 Thioketens** | **196** |
| **5 Sulphines** | **197** |
| **6 Sulphenes** | **199** |
| **7 Thioamides** | **201** |
| Synthesis | 201 |
| Reactions | 204 |
| The Thio-Claisen Rearrangement | 207 |

*Contents*

| | |
|---|---|
| **8 Thioureas** | 209 |
| Synthesis | 209 |
| Reactions | 210 |
| **9 Thiosemicarbazides and Derivatives** | 217 |
| Synthesis | 217 |
| Reactions | 218 |
| **10 Thionocarboxyclic and Dithiocarboxylic Acids, and Derivatives** | 220 |
| Synthesis | 220 |
| Reactions | 222 |
| **11 Thionocarbonates, Thionodithiocarbonates, and Trithiocarbonates** | 224 |
| Synthesis | 224 |
| Reactions | 225 |
| **12 Thionocarbamic and Dithiocarbamic Acids, and Derivatives** | 228 |
| Synthesis | 228 |
| Reactions | 230 |
| **13 Selenocarbonyl and Tellurocarbonyl Compounds** | 233 |
| Synthesis | 233 |
| Reactions | 234 |
| **14 Physical Properties** | 234 |
| Structure | 235 |
| Enethiol Tautomerism | 235 |
| Ring–Chain Isomerism | 237 |
| Polarisation Effects and Restricted Rotation | 238 |
| Spectra | 240 |
| U.v. spectra | 240 |
| I.r. spectra | 242 |
| N.m.r. spectra | 244 |
| Mass spectra | 245 |
| Other Physical Properties | 245 |
| Polarography | 245 |
| Dipole Moments | 246 |
| Acidity and Basicity Measurements | 246 |

**Chapter 6 Ylides of Sulphur, Selenium, Tellurium and Related Structures**
*By A. William Johnson*

| | |
|---|---|
| **1 Introduction** | 248 |

|   |   |
|---|---|
| 2 Sulphonium Ylides | 249 |
| Methods of Synthesis | 249 |
| Physical Properties | 252 |
| Chemical Stability | 257 |
| Reactions of Sulphonium Ylides | 267 |
| Sulphonium Ylides as Reaction Intermediates | 275 |
| 3 Oxysulphonium Ylides | 276 |
| Synthesis | 276 |
| Chemical Stability | 278 |
| Reactions of Oxysulphonium Ylides | 279 |
| 4 Sulphinyl Ylides | 287 |
| 5 Sulphonyl Ylides | 289 |
| 6 Sulphenes and Sulphines | 290 |
| Preparation of Sulphenes | 290 |
| Evidence for the Existence of Sulphenes | 291 |
| Reactions of Sulphenes | 292 |
| Sulphines | 296 |
| 7 Sulphur Imines | 299 |
| Iminosulphuranes | 299 |
| Imino-oxysulphuranes | 301 |
| Sulphurdi-imines | 303 |
| Miscellaneous Properties of Sulphur Imines | 304 |
| 8 Sulphonyl and Sulphinyl Amines | 306 |
| 9 Selenonium Ylides | 308 |

Chapter 7 Heterocyclic Compounds of Quadricovalent Sulphur
*By D. H. Reid*

|   |   |
|---|---|
| 1 Introduction | 309 |
| 2 Thiabenzenes | 309 |
| 3 Thiabenzene-1-oxides | 312 |
| 4 2-Thiaphenalenes | 314 |
| 5 Thieno[3,4-*c*]-thiophens | 318 |
| 6 Thieno[3,4-*c*]-1,2,5-thiadiazoles | 319 |

Chapter 8 6*a*-Thiathiophthenes and Related Compounds
*By D. H. Reid*

|   |   |
|---|---|
| 1 Introduction | 321 |
| 2 Structural Studies | 321 |

|   |   |
|---|---|
| 3 Bonding | 326 |
| 4 Formation and Syntheses | 327 |
| 5 Reactions | 334 |

## Chapter 9 1,2- and 1,3-Dithioles
*By R. J. S. Beer*

|   |   |
|---|---|
| 1 1,2-Dithiole-3-thiones and 1,2-Dithiole-3-ones | 336 |
| 2 1,2-Dithiolium Salts | 339 |
| 3 Derivatives of 1,3-Dithiole | 343 |

## Chapter 10 Thiopyrans
*By R. J. S. Beer*

|   |   |
|---|---|
| 1 $2H$-Thiopyrans and Related Compounds | 346 |
| 2 $4H$-Thiopyran-4-ones and $4H$-Thiopyran-4-thiones | 349 |
| 3 $2H$-Thiopyran-2-ones, $2H$-Thiopyran-2-thiones, and Related Compounds | 351 |
| 4 Complex Thiopyran and Selenopyran Derivatives | 354 |
| 5 'Quadrivalent' Sulphur Compounds | 356 |

## Chapter 11 Thiepines and Dithiins
*By D. H. Reid*

|   |   |
|---|---|
| 1 Thiepines | 358 |
| 2 Dithiins | 364 |
|     1,2-Dithiins | 364 |
|     1,4-Dithiins | 365 |
|     Benzo- and Dibenzo-1,4-dithiins | 366 |

## Chapter 12 Isothiazoles
*By F. Kurzer*

|   |   |
|---|---|
| 1 Introduction | 369 |
| 2 Syntheses | 369 |
| 3 Properties | 370 |
|     Alkylation | 370 |

| | |
|---|---|
| Acylation | 372 |
| Ring-cleavage | 373 |
| Ring-cleavage with Ring-expansion | 374 |
| Other Reactions | 374 |
| **4 Condensed Ring Systems** | 376 |
| Isothiazolo[2,3-*a*]pyridines | 376 |
| Isothiazolo[5,4-*d*]pyrimidines | 377 |

## Chapter 13 Thiazoles and Related Compounds
### By F. Kurzer

| | |
|---|---|
| **1 Introduction** | 378 |
| **2 Thiazoles** | 378 |
| Syntheses | 379 |
| Hantzsch's Synthesis: (Type A, C—C+S—C—N) | 379 |
| (i) Mechanism | 379 |
| (ii) Applications | 380 |
| (iii) Labelled Thiazoles | 383 |
| From 1-Thiocyanato-2-isothiocyanatoethylenes: (Type A, C—C+S—C—N) | 383 |
| From Amino-acids: (Type B, C—S+N—C—C) | 384 |
| Physical Properties | 385 |
| Chemical Properties | 387 |
| Radical Reactions | 387 |
| Nitration and Halogenation | 387 |
| Other Substitution Reactions | 388 |
| Alkylthiazoles | 390 |
| Complexes with Metals | 391 |
| Biochemical Aspects | 391 |
| Mesoionic Thiazoles | 392 |
| **3 2-Thiazolines** | 393 |
| Syntheses | |
| From *N*-($\beta$-Halogenoalkyl)imidohalides: (Type C, C—C—N—C+S) | 394 |
| Condensation of $\alpha$-Thiolo-acids and Cyanogen: (Type D, C—C—S+C—N) | 395 |
| From *N*-Thiobenzoyl-$\alpha$-amino-acids: (Type F, C—C—N—C—S) | 396 |
| From Chloroacyl Isothiocyanates: (Type F, C—C—N—C—S) | 396 |
| From Aziridines | 397 |
| Properties | 398 |
| Metal Complexes | 399 |

## 4 4-Thiazolines — 399
### Syntheses — 399
Condensation of α-Thioloketones and Schiff's Bases: (Type D, C—C—S+N—C) — 399
Condensation of 3-Aroylaziridines and Aryl Isothiocyanates: (Type C—N—C+S—C) — 400
Direct Cyclisation: (Type C—N—C—C—S) — 401
### Chemical Reactions — 401

## 5 Thiazolidines — 402
### Syntheses — 402
Condensations Involving Thioamido- or Thiocyanato-compounds: (Type A, C—C+S—C—N) — 402
From Ethanolamines: (Type B, S—C+N—C—C) — 403
From a Sulphur Dioxide–Keten Adduct: (Type D, C—C—S+C—N) — 404
From Ethyl Isothiuronium Compounds: (Type C—S—C—C—N, or Type A) — 405
Intramolecular Radical Addition: (Type C—N—C—C—S) — 405
### Properties — 408
### Biological Activity — 409

# Chapter 14 Benzothiazoles
### By F. Kurzer

## 1 Introduction — 410

## 2 Synthesis — 410
Synthesis from *o*-Aminothiophenols — 410
Jacobson–Hugershoff Synthesis — 411
Synthesis from Benzothiazines — 412

## 3 Physical Properties of Benzothiazoles — 414

## 4 Chemical Properties of Benzothiazoles — 414
Reduction and Oxidation — 414
2-Methylenebenzothiazoles — 415
Photochromic Benzothiazole Spiranes — 416
Polymers — 417
Metal Complexes — 417
Biochemical Aspects: Phaeomelanins — 419

## 5 Chemical Properties of Benzothiazolines — 420
Alkylation of Benzothiazoline-2-thiones — 420

| | |
|---|---:|
| Thionation of Benzothiazolin-2-ones | 421 |
| Miscellaneous Reactions | 421 |
| Uses | 422 |

## Chapter 15 Condensed Ring Systems Incorporating Thiazole
*By F. Kurzer*

| | |
|---|---:|
| **1 Introduction** | 424 |
| **2 Structures Comprising Two Five-membered Rings (5,5)** | 425 |
| Thiazolo[3,2-*d*]tetrazoles | 425 |
| Imidazo[2,1-*b*]thiazoles | 425 |
| Thiazolo(or Oxazolo)[2,3-*b*]thiazole | 425 |
| Thiazolidino[4,5-*d*]thiazoles | 426 |
| **3 Structures Comprising One Five-membered and One Six-membered Ring (5,6)** | 426 |
| Thiazolo[3,2-*a*]-*s*-triazines | 426 |
| Thiazolo[3,2-*a*(and *c*)]pyrimidines | 427 |
| Thiazolo[4,5-*d*]pyrimidines | 429 |
| Thiazolo[4,5-*d*]pyridazine | 429 |
| Thiazolo[3,2-*a*]pyridines | 429 |
| Thiazolo[3,4-*a*]pyridines | 434 |
| **4 Structures Comprising Two Five-membered and One Six-membered Ring (5,5,6)** | 434 |
| Tetrazolo[5,1-*b*]benzothiazoles | 434 |
| Bisthiazolo[3,2-*a*; 5′,4′-*e*]pyrimidines | 435 |
| Thiazolo[2,3-*b*]benzothiazoles | 436 |
| Benzobisthiazoles | 436 |
| Thiazolo[3,2-*a*]benzimidazoles | 437 |
| Imidazo[2,1-*b*]benzothiazoles | 437 |
| **5 Structures Comprising One Five-membered and Two Six-membered Rings (5,6,6)** | 438 |
| Pyrido[3,2-*d*]thiazolo[3,2-*b*]pyridazines and Related Ring Systems | 438 |
| Thiazolo[3,2-*a*]quinazoline | 439 |
| Piperidino[5,4-*c*]thiazoline-spiro-piperidines | 440 |
| Naphtho[1,2-*d*]thiazoles | 441 |
| **6 Structures Comprising Four Rings** | 441 |
| Thiazolo[3,2-*a*]pyrimidines | 441 |
| Benzothiazolo[2,3-*b*]quinazolines | 441 |
| Thiazolo[3,2-*b*][2,4]- and Thiazolo[2,3-*b*][1,3]-benzodiazepines | 442 |

## Chapter 16 Thiadiazoles
### By F. Kurzer

| | |
|---|---|
| 1 Introduction | 444 |
| 2 1,2,3-Thiadiazoles | 444 |
|    1,2,3-Thiadiazolo[5,4-*d*]pyrimidines | 445 |
| 3 1,2,4-Thiadiazoles | 446 |
| 4 1,3,4-Thiadiazoles | 449 |
|    Synthesis | 449 |
|    Analytical | 451 |
|    Mesoionic 1,3,4-Thiadiazoles | 451 |
|    1,3,4-Thiadiazolo[2,3-*a*]isoquinolines | 452 |
| 5 1,2,5-Thiadiazoles | 453 |
|    2,1,3-Benzothiadiazole | 453 |

## Chapter 17 Thiazines and Thiazepines
### By D. H. Reid

| | |
|---|---|
| 1 1,2-Thiazines | 454 |
| 2 1,3-Thiazines | 455 |
|    Simple 1,3-thiazines | 455 |
|    Benzo-1,3-thiazines | 458 |
| 3 1,4-Thiazines | 462 |
|    Monocyclic 1,4-Thiazines | 462 |
|    Benzo-1,4-thiazines | 465 |
|    Phenothiazines (dibenzo-1,4-thiazines) and Related Compounds | 466 |
| 4 Phosphorus Analogues of Thiazines | 469 |
| 5 1,4-Thiazepines | 470 |

## Chapter 18 Thiadiazines
### By D. H. Reid     473

Author Index     477

# *Abbreviations*

The following abbreviations have been used:
| | |
|---|---|
| c.d. | circular dichroism |
| CIDNP | chemically induced dynamic nuclear polarisation |
| DMF | $NN$-dimethylformamide |
| DMSO | dimethyl sulphoxide |
| endor | electron nuclear double resonance |
| e.s.r. | electron spin resonance |
| g.l.c. | gas–liquid chromatography |
| i.r. | infrared |
| LAH | lithium aluminium hydride |
| MHz | megaHertz |
| MO | molecular orbital |
| NBS | $N$-bromosuccinimide |
| nm | nanometres |
| n.m.r. | nuclear magnetic resonance |
| o.r.d. | optical rotatory dispersion |
| p.p.m. | parts per million |
| PPP | Pariser–Parr–Pople |
| py | pyridine |
| SCF | self-consistent field |
| THF | tetrahydrofuran |
| t.l.c. | thin-layer chromatography |
| Tos | toluene-$p$-sulphonyl |
| u.v. | ultraviolet |

# 1
# Theoretical Aspects of Organosulphur, Organoselenium, and Organotellurium Compounds

## 1 Introduction

This report covers the period April 1969—March 1970. However, for the sake of completeness some earlier and later work is also discussed.

The title is to some extent misleading, since there has been little of theoretical significance published concerning organo-selenium or -tellurium compounds, and indeed it seems unlikely that the situation will improve dramatically in the near future, at least for the latter. For the most part, therefore, we shall be concerned with organosulphur compounds. Fortunately there have been a few important developments in this area.

Although good SCF wave-functions have been available for various electronic states of the sulphur, selenium, and tellurium atoms for some considerable time, even as late as 1966 the only non-empirical calculations on molecules were restricted to $H_2S$ and COS.[1] Even semi-empirical PPP $\pi$-electron calculations on sulphur-containing heterocycles were few and far between at that stage, and discussions of bonding in organosulphur compounds were mostly restricted to Hückel treatments (for a review see ref. 2).

In the past two years in particular, however, the situation has improved dramatically as far as our understanding of bonding in organosulphur compounds is concerned, although the situation with regard to selenium and tellurium compounds has remained roughly static. Before discussing in some detail the results, it is worth digressing on the reason for this relatively sudden improvement.

The past decade has seen a tremendous change in the level of sophistication of theoretical treatments of molecules of interest to organic chemists.

---

[1] For a review see D. W. J. Cruickshank and B. Webster, Chapter 2 in 'Inorganic Sulphur Chemistry', ed. G. Nickless, Elsevier, 1968.
[2] R. Zahradnik, 'Advances in Heterocyclic Chemistry', ed. A. R. Katritzky, vol. 5, p. 1, 1965.

Thus, even as late as 1960, *ab initio* treatments of organic molecules were restricted to those containing just one carbon atom.[3] With the lack of any semi-empirical treatments to include $\sigma$-electrons as well, organic chemists had to be satisfied with the vast body of $\pi$-electron calculations within the Hückel or Pariser–Parr–Pople (PPP) SCF formalism. However, in the past four years in particular, the situation has changed dramatically.

The extension of Hückel theory to include *all* valence electrons [4] and the development of semi-empirical SCF all-valence-electron treatments [5–8] has transformed the situation, and semi-empirical all-valence-electron calculations on a wide variety of medium-sized organic molecules have appeared in the literature. The spectacular advance in computer capability, coupled with the use of Gaussian Type (GTF) [9] instead of Slater Type (STF) [9] basis functions in the expansion method, has seen the development of non-empirical all-electron treatments on large numbers of polyatomic molecules.[10]

Not unnaturally, the increased problems of dealing with second-row atoms has meant that there has been much less attention paid to organo-sulphur compounds than to, say, organo-nitrogen or -oxygen compounds. With regard to non-empirical calculations, whilst optimized orbital exponents for Slater type orbitals for sulphur have been available [11] for some time, calculations on polyatomic sulphur-containing molecules have been limited by the lack of efficient programs for the evaluation of three- and four-centre two-electron integrals. The situation is already improving, however, and good calculations have appeared on hydrogen sulphide [12,13] and thiiran.[14] In contrast, whilst efficient programs for evaluation of multicentre integrals over Gaussian functions have been available for some time, it is only within the past year or so that optimised exponents have become available,[15–18] and good calculations on a wide variety of sulphur com-

---

[3] L. C. Allen and A. M. Karo, *Rev. Mod. Phys.*, 1960, **32**, 275.
[4] R. Hoffman, *J. Chem. Phys.*, 1963, **39**, 1397.
[5] J. A. Pople, D. P. Santry, and G. A. Segal, *J. Chem. Phys.*, 1965, **43**, 3129.
[6] J. A. Pople and G. A. Segal, *J. Chem. Phys.*, 1965, **43**, 5136.
[7] J. A. Pople, D. L. Beveridge, and P. A. Dobish, *J. Chem. Phys.*, 1967, **47**, 2026.
[8] M. J. S. Dewar and G. Klopman, *J. Amer. Chem. Soc.*, 1967, **89**, 3089, and subsequent papers.
[9] *Cf.* definitions, S. Huzinaga, *J. Chem. Phys.*, 1965, **42**, 1293.
[10] *Cf.* E. Clementi, *Chem. Rev.*, 1968, **68**, 341.
[11] E. Clementi and D. L. Raimondi, *J. Chem. Phys.*, 1963, **38**, 2686.
[12] F. P. Boer and W. N. Lipscomb, *J. Chem. Phys.*, 1969, **50**, 989.
[13] S. Polezzo, M. P. Stabilini, and M. Simonetta, *Mol. Phys.*, 1969, **17**, 609.
[14] R. Bonaccorsi, E. Scrocco, and J. Tomasi, *J. Chem. Phys.*, 1970, **52**, 5270.
[15] A. Rauk and I. G. Csizmadia, *Canad. J. Chem.*, 1968, **46**, 1205.
[16] *a* S. Huzinaga and Y. Sakai, *J. Chem. Phys.*, 1969, **50**, 1371; *b* A. Veillard, *Theor. Chim. Acta*, 1968, **12**, 405.
[17] J. D. Petke, J. L. Whitten, and A. W. Douglas, *J. Chem. Phys.*, 1969, **51**, 256.
[18] *a* B. Roos and P. Siegbahn, 'Proceedings of the Seminar on Computational Problems in Quantum Chemistry, Strassburg, 1969', p. 29; *b* *ibid.* p. 54; *c* B. Roos and P. Siegbahn, *Theor. Chim. Acta*, 1970, **17**, 199.

pounds have appeared.[19-23] Whilst the situation looks very hopeful for improving our knowledge of bonding in organosulphur compounds over the next few years, it seems unlikely that good non-empirical wave-functions for organo-selenium and -tellurium compounds will become available. The main hope in these cases must be in the development of adequate approximate SCF MO treatments. In this connection, the experimental results obtained by photoelectron spectroscopy, both $X$-ray and u.v., are likely to be of considerable importance.

## 2 $d$-Orbital Participation in the Ground States of Organosulphur Compounds

Reviews relevant to this section are given in refs. 24—27.

The role of $d$-orbital participation in the ground state of organosulphur compounds has interested chemists for many years. Despite numerous theoretical[28] and experimental investigations, it is only now that a clear picture of the situation is emerging. Most theoretical investigations until 1968 had concentrated on $\pi$-bonding between divalent sulphur and carbon; the interesting feature being that there was little measure of agreement between various workers as to the importance of $d$-orbital participation.[28] This is not too surprising, since both the Hückel and PPP $\pi$ SCF MO procedures are unsuitable for attempting to establish the extent to which the $d$-orbitals are involved in bonding, because of the large number of parameters whose values must be estimated but which are decisive in deriving orbital occupation numbers.

The question of $d$-orbital participation can really only adequately be dealt with by non-empirical quantum mechanical treatments, and perhaps the most significant theoretical advance has been in this area. We start off with a detailed discussion of recent work on $H_2S$, which for the purist, no doubt, is classified as an inorganic molecule. Nonetheless, a number of very important points are illustrated in this work.

**Calculations with Slater-type Basis Functions.**—Non-empirical LCAO MO SCF calculations using a minimal Slater basis set have been

---

[19] [a] S. Wolfe, A. Rauk, and I. G. Csizmadia, *J. Amer. Chem. Soc.*, 1967, **89**, 5710; [b] S. Wolfe, A. Rauk, and I. G. Csizmadia, *Canad. J. Chem.*, 1969, **47**, 113; [c] S. Wolfe, A. Rauk, and I. G. Csizmadia, *J. Amer. Chem. Soc.*, 1969, **91**, 1567; [d] S. Wolfe, A. Rauk, and I. G. Csizmadia, *Chem. Comm.*, 1970, 96.
[20] [a] D. T. Clark, 'Quantum Aspects of Heterocyclic Compounds in Chemistry and Biochemistry', Proceedings of the 2nd International Jerusalem Symposium, Israel Academy of Science and Humanities, Jerusalem, 1970, p. 238; [b] D. T. Clark, *Theor. Chim. Acta*, 1969, **15**, 225.
[21] I. H. Hillier and V. R. Saunders, *Chem. Phys. Letters*, 1969, **4**, 163.
[22] D. T. Clark and D. R. Armstrong, *Chem. Comm.*, 1970, 319.
[23] U. Gelius, B. Roos, and P. Siegbahn, *Chem. Phys. Letters*, 1970, **4**, 471.
[24] W. G. Salmond, *Quart. Rev.*, 1968, **22**, 253.
[25] C. A. Coulson, *Nature*, 1969, **221**, 1106.
[26] J. I. Musher, *Angew. Chem. Internat. Edn.*, 1969, **8**, 54.
[27] K. A. R. Mitchell, *Chem. Rev.*, 1969, **69**, 157.
[28] Ref. 19b lists a considerable number of early attempts.

reported by two groups of workers,[12,13] and the results are shown in Table 1. (The discrepancy between the atomic populations and dipole

**Table 1** Calculated properties for $H_2S$

|  |  | Boer and Lipscomb[12] |  | Pollezo, Stabilini, and Simonetta[13] |
|---|---|---|---|---|
|  |  | 1 | $2^a$ |  |
| Total energy (in a.u.) |  | −397·7881 | −397·8415 | −397·905$^b$ |
| Gross atomic populations | S | 16·186 | 16·352 | 16·144 |
|  | H | 0·907 | 0·824 | 0·928 |
| Total $d$-orbital population |  | — | 0·147 | 0·104 |
| Dipole moment (Debye) |  | 1·84$^c$ | 0·74$^c$ | −0·054 |

$^a$ Including 3$d$ orbitals on sulphur, optimized exponent 1·7077.
$^b$ Including 3$d$ orbitals on sulphur, optimized exponent 1·6. Two-electron three-centre integrals involving 3$d$ orbitals estimated from Mulliken's approximation.
$^c$ E. Scrocco, personal communication.

moments for the two calculations including 3$d$ orbitals is somewhat puzzling.) However, the striking feature is the minute effect of including $d$-orbitals on the total energy. This can be contrasted with the change in the population analyses and computed dipole moments. Here we have an apparent dilemma: the total energies show quite clearly that $d$-orbital participation is negligible, but some computed properties would tend to give weight to an opposite view. This situation is a direct consequence of using a minimal basis set and the variational principle for determining the best energy and corresponding coefficients. If we added 4$f$ functions on sulphur and, say, 3$d$ functions on the hydrogens, the change in energy would be very small but there would still be small electron populations in these orbitals, and, what is more, the population of the 3$d$ orbitals on sulphur would almost certainly have changed. This emphasizes the fact that no sound chemical conclusion can properly be drawn from such small orbital populations, but if they turn out to be large, then the situation is quite different. If the calculations had shown a large change in energy and orbital populations, then the situation is still not necessarily clear-cut because it could mean that the basis set describing the $s$ and $p$ basis set (*i.e.* H1$s$, S1$s$,2$s$,3$s$,2$p$,3$p$) was so inadequate that the functions we were adding were making up for the deficiencies. In carrying out calculations, therefore, it is important to ensure that the basis set of functions (either Slater or Gaussian) is adequate and gives a physically balanced, and as far as possible a formally balanced, wave-function.[29]

A physically unbalanced basis set gives erroneous results for well-defined physical quantities (dipole moments *etc.*), while formal unbalance gives unreasonable results for charges on atoms. Clearly, the main effect of the 3$d$ orbitals on sulphur in $H_2S$ is not to build hybrids in the normal chemical sense, but to polarize $s$ and $p$ valence-orbitals.

[29] *Cf.* ref. 18, p. 37.

This discussion illustrates an important point. Inclusion of $d$-orbitals on sulphur (for formally dicovalent sulphur compounds see later) makes very little difference to the total energy, and when organic chemists have invoked $d$-orbital participation it is usually this aspect which has been under consideration, *e.g.* in discussing the relative energies of two molecules. This shows that $3d$ orbitals on sulphur are acting like polarization functions, much as, say, a $2p$ function on hydrogen. Inclusion of polarization function will have a significant effect, however, on such properties as gross atomic population and hence charge on an atom. The difficulty of defining the latter, and its detailed dependence on basis set, point out the difficulty of using 'charge densities' in discussing the chemistry of organic molecules in general, as organic chemists have been wont to do. Whatever other theoretical arguments may be levelled at using this concept, for discussing reactivities for example, this factor alone should caution organic chemists against the indiscriminate use of charge densities.

**Calculations with Gaussian-type Basis Functions.**—A number of workers have published exponents for Gaussian basis sets for sulphur in the past year.[16-18] Following the pioneering work of Csizmadia and Rauk [15] with a very limited number of functions ($5s, 2p, 2d$), Huzinaga [16a] and Veillard [16b] have published optimized $17s, 12p$ and $12s, 9p$ Gaussian basis sets respectively for sulphur. An extended Gaussian lobe basis set for sulphur has also appeared;[17] however, as yet no calculations have been carried out on organosulphur compounds using such a basis set. Probably the most significant work, however, has been that of Roos and Siegbahn, who have published [18a, b, c] optimized exponents for a $9s, 5p$ and $10s, 6p$ basis set for sulphur and investigated in considerable detail the inclusion of $d$-orbitals on sulphur. It is instructive to compare the inclusion of $d$-orbitals in calculations on water and hydrogen sulphide, since this illustrates a number of important points. Comparative calculations on the extent of $d$-orbital participation in $H_2O$ and $H_2S$ have been carried out by two groups of workers.[18, 21] The results are shown in Tables 2 and 3. On the basis of comparison between calculated and experimentally determined first ionization potentials and dipole moments, Hillier and Saunders [21] concluded that $3d$ orbitals on sulphur are important in discussing bonding in $H_2S$. On the basis of total energy this is clearly not the case, and it is evident that the $3d$ orbitals on sulphur are acting in much the same manner as $3d$ orbitals on oxygen, *i.e.* as polarization functions (*cf.* Tables 2 and 3). This is shown up very nicely in the detailed investigation of Roos and Siegbahn, and emphasizes the fact that it is extremely difficult to say anything about the extent of $d$-orbital participation in bonding by comparing such properties as dipole moments and ionization potentials; for the former property, because of the problems in ensuring that a physically balanced basis set does not become unbalanced by addition of polarization functions to just one atom, and for the latter, because of the approximations inherent in Koopmans' theorem.

**Table 2** Calculated properties for $H_2O$

| | Hillier and Saunders [21] | | | Roos and Siegbahn [18a] | |
|---|---|---|---|---|---|
| [a]Basis set GTO<br>CGTO | (O/6,3)[b] (H/3)<br>(O/2,1) (H/1) | (O/6,3,3) (H/3)<br>(O/2,1,1) (H/3) | (O/7,3) (H/4)<br>(O/4,2) (H/2) | (O/7,3,1)[c] (H/4)<br>(O/4,2,1) (H/4) | (O/7,3) (H/4,1)[d]<br>(O/4,2) (H/2,1) |
| Total energy in a.u. | −74.951 | −75.002 | −75.875 | −75.896 | −75.914 |
| Dipole moment (Debye).<br>Experimental 1·84 | 1·74 | 0·66 | 2·415 | 2·205 | 2·115 |
| First ionization potential (a.u.).<br>Experimental 0·463 | 0·401 | 0·406 | 0·486 | 0·483 | 0·479 |
| Gross atomic populations $\begin{cases} O \\ H \end{cases}$ | Not given | | 8·688<br>0·656 | 8·704<br>0·648 | 8·280<br>0·860 |

[a] Nomenclature of ref. 18a [e.g. (O/6,3,1) GTO means an uncontracted basis set of 6s-type, 3p-type, and 1d-type Gaussian functions].
[b] Corresponds to minimal Slater basis set, 3 Gaussians used to approximate each member: d-orbital exponent (for STO 1·66).
[c] d-Orbital exponent (for GTO 1·325).
[d] p-Orbital exponent (for GTO 0·80).

**Table 3** Calculated properties for $H_2S$[a]

| | Hillier and Saunders[21] | | | Roos and Siegbahn[18a] | |
|---|---|---|---|---|---|
| Basis set GTO<br>CGTO | (S/9,6)[b] (H/3)<br>(S/3,2) (H/1) | (S/9,6,3) (H/3)<br>(S/3,2,1) (H/1) | (S/9,5) (H/4)<br>(S/6,4) (H/2) | (S/9,5,1)[c] (H/4)<br>(S/6,4,1) (H/1) | (S/9,5) (H/4,1)[d]<br>(S/6,4) (H/2,1) |
| Total energy in a.u. | −394·463 | −394·516 | −398·186 | −398·228 | −398·212 |
| Dipole moment (Debye).<br>Experimental 0·92 | 1·74 | 0·66 | 1·48 | 0·88 | 1·190 |
| First ionization potential (a.u.).<br>Experimental 0·384 | 0·356 | 0·343 | 0·355 | 0·343 | 0·352 |
| Gross atomic populations { S<br>{ H | Not given | | 16·133<br>0·933 | 16·242<br>0·879 | 15·927<br>1·036 |

[a] Nomenclature of 18a [e.g. (S/9,5,1) GTO means an uncontracted basis set of 9s-type, 5p-type, and 1d-type Gaussian functions].
[b] Corresponds to minimal Slater basis set, 3 Gaussians used to approximate each member: d-orbital exponent (for STO 1·7077).
[c] d-Orbital exponent (for GTO 0·541).
[d] p-Orbital exponent (for GTO 0·80).

The $s$ and $p$ basis sets of contracted Gaussian functions of Roos and Siegbahn [18a] are considerably better than those of Hillier and Saunders,[21] the total energies being lower by some 0·924 a.u. and 3·723 a.u. for $H_2O$ and $H_2S$ respectively. This has the effect of giving a better estimate of the extent of $d$-orbital participation as such, since there is always the problem of making up for deficiencies in the $s$ and $p$ basis set. For example, in calculations on sulphate anion, where $d$-orbitals are of some importance in bonding, the energy lowerings on inclusion of $d$-orbitals in two calculations are 0·838 a.u.[30] and 0·232 a.u.[31] respectively. The much larger value for the former almost certainly arises for this reason.

**Table 4** *Radius of maximal charge density for oxygen 2p,3d and sulphur 3p,3d orbitals*

| Atom | $r_{max}(np)^a$ | $r_{max}(3d)^b$ | $3d^c$ |
|---|---|---|---|
| O | 0·84 | 1·06 | 1·33 |
| S | 1·59 | 1·67 | 0·54 |

<sup>a</sup> Radius of maximum charge density for $2p$ and $3p$ orbitals, respectively (in a.u.).
<sup>b</sup> Corresponding radius of maximum charge density for $3d$ orbitals.
<sup>c</sup> $3d$ Orbital exponents (GTF) (ref. 18a).

The effect of adding $3d$ orbitals on oxygen parallels that for sulphur, and for both $H_2O$ and $H_2S$ the inclusion of polarization functions on hydrogen is of comparable importance. Roos and Siegbahn have optimized the $d$-orbital exponents, and it is of interest to compare the radial maxima for their functions with those for the valence $p$-orbitals. Their results are shown in Table 4. Since the main effect of the $3d$ orbitals is to polarize the valence orbitals, the former should have charge densities which overlap strongly with the $2p$ and $3p$ orbitals for O and S respectively. The charge density maxima for the $3d$ orbitals lie on the outer side of the maxima for the $2p$ and $3p$ orbitals, (1·06 a.u., 0·84 a.u.) and (1·67 a.u., 1·59 a.u.) respectively.

It is worth noting that the optimized S $3d$ exponent for a Slater type orbital of 1·707 (corresponding GTO exponents 0·49) found by Lipscomb and Boer [12] corresponds very closely with the optimized GTO exponent of Roos and Siegbahn of 0·54.[18a]

**Organosulphur Compounds.**—The question of $d$-orbital participation in organosulphur compounds has received considerable attention either directly or indirectly, and non-empirical calculations have appeared on $CS$,[21] $COS$,[23] $CS_2$,[23] thiiran,[14, 20] hydrogen methyl sulphoxide and its anion,[19a, b] dimethyl sulphoxide and its anion,[19a, b] the carbanions derived from hydrogen methyl sulphone [19c] and sulphide,[19d] dimethyl sulphide dianion,[19d] and thiophen.[22, 23] The results of these investigations will be

[30] I. H. Hillier and V. R. Saunders, *Chem. Comm.*, 1969, 1181.
[31] U. Gelius, personal communication.

discussed in some detail in the next section, and at this stage evidence will be considered relevant to the contribution of $3d$ orbitals on sulphur to bonding in these molecules.

*Carbon Monosulphide*: CS. Calculations have been carried out in a limited Gaussian basis set,[21] and it is of interest to compare the results with that for carbon monoxide.[21] Inclusion of $3d$ orbitals on the sulphur and oxygen atoms in the two molecules results in energy lowering of 0·062 a.u. and 0·081 a.u. respectively, indicating that the $3d$ orbitals are acting as polarization functions for both oxygen and sulphur.

*Carbonyl Sulphide and Carbon Disulphide*: COS and $CS_2$. Calculations carried out with an extended contracted Gaussian-type basis set have been reported by Gelius, Roos, and Siegbahn.[23, 31] For COS, calculations were performed in which $3d$ orbitals were included on sulphur, and $\pi$-type $d$-orbitals on carbon, oxygen, and sulphur. The results are shown in Table 5. Again it is clear that the S $3d$ orbitals are acting as polarization functions. This is also true for $CS_2$, the energy lowering on inclusion of $3d$ orbitals on sulphur being 0·075 a.u. in this case.[18, 31]

**Table 5** *Comparative calculations on* COS [23, 31]

|  | Energy (a.u.) | Change in energy (a.u.) | |
| --- | --- | --- | --- |
| $s,p$ basis set | 509·917581 | (0·0) | |
| $+3d$ on S | 509·955657 | 0·038 | (0·0) |
| $s,p$ basis set with | | | |
| $\pi$-type $3d$ on C,O,S | 510·024663 | 0·107 | 0·069 |

*Thiiran*: $C_2H_4S$. Calculations employing a minimal Slater basis set have been carried out by Scrocco *et al.*[14] on the formally dicovalent sulphur compound thiiran. The energy lowering on inclusion of $d$-orbitals amounts to 0·040 a.u., again indicating the role of $3d$ orbitals as polarization functions.

*Thiophen*: $C_4H_4S$. Thiophen, above all organosulphur compounds, has been the most studied. The pioneering Pariser–Parr–Pople $\pi$-electron and CNDO all-valence-electron semi-empirical LCAO MO SCF treatments, in which $3d$ orbitals on sulphur were first explicitly considered,[32, 33] indicated that inclusion of the latter did not significantly affect the total energy. However, the large number of approximations inherent in these treatments inevitably means that the results obtained are not as clear-cut as one would like. One of the most significant advances in theoretical aspects of organosulphur compounds in the past year has been the extension of non-empirical treatments to molecules the size of thiophen. Although still a small molecule in the chemical sense, the 44-electron system presents considerable computational problems. Two groups of

---

[32] M. J. Bielefeld and D. D. Fitts, *J. Amer. Chem. Soc.*, 1966, **88**, 4804.
[33] D. T. Clark, *Tetrahedron*, 1968, **24**, 2663.

workers have now published preliminary communications on 'ab initio' calculations on thiophen and the extent of $d$-orbital participation therein.[22, 23, 31] The different approaches both have their merits and give the same answer; namely that $d$-orbitals are involved to only a minor extent in bonding on thiophen, thus settling a question which has intrigued chemists for some 30 years. The approaches differ in choice of basis set of GTF and contracted GTF, and the method of dealing with the $d$-orbitals. With the usual definition of GTF there are six $d$-type functions ($d_{x^2}$, $d_{y^2}$, $d_{z^2}$, $d_{xy}$, $d_{xz}$, $d_{yz}$) instead of the usual five. It is possible to effect a linear transformation of this set to a $3s$-type ($d_{x^2+y^2+z^2}$) and five $3d$-type ($d_{x^2-y^2}$, $d_{3z^2-r^2}$, $d_{xy}$, $d_{xz}$, and $d_{yz}$)[15] functions before carrying out the SCF procedure. At the same time, the $3s$-type function can be deleted, and this simplifies the interpretation of the results. However, in doing this a valuable degree of variational freedom is lost. By allowing the coefficients of the six $d$-type functions to be determined in the SCF procedure, a better wavefunction should be obtained since distorted $s$- and $d$-type orbitals can be accommodated. The results, however, are correspondingly more difficult to interpret.

The basis sets used and the results of the calculations are given in Tables 6 and 7. The S basis set used by Clark and Armstrong[22] is considerably larger than that used by Gelius et al.,[23] but is more heavily contracted. The energy lowering on introducing $3d$ orbitals on S amounts to 0·0527 a.u.,[31] and it is interesting to compare this with the effect of introducing polarization functions on hydrogen. The energy lowering then amounts to 0·0224 a.u., and this emphasizes that the $3d$ orbitals on S are indeed polarization functions. This is also shown by the other calculation,[22] but in this case the small but significant energy lowering of 0·1179 a.u. on introduction of $d$-orbitals on sulphur arises almost solely from the increased variational freedom in the S basis set due to the added $3s$-type function (the contribution is actually from a distorted orbital). This is most readily appreciated by inspection of the eigenfunctions and by calculations on the spherically symmetric sulphide ion ($S^{2-}$). The total energies for the latter with and without inclusion of $d$ functions are $-396·8185$ a.u. and $-396·7094$ a.u. In this case, the energy lowering of 0·1091 a.u. arises solely from the added $3s$-type function. The small degree of participation in bonding of the five $d$-type orbitals in thiophen is also illustrated by the gross atomic populations given in Table 6. It is clear that the major difference in atomic population occurs from the $3s$ function of sulphur. The lower contribution in the case of the calculation including $d$-orbitals is offset to a considerable extent by the population of the $3s$-type orbital formed from the $3d$ functions (actually a distorted $3s$-type function).

It will be of interest now to repeat the SCF procedure, deleting the $3s$-type function to compare the results more directly with the other calculation, and also to compute the density contour maps to investigate the wavefunctions more closely.

# Organosulphur, Organoselenium, and Organotellurium

**Table 6** *Non-empirical LCAO MO SCF calculations on thiophen* [23, 31]

|  | 1 | 2 |
|---|---|---|
| Basis set GTO | (S/10,6) (C/7,3) (H/4) | (S/10,6) (C/7,3) (H/4,1) |
| CGTO | (S/6,4) (C/4,2) (H/2) | (S/6,4) (C/4,2) (H/2,1) |
| Total energy in a.u. | −550·92339 | −550·94583 |

|  | 3 | 4 |
|---|---|---|
| Basis set GTO | (S/10,6,1) (C/7,3) (H/4) | (S/10,6,1) (C/7,3) (H/4,1) |
| CGTO | (S/6,4,1) (C/4,2) (H/2) | (S/6,4,1) (C/4,2) (H/2,1) |
| Total energy in a.u. | −550·97612 | −550·99903 |


**Table 7** *Non-empirical LCAO MO SCF calculations on thiophen* [22]

|  |  | 5 | 6 |
|---|---|---|---|
|  | Basis set GTO | (S/12,9) (C/7,3) (H/3) | (S/12,9,2) (C/7,3) (H/3) |
|  | CGTO | (S/4,2) (C/3,1) (H/1) | (S/4,2,2) (C/3,1) (H/1) |
|  | Total energy | −550·4174 a.u. | −550·5353 a.u. |
| Atomic populations |  |  |  |
| S | $3s$ | 1·716 | 1·140 |
|  | $3p_x$ | 0·872 | 0·913 |
|  | $3p_y$ | 1·347 | 1·317 |
|  | $3p_z$ | 1·680 | 1·684 |
|  | $3d_{x^2}$ | — | 0·172 |
|  | $3d_{y^2}$ | — | 0·184 |
|  | $3d_{z^2}$ | — | 0·184 |
|  | $3d_{xy}$ | — | 0·108 |
|  | $3d_{xz}$ | — | 0·052 |
|  | $3d_{yz}$ | — | 0·018 |
| Total |  | 15·616 | 15·773 |
| C-1(C-4) | $2s$ | 1·179 | 1·139 |
|  | $2p_x$ | 1·114 | 1·094 |
|  | $2p_y$ | 0·997 | 0·995 |
|  | $2p_z$ | 1·106 | 1·074 |
| Total |  | 6·396 | 6·302 |
| C-2(C-3) | $2s$ | 1·096 | 1·093 |
|  | $2p_x$ | 1·023 | 1·022 |
|  | $2p_y$ | 1·076 | 1·077 |
|  | $2p_z$ | 1·054 | 1·049 |
| Total |  | 6·249 | 6·242 |
| H-1(H-4) |  | 0·769 | 0·784 |
| H-2(H-3) |  | 0·778 | 0·787 |

**Table 9** Calculated properties for COS and $CS_2$ [23, 31]

|  | COS | | | $CS_2$ | |
|---|---|---|---|---|---|
|  | i | ii | iii | i | ii |
| [a]Basis set GTO | (S/10,6) | (S/10,6,1) | (S/10,6,1)[b] | (S/10,6) | (S/10,6,1) |
| CGTO | (S/6,4) | (S/6,4,1) | (S/6,4,1) | (S/6,4) | (S/6,4,1) |
| Total energy (in a.u.). | −509.91758 | −509.95566 | −510.02466 | −832.52983 | −832.60500 |
| Highest occupied orbital energies (in eV) | −21.05 $\sigma$ | −21.82 $\sigma$ | −20.98 $\sigma$ | −18.21 $\sigma_g$ | −18.37 $\sigma_g$ |
|  | −18.21 $\pi$ | −17.96 $\pi$ | −17.70 $\pi$ | −15.65 $\sigma_u$ | −15.74 $\sigma_u$ |
|  | −16.99 $\sigma$ | −17.23 $\sigma$ | −17.03 $\sigma$ | −14.62 $\pi_u$ | −14.28 $\pi_u$ |
|  | −11.29 $\pi$ | −11.20 $\pi$ | −11.20 $\pi$ | −10.07 $\pi_g$ | −9.87 $\pi_g$ |
| Gross atomic populations C | 5.817 | 5.340 | 5.492 |  |  |
| O | 8.500 | 8.493 | 8.379 |  |  |
| S | 15.683 | 16.167 | 16.130 | 15.659 | 16.117 |
|  |  |  |  | 6.682 | 5.766 |

[a] C, O GTO 7,3; CGTO 4,2
[b] Includes $\pi$-type $d$-functions on C, O, and S.

The He I lamp photoelectron spectra of both COS [34] and $CS_2$ [34, 35] have recently been investigated, and it is of interest to compare the energy levels with those calculated using Koopmans' theorem. This is shown in Table 10.

**Table 10** *Calculated and experimental ionization potentials (in eV)*

| | COS | | | |
|---|---|---|---|---|
| | $\pi$ | $\sigma$ | $\pi$ | $\sigma$ |
| i | 11·29 | 16·99 | 18·21 | 21·05 |
| ii | 11·20 | 17·23 | 17·96 | 20·82 |
| iii | 11·20 | 17·03 | 17·70 | 20·98 |
| Experimental [34] | 11·2 | 16·0 | 15·5 | 18·0 |

| | $CS_2$ | | | |
|---|---|---|---|---|
| | $\pi_g$ | $\pi_u$ | $\sigma_u$ | $\sigma_g$ |
| i | 10·07 | 14·62 | 15·65 | 18·21 |
| ii | 9·87 | 14·28 | 15·74 | 18·37 |
| Experimental [34, 35] | 10·09 | 12·69 | 14·47 | 16·19 |

For carbonyl sulphide, the ordering of the first $\sigma$ and second $\pi$ states is reversed compared with the experimental order, but is in much closer agreement than Clementi's calculation.[36]

The interesting feature evident in the spectrum of $CS_2$ is a broad band centred at 17·1 eV. Brundle and Turner [34] have suggested that this corresponds to ionization from the strongly bonding $\sigma_u$ orbital. However, both calculations [31] put the energy of this orbital at $>27$ eV, so that this explanation seems unlikely.

*Thiiren and Thiiran*: △S and △S. Calculations using a small uncontracted Gaussian basis set have been reported [20a, b] on the as-yet-unknown anti-aromatic thiiren and the corresponding saturated compound, thiiran. A minimal basis set calculation using Slater type orbitals has also been carried out [14] for the latter. The Gaussian basis set consisted of a 5s,2p atomic set for C and S and 2s atomic set for H. However, a 5s,2p basis set on sulphur is minimal and cannot be expected to give a good wavefunction. Nonetheless, for energy differences between species the calculations should be qualitatively correct. The calculations on thiiren were aimed at discussing the likely chemistry of this as-yet-unknown molecule and the related oxiren and 1- and 2-azirines.

A possible synthetic route for this molecule would be the reaction of atomic sulphur ($^3P$) with acetylene. However, the calculated enthalpy changes in this reaction [20a] would seem to preclude this. Other calculated properties were heats of isomerization to thioketen, decomposition to carbon monosulphide and $^1A_1$ methylene, and relative heat of hydrogenation with respect to cyclopropene.

[34] C. R. Brundle and D. W. Turner, *Internat. J. Mass Spectrometry Ion Phys.*, 1969, **2**, 195.
[35] J. H. D. Eland and G. J. Danby, *Internat. J. Mass Spectrometry Ion Phys.*, 1968, **1**, 111.
[36] E. Clementi, *J. Chem. Phys.*, 1962, **36**, 750.

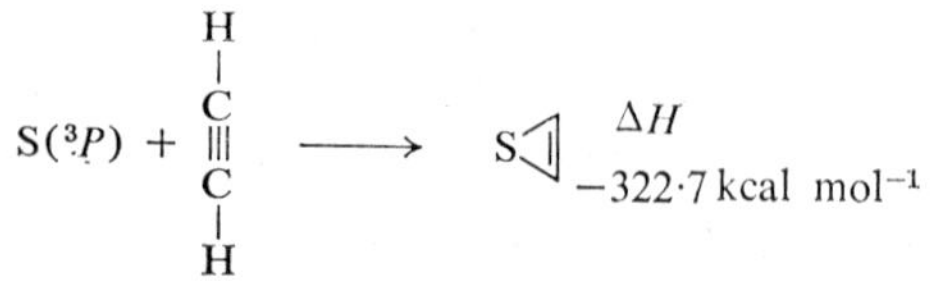

$$S(^3P) + \begin{array}{c} H \\ | \\ C \\ ||| \\ C \\ | \\ H \end{array} \longrightarrow \text{S}\triangleleft \quad \begin{array}{l} \Delta H \\ -322 \cdot 7 \text{ kcal mol}^{-1} \end{array}$$

$$\triangleright\text{S}\begin{array}{c} \nearrow CH_2=C=S \\ \searrow \\ {}^1A_1CH_2 + CS \end{array} \quad \begin{array}{l} \Delta H \\ -122 \cdot 1 \text{ kcal mol}^{-1} \\ \\ 34 \cdot 45 \text{ kcal mol}^{-1} \end{array}$$

$$\triangleright X \xrightarrow{H_2} \triangleright X \quad \begin{array}{l} \Delta H_{rel} \\ X = CH_2 \quad (0 \cdot 0) \\ X = S \quad 20 \cdot 9 \text{ kcal mol}^{-1} \end{array}$$

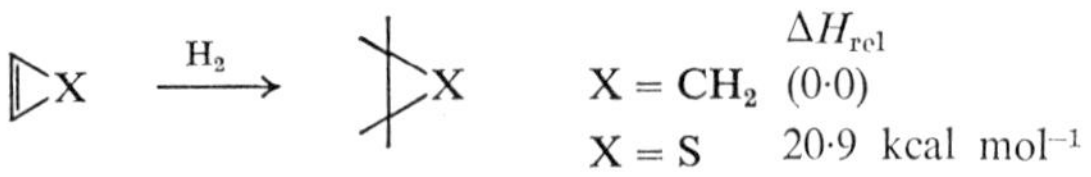

Comparison has also been drawn with the corresponding nitrogen and oxygen heterocycles, and the corresponding saturated derivatives, thiiran, aziridine, and oxiran.[20a, b]

A minimal basis set of STO's gives a much lower energy and better wavefunction than a minimal Gaussian basis set. For example, the total energies calculated for thiiran in the two basis sets are 456·0016 [20b] and 474·4759 a.u. respectively. The deficiencies of a $5s,2p$ GTF basis set on sulphur are clearly evident. Some of the calculated properties [14] for thiiran are shown in Table 11. Of particular interest is the minor effect of including $d$-orbitals on the second moment of electronic charges, and the considerable effect on dipole and quadrupole moments. A discussion has also been given in terms of localized orbitals, and from these the 'bent bonds' on the rings are clearly evidenced. A feature of the chemistry of the three-membered ring heterocycles is their ease of ring opening. For example, in acid media:

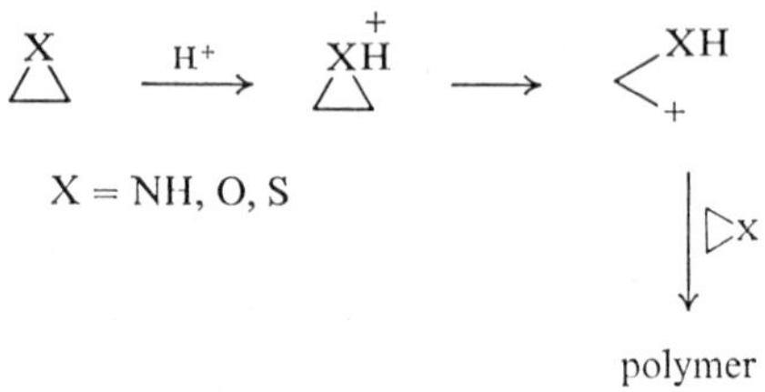

$$\overset{X}{\triangle} \xrightarrow{H^+} \overset{\overset{+}{X}H}{\triangle} \longrightarrow \overset{XH}{\underset{+}{<}}$$

$X = NH, O, S$

$$\downarrow \triangleright X$$

polymer

The initial stage in this reaction is approach of a proton to the heterocycle. In a very elegant study, Scrocco and co-workers [14] have investigated a prototype potential energy surface for the first step in the reaction. Using the molecular wave-function for thiiran (including $d$-orbitals on S) the electrostatic potential produced in the neighbouring space by the nuclear and electronic charges has been calculated, by evaluating the function

$$V(r_i) = -\int \frac{\rho(1)}{r_{1i}} d\tau_1 + \sum_{\alpha}^{Nucl} \frac{Z_\alpha}{r_{\alpha i}}$$

**Table 11** *Calculated properties for thiiran* [14]

|  |  | 1 | $2^a$ | Experimental |
|---|---|---|---|---|
| Total energy (a.u.) | | −474·4759 | −474·5159 | −477·846 |
| Ionization potential (eV) | | 9·01 | 8·75 | 8·87 |
| Dipole moment (Debye) | | 1·56 | 0·84 | 1·84 |
| Quadrupole moment | $xx$ | −1·920 | −0·282 | |
| | $yy$ | 0·407 | 0·585 | |
| | $zz$ | 1·513 | 0·867 | |
| Second moment of | $x^2$ | 106·4376 | 105·2215 | |
| electronic charges (a.u.) | $y^2$ | 30·7013 | 30·7890 | |
| | $z^2$ | 64·3492 | 64·2653 | |

$^a$ Basis set containing $d$-orbitals.

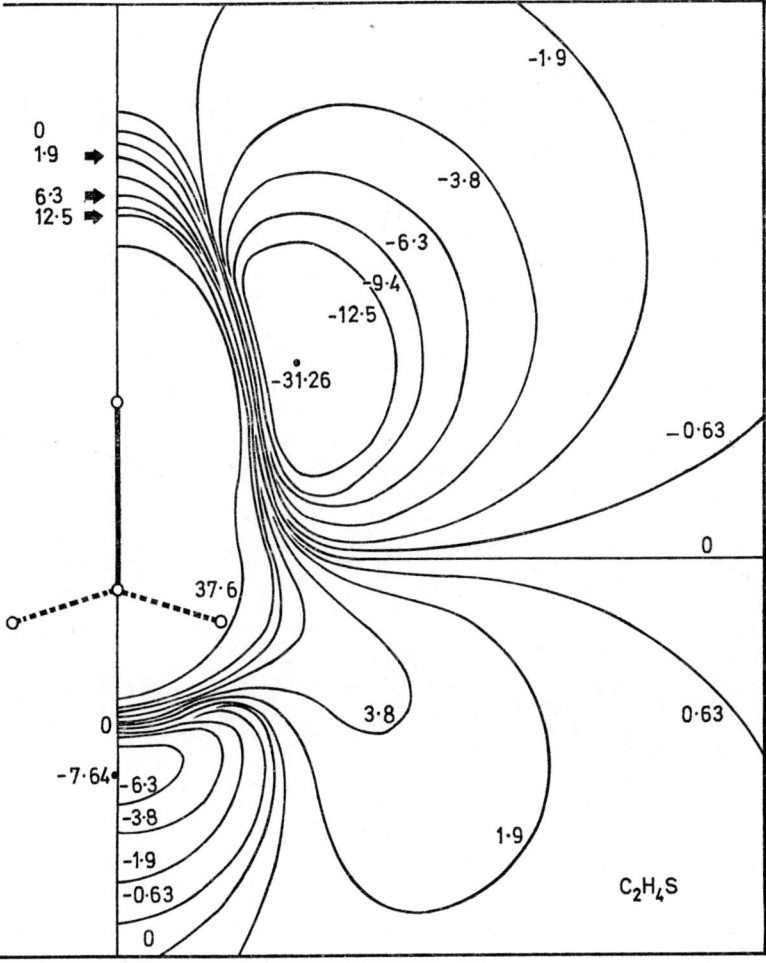

**Figure 1** *Electrostatic potential energy map for thiiran. (Energy in kcal mol$^{-1}$, plane perpendicular to ring plane)*

point by point, where $\rho(1)$ is the molecular monoelectronic density function and $Z_\alpha$ the charge of the nucleus $\alpha$. The distances of the electronic charge element $\rho(1)d\tau_1$ and of the nucleus $\alpha$ from the point considered, $i$, where the potential $V$ is calculated, are respectively $r_{1i}$ and $r_{\alpha i}$. The contour maps, Figures 1 and 2 (energies in kcal mol$^{-1}$) show the energies experienced by a unit positive charge (prototype for a proton).

In the region corresponding to the sulphur lone pairs, two minima are present ($-31 \cdot 26$ kcal mol$^{-1}$), and there is a secondary minimum near the

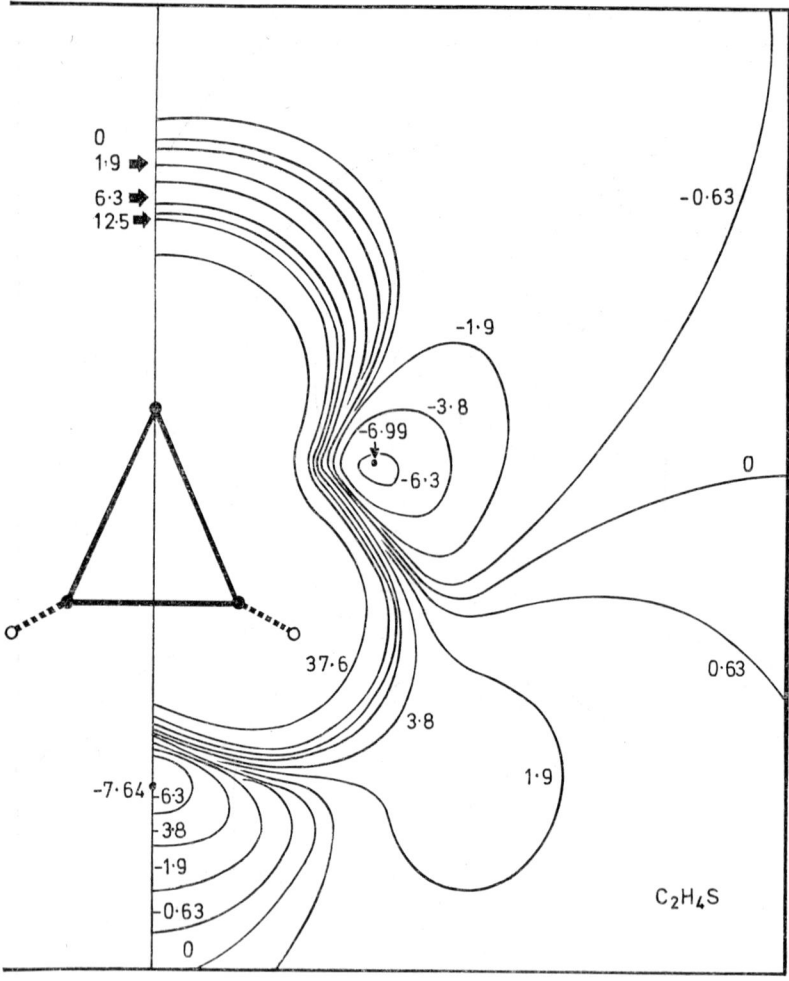

**Figure 2** *Electrostatic potential energy map for thiiran. (Energy in kcal mol$^{-1}$, ring plane)*

C—C bond (energy $-7.64$ kcal mol$^{-1}$). The corresponding minima for the calculation without $d$-orbitals on sulphur are $-31.93$ kcal mol$^{-1}$ and $-4.74$ kcal mol$^{-1}$ respectively. This simple electrostatic model indicates that the sulphur is more easily subject to electrophilic attack than the $CH_2$—$CH_2$ region (cf. cyclopropane itself). Similar calculations for the corresponding oxygen heterocycle, oxiran, give deep minima of $-44$ kcal mol$^{-1}$ close to the lone pairs on oxygen and $-17$ kcal mol$^{-1}$ near the C—C bond. This approach therefore predicts that the proton affinity of the oxygen heterocycle is greater than that for the sulphur compound.

*Hydrogen Methyl Sulphoxide and its Anion*: $CH_3$—SO—H (*HMSO*) and C$^-$H$_2$—SO—H (*HMSO$^-$*). Csizmadia and co-workers [15, 19a, b] have made extensive series of calculations on HMSO and HMSO$^-$ with a small uncontracted Gaussian basis set. For HMSO the rotation about the C—S bond was investigated, and for HMSO$^-$ a conformational energy surface (total energy as a function of rotation about the C—S bond and inversion of the carbanion angle) determined. The whole investigation had as its object interpretation and prediction of the course of proton exchange in sulphoxides. The detailed application of the theory to the experimental results will be dealt with elsewhere in this Report, and we will concentrate here on the theoretical results themselves. The calculations used a basis set of C, O $3s,1p$, S $5s,2p,1d$, and H $1s$. Although this basis set is extremely limited and gives a poor total energy, this does not necessarily invalidate the results if one is just interested in relative energies of species. As it turns out, one of the most readily calculable properties of a molecule, even using a very limited basis set, is its geometry.[15] The work of Csizmadia, Rauk, and Wolfe represents a very good example of the cross-fertilization of theoretical and experimental approaches to a given problem.

Figure 3 shows the variation of total energy as a function of rotation about the C—S bond, calculated for hydrogen methyl sulphoxide. The positions of energy minima and maxima correspond to angles of rotation $\theta = 21°$ and $81°$ respectively, with a barrier to rotation of $2.68$ kcal mol$^{-1}$. An analysis of the barrier in terms of the component one- and two-electron contributions has been presented, and also a discussion of orbital energies, gross atomic populations, and bond overlap populations. The limitation of the basis set employed in the calculations limits the usefulness of these results. However, as mentioned previously, there is little evidence for the importance of $d$-orbital participation.

The conformation of HMSO$^-$ was defined by two independent parameters, the rotational angle, $\theta$, about the C—S bond, and the anion angle, $\phi$, where the sign of the angle $\phi$ is defined with respect to a vertical plane through the C—S bond as drawn, i.e. for H$_6$ and H$_7$ in Figure 4 reflected across this plane, $\phi$ becomes negative. A projection of the computed three-dimensional P.E. surface is given in Figure 5.

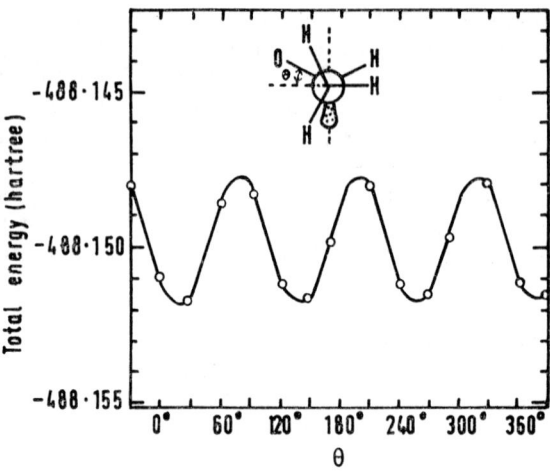

Figure 3

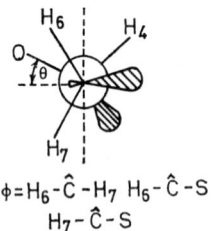

Figure 4

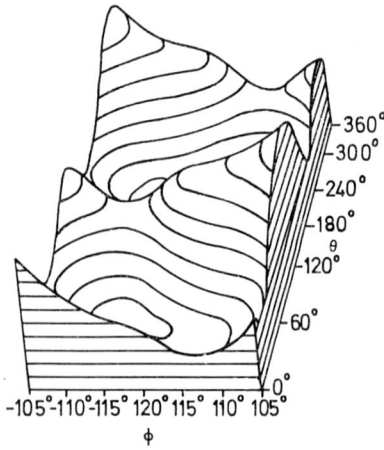

Figure 5

The energy minimum corresponds to $\theta = 240°$, $\phi = 115°$, and a Newman projection of this conformation is shown in Figure 6.

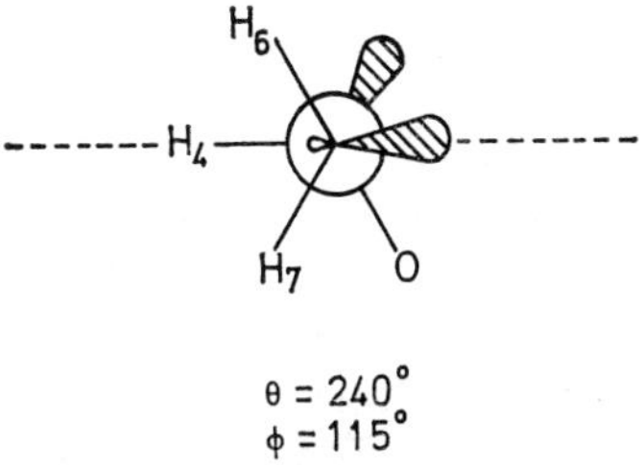

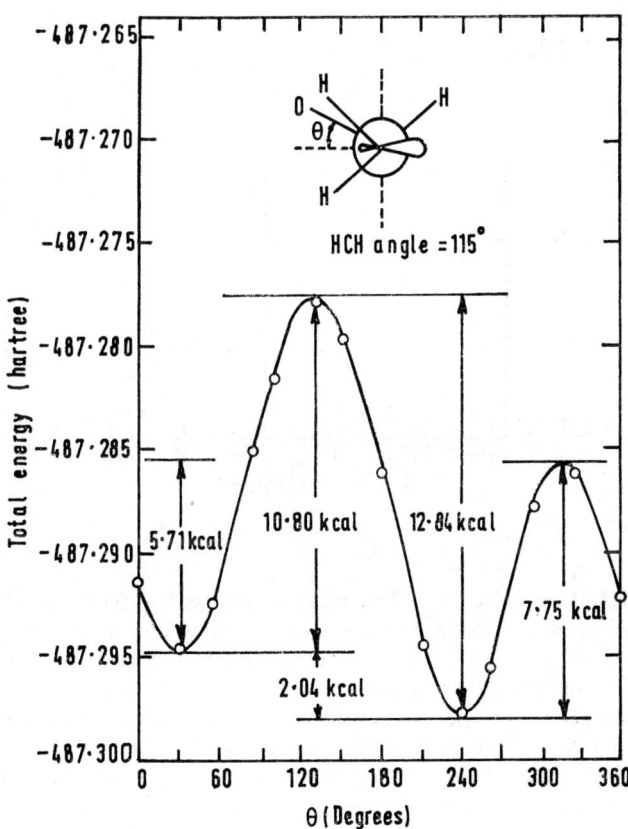

**Figure 6**

**Figure 7**

Cross-sections of the energy surface of HMSO⁻ at constant $\theta = 240°$ and $\phi = 115°$ have previously been published,[19a] and are reproduced in Figures 7 and 8. The barrier to rotation is much higher than in HMSO, and in the energy minimum the carbanion lone pair is situated in the plane of the bisector of the O—S—lone pair angle. This is in agreement with the

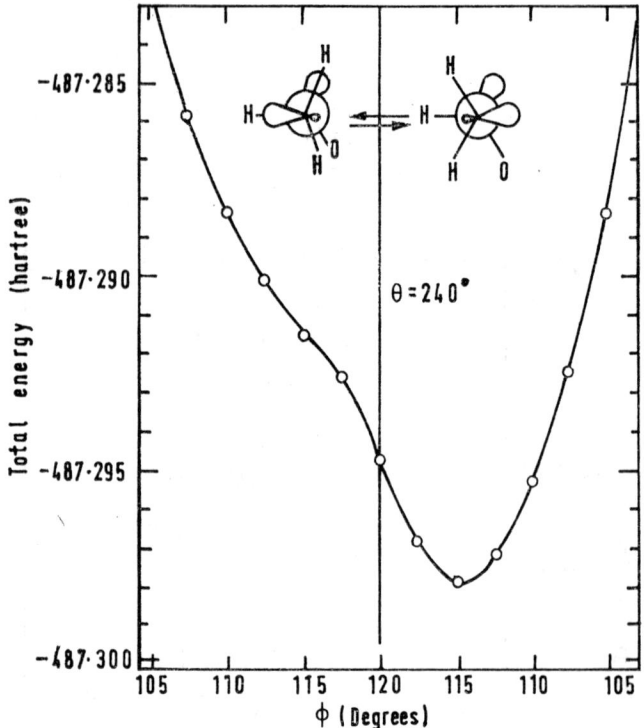

**Figure 8**

experimental evidence.[19a, b] The maximum in energy corresponds to a *trans* arrangement of the carbanion and sulphur lone pairs. There are secondary minima and maxima corresponding to the carbanion lone pair *trans* to oxygen, and nearly eclipsed with sulphur lone pair respectively. An investigation of the component energy terms as a function of $\theta$ ($\phi = 115°$) shows that there is no simple explanation for the lower energy of the preferred conformer. Reference to Figure 8 suggests that a carbanion generated on the side opposite to oxygen and the lone pair would undergo spontaneous inversion. These results have been used to discuss in detail the stereochemistry of proton exchange in sulphoxides.

Figure 9 shows the variation of the overlap population of the CS and SO bonds as a function of $\theta$ for HMSO (I) and HMSO⁻ (II) ($\phi = 115°$). Whilst the limited basis set employed means little quantitative reliance can be placed on the absolute magnitude of the numbers, the trends should be

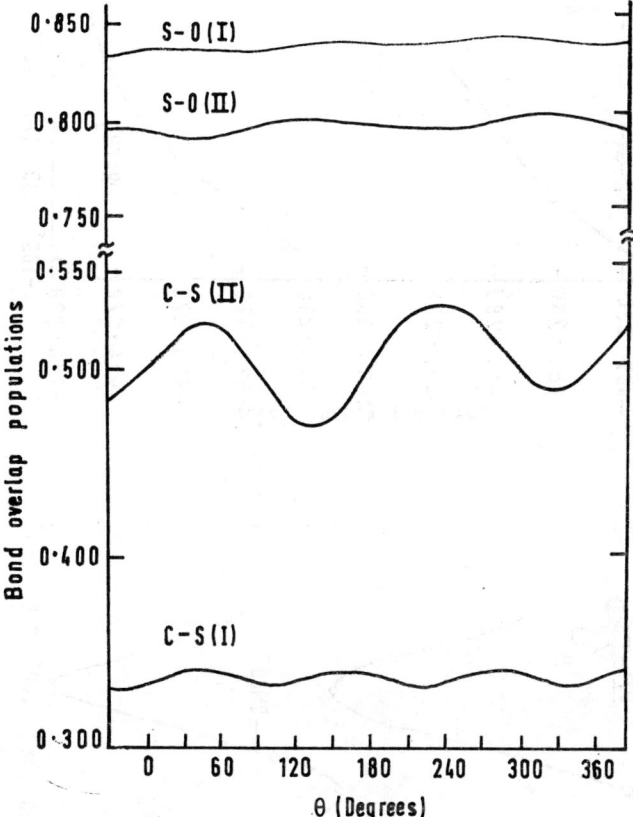

Figure 9

reliably produced. The overlap population of the C—S bond increases, and that of the S—O bond decreases, in going from the neutral molecule to the anion; there is also a significant dependence on $\theta$. The increased C—S bond overlap population for the anion could be explained by invoking $d$-orbital participation on sulphur. However, inspection of the coefficient matrix of the MO's reveals little evidence for this. It is not necessary to invoke $d$-orbital participation, therefore, to explain the asymmetry of sulphinyl carbanions.

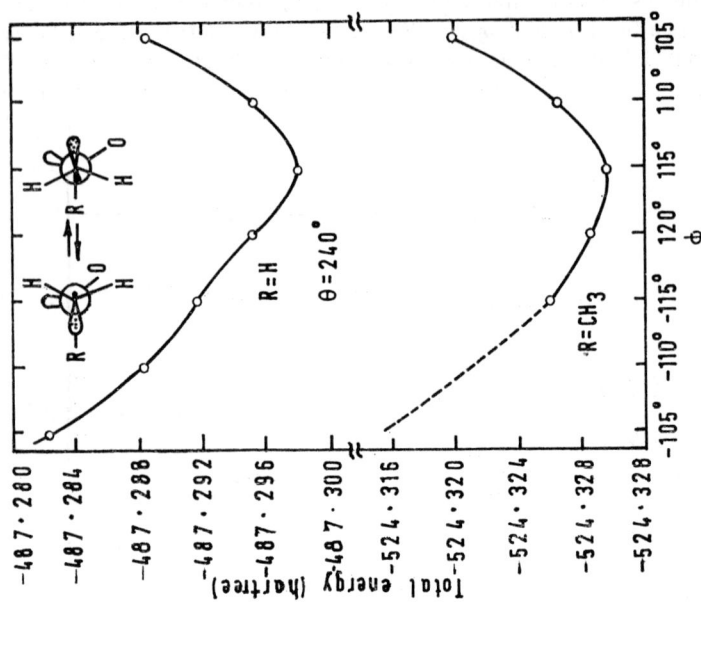
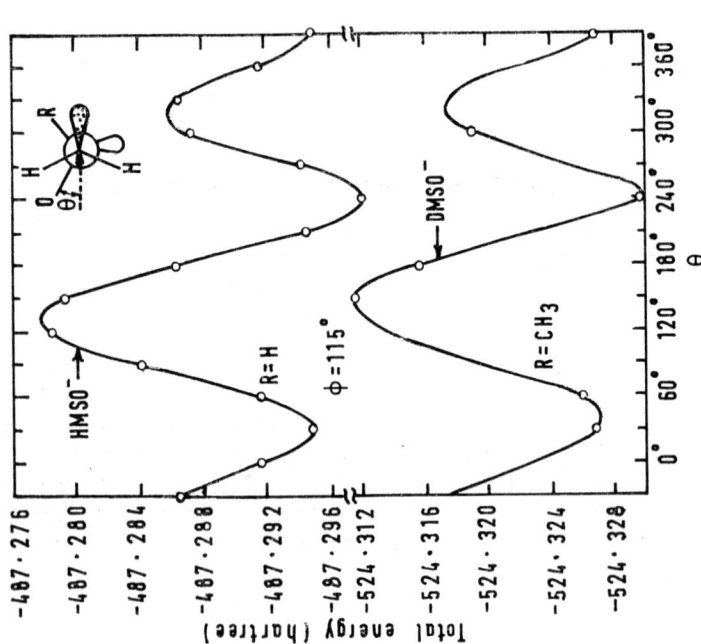

Figure 10

*Dimethyl Sulphoxide and its Anion*: $CH_3-SO-CH_3$ (*DMSO*) and $CH_3-SO-CH_2^-$ (*DMSO*⁻). There are, of course, dangers in applying the results of calculations on hypothetical molecules to real situations. To cover this eventuality, Csizmadia *et al.*[19b] also carried out a limited number of calculations on DMSO and DMSO⁻. The energy cross-sections for rotation and inversion about the carbanionic centre in DMSO⁻ parallel very closely the results for HMSO⁻ (*cf.* Figure 10) and lend considerable support to the view that substitution of a simple alkyl group for hydrogen in HMSO⁻ will not qualitatively alter the overall features of the P.E. surface.

*Hydrogen Sulphonyl Carbanion*: $H-SO_2-CH_2^-$. Conformationally unconstrained α-sulphonyl carbanions are almost unique among reactive intermediates in their ability to retain asymmetry under a wide variety of experimental conditions. Structural discussions have followed two main lines: either that the carbanion is pyramidal and there is a high barrier to inversion; or that the carbanion is planar and there is a high barrier to rotation. The structure of the simplest prototype for an α-sulphonyl carbanion has now been investigated in some detail by Csizmadia, Rauk, and Wolfe.[19c] The Gaussian basis set was identical to that used for the investigation of hydrogen methyl sulphoxide described above.[19a,b]

The structure of hydrogen sulphonyl carbanion was investigated as a function of angle of rotation, $\theta$, about the C—S bond and the anion angle $\phi$. The calculated P.E. surface is shown in Figure 11. Figure 12 shows

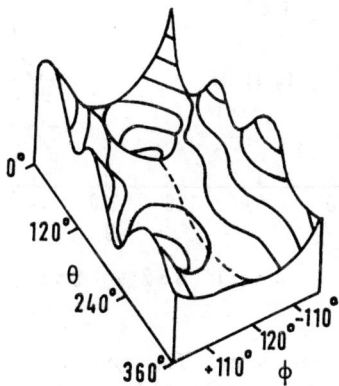

**Figure 11**

sections through the P.E. surface for constant carbanion angles of 115° (pyramidal) and 120° (planar). The energy minimum is clear-cut and corresponds to a pyramidal carbanion ($\phi = 115°$) in which the lone pair on carbon bisects the O—S—O angle. This is some 5 kcal lower in energy than the corresponding conformer with the lone pair on carbon eclipsing hydrogen. The most stable planar conformation has the carbon $2p$ lone pair bisecting the O–$\widehat{S}$–O angle: however, this is still some 2·5 kcal mol$^{-1}$ higher in energy than the most stable conformer. The planar conformer with H–C–H plane bisecting the O–$\widehat{S}$–O angle is some

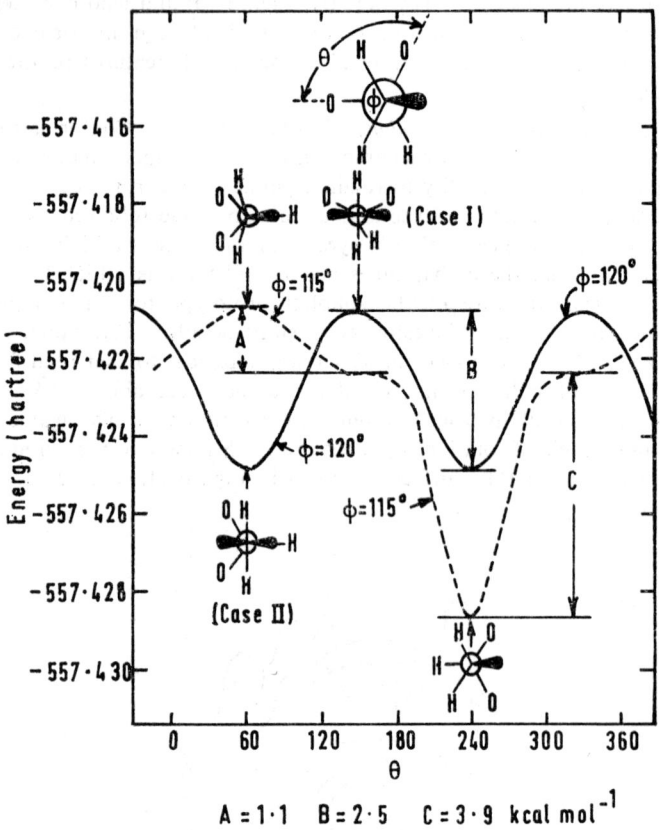

**Figure 12**

2·5 kcal mol$^{-1}$ higher in energy. This constrasts with the expectation from qualitative arguments set forward by Koch and Moffitt.[37]

As with the α-sulphinyl carbanions discussed in the preceding section, there is no significant contribution from the $3d$ orbitals on sulphur.

*Methanethiol*: CH$_3$–S–H. Calculations have been performed[19d] on the carbanion derived from methanethiol (HMS$^-$) to complete the series sulphide, sulphoxide, sulphone. The basis set was identical to that

[37] H. P. Koch and W. E. Moffitt, *Trans. Faraday Soc.*, 1951, **47**, 7.

described for hydrogen methyl sulphoxide. The same parameters were investigated, namely rotation about the C—S bond and inversion about the carbanion centre. A cross-section of the calculated energy surface of HMS⁻ for rotation about the C—S bond ($\theta$) through the energy minima $\phi = 115°$ is shown in Figure 13. The close similarity with that for the

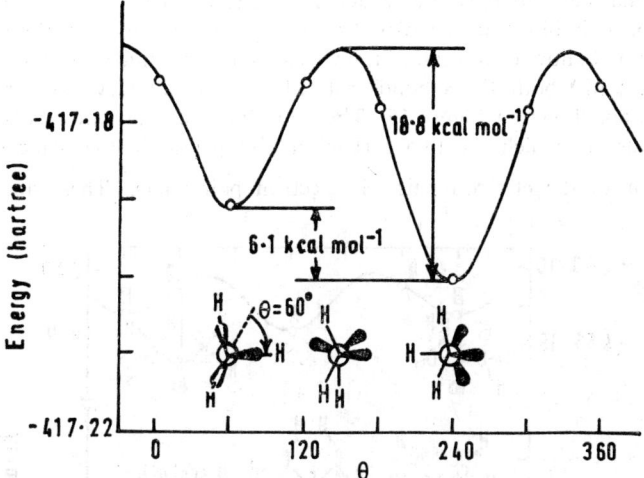

Figure 13

corresponding sulphoxide is striking. Again the most stable conformer is pyramidal, with the carbanion lone pair bisecting the lone pair—S—lone pair angle. There is a high barrier to internal rotation of 18·8 kcal mol⁻¹. Inspection of the MO coefficient matrix again shows little evidence for *d*-orbital participation on sulphur.

*Dimethyl Sulphide Dicarbanion*: $\bar{C}H_2$—S—$\bar{C}H_2$. Calculations have been reported on dimethyl disulphide carbanion,[19d] as a prototype for a system in which sulphur is adjacent to two carbanionic centres, the motivation for this study being the experimental observation of the ready formation of the dicarbanion of dimethyl tetrathia-adamantane (I → II, Figure 14) and the interpretation of this in terms of *d*-orbital participation by the

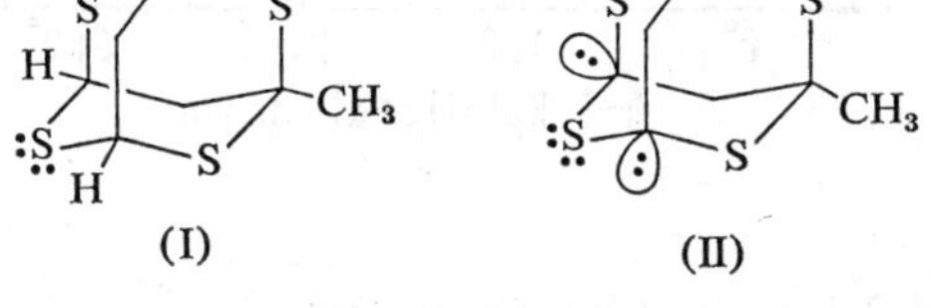

Figure 14

intervening sulphur atom.[38] In the dianion the carbanion lone pairs bisect the electron pair—$\widehat{S}$—electron pair angle of two sulphurs, and hence should represent a particularly favourable arrangement. To confirm this, calculations were carried out on dimethyl sulphide carbanion with pyramidal ($\phi = 115°$) geometry about the carbanion centre. Three types of rotational behaviour were considered: (i) one C—S bond fixed, with the electron pair bisecting the two electron pairs of sulphur and the second C—S bond allowed to rotate; (ii) both C—S bonds rotated in a disrotatory manner; (iii) both C—S bonds rotated in a conrotatory manner. The results are shown in Figure 15. The three stable conformations (A, B, C) are those in which the two carbanions are placed in the plane of the bisector of the electron pair—$\widehat{S}$—electron pair angle. These are some

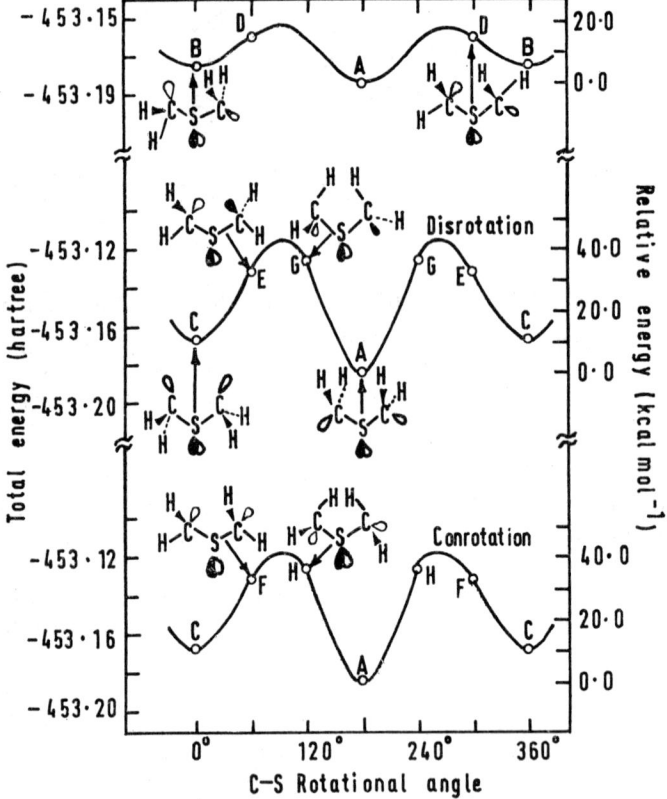

Figure 15

[38] K. C. Bank and D. L. Coffen, *Chem. Comm.*, 1969, 8.

3·8—14·8 kcal mol$^{-1}$ more stable than a conformation (D) in which one carbanion is in the plane of the bisector and the other is not; and this conformation is at least 18 kcal mol$^{-1}$ more stable than any conformation (E, F, G, H) in which neither carbanion is in the plane of the bisector. Again there is little evidence for $d$-orbital participation, and this should caution organic chemists in particular against the indiscriminate use of pictorial explanations invoking $d$-orbitals. These may well look very nice on paper, but, as the work summarized in the above sections of this Report shows, they often have little scientific foundation. The most significant feature of Csizmadia, Rauk, and Wolfe's [19a–d] work is the demonstration of the importance of electron pair–electron pair or electron pair–polar bond interactions in deciding the structures of carbanions generated next to sulphur, sulphoxide, and sulphone.

*Thiophen*: ⟨_S_⟩. Preliminary communications of calculations on thiophen [22, 23, 31] have appeared. The results are shown in Tables 12 and 13.

**Table 12** *Gross atomic populations for thiophen*

|  | Gelius, Roos, and Siegbahn [23, 31] | | | | | Clark and Armstrong [22] | |
|---|---|---|---|---|---|---|---|
|  | 1[a] | 2 | 3 | 4 | | 5 | 6 |
| S | 15·518 | 15·509 | 15·984 | 15·999 |  | 15·616 | 15·773 |
|  |  |  |  |  | $\sigma$ | 13·936 | 14·019 |
|  |  |  |  |  | $\pi$ | 1·680 | 1·754 |
| C-1(C-4) | 6·568 | 6·444 | 6·245 | 6·108 |  | 6·396 | 6·302 |
|  |  |  |  |  | $\sigma$ | 5·290 | 5·228 |
|  |  |  |  |  | $\pi$ | 1·106 | 1·074 |
| C-2(C-3) | 6·176 | 6·056 | 6·244 | 6·119 |  | 6·249 | 6·242 |
|  |  |  |  |  | $\sigma$ | 5·195 | 5·193 |
|  |  |  |  |  | $\pi$ | 1·054 | 1·049 |
| H-1(H-4) | 0·733 | 0·853 | 0·751 | 0·877 |  | 0·769 | 0·784 |
| H-2(H-3) | 0·764 | 0·892 | 0·768 | 0·897 |  | 0·778 | 0·787 |

[a] Numbering corresponds to Tables 6 and 7.

The total energies have previously been commented on, and the discussion here will be centred on orbital energies and gross atomic populations. A full comparison can be made when the full papers are published. However, it is worthwhile at this stage making a few preliminary observations. The calculations without polarization functions (1, 5) show many common features. Thus the sulphur is overall electron-deficient and the carbons electron-rich, with the electron population on C-1(C-4) being greater than C-2(C-3). That this arises predominantly from the $\sigma$ system is evident from the $\sigma$- and $\pi$-electron populations (5). The hydrogens are electron-deficient, with H-1(H-4) carrying the greater positive charge. The electron populations on the hydrogens are in fact very similar for the two calculations (1, 5), the main difference being in the populations on the carbons. The main effect of introducing polarization functions on hydrogen is to

**Table 13** Calculated orbital energies (in a.u.) for thiophen

|      |      | 1[a]      | 2         | 3         | 4         | 5         | 6         |
|------|------|-----------|-----------|-----------|-----------|-----------|-----------|
|      | S1s  | −91.9813  | −91.9802  | −91.9749  | −91.9742  | −91.7793  | −91.8679  |
|      | C1s  | −11.2775  | −11.2785  | −11.2682  | −11.2690  | −11.4176  | −11.4098  |
|      | C1s  | −11.2526  | −11.2530  | −11.2473  | −11.2479  | −11.4169  | −11.4061  |
|      | S2s  | −8.9879   | −8.9870   | −8.9767   | −8.9762   | −8.9239   | −8.9360   |
|      | S2p  | −6.6785   | −6.6776   | −6.6682   | −6.6677   | −6.6907   | −6.7133   |
| A1σ  |      | −1.1983   | −1.1994   | −1.1833   | −1.1846   | −1.2589   | −1.2389   |
|      |      | −0.9974   | −0.9970   | −0.9862   | −0.9859   | −1.0523   | −1.0404   |
|      |      | −0.7670   | −0.7657   | −0.7634   | −0.7620   | −0.8103   | −0.8038   |
|      |      | −0.7087   | −0.7070   | −0.7030   | −0.7015   | −0.7501   | −0.7405   |
|      |      | −0.5455   | −0.5444   | −0.5412   | −0.5402   | −0.6081   | −0.6003   |
|      |      | −0.4710   | −0.4705   | −0.4755   | −0.4754   | −0.5189   | −0.5325   |
| A2π  |      | −0.3389   | −0.3377   | −0.3333   | −0.3322   | −0.4106   | −0.3981   |
|      | S2p  | −6.6749   | −6.6740   | −6.6656   | −6.6651   | −6.6880   | −6.7112   |
| B1π  |      | −0.5310   | −0.5303   | −0.5242   | −0.5237   | −0.5986   | −0.5864   |
|      |      | −0.3501   | −0.3496   | −0.3421   | −0.3418   | −0.4152   | −0.4056   |
|      | C1s  | −11.2776  | −11.2785  | −11.2683  | −11.2691  | −11.4178  | −11.4098  |
|      | C1s  | −11.2519  | −11.2523  | −11.2466  | −11.2471  | −11.4170  | −11.4062  |
|      | S2p  | −6.680    | −6.6791   | −6.6693   | −6.6688   | −6.6925   | −6.7140   |
| B2σ  |      | −1.001    | −1.001    | −0.9953   | −0.7949   | −1.0529   | −1.0420   |
|      |      | −0.7552   | −0.7539   | −0.7519   | −0.7504   | −0.7982   | −0.7937   |
|      |      | −0.5903   | −0.5899   | −0.5833   | −0.5830   | −0.6352   | −0.6248   |
|      |      | −0.5229   | −0.5223   | −0.5239   | −0.5234   | −0.5749   | −0.5758   |

[a] Numbering corresponds to Tables 6 and 7.

shift electron density from carbon towards the hydrogen. Comparing now (3) and (6), the calculations including $3d$ orbitals on sulphur, the overall trends are similar for the two calculations. Electron density is built up on sulphur mostly at the expense of C-1(C-4). This is much more marked in (3) than in (5). There is very little change in the populations on the hydrogens. These effects are also clear in comparing (2) and (4). The results in Table 12 should act further as a cautionary note to the indiscriminate use of electron densities in discussing reactivities of molecules, and emphasizes the fact that a Mulliken population analysis is a very crude guide to the actual electron distribution within a molecule and depends markedly on the basis set employed in the calculations. The calculated orbital energies are given in Table 13. An interesting difference between the calculations of the different workers arises for the C1$s$ core levels. The calculations by Gelius *et al.*[23, 31] predict a split of $\sim 0.7$ eV, whereas Clark and Armstrong's [22] calculations would indicate a negligible split. It will be of considerable interest, therefore, to investigate the C1$s$ levels of thiophen experimentally. The ordering of the first few energy levels is the same for all calculations, namely $a_{2\pi}$, $b_{1\pi}$, $a_{1\sigma}$, $b_{2\sigma}$, $b_{1\pi}$, $a_{1\sigma}$, $b_{2\sigma}$, $a_{1\sigma}$. Of particular note is the deep-lying $b_{1\pi}$ level, and this may be compared with the predictions based on semi-empirical treatments (see later). The He I lamp photoelectron spectrum of thiophen has been measured,[39, 40] and reveals all eight energy levels predicted by the calculations. The approximations inherent in Koopmans' theorem mean that in absolute magnitude the calculated orbital energies do not correlate particularly well with the experimentally measured ionization potentials. However, trends should be recognisable. In Figure 16 the lowest occupied orbital energies are plotted for the calculations, together with the experimental results. The correlation is surprisingly good, the trends and energy gaps being reproduced rather better by calculations (5) and (6) compared with the others.

*Molecular Core Binding Energies.* By implicitly treating all electrons in an *ab initio* molecular orbital method, not only are the energy levels of valence electrons computed but also the core levels. It is now known that the binding energy of a core electron is a sensitive function of the electronic environment of a given atom within a molecule.[41a, b] The exciting new field of $X$-ray photoelectron spectroscopy (ESCA)[41a, b] offers the possibility of direct measurement of molecular core binding energies, and in

---

[39] J. H. D. Eland, *Internat. J. Mass Spectrometry Ion Phys.*, 1969, **2**, 471.
[40] A. D. Baker and D. Betteridge, *Analyt. Chem.*, 1970, in press.
[41] [a] K. Siegbahn, C. Nordling, A. Fahlman, R. Nordberg, K. Hamrin, J. Hedman, G. Johansson, T. Bergmark, S. E. Karlsson, I. Lindgran, and B. Lindberg, 'ESCA Atomic, Molecular, and Solid State Structure Studied by Means of Electron Spectroscopy', Nova Acta Regiae Soc. Sci. Upsaliensis, Ser IV, Vol. 20, 1967; [b] K. Siegbahn, C. Nordling, G. Johansson, J. Hedman, P. F. Heden, K. Hamrin, U. Gelius, T. Bergmark, L. O. Werme, R. Manme, and Y. Baer, 'ESCA Applied to Free Molecules', North Holland, Amsterdam, 1969.

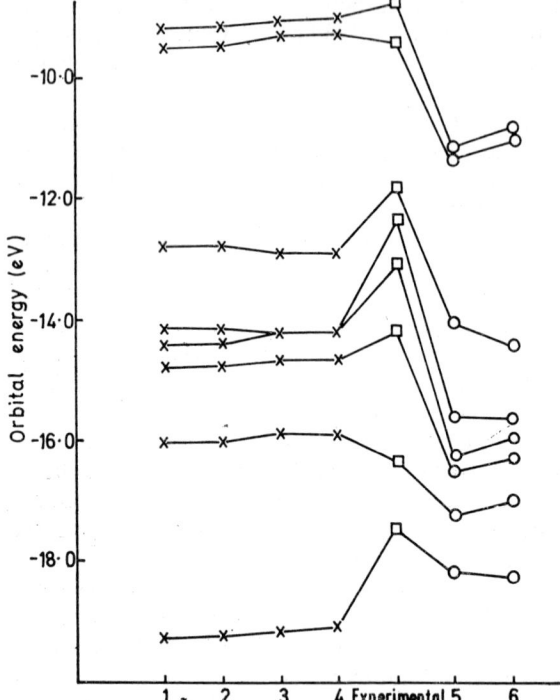

**Figure 16** (*Numbering corresponds to Tables 6 and 7*)

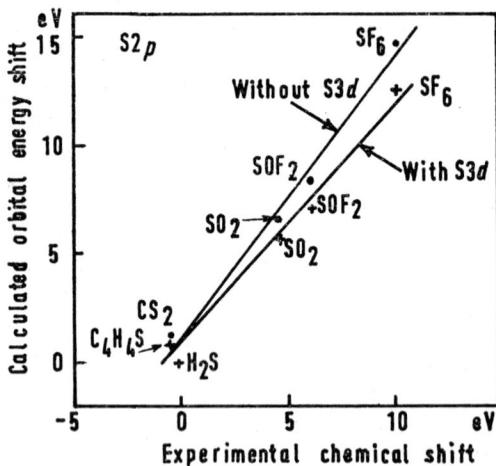

Figure 17

the past few years a considerable body of experimental measurements on both organic and inorganic sulphur compounds has been accumulated (mainly on the $2p_{\frac{1}{2}}$ $2p_{\frac{3}{2}}$ spin doublet).[41a, b, 42] It is of interest, therefore, to see how these energy levels compare with those calculated (using Koopmans' theorem).

Gelius, Roos, and Siegbahn [23] have studied $SO_2$, $SOF_2$, and $SF_6$ in addition to the organosulphur compounds previously mentioned in this report. The correlation between the calculated sulphur $2p$ orbital energy shift (with respect to hydrogen sulphide) and experimental shift is shown in Figure 17. It is interesting to note that the slope decreases from 1·29 to 1·09 on including $d$-orbitals largely because of the reduction in shift for $SOF_2$ and $SF_6$, where $d$-orbital participation is likely to be enhanced.

**Semi-empirical All-valence-electron Treatments.**—(*a*) *Introduction.* There have been very few semi-empirical all-valence-electron calculations published on organosulphur compounds.[43-46] This to some extent reflects the difficulties in parametrizing calculations on molecules containing second-row atoms. The calculations fall into two classes; extended Hückel treatments and SCF treatments in the CNDO approximation. For both types of calculation spectroscopic data are required for the one- and two-centre one-electron integrals. For the CNDO treatments spectroscopic data may also be required for the two-electron integrals, depending on the particular method of parametrization. The off-diagonal elements of the H (EHT) or F (CNDO) matrices also require the evaluation of overlap integrals over basis functions. In parametrizing semi-empirical treatments, therefore, there are at least two distinct major problems; of estimating one-electron core integrals from spectroscopic data, and of estimating orbital exponents for calculating overlap integrals (the basis sets used have been exclusively STO's). Papers relevant to this discussion but not specifically involving calculations on organosulphur compounds are given in references 47—51.

The one-centre one-electron core integral involving the $3s$ and $3p$ orbitals on sulphur may be evaluated from valence state ionization potentials, and there is a considerable measure of agreement in the values to be used in a

[42] B. J. Lindberg, K. Hamrin, G. Johansson, U. Gelius, A. Fahlman, C. Nordling, and K. Siegbahn, *Physica Scripta*, 1970, **1**, in the press.
[43] D. T. Clark, *Tetrahedron*, 1968, **24**, 2663.
[44] R. Gleiter and R. Hoffman, *Tetrahedron*, 1968, **24**, 5899.
[45] D. C. Owsley, G. K. Helmkamp, and M. F. Rettig, *J. Amer. Chem. Soc.*, 1969, **91**, 5239.
[46] T. Yonezawa, H. Konishi, and H. Kato, *Bull. Chem. Soc. Japan*, 1969, **42**, 933.
[47] D. W. J. Cruickshank, B. C. Webster, and M. A. Spinnler, *Internat. J. Quantum Chem.,* 1967, **1**, 225.
[48] P. G. Perkins and K. A. Levison, *Theor. Chim. Acta*, 1969, **14**, 206.
[49] W. W. Fogleman, D. J. Miller, H. B. Jonassen, and L. C. Cusachs, *Inorg. Chem.*, 1969, **8**, 1209.
[50] L. C. Cusachs, D. J. Miller, and C. W. McMurdy jun., *Spectroscopy Letters*, 1969, **2**, 141.
[51] R. D. Brown and J. B. Peel, *Austral. J. Chem.*, 1968, **21**, 2589.

particular case (Table 14). For the 3d orbitals, however, the difficulty arises in that the 3d orbitals of neutral sulphur atoms are formally unoccupied. This can be circumvented by making use of spectroscopic data on excited configurations, but the data are incomplete. However, even for the 3d and, for that matter, the 4s orbital on sulphur, there is a considerable measure of agreement in the literature as to the values to be used (Table 14).

**Table 14** Valence state ionization potentials used for sulphur (bicovalent) ($-I_\mu$ in eV)

| Reference | $3s$ | $3p$ | $3d^a$ | $4s^b$ | $4p^c$ |
|---|---|---|---|---|---|
| 43 | 20·77 | 11·98 | 2·00 | 3·75 | 2·40 |
| 44 | 20·0 | 13·3 | 4·0 | — | — |
| 45 | 20·08 | 13·32 | — | — | — |
| 47 | — | — | 1·8 | — | — |
| 48 | 20·52 | 10·78 | 3·06 | — | — |
| 50 | 21·29 | 12·00 | 2·00 | 3·66 | — |

$^a$ For process $3s^23p^33d^1 \rightarrow 3s^23p^3$.
$^b$ For process $3s^23p^34s^1 \rightarrow 3s^23p^3$.
$^c$ For process $3s^23p^34p^1 \rightarrow 3s^23p^3$.

Both EHT and CNDO SCF treatments depend markedly on the values assigned to overlap integrals, and hence on the exponents given to the Slater-type orbitals. Too little attention has been paid to this particular aspect in the literature. If Slater type orbitals are being used to obtain a reliable estimate of overlap integrals (particularly involving orbitals with principal quantum number $\geqslant 3$), then exponents must be selected to give a good fit to an atomic SCF wave-function at distances from the nucleus appropriate for bonding. In general, therefore, exponents defined by, say, Slater's rules are not adequate, since they are defined to give orbitals with good energies, and hence a good fit to an accurate wave-function, close to the nucleus. However, the fit at distances further from the nucleus may well be poor. This problem has been specifically investigated by Burns,[52] who has proposed a modified set of rules which define orbital exponents and which produce overlap integrals very similar to those obtained using atomic SCF orbitals. This aspect is particularly striking when the 3d orbitals are considered on sulphur. From Slater's rules, a 3d orbital exponent of zero is calculated for the configuration $3s^23p^33d^1$, whereas Burn's rules give an exponent of 1·0. On the basis of this, Clark[43] has shown that d-orbital participation on sulphur in thiophen cannot be excluded on the grounds of the overlap integrals being too small. In Table 15 are given some of the orbital exponents which have been used for sulphur.

(b) *CNDO SCF MO Calculations: Thiophen* (1), *2,3-Thiophyne* (2), *and 3,4-Thiophyne* (3). Since the original calculation on thiophen,[43] there has

[52] G. Burns, *J. Chem. Phys.*, 1964, **41**, 1521.

**Table 15** Orbital exponents (STO) used for sulphur (bicovalent)

| Reference | 3s | 3p | 3d | 4s | 4p |
|---|---|---|---|---|---|
| 43 | 1·967 | 1·817 | 1·0 | 0·65 | 0·325 |
| 44 | 2·122 | 1·827 | 0·8 | — | — |
| 46 | 2·122 | 1·827 | — | — | — |
| 48 | — | — | 1·0 | — | — |
| 50 | — | — | 1·70 | 1·35 | — |

been only one further investigation in the CNDO approximation. Yonezawa, Konishi, and Kato have carried out calculations on the ground and excited states of thiophen (1), 2,3-thiophyne (2), and 3,4-thiophyne (3). It is of some interest to compare the two calculations on thiophen and the *ab initio* results reported in the previous section.

(1)          (2)          (3)

The objectives of the two investigations were somewhat different in scope. Clark's investigation was primarily concerned with assessing the importance of *d*-orbital participation in the ground state of thiophen, using trends in calculated total energy, dipole moments, and $^1H-^1H$ and directly bonded $^{13}C-^1H$ coupling constants, whereas Yonezawa, *et al.* have been concerned with calculated electronic transitions in (1), (2), and (3) and the relative energies and geometries of the latter.

The methods of calculation are closely similar, the major difference being the retention [46] of one-centre exchange terms by Yonezawa *et al.*, which makes their calculations rather similar to those of the INDO [53] method. 3*d* Orbitals were not explicitly considered in the latter calculations, although both the non-empirical [22,23] and semi-empirical calculations indicate significant contribution of the 3*d* orbitals to some of the lower-lying virtual orbitals, implying they may be of some importance in discussions of excited states (*cf.* ref. 32).

Exponents defined by Burn's rules have been used by Clark,[43] and, as the previous discussion has indicated, in a semi-empirical treatment these are better for calculating overlap integrals than the best atom values used by Yonezawa *et al.*[46] The first terms in the off-diagonal elements ($\beta$'s) of the F matrix have been evaluated from the Wolfsberg–Helmholtz equation,[54] the one-centre two-electron integrals from spectroscopic data, and the two-centre two-electron integrals from the refined Mataga treatment.[55a,b] This differs considerably from the parametrization in the

---

[53] J. A. Pople, D. L. Beveridge, and P. A. Dobosh, *J. Chem. Phys.*, 1967, **47**, 2026.
[54] M. Wolfsberg and L. Helmholtz, *J. Chem. Phys.*, 1952, **20**, 837.
[55] [a] N. Mataga and K. Nishimoto, *Z. phys. Chem. (Frankfurt)*, 1957, **13**, 140; [b] K. Ohno, *Theor. Chim. Acta*, 1964, **2**, 219.

original CNDO/2 treatment but probably gives a better wave-function and allows a reasonable discussion of the excited states. One pitfall to be avoided, however, is in the calculation of the nuclear repulsion terms. Since the two-electron integrals calculated from spectroscopic data are much smaller in absolute magnitude than the theoretically calculated values, the electronic energy is much smaller for the modified CNDO treatments. If the nuclear repulsion energy is calculated, therefore, from the point charge approximation as for the original CNDO/2 treatment, then the total energy is dominated by this term and can give erroneous results when calculating energies between molecules. It is therefore better to calculate the nuclear repulsion energy from the two-centre two-electron integrals. (For a discussion see ref. 56.)

Considering first the calculated energies for thiophen, the estimated change in total valence electron energy on inclusion of $3d$, $4s$, $4p$ orbitals on sulphur in thiophen in the semi-empirical treatment [43] is in good agreement with the non-empirical calculations reported in the previous sections, and indicates that the parametrization is reasonable. Table 16 shows the

**Table 16** Thiophen: Semi-empirical all-valence-electron SCF calculated orbital energies (in eV)

| 1[a] | 2[b] |
|---|---|
| $-10.04\ a_{2\pi}$ | $-11.00\ a_{2\pi}$ |
| $-10.51\ b_{1\pi}$ | $-11.44\ b_{1\pi}$ |
| $-10.87\ a_{1\sigma(n)}$[c] | $-12.78\ a_{1\sigma(n)}$[c] |
| $-12.43\ b_{2\sigma}$ | $-14.11\ a_{1\sigma}$ |
| $-12.79\ a_{1\sigma}$ | $-14.13\ b_{2\sigma}$ |
| $-13.81\ b_{2\sigma}$ | $-14.91\ b_{1\pi}$ |
| $-18.81\ b_{1\pi}$ | $-15.43\ b_{2\sigma}$ |
| $-20.54\ a_{1\sigma}$ | $-17.47\ a_{1\sigma}$ |
| $-23.04\ a_{1\sigma}$ | $-20.75\ a_{1\sigma}$ |
| $-23.23\ b_{2\sigma}$ | $-22.34\ b_{2\sigma}$ |
| $-31.03\ a_{1\nu}$ | $-27.37\ a_{1\sigma}$ |
| $-34.71\ b_{2\sigma}$ | $-30.37\ b_{2\sigma}$ |
| $-48.34\ a_{1\sigma}$ | $-35.35\ a_{1\sigma}$ |

[a] Reference 43, unpublished data, basis set without $d$-orbitals.
[b] Reference 46.
[c] $n$ indicates that this orbital has considerable lone pair characteristics (on S).

calculated orbital energies for the two calculations. Comparing these with one another and with the non-empirical and experimental results, the following observations may be made. In the energy range $<21$ eV all calculations predict eight levels, in agreement with experiment. The ordering of the first three energy levels is the same for all calculations (including $3d$ orbitals on S in the CNDO treatment [43] reverses the assignment to $b_{1\pi}$, $a_{1\sigma}$, $a_{2\pi}$). A major difference arises, however, in the energy gap between the $\pi$- and highest-energy $\sigma$-orbital. The non-empirical calculations

[56] D. T. Clark and G. Smale, *Tetrahedron*, 1969, **25**, 13.

give a clear-cut gap of ~3 eV, whilst the semi-empirical treatments give much smaller differences of 0·36 eV [43] and 1·34 eV [46] respectively. All calculations indicate a deep-lying $\pi$-orbital, although there is disagreement between the non-empirical calculations (which all agree) on the one hand and the two semi-empirical calculations on the other as to the exact location of this level.

For 2,3- and 3,4-thiophynes, Yonezawa et al.[46] used an interesting procedure for estimating geometries. This depends on relationships established between theoretically calculated bond energies (C—C and C—S) and experimental bond lengths for a set of reference molecules. As initial guesses, the geometries of thiophen with hydrogen removed from the 2,3 or 3,4 positions were taken. At each iteration new estimates were made of the bond lengths until self-consistency was obtained. The calculated bond lengths are shown in Figure 18, and they seem entirely reasonable.

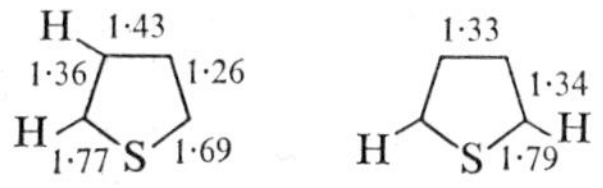

**Figure 18** *Bond lengths in Å*

The relative energies of (2) and (3) were not reported. However, the authors do comment that, in contrast with experiment,[57] (3) is predicted to be more stable than (2), and this almost certainly arises from the overemphasis of the nuclear repulsion terms due to the point charge approximation.

Because of their relevance to calculation in the previous and subsequent sections, it is worthwhile considering the orbital populations and charge distribution on thiophen. In a CNDO treatment we are effectively assuming that we are dealing with a basis set of orthogonalized (Lowdin) orbitals, and hence no problems arise concerning overlap densities, as in a non-empirical treatment. This assumption, however, has been severely questioned.[58] The orbital populations and hence charge distribution are, therefore, defined differently in a CNDO treatment than in a non-empirical treatment, and the two are not strictly comparable. Nonetheless, trends should be discernible common to both. Table 17 shows the orbital populations for thiophen. The overall trend in charge distribution is much closer to the non-empirical calculations for reference 43 than reference 46, a fact reflected in the much poorer dipole moment of 1·90 Debye [46] compared with experimental 0·55 and 0·90 Debye.[43] All calculations predict substantial migration of $\pi$-electrons from sulphur coupled with a $\sigma$-migra-

[57] G. Wittig, *Angew Chem.*, 1962, **74**, 479.
[58] D. B. Cook P. C. Hollis, and R. McWeeny, *Mol. Phys.*, 1967, **13**, 553.

tion in the reverse direction. This raises the question of a rigid non-polarizable σ-framework usually assumed in PPP SCF MO π calculations on organosulphur compounds. Attention is also drawn to the difference in the assignment of the site of highest π-electron density between the calculations: this should come as a salutary lesson to those unwise enough to use charge densities to discuss reactivities.

**Table 17** Thiophen: Semi-empirical all-valence-electron SCF calculated orbital populations

|  |  |  | 1[a] | 2[b] |
|---|---|---|---|---|
| H-1(H-4) |  | 1s | 0·966 | 0·974 |
| H-2(H-3) |  | 1s | 0·972 | 0·961 |
| C-1(C-4) |  | 2s | 1·017 | 1·248 |
|  |  | $2p_x$ | 0·950 | 0·835 |
|  |  | $2p_y$ | 1·006 | 0·788 |
|  |  | $2p_z$ | 1·085 | 1·045 |
| Total charge |  | σ | +0·027 | +0·129 |
|  |  | π | −0·085 | −0·045 |
| C-2(C-3) |  | 2s | 1·015 | 1·246 |
|  |  | $2p_x$ | 0·991 | 0·883 |
|  |  | $2p_y$ | 0·995 | 0·876 |
|  |  | $2p_z$ | 1·071 | 1·078 |
| Total charge |  | σ | −0·004 | −0·005 |
|  |  | π | −0·071 | −0·078 |
| S |  | 3s | 1·569 | 1·702 |
|  |  | $3p_x$ | 1·143 | 1·123 |
|  |  | $3p_y$ | 1·466 | 1·553 |
|  |  | $3p_z$ | 1·688 | 1·755 |
| Total charge |  | σ | −0·178 | −0·378 |
|  |  | π | +0·312 | +0·245 |

[a] Reference 43 calculation without d-orbitals.
[b] Reference 46.

The excited states of thiophen, 2,3- and 3,4-thiophynes have also been discussed.[46] The results are reproduced in Table 18. For thiophen, the lower energy transitions and the lowest triplet state are in good agreement with experiment. The first σ–π* transition in thiophen (roughly n on S–π*) is predicted to occur at 6·37 eV, and with the low oscillator strength will be masked by the much stronger π–π* transition at 6·39 eV. The two thiophynes are both predicted to have singlet ground states with relatively low-lying triplet states. In each case the first absorption band is predicted to be σ–σ*.

(c) *Extended Hückel Calculations*. (i) *Thiathiophthene*: As part of a general discussion on stabilization of centres with six valence electrons through electron-rich three-centre bonding, Gleiter and Hoffman[44] have carried out calculations on the thiathiophthene ring system.

**Table 18** Excited states of thiophen, 2,3-thiophyne, and 3,4-thiophyne [46]

| | Thiophen | | | | 2,3-Thiophyne | | | | 3,4-Thiophyne | |
|---|---|---|---|---|---|---|---|---|---|---|
| Transition | State | Energy | f | Transition | State | Energy | f | Transition | State | Energy | f |
| $\pi\rightarrow\pi^*$ | $^1B_2$ | 5.81 | 0.41 | $\sigma\rightarrow\sigma^*$ | $^1A'$ | 2.80 | 0.11 | $\sigma\rightarrow\sigma^*$ | $^1B_2$ | 2.91 | 0.18 |
| $\pi\rightarrow\pi^*$ | $^1A_1$ | 6.39 | 0.62 | $\pi\rightarrow\pi^*$ | $^1A'$ | 5.87 | 0.38 | $\pi\rightarrow\pi^*$ | $^1B_2$ | 5.54 | 0.49 |
| $\pi\rightarrow\pi^*$ | $^1A_1$ | 7.39 | 0.47 | $\pi\rightarrow\pi^*$ | $^1A'$ | 6.57 | 0.55 | $\pi\rightarrow\pi^*$ | $^1A_1$ | 6.30 | 0.51 |
| $\pi\rightarrow\pi^*$ | $^1B_2$ | 7.69 | 0.17 | $\sigma\rightarrow\sigma^*$ | $^1A''$ | 6.75 | 0.003 | $\sigma\rightarrow\sigma^*$ | $^1B_1$ | 6.19 | 0.005 |
| $\sigma\rightarrow\pi^*$ | $^1B_2$ | 6.37 | 0.008 | $\sigma\rightarrow\pi^*$ | $^3A'$ | 0.85 | 0 | $\sigma\rightarrow\sigma^*$ | $^3B_2$ | 0.86 | 0 |
| $\sigma\rightarrow\pi^*$ | $^1A_2$ | 8.58 | 0 | $\pi\rightarrow\pi^*$ | $^3A'$ | 3.99 | 0 | $\pi\rightarrow\pi^*$ | $^3B_2$ | 3.28 | 0 |
| $\pi\rightarrow\pi^*$ | $^3B_2$ | 3.81 | 0 | $\sigma\rightarrow\pi^*$ | $^3A''$ | 6.57 | 0 | $\sigma\rightarrow\pi^*$ | $^3B_1$ | 6.01 | 0 |
| $\pi\rightarrow\pi^*$ | $^3A_1$ | 4.84 | 0 | | | | | | | | |
| $\sigma\rightarrow\pi^*$ | $^3B_1$ | 6.18 | 0 | | | | | | | | |
| $\sigma\rightarrow\pi^*$ | $^3A_2$ | 8.45 | 0 | | | | | | | | |

Thiathiophthene

Starting from a structure with the $S^1$–$S^{6a}$ and $S^{6a}$–$S^6$ bond lengths equal, calculations were carried out both with and without *d*-orbitals on sulphur, corresponding to displacement of the central sulphur atom from its symmetrical location. The results are shown in Figure 19. For both sets of calculations an unsymmetrical structure is favoured, although it is not clear that the calculation without 3*d* orbitals will give a minimum. For the calculation including *d*-orbitals on sulphur, the results with a small

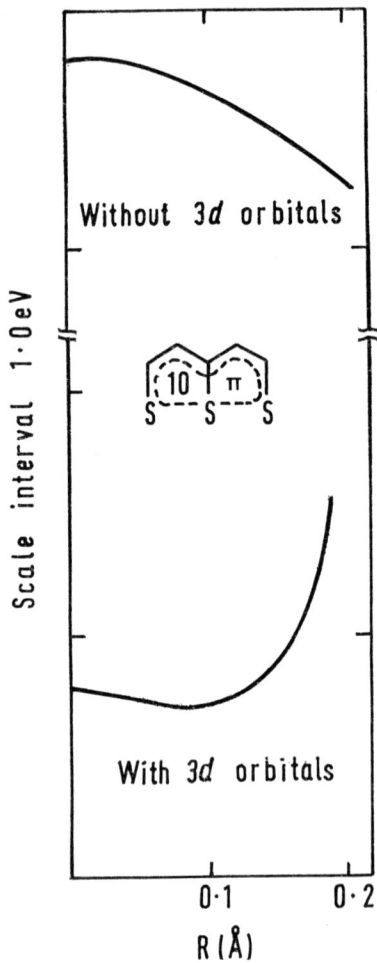

Figure 19

energetic preference for the unsymmetrical structure are in good agreement with the experimental data.[59] This should not be taken as evidence for the importance of $d$-orbital participation on sulphur, however, since in EHT calculated geometries are determined almost solely by the magnitude of overlap integrals (via the off-diagonal elements of the H matrix), and these will be sensitive functions of the orbital exponents. It is probably true to say, therefore, that by a judicious choice of parameters any desired answer could be produced.

(ii) *Episulphonium cation*: $\begin{bmatrix} CH_2 \diagdown \diagup CH_3 \\ | \quad\; S\!:\! \\ CH_2 \diagup \end{bmatrix}^+$ . Calculations have been carried

out [45] on the episulphonium cation and hypothetical precursor 1-chloro-1-methyl-1-thiacyclopropane. For the latter, two possible geometries about sulphur were considered; trigonal-bipyramidal (III) and square-pyramidal (IV).

$$\underset{(III)}{\overset{\overset{\displaystyle CH_3}{|}}{\underset{\underset{\displaystyle Cl}{|}}{\triangle\!S\!:}}} \qquad\qquad \underset{(IV)}{\triangleright\!\!\overset{\displaystyle \ddot{S}\!\diagdown\! Cl}{\diagdown\! CH_3}}$$

Calculations with and without $d$-orbitals on sulphur were carried out, and $d$-orbital participation found to be negligible. (IV) Was calculated to be some 12 eV lower in energy than (III).

(*d*) *Semi-empirical π-Electron Treatments.* There have been a considerable number of papers published of calculations in the PPP SCF MO formalism.[60-79] With one or two notable exceptions [63, 79] these have been

[59] E. Klingsberg, *Quart. Rev.*, 1969, **23**, 537.
[60] J. Fabian, *Z. Chem.*, 1968, **8**, 275.
[61] D. T. Clark, *J. Mol. Spectroscopy*, 1968, **26**, 181.
[62] J. Fabian, A. Mehlhorn, and R. Zahradnik, *J. Phys. Chem.*, 1968, **72**, 3975.
[63] D. T. Clark, *Tetrahedron*, 1968, **24**, 2567.
[64] J. Fabian, *Theor. Chim. Acta*, 1968, **12**, 200.
[65] J. Fabian, A. Mehlhorn, and R. Zahradnik, *Theor. Chim. Acta*, 1968, **12**, 247.
[66] P. Markov and P. N. Skanche, *Acta Chem. Scand.*, 1968, **22**, 2052.
[67] C. B. Choudhury and R. Basu, *J. Indian Chem. Soc.*, 1969, **46**, 779.
[68] J. Metzger, *Z. Chem.*, 1969, **9**, 99.
[69] J. Fabian, *Z. Chem.*, 1969, **9**, 272.
[70] J. Fabian and H. Hartmann, *Tetrahedron Letters*, 1969, 239.
[71] J. Fabian and H. Hartmann, *Chem. Ber.*, 1969, **73**, 107.
[72] J. Fabian and A. Mehlhorn, *Z. Chem.*, 1969, **9**, 271.
[73] J. Fabian and G. Laban, *Tetrahedron*, 1969, **25**, 1441.
[74] T. Zwkovic and N. Trinajstic, *Canad. J. Chem.*, 1969, **47**, 697.
[75] R. A. W. Johnstone and S. D. Ward, *Tetrahedron*, 1969, **25**, 5485.
[76] R. A. W. Johnstone and S. D. Ward, *Theor. Chim. Acta*, 1969, **14**, 420.
[77] M. Bossa, *J. Chem. Soc. (B)*, 1969, 1182.
[78] L. Paolini, M. Cignitti, and M. L. Tosato, 'Quantum Aspects of Heterocyclic Compounds in Chemistry and Biochemistry', Proceedings of the 2nd International Jerusalem Symposium, Israel Academy of Science and Humanities, Jerusalem, 1970, p. 324.
[79] M. J. S. Dewar and N. Trinajstic, *J. Amer. Chem. Soc.*, 1970, **92**, 1453.

almost exclusively directed towards understanding the electronic spectra associated with the $\pi$-electronic structure of organosulphur compounds. In this connection the work of Fabian and Zahradnik and their collaborators is of considerable importance since it systematizes a large amount of experimental data. This emphasizes perhaps the remaining role for PPP $\pi$-only SCF calculations, with the development of the more adequate all-valence-electron SCF MO treatments.

Table 19 lists the material covered in each paper, with an indication of the class of compound studied in each, since the numbers of individual compounds actually studied are too numerous to list individually.

Before discussing a few of the more important of these results, a few points are worth commenting on. The results of both the non-empirical all-electron and semi-empirical all-valence-electron calculations call into question the normal assumption in $\pi$-only treatments of organosulphur compounds, of a rigid non-polarizable $\sigma$ core. This is particularly true for molecules in which other heteroatoms are involved, e.g. thiazole. Also, the finding that there are relatively low-lying $\sigma$ levels, in thiophen for example, which in turn can give rise to low-lying excited states also needs to be borne in mind when considering the interpretation of electronic spectra in the PPP formalism. With intelligent parametrization, PPP SCF MO calculations can still be useful in calculating $\pi$-energy differences between molecules. However, properties such as $\pi$-electron distributions are much too sensitive to small changes in parameter values to be of any real value. In discussing reactivities, therefore, it is much safer to use calculated localization energies than charge distribution. It is instructive to consider the numerous calculations which have been published on the thienothiophens for example (Table 20).

Comparing the $\pi$-electron distributions should act as a warning to those unwise enough to use such numbers to interpret their experimental results. The one advantage, however, which the author is prepared to concede is that no matter what the experimental results may be, it should be possible to pick a particular set of numbers to 'explain' them.

It is clear from Table 19 that a small range of parameters ($U_C$, $U_S$, $\beta_{CC}$, $\beta_{CS}$, $\gamma_{CC}$, $\gamma_{SS}$, etc.) is sufficient to give an adequate systemization of a large number of different types of organosulphur $\pi$-systems. In semi-empirical PPP calculations the treatment of ground-state as opposed to excited-state properties requires different parametrization (mainly the $\beta$'s). There have been only two investigations specifically aimed at discussing ground state properties.[63,79] The extensive calculations of Dewar and Trinajstic.[79] covering heats of atomization, resonance energies, ionization potentials, and bond lengths of a wide variety of molecules, overlap considerably with the much earlier work of Clark.[63] It is instructive to compare the results for thiophen and the thienothiophens. The key compound in both investigations is thiophen, and parametrization of the calculations depends on empirical estimates of heats of atomization of the $\sigma$

**Table 19** PPP SCF MO calculations on organosulphur compounds

| Ref. | Compounds | Properties studied | One-electron integrals | Two-electron integrals |
|---|---|---|---|---|
| 60 | 2-Dimethylaminothiophen | Electronic spectrum, $\pi$-electron distribution | $U_C$ 11·42 eV, $U_S$ 20·00 eV<br>$U_N$ 22·50 eV<br>$\beta_{CC}$ 2·318 eV<br>$\beta_{CS}$ 1·623 eV<br>$\beta_{CN}$ 2·434 eV | $\gamma_{CC} = \gamma_{SS} = 10·84$ eV<br>$\gamma_{NN} = 14·50$ eV<br>Two-centre Mataga[a] |
| 61 | Thienothiophens | Electronic spectra, bond lengths change on excitation, $\pi$-electron distribution | $U_C$ 11·16 eV, $U_S$ 22·88 eV<br>$\beta_{CS}$ 1·45 eV<br>$\beta_{CC}$ 2·37 eV | $\gamma_{CC}$ 11·13 eV,<br>$\gamma_{SS}$ 11·90 eV<br>Two-centre uniformly charged sphere[b] |
| 62 | Alternant and non-alternant hydrocarbons in which CH=CH group has formally been replaced by sulphur, benzene, acenes, phenanthrene, benzphenanthrenes, tropylium, azulene, heptalene, cyclohepta[d,e]naphthalene, cyclopenta[c,f]heptalene, acenaphtho[1,2-j]fluoranthene | Electronic spectrum, ionization potentials, electron affinities, dipole moment, $\pi$-electron distribution | $U_C$ 11·42 eV, $U_S$ 20·0 eV<br>$\beta_{CS} = 0·7\beta_{CC}$<br>$\beta_{CC}$ 2·318 eV | $\gamma_{SS} = \gamma_{CC} = 10·84$ eV<br>Two-centre Mataga[a] |
| 63 | Thiophen, thienothiophens | $\pi$-Bonding energies, heats of formation, localization energies for electrophilic, nucleophilic, and free radical substitution | $U_C$ 11·16 eV, $U_S$ 22·88 eV<br>$\beta_{CS}$ 1·07 eV, $\beta_{CC}$ 1·75 eV | $\gamma_{SS}$ 11·90 eV,<br>$\gamma_{CC}$ 11·13 eV<br>Two-centre uniformly charged sphere[b] |
| 64 | Thiocarbonic acid derivatives | Electronic spectra | As in ref. 62 | |
| 65 | As in ref. 62 | Electronic spectra, iso-electronic O, N, Se compounds also studied | $U_C$ 11·42 eV, $U_S$ 20·27 eV<br>$\beta_{CC}$ various values<br>$\beta_{CS} = 0·7\beta_{CC}$ | $\gamma_{SS}$ 9·80 eV<br>$\gamma_{CC}$ 10·84 eV<br>Two-centre Mataga[a] |

**Table 19** (cont.)

| Ref. | Compounds | Properties studied | One-electron integrals | Two-electron integrals |
|---|---|---|---|---|
| 66 | 1,3,4-Thiadiazole, 1,2,5-thiadiazole | Charge densities, bond orders, ionization potentials, bond lengths | $U_C$ 11·54 eV, $U_S$ 22·30 eV<br>$U_N$ 14·32 eV<br>$\beta_{CC}$ 2·39 eV<br>$\beta_{CX} = (S_{CX}/S_{CC}).\beta_{CC}$ eV | $\gamma_{SS}$ 9·80 eV<br>$\gamma_{CC}$ 10·53 eV<br>$\gamma_{NN}$ 12·73 eV<br>Two-centre uniformly charged sphere[b] |
| 67 | Thiophen, thiazole, 1,2,5-thiadiazole, 1,3,4-thiadiazole | Electronic spectra | $U_C$ 11·54 eV, $U_S$ 22·30 eV<br>$U_N$ 14·63 eV, $\beta_{CC}$ 2·39 eV<br>$\beta_{CS}$ 2·14 eV, $\beta_{CN}$ 2·57 eV | $\gamma_{CC}$ 11·08 eV<br>$\gamma_{SS}$ 9·80 eV<br>$\gamma_{NN}$ 11·27 eV |
| 68 | Thiazole | Electronic spectrum, σ system also considered, electron distribution | | |
| 69 | Naphthothiopyran, naphtho[1,8-c,d]thiopyran | Electronic spectra, charge density, bond order diagrams for lowest energy singlet and triplet states | $U_C$ 11·42 eV, $U_S$ 20·0 eV<br>$\beta_{CS} = 0.7\ \beta_{CC}$<br>$\beta_{CC}$ 2·318 eV | $\gamma_{SS} = \gamma_{CC} = 10.84$ eV<br>Two-centre Mataga[a] |
| 70 | Thieno[b]thiapyrylium ions, thieno[b]tropylium ions | Electronic spectra, charge densities, bond orders | As in ref. 62 | |
| 71 | Thiophenol, thioanisole | Electronic spectra (also studied by MIM method) | $U_C$ 11·42 eV, $U_{SH}$ 21·00 eV<br>$U_{SCH_3}$ 20·40 eV<br>$\beta_{CC}$ 2·318 eV<br>$\beta_{CS} = 0.5\ \beta_{CC}$ | $\gamma_{SS} = \gamma_{CC} = 10.84$ eV<br>Two-centre Mataga[a] |
| 72 | Thioformaldehyde | Electronic spectrum $n \to \pi^*$ considered | | |
| 73 | 2H-Thiopyran-2-thiones, 4H-thiopyran-4-thione | Electronic spectra | As in ref. 62 | |

**Table 20** Sites of substitution predicted for electrophilic, nucleophilic, and free radical substitution in thienothiophens

| Molecule | Electrophilic | | Nucleophilic | | Free radical | |
|---|---|---|---|---|---|---|
| | Electron density | Localization energy | Electron density | Localization energy | Free valence | Localization energy |
| 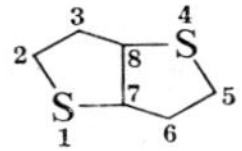 (structure 1) | 3(6) [80]<br>2(5) [63]<br>2(5) [79] | 2(5) [63] | 2(5) [80]<br>3(6) [63]<br>3(6) [79] | 2(5) [63] | 2(5) [80] | 2(5) [63] |
| 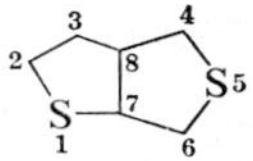 (structure 2) | 6 > 4 [80]<br>6 > 4 [63]<br>6 > 4 [79] | 4 > 6 [63] | 2 > 3 [80]<br>3 > 2 [63]<br>2 > 3 [79] | 4 > 6 [63] | 4 > 6 [80] | 4 > 6 [63] |
| 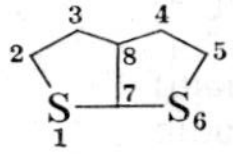 (structure 3) | 2(5) [80]<br>2(5) [63]<br>2(5) [79] | 2(5) [63] | 3(4) [80]<br>3(4) [63]<br>3(4) [79] | 2(5) [63] | 2(5) [80] | 3(4) [63] |

[80] N. Trinajstic and Z. Majerski, *Z. Naturforsch.*, 1967, **22a**, 1475.

Table 19 (cont.)

| Ref. | Compounds | Properties studied | One-electron integrals | Two-electron integrals |
|---|---|---|---|---|
| 74 | 2,2′-Bithienyl, thieno[3,2-b]thiophen | C—C, C—S bond lengths | | |
| 75 | Thiophen, benzo[b]thiophen, dibenzothiophen, thieno[2,3-b]thiophen, thieno[3,2-b]thiophen, trithione | Electronic spectra, ionization potentials | $U_S - U_C$ in range 8·0—18·0 eV<br>$\beta_{CS}$ 0·9—2·90<br>Best fit $U_S - U_C$ 8—10 eV<br>$\beta_{CC} = 2524 \exp(-5r)$ eV<br>$\beta_{CS}$ 1·2—1·5 eV | $\gamma_S$ 9·79, 11·9, 12·14<br>Two-centre Mataga[a] and uniformly charged sphere[b] |
| 76 | 6a-Thiathiophthenes | Electronic spectra, ionization potentials | $U_S - U_C$ 9—10·5 eV<br>$U_S - U_C$ 1·0—0·0 eV<br>$\beta_{SS}$ 0·4 eV<br>$\beta_{CS}$ 1·1 eV and 0·8 eV | As in ref. 75 |
| 77 | Xanthates, thiocarbonates | Electronic spectra, charge densities, and bond orders; VESCF also investigated | $U_S$ 23·74 eV, 23·34 eV<br>$U_S$ 12·70 eV, 11·83 eV<br>$U_C$ 11·16 eV, 11·73 eV<br>$\beta_{CS}$ 1·40 eV<br>$\beta_{CC}$ 6·442 exp(−5·686r) eV | $\gamma_{CC}$ 11·13 eV<br>$\gamma_{SS}$ 9·94 eV<br>Two-centre Mataga[a] |
| 78 | Isomers of trimethyl (N- and S-substituted) trithiocyanuric acid | Electronic spectra, relative π energies, dipole moments, charge densities, bond orders | $U_C$ 11·16 eV, $U_N$ 14·12 eV<br>$U_N$ 28·71 eV, $U_S$ 12·70 eV<br>$U_S$ 23·74 eV | $\gamma_{CC}$ 11·13 eV<br>$\gamma_N$ 12·34 eV, 16·75 eV<br>$\gamma_S$ 9·94 eV, 12·09 eV |
| 79 | Thiophen, thiophenol, benzothiophens, thienothiophens, dibenzothiophen, bithienyls, thiepin, benzo[d]thiepin, | Heats of atomization, ionization potentials, resonance energies, bond lengths, electron densities | $U_S$ 22·88 eV, $U_C$ 11·16 eV<br>$\beta_{CC}$ 6·927$S_{CC}$<br>$\beta_{CS}$ 15·7265$S_{CS}$ | $\gamma_{SS}$ 11·90 eV<br>$\gamma_{CC}$ 11·13 eV<br>Two-centre Mataga[a] |

framework which are in striking agreement for the two calculations. The relative heats of atomization for the thienothiophens are given in Table 21.

**Table 21** Calculated differences in $\pi$-bonding energies for the isomeric thienothiophens

| Molecule | Relative $\pi$-bonding energies (in eV) | |
|---|---|---|
| | Ref. 63 | Ref. 79 |
| Thieno[3,2-b]thiophen | (0·0) | (0·0) |
| Thieno[2,3-b]thiophen | 0·054 | 0·035 |
| Thieno[3,4-b]thiophen | 0·313 | 0·236 |
| Thieno[3,4-c]thiophen | 2·00[a] | 1·961 |

[a] For singlet state, predicted to be 0·11 eV higher in energy than triplet ground state.

Again the results are in good agreement. For thieno[3,4-c]thiophen, the more extensive calculations of Clark[63] indicate a triplet ground state, although the singlet–triplet separation is small, ~0·11 eV. A dimethyl derivative has been shown[81a] to have a transitory existence and undergoes facile addition reactions. (The tetraphenyl-substituted compound is somewhat more stable.[81b])

Localization energies for electrophilic, nucleophilic, and free radical substitution in the thienothiophens have also been calculated[63] and are in satisfactory agreement with experiment. Resonance energies for the compounds studied by Dewar and Trinajstic[79] are given in Table 22, and

**Table 22** Calculated resonance energies for organosulphur $\pi$-electron systems[79]

| Compound | Resonance energy (kcal mol$^{-1}$) |
|---|---|
| Benzene | 22·6 |
| Thiophen | 6·5 |
| Thiophenol | 22·1 |
| Benzo[b]thiophen | 24·8 |
| Benzo[c]thiophen | 9·3 |
| Thieno[3,2-b]thiophen | 11·3 |
| Thieno[2,3-b]thiophen | 10·5 |
| Thieno[3,4-b]thiophen | 5·9 |
| Thieno[3,4-c]thiophen | −33·9 |
| Dibenzothiophen | 44·6 |
| 2,2′-Bithienyl | 12·7 |
| 2,3′-Bithienyl | 12·1 |
| 3,3′-Bithienyl | 11·0 |
| Thiepin | −1·45 |
| Benzo[d]thiepin | 19·9 |
| Thieno[c,d]thiepin | 3·5 |

[81] [a] M. P. Cava and N. M. Pollack, *J. Amer. Chem. Soc.*, 1967, **89**, 3639; [b] M. P. Cava and G. E. M. Husbands, *J. Amer. Chem. Soc.*, 1969, **91**, 3952.

for comparison the value calculated for benzene is also included. A few points are worth considering in more detail. The calculated resonance energies of the isomeric benzothiophens differ considerably and reflect the difference in stability of the two systems. For the bithienyls, little interaction is predicted between the rings, the calculated resonance energies being roughly twice that for thiophen. The results for thiepin, benzo[*d*]-thiepin, and thieno[*cd*]thiepin show that the molecules are not likely to be aromatic.

## 4 Semi-empirical PPP SCF MO Calculations on Organoselenium Compounds

The theoretical treatment of organoselenium compounds is still in its infancy, and the limited work so far published has concerned the interpretation of electronic spectra in the PPP formalism. Figure 20 indicates the

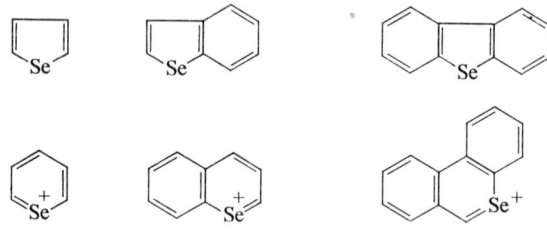

**Figure 20**

compounds so far studied by Fabian, Zahradnik, and co-workers.[65] The final parameters for selenium ($U_{Se}$ 19·17 eV, $\gamma_{SeSe}$ 9·29 eV, $\beta_{CSe} = 0.6 \beta_{CC}$) give a surprisingly good fit to the experimental spectra, and would suggest that it should be possible to systematize the electronic spectra of organoselenium $\pi$-systems in much the same manner as has proved possible with organosulphur compounds. A detailed understanding of the electronic structure of organoselenium compounds in general, however, would still appear to be someways off.

# 2
# Aliphatic Organo-sulphur Compounds, Compounds with Exocyclic Sulphur Functional Groups, and Their Selenium and Tellurium Analogues

## 1 General

This Chapter, based on more than 500 papers published mostly within a period of twelve months, discloses few examples of novel sulphur-containing functional groups; indeed, most of the preparative methods described in a 1943 review [1] remain standard procedure. However, no single year's literature is merely a précis of work of preceding years, and advances over a broad front have been made in the period under review.

Recent text-book coverage includes sulphonation,[2] and information [3] on representative syntheses of thiocyanates, isothiocyanates, thiols, sulphides, disulphides, sulphoxides, sulphones, and oxy-sulphur acids. The Proceedings of two 1966 Conferences [4,5] are valuable source material; 'Organic Chemistry of Bivalent Sulphur' has reached six volumes,[6] and Mechanisms of Reactions of Sulfur Compounds,[7] Chemistry of Organic Sulphur Compounds,[8] and *Quarterly Reports on Sulfur Chemistry* [9] are continuing series.

Volume I of *Intra-Science Chemistry Reports* [10] comprises a bibliography of reviews on sulphur chemistry for 1961—67;[10a] selected studies of

[1] R. Connor, in 'Organic Chemistry: an Advanced Treatise', ed. H. Gilman, 2nd edn., J. Wiley & Sons, New York, vol. I, 1943, pp. 835—943.
[2] 'Mechanistic Aspects in Aromatic Sulphonation and Desulphonation', H. Cerfontain, Interscience, New York, 1968.
[3] 'Organic Functional Group Preparations', S. R. Sandler and W. Karo, Academic Press, New York, 1968.
[4] 'Chemistry of Sulphides', ed. A. V. Tobolsky, Interscience, New York, 1968.
[5] 'Organo-sulphur Chemistry', ed. M. J. Janssen, Interscience, New York, 1967.
[6] E. E. Reid, Chemical Publishing Co., Inc., New York, vol. VI, 1966.
[7] Ed. N. Kharasch, B. S. Thyagarajan, and A. I. Khodair, Intra-Science Research Foundation, Santa Monica, California; vol. I, 1966; vol. II, 1968; vol. III, in the press (*Tetrahedron Letters*, 1970, 31 refers.).
[8] Ed. N. Kharasch and C. Meyers, Pergamon Press, Oxford, vol. II, 1966.
[9] Intra-Science Research Foundation, Santa Monica, California; vol. IV, 1969.
[10] *Intra-Sci. Chem. Reports*, 1967, **1**, No. 4, [a] 297, J. L. Day; [b] 265, A. I. Khodair; [c] 225, F. A. Billig; [d] 179, B. S. Thyagarajan and N. Kharasch; [e] 337, N. Kharasch and Z. S. Ariyan.

reaction mechanisms of organic sulphur compounds;[10b] i.r. spectra of sulphur compounds reported 1965—67;[10c] chemistry of dimethyl sulphoxide 1966—67;[10d] review of sulphenyl halides and chlorodithio-compounds.[10e] Other reviews include: the literature of the last five years on the preparation of sulphonic acids, sulphonyl halides, and alkyl sulphonates;[11] polyhalogeno-alkyl derivatives of sulphur;[12] i.r. spectra of sulphinyl compounds;[13] the chemistry of acyl and sulphonyl isothiocyanates and isoselenocyanates;[14] nucleophilic, electrophilic, and radical substitution reactions of sulphur compounds;[15] nucleophilic substitution at sulphur;[16] an integrated treatment of sulphoxides, sulphones, sulphinic acids, and sulphonic acids;[17] acid-catalysed rearrangements of sulphoxides;[18a] rearrangements of α-halogenosulphones.[18b]

Sulphur compounds are included in reviews of radicals;[19,20] recent work by Modena's research school at the University of Bari, on the oxidation of sulphides and reactions of sulphenyl chlorides,[21] and by Montanari's group,[22] on α-sulphonyl carbanions, neighbouring group participation by sulphoxide oxygen, asymmetric synthesis, absolute configuration, and o.r.d. of sulphoxides, has been reviewed briefly.

Selenium and tellurium may be estimated in plant samples at low concentrations, Te by neutron activation analysis [23] down to 350 parts per billion, and Se by neutron activation analysis ($\pm 6\%$ at 3 $\mu$g level [24]), at $\geqslant 0.005$ p.p.m.,[25] or by fluorimetry.[26]

## 2 Thiols

**Preparation.**—Addition of $H_2S$ to methyl α-chloroacrylate and the corresponding nitrile, in the presence of $Me_3N$, gives [27] β-mercaptopropionic acid derivatives, together with corresponding di- and tri-sulphides and

[11] E. E. Gilbert, *Synthesis*, 1969, **1**, 3.
[12] R. D. Dresdner and T. R. Hooper, *Fluorine Chem. Rev.*, 1969, **4**, 1.
[13] R. Steudel, *Z. Naturforsch.*, 1970, **25b**, 156.
[14] M. O. Lozinskii, *Russ. Chem. Rev.*, 1968, 363.
[15] A. Fava, *Stereochim. Inorg., Accad. Naz. Lincei, Corso Estivo Chim. 9th*, 1965, Accad. Naz. Lincei, Rome, 1967, p. 333.
[16] E. Ciuffarin and A. Fava, *Progr. Phys. Org. Chem.*, 1968, **6**, 81.
[17] Y. A. Kolesnik, *Russ. Chem. Rev.*, 1968, 519.
[18] [a] L. A. Paquette, in 'Mechanisms of Molecular Migrations', ed. B. S. Thyagarajan, vol. I, 1968, pp. 121—156; [b] G. A. Russell and G. J. Mikol, pp. 157—207.
[19] 'Organic Chemistry of Stable Free Radicals', A. R. Forrester, J. M. Hay, and R. H. Thomson, Academic Press, London, 1968.
[20] 'Radical Ions', ed. E. T. Kaiser and L. Kevan, Interscience, New York, 1968.
[21] G. Modena, *Corsi Semin. Chim.*, 1968, 104.
[22] F. Montanari, *Corsi Semin. Chim.*, 1968, 112.
[23] A. Abu-Samra and G. W. Leddicotte, *Nat. Bur. Stand. (U.S.A.), Spec. Publ.*, 1969, No. 312, p. 134.
[24] J. P. F. Lambert, O. Levander, L. Argrett, and R. E. Simpson, *J. Assoc. Offic. Analyt. Chemists*, 1969, **52**, 915; J. Leibetseder, *Chem. Abs.*, 1969, **71**, 109,711 (*Trace Miner. Stud. Isotop. Dom. Anim., Proc. Panel*, 1968, *Int. At. Energy Agency*, 1969, 77).
[25] R. J. Hall and P. L. Gupta, *Analyst*, 1969, **94**, 292.
[26] L. Karelina and R. Salmane, *Opred. Mikroelem. Biol. Obektakh*, 1968, 151 (*Chem. Abs.*, 1969, **71**, 109,741).
[27] W. H. Mueller, *J. Org. Chem.*, 1969, **34**, 2955.

sulphur. The reaction of benzylidene-acetone, $H_2S$, and $NH_3$ has been reinvestigated;[28] basic products are two stereoisomers of 1,5-dimethyl-3,7-diphenyl-9-aza-2,6-dithiabicyclo[3,3,1]nonane (1), rather than bis(α-mercapto-α-methylcinnamyl)amine stereoisomers (2) as earlier claimed,[29]

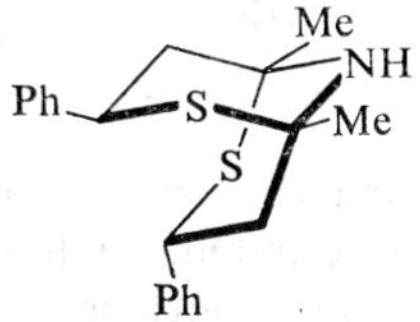

and neutral products are 4-mercapto-4-phenylbutan-2-one and 5-acetyl-2,6-diphenyl-4-hydroxy-4-methyltetrahydrothiapyran (3). Dibenzylideneacetone gives [28] a dispirotrithiolan, via a tetrahydro-1-thiapyranone, with $H_2S$ and $NH_3$. D-Glucose gives 1-thio-D-glucitol by treatment with $H_2S$ in aqueous solution at 150 °C, followed by reduction with zinc;[30] thiol analogues of monosaccharides are more usually prepared by nucleophilic displacement reactions [31, 32] or addition [33] of thiolate or thioacylate anions, followed by reductive or hydrolytic cleavage to unmask the thiol function. Protected O-(toluene-p-sulphonyl)-L-serine (4) gives [34] L-selenocysteine derivatives (5) with $SeH^-$, supplementing an earlier study leading to the

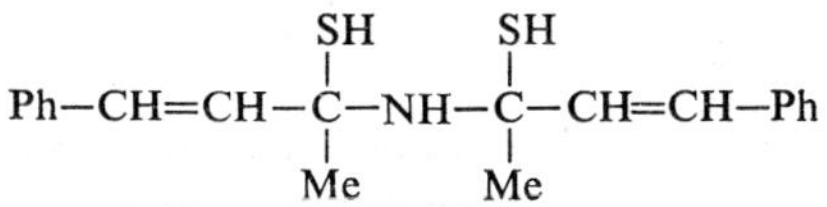

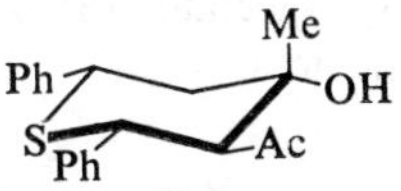

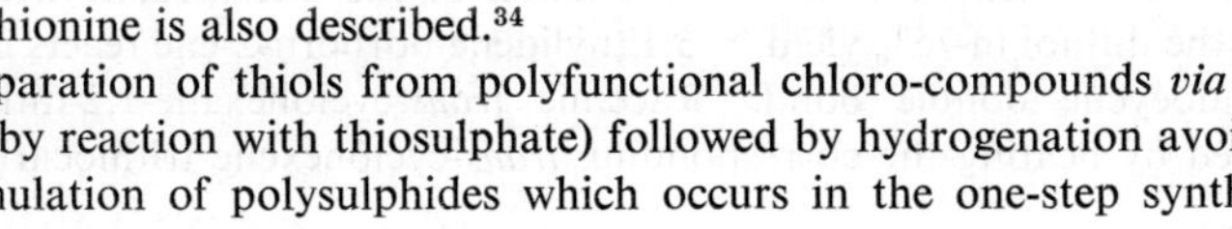

Se-benzyl analogue; conversion of (5) into Se analogues of L-cystine and L-lanthionine is also described.[34]

Preparation of thiols from polyfunctional chloro-compounds via Bunté salts (by reaction with thiosulphate) followed by hydrogenation avoids the accumulation of polysulphides which occurs in the one-step synthesis.[35]

---

[28] G. C. Forward and D. A. Whiting, J. Chem. Soc. (C), 1969, 1647.
[29] E. Fromm and F. Haas, Annalen, 1912, 394, 290.
[30] A. R. Procter and R. H. Wiekenkamp, Carbohydrate Res., 1969, 10, 459.
[31] J. M. Heap and L. N. Owen, J. Chem. Soc. (C), 1970, 707, 712.
[32] R. L. Whistler, U. G. Nayak, and A. W. Perkins, J. Org. Chem., 1970, 35, 519.
[33] R. L. Whistler and R. E. Pyler, Carbohydrate Res., 1970, 12, 201.
[34] J. Roy, W. Gordon, I. L. Schwartz, and R. Walter, J. Org. Chem., 1970, 35, 510.
[35] R. T. Wragg, J. Chem. Soc. (C), 1969, 2582.

The well-known thiourea alkylation procedure for the synthesis of thiols has been extended in novel ways,[36–39] e.g.[36] (6) → (7); o-chloroaniline is

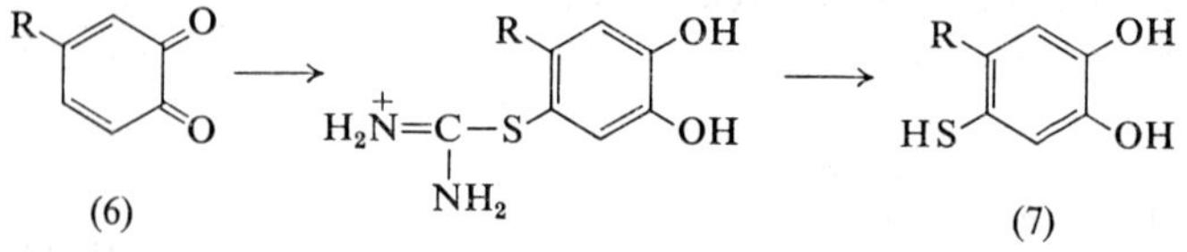

converted into o-chlorothiophenol, the key step being reaction of the derived diazonium tetrafluoroborate with thiourea.[37] Pyrrole and indole thiols [38, 39] are obtainable from the heterocycles themselves via thiourea substitution products, corresponding thiazole-5-thiols being accessible [40] through hydrolysis of corresponding S-acetates (8) obtained by condensation of α-amino-acids with thioacetic acid.

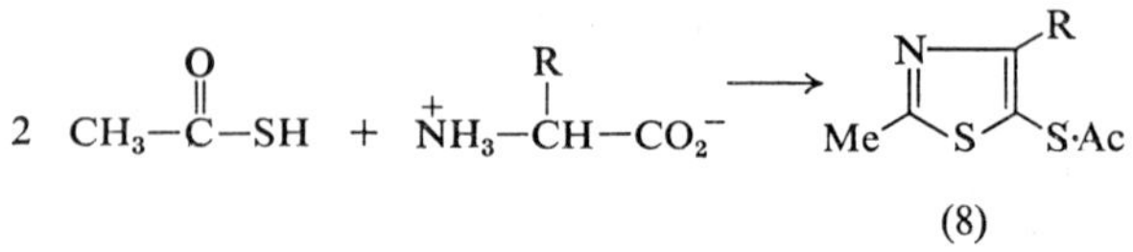

Free-radical addition of $H_2S$, Me·SH, Ac·SH, or Ph·$CH_2$·SH to 1-chloro-, 2-chloro-, 1-methyl-, or 2-methyl-4-t-butylcyclohex-1-ene, initiated by u.v.-irradiation or by added azo-bisisobutyronitrile, gives predominantly trans-diaxial adducts, conformational factors apparently favouring axial attack by a thiyl radical.[41] It is suggested [41] that additions involving thiyl radicals are less stereoselective than, e.g., additions of HBr, since the radical addition step with R·S· may be reversed more easily, reversibility being greatest for $H_2S$ addition.

Substituted phenyl-lithium compounds from 1,3,5-triadamantylbenzene give 2-butyl-thio-compounds with equivalents of sulphur and BuBr;[42] and symmetrical disulphides with two equivalents of sulphur. The thiols obtained on reductive cleavage yield radicals with $PbO_2$ and become violet in colour on u.v.-irradiation in the solid state.[42]

cis-Dithiols result from treatment of norborn-2-ene (9) with sulphur and ammonia in dimethylformamide, reduction of the intermediate trithiolan giving the dithiol in 75% yield.[43] 5-Ethylidene norborn-2-ene reacts only at the endocyclic double bond. Racemic trans-cyclohexane-1,2-dithiol is obtained by heating the corresponding trans-cyclohexene trithiocarbonate

[36] J. Daneke, U. Jahnke, B. Pankow, and H.-W. Wanzlick, Tetrahedron Letters, 1970, 1271.
[37] B. V. Kopylova and M. N. Khasanova, Izvest. Akad. Nauk S.S.S.R., Ser. khim., 1969, 2619.
[38] R. L. N. Harris, Tetrahedron Letters, 1969, 4465.
[39] R. L. N. Harris, Tetrahedron Letters, 1968, 4045.
[40] G. C. Barrett, A. R. Khokhar, and J. R. Chapman, Chem. Comm., 1969, 818.
[41] N. A. LeBel, R. F. Czaja, and A. DeBoer, J. Org. Chem., 1969, 34, 3112.
[42] W. Rundel, Chem. Ber., 1969, 102, 1649.
[43] T. C. Shields and A. N. Kurtz, J. Amer. Chem. Soc., 1969, 91, 5415.

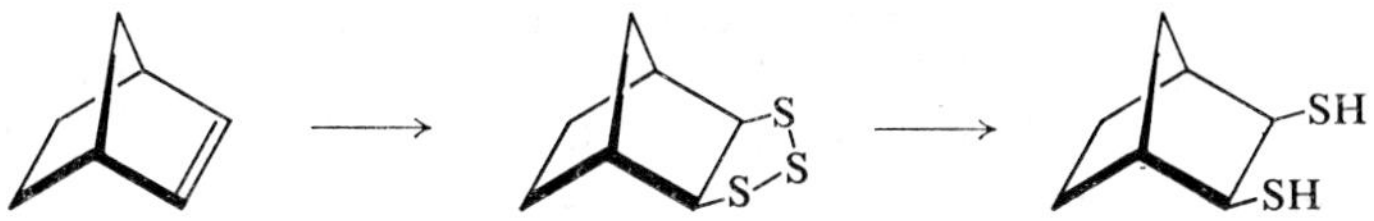

(9) → → 

with one equivalent of 2-aminoethanol at 80—120 °C.[44] Thiols are obtained similarly from dithiolcarbonates, xanthates, and thiol esters,[44] this general route avoiding oxidation and polymerisation side-reactions (see also ref. 45).

Ring-opening of (10) with sodium sulphide in dimethyl sulphoxide gives a deep red solution containing the dithiolate anion (11);[46] dilution with

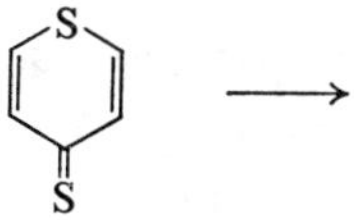

(10) ⟶ $^-$S—CH=CH—CS—CH=CH—S$^-$
(11)

water reverses the reaction. 2-Phenyl-4$H$-thiapyran-4-thione gives [46] HO·CH:CH·CS·CH:C(Ph)·S$^-$ with NaOH in aqueous dimethylformamide. Ph·C⋮C·SH (⇌ Ph·CH:C:S) is evidently formed as an intermediate in the reaction of the corresponding methyl sulphide with secondary amines, to give thioamides Ph·CH$_2$·CS·NR$_2$, since 2-methylthio-3,4-diphenylthiophen is a minor product.[47] 1-Alkenethiolates are formed from 1-alkenyl alkyl sulphides by their treatment with an alkali-metal in ammonia,[48] giving di(1-alkenyl)disulphides on I$_2$ oxidation and 1-alkenyl alkyl disulphides in good yield by reaction with R·SO$_2$·S·R or R·SCN. 1-Alkenyl- or -alkynyl thiolates and selenates, prepared similarly (Li in NH$_3$), may be $S$- or $Se$-alkylated with an alkyl bromide.[49]

**Addition of Thiols to Olefins, Carbonyl Compounds, and Epoxides.**—Conversion of $N$-$\beta$-(3-indolyl)acryloylimidazole (12) into the corresponding thiolester by displacement of imidazole by cysteamine or $NN$-dimethyl-

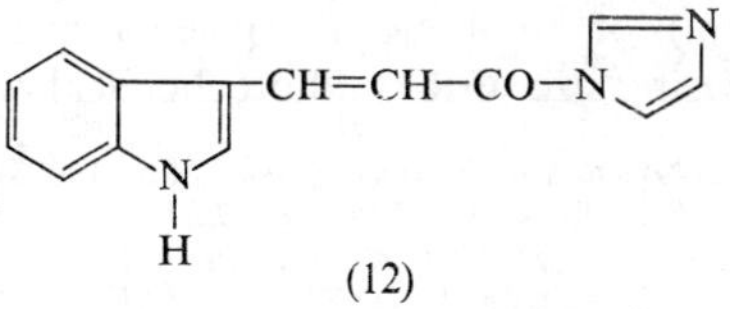

(12)

---

[44] T. Taguchi, Y. Kiyoshima, O. Komori, and M. Mori, *Tetrahedron Letters*, 1969, 3631.
[45] K. Mori and Y. Nakamura, *J. Org. Chem.*, 1969, **34**, 4170.
[46] J. G. Dingwall, D. H. Reid, and J. D. Symon, *Chem. Comm.*, 1969, 466.
[47] M. L. Petrov, B. S. Kupin, and A. A. Petrov, *Zhur. org. Khim.*, 1969, **5**, 1759.
[48] H. E. Wijers, H. Boelens, A. Van der Gen, and L. Brandsma, *Rec. Trav. chim.*, 1969, **88**, 519.
[49] L. Brandsma and P. J. W. Schuijl, *Rec. Trav. chim.*, 1969, **88**, 513.

cysteamine conceals a reaction sequence (conjugate addition of $R \cdot S^-$, second nucleophilic addition to CO followed by $\beta$-elimination) which is not followed by simple O- or N-nucleophiles; these add at the carbonyl group.[50] Ring-opening of 1,2-epoxy-2-methylbut-3-ene (13) by thiols gives products dependent upon reaction conditions, free-radical conditions resulting in 60% 1,4-addition and giving also 40% (14).[51]

$$CH_2=CH-\underset{SR}{\underset{|}{\overset{Me}{\overset{|}{C}}}}-CH_2OH$$

$BF_3$ / excess R·SH

$$CH_2=CH-\underset{O}{\overset{Me}{\underset{\diagdown\diagup}{C}}}-CH_2 \xrightarrow{R \cdot SH \text{ or } R \cdot S^- Na^+} CH_2=CH-\underset{OH}{\underset{|}{\overset{Me}{\overset{|}{C}}}}-CH_2-SR$$

(13)  (14)

Other epoxide-opening reactions reported include the conversion of ethyl 3,4-anhydro-$\beta$-L-ribopyranoside into ethyl S-benzyl-4-thio-$\alpha$-D-lyxopyranoside by $Ph \cdot CH_2 \cdot S^- Na^+$ ($-20\ °C$);[52] 1,2-epoxyalkenes with $Pr \cdot SeH$ or $Ph \cdot CO \cdot SeH$;[53] oxepin (15) with $Na_2S,H_2O$ (0 °C; 45 min);[54] and epoxyalkanes with $H_2S$.[55, 56]

(15)  (16)

Photo-addition of Me·SH to acetylene and to butadiene [57] gives results consistent with the previously postulated mechanism for olefins, except that the rate of dehydrogenation of thiol by the primary adduct radical is comparable with the rate of radical decomposition in the case of acetylene. Additions of simple thiols to activated olefins [58] and to conjugated di-ynes [59] (1,4-addition gives $R^1S \cdot CR^2 : CH \cdot CH : CR^3 \cdot SR$) have been studied; photolysis (253·7 nm) of thiols R·SH gives radicals whose reactivity increases with increasing size of R.[60] Photochemical addition of Ph·SH to

[50] M. F. Dunn and S. A. Bernhard, *J. Amer. Chem. Soc.*, 1969, **91**, 3274.
[51] G. I. Zaitseva and V. M. Albitskaya, *Zhur. org. Khim.*, 1969, **5**, 612.
[52] J. P. H. Verheyden and J. G. Moffatt, *J. Org. Chem.*, 1969, **34**, 2643.
[53] L. A. Khazemova and V. M. Albitskaya, *Zhur. org. Khim.*, 1969, **5**, 1926.
[54] R. M. De Marinis and G. A. Berchtold, *J. Amer. Chem. Soc.*, 1969, **91**, 6525.
[55] W. Umbach, R. Mehren, and W. Stein, *Fette, Seifen, Anstrichm.*, 1969, **71**, 199.
[56] E. Tobler, *Ind. and Eng. Chem. (Product Res. and Development)*, 1969, **8**, 415.
[57] D. M. Graham and J. F. Soltys, *Canad. J. Chem.*, 1969, **47**, 2529.
[58] P. Bravo, G. Gaudiano, and T. Salvatori, *Gazzetta*, 1968, **98**, 1046.
[59] W. Schroth, F. Billig, and A. Zschunke, *Z. Chem.*, 1969, **9**, 184.
[60] B. G. Dzantiev, A. V. Shishkov, and M. S. Unukovich, *Khim. vysok. Energii*, 1969, **3**, 111.

norborn-2-ene at 0 °C gives 99·5% *exo*-2-norbornyl phenyl thioether;[61] and the 7,7-dimethyl analogue (*apo*-born-2-ene), although slower reacting, also gives predominantly (95%) *exo*-addition; this is against steric reasoning, and, although a two-stage addition mechanism may explain the result, further investigation is promised.[61] The predominance of *exo* thiol addition was established earlier [62] in this series; *S*-deuteriobenzenethiol gives the *exo-cis*-adduct (17) with norborn-2-ene under free-radical conditions,[63] indicating that chain transfer of intermediate radicals takes place from the *exo*-direction. Additional data are presented [63] for Ph·SH, Bu$^t$·SH, Me·SH, and Ac·SH as addenda; with 2-methylnorborn-2-ene, Ph·S· attacks the more accessible 3-position, leading to the more highly-substituted intermediate radical.[63] Radical addition [azo-bisisobutyronitrile (15 hr; 75—80 °C)] of Bu$^t$·SH to hexamethylnorborna-2,5-diene (18) gives mainly (51%) the 5-*endo*-adduct (19), together with 45% rearranged product (20) and only 4% 5-*exo*-adduct.[64]

(17)   (18)   (19)   (20)

Enhanced ring strain in norbornenes compared with bicyclo[2,2,2]octenes is shown [65] in the greater reactivity of the former towards thiol addition; whereas thiols give 1,2-adducts with α- or β-pinene, addition of other free radical species is accompanied by cyclobutane ring-opening;[66] thuj-4(10)-ene, however, suffers ring-opening with thiyl radicals, determined by the slower chain transfer by the first-formed intermediate radical in this case.[66]

Allyl alcohol with Bu$^t$·SH gives Bu$^t$·S·CH(CH$_3$)·CH$_2$OH in the presence of sulphur, though under free radical conditions, Bu$^t$·S·(CH$_2$)$_3$·OH is obtained;[67] i.r. spectra (4000—400 cm$^{-1}$) of these compounds, and of Me·S(CH$_2$)$_3$S·Me, prepared from allyl chloride (Me·S· addition, Me·S$^-$ substitution), have been analysed.[67] The vinyl ether (CH$_2$:CH·O·CH$_2$·CH$_2$·S·)$_2$ is cleaved into the corresponding thio by a dialkyl phosphonate, in 90—95% yield, the product undergoing polymerisation to (·CH$_2$·CH$_2$·O·CH$_2$·CH$_2$·S·)$_{10-15}$ and cyclisation to

---

[61] H. C. Brown and J. H. Kawakami, *J. Amer. Chem. Soc.*, 1970, **92**, 201.
[62] D. I. Davies and P. J. Rowley, *J. Chem. Soc.* (*C*), 1968, 1832.
[63] D. I. Davies, L. T. Parfitt, C. K. Alden, and J. A. Claisse, *J. Chem. Soc.* (*C*), 1969, 1585.
[64] E. N. Prilezhaeva, V. A. Asovskaya, G. U. Stepanyanz, D. Mondeshka, and R. J. Shekhtman, *Tetrahedron Letters*, 1969, 4909.
[65] D. I. Davies and L. T. Parfitt, *J. Chem. Soc.* (*C*), 1969, 1401.
[66] J. A. Claisse, D. I. Davies, and L. T. Parfitt, *J. Chem. Soc.* (*C*), 1970, 258.
[67] K. S. Boustany and A. Jacot-Guillarmod, *Chimia* (*Switz.*), 1969, **23**, 331.

2-methyl-oxathiolan;[68] polymerisation is not inhibited by $K_2CO_3$ and is therefore taken to be a free-radical process, in which case it constitutes the first reported instance of the normal addition of a thiol to a vinyl ether under free-radical initiation.[68]

Equilibria between Pr·SH and simple carbonyl compounds have been studied in $CH_2Cl_2$;[69] the resulting α-hydroxysulphides may be converted into mercaptals or mercaptoles where equilibrium constants are less than $10^2$, by addition of an acid catalyst ($BF_3$ or HCl). Rates and equilibrium constants referring to the formation and breakdown of hemithioacetals (acetaldehyde + Ph·SH, or Ac·SH, or $p$-$NO_2$·$C_6H_4$·SH) reveal a diffusion-controlled rate-determining step, with proton transfer in some sense concerted with cleavage and formation of the C—S bond.[70]

Addition of thiophenols to phenylacetylene, and radical isomerisation of vinyl sulphides such as Ph·CH:CH·S·Ar, is inhibited by $Fe(CO)_5$ and $[Fe(CO)_3SEt]_2$ but not by $Mo(CO)_6$ and $W(CO)_6$.[71]

**Nucleophilic Substitution Reactions of Thiols.**—Thiophenol reacts faster than its anion with a bromopyridine, in methanol, due to a rapid acid–base pre-equilibrium in which the pyridine is protonated;[72] an $o$-MeO-substituent accelerates the replacement of Br, and a small rate increase is noted on going from MeOH to dimethyl sulphoxide as solvent. Pentafluoropyridine is substituted in the 4-position by $HS^-$, $Ph·S^-$, $Ph·SO_2^-$, or $Na_2SO_3$ in aprotic solvents;[73] the 4-sulphonate yields the 4-cyano-compound with $CN^-$. Pentachloropyridine reacts similarly with $P_2S_5$ or $HS^-$.[74] A spectroscopic study of 1:1-adduct formation between 1,3,5-trinitrobenzene and nucleophiles sets up the order of relative nucleophilicities $Et·S^- > Me·O^- > Ph·S^- > Ph·O^-$.[75] Thiophenol reacts with 2,4-dinitrofluorobenzene with no evidence of acid catalysis; the first step (nucleophilic addition) is rate-limiting,[76] contrary to other claims.[77]

The 3-SH group in 3,5-dimercapto-4-cyanoisothiazole is alkylated most readily,[78] and exclusive $S$-tritylation is observed with reactions of $o$-, $m$-, or $p$-aminothiophenol with $Ph_3C^+ClO_4^-$ or $Ph_3C·OH$–AcOH.[79] Reactions of flavanoids with mercaptoacetic acid [80] give further examples of benzylic

---

[68] B. A. Trofimov, A. S. Atavin, A. V. Gusarov, S. V. Amosova, and S. E. Korostova, *Zhur. org. Khim.*, 1969, **5**, 816.
[69] L. Field and B. J. Sweetman, *J. Org. Chem.*, 1969, **34**, 1799.
[70] R. E. Barnett and W. P. Jencks, *J. Amer. Chem. Soc.*, 1969, **91**, 6758.
[71] I. I. Kandror, R. G. Petrova, P. V. Petrovskii, and R. K. Freidlina, *Izvest. Akad. Nauk S.S.S.R., Ser. khim.*, 1969, 1621.
[72] R. A. Abramovitch, F. Helmer, and M. Liveris, *J. Org. Chem.*, 1969, **34**, 1730.
[73] R. E. Banks, R. N. Haszeldine, D. R. Karsa, F. E. Rickett, and I. M. Young, *J. Chem. Soc. (C)*, 1969, 1660.
[74] E. Ager, B. Iddon, and H. Suschitzky, *J. Chem. Soc. (C)*, 1970, 193.
[75] M. R. Crampton, *J. Chem. Soc. (B)*, 1968, 1208.
[76] J. F. Bunnett and N. S. Nudelman, *J. Org. Chem.*, 1969, **34**, 2038.
[77] K. C. Ho, J. Miller, and K. W. Wong, *J. Chem. Soc. (B)*, 1966, 310.
[78] G. A. Hoyer and M. Kless, *Tetrahedron Letters*, 1969, 4265.
[79] G. Chuchani and K. S. Heckmann, *J. Chem. Soc. (C)*, 1969, 1436; M. K. Eberhardt and G. Chuchani, *Tetrahedron*, 1970, **26**, 955.
[80] M. J. Betts, B. R. Brown, and M. R. Shaw, *J. Chem. Soc. (C)*, 1969, 1178.

substitution by mercaptide ions; both 2- and 4-benzylic positions can be substituted, though the 4-position is attacked exclusively at high pH. Polymerisation of 2-mercaptoalkanols is acid-catalysed at the primary cyclodehydration step leading to episulphide,[81] which undergoes self-condensation to poly(ethylene sulphides).

**Reactivity of —SH Groups in Proteins.**—The p$K$ of the thiol grouping is sensitive to local structure, in obvious ways through inductive effects, but also through 'medium effects'. For peptides, the —SH group of a terminal cysteine residue, or of a cysteine residue surrounded by basic amino-acid side-chains is considerably more acidic than the —SH group in a $C$-terminal cysteine residue, or in a cysteine residue surrounded by acidic amino-acid side-chains.[82] Such factors are crucial in determining the function of thiol groupings in enzyme action, and lead to necessarily loose phraseology ('the role of different types of thiol group'[83]); relevant aspects were discussed at a 1970 Conference [84] and striking demonstration is given in the fact that rates of alkylation of essential thiol groups of several enzymes range over five orders of magnitude.[85]

Typical reagents for locating thiol groups in proteins include iodoacetamide, $N$-ethylmaleimide, and 5,5′-dithiobis(2-nitrobenzoic acid), whose rates of reaction reveal relative reactivities of the thiol groups concerned [86] (see also ref. 509). Detailed studies of aldolase [86, 87] using such methods have been reported, and stopped-flow spectroscopic techniques with very rapidly reacting chromophoric organomercurial compounds have been reported [83] (see also ref. 510). $\beta$-Nitrostyrene adds to —SH groups at rates determined by relative reactivities of the groups concerned, and has been applied as a reagent for this purpose,[88] in comparison with Ph·Hg·Cl and 5,5′-dithiobis(2-nitrobenzoic acid). 7-Chloro-4-nitrobenzo-2-oxa-1,3-diazole is proposed as a specific probe to assess the reactivity of model compounds and proteins containing thiol groups,[89] since fluorescent products with characteristic visible absorption (400—500 nm) are obtained ($N$-acetylcysteine gives a product with $\lambda_{max}$ 425 nm).

**Estimation and Identification of Thiol Groups.**—Standard techniques [90] have been supplemented considerably very recently, owing to current

---

[81] A. D. B. Sloan, *J. Chem. Soc.* (*C*), 1969, 1252.
[82] S. J. Rogers, *J. Chem. Educ.*, 1969, **46**, 239.
[83] H. Gutfreund and C. H. McMurray, *Biochem. J., Proceedings* 1970, in the press.
[84] 'Chemical Reactivity and Biological Role of Functional Groups in Enzymes', *Biochem. J., Proceedings* 1970, in the press.
[85] K. Wallenfals and B. Eisele, in ref. 84.
[86] P. J. Anderson and R. N. Perham, in ref. 84.
[87] P. A. M. Eagles, L. N. Johnson, M. A. Joynson, C. H. McMurray, and H. Gutfreund, *J. Mol. Biol.*, 1969, **45**, 533.
[88] R. T. Louis-Ferdinand and G. C. Fuller, *J. Pharm. Sci.*, 1969, **58**, 1155.
[89] D. J. Birkett, N. C. Price, G. K. Radda, and A. G. Salmon, *F.E.B.S. Letters*, 1970, **6**, 346.
[90] R. Cecil, in 'The Proteins: Composition, Structure, and Function', ed. H. Neurath, 2nd edn., Academic Press, New York and London, 1963, vol. 1, pp. 379—476.

interest in the role of thiols in enzyme action, and some of the reagents used are named in the preceding paragraph. Saville's method [91] using colorimetric quantitation has been extended [92] to proteins. U.v. spectrometric determination of the substituted pyridine-2-thione produced by cleavage by thiols of the corresponding 2,2′-dipyridyl disulphide has been used for quantitative assessment [93] and β-hydroxyethyl-2,4-dinitrophenyl disulphide has been used similarly.[94] Tetranitromethane yields nitroformate ($\lambda_{max}$ 350 nm, $\varepsilon$ 14,400) in oxidising thiols to disulphides, or sulphinic acids; in oxidising thiol groups in proteins to disulphides, oxidation of other residues may occur and caution is required in use of the reagent for quantitative thiol analysis.[95] Quantitative conversion of cysteine into S-sulphocysteine may be achieved [96] using sodium tetrathionate, offering an alternative to carboxymethylation or conversion into cysteic acid ($\cdot SH \rightarrow \cdot SO_3H$) for accurate analysis of cysteine in protein hydrolysates. However, sodium tetrathionate can introduce $\cdot S \cdot SO_3H$ groups into complex molecules [97] and this should be borne in mind when using this procedure more generally.

2-Mercaptoethanol (but not $Pr^n \cdot SH$, 2,2′-dithioethanol, cysteine, or dithiothreitol) gives a soluble complex with $FeCl_3 + Na_2S$ in aqueous solution at pH 9, with an absorption spectrum ($\lambda\lambda_{max}$ 318, 412 nm) similar to that of non-haem iron proteins, suggesting that an Fe—S bond is involved with a correctly-located OH group as chromophore in these compounds.[98]

**Protection of Thiol Groups in Synthesis.**—Nucleophilic addition and substitution reactions of thiols have been utilised in its reversible protection. S-Di-ethoxycarbonylethyl substitution of cysteine gives protection compatible with peptide synthesis procedures.[99] The group, introduced by reaction with diethyl methylenemalonate, $(EtO_2C)_2C:CH_2$, is stable to HBr-AcOH and to trifluoroacetic acid, and is removed by treatment with N-alcoholic KOH at room temperature within a few minutes. An earlier study [100] of the use of S-isobutyloxymethyl derivatives of cysteine in peptide synthesis has been continued.[101] Its use is compatible with N-phthaloyl, N-trityl, C-t-butyl protection; deprotection is effected with thiocyanogen or a sulphenyl thiocyanate, as for S-trityl analogues, and is

---

[91] B. Saville, *Analyst*, 1958, **83**, 670.
[92] M. Gronow and P. Todd, *Analyt. Biochem.*, 1969, **29**, 540.
[93] D. R. Grassetti and J. F. Murray, *Analyt. Chim. Acta*, 1969, **46**, 139; J. N. Mehrishi and D. R. Grassetti, *Nature*, 1969, **224**, 563.
[94] J. Witwicki and K. Zakrzewski, *European J. Biochem.*, 1969, **10**, 284.
[95] M. Sokolovsky, D. Harell, and J. F. Riordan, *Biochemistry*, 1969, **8**, 4740.
[96] A. S. Inglis and T.-Y. Liu, *J. Biol. Chem.*, 1970, **245**, 112.
[97] J. T. Edward and J. Whiting, *Canad. J. Chem.*, 1970, **48**, 337.
[98] C. S. Yang and F. M. Huennekens, *Biochem. Biophys. Res. Comm.*, 1969, **35**, 634.
[99] T. Wieland and A. Sieber, *Annalen*, 1969, **722**, 222; *ibid.*, **727**, 121.
[100] G. T. Young, P. J. E. Brownlee, M. E. Cox, B. O. Handford, and J. C. Marsden, *J. Chem. Soc.*, 1964, 3832.
[101] R. G. Hiskey and J. T. Sparrow, *J. Org. Chem.*, 1970, **35**, 215.

not accompanied by thiol–disulphide interchange. $S$-Ethylthio-protection of cysteine residues has been used in a Merrifield synthesis of insulin A-chain,[102] final treatment with $Na_2SO_3 + Na_2S_4O_6$ in tris buffer giving the $S$-sulphonate of the required peptide. Base-induced $\beta$-elimination reactions during deprotection of cysteine derivatives can be troublesome; their intervention during manipulation of $S$-benzhydryl-cysteine peptides has been assessed [103] by estimation of pyruvic acid released on acid hydrolysis. Dehydroalanine formation resulting from elimination is greatest when polar solvents are used to mediate deprotection.

Thiobenzoylation of thiol groups in thioglycerols may be achieved [104] without masking —OH groups (using Ph·CS·S·Me as reagent), unmasking being brought about by hydrolysis or aminolysis.

**Other Reactions of Thiols.**—Displacement by halogens has been noted; cysteine ethyl ester hydrochloride gives $\beta$-chloroalanine ethyl ester hydrochloride (overall ·$CH_2SH \rightarrow$ ·$CH_2$·Cl) on treatment with $Cl_2$ at $-20\,°C$ during 2 days,[105] and treatment of trimethylsilyl derivatives of 1-thio-glycopyranosides with $Br_2$ gives corresponding glycosyl 1-bromides.[106] Cysteine, cystine, and methionine are partially desulphurised on treatment with triethyl phosphite during several days;[107] de-gassed Raney nickel desulphurises cyclo-octyl thiol to a 4:1 mixture of cyclo-octane and cis-cyclo-octene, and the corresponding sulphide to a 7:93 mixture of the same compounds,[108] though it is not obvious that a conventional free-radical intermediate is involved.

Thiophenol is useful as a free-radical trap, and has been used as such in a study of the oxidation of hydrazones to hydrazonyl free radicals.[109] Thiols have no effect on the singlet excited state of photosensitized diphenyl-pyrazoline, which is quenched by indoles or phenols,[110] though thiyl radicals can be shown to be present in reaction mixtures (e.s.r. study). $X$-Irradiation of cysteine [111] in aqueous solution (pH 2·2—3·25), or radiolysis (pH 7),[112] gives the corresponding disulphide (cystine) and $H_2O_2$. $\gamma$-Radiolysis of alkanethiols, $C_nH_{2n+1}SH$, at 77 K produces decreasing amounts of radicals (e.s.r. study) with increasing $n$.[113]

---

[102] U. Weber, *Z. physiol. Chem.*, 1969, **350**, 1421.
[103] R. G. Hiskey, R. A. Upham, G. M. Beverly, and W. C. Jones, *J. Org. Chem.*, 1970, **35**, 513.
[104] E. J. Hedgley and N. H. Leon, *J. Chem. Soc. (C)*, 1970, 467.
[105] B. E. Norcross and R. L. Martin, *J. Org. Chem.*, 1969, **34**, 3703.
[106] W. Hengstenberg and K. Wallenfals, *Carbohydrate Res.*, 1969, **11**, 85.
[107] C. Ivanov and O. C. Ivanov, *Doklady Bolg. Akad. Nauk*, 1969, **22**, 49.
[108] A. C. Cope and J. E. Engelhart, *J. Org. Chem.*, 1969, **34**, 3199.
[109] C. Wintner and J. Wiecko, *Tetrahedron Letters*, 1969, 1595.
[110] I. H. Leaver, G. C. Ramsay, and E. Suzuki, *Austral. J. Chem.*, 1969, **22**, 1891; I. H. Leaver and G. C. Ramsay, *ibid.*, 1969, **22**, 1899.
[111] T. C. Owen and M. T. Brown, *J. Org. Chem.*, 1969, **34**, 1161.
[112] J. E. Packer and R. V. Winchester, *Canad. J. Chem.*, 1970, **48**, 417.
[113] A. D. Bichiashvili, E. M. Nanobashvili, and R. G. Barsegov, *Soobshch. Akad. Nauk. Gruz. S.S.S.R.*, 1969, **53**, 337.

Sodium selenophenolate, Ph·Se⁻Na⁺, reacts faster than Ph·S⁻Na⁺ in dealkylation of alkaloidal quaternary ammonium salts.[114]

### 3 Sulphides

**Preparation.**—Symmetrical diaryl sulphides may be prepared by direct methods. Typically, azulene and $SCl_2$ at $-78\,°C$ gives di-(1-azulyl) sulphide;[115] selenionyl chloride ($SeOCl_2$) and 2-t-butyl-*p*-cresol ($CHCl_3$, room temperature) gives (21).[116]

Me      Me

Bu^t — [ring] — Se — [ring] — Bu^t
      OH      OH

(21)

Phenol reacts with sulphur to give thiobisphenols,[117] mixtures of 2,2'-, 2,4'-, and 4,4'-isomers being obtained in proportions 45:45:10 at 180 °C. Separated products disproportionate into the same mixture when heated to 180 °C in phenol in the presence of NaOH, through heterolytic cleavage into monothiobenzoquinones, a mechanism supported by the facts that corresponding disulphides lose $H_2S$ in giving the same product mixture, and that the 3,3'-dithiobisphenol is stable to these conditions.[117]

Sodium methyl acetylide, treated with selenium or tellurium followed by MeI, gives Me·C⫶C·Se·Me or its Te analogue.[118] While the latter gives back methylacetylene on attempted hydration, the selenoether gives Me·CH₂·CO·Se·Me with $H_2O$–$HgSO_4$.

Syntheses of sulphides from thiols have been surveyed in the preceding section, and transformations of other sulphur functional groups give alternative general routes: de-oxygenation of sulphoxides may be brought about by heating at 130—135 °C in diglyme or $Bu_2O$ with $Fe(CO)_5$, though $Mo(CO)_6$ is ineffective;[119] reduction of sulphoxides with $Si_2Cl_6$ [120] or hydrogenation in the presence of $Rh^{III}$ complexes [121] offers mild conditions, and deoxygenation with $PCl_3$ avoids chlorination which accompanies the use of $PCl_5$ or $POCl_3$ for the purpose.[122]

3-Methylbut-2-enylthiolation of 2-lithio-4,5-dimethylthiazole is brought about using the corresponding disulphide.[123] Dimethyl disulphide gives

---

[114] V. Simanek and A. Klasek, *Tetrahedron Letters*, 1969, 3039.
[115] L. L. Replogle, G. C. Peters, and J. R. Maynard, *J. Org. Chem.*, 1969, **34**, 2022.
[116] S. Korcek, S. Holorik, J. Lesko, and V. Vesely, *Chem. Zvesti*, 1969, **23**, 281.
[117] A. J. Neale, P. J. S. Bain, and T. J. Rawlings, *Tetrahedron*, 1969, **25**, 4583, 4593.
[118] Y. A. Boiko, B. S. Kupin, and A. A. Petrov, *Zhur. org. Khim.*, 1969, **5**, 1553.
[119] H. Alper and E. C. H. Keung, *Tetrahedron Letters*, 1970, 53.
[120] K. Naumann, G. Zon, and K. Mislow, *J. Amer. Chem. Soc.*, 1969, **91**, 7012.
[121] B. R. James, F. T. T. Ng, and G. L. Rempel, *Canad. J. Chem.*, 1969, **47**, 4521.
[122] I. Granoth, A. Kalir, and Z. Pelah, *J. Chem. Soc. (C)*, 1969, 2424.
[123] I. Fleming, *Chem. and Ind.*, 1969, 1661.

all-*exo*-2,3,5,6-tetrakis(methylthio)norbornane (22) with norbornadiene–
$Cr(CO)_4$ in boiling methylcyclohexane;[124] mono-desulphuration o

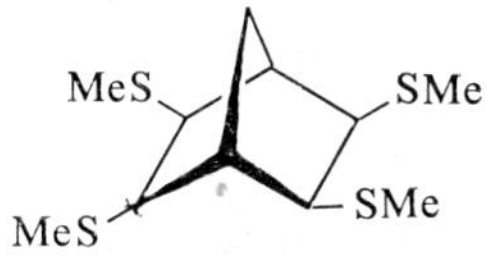

(22)

disulphides $R^1 \cdot S \cdot S \cdot R^2$ with tris(diethylamino)phosphine in refluxing benzene gives sulphides with no 'crossed' products.[125] Thiolcarbonate esters give sulphides on heating with trioctylphosphine, or $(Et_4N^+)F^-$ in dimethylformamide or $Me \cdot CN$; xanthate esters can react similarly but side-reactions make their use unprofitable.[126] Dephenylation of triphenylsulphonium tetrafluoroborate with vinyl-lithium proceeds *via* five-co-ordinate sulphur in giving diphenyl sulphide and styrene, in tetrahydrofuran at $-78\,°C$.[127] $\beta\gamma$-Unsaturated sulphonium salts give sulphides through [3,3]-sigmatropic rearrangements.[128]

Vinyl sulphides are accessible through the reaction of alkylidene-triphenylphosphoranes with thiolesters,[129, 130] and from acetylenes by addition of aryl sulphenyl halides [$Me \cdot C \vdotdbl C \cdot CO_2Et$ gives $Me \cdot C(X) \colon C(S \cdot Ar)CO_2Et$][131] (see also ref. 359) or Se analogues.[132] Addition of phosgene to 1-alkynyl sulphides gives $R \cdot S \cdot C(Cl) \colon C(CO \cdot Cl)R$.[133] An adaptation of the phosphonate olefin synthesis leads to vinyl sulphides,[134] hydrolysis of which (in the presence of $Hg^{2+}$) gives ketones; overall conversion (Scheme 1) is $R^1 \cdot CO \cdot R^2$ to $R^1R^2CH \cdot CO \cdot R$ in good yield.

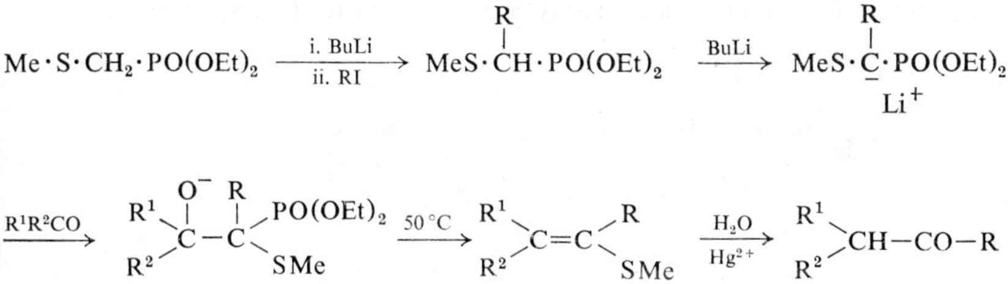

**Scheme 1**

[124] R. B. King, *J. Org. Chem.*, 1970, **35**, 274.
[125] D. N. Harpp, J. G. Gleason, and J. P. Snyder, *J. Amer. Chem. Soc.*, 1968, **90**, 4181.
[126] F. N. Jones, *J. Org. Chem.*, 1968, **33**, 4290.
[127] B. M. Trost, R. LaRochelle, and R. C. Atkins, *J. Amer. Chem. Soc.*, 1969, **91**, 2175.
[128] J. E. Baldwin, R. E. Hackler, and D. P. Kelly, *Chem. Comm.*, 1968, 537, 538, 899, 1083.
[129] T. Kumamoto, K. Hosoi, and T. Mukaiyama, *Bull. Chem. Soc. Japan*, 1968, **41**, 2742.
[130] H. Saikachi and S. Nakamura, *Yakugaku Zasshi*, 1969, **89**, 1446 (*Chem. Abs.*, 1970, **72**, 12,453).
[131] M. Verny and R. Vessiere, *Bull. Soc. chim. France*, 1970, 746.
[132] L. M. Kataeva, E. G. Kataev, and T. G. Mannafov, *Zhur. strukt. Khim.*, 1969, **10**, 830.
[133] G. Van den Bosch, H. J. T. Bos, and J. F. Arens, *Rec. Trav. chim.*, 1970, **89**, 133.
[134] E. J. Corey and J. I. Shulman, *J. Org. Chem.*, 1970, **35**, 777.

Copper acetylides give alkynyl aryl sulphides with aryl sulphenyl chlorides;[135] ynamine thioethers, $R^1{}_2N \cdot C \vdots C \cdot S \cdot R^2$, result from reaction of *trans*-1,2-dichlorovinyl sulphides with lithium salts of secondary amines.[136] Preparation of β-chloroalkyl sulphides by the addition of sulphenyl chlorides to alkenes [137] is well represented in the recent literature; 2-chloroethyl sulphenyl chloride adds to crotonic acid derivatives (23) under kinetic control, the adduct (24) rearranging into the more stable isomer (25), possibly *via* a thi-iranium intermediate.[138]

$$\text{Me}-\text{CH}=\text{CH}-\text{CO}-\text{X} \longrightarrow \underset{\underset{\text{S}-\text{CH}_2-\text{CH}_2-\text{Cl}}{|}}{\text{Me}-\text{CH}-\overset{\overset{\text{Cl}}{|}}{\text{CH}}-\text{CO}-\text{X}}$$

(23)     (24)

$$\underset{\underset{\text{Cl}}{|}}{\text{Me}-\text{CH}}-\overset{\overset{\text{S}-\text{CH}_2-\text{CH}_2-\text{Cl}}{|}}{\text{CH}}-\text{CO}-\text{X}$$

(25)

Phenyl sulphides $Ph \cdot S \cdot CH_2 \cdot CH_2 \cdot NCO$, and corresponding isothiocyanate and *N*-sulphinylamine, are obtained by thermal isomerisation of *N*-substituted aziridines $Ph \cdot S \cdot X \cdot N(CH_2)_2$, where X = CO, CS, or SO.[139]

**Reactions.**—Photochemical demethylthiolation of solid 2-methylthio-4,6-bis(ethylamino)-*s*-triazine (310 nm, 72 hr in air) does not involve sulphinyl or sulphonyl intermediates.[140] 2-Methylthio-indan-1-one behaves similarly,[141] giving indan-1-one and (presumably) thioformaldehyde. Reaction of certain allenic and alkynyl sulphides with MeI similarly results

$$\underset{\text{Me}}{\overset{\text{Ph}}{\diagdown}}\text{C}=\text{C}=\text{CH}-\text{SMe} \xrightarrow{\text{MeI}} \text{CH}_2=\overset{\overset{\text{Ph}}{|}}{\text{C}}-\underset{\underset{\text{I}}{|}}{\text{C}}=\text{CH}_2$$

(26)

+

$$\text{CH}_2=\underset{\underset{\text{I}}{|}}{\text{C}}-\underset{\underset{\text{Ph}}{|}}{\text{C}}\overset{cis}{=}\underset{\underset{\text{H}}{|}}{\text{C}}-\text{C}\equiv\text{C}-\text{Ph}$$

[135] C. E. Castro, R. Havlin, V. K. Honwad, A. Malte, and S. Mojé, *J. Amer. Chem. Soc.*, 1969, **91**, 6464.
[136] S. Y. Delavarenne and H. G. Viehe, *Tetrahedron Letters*, 1969, 4761.
[137] W. H. Mueller and P. E. Butler, *J. Amer. Chem. Soc.*, 1968, **90**, 2075.
[138] T. Pranskiene, Z. Stumbreviciute, L. Rasteikiene, M. G. Linkova, and I. L. Knunyants, *Izvest. Akad. Nauk S.S.S.R., Ser. khim.*, 1969, 2063.
[139] D. A. Tomalia, D. P. Sheetz, and G. E. Ham, *J. Org. Chem.*, 1970, **35**, 47.
[140] J. R. Plimmer, P. C. Kearney, and U. I. Klingebiel, *Tetrahedron Letters*, 1969, 3891.
[141] R. H. Fish, L. C. Chow, and M. C. Caserio, *Tetrahedron Letters*, 1969, 1259.

in loss of the sulphur grouping;[142] e.g. (26) gives two products, but (Ph·C⋮C·CH$_2$·)$_2$S+MeI gives (27). Irradiation of allyl methallyl sulphide [142a] gives a 1:2:1 mixture of hexa-1,5-diene, 2-methylhexa-1,5-diene, and 2,5-dimethylhexa-1,5-diene, in organophosphorus solvents.

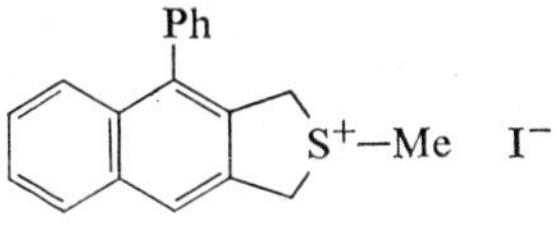

(27)

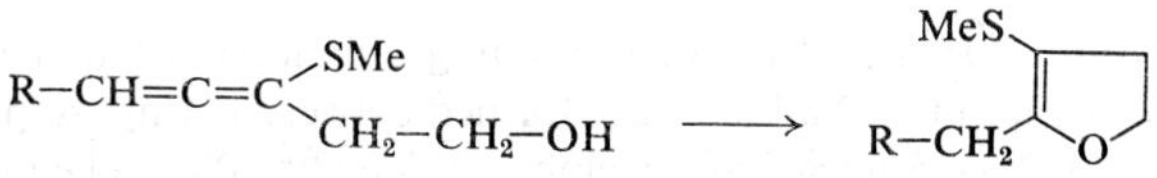

(28) → (29)

While β-hydroxyethyl-1-methylthio-allenes (28) give corresponding dihydrofurans (29), the oxygen analogues give mixtures of (29; O in place of S) and 2-alkylidene 3-methoxytetrahydrofuran.[143] This well-known property of sulphur in stabilising an adjacent double bond appears also in base-promoted rearrangements of aryl allyl selenides [144] [Ph·Se·CH(Me)·CH:CH$_2$ gives Ph·Se·C(Me):CH·Me in dimethyl sulphoxide, though but-2-enyl and phenylallyl analogues resist isomerisation], comparable rearrangements of sulphides being well established.[145] In contrast to oxygen analogues, a sulphur atom stabilises an α-radical by electron-sharing conjugation [146] (Ph—S̈—ĊMe$_2$ ←→ Ph—S̈=CMe$_2$) shown by decomposition of Ph·S·CMe$_2$·N:N·CMe$_2$·S·Ph in Ph$_2$O at 120 °C in giving 37% Ph·S·CHMe$_2$, 15% (Ph·S·CMe$_2$·)$_2$, 10% Ph·S·C(Me):CH$_2$, and 8% Ph·S·S·Ph with small enthalpy and entropy of activation compared with the oxygen analogue, which gives 52% acetophenone+CH$_4$. Firestone [147] suggests that radical stabilisation by adjacent sulphur involves saturated species (R—S̈—Ċ= ←→ R—S̈$^{\frac{1}{2}+}$—Ċ$^{\frac{1}{2}-}$=), on the basis of Linnett theory. Aliphatic sulphides possess a remarkable ability to quench photo-excited benzophenone (1 mole Bu$_2$S prevents photo-reduction of 23 moles Ph·CO·Ph).[148] Deuterium exchange rates of aryl deuteriomethyl sulphides in 0·02N-KNH$_2$–NH$_3$, compared with those of oxygen analogues, suggest

[142] A. Terada and Y. Kishida, Chem. and Pharm. Bull. (Japan), 1969, 17, 974.
[142a] E. J. Corey and E. Block, J. Org. Chem., 1969, 34, 1233.
[143] S. Hoff, L. Brandsma, and J. F. Arens, Rec. Trav. chim., 1969, 88, 609.
[144] E. G. Kataev, G. A. Chmutova, A. A. Musina, and E. G. Yarkova, Doklady Akad. Nauk S.S.S.R., 1969, 187, 1308.
[145] D. S. Tarbell and W. E. Lovett, J. Amer. Chem. Soc., 1956, 78, 2259.
[146] A. Ohno and Y. Ohnishi, Tetrahedron Letters, 1969, 4405.
[147] R. A. Firestone, J. Org. Chem., 1969, 34, 2621.
[148] J. Guttenplan and S. G. Cohen, Chem. Comm., 1969, 247.

electron release through vacant $d$-orbitals of the sulphur atom in the sulphides.[149]

Vinyl sulphides are effective dienophiles,[150] tetrachlorocyclopentadiene giving (30) and small amounts of the *exo*-isomer, with methyl vinyl sulphide

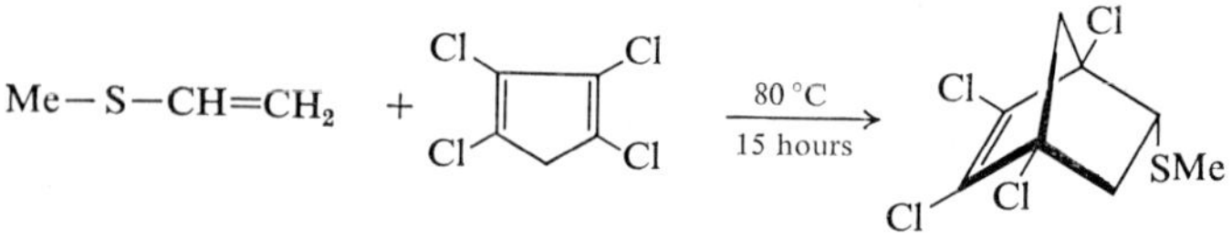

$$Me-S-CH=CH_2 + \underset{Cl}{\overset{Cl}{\diagup}}\underset{Cl}{\overset{Cl}{\diagdown}} \xrightarrow[15 \text{ hours}]{80\,°C}$$

(30)

at 80 °C during 15 hr. The corresponding sulphone behaves similarly; mixtures of 1 : 1 and 1 : 2 adducts are formed between divinyl sulphone and 1,1-dimethoxy-2,3,4,5-tetrachlorocyclopentadiene.[150] Photochemical cyclisation of $(Ph \cdot CH : CH \cdot)_2 S$ in $Et_2O$ gives the first cyclobutene episulphide, *trans*-1,2-diphenylcyclobut-3-ene episulphide (31), together with the isomeric 3,4-diphenyl-2,3-dihydrothiophen;[151] vinyl sulphides add to cyclohex-2-en-1-one under u.v. irradiation, giving a simple entry to the bicyclo-octane ring system (32).[152] Photochemical studies with cyclic

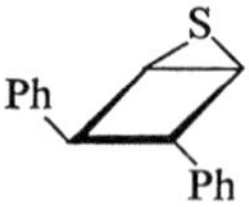

(31) 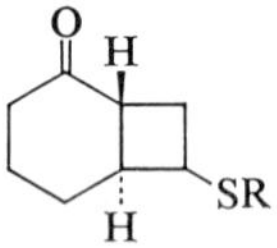   (32)

$\gamma$-ketosulphides reveal a charge-transfer interaction which is not evident in acyclic analogues.[153]

*cis*-2-Phenylthio-vinyl ketones, $Ph \cdot S \cdot CH : CH \cdot CO \cdot R$, are more stable than their *trans*-isomers, apparently [154] owing to favourable electrostatic interaction between the sulphur and CO groupings since the reverse stability relationship holds for corresponding sulphones. Related observations touching upon differences between related functional groups have been reported: 2-alkylthiothiophens are scarcely reactive towards BuLi whereas corresponding selenoethers give thienyl-lithium more readily;[155] alkylthio-quinoxalines and sulphinyl and sulphonyl analogues show decreasing reactivity towards $MeO^-$, with increasing size of the alkyl grouping through the series Me, Et, $Pr^i$, $Bu^t$, the effect being smallest for the sulphides

---

[149] E. A. Gvozdeva and A. I. Shatenshtein, *Teor. i eksp. Khim.*, 1969, **5**, 191.
[150] V. A. Azovskaya, E. N. Prilezhaeva, and G. U. Stepanyants, *Izvest. Akad. Nauk S.S.S.R., Ser. khim.*, 1969, 662.
[151] E. Block and E. J. Corey, *J. Org. Chem.*, 1969, **34**, 896.
[152] J. Y. Vanderhoek, *J. Org. Chem.*, 1969, **34**, 4184.
[153] P. Y. Johnson and G. A. Berchtold, *J. Org. Chem.*, 1970, **35**, 584.
[154] B. Cavalchi, D. Landini, and F. Montanari, *J. Chem. Soc. (C)*, 1969, 1204.
[155] Y. L. Goldfarb, V. P. Litvinov, and A. N. Sukiasyan, *Doklady Akad. Nauk S.S.S.R.*, 1968, **182**, 340.

(rate decrease 1 : 8) and largest for the sulphoxides (1 : 140);[156] decarboxylation rates of $Me_2C:C(R) \cdot CH(CO_2H)_2$ in 4 : 1 $(CD_3)_2CO-D_2O$ decrease in the order R = $Ph \cdot CH_2 \cdot SO \cdot$, $Ph \cdot CH_2 \cdot SO_2 \cdot$, $Ph \cdot CH_2 \cdot S \cdot$ (53·5 : 3·1 : 1).[157] Chloromethyl methyl sulphide adds to vinyl chloride in $H_2SO_4$, giving $Me \cdot S \cdot CH_2 \cdot CH_2 \cdot CHO$ after dilution with water; the methyl thiomethyl cation is an intermediate.[158] Bis(chloromethyl)sulphide undergoes self-condensation in the presence of sodium methoxide, giving a series of chlorine-free sulphides;[159] thermal degradation of poly(alkylene sulphides) at $>200\ °C$ gives $>95\%$ conversion into $H_2S$, olefins, and dimers.[160] The fate of every atom in $^5CH_3 \cdot S \cdot {^4CH_2} \cdot {^3CH_2} \cdot {^2CO} \cdot {^1CO_2H}$ during peroxidase degradation in the presence of $Mn^{2+}$, $SO_3^{2-}$ and phenol, in giving $^5CH_3 \cdot S \cdot S \cdot {^5CH_3}$, $^4CH_2 : {^3CH_2}$, $^2CO_2$, and $^1CO_2$, has been determined;[161] methional $(CH_3 \cdot S \cdot CH_2 \cdot CH_2 \cdot CHO)$ gives dimethyl disulphide, ethylene, and formic acid.

Neighbouring group participation by sulphur (further examples are given later in this Chapter) is invoked to account for differing rates of solvolysis through a series $p\text{-}Me \cdot C_6H_4 \cdot S \cdot (CH_2)_n \cdot Br$, $n = 3$—6,[162] and also to account for the larger $\rho$-value noted for $p$-substituted phenyl 3-bromopropyl sulphides in comparison with oxygen analogues in $E2$-elimination by $Bu^tO^-$, the transition state (33) representing assistance by sulphur through a vacant $d$-orbital.[163]

$$X-\langle\phantom{O}\rangle-S\cdots\overset{H}{\underset{CH_2}{CH}}=CH_2\cdots Br$$

(33)

Cumulene thioethers (34) and a 1-(alkylthio)vinylacetylene (35) are formed on treatment of a 1,4-di(alkylthio)but-2-yne (36) with two equivalents of BuLi in ether;[164–166] the cumulene suffers polymerisation or explosive decomposition.[164] 1,3-Bis(methylthio)allyl-lithium, dark red in colour, is valuable as a nucleophilic equivalent of the $\cdot CH:CH \cdot CHO$ grouping, and has been used in a prostaglandin synthesis,[167] (Scheme 2).

[156] G. B. Barlin and W. V. Brown, *J. Chem. Soc.* (B), 1969, 333.
[157] S. Allenmark and O. Bohman, *Arkiv Kemi*, 1969, **31**, 305.
[158] T. Ichikawa, H. Owatari, and T. Kato, *J. Org. Chem.*, 1970, **35**, 344.
[159] B. J. Christ and A. N. Hambly, *Austral. J. Chem.*, 1969, **22**, 2471.
[160] R. T. Wragg, *J. Chem. Soc.* (B), 1970, 404.
[161] S.-F. Yang, *J. Biol. Chem.*, 1969, **244**, 4360.
[162] A. C. Knipe and C. J. M. Stirling, *J. Chem. Soc.* (B), 1968, 1218.
[163] Y. Yano and S. Oae, *Tetrahedron*, 1970, **26**, 67.
[164] R. Mantione, A. Alves, P. P. Montijn, H. J. T. Bos, and L. Brandsma, *Tetrahedron Letters*, 1969, 2483.
[165] R. Mantione, A. Alves, P. P. Montijn, G. A. Wildschut, H. J. T. Bos, and L. Brandsma, *Rec. Trav. chim.*, 1970, **89**, 97.
[166] L. Brandsma, H. J. T. Bos, and J. F. Arens, in 'Chemistry of Acetylenes', ed. H. G. Viehe, Dekker, New York, 1969.
[167] E. J. Corey and R. Noyori, *Tetrahedron Letters*, 1970, 311.

$CH_2=C=C=CH-SR$ + $CH_2=CH-C\equiv C-SR$
(34) (35)

↑

$RS-CH_2-C\equiv C-CH_2-SR$
(36)

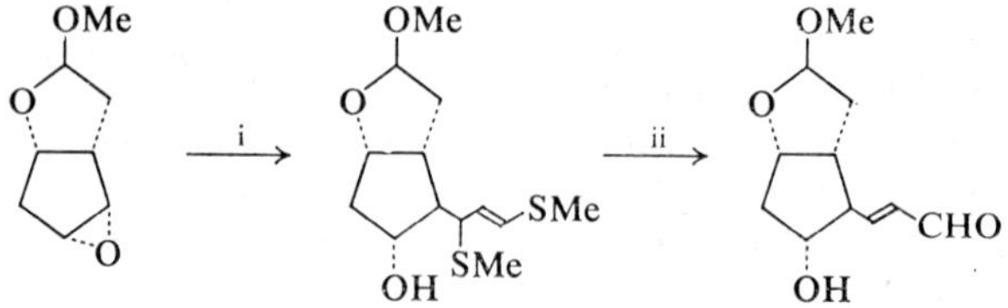

Reagents: i, 1·05 equiv. MeS·CHLi·CH:CH·SMe/−78 °C; ii, HgCl$_2$–CaCO$_3$ in MeCN–H$_2$O (4:1)/50 °C/3 hr) (epoxide-opening gives two isomers, but only one is shown)

**Scheme 2**

**Rearrangements of Sulphides.**—Allyl sulphides suffer carbene insertion with rearrangement, as well as carbene addition at the double bond, on reaction with di(methoxycarbonyl)carbene, $R\cdot S\cdot CH_2\cdot CR^1:CHR^2$ giving $R\cdot S\cdot C(CO_2Me)_2\cdot CHR^2\cdot CR^1:CH_2$. The relative proportions of the two products depend upon the mode of *in situ* carbene formation (from dimethyl diazomalonate).[168] Little-studied thio-Claisen rearrangements of allyl aryl sulphides are now well understood as a result of significant work during the present period. The intermediate thione in the rearrangement of an allyl 4-quinolyl sulphide (37) to a 2,3-dihydrothieno[3,2-c]quinoline (38) has

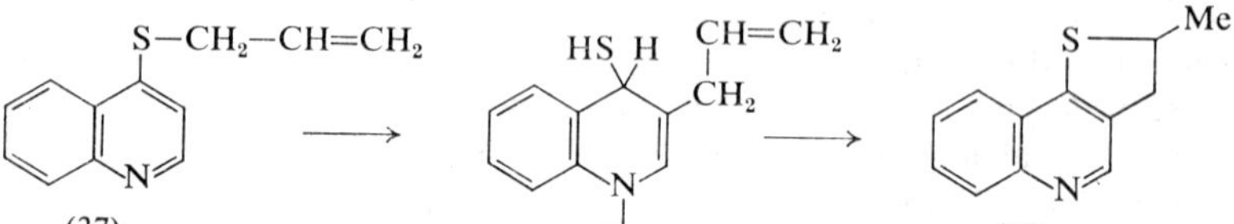

been trapped as its *S*-butyryl derivative, dispensing with alternative mechanisms;[169] 4-propargylthio-quinoline behaves similarly (200 °C; 2 hr in dimethylaniline), giving a thienoquinoline *via* a 3-allenylquinoline-4-thione,[169] a novel [3,3]-sigmatropic rearrangement exemplified also in rearrangements of 2- or 3-propynylthiothiophens,[170] *e.g.* (39) → (40) + (41), ratios dependent upon solvent. E.s.r.-monitoring shows the intervention

[168] W. Ando, K. Nakayama, K. Ichibori, and T. Migita, *J. Amer. Chem. Soc.*, 1969, **91**, 5164.
[169] Y. Makisumi and A. Murabayashi, *Tetrahedron Letters*, 1969, 1971.
[170] L. Brandsma and D. Schuijl-Laros, *Rec. Trav. chim.*, 1970, **89**, 110.

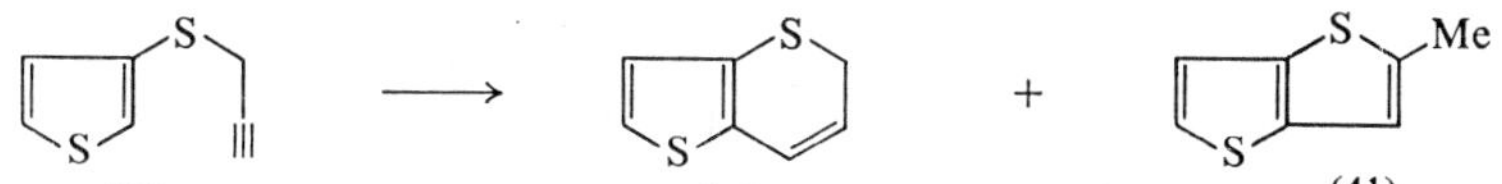

(39) → (40) + (41)

of free-radical intermediates in the analogous rearrangement of allyl 3-quinolyl sulphide (42), (43) predominating in the absence of radical initiators and (44) being formed exclusively under free-radical initiation.[171]

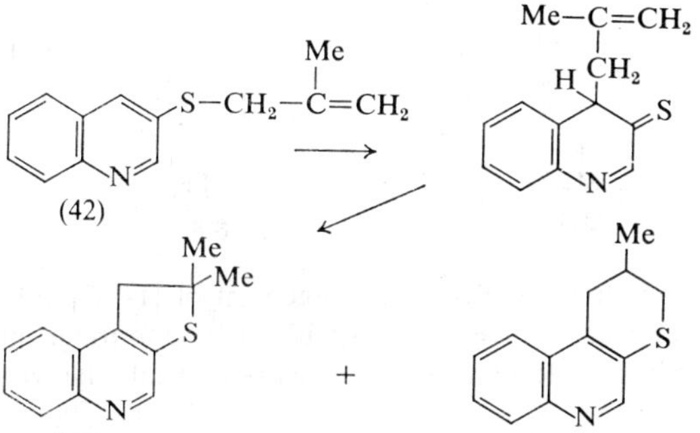

(42) → (intermediate) ← (44)
(43) +

Allyl 2-quinolyl sulphide gives 75% propenyl isomer (*i.e.* double-bond migration towards sulphur), di(2-quinolyl) sulphide and disulphide, and 1% *N*-allyl-thiocarbostyril; pure samples of the latter were shown to undergo retro-thio-Claisen rearrangement to allyl 2-quinolyl sulphide at 200 °C, and its presence in only trace amounts in product mixtures should not be taken to show that its formation is particularly disfavoured.[172]

Rearrangements of 3-substituted *N*-methyl(2-allylthio)indoles, and propynyl analogues, in refluxing toluene, *e.g.* (45) → (46), (47) → (48), show activation energies consistent with concerted pathways;[173] the products show no tendency towards further cyclisation under rearrangement conditions.

*S*-(2-Aminoethyl)-2-mercaptopyrimidines undergo reversal of the 2-substituent in solution at pH 7 ($\cdot$S$\cdot$CH$_2\cdot$CH$_2\cdot$NH$_2$ → $\cdot$NH$\cdot$CH$_2\cdot$CH$_2\cdot$S$\cdot$S$\cdot$CH$_2\cdot$CH$_2\cdot$NH$\cdot$);[174] 2-(alkylthio)malonates give rearranged products at elevated temperatures,[175] Et$\cdot$S$\cdot$CH(CO$_2$Me)$_2$ giving Et$\cdot$S$\cdot$CH$_2\cdot$CO$_2$Me and CH$_2$(CO$_2$Me)$_2$ in proportions 5:4, ethylene sulphide being a presumed third product, after 5 hr at 230 °C; formation of Ph$\cdot$CH$_2\cdot$CH(S$\cdot$Me)$\cdot$CO$_2$Me from the corresponding malonate involves a similar but essentially different seven-membered

---

[171] Y. Makisumi and A. Murabayashi, *Tetrahedron Letters*, 1969, 2449, 2453.
[172] Y. Makisumi and T. Sasatani, *Tetrahedron Letters*, 1969, 1975.
[173] B. W. Bycroft and W. Landon, *Chem. Comm.*, 1970, 168.
[174] L. V. Pavlova and F. Y. Rachinskii, *Khim. geterotsikl. Soedinenii*, 1969, 533.
[175] W. Ando, H. Matsuyama, S. Nakaido, and T. Migita, *Tetrahedron Letters*, 1969, 3825.

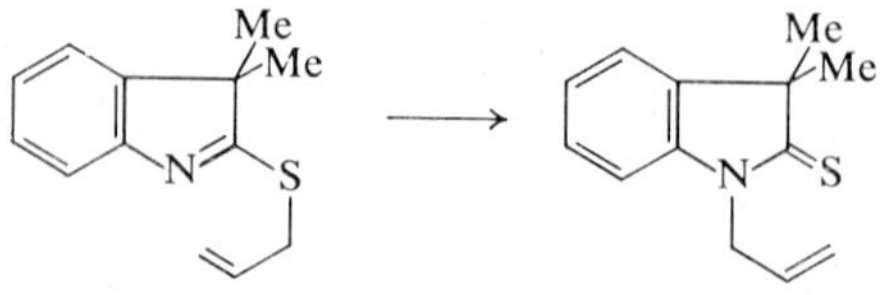

(45) → (46)

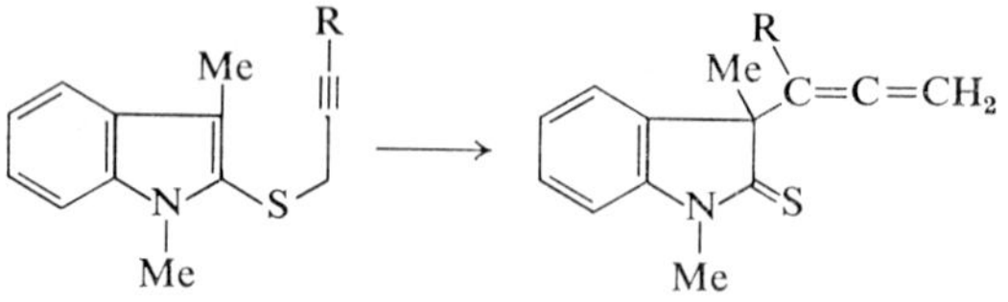

(47) → (48)

cyclic transition state. Skeletal rearrangement of [1-$^{13}$C]diphenyl sulphide on electron impact involves ring-expansion;[176] a nitrene intermediate is implicit in the intramolecular rearrangement of o-azidophenyl aryl sulphides (49) → (50).[177]

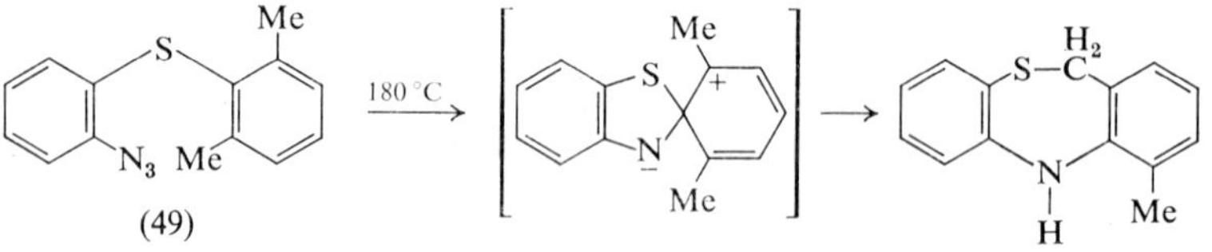

(49) →[180 °C] → (50)

**Naturally-occurring Acyclic Sulphides (excluding Methionine Derivatives).**—Sporidesmin D (51; $R^1 = R^3 = Me·S·$; $R^2 = Me$) and Sporidesmin F (51; $R^3 = Me·S·$; $R^1R^2 = CH_2$) have been isolated as biologically-inactive metabolites of *Pithomyces chartarum*.[178] *Se*-Methyl seleno-L-cysteine and its *N*-(γ-L-glutamyl) derivative have been isolated from the leaves and seeds, respectively, of *A. bisulcatus*,[179] a species which tolerates massive doses of selenium.

Penicillin V, with Cl-CO$_2$Me and NEt$_3$ in dimethylformamide, gives the azlactone (52) and the corresponding symmetrical sulphide.[180] Synthesis of lincomycin (53) through a route incorporating the methylthio-substituent at an early stage (D-galactose + Me·SH + HCl gives methyl-1-thio-α-D-galactopyranoside) has been reported.[181]

---

[176] J. D. Henion and D. G. I. Kingston, *Chem. Comm.*, 1970, 258.
[177] J. I. G. Cadogan and S. Kulik, *Chem. Comm.*, 1970, 233.
[178] W. D. Jamieson, R. Rahman, and A. Taylor, *J. Chem. Soc. (C)*, 1969, 1564.
[179] S. N. Nigam and W. B. McConnell, *Biochim. Biophys. Acta*, 1969, **192**, 185.
[180] S. Kukolja, R. D. G. Cooper, and R. B. Morin, *Tetrahedron Letters*, 1969, 3381.
[181] B. J. Magerlein, *Tetrahedron Letters*, 1970, 33.

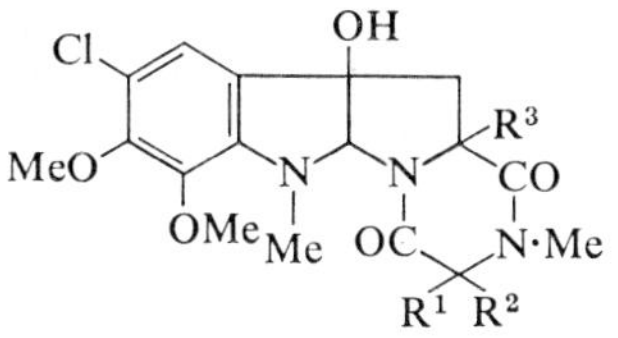

## 4 Thioacetals, Dithioacetals, Orthothioformates, and Orthothiocarbonates

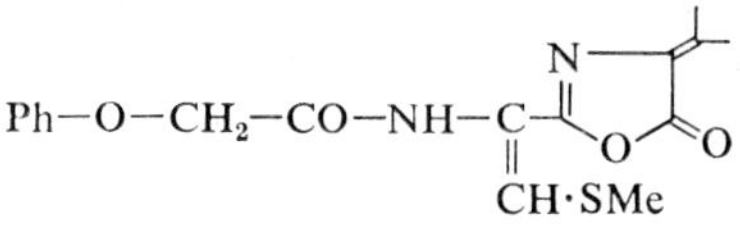

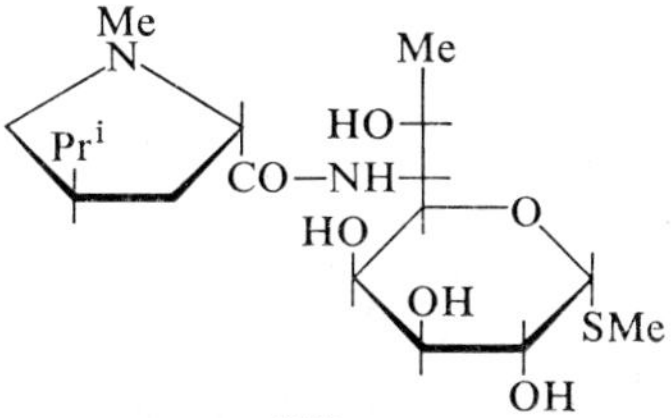

**Preparation.**—Methanedithiol is easily prepared from formaldehyde or $(HO \cdot CH_2)_2S$, with $H_2S$.[182] 2-(Alkylthio)tetrahydropyran,s obtained from the 2-hydroxy-compounds by reaction with $R \cdot SH$–dil. HCl, give 1,1-di(alkylthio)-5-hydroxypentanes on treatment with $R \cdot SH$–conc. HCl.[183] Dithioacetals result from treatment of 1,1-dimorpholides, *e.g.* $Ph \cdot CH(NR_2)_2$, with thiols in the presence of HCl (30 min, > 40 °C);[184] one equivalent of thiol gives the unstable $S,N$-acetal. Electrophilic substitution of the CH proton in keten $S,N$-acetals, $R^1 \cdot CH : C(SR^2)NR^3{}_2$, prepared by alkylation of corresponding thioamides, has been studied.[185] Amide mercaptals are discussed as part of a broader review of amide acetals.[186] Thioacetals of 3-methylpiperid-4-ones form more easily than corresponding acetals (n.m.r. study).[187] Diethyl dithioacetals of peracetylated aldoses have been studied by 100 MHz n.m.r. to provide information on preferred conformations of acyclic forms of monosaccharides.[188]

**Reactions.**—Among other points of interest in the chemistry of α-chloro-β-ketosulphides (54) shown in Scheme 3 is a convenient synthesis of α-hydroxy-aldehydes;[189] the Favorski-type rearrangement of α-hydroxy-dithioacetals shown in Scheme 3[189] is quite general,[190,191] careful study of

---

[182] *Ger., Offen.*, 1,921,776; *Chem. Abs.*, 1970, **72**, 31,218.
[183] L. Bassery, C. Leroy, and M. Martin, *Compt. rend.*, 1969, **269**, C, 1131.
[184] Y. Le Floc'h, A. Briault, and M. Kerfanto, *Compt. rend.*, 1969, **268**, C, 1718.
[185] R. Gompper and W. Elser, *Annalen*, 1969, **725**, 64, 73.
[186] J. Gloede, L. Haase, and H. Gross, *Z. Chem.*, 1969, **9**, 201.
[187] R. F. Branch and A. F. Casy, *J. Chem. Soc. (B)*, 1968, 1087.
[188] D. Horton and J. D. Wander, *Carbohydrate Res.*, 1969, **10**, 279.
[189] G. A. Russell and L. A. Ochrymowycz, *J. Org. Chem.*, 1969, **34**, 3618.
[190] B. A. Trofimov, A. S. Atavin, A. I. Mikhaleva, G. A. Kalabin, and N. P. Vasilev, *Zhur. org. Khim.*, 1969, **5**, 1886.
[191] G. A. Russell and L. A. Ochrymowycz, *J. Org. Chem.*, 1970, **35**, 764.

products of acid treatment of $(Bu^i \cdot S)_2 C(Me) \cdot CH_2 \cdot OH$ showing 50—60% conversion into $Bu^i \cdot S \cdot C(Me):CH \cdot S \cdot Bu^i$, but $Bu^i \cdot S \cdot CH(Me) \cdot CHO$, $Bu^i \cdot S \cdot CH_2 \cdot CO \cdot Me$, and $Bu^i \cdot SH$ [190] are also formed.

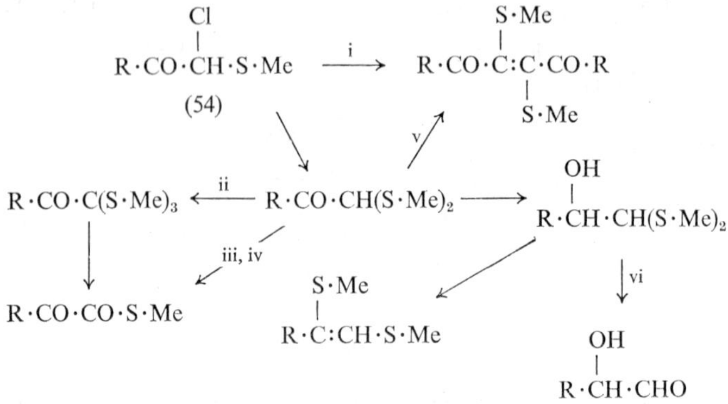

Reagents: i, pyrolysis; ii, Me·S·Cl; iii, $Br_2$; iv, $H_2O$; v, $Br_2$; vi, $I_2$

**Scheme 3**

Alkylthiolation by disulphides or thiolsulphonates, of anions derived from acetylenic dithioacetals, $R \cdot C:C \cdot CH(S \cdot Et)_2$, gives corresponding allenes $R^1 \cdot C(S \cdot R^2):C:C(S \cdot Et)_2$, in good yield.[192, 165] These ($R^1 = Me$) rearrange to butadienes $CH_2:C(S \cdot R^2) \cdot CH:C(S \cdot Et)_2$ on warming with acid; methylthio-alkynyl dithioacetals $Me \cdot S \cdot C:C \cdot CH(S \cdot Et)_2$ give $(Me \cdot S)_2 C:C:C(S \cdot Et)_2$ with $Me_2S_2$,[192] and keten dithioacetals $R \cdot CH_2 \cdot CR^2:C(S \cdot Et)_2$ give ethyl orthothioesters $R^1 \cdot CH:CR^2 \cdot C(S \cdot Et)_3$ with Et·SH.

Novel thio-Claisen rearrangements of dithioacetals, leading to dithio-esters, have been reported,[49, 193, 194] (55) → (56), (57) → (58).

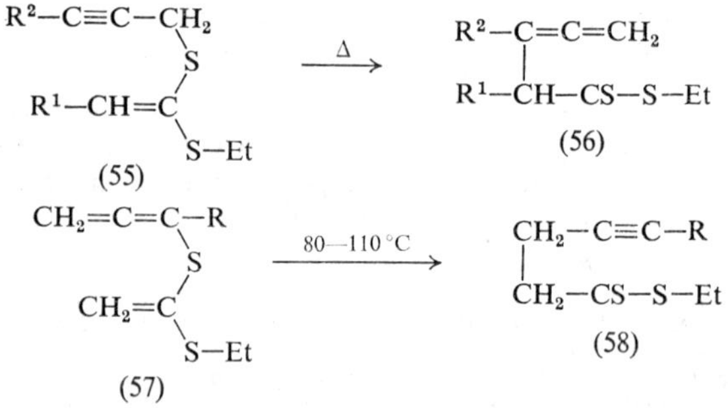

[192] G. A. Wildschut, L. Brandsma, and J. F. Arens, *Rec. Trav. chim.*, 1969, **88**, 1132.
[193] P. J. W. Schuijl, H. J. T. Bos, and L. Brandsma, *Rec. Trav. chim.*, 1969, **88**, 597.
[194] P. J. W. Schuijl and L. Brandsma, *Rec. Trav. chim.*, 1969, **88**, 1201.

Di(methylthio)carbene, $(Me \cdot S)_2C:$, is formed [195] from $(Me \cdot S)_2CH^+BF_4^-$ in $CH_2Cl_2$ or sulpholane at $-10$ °C, in the presence of $Pr^i_2NH$, since $(Me \cdot S)_2C:C(S \cdot Me)_2$ may be isolated; a similar intermediate is involved [196] in the formation of tetra(phenylthio)ethylene from $(Ph \cdot S)_3C^-Li^+$ at 40 °C, and may be implicated [197] in the conversion $R \cdot CO \cdot CHN_2 \rightarrow R \cdot CO \cdot CH(S \cdot Me) \cdot CH(S \cdot Me)_2$ with $CH(S \cdot Me)_3$–$BF_3$ (R = $o$-methoxycarbonylphenyl). Dithioacetal ylides $Me_2S:CH \cdot S \cdot Ph$ behave as sources of thiocarbenes $:CH \cdot S \cdot Ph$, in giving cyclopropanes with ethylene derivatives.[198] Temperature-dependent e.s.r. spectra of cation radicals derived from $(R \cdot S)_2C:C(S \cdot R)_2$ have been studied.[199]

Tetramethyl orthothiocarbonate, $C(S \cdot Me)_4$, gives spiro-compounds (59) with alkanedithiols.[200]

(59)

## 5 Sulphoxides and Sulphones

Chemistry of sulphoxides reported in the period 1961—66 has been surveyed,[201] dealing with syntheses, reactions, hydrogen bonding, complex formation, and physical properties.

**Preparation.**—Sulphides give sulphoxides with photochemically-generated singlet oxygen,[202] and with anhydrous $H_2O_2$ in $Bu^t \cdot OH$ at $-45$ °C with $V_2O_5$ catalysis (at 15—20 °C, sulphones are formed).[203] The latter procedure is valuable in the partial oxidation of labile α-chloro- or α-acetoxy-sulphides, and is a well-researched modification of the simple laboratory preparation of sulphoxides; acid-catalysed peroxide oxidation of sulphides involves hydrogen-bonded intermediates.[204] Tetracovalent sulphur intermediates have been identified as their mercuric chloride adducts $R_2S^+ \cdot O \cdot Bu^t, HgCl_3^-$ in the oxidation of sulphides to sulphoxides by $Bu^tOCl$.[205] Oxidation of chiral esters (60) of $o$-(methylthio)benzoic acid with achiral peracids gives a preponderance of the sulphoxide with the

---

[195] R. A. Olofson, S. W. Walinsky, J. P. Marino, and J. L. Jernow, *J. Amer. Chem. Soc.*, 1968, **90**, 6554.
[196] D. Seebach and A. K. Beck, *J. Amer. Chem. Soc.*, 1969, **91**, 1540.
[197] A. Schoenberg, J. Kohtz, and K. Praefcke, *Chem. Ber.*, 1969, **102**, 1397.
[198] Y. Hayasi, M. Takaku, and H. Nozaki, *Tetrahedron Letters*, 1969, 3179.
[199] D. H. Geske and M. V. Merritt, *J. Amer. Chem. Soc.*, 1969, **91**, 6921.
[200] D. L. Coffen, *J. Heterocyclic Chem.*, 1970, **7**, 201.
[201] C. R. Johnson and J. C. Sharp, *Quart. Reports Sulfur Chem.*, 1969, **4**, 1.
[202] K. Gollnick, *Adv. Photochem.*, 1968, **6**, 1.
[203] F. E. Hardy, P. R. H. Speakman, and P. Robson, *J. Chem. Soc. (C)*, 1969, 2334.
[204] R. Curci, R. DePrett, J. D. Edwards, and G. Modena, in 'Hydrogen-bonded Solvent Systems', ed. A. K. Covington and P. Jones, Taylor and Francis, London, 1968, pp. 303—321.
[205] C. R. Johnson and J. J. Rigau, *J. Amer. Chem. Soc.*, 1969, **91**, 5398.

($R$)-configuration (61).[206] Photo-oxidation of methionine to the sulphoxide

(60) → (61)  [Ar—CO—OOH]

is specifically brought about in the presence of haematoporphyrin both in neutral aqueous solution and in acetic acid, in the presence of other photo-oxidisable groupings.[207] Periodate oxidation gives the corresponding sulphoxide and sulphone from di(1-azulyl) sulphide;[115] selenides give selenoxides by oxidation with periodate or $Ph \cdot ICl_2$ below room temperature,[208] or with ozone.[209] Acetylenic sulphones $R \cdot SO_2 \cdot C\!:\!CH$ have been prepared from the corresponding sulphide without difficulty, for polarographic study.[210]

Sulphonyl fluorides or sulphonate esters, with a $\cdot CH_2 \cdot$ group adjacent to the sulphur atom, give sulphones with $Ph \cdot Li$, the formation of $Ph \cdot CH_2 \cdot SO_2 \cdot CH(Ph) \cdot SO_2 \cdot OPh$ from $Ph \cdot CH_2 \cdot SO_2 \cdot OPh$ at $-60\ °C$ in 85% yield being the first example of a Claisen-type condensation of a simple sulphonate ester.[211] Moderate yields of sulphones may be obtained from alkyl or aryl sulphonyl chlorides and aryl trimethyl stannanes in the presence of $AlCl_3$; $Ar \cdot SiMe_3$ behaves similarly though some isomerisation occurs ($p$-$Me \cdot C_6H_4 \cdot SiMe_3$ gives $o$- and $p$-$Me \cdot C_6H_4 \cdot SO_2 \cdot Ph$ with $Ph \cdot SO_2Cl$).[212] Sulphines ($R \cdot CH\!:\!S\!:\!O$) react with organolithium reagents giving sulphoxides;[213] dienes give mixtures of sulphones with sulphene ($CH_2\!:\!SO_2$).[214] Trityl sulphones result from $SO_2$ insertion into azo-compounds, e.g. $Ph \cdot N\!:\!N \cdot CPh_3$ gives $Ph \cdot SO_2 \cdot CPh_3$ in refluxing benzene, the free-radical nature of the reaction being shown by its suppression in the presence of a radical scavenger such as $Bu \cdot SH$.[215] Addition of arylsulphinate anions to $Br \cdot CH\!:\!CH \cdot CN$ gives $Ar \cdot SO_2 \cdot CHBr \cdot CH_2 \cdot CN$,[216] while addition of organo-magnesium, -zinc, or -cadmium compounds to sulphinate esters gives sulphoxides. $X$-Ray crystal analysis, like an earlier o.r.d./c.d. study, indicates inversion of configuration at the sulphinyl

[206] G. Barbieri, V. Davoli, I. Moretti, F. Montanari, and G. Torre, *J. Chem. Soc.* (*C*), 1969, 731.
[207] G. Jori, G. Galiazzo, and E. Scoffone, *Biochemistry*, 1969, **8**, 2868.
[208] M. Cinquini, S. Colonna, and R. Giovini, *Chem. and Ind.*, 1969, 1737.
[209] R. Paetzold and G. Bochmann, *Z. anorg. Chem.*, 1968, **360**, 293.
[210] V. I. Laba, M. K. Polievktov, E. N. Prilezhaeva, and S. G. Mairanovskii, *Izvest. Akad. Nauk S.S.S.R., Ser. khim.*, 1969, 2149.
[211] T. Shirota, T. Nagai, and N. Tokura, *Tetrahedron*, 1969, **25**, 3193.
[212] S. N. Bhattacharya, C. Eaborn, and D. R. M. Walton, *J. Chem. Soc.* (*C*), 1969, 1367.
[213] A. G. Schultz, unpublished work quoted in C. N. Skold and R. H. Schlessinger, *Tetrahedron Letters*, 1970, 791.
[214] L. A. Paquette and R. W. Begland, *J. Org. Chem.*, 1969, **34**, 2896.
[215] H. Takeuchi, T. Nagai, and N. Tokura, *Tetrahedron*, 1969, **25**, 2987.
[216] V. N. Mikhailova, N. H. Cao, and A. D. Bulat, *Zhur. org. Khim.*, 1969, **5**, 1459.

centre (62) → (63)[217] (see also ref. 398). Similar reaction of a Grignard reagent with a sulphonate ester gives a sulphone (Scheme 4),[218] with the inversion at sulphur which is a feature of all nucleophilic substitution reactions at sulphur which have been studied so far,[218] though there are recent exceptions to this generalisation (see refs. 244, 245 later).

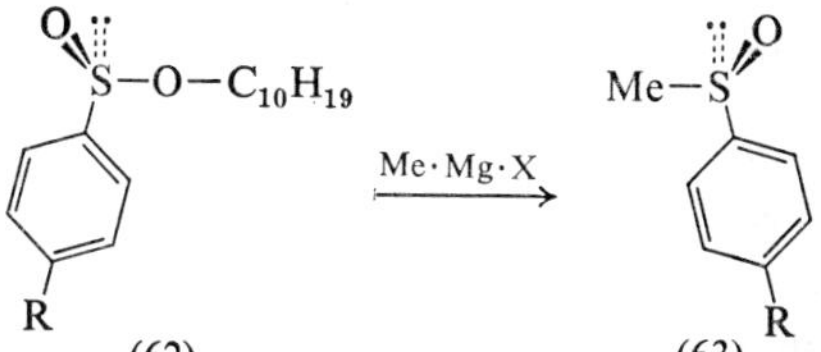

$[C_{10}H_{19} = (-)\text{-menthyl}]$

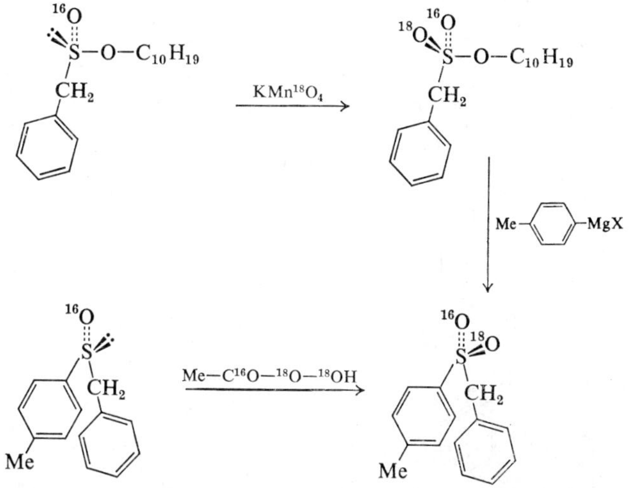

**Scheme 4**

Rearrangements of allylic sulphenates into allyl sulphoxides are [2,3]-sigmatropic processes which have been briefly discussed in the general context of the Claisen rearrangement,[219] and demonstrated with the simplest representative compounds.[220-222]

---

[217] H. Hope, U. De La Camp, G. D. Homer, A. W. Messing, and L. H. Sommer, *Angew. Chem., Internat. Edn.*, 1969, **8**, 612.
[218] M. A. Sabol and K. K. Andersen, *J. Amer. Chem. Soc.*, 1969, **91**, 3603.
[219] A. Jefferson and F. Scheinmann, *Quart. Rev.*, 1968, **22**, 419.
[220] N. S. Zefirov and F. A. Abdulvaleeva, *Vestnik Moskov Univ., Khim.*, 1969, **24**, 135 (*Chem. Abs.*, 1969, **71**, 112,345).
[221] D. J. Abbott and C. J. M. Stirling, *J. Chem. Soc. (C)*, 1969, 818.
[222] V. Rautenstrauch, *Chem. Comm.*, 1970, 526.

**Physical Properties of Sulphoxides and Sulphones.**—The preferred conformation of aryl isopropyl sulphoxides is (64);[223] the *p*-substituent influences the chemical shift of the aliphatic protons and the degree of association with trifluoroacetic acid at the sulphoxide group in (64).[223] Stereospecific transmission of electronic effects across a sulphoxide grouping is seen also in chemical shifts of (*R*)-α-methylbenzyl *p*-substituted-(*S*)-sulphoxides (65).[224] Corresponding sulphones and sulphides provide comparative data, shifts of $-0.29$ p.p.m. accompanying change of *p*-substituent from Me to $NO_2$ in (65)-sulphide (: in place of O), $-0.22$ for (65), $-0.05$ for the epimer of (65), and $-0.09$ for (65)-sulphone (O in place of :),[224] confirming that sulphenyl sulphur is more capable of transmitting effects of electron withdrawal in these compounds than ·SO· or ·$SO_2$· groupings.[224] Chemical shift of $H_a$ (gauche to the lone pair) of (66) is more sensitive to substituent

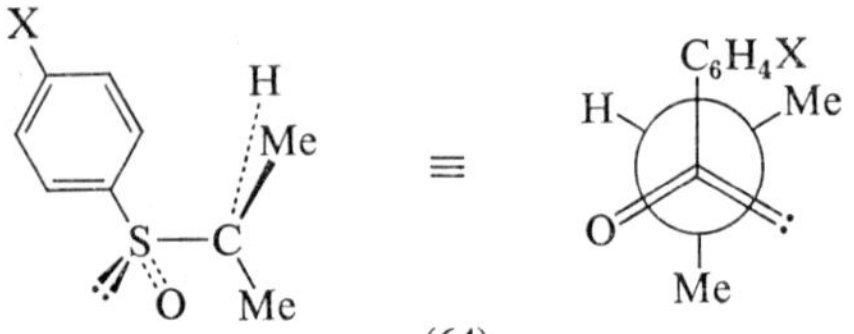

(64)

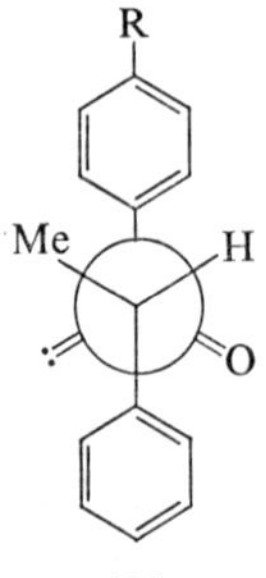

(65) 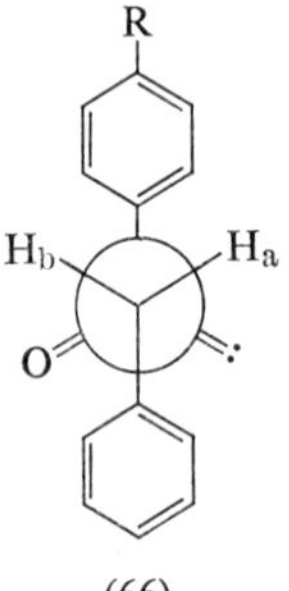  (66)

changes in the substituted phenyl group, tentatively attributed to 'the conformational preference of the phenyl group',[224] and which may be exploited for configurational assignments to diastereoisomeric sulphoxides of type (65).[225] Further applications of n.m.r. for stereochemical assignments to sulphoxides are reported;[226–228] self-association of these com-

[223] F. Taddei, *Congr., Conv. Simp. Sci.*, 1967, **11**, 106 (*Chem. Abs.*, 1969, **71**, 112,253); *J. Chem. Soc. (B)*, 1970, 653.
[224] M. Nishio, *Chem. Comm.*, 1969, 560.
[225] M. Nishio, *Chem. and Pharm. Bull. (Japan)*, 1969, **17**, 262, 274; *Chem. Comm.*, 1969, 51.
[226] A. L. Ternay, *Quart. Reports Sulfur Chem.*, 1968, **3**, 145.
[227] S. Allenmark and O. Bohman, *Acta Chem. Scand.*, 1968, **22**, 3326.
[228] C. Y. Meyers and A. M. Malte, *J. Amer. Chem. Soc.*, 1969, **91**, 2123.

pounds is demonstrated by n.m.r., and further examples of non-equivalence of methylene protons adjacent to the sulphinyl grouping in (91) [388] and in sulphites [229] have been reported.

The basicity of representative aliphatic and aromatic sulphoxides towards aqueous perchloric and sulphuric acids does not follow the $H_0$-function;[230] n.m.r. and u.v. data give $pK_a$ values varying from $-1.8$ for $Me_2SO$ to $-2.9$ for methyl $p$-nitrophenyl sulphoxide, and $\phi$ values within the range 0·4—0·6, similar to values reported for amides. The $H_A$-function is as satisfactory for representing the protonation behaviour of sulphoxides as for $\alpha\beta$-unsaturated carbonyl compounds.[230] The $pK_{BH^+}$ values of nine $m$- and $p$-substituted diphenyl sulphoxides ($-3.67$ to $-5.74$) show good correlation with Brown–Okamoto's $\sigma^+$ values ($\rho = -2.00$) rather than $\sigma$ values, and considerable conjugative interaction between the sulphinyl group and phenyl residues is indicated.[231] Sulphoxides cannot be considered to be Hammett bases,[232] a u.v.-spectrometric study aimed at determining p$K$ values requiring cautious interpretation in this area.[232]

**Stereochemistry of Sulphinyl Compounds.**—An earlier review of the stereomutation of sulphoxides remains excellent background reading.[233]

Standard methods are applicable for resolution of sulphoxides in which asymmetry is centred on sulphur [cf. (54) and non-superimposable mirror-image]; examples this year include $Ph \cdot CH_2 \cdot SO \cdot CEt:C(CO_2H)_2$,[234] $Ar \cdot SO \cdot CH_2 \cdot CO_2H$,[235, 236] and $o$-Me·SO-benzoic acid.[237] Partial reduction of $dl$-methyl alkyl sulphoxides with an ($R$)-$O$-alkyl ethylphosphonothioic acid during several weeks is more rapid with the ($R$)-isomer, i.e. unreacted starting material is enriched in the ($S$)-sulphoxide.[238]

($-$)-2,2,2-Trifluoro-1-phenylethanol causes n.m.r.-enantiomeric spectral non-equivalence of adjacent methylene protons in aliphatic sulphinyl compounds $R^1 \cdot X \cdot SO \cdot Y \cdot R^2$, n.m.r. spectra in optically-active solvents permitting correlation of absolute configurations and assessment of optical purity in this series.[239, 240] Ethylthio-L-galactitol gives corresponding ($R$)- and ($S$)-sulphoxides in unequal amounts (41% and 14%) on controlled oxidation with $H_2O_2$, the ($S$)- configuration being assigned on the basis of a positive Cotton effect near 220 nm shown by the major product and by the

[229] R. E. Lack and L. Tarasoff, *J. Chem. Soc.* (*B*), 1969, 1095.
[230] D. Landini, G. Modena, G. Scorrano, and F. Taddei, *J. Amer. Chem. Soc.*, 1969, **91**, 6703.
[231] S. Oae, K. Sakai, and N. Kuneida, *Bull. Chem. Soc. Japan*, 1969, **42**, 1964.
[232] N. C. Marziano, G. Cimino, U. Romano, and R. C. Passerini, *Tetrahedron Letters*, 1969, 2833.
[233] K. Mislow, *Rec. Chem. Progr.*, 1967, **28**, 217.
[234] O. Bohman and S. Allenmark, *Arkiv Kemi*, 1969, **31**, 299.
[235] M. Janczewski and M. Podgorski, *Roczniki Chem.*, 1969, **43**, 683, 1479.
[236] M. Janczewski and H. Maziarczyk, *Roczniki Chem.*, 1968, **42**, 657.
[237] U. Folli, D. Iarossi, and G. Torre, *Ricerca Sci.*, 1968, **38**, 914.
[238] M. Mikolajczyk and M. Para, *Chem. Comm.*, 1969, 1192.
[239] W. H. Pirkle and S. D. Beare, *J. Amer. Chem. Soc.*, 1968, **90**, 6250.
[240] W. H. Pirkle, S. D. Beare, and R. L. Muntz, *J. Amer. Chem. Soc.*, 1969, **91**, 4575.

(+)-Et·SO·CH$_2$·CHO obtained on periodate cleavage.[241] The correlation of sign of Cotton effect with absolute configuration follows well-established precedent; however, the (R)-isomer of S-allyl-L-cysteine-S-oxide also shows a positive Cotton effect, though the (R)-S-propyl analogue falls in line with negative o.r.d.–c.d. behaviour; this observation illustrates interaction of a neighbouring unsaturated centre with the asymmetrically-perturbed sulphoxide chromophore, and prescribes caution in the use of o.r.d.–c.d. data for assignment of absolute configuration in such circumstances.[242]

Chiral sulphoxides of known absolute configuration can be correlated with secondary alcohols, permitting assignment of configuration; thus, (R)-(+)-methyl p-tolyl sulphoxide predominates in the mixture resulting from treatment of the di(toluene-p-sulphinate) of (+)-HO·CH$_2$·CH(Me)·OH with MeMgI, and the alcohol is therefore (S)[243] (cf. 62 → 63).

Conversion of a sulphoxide into a sulphimine exemplifies nucleophilic substitution at sulphur with retention of configuration, a conclusion based on the formation of predominantly (R)-Me·(O)S(NH)·CH$_2$·CH$_2$·CH(NH$_2$)·CO$_2$H from L-methionine-(S)-sulphoxide via ·SO· + p-Me·C$_6$H$_4$·SO$_2$·N:S:O → ·S(N·Tos)· → ·(O)S(N·Tos)·.[244] Retention is suggested from results of other reactions with sulphoxides,[245] though the general rule is inversion of configuration through nucleophilic substitution at sulphur.[217] $^{18}$O-Labelled diaryl sulphoxides undergo concurrent oxygen-exchange and racemisation via an $S_N$1-like path in conc. H$_2$SO$_4$, though with inversion in 70% H$_2$SO$_4$ at 30 °C;[246] and much remains to be established concerning the topology of the transition state in nucleophilic substitution at sulphur.

Properties of sulphinyl substituents in condensed alicyclic systems have given detailed stereochemical information through o.r.d./c.d.–structure relationships and syn-elimination reactions and neighbouring-group participation by sulphinyl oxygen. Steroidal methyl[247] sulphoxides and alkyl analogues (Et, Pr$^n$, Pr$^i$, Bu$^n$, Bu$^t$)[248] and benzyl[249] analogues have been compared in relation to preferred conformations [rotational isomers; cf. (67)] and absolute configurations at sulphur. The Cotton effects near 230 nm in 6- or 7-alkylsulphinyl-5α-cholestanes are associated with an inflection in the isotropic absorption spectrum ($\lambda_{inf}$ 227—236 nm in hexane) which shifts to shorter wavelengths with increase in solvent polarity and is therefore associated with an $n \to \pi^*$ transition; these compounds show

[241] R. J. Ferrier and D. T. Williams, *Carbohydrate Res.*, 1969, **10**, 157.
[242] P. D. Henson and K. Mislow, *Chem. Comm.*, 1969, 413.
[243] C. Fuganti and D. Ghiringhelli, *Gazzetta*, 1969, **99**, 316.
[244] B. W. Christensen and A. Kjaer, *Chem. Comm.*, 1969, 934.
[245] S. Oae, M. Yokoyama, M. Kise, and N. Furukawa, *Tetrahedron Letters*, 1968, 4131.
[246] N. Kunieda and S. Oae, *Bull. Chem. Soc. Japan*, 1969, **42**, 1324.
[247] D. N. Jones, M. J. Green, and R. D. Whitehouse, *J. Chem. Soc.* (C), 1969, 1166.
[248] D. N. Jones, D. Mundy, and R. D. Whitehouse, *J. Chem. Soc.* (C), 1969, 1668.
[249] D. N. Jones and W. Higgins, *J. Chem. Soc.* (C), 1969, 2159.

## Aliphatic Organo-sulphur Compounds etc.

irregular Cotton effect–configuration relationships (absolute configurations being deduced from pyrolytic elimination rates and n.m.r. data) which could not be rationalised in terms of solvational or conformational equilibria.[247] The benzylsulphinyl analogues were studied to test this aspect further, (*R*)-6β-benzylsulphinyl-5α-cholestane (67) showing a positive

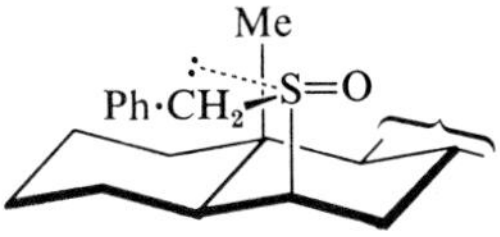

(67)

Cotton effect near 230 nm (EtOH), with a negative Cotton effect below 220 nm which is of much larger amplitude than that of alkylsulphinyl analogues.[249] Steroidal phenyl selenoxides chiral at selenium (the first resolution of this class of compound) have been studied;[250] 6α-methanesulphonyloxy-5α-cholestane gives the 6β-Ph·Se-analogue with Ph·Se$^-$, thence to corresponding phenylselenoxides in the ratio (*R*) : (*S*) = 2 : 1 by treatment with ozone at $-78$ °C. The (*S*)-configuration was assigned to the isomer which decomposed fastest in hexane at 0 °C into Ph·SeO$_2$H and 5α-cholest-6-ene; other differences between the diastereoisomers lay in absorption spectra, (*R*) showing $\lambda_{max}$ 265 nm, with (*S*) showing $\lambda_{max}$ 245 nm, and in practically mirror-image relationship o.r.d. curves. The (*R*)-isomer shows a positive Cotton effect, like its sulphur analogue, and configurational correlations between sulphur and selenium compounds of this series should be straightforward. The stereospecific elimination reaction is brought about under much milder conditions with selenoxides since they are more basic than sulphoxides.[250]

2,3-Di(phenylsulphinyl)norborn-5-enes, obtained by Diels–Alder addition of *cis*- or *trans*-1,2-di(phenylsulphinyl)ethylene to cyclopentadiene, have provided clear evidence of participation by sulphinyl oxygen.[251] Addition of I$_2$–H$_2$O to the adduct (68) from the *cis*-ethylene gives an iodhydrin with

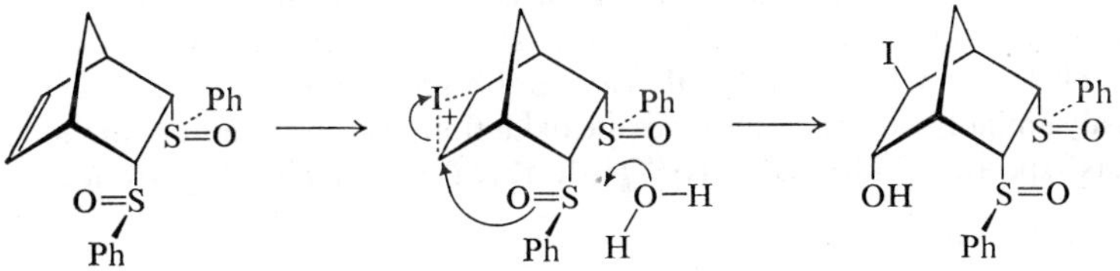

(68)         (69)

inversion at one sulphinyl centre (69), since acetylation followed by reduction (Zn–py) gives an epimeric 2,3-di(phenylsulphinyl)norborn-5-ene which does not undergo iodhydrin formation under the same conditions.[251]

---

[250] D. N. Jones, D. Mundy, and R. D. Whitehouse, *Chem. Comm.*, 1970, 86.
[251] M. Cinquini, S. Colonna, and F. Montanari, *J. Chem. Soc. (C)*, 1970, 572.

This splendid essay in sulphur stereochemistry includes demonstration of a Walden inversion cycle at sulphur with the Diels–Alder adducts.[251] All four (racemic) isomers of 2-phenylsulphinylcyclopentane-1-carboxylic acid have been prepared from the corresponding olefin;[252] reduction of the *cis*-isomers may be brought about under conditions to which *trans*-isomers do not respond, ascribed to anchimeric assistance by sulphinyl oxygen.[252]

**Reactions of Sulphoxides and Sulphones.**—Sulphinyl and sulphonyl compounds resemble carbonyl analogues in their base-catalysed reactions at α-methylene and -methine groups; a Claisen condensation of a sulphonate[211] and haloform reaction of methyl sulphones[253] ($R \cdot SO_2 \cdot CH_3 + CCl_4$–$KOH \rightarrow R \cdot SO_2 \cdot CCl_3 \rightarrow CHCl_3 + R \cdot SO_3^-$) being striking examples. Conversion of a sulphide into a sulphone in a complex molecule is accompanied by activation at more remote positions, too, and the potential of sulphur functional groups in synthesis is becoming more widely appreciated. Sulphoxides contribute their redox property for consideration for synthetic exploitation, and this feature causes notable differences in their reactions in comparison with sulphones and sulphides; recent work exposes the side-reactions to be expected with sulphoxides, though their potential utility in synthesis encourages further work aimed at understanding and taming their more capricious excesses.

Allyl aryl sulphones are isomerised into prop-1-enyl isomers with $NEt_3$.[254] Whereas phenylsulphonylcyclopropane results from Simmons–Smith methylenation of $Ph \cdot SO_2 \cdot CH:CH_2$, $Ph \cdot S \cdot CH:CH_2$ fails to react; also, $Ph \cdot SO_2 \cdot (CH_2)_3 \cdot Cl$ with $Bu^tO^-$ gives the cyclopropane but $Ph \cdot S \cdot (CH_2)_3 \cdot Cl$ gives allyl phenyl sulphide.[255] Differences in properties of sulphones compared with sulphides are exploited in the use of 2-(alkylthio)ethyl esters for carboxy-group protection in peptide synthesis,[256, 257] oxidation to the sulphone labilising the ester group towards saponification under very mild conditions. A recent study[258] of the 2-(*p*-nitrophenylthio)ethyl group for the same purpose uses molybdate-catalysed peroxide oxidation in acetone solution (*cf.* ref. 259), a mixture of $H_2O_2$ with acetone risking violent explosions.[260] Activation of this type is used in a new modification of Merrifield solid-phase peptide synthesis;[261] *N*-protected amino-acids are coupled to $P \cdot CH_2 \cdot S \cdot C_6H_4 \cdot OH$ (P = polystyrene) giving esters which are easily aminolysed only after peroxide oxidation to corresponding sulphones. Mass spectra of $Ph \cdot CO \cdot CH_2 \cdot SO \cdot Me$ reveal a McLafferty rearrangement

[252] S. Allenmark and H. Johnsson, *Acta Chem. Scand.*, 1969, **23**, 2902.
[253] C. Y. Meyers, A. M. Malte, and W. S. Matthews, *J. Amer. Chem. Soc.*, 1969, **91**, 7511.
[254] V. N. Mikhailova and A. D. Bulat, *Zhur. org. Khim.*, 1969, **5**, 1263.
[255] S. M. Shostakovskii and A. V. Bobrov, *Zhur. org. Khim.*, 1969, **5**, 908.
[256] A. W. Miller and C. J. M. Stirling, *J. Chem. Soc.* (C), 1968, 2612.
[257] M. J. S. A. Amaral, G. C. Barrett, H. N. Rydon, and J. E. Willett, *J. Chem. Soc.* (C), 1966, 807.
[258] M. J. S. A. Amaral, *J. Chem. Soc.* (C), 1969, 2495.
[259] P. M. Hardy, H. N. Rydon, and R. C. Thompson, *Tetrahedron Letters*, 1968, 2525.
[260] C. J. M. Stirling, *Chem. in Britain*, 1969, **5**, 36.
[261] D. L. Marshall and I. E. Liener, *J. Org. Chem.*, 1970, **35**, 867.

into $CH_2 \cdot SO$ and $Ph \cdot CO \cdot Me$ which is not shown by the corresponding sulphone.[262] Several papers have appeared on the Smiles rearrangement of diaryl sulphones, into sulphinic acids [*e.g.* (70) → (71)].[263]

(70)  (71)

Vinyl sulphones $p\text{-}X \cdot C_6H_4 \cdot SO_2 \cdot CH:CH \cdot Ph$ give epoxides in high yields, directly $(H_2O_2\text{-}OH^-)$;[264] the same product is accessible in other ways [$Tos \cdot CH_2Cl + Ph \cdot CHO$ in the presence of NaH; and from $R \cdot SO_2 \cdot CHCl \cdot CH(OH)Ph$].[265] The double bond in vinyl sulphones is relatively unreactive,[216] addition of HBr requiring prolonged reflux. α-Chlorination of dialkyl- and arylalkyl-sulphoxides (and sulphides if water is present) with $Ph \cdot ICl_2$ in pyridine at $-40\,°C$ is highly stereospecific; sulphones do not react. Since a large isotope effect is observed, an α-carbanion intermediate is involved.[266] The benzylic proton *anti* to the sulphoxide lone pair [*cf.* (66)] is substituted under these conditions in methyl benzyl sulphoxide.

An intense e.s.r. spectrum consisting of two quartets (one for Me·, the other corresponding to $Me \cdot SO_2\cdot$) is observed for a solution of dimethyl sulphoxide (0·5%) in water containing $Ti^{2+}\text{-}H_2O_2$ (a source of HO·);[267] photochemically-generated HO· is reported to yield alkyl radicals from aliphatic sulphoxides but not from sulphones.[268]

Arylation of sulphones with $Ph \cdot N_2^+ \; BF_4^-$ gives aryloxysulphoxonium salts, $R^1 \cdot (PhO)\overset{+}{S}(O) \cdot R^2 \; BF_4^-$;[269] the reaction fails for sulphonate esters, and if $R^1$ or $R^2$ contain $NO_2$ or CO groups.

Sulphinyl or sulphonyl groups can suffer base-catalysed displacement in reactions of synthetic utility; *E*1cb-elimination of $-SO \cdot Me$ from $Ar \cdot (CH_2)_2 \cdot SO \cdot Me$ has been studied from a kineticist's viewpoint,[270] leading to corresponding aryl cyclopropanes by 1,3-elimination. Competing

[262] W. R. Oliver and T. H. Kinstle, *Tetrahedron Letters*, 1969, 4317.
[263] V. N. Drozd, L. A. Nikonova, and L. I. Zefirova, *Izvest. Timiryazev. Selskokhoz. Akad.*, 1969, 179, 208; V. N. Drozd, K. A. Pak, and Y. A. Ustynyuk, *Zhur. org. Khim.*, 1969, 5, 1267, 1446; V. N. Drozd, L. I. Zefirova, K. A. Pak, and Y. A. Ustynyuk, *Zhur. org. Khim.*, 1969, 5, 933.
[264] B. Zwanenburg and J. ter Wiel, *Tetrahedron Letters*, 1970, 935.
[265] F. Bohlmann and G. Haffer, *Chem. Ber.*, 1969, 102, 4017.
[266] M. Cinquini, S. Colonna, and F. Montanari, *Chem. Comm.*, 1969, 607.
[267] W. Damerau, G. Lassmann, and K. Lohs, *Z. Chem.*, 1969, 9, 343.
[268] C. Lagercrantz and S. Forshult, *Acta Chem. Scand.*, 1969, 23, 811.
[269] G. R. Chalkley, D. J. Snodin, G. Stevens, and M. C. Whiting, *J. Chem. Soc. (C)*, 1970, 682.
[270] R. Baker and M. J. Spillett, *J. Chem. Soc. (B)*, 1969, 481, 581, 880.

concerted 1,2-elimination gives *trans*-ethylenes, and this route is favoured with a bulky substituent at the central methylene group; dimethyl sulphoxide as solvent favours the 1,3-elimination route. Base-catalysed (Bu$^t$O$^-$–Bu$^t$OH–dimethylformamide) γ-elimination from δ-oxo-sulphones gives cyclopropanes, first results suggesting that one particular stereoisomer is favoured [270, 271] [*e.g.* (72) → (73)].[271] Kinetics of hydrolysis and n-pentylaminolysis of 2-arylsulphonyl-pyrimidines and -purines, and related sulphoxides and sulphones, have been reported.[272]

Sulphoxides react with Grignard reagents at elevated temperatures to give $R^1 \cdot \overset{+}{S}(OMgX) \cdot \bar{C}H \cdot R^2$, which suffers decomposition into sulphides and disulphides, with hydrocarbons $R^3 \cdot R^3$ and $R^3 \cdot H$ derived from the organic residue of the Grignard reagent.[273] Oxidation of thiols to disulphides is efficiently brought about using dimethyl sulphoxide in boiling acetic acid,[274] and the formation of aldehydes from $\cdot SO \cdot CH_2 \cdot$ compounds in this medium is notable (Ph·SO·CH$_2$·CO$_2$H → Ph·S·S·Ph + OCH·CO$_2$H); more broadly, no single mechanism can account for all redox reactions of sulphoxides,[274] and the term 'Pummerer reaction' should be restricted to reactions of the type Ph·SO·CH$_2$·CO$_2$H + Ac$_2$O → Ph·S·CH(OAc)·CO$_2$H. The formation of the sulphone-sulphide R·CO·CH(SMe)SO$_2$Me from R·CO·CH$_2$·SO·Me and MeSOCl is an additional example.[275] Treatment of Ph·SO·CH$_2$·CO·CH$_2$Cl with KOAc–AcOH gives Ph·S·S·Ph and Ph·S·CH(OAc)·CO·Me and other products resulting from rearrangement of the sulphide;[276, 277] in the absence of KOAc, the Pummerer product is obtained. Reductive amination of representative sulphoxides with Ti(NMe$_2$)$_4$ has been reported (Me·SO·R gives Me$_2$N·CH$_2$·S·R).[278]

Dimethyl sulphoxide gives only 2% dimethyl sulphone in causing de-oxygenation of pyridine-*N*-oxides, indicating that a mechanism involving initial co-ordination complex formation followed by attack of nucleophilic oxygen on sulphur, though acceptable for Ph·SO·Ph as reagent, does not apply to the dimethyl analogue.[279] Deoxygenation of 4-methylpyridine-*N*-oxide with five equivalents of dimethyl sulphoxide under reflux gives 35% of the expected product and 1% 4-formylpyridine; the 4-benzyl

---

[271] W. L. Parker and R. B. Woodward, *J. Org. Chem.*, 1969, **34**, 3085.
[272] D. J. Brown and P. W. Ford, *J. Chem. Soc.* (C), 1969, 2720.
[273] P. Manya, A. Sekera, and P. Rumpf, *Tetrahedron*, 1970, **26**, 467.
[274] D. A. Davenport, D. B. Moss, J. E. Rhodes, and J. A. Walsh, *J. Org. Chem.*, 1969, **34**, 3353.
[275] G. A. Russell and E. T. Sabourin, *J. Org. Chem.*, 1969, **34**, 2336.
[276] V. Rosnati, F. Sannicolo, and G. Zecchi, *Gazzetta*, 1969, **99**, 651.
[277] V. Rosnati, F. Sannicolo, and G. Zecchi, *Tetrahedron Letters*, 1970, 599.
[278] G. Chandra, T. A. George, and M. F. Lappert, *J. Chem. Soc.* (C), 1969, 2565.
[279] M. E. C. Biffin, J. Miller, and D. B. Paul, *Tetrahedron Letters*, 1969, 1015.

analogue gives 29% 4-benzylpyridine but also 41% 4-benzoylpyridine and 9% Ph·CHR·CHR·Ph (R = 4-pyridyl), the formation of the dimeric product suggesting intervention of radicals.[280] Two Me·S·CH$_2$· substituents are introduced into $p$-hydroxybenzaldehyde by refluxing dimethyl sulphoxide, entering the positions *ortho* to the —OH group; vanillin gives 61% (74).[281]

Dimethyl sulphoxide now has considerable applications in organic synthesis,[10d, 282, 283] but its side-reactions are well illustrated in a number of recent papers. Products from the reaction of (75) with two equivalents of Ph·NCO in Me$_2$SO have been rationalised on the basis of solvent involvement; acyloxysulphonium salts are to be expected as intermediates where anhydrides are present [the carboxy-group of (75) gives ·CO·O·CO·NH·Ph, reaction with Me$_2$SO giving Ph·NH·CO·O·SMe$_2$+].[284] Schiff's bases give condensation products with

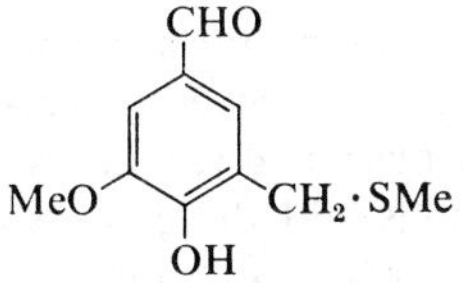

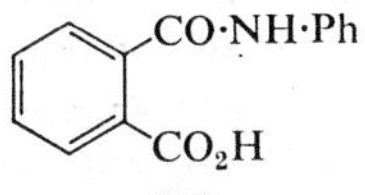

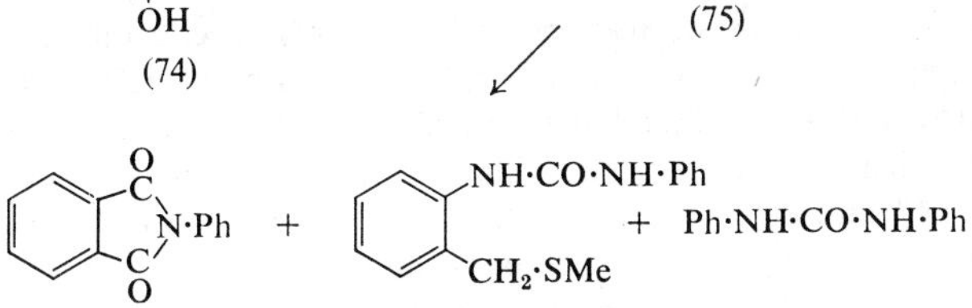

Me$_2$SO or Me$_2$SO$_2$ in dimethylformamide [Ph·N:CH·Ph gives Ph·C(NH·Ph):CH·C(Ph):N·Ph by an addition–$\beta$-elimination sequence],[285] and other sulphinyl carbanion reactions with Schiff's bases [286,287]

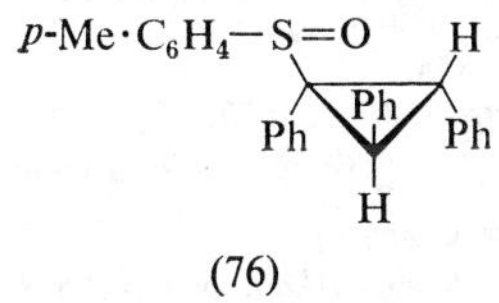

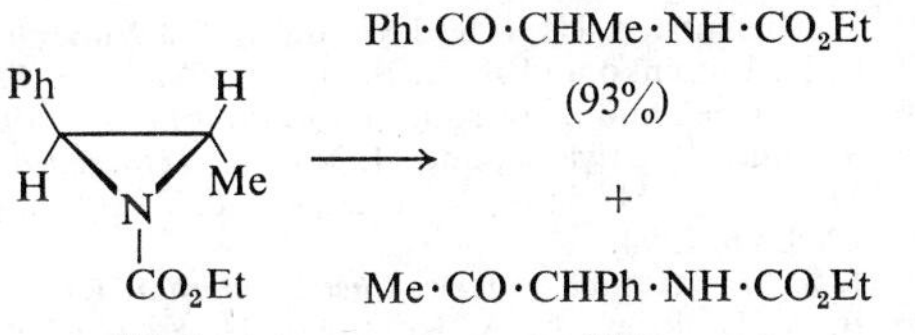

[280] V. J. Traynelis and K. Yamauchi, *Tetrahedron Letters*, 1969, 3619.
[281] J. Doucet, D. Gagnaire, and A. Robert, *Compt. rend.*, 1969, **268**, *C*, 1700.
[282] A. Durst, in 'Advances in Organic Chemistry: Methods and Results', vol. VI, ed. E. C. Taylor and H. Wynberg, Interscience, New York, 1969, p. 285.
[283] W. S. MacGregor, *Quart. Reports Sulfur Chem.*, 1968, 3, 149.
[284] P. S. Carleton and W. J. Farrissey, *Tetrahedron Letters*, 1969, 3485.
[285] H. D. Becker, *J. Org. Chem.*, 1969, **34**, 4162.
[286] J.-C. Richer and D. Perelman, *Canad. J. Chem.*, 1970, **48**, 570.
[287] A. Nudelman and D. J. Cram, *J. Org. Chem.*, 1969, **34**, 3659.

include an interesting cyclopropane synthesis [(76)+Ph·NH·NH·Ph formed from Ph·N:CH·Ph and $p$-Me·$C_6H_4$·SO·$CH_2$·Ph in the presence of BuLi].[287]

An activated methylene grouping may be dimethylthiolated with dimethyl sulphoxide [R·$SO_2$·$CH_2$·$SO_2$·R → R·$SO_2$·C·($SMe_2^+$)·$SO_2$·R];[288, 289] an investigation of an explosion following an apparently unexceptionable experiment led to the discovery of the acid-catalysed self-reaction of dimethyl sulphoxide in the presence of inorganic salts [290] ($Me_2S^+OH$ + $Me_2SO$ → $Me_2S^+$·OMe + Me·SOH;   Me·SOH + 2$Me_2SO$ → Me·$SO_3H$ + 2$Me_2S$; $Me_2S^+$·OMe → $Me_2S$ + H·CHO + $H^+$) though the role of the salt is not understood.

Deliberate exploitation of the mild oxidative properties of dimethyl sulphoxide continues. This feature is implicit in a number of the reactions discussed in the preceding paragraphs and in oxidative ring-opening of (76a); the key step in this reaction, a useful general method for the preparation of α-alkoxycarbonylamino-ketones, is proton transfer and loss of $Me_2S$.[291] Dimethyl sulphoxide is more usually employed in conjunction with another reagent in the conversion of primary and secondary alcohols into aldehydes and ketones under mild conditions, with dicyclohexylcarbodi-imide,[292-294] with phosphorus pentoxide,[293, 241] with acetic anhydride,[293-297] with sulphur trioxide or pyridine–$SO_3$,[298] with ketenimines,[299] with mercury(II) acetate;[300] dimethyl sulphoxide with iodine, at 140—160 °C, is a convenient reagent for the oxidation of 2-picolines to corresponding aldehydes,[301] and elevated temperatures are involved in a number of the other procedures also.[297, 300] Refluxing dimethyl sulphoxide, alone [291] or with an air stream,[295] is also an effective oxidant for preparation of ketones.

Reagents of this type have been widely adopted in carbohydrate chemistry,[293, 302] though side-reactions can occur; whereas a ·$CH_2OH$ group of a sugar is oxidised cleanly by dimethyl sulphoxide–dicyclohexylcarbodi-

---

[288] H. Diefenbach, H. Ringsdorf, and R. E. Williams, *Chem. Ber.*, 1970, **103**, 183.
[289] L. N. Lutsenko and B. G. Boldyrev, *Zhur. org. Khim.*, 1969, **5**, 1246.
[290] C.-T. Chen and S.-J. Yan, *Tetrahedron Letters*, 1969, 3855.
[291] S. Fujita, T. Hiyama, and H. Nozaki, *Tetrahedron Letters*, 1969, 1677.
[292] M. Matsui, M. Saito, M. Okada, and M. Ishidate, *Chem. and Pharm. Bull. (Japan)*, 1968, **16**, 1294.
[293] J. S. Brimacombe, *Angew. Chem., Internat. Edn.*, 1969, **8**, 401.
[294] B. A. Dmitriev, A. A. Kost, and N. K. Kochetkov, *Izvest. Akad. Nauk S.S.S.R., Ser. khim.*, 1969, 903.
[295] W. H. Clement, T. J. Dangieri, and R. W. Tuman, *Chem. and Ind.*, 1969, 755.
[296] K. Antonakis, *Bull. Soc. chim. France*, 1968, 2972.
[297] M. I. Sayed and D. A. Wilson, *J. Chem. Soc. (C)*, 1969, 2168.
[298] J. R. Parikh and W. von E. Doering, *U.S. Pat.* 3,444,216 (*Chem. Abs.*, 1969, **71**, 50,375); *J. Amer. Chem. Soc.*, 1967, **89**, 5505.
[299] R. E. Harmon, C. V. Zenarosa, and S. K. Gupta, *Chem. and Ind.*, 1969, 1428; R. E. Harmon, C. V. Zenarosa, and S. K. Gupta, *Tetrahedron Letters*, 1969, 3781.
[300] J. M. Tien, H.-J. Tien, and J.-S. Ting, *Tetrahedron Letters*, 1969, 1483.
[301] A. Markovac, C. L. Stevens, A. A. Ash, and B. E. Hackley, *J. Org. Chem.*, 1970, **35**, 841.
[302] A. K. Mallams, *J. Amer. Chem. Soc.*, 1969, **91**, 7505.

imide (the 'Pfitzner–Moffatt reagent'), $Me_2SO–Ac_2O$ (the 'Albright–Goldman reagent') gives three products, one of which is the $\cdot CH_2 \cdot OMe$ analogue (see also ref. 297). Acetates [294, 302, 303] and methylthiomethyl ethers [294] may also be formed by $Me_2SO–Ac_2O$, though this reagent is useful in causing no cleavage while oxidising cleavage-sensitive secondary alcohols:[295] it is also valuable for oxidative modification of quinones.[304] Tertiary alcohols in which the $-OH$ group lies $\beta$ to a nitro-group are converted into nitromethylene analogues, $>C:CH\cdot NO_2$, with $Me_2SO–Ac_2O$, in quantitative yield and with a preponderance of one geometrical isomer.[305] $Me_2SO$–Dicyclohexylcarbodi-imide does not oxidise a secondary alcohol in which the $-OH$ group is *cis* to an $-NMe_2$ grouping;[302] such compounds give only *O*-acetates with $Me_2SO–Ac_2O$.[302, 303]

An attempted intramolecular oxidation of ethylsulphinyl-L-galactitol, in solution with dicyclohexylcarbodi-imide, $Ac_2O$, $H_3PO_4$, $P_2O_5$, or a cation exchanger, was unsuccessful.[241]

## 6 Thiocyanates and Isothiocyanates; Sulphonyl Cyanides

**Preparation.**—Simple alkyl thiocyanates and isothiocyanates are obtained from corresponding benzoate esters in fused KSCN–NaSCN; $R \cdot CO_2Me$ (R = 3,5-dihydroxyphenyl) gives 75% $Me \cdot SCN$ and 6% $Me \cdot NCS$.[306] Aryl nitro-compounds give isothiocyanates in better yields (50—84%) through treatment with $CS_2$ and an alkyl or aryl sulphide (150—170 °C) than by $CS_2$–$RO^-Na^+$–150 to 170 °C or COS–aryl sulphide–150 to 170 °C;[307] *m*-dinitrobenzene gives *m*-di(isothiocyanato)benzene as minor product. Treatment of sulphenyl anions with BrCN gives thiocyanates, though a cyano-group enters the phenyl grouping too in the reaction with $(Ph\cdot N:CR\cdot S^-)_2Hg^{2+}$.[308] Thiocyanogen, $(SCN)_2$, has been used for thiocyanation of thiophen in the presence of $AlCl_3$;[309] stereospecific addition of thiocyanogen to alkenes requires the presence of a radical inhibitor, when *cis*-2,5-dimethylhex-3-ene gives ($\pm$)-3,4-dithiocyanato- and *threo*-3-thiocyanato-4-isothiocyanato-2,5-dimethylhexanes, and the *trans*-isomer gives the corresponding *meso*- and *erythro*-adducts.[310] Hex-1-ene gives 1,2-dithiocyanato-, 1-thiocyanato-2-thiocyanato-, and 1-isothiocyanato-2-thiocyanato-hexanes, *trans*-2-octalin gives the diaxial adduct; but addition to styrenes is non-stereospecific and gives only Markownikoff-oriented adducts.[310]

---

[303] Y. Ali and A. C. Richardson, *J. Chem. Soc.* (C), 1969, 320.
[304] H. W. Moore and R. J. Wikholm, *Tetrahedron Letters*, 1968, 5049.
[305] H. P. Albrecht and J. G. Moffatt, *Tetrahedron Letters*, 1970, 1063.
[306] E. M. Wadsworth and T. I. Crowell, *Tetrahedron Letters*, 1970, 1085.
[307] G. Ottmann and E. H. Kober, *Angew. Chem., Internat. Edn.*, 1969, **8**, 760.
[308] W. Ried and W. Merkel, *Tetrahedron Letters*, 1969, 4421.
[309] F. M. Stoyanovich, G. I. Gorushkina, and Ya. L. Goldfarb, *Izvest. Akad. Nauk S.S.S.R., Ser. khim.*, 1969, 387.
[310] R. G. Guy, R. Bonnett, and D. Lanigan, *Chem. and Ind.*, 1969, 1702.

**Reactions.**—Trimethylstannyl isothiocyanate is obtained in 83% yield from Ph·CO·NCS and Me$_3$Sn·NMe$_2$, though the corresponding bis(trimethylstannyl)methylamine reacts differently in giving (Me$_3$Sn)$_2$S and Ph·CO·N:C:N·Me.[311] Benzoyl isothiocyanate gives Ph·CO·N:C(SR)$_2$ by successive treatment with R·SH and R·I;[312] isothiocyanates react similarly with alkylidenetriphenylphosphoranes giving thioamides,[313, 314] and sulphonyl isothiocyanates are alkylated in a similar way by enamines.[315] The addition of amines at the same point is an essential step in the Edman stepwise degradation of polypeptides; further studies with Me·NCS in place of phenyl isothiocyanate have been reported,[316] using isotope dilution techniques to aid quantitative analysis.

New cycloaddition reactions of isothiocyanates have been reported, thiamine giving (77),[317] and 3-aroylaziridines giving thiazolines.[318]

Hydrosulphite reduction of selenocyanates and reaction with Cl·CH$_2$·CO$_2$H gives carboxymethyl selenides,[319] and 2-anilino-2-(alkylthio)malonates are obtained through treatment of malonate anion with isothiocyanates.[320] Carbamoyl isothiocyanates show a range of reactions consistent with equilibration with thiocarbamoyl isocyanates, R$_2$N·CS·NCO, in reaction mixtures.[321] Cyclisation of desyl thiocyanate (78) to (79) and cis-dibenzoylstilbene episulphide by NaH is best explained

in terms of Ph·CHS as an intermediate.[322] 1-Thiocyanato-2-(isothiocyanato)naphthalene gives dithiazenes with amines;[323] halogenated 2-chlorophenyl isothiocyanates give corresponding 2-chlorobenzothiazoles

---

[311] K. Itoh, I. Matsuda, and Y. Ishii, *Tetrahedron Letters*, 1969, 2675.
[312] B. W. Nash, R. A. Newberry, R. Pickles, and W. K. Warburton, *J. Chem. Soc. (C)*, 1969, 2794.
[313] H. Saikachi and K. Takai, *Yakugaku Zasshi*, 1969, **89**, 1401 (*Chem. Abs.*, 1970, **72**, 12,452).
[314] H. J. Bestmann and S. Pfohl, *Angew. Chem., Internat. Edn.*, 1969, **8**, 762.
[315] J. Goerdeler and U. Krone, *Chem. Ber.*, 1969, **102**, 2273.
[316] F. F. Richards, W. T. Barnes, R. E. Lovins, R. Salomone, and M. D. Waterfield, *Nature*, 1969, **221**, 1241.
[317] A. Takamizawa, K. Hirai, S. Matsumoto, S. Sakai, and Y. Nakagawa, *Chem. and Pharm. Bull. (Japan)*, 1969, **17**, 910.
[318] J. W. Lown, G. Dallas, and T. W. Maloney, *Canad. J. Chem.*, 1969, **47**, 3557.
[319] L. Christiaens and M. Renson, *Bull. Soc. chim. belges*, 1968, **77**, 153.
[320] I. T. Kay and P. J. Taylor, *J. Chem. Soc. (C)*, 1968, 2656.
[321] J. Goerdeler and D. Wobig, *Annalen*, 1970, **731**, 120.
[322] G. E. Kuhlmann and D. C. Dittmer, *J. Org. Chem.*, 1969, **34**, 2006.
[323] R. Pohloud ek-Fabini and M. Selchau, *Arch. Pharm.*, 1969, **302**, 527.

at 330—380 °C,[324] while 2-thiocyanatobenzoyl chlorides and selenium analogues give benzothiazinones and selenazinones with HCl.[325]

Although $p$-hydroxybenzyl isothiocyanate is hydrolysed cleanly at $-$NCS, allyl isothiocyanate decomposes in aqueous solution into allylamine, also to the dithiocarbamate which gives tetra- and penta-sulphides, eventually hydrocarbons and sulphur.[326] Optically active thiocyanates $R^1R^2CH \cdot SCN$ can be isomerised into isothiocyanates with net retention of configuration;[327, 330] other ion-pair return studies deal with *exo*-2-norbornyl[328] and furfuryl[329] thiocyanates, whose isomerisations depend critically on solvent. Diphenylmethyl thiocyanates give thiobenzophenones and $CN^-$ *via* a concerted $E$1cb-like mechanism on treatment in isopropanol with $Pr^i \cdot O^-Na^+$,[330] though their solvolysis, exchange, and isomerisation reactions involve carbonium ions in 95% aqueous acetone.[331] Thiocyanate ion, in aqueous solution, reacts with HO· (produced by pulse radiolysis of solvent) giving the thiocyanate radical–solute complex $(CNS)_2^-$; this species reacts specifically with the indole grouping of tryptophan residues in lysozyme, resulting in enzyme inactivation, the visible absorption of the species ($\lambda_{max}$ 475 nm) enabling direct monitoring of the reaction.[332]

Isothiocyanates $SCN \cdot CHR \cdot CO_2Et$ may be used in peptide synthesis, though their coupling with *N*-benzyloxycarbonyl-glycyl-L-phenylalanine requires conditions (60—150 °C; 1·5—5 hr; toluene, pyridine, or dioxan) resulting in extensive racemisation.[333] Cotton effects at 340, 260, and 205 nm are displayed by these derivatives, and corresponding chiral isothiocyanatoalkanols,[334] though isotropic absorption features associated with the longest-wavelength Cotton effect are apparently not detectable.

**Sulphonyl Cyanides.**—Oxidation of aryl thiocyanates with *m*-chloroperbenzoic acid, in hexane at 60 °C, gives good yields of corresponding sulphonyl cyanides,[335] and an alternative route, treatment of an aryl sulphinate with $Cl \cdot CN$ in aqueous solution at 20 °C, is similarly efficient.[336] Recent studies[337] of nucleophilic displacement of $CN^-$, and demonstration of a cycloaddition of a type not shown by simple cyanides (80) → (81), promise a busy future for these novel compounds.

[324] E. Degener, G. Beck, and H. Holtschmidt, *Angew. Chem., Internat. Edn.*, 1970, **9**, 65.
[325] G. Simchen and J. Wenzelburger, *Chem. Ber.*, 1970, **103**, 413.
[326] S. Kawakishi and M. Namiki, *Agric. and Biol. Chem. (Japan)*, 1969, **33**, 452.
[327] U. Tonellato, O. Rossetto, and A. Fava, *J. Org. Chem.*, 1969, **34**, 4032.
[328] L. A. Spurlock and T. E. Parks, *J. Amer. Chem. Soc.*, 1970, **92**, 1279.
[329] L. A. Spurlock and R. G. Fayter, *J. Org. Chem.*, 1969, **34**, 4035.
[330] A. Ceccon, U. Miotti, U. Tonellato, and M. Padovan, *J. Chem. Soc. (B)*, 1969, 1084.
[331] A. Ceccon, A. Fava, and I. Papa, *J. Amer. Chem. Soc.*, 1969, **91**, 5547.
[332] J. E. Aldrich, R. B. Cundall, G. E. Adams, and R. L. Willson, *Nature*, 1969, **221**, 1049.
[333] J. H. Jones and R. Fairweather, *Makromol. Chem.*, 1969, **128**, 279.
[334] B. Halpern, W. Patton, and P. Crabbé, *J. Chem. Soc. (B)*, 1969, 1143.
[335] R. G. Pews and F. P. Corson, *Chem. Comm.*, 1969, 1187.
[336] J. M. Cox and R. Ghosh, *Tetrahedron Letters*, 1969, 3351.
[337] A. M. van Leusen and J. C. Jagt, *Tetrahedron Letters*, 1970, 967, 971.

Me—⟨C₆H₄⟩—SO₂—CN + CH₂N₂ $\xrightarrow[-30\,°C]{Et_2O}$ Me—⟨C₆H₄⟩—SO₂—N(NH)=N
(80)                                                    (81)

## 7 Sulphenic Acids and their Derivatives

Sulphenic acids, R·S·OH, lie on the reduction pathway from more stable oxysulphur acids and are briefly mentioned in this respect elsewhere in this Chapter. Stable nitroxyl radicals derived from 2,2,6,6-tetramethyl-4-piperidone oxidise aryl thiols to mixtures of disulphide and sulphenic acid, together with the $N$-(arylsulphenyl)piperidone.[338] Treatment of bis(1-β-D-ribofuranosyl-4-thiouracil) disulphide with aqueous alkali gives a mixture of the corresponding thione and sulphenic acid (isolated as the Ag salt), though aqueous acid gives thione and uracil derivatives.[339]

Alkyl arenesulphenates have been used in studies concerned more with the fate of the alkyl residue during chlorinolysis in acetic acid, which with Ar·S·O·CH($^{14}$CH₃)·CH(CH₃)Ph gives alkyl chloride and acetate, together with products of group migration, and (presumably) Ar·SO·Cl (Ar = 2,4-dinitrophenyl).[340] Pyrolysis of a diaryl sulphenate gives a mixture of $o$-hydroxyphenyl aryl sulphide and disulphide;[341] t-butyl 2-nitrophenyl-sulphenate decomposes in anisole through an $S_N1$ pathway (via Ar·S⁺ + Bu$^t$O⁻) since t-butanol can be identified in the products, which include 2-nitrophenyl methoxyphenyl sulphides formed by attack of the sulphenyl cation on the solvent.[342] Allyl sulphenates, obtained [220] from allyl alcohol and a sulphenyl halide or SCl₂, rearrange into corresponding allyl sulph-oxides;[219-222] aryl allyl sulphoxides show reactions which may be rationalised on the basis of competition between isomers Ar·S·O·CH₂·CH:CH₂ ⇌ Ar·SO·CH₂·CH:CH₂ ⇌ Ar·SO·CH:CH·Me,[221] formation of $N$-arylsulphenylpiperidine by addition of piperidine being shown to be derived from the sulphenate rather than from the arylsulphinylpiperidine.

**Sulphenyl Halides.**—Interest in these compounds, readily prepared by halogen cleavage of disulphides,[10e] has been sustained. Chlorinolysis of carbohydrate disulphides containing the grouping ·O·CS·S·S·CS·O· is less prominent than chlorine addition, giving ·O·CCl(S·Cl)·S·S·CCl(S·Cl)·O·.[343] Chlorinolysis of dimethyl diselenide gives crystalline Me₂SeCl₃, which is mainly dimeric in CH₂Cl₂;[344] the thermal

[338] K. Murayama and T. Yoshioka, *Bull. Chem. Soc. Japan*, 1969, **42**, 1942.
[339] B. C. Pal, M. Uziel, D. G. Doherty, and W. E. Cohn, *J. Amer. Chem. Soc.*, 1969, **91**, 3634.
[340] H. Kwart, E. N. Givens, and C. J. Collins, *J. Amer. Chem. Soc.*, 1969, **91**, 5532.
[341] D. R. Hogg, J. H. Smith, and P. W. Vipond, *J. Chem. Soc. (C)*, 1968, 2713.
[342] D. R. Hogg and P. W. Vipond, *J. Chem. Soc. (C)*, 1970, 60.
[343] B. S. Shasha, W. M. Doane, C. E. Russell, and C. E. Rist, *J. Org. Chem.*, 1969, **34**, 1642.
[344] K. J. Wynne and J. W. George, *J. Amer. Chem. Soc.*, 1969, **91**, 1649.

*Aliphatic Organo-sulphur Compounds etc.*

stability of tetracovalent halides increases in the order $MeSCl_3$, $MeSeCl_3$, $MeTeCl_3$.[345] Controlled chlorination of pyrazole-4,4′-disulphides in moist acetic acid at 15 °C gives the 4-sulphonyl chloride;[346] an improved synthesis of trichloromethanesulphonyl chloride from $Cl_3C \cdot S \cdot Cl$, using HOCl at 60 °C,[347] probably involves the sulphine $Cl_2CSO$,[348] and the high yield reported [347] (90%) is surprising in view of the instability of the sulphine, and the hydrolysis of $Cl_3C \cdot S \cdot Cl$ to $Cl \cdot CO \cdot S \cdot Cl$.[348, 350]

Novel sulphenyl fluorides are obtained from $CF_3 \cdot S \cdot Cl$ by treatment with $HgF_2$, first-formed $CF_3 \cdot S \cdot F$ existing in equilibrium with $CF_3 \cdot SF_2 \cdot S \cdot CF_3$ below 0 °C, which suffers cleavage into $CF_3 \cdot S \cdot S \cdot CF_3 + CF_3 \cdot SF_3$ by $CF_3 \cdot S \cdot F$.[349]

The stability of azobenzene-2-sulphenyl halides in aqueous or ethanolic solution, in contrast with other sulphenyl halides, has been ascribed to reversible cyclisation (82) → (83); further u.v.-spectroscopic studies have

been reported.[351] N.m.r.-spectroscopic studies [352] of the spontaneous decomposition of $Me \cdot S \cdot Cl$ (complete within a few hours at room temperature) show formation of a mixture of alkyl and sulphenyl chlorides and polysulphides; e.s.r. monitoring provides no evidence for intervention of free radicals. Crystalline $MeSCl_3$ decomposes rapidly at room temperature,[345, 352] giving 60% $Cl \cdot CH_2 \cdot S \cdot Cl$ and 20% $Me \cdot S \cdot Cl$ with traces of polysulphides. Nuclear quadrupole resonance spectra of aliphatic sulphenyl chlorides have been reported.[353] Unusual features of the thermal decomposition of sulphenyl chlorides have been discussed;[354] assuming well-documented *trans*-addition, the sulphenyl chloride obtained from *trans*-stilbene and $SCl_2$ is (84), and formation of (85) on heating, as the major product, requires the mechanism shown or its ion-pair equivalent.

Addition reactions of $Me \cdot S \cdot Cl$ and $SCl_2$ to alkenes have been reviewed;[355] closely-related compounds $R_2N \cdot S \cdot Cl$ and $Cl_2C:N \cdot S \cdot Cl$ (from thiocyanogen and $Cl_2$) show similar addition reactions.[350, 355] Substituent effects

---

[345] K. J. Wynne and P. S. Pearson, *Inorg. Chem.*, 1970, **9**, 106.
[346] R. J. Alabaster and W. J. Barry, *J. Chem. Soc. (C)*, 1970, 78.
[347] E. Dykman and P. Bracha, *Israel J. Chem.*, 1969, **7**, 589.
[348] J. Silhanek and M. Zbirovsky, *Chem. Comm.*, 1969, 878.
[349] F. Seel and W. Gombler, *Angew. Chem., Internat. Edn.*, 1969, **8**, 773.
[350] G. Zumach and E. Kühle, *Angew. Chem., Internat. Edn.*, 1970, **9**, 54.
[351] A. Chaudhuri and P. K. Sarma, *J. Chem. Soc. (C)*, 1969, 1828.
[352] I. B. Douglass, R. V. Norton, R. L. Weichmann, and R. B. Clarkson, *J. Org. Chem.*, 1969, **34**, 1803.
[353] T. A. Babushkina, V. S. Levin, M. I. Kalinkin, and G. K. Semin, *Izvest. Akad. Nauk S.S.S.R., Ser. khim.*, 1969, 2340.
[354] T. J. Barton and R. G. Zika, *Tetrahedron Letters*, 1970, 1193.
[355] W. H. Mueller, *Angew. Chem., Internat. Edn.*, 1969, **8**, 482.

in the electrophilic addition of 4-substituted 2-nitrophenyl sulpheny bromides to cyclohexene in acetic acid at 25 °C have been assessed;[356] the electrophilic addition step is rate-limiting. Kinetic control in the addition of 4-chlorobenzenesulphenyl chloride to butadienes in non-polar solvents gives 1,2-adducts, which rearrange to 1,4-adducts in polar solvents [357] (cf. ref. 417). Ethanesulphenyl chloride gives only one isomer (86) with 2-methoxy-

$\Delta^3$-dihydropyran.[358] Peroxide-induced addition of toluene-p-sulphenyl chloride to alkynes gives mixtures of isomeric 2-chlorovinyl sulphides.[359, 131] Arylsulphenylation of ketones proceeds via the enol tautomer, optically-active 2-methyl-indan-1-one giving a racemic adduct.[360] $SF_5Cl$ [361] undergoes reversible addition ($\cdot SF_5$ radicals) to trifluoroethylene in the gas phase giving medium-controlled ratios of $SF_5 \cdot CHF \cdot CF_2Cl$ and $SF_5 \cdot CF_2CHFCl$.[362] Ethane-1,2-di(sulphenyl chloride) gives p-dithians by reaction with acetylenes.[363]

Substitution reactions of sulphenyl halides are well represented; $Me \cdot S \cdot Cl$ reacts with ethyl acetoacetate to give only traces of the 2-methylthio-substitution product, but a good yield of this compound may be obtained from ethyl α-chloroacetoacetate with $Me \cdot S \cdot Cl$ (prepared in situ from $Me \cdot S \cdot S \cdot Me$ and $SO_2Cl_2$ at $-40$ °C) followed by zinc–acetic acid reduction at $-5$ °C.[364] Phenyl sulphenyl chloride, or the sulphonyl chloride, reacts with $NN'$-diphenyl-diketopiperazines through chain reactions (in boiling pyridine) to give new meso-ionic compounds (87), which undergo expected 1,3-dipolar cycloaddition reactions.[365] Unstable S-substitution products $R^1 \cdot C(:N \cdot R^2) \cdot S \cdot S \cdot R^3$ result from thioamides $R^1 \cdot CS \cdot NH \cdot R^2$ and $R^3 \cdot S \cdot Cl$,

---

[356] C. Brown and D. R. Hogg, J. Chem. Soc. (B), 1969, 1054.
[357] G. Kresze and W. Kosbahn, Annalen, 1970, 731, 67.
[358] M. J. Baldwin and R. K. Brown, Canad. J. Chem., 1969, 47, 3553.
[359] V. Calo, G. Scorrano, and G. Modena, J. Org. Chem., 1969, 34, 2020.
[360] R. Gustafsson, C. Rappe, and J. O. Levin, Acta Chem. Scand., 1969, 23, 1843.
[361] R. E. Banks, R. N. Haszeldine, and W. D. Morton, J. Chem. Soc. (C), 1969, 1947.
[362] H. W. Sidebottom, J. M. Tedder, and J. C. Walton, Chem. Comm., 1970, 253.
[363] W. H. Mueller and M. B. Dines, J. Heterocyclic Chem., 1969, 6, 627.
[364] I. T. Kay, D. J. Lovejoy, and S. Glue, J. Chem. Soc. (C), 1970, 445.
[365] J. Honzl and M. Sorm, Tetrahedron Letters, 1969, 3339.

Ph
|
Ph·S\_N+/S·Ph
⁻O/N+\O⁻
|
Ph

(87)

the products giving $R^1 \cdot CN + Ph \cdot S \cdot SH$ when $R^2 = H$, $R^3 = Ph$.[366] Mixtures of sulphides and disulphides are obtained from a 2-nitrophenylthiosulphenyl chloride–anisole–iron powder reaction mixture, the major product being *p*-methoxyphenyl 2-nitrophenyl disulphide (61%).[367] The ready substitution of indoles by aryl sulphenyl chlorides is exploited in the use of 2,4-dinitrophenyl-1,5-disulphenyl chloride for cross-linking peptide chains at tryptophan residues.[368]

A wide range of products can result from the decomposition of Me·S·Cl in the presence of secondary amines at $-60$ °C, major products (Me·S·NR$_2$, R$_2$N·CH$_2$·NR$_2$, Me$_2$S$_2$, Me$_2$S$_3$) being consistent with the primary formation of thioformaldehyde through either free-radical or 1,2-elimination pathways.[369] The reaction of triphenylmethane sulphenyl chloride with Bu·NH$_2$ proceeds at a rate proportional to the square of the amine concentration, and is accelerated by added base or salt.[370] Ethanesulphenyl chloride (from the disulphide and SO$_2$Cl$_2$) gives diethyl disulphide on treatment with NEt$_3$ in CHCl$_3$ at 30 °C, but with 1,4-bis(dimethylamino)benzene it gives a blue coloration (tetrahydrofuran as solvent) followed by precipitation of the cation-radical of the amine chloride, characterised by a strong e.s.r. signal.[371]

**Sulphenamides.**—Aminolysis of a thiolsulphonate, $R^1 \cdot S \cdot SO_2 \cdot R^2$, gives a sulphenamide and an ammonium sulphinate;[372–374] other preparations of sulphenamides are noted earlier in this section. *N*-(*o*-Nitrophenylsulphenyl)amino-acids continue to be widely used as *N*-protected components in peptide synthesis;[375] a novel variant involves concurrent deprotection and coupling, an acylamino-acid thiophenyl ester reacting with an *N*-(*o*-nitrophenylsulphenyl)amino-acid or peptide in the presence of imidazole to give corresponding terminal *N*-acyl di- or oligo-peptides respectively in high yields.[376] Unmasking of an *o*-nitrophenylsulphenyl-protected amine is brought about with two equivalents of HCl in MeOH.[375]

[366] W. Walter and P. M. Hell, *Annalen*, 1969, **727**, 22.
[367] T. Fujisawa, T. Kobori, and G. Tsuchihashi, *Tetrahedron Letters*, 1969, 4291.
[368] F. M. Veronese, E. Boccu, and A. Fontana, *Internat. J. Protein Res.*, 1970, **2**, 67.
[369] D. A. Armitage and M. J. Clark, *Chem. Comm.*, 1970, 104.
[370] E. Ciuffarin and G. Guaraldi, *J. Amer. Chem. Soc.*, 1969, **91**, 1745.
[371] N. T. Ioffe, M. I. Kalinkin, I. N. Rozhkov, and I. L. Knunyants, *Izvest. Akad. Nauk S.S.S.R., Ser. khim.*, 1969, 2070.
[372] B. G. Boldyrev and S. A. Kolesnikova, *Zhur. org. Khim.*, 1969, **5**, 1088.
[373] B. G. Boldyrev and L. E. Kolmakova, *Zhur. org. Khim.*, 1969, **5**, 1669.
[374] J. E. Dunbar and J. H. Rogers, *J. Org. Chem.*, 1970, **35**, 279.
[375] B. J. Johnson, *J. Chem. Soc. (C)*, 1969, 1412.
[376] H. Faulstich, *Chimia (Switz.)*, 1969, **23**, 150.

The rotation barrier about the S—N bond in sulphenamides can be as large as ca. 20 kcal mol$^{-1}$ (in $p$-R·C$_6$H$_4$·S·NPr$^{\text{i}}$·SO$_2$·Ph),[377] electronic factors as well as the more obvious lone-pair repulsions in the sulphe amide unit determining the height of the rotation barrier. N.m.r. data give this information; chemical shift non-equivalence (AB quartets at low temperatures) of the benzyl protons in Cl$_3$C·S·NR·CH$_2$·Ph indicates a substantial barrier,[378] but the flexing of these molecules (Scheme 5) includes inversion

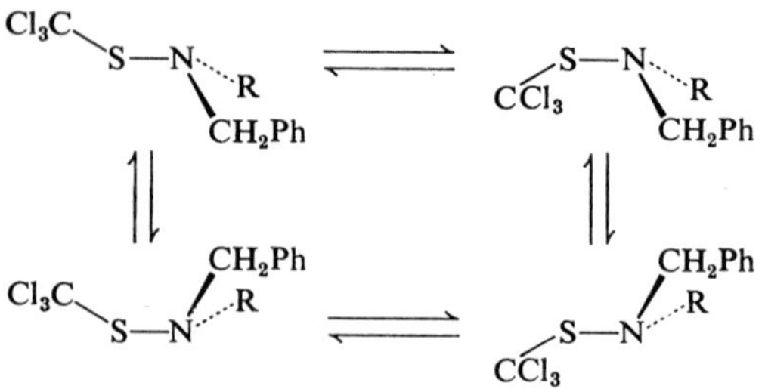

**Scheme 5**

of the nitrogen pyramid. The rate-determining step in the overall conformational change is the notably 'slow' rotation about the S—N bond.[378] A more general review of the use of n.m.r. spectroscopy in assessing hindrance to rotation about single bonds offers further detail.[379]

New reactions of sulphenamides include CS$_2$ insertion, giving amine carbotrithioates R·S·S·CS·NR$_2$;[374] cyclisation of N-acyl trichloromethane sulphenamides to give (88), in aqueous acid or alkali;[380] conversion of menthyl methylphosphinate into its methylthio-analogue C$_{10}$H$_{19}$·O·(O)P(Me)·S·Me with Me·S·NEt$_2$ at 130 °C as an alternative to successive treatment with sulphur and MeI.[381]

Rearrangement of 2-nitrophenyl sulphenanilides (89) into corresponding $o$-arylazophenylsulphinates (90) in aqueous alcoholic NaOH is first order in sulphenanilide and OH$^-$, $^{18}$O-labelling showing that transfer of both oxygen atoms from nitrogen to sulphur is involved;[382] no rearrangement occurs when R = $p$-nitrophenyl, though (90; R = $p$-methoxyphenyl) rearranges slightly faster than the phenyl analogue. Speculation that (90; R = Ph) is formed from (82), rather than (83) as claimed,[351] in aqueous solution, ends on referring to literature cited in reference 351.

---

[377] M. Raban and F. B. Jones, *J. Amer. Chem. Soc.*, 1969, **91**, 2180.
[378] M. Raban, F. B. Jones, and G. W. J. Kenney, *Tetrahedron Letters*, 1968, 5055; *J. Amer. Chem. Soc.*, 1969, **91**, 6677.
[379] H. Kessler, *Angew. Chem.*, 1970, **82**, 237.
[380] F. Becke and J. Gnad, *Annalen*, 1969, **726**, 110.
[381] H. P. Benschop, D. H. J. M. Platenburg, F. H. Meppelder, and H. L. Boter, *Chem. Comm.*, 1970, 33.
[382] C. Brown, *J. Amer. Chem. Soc.*, 1969, **91**, 5832.

CCl$_3$·S·NH·CO·R  ⟶  (88) [structure: N—S heterocycle with O and =O]

(89) [o-nitrophenyl S·NH·R]   OH⁻ ⟶   (90) [o-(N=N-R)phenyl SO$_2^-$]

## 8 Sulphinic Acids and their Derivatives

Hydrogenolysis of aryl sulphones, sulphonic acid esters, and sulphonamides at a mercury cathode gives sulphinic acids (Ar·SO$_2$·R + 2e + 2H$^+$ → Ar·SO$_2$H + R·H), via radical anions which are often relatively long-lived (e.s.r.);[383] NN-diphenyl toluene-p-sulphonamide gives a red-brown radical anion. Reduction (Na–NH$_3$) of benzylsulphonyl compounds derived from cysteine and related thiols gives near-quantitative conversion into corresponding sulphinic acids (·SH → ·S·CH$_2$Ph → ·SO$_2$·CH$_2$Ph → ·SO$_2$H);[384] purine-6-sulphonic acid gives the corresponding 6-sulphinic acid on electrolysis at low pH, though further reduction to 6-thiopurine takes place rapidly.[385]

Chlorosulphinyl compounds may be prepared from hydrocarbons by treatment with SOCl$_2$ and AlCl$_3$, a recent example being 1-chlorosulphinyl-adamantane [386] which has been elaborated into the corresponding acid and its amides and esters. Ethane-1,2-disulphinic acid, m.p. 109—111 °C, has been prepared from the corresponding anhydride by treatment with boiling water during 1 min.[387] The anhydride is prepared by careful hydrolysis of the known 1,2-disulphinyl chloride, in moist tetrahydrofuran;[387] the anhydride of β-carboxyethanesulphinic acid (91) results from treatment of (HO$_2$C·CH$_2$·CH$_2$·S·)$_2$ with 2 equivalents of Cl$_2$,[388] and, like other sulphinyl compounds,[240, 228] shows chemical shift non-equivalence of the ·CH$_2$·SO· protons. New 2-pyridylsulphinic acids, prepared from thiolsulphonate esters py·SO$_2$·S·Ar by alkaline cleavage,[389] are not stable except in the form of their salts, and this observation supports a general assumption of the lability of this class of compound; the obvious stability of ethanedisulphinic acid [387] is a remarkable contrast. Benzenesulphinic acid disproportionates in benzene solution under N$_2$ into Ph·SH + Ph·SO$_3$H;[390] oxygenated

[383] L. Horner and R.-J. Singer, *Tetrahedron Letters*, 1969, 1545.
[384] D. B. Hope, C. D. Morgan, and M. Wälti, *J. Chem. Soc.* (C), 1970, 270.
[385] G. Dryhurst, *J. Electrochem. Soc.*, 1969, **116**, 1357.
[386] H. Sletter, M. Krause, and W. D. Last, *Chem. Ber.*, 1969, **102**, 3357.
[387] W. H. Mueller and M. B. Dines, *Chem. Comm.*, 1969, 1205.
[388] Y. H. Chiang, J. S. Luloff, and E. Schipper, *J. Org. Chem.*, 1969, **34**, 2397.
[389] W. Walter and P. M. Hell, *Annalen*, 1969, **727**, 35.
[390] Z. Yoshida, H. Miyoshi, and K. Kawamoto, *Kogyo Kagaku Zasshi*, 1969, **72**, 1295 (*Chem. Abs.*, 1969, **71**, 101,463).

solutions give $Ph \cdot SO_3H$, $Ph \cdot S \cdot S \cdot Ph$, $Ph \cdot S \cdot SO_2 \cdot Ph$, and $Ph \cdot SO_2 \cdot SO_2 \cdot Ph$, while $Ph \cdot SO_2Et$ is the main product in aerated ethanol. Theoretical assessment of the acidity of $Ph \cdot SO_2H$ and $Ph \cdot SeO_2H$ suggests that conjugation between the benzene ring and its substituent is impeded in these compounds.[391]

Sulpholane, heated at 240—250 °C during 2—3 hr with NaOEt, gives butadiene and ca. 25% $CH_2\!:\!CH \cdot CH\!:\!CH \cdot SO_2^- Na^+$, longer periods of heating giving $Na_2SO_3$ and a mixture of six isomeric octadienes;[392] 2,5-dihydrothiophen 1,1-dioxide and related heterocyclic compounds undergo electrocyclic ring-opening with $KNH_2$–$NH_3$ at $-40$ to $-60$ °C,[393] the dioxide giving the same sulphinate anion as obtained from sulpholane.

Enzymic decarboxylation of L-homocysteine sulphinic acid gives 3-aminopropanesulphinic acid (homohypotaurine);[394] preparative and analytical aspects of the chemistry of amino-sulphonic and -sulphinic acids are discussed,[394] use of iodoplatinate offering rapid quantitative determination of the latter.

Addition of benzenesulphinic acid to p-benzoquinone, giving phenyl 2,5-dihydroxyphenyl sulphone in aqueous solution, pH < 6, follows a similar mechanism (nucleophilic addition followed by deprotonation) to that of the addition of thiosulphate.[395]

A paper giving entry to the small volume of literature on sulphinamides describes[388] straightforward preparations from other sulphinic acid derivatives. While 2-propanesulphinamide possesses mild oxidising properties towards thiols, sodium hydroxymethyl sulphinate ('Rongalite C') continues to be used as a mild reducing agent, giving dihydropyridines.[396] Carboxymethyl sulphenyl chloride gives a sulphinamide with morpholine, the unusual u.v. absorption ($\lambda_{max}$ 273 nm; $\varepsilon$ 14,000; unaffected by alkali) compared with other aliphatic analogues prompting its formulation as the resonance-stabilised betaine (92);[388] aminolysis of the mixed anhydride (91) gives $(R \cdot NH \cdot CO \cdot CH_2 \cdot CH_2)_2SO_2$.[388]

Sulphinic acid esters adopt the *gauche* conformation (93) rather than (94), as judged by dipole moment measurements,[397] lone-pair repulsion

(91)    (92)

---

[391] D. De Filippo and F. Momicchioli, *Tetrahedron*, 1969, **25**, 5733.
[392] V. I. Dronov and A. U. Baisheva, *Khim. seraorg. Soedinenii, Soderzh. Neftyakh Nefteprod.*, 1968, **8**, 144 (*Chem. Abs.*, 1969, **71**, 80,599).
[393] H. Kloosterziel, J. A. A. Van Drunen, and P. Galama, *Chem. Comm.*, 1969, 885.
[394] B. Jolles-Bergeret, *European J. Biochem.*, 1969, **10**, 569.
[395] Y. Ogata, Y. Sawacki, and M. Isono, *Tetrahedron*, 1969, **25**, 2715; *ibid.*, 1970, **26**, 731.
[396] K. Wallenfals and W. Haustein, *Annalen*, 1970, **732**, 139.
[397] O. Exner, P. Dembech, and P. Vivarelli, *J. Chem. Soc. (B)*, 1970, 278.

(93)    (94)

being more important than steric considerations. Pyrolysis of sulphinates derived from 5α-cholestane [6β-SH or disulphide gives mixtures of 6-oxocholestanes and (R)- and (S)-6α- and -6β-methyl sulphinates with Pb(OAc)$_4$ and MeOH–CHCl$_3$] gives 5α-cholest-6-ene at a rate dependent upon the the chirality at sulphur; the reaction proceeds through a five-membered cyclic transition state as for other *syn*-eliminations of *N*- or *S*-mono-oxides, though less readily, the more reactive isomer (giving 70% olefin after 16 hr reflux in decalin) being the (R)-isomer (67; MeO in place of Ph·CH$_2$) since severe non-bonded interactions present in the transition state for the (S)-6β-sulphinate would be expected to diminish the reactivity of this diastereoisomer.[398] 6α-Sulphinates react with MeMgI, to give 6α-methylsulphinyl analogues, but the 6β-isomers do not react and may be isolated from mixtures of 6α- and 6β-isomers after treatment with MeMgI. In showing a positive Cotton effect (in hexane) centred at 230 nm, the (R)-isomer of the 6β-sulphinate behaves as does menthyl (R)-butane-1-sulphinate, whose absolute configuration had been established earlier in a different way.

## 9 Sulphonic Acids and their Derivatives

**Preparation.**—Sulphur dioxide insertion reactions into organometallic compounds can give products with $\geq$C·SO·O· groupings; this is a fast-growing subject, and entry to its literature is given in references 399—401. Typically, C$_5$H$_5$Mo(CO)$_3$R → ·O·SO·R;[399] [C$_5$H$_5$Fe(CO)$_2$]$_2$SnPh$_2$ → ·(O·SO·Ph)$_2$;[400] (Ph·CH$_2$)$_2$Hg → (Ph·CH$_2$·SO$_2$)Hg·CH$_2$·Ph;[401] SO$_2$-insertion into the C—Si bond of silylacetylenes is effected with Cl·SO$_3$·SiMe$_3$, giving R·C⦂C·SO$_3$·SiMe$_3$, readily hydrolysed (aq. Na$_2$CO$_3$) into R·CO·CH$_2$·SO$_3^-$Na$^+$,[402] though sulphonic acids would also be expected as products of mild oxidative hydrolysis of the SO$_2$-insertion products. Oxidation of thiols to sulphonic acids is exemplified in new modifications; photo-oxidation of cysteine to cysteic acid in acetic acid is sensitized by Crystal Violet;[3] a relatively simple new preparation of trifluoromethane-

---

[398] D. N. Jones and W. Higgins, *J. Chem. Soc. (C)*, 1970, 81.
[399] M. Graziani, J. P. Bibler, R. M. Montesano, and A. Wojcicki, *J. Organometallic Chem.*, 1969, **16**, 507.
[400] R. C. Edmondson and M. J. Newlands, *Chem. Comm.*, 1968, 1219.
[401] W. Kitching, B. Hegarty, S. Winstein, and W. G. Young, *J. Organometallic Chem.*, 1969, **20**, 253.
[402] R. Calas and P. Bourgeois, *Compt. rend.*, 1969, **268**, *C*, 1525.
[403] G. Jori, G. Galiazzo, and E. Scoffone, *Internat. J. Protein Res.*, 1969 **1**, 289.

sulphonic acid involves $H_2O_2$ oxidation (100 °C; 4 hr) of $Hg(S \cdot CF_3)_2$, produced by heating $CS_2$ with $HgF_2$ at 250 °C; 50 atm; 4 hr.[404]

Sulphonation[2] with $H_2SO_4$ is well documented, though studies of isomerisation of sulphonic acids prepared in this medium continue to be reported; isomerisation of monosulphonic acids of *o*- and *m*-xylene in aq. $H_2SO_4$ has been studied further;[405] pyrrolo[1,2-*a*]quinoxaline gives the 2-sulphonic acid in conc. $H_2SO_4$ at 130 °C, but the 3-isomer on sulphonation at room temperature.[406] Interconversion of isomers of $C_3$-sulphonic acids and related hydroxyalkanesulphonic acids in aq. NaOH or conc. $H_2SO_4$ has been studied by n.m.r. spectroscopy.[407]

The bisulphite addition product of D-glucose gives the corresponding acylic hexa-acetyl sulphonic acid on acetylation with acetic anhydride–pyridine;[408] sulphitolysis of aryl methyl ethers in neutral solution involves $S_N2$ attack on the methyl carbon atom, giving $MeSO_3H$,[409] and has been compared with demethylation reactions of $^-SH$ and $S^{2-}$. Nucleophilic addition of sulphite to the thione grouping of 4-thiouridine precedes displacement of $HS^-$, giving uridine-4-sulphonate (95);[410] uracil (95; H in

(95)

place of β-D-ribofuranosidyl, OH in place of $SO_3^-Na^+$) gives the 6-sulphonate with $NaHSO_3$ in aqueous solution at pH 6,[411] and cytosine (95; H in place of β-D-ribofuranosidyl, $NH_2$ in place of $SO_3^-Na^+$) suffers deamination in giving the same product.[411] Reversible addition of $HSO_3^-$ at the 6-position explains the bisulphite-catalysed decarboxylation of 5-carboxy-uracil[412] and the deamination of cytosine derivatives.[413] Bisulphites, and $SO_2$, should be considered genetically hazardous in view of these results,[413] though they are widespread in the urban environment. Aeruginosin B was, in 1964, the first example of a naturally-occurring

[404] M. Schmeisser, P. Sartori, and B. Lippsmeier, *Chem. Ber.*, 1970, **103**, 868.
[405] A. Koeberg-Telder, A. J. Prinsen, and H. Cerfontain, *J. Chem. Soc. (B)*, 1969, 1004.
[406] G. W. H. Cheeseman and P. D. Roy, *J. Chem. Soc. (C)*, 1969, 856.
[407] D. M. Brouwer and J. A. Van Doorn, *Rec. Trav. chim.*, 1969, **88**, 1041.
[408] D. L. Ingles, *Chem. and Ind.*, 1969, 50.
[409] J. Gierer and B. Koutek, *Acta Chem. Scand.*, 1969, **23**, 1343.
[410] H. Hayatsu, *J. Amer. Chem. Soc.*, 1969, **91**, 5693.
[411] H. Hayatsu, Y. Wataya, and K. Kai, *J. Amer. Chem. Soc.*, 1970, **92**, 724.
[412] K. Isono, S. Suzuki, M. Tanaka, T. Nanbata, and K. Shibuya, *Tetrahedron Letters*, 1970, 425.
[413] R. Shapiro, R. E. Servis, and M. Welcher, *J. Amer. Chem. Soc.*, 1970, **92**, 422.

# Aliphatic Organo-sulphur Compounds etc. 95

sulphonic acid; full details have now been published supporting its formulation as a substituted phenazinium-8-sulphonate betaine.[414]

**Reactions of Sulphonic Acids.**—Reductive acylation of aryl sulphonic anhydrides or thiolsulphonates by carboxylic acids in the presence of $P + I_2$ gives thiolesters $(Ar \cdot SO_2 \cdot OR^1 + R^2 \cdot CO_2H \rightarrow Ar \cdot S \cdot CO \cdot R^2)$.[415] Mixed sulphonic carboxylic anhydrides are prepared from a silver sulphonate with $R \cdot CO \cdot Cl$.[416] Silver 2,4,6-trinitrobenzenesulphonate gives methanesulphenyl 2,4,6-trinitrobenzenesulphonate with MeSBr; the salt provides the methanesulphenyl cation in condensation reactions with butadienes.[417] Trinitrobenzenesulphonic acid has been proposed as a specific probe for primary amino-groups in proteins,[418, 419] rates of trinitrophenylation, followed spectrophotometrically, revealing relative reactivities of different amino-groups (the ε-amino-groups in tobacco mosaic virus protein are relatively unreactive [418]). Methanesulphonic acid, like $H_2SO_4$, is a valuable solvent for far-u.v. and c.d. studies to 185 nm; polypeptides adopt random-coil conformations in $MeSO_3H$.[420] Mercury trifluoromethanesulphonate gives trichloromethyl ester with $CXCl_3$ (X = Cl, Br, I);[421] $(CF_3 \cdot SO_2 \cdot O)_2CCl_2$ is also formed with $ClCl_3$ and is readily transformed into $(CF_3 \cdot SO_2)_2O + COCl_2$. Vinyl trifluoromethanesulphonates result from addition of $CF_3 \cdot SO_3H$ to alkynes.[422a] Replacement of $\cdot SO_3H$ by $\cdot CN$, or by $\cdot Cl$ (using $PCl_5$), has been reported for aliphatic sulphonic acids.[408]

**Sulphonate Esters.**—Methyl p-nitrobenzenesulphonate specifically methylates residue 57 (histidine) in chymotrypsin, but does not modify trypsin or subtilisin.[423] Methoxymethyl methanesulphonate, $Me \cdot SO_2 \cdot O \cdot CH_2 \cdot O \cdot Me$, is a reactive oxyalkylating agent; it is prepared from the mixed anhydride $Me \cdot SO_2 \cdot OAc$ and $CH_2(OMe)_2$.[424] A novel displacement of benzyl sulphonate from residue 195 (serine) of chymotrypsin by $H_2O_2$ gives the hydroperoxy-enzyme.[425]

New syntheses of enol sulphonates have been reported; pyrazol-5-one gives the 5-sulphonate with a sulphonyl chloride,[426] and vinyl sulphonates result from treatment of ketones with free α-positions with $MeSO_2Cl-NEt_3$

---

[414] R. B. Herbert and F. G. Holliman, *J. Chem. Soc.* (C), 1969, 2517.
[415] J. Morgenstern and R. Mayer, *Z. Chem.*, 1969, **9**, 182.
[416] E. Sarlo and T. Lanigan, *Org. Prep. Procedures*, 1969, **1**, 157.
[417] G. K. Helmkamp, D. C. Owsley, and B. R. Harris, *J. Org. Chem.*, 1969, **34**, 2763.
[418] R. B. Scheele and M. A. Lauffer, *Biochemistry*, 1969, **8**, 3597.
[419] R. B. Freedman and G. K. Radda, *Biochem. J.*, 1969, **114**, 611.
[420] E. Peggion, L. Strasorier, and A. Cosani, *J. Amer. Chem. Soc.*, 1970, **92**, 381.
[421] M. Schmeisser, P. Sartori, and B. Lippsmeier, *Chem. Ber.*, 1969, **102**, 2150.
[422] *a* P. J. Stang and R. Summerville, *J. Amer. Chem. Soc.*, 1969, **91**, 4600; *b* P. E. Peterson and J. M. Indelicato, *J. Amer. Chem. Soc.*, 1969, **91**, 6194.
[423] Y. Nakagawa and M. L. Bender, *J. Amer. Chem. Soc.*, 1969, **91**, 1566.
[424] M. H. Karger and Y. Mazur, *J. Amer. Chem. Soc.*, 1969, **91**, 5663.
[425] M. J. Gibian, D. L. Elliott, and W. R. Hardy, *J. Amer. Chem. Soc.*, 1969, **91**, 7528.
[426] G. A. Galoyan, S. G. Agbalyan, and G. T. Esayan, *Armyan. khim. Zhur.*, 1969, **22**, 430 (*Chem. Abs.*, 1969, **71**, 70,535).

(a source of sulphene, $CH_2:SO_2$), perhalogenated aliphatic ketones giving $\beta$-sultones.[427] This route is claimed to be better than that using a sulphonic anhydride,[428] only 5—40% yields of steroid enol toluene-$p$-sulphonates being obtained on treatment of a steroid ketone with toluene-$p$-sulphonic anhydride in dimethylformamide at 140 °C during 30—120 min.[428] Base-induced elimination from $Ph \cdot CH:CMe \cdot O \cdot Tos$ with $Bu^tO^--Bu^tOH$ at room temperature gives $Ph \cdot CH:C:CH_2$ in high yield,[428] though this reaction is found to give back the parent ketone more generally.[428] Solvolysis of enol sulphonates in hydroxylic solvents proceeds *via* the vinyl cation,[422] where *cis*-isomers are concerned, though *via* an E2-like transition state with *trans*-isomers, ultimate products being $R^1 \cdot C \vdots C \cdot R^2$ and $R^1 \cdot CH_2 \cdot CO \cdot R^2$ from $R^1 \cdot CH:CR^2 \cdot O \cdot SO_2 \cdot CF_3$.[422a] Reaction of dimethylformamide with toluene-$p$-sulphonic anhydride, or the sulphonyl chloride, gives $(Me_2\overset{+}{N}:CH \cdot O \cdot Tos)X^-$, which gives $(Me_2\overset{+}{N}:CH \cdot NMe_2)X^-$ with excess dimethylformamide.[429] Vinyl sulphonates and corresponding sulphonyl chlorides react with amines, giving taurine derivatives and corresponding betaines, typically $CH_2:CH \cdot SO_3R + R \cdot NH_2 \rightarrow R \cdot NH \cdot CH_2 \cdot CH_2 \cdot SO_3R$.[430]

Sulphonate esters have been used in a number of studies of solvolysis reactions aimed at refining knowledge of the $S_N$ transition state. Data on the trifluoroacetolysis of a series of alkyl esters of toluene-$p$-sulphonic acid may be separated into rate constants $k_\Delta$ (anchimerically-assisted ionisation) and $k_s$ (anchimerically-unassisted ionisation), whose ratio is largely determined by solvent characteristics ($k_\Delta/k_s$ increases through the series $H \cdot CO_2H < CF_3 \cdot CO_2H < F \cdot SO_3H$).[431–434] The small upward drift in the rate of acetolysis of alkyl toluene-$p$-sulphonates is caused by acid catalysis by released toluene-$p$-sulphonic acid rather than by salt effects.[435] Acetolysis rates of arylsulphonate esters $Ar \cdot SO_2 \cdot O \cdot R$ (R = 3- or 4-tetrahydropyranyl) are slower than those of open-chain analogues;[436] steric factors and polar influences governing this observation,[436] and rates of substitution of carbohydrate sulphonates [437, 438] have been reviewed. Acetolysis of trifluoromethanesulphonates ('triflates') of norbornyl and related alicyclic systems follows good first-order kinetics, with rates $10^{4.3}$—$10^{5.2} \times$ faster than those for corresponding toluene-$p$-sulphonates.[439]

[427] W. E. Truce and L. K. Liu, *Chem. and Ind.*, 1969, 457; *Tetrahedron Letters*, 1970, 517.
[428] N. Frydman, R. Bixon, M. Sprecher, and Y. Mazur, *Chem. Comm.*, 1969, 1044.
[429] H. Schindlbauer, *Monatsh.*, 1969, **100**, 1590.
[430] A. Le Berre, A. Etienne, and B. Dumaitre, *Bull. Soc. chim. France*, 1970, 946, 954.
[431] I. L. Reich, A. Diaz, and S. Winstein, *J. Amer. Chem. Soc.*, 1969, **91**, 5635.
[432] P. C. Myhre and E. Evans, *J. Amer. Chem. Soc.*, 1969, **91**, 5641.
[433] P. C. Myhre and K. S. Brown, *J. Amer. Chem. Soc.*, 1969, **91**, 5639.
[434] A. Diaz, I. L. Reich, and S. Winstein, *J. Amer. Chem. Soc.*, 1969, **91**, 5637.
[435] M. L. Sinnott, *J. Org. Chem.*, 1969, **34**, 3638.
[436] D. S. Tarbell and J. R. Hazen, *J. Amer. Chem. Soc.*, 1969, **91**, 7657.
[437] A. C. Richardson, *Carbohydrate Res.*, 1969, **10**, 395.
[438] D. H. Ball and F. W. Parrish, *Adv. Carbohydrate Chem.*, 1968, **23**, 233.
[439] T. M. Su, W. F. Sliwinski, and P. Von R. Schleyer, *J. Amer. Chem. Soc.*, 1969, **91**, 5386.

## Aliphatic Organo-sulphur Compounds etc.

Aryl triflates resist solvolysis.[439] Thiolysis of *p*-nitrophenyl benzenesulphonate by Ph·S⁻ is slower than lysis by OH⁻.[440] In compounds containing both sulphonate and carboxylate groupings, *e.g.* diphenyl sulpho-acetate Ph·O·CO·CH$_2$·SO$_2$·O·Ph, hydrolysis in aqueous acid involves primarily the carboxylate grouping, but in pyridine, hydrolysis takes place at the sulphonate grouping exclusively;[441] this may be due to intermediate sulphene formation,[441] though such eliminations usually involve a better leaving group than ·O·Ph. Kinetics of hydrolysis of mixed carboxylic-sulphonic anhydrides have been reported.[442]

Acetolysis rates of toluene-*p*-sulphonates (96)[443] and (98)[444] are greatly enhanced relative to those of analogous compounds in which neighbouring group participation by sulphur is not possible [(97) and (99), respectively].

(96)        (98)        (99)

(97) = *p*-isomer of (96)

Very mild conditions (1 equivalent KOH during 20 min at room temperature) are sufficient for β-elimination of D-glucopyranosiduronate 4-sulphonates to Δ$^{4,5}$-4-deoxy-analogues;[445] other examples[445] bear out the ease with which axial–equatorial eliminations of this type can be brought about.

Rearrangement of *N*-toluene-*p*-sulphonyloxy-isocarbostyril into 3-toluene-*p*-sulphonyloxy-isocarbostyril has been studied;[446] intriguing mechanistic possibilities are discussed but, with the help of isotopic labelling of the sulphonate oxygen atoms, migration is concluded to involve mainly a solvent-separated ion-pair path. ³⁵S-Studies of the equilibration of 1-naphthyl sulphamic acid into 1-naphthylamine-4-sulphonic acid in H$_2$SO$_4$–dioxan suggest the intervention of multiply-sulphonated naphthylamine intermediates; possibly, more than one of the suggested mechanisms is involved.[447]

**Sulphonyl Halides.**—The unusual stability of tetrachloropyridine-4-sulphonyl chloride (unchanged during 16 hr in AcOH at 100 °C) is ascribed to steric

---

[440] W. Tagaki, T. Kurusu, and S. Oae, *Bull. Chem. Soc. Japan*, 1969, **42**, 2894.
[441] B. E. Hoogenboom, M. S. El-Faghi, S. C. Fink, E. D. Hoganson, S. E. Lindberg, T. J. Lindell, C. J. Linn, D. J. Nelson, J. O. Olson, L. Rennerfeldt, and K. A. Wellington, *J. Org. Chem.*, 1969, **34**, 3414.
[442] R. M. Laird and M. J. Spence, *J. Chem. Soc. (B)*, 1970, 388.
[443] M. Hojo, T. Ichi, Y. Tamaru, and Z. Yoshida, *J. Amer. Chem. Soc.*, 1969, **91**, 5170.
[444] L. A. Paquette, G. V. Meehan, and L. D. Wise, *J. Amer. Chem. Soc.*, 1969, **91**, 3231.
[445] J. Kiss, *Carbohydrate Res.*, 1969, **10**, 328.
[446] K. Ogino, S. Kozuka, and S. Oae, *Tetrahedron Letters*, 1969, 3559.
[447] F. L. Scott, C. B. Goggin, and W. J. Spillane, *Tetrahedron Letters*, 1969, 3675.

hindrance;[448] pyridine-2- and -4-sulphonyl chlorides decompose by an $S_N i$ mechanism into $SO_2$ and the 2- and 4-chloropyridines, at room temperature. The reaction between pyridine and ethanesulphonyl chloride in AcOH is claimed to give $N$-(2-sulphoethyl)pyridinium betaine, $py^+ \cdot CH_2 \cdot CH_2 \cdot SO_3^-$ after 30 min at room temperature.[449] Alkanesulphonyl chlorides react as sulphenes in the presence of a tertiary amine at low temperatures, 2-cyano-enamines and -ynamines giving (methanesulphonyl)methyl vinyl sulphones with $MeSO_2Cl–NEt_3$, e.g. $(Et_2N)_2C:CH \cdot CN \rightarrow (Et_2N)_2C:C(CN) \cdot SO_2 \cdot CH_2 \cdot SO_2Me$.[450] Elimination of axial 3-chloro-sulphonyl-*trans*-decalin proceeds faster than with the equatorial isomer, consistent with an $E2$-mechanism for sulphene formation.[451]

Perfluoroalkyl sulphonyl fluorides convert tetra-alkoxysilanes into tri-alkoxyfluorosilanes in the presence of a tertiary amine.[452]

Rates of $S_N 2$-solvolysis of methanesulphonyl chloride and the phenyl analogue in $H_2O$ and $D_2O$ have been determined with considerable accuracy.[453]

**Sulphonamides.**—Birch reduction of benzenesulphonamide gives 65% $Ph \cdot SH$ and 9% $Ph \cdot S \cdot S \cdot Ph$;[454] benzenesulphonic acid does not react when treated similarly ($Li–NH_3$). Cleavage of aryl sulphonamides, $Ar \cdot SO_2 \cdot NR_2$, and sulphonate esters by arene anion radicals (from e.g. naphthalene + Na in an ether solvent) requires exactly 2 equivalents of reagent at $-60\ °C$, giving corresponding amide and arylsulphinate anions.[455, cf. 383] The sulphinate suffers further reduction into $Ar \cdot H$, sulphite, thiosulphate, and sulphide salts, though $Ar \cdot SH$ is not formed. The toluene-*p*-sulphonyl radical appears on the photolysis route followed by toluene-*p*-sulphonyl-iminodiphenyl cyclopropene (100) on irradiation in benzene solution;

Ph
 \
  >=N—Tos    ⟶    Ph\C=C=N—Tos
 /                    / ·
Ph                  Ph
 (100)

expulsion of Tos· from the first-formed diradical gives the relatively more stable radical $Ph \cdot C(CN):\dot{C} \cdot Ph$, and toluene-*p*-sulphonamide and $(Tos)_2S$ are present in the reaction product mixture.[456]

---

[448] E. Ager and B. Iddon, *Chem. Comm.*, 1970, 118.
[449] A. Etienne, A. Le Berre, and B. Dumaitre, *Fr. Pat.*, 1,529,883 (*Chem. Abs.*, 1969, **71**, 70,507).
[450] T. Sasaki and A. Kojima, *J. Chem. Soc.* (C), 1970, 476.
[451] J. F. King and T. W. S. Lee, *J. Amer. Chem. Soc.*, 1969, **91**, 6524.
[452] V. Beyl, H. Niederprüm, and P. Voss, *Annalen*, 1970, **731**, 58.
[453] R. E. Robertson, B. Rossall, S. E. Sugamori, and L. Treindl, *Canad. J. Chem.*, 1969, **47**, 4199.
[454] C. A. Matuszak and T. P.-F. Niem, *Chem. and Ind.*, 1969, 952.
[455] W. D. Closson, S. Ji, and S. Schulenberg, *J. Amer. Chem. Soc.*, 1970, **92**, 650.
[456] N. Obata, A. Hamada, and T. Takizawa, *Tetrahedron Letters*, 1969, 3917.

*Aliphatic Organo-sulphur Compounds etc.* 99

$N$-Acyl- and $N$-sulphonyl-sulphonamides are easily prepared from sulphonamides by way of the $N$-sulphinyl compound $R \cdot SO_2 \cdot N:S:O$, obtained by refluxing with excess $SOCl_2$; $R \cdot SO_2 \cdot NSO$ heated with a sulphonic acid or a carboxylic acid gives the $N$-substituted sulphonamides.[457] Toluene-$p$-sulphonamide gives $(Tos \cdot NH)_2CH_2$ with formaldehyde in neutral or alkaline solution, not $Tos \cdot NH \cdot CH_2OH$ as claimed earlier; the corresponding cyclic trimer (101) and polymers are formed in aqueous acid solutions.[458]

$$\begin{array}{c} Tos \\ | \\ N \\ \diagup \ \diagdown \\ Tos-N \diagdown \diagup N-Tos \end{array}$$

(101)

New reactions of $NN$-dichloro-toluene-$p$-sulphonamide and the corresponding sodium-chlorosulphonamide (Chloramine-T) give access to further $N$-substituted sulphonamides [$\cdot SO_2 \cdot NCl_2$ or $\cdot SO_2 \cdot N(Na)Cl +$ $R \cdot S^-Na^+ \rightarrow \cdot SO_2 \cdot N:S(R) \cdot NH \cdot SO_2 \cdot$;[459, 460] $+ Me_2SO \rightarrow$ $\cdot SO_2 \cdot N:S(O)Me_2$;[461] $+ R_3P$ followed by treatment with Cu powder $\rightarrow$ $\cdot SO_2 \cdot N:PR_3$ [461]]. $N$-($p$-Nitrobenzenesulphonyloxy)benzenesulphonamide similarly gives $Ph \cdot SO_2 \cdot N:S(O)Me_2$ with dimethyl sulphoxide and $NEt_3$, though in low yield,[462] and corresponding $\cdot SO_2 \cdot N:SMe_2$ derivatives with $Me_2S-NEt_3$.[462] Displacement of arene-sulphonate in the latter reactions could be taken to be the mechanism, though a nitrene intermediate is more likely.[462] The thermal decomposition of toluene-$p$-sulphonyl azide in solution is a first-order kinetic reaction, giving $N_2$ and nitrene; aliphatic sulphonyl azides behave similarly but an alternative radical chain reaction involving loss of $SO_2$ competes.[463] Thermolysis of mesitylene-2-sulphonyl azide in dodecane gives six identified products,[464] two of which (mesidine in 20.7% yield and the corresponding azobenzene in 0.35% yield) result from a Curtius-type rearrangement of the sulphonyl nitrene. Further studies of aromatic substitution by methanesulphonyl nitrene, generated thermally from the azide, are reported;[465] products are often complex $(Me \cdot SO_2 \cdot N_3 + Ph \cdot NO_2 \rightarrow Me \cdot SO_2 \cdot NH \cdot C_6H_4NO_2 + Ph \cdot NH \cdot SO_2 \cdot Me + Ph \cdot O \cdot SO_2Me + Me \cdot SO_2 \cdot NH_2)$ and tar is the major product, though

[457] F. Bentz and G.-E. Nischk, *Angew. Chem., Internat. Edn.*, 1970, **9**, 66.
[458] C. D. Egginton and A. J. Lambie, *J. Chem. Soc.* (C), 1969, 1623.
[459] M. M. Kremlev and I. V. Koval, *Zhur. org. Khim.*, 1969, **5**, 2014.
[460] M. M. Kremlev, G. F. Kodachenko, and V. F. Baranovskaya, *Zhur. org. Khim.*, 1969, **5**, 914.
[461] A. Schoenberg and E. Singer, *Chem. Ber.*, 1969, **102**, 2557.
[462] M. Okahara and D. Swern, *Tetrahedron Letters*, 1969, 3301.
[463] D. S. Breslow, M. F. Sloan, N. R. Newburg, and W. B. Renfrow, *J. Amer. Chem. Soc.*, 1969, **91**, 2273.
[464] R. A. Abramovitch and W. D. Holcomb, *Chem. Comm.*, 1969, 1298.
[465] R. A. Abramovitch, G. N. Knaus, and V. Urna, *J. Amer. Chem. Soc.*, 1969, **91**, 7532.

results are consistent with the intervention of both singlet and triplet nitrene.

2- and 4-(p-Nitrobenzene)sulphonamido-pyrimidines suffer elimination of $SO_2$ by treatment with strong bases, giving p-nitrophenylamino-pyrimidines (102). Judicious $^{15}N$-labelling shows that ring-opening and

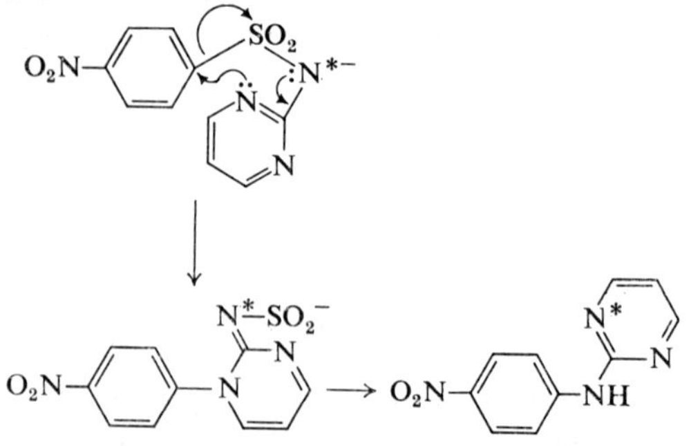

(102)

cyclisation in the sense of a Dimroth rearrangement take place, and the failure of the 5-substituted compound to react in the same way supports the mechanism.[466]

o-Benzenedisulphonimide is completely ionised in water, implying that the sulphonimide anion is highly resonance-stabilised.[467] The acidity of $(Ph \cdot SO_2)_2NH$ is remarkably high; with $pK_a = 1.45$, this compound is roughly as strong an acid as $H_3PO_4$.[468] Crystal structure analysis of di(benzenesulphonyl)imine shows no undue shortening of the S—N bonds, which might have been expected on the basis of increased $d_\pi$–$p_\pi$ interaction resulting from delocalisation of negative charge in the anion, and the acidity must therefore be ascribed solely to the combined electron-withdrawing power of the two benzenesulphonyl groups.[468] Primary aliphatic amines may be converted into alkyl halides via their toluene-p-sulphonimide derivatives, which are susceptible to nucleophilic displacement ($R \cdot NH_2 \rightarrow R \cdot NTos_2 \rightarrow R \cdot I$ with $I^-$ in dimethylformamide).[469]

N-Phenacylsulphonyl derivatives of amines may have uses as N-protected intermediates in synthesis (deprotection through zinc–acetic acid reduction to the N-sulphinate, which decomposes with loss of $SO_2$) and in the preparation of secondary amines, since the derivatives are readily N-alkylated; the latter process can be wasteful with respect to the alkylating agent, since the phenacyl methylene group is also alkylated.[470]

[466] G. Malewski, H. Walther, H. Just, and G. Hilgetag, *Tetrahedron Letters*, 1969, 1057.
[467] J. B. Hendrickson, S. Okano, and R. K. Bloom, *J. Org. Chem.*, 1969, **34**, 3434.
[468] F. A. Cotton and P. F. Stokely, *J. Amer. Chem. Soc.*, 1970, **92**, 294.
[469] P. J. De Christopher, J. P. Adamek, G. D. Lyon, J. J. Galante, H. E. Haffner, R. J. Boggio, and R. J. Baumgarten, *J. Amer. Chem. Soc.*, 1969, **91**, 2384.
[470] J. B. Hendrickson and R. Bergeron, *Tetrahedron Letters*, 1970, 345.

## Aliphatic Organo-sulphur Compounds etc.

**Sulphonyl Isocyanates.**—Toluene-$p$-sulphonyl isocyanate is an acylating agent under Friedel–Crafts reaction conditions,[471] like other heterocumulenes. Triphenylmethanol gives $N$-trityl toluene-$p$-sulphonamide and $CO_2$ with Tos·NCO;[472] substituent effects,[473] and the effects of added pyridine, on the rate of release of $CO_2$, have been discussed.[472]

Cycloaddition reactions of sulphonyl isocyanates, particularly of Cl·$SO_2$·NCO [474] and F·$SO_2$·NCO,[475] have merited a considerable number of papers because of the variety of structures which may be obtained. Toluene-$p$-sulphonyl isocyanate gives $\beta$-lactams (103) and (104) with 1,1-di(cyclopropyl)ethylene [476] and diazofluorene [477] respectively; a wide range of ethylenes gives corresponding products but also $N$-sulphonyl-acrylamides $R^1$·CH:$CR^2$·CO·NH·$SO_2R^3$.[478] Tos·NCO gives (105) with diazomethane, with evolution of $N_2$.[477]

In the presence of radical initiators, Cl·$SO_2$·NCO gives Cl·$CH_2$·$CH_2$·$SO_2$·NCO and (106) with ethylene.[479]

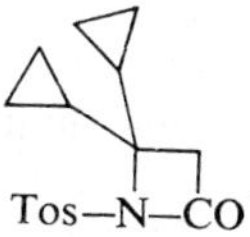

Tos—N—CO
(103)

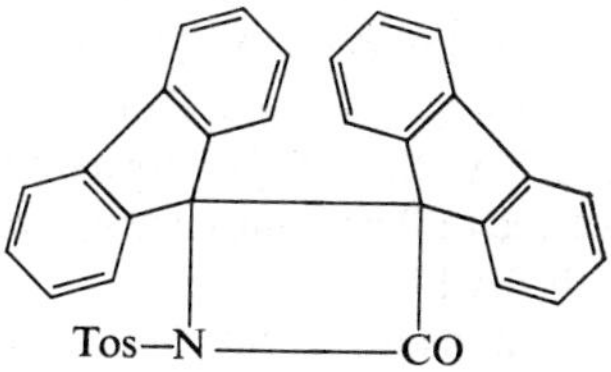

Tos—N————CO
(104)

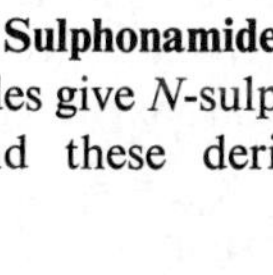

$$R-\overset{O}{\underset{O}{S}}=N-\underset{H_2}{C}-CO$$
(105)

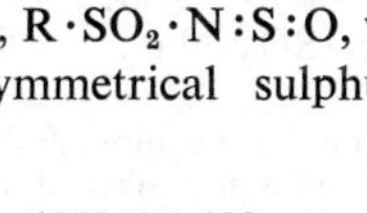

(106)

**$N$-Sulphinyl Sulphonamides, and $NN'$-Disulphonyl Sulphur Di-imides.**—Sulphonamides give $N$-sulphinyl derivatives, R·$SO_2$·N:S:O, with refluxing $SOCl_2$,[457] and these derivatives yield symmetrical sulphur di-imides,

---

[471] J. W. McFarland and L. Y. Yao, *J. Org. Chem.*, 1970, **35**, 123.
[472] J. W. McFarland, D. Green, and W. Hubble, *J. Org. Chem.*, 1970, **35**, 702.
[473] J. W. McFarland and D. J. Thoennes, *J. Org. Chem.*, 1970, **35**, 704.
[474] L. A. Paquette and G. R. Krow, *J. Amer. Chem. Soc.*, 1969, **91**, 6107; R. Askani, *Angew. Chem.*, 1970, **82**, 176; L. A. Paquette, J. R. Malpass, and T. J. Barton, *J. Amer. Chem. Soc.*, 1969, **91**, 4714; E. J. Moriconi, J. G. White, R. W. Franck, J. Jansing, J. F. Kelly, R. A. Salomone, and Y. Shimakawa, *Tetrahedron Letters*, 1970, 27.
[475] K. Clauss and H. Jensen, *Tetrahedron Letters*, 1970, 119; K.-D. Kampe, *ibid.*, p. 123.
[476] F. Effenberger and W. Podszun, *Angew. Chem., Internat. Edn.*, 1969, **8**, 976.
[477] G. Lohaus, *Tetrahedron Letters*, 1970, 127.
[478] R. Lattrell, *Annalen*, 1969, **722**, 132.
[479] D. Gunther and F. Soldan, *Chem. Ber.*, 1970, **103**, 663.

R·SO$_2$·N:S:N·SO$_2$·R with base,[480, 481] and sulphonyl isocyanates with phosgene.[482] NN'-Disulphonyl sulphur di-imides give N-sulphonyl-N'-sulphenyl analogues R$^1$·SO$_2$N:S:N·S·R$^2$ on treatment with a sulphenamide, or its N-sulphinyl derivative R$^2$·S·N:S:O,[480] NN'-disulphenyl analogues (o-NO$_2$·C$_6$H$_4$·S·N:)$_2$S being formed from N-toluene-p-sulphonyl-N'-o-nitrophenylsulphenyl sulphur di-imide on treatment with a primary or secondary amine.[480] Studies of cycloaddition reactions of these heterocumulenes would be of interest, complementing recent work.[483]

**10 Disulphides and Polysulphides; their Oxy-sulphur Analogues**

**Preparation.**—Simple syntheses of mixed or symmetrical disulphides, diselenides, and ditellurides start with the addition of an alkyl iodide to a solution of sulphur, selenium, or tellurium with one equivalent of sodium, in liquid ammonia;[484] successive addition of further equivalents of sodium and an alkyl halide completes the procedure. 1,2,4-Trichlorobenzene gives the 5,5'-trisulphide on treatment with S$_2$Cl$_2$ and Al during 2 hr at 60 °C,[485] while pyrrole disulphides are obtained with S$_2$Cl$_2$,[39] and 4,4'-dithiobis(2,6-di-t-butylphenol) is obtained from 2,6-di-t-butylphenol with S$_2$Cl$_2$ and I$_2$.[486] Practical yields of β-chloroalkyl disulphides are obtained from reactions between alkenes (in excess) with S$_2$Cl$_2$,[487] and similar results are obtained with SeCl.[488]

Sulphonyl chlorides yield disulphides when treated with 1·1—1·3 equivalents of Mo(CO)$_6$ in tetramethylurea,[489] but thiolsulphonates are obtained if Fe(CO)$_5$ is used.[490]

New mild oxidation procedures for conversion of thiols into disulphides have been reported; flavins anaerobically oxidise excess thiol to disulphide (Ph·CH$_2$·SH → 81% Ph·CH$_2$·S·S·CH$_2$·Ph) with concomitant reduction to dihydroflavin;[491] 1,2-dithian-SS-tetroxide quantitatively oxidises a thiolate to its disulphide [with concomitant reduction to Na$^+$ $^-$O$_2$S·(CH$_2$)$_4$·SO$_2$$^-$ Na$^+$], a procedure proposed to be of potential value in biological systems,[492] though the corresponding cyclic thiolsulphonate suffers ring-opening with R·S$^-$Na$^+$, giving R·S·S·(CH$_2$)$_4$·SO$_2$$^-$Na$^+$. 'Diamide' [diazenecarboxylic acid bis(NN-

---

[480] H. C. Buchholt, A. Senning, and P. Kelly, *Acta Chem. Scand.*, 1969, **23**, 1279.
[481] H. H. Hoerhold and J. Beck, *J. prakt. Chem.*, 1969, **311**, 621.
[482] H. Ulrich, B. Tucker, and A. A. R. Sayigh, *J. Org. Chem.*, 1969, **34**, 3200.
[483] H. H. Hoerhold and H. Eibisch, *Tetrahedron*, 1969, **25**, 4277; *Chem. Ber.*, 1968, **101**, 3567; C. D. Campbell and C. W. Rees, *J. Chem. Soc.* (C), 1969, 748.
[484] G. M. Bogolyubov, Y. N. Shlyk, and A. A. Petrov, *Zhur. obshchei Khim.*, 1969, **39**, 1804.
[485] A. Sakic, A. F. Damanski, and Z. J. Binenfeld, *Compt. rend.*, 1969, **268**, C, 1779.
[486] R. J. Laufer, *U.S. Pat.* 3,479,407 (*Chem. Abs.*, 1970, **72**, 31,445).
[487] F. K. Lautenschlaeger and N. V. Schwartz, *J. Org. Chem.*, 1969, **34**, 3991.
[488] F. K. Lautenschlaeger, *J. Org. Chem.*, 1969, **34**, 4002.
[489] H. Alper, *Angew. Chem., Internat. Edn.*, 1969, **8**, 677.
[490] H. Alper, *Tetrahedron Letters*, 1969, 1239.
[491] M. J. Gibian and D. V. Winkelman, *Tetrahedron Letters*, 1969, 3901.
[492] L. Field and R. B. Barbee, *J. Org. Chem.*, 1969, **34**, 1792.

*Aliphatic Organo-sulphur Compounds etc.* 103

dimethylamide)] stoicheiometrically oxidises reduced glutathione within the human red blood cell, without interfering with cell function.[493] The use of Ph·N:N·CO$_2$Me for thiol oxidation in biochemistry has been reviewed.[494] Trialkyl arylthiostannanes, R$_3$Sn·SAr, give mixed aryl disulphides in high yields with aryl sulphenyl halides;[495] the unstable *S*-substitution product resulting from a thioamide and a sulphenyl halide disproportionates into a hydrodisulphide and a nitrile (R·CN + Ph·S·SH from R·CS·NH$_2$ + Ph·S·Cl).[366]

U.v.-irradiation of 1,2-dithio-oxalates and trithiocarbonates gives disulphides (Ar·S·CO·CO·S·Ar → Ar·S·S·Ar + Ar·S·CO·S·Ar + CO);[496] photolysis of thioacetates similarly gives high yields of aryl disulphides (toluene-4-thioacetate gives 77% di-*p*-tolyl disulphide and 7% di-*p*-tolyl sulphide) by dimerisation of intermediate arylthiyl radicals.[497] Homolytic cleavage of cyclohexene-1-thioacetate gives a mixture of *cis*- and *trans*-2-acetylcyclohexane thioacetate, H$_2$S, and 15% 2,3:4,5-bis(tetramethylene)-thiophen formed through a radical dimerisation path.[497] Cysteine is oxidised to cystine by *N*-nitroso-*N*-methyl toluene-*p*-sulphonamide,[498] and thiophenols similarly give good yields of disulphides with stable nitroxyl radicals;[338] oxidation of thiols to sulphonic acids by nitroxyl radicals has been reported, however,[499] and clearly, the reaction conditions used determine the level of oxidation. Trifluoromethyl thionitrite, CF$_3$·S·N:O, a red gas prepared from the thiol by reaction with NOCl or an alkyl nitrite, or from (CF$_3$·S)$_2$Hg and NOCl, is less stable even than alkyl analogues, and disproportionates into the disulphide and nitric oxide.[500]

The challenge offered by non-random synthesis of peptides carrying several disulphide links within each molecule has been taken up during the 1960's; a recent paper, utilising successfully the sulphenyl isothiocyanate method for disulphide bond formation, refers back to the broader literature on this topic.[501]

Kinetics of the reaction of cystine with Na$_2$S, giving polysulphides, have been determined by polarographic and spectroscopic methods (R·S·S$^-$; $\lambda_{max}$ 335 nm), revealing a rate-determining first step,[502] at pH 13·8. Diphenyl disulphide gives a mixture of polysulphides, Ph·S$_n$·Ph ($n$ = 3—6), with H$_2$S$_2$;[503] cystine does not react with H$_2$S$_2$, neither does benzyl di-

---

[493] N. S. Kosower, E. M. Kosower, B. Wertheim, and W. S. Correa, *Biochem. Biophys. Res. Comm.*, 1969, **37**, 593.
[494] E. M. Kosower and N. S. Kosower, *Nature*, 1969, **224**, 117.
[495] J. L. Wardell and D. W. Grant, *J. Organometallic Chem.*, 1969, **20**, 91.
[496] H. G. Heine and W. Metzner, *Annalen*, 1969, **724**, 223.
[497] J. R. Grunwell, *Chem. Comm.*, 1969, 1437.
[498] U. Schulz and D. R. McCalla, *Canad. J. Chem.*, 1969, **47**, 2021.
[499] J. D. Morrisett and H. R. Drott, *J. Biol. Chem.*, 1969, **244**, 5083.
[500] J. Mason, *J. Chem. Soc. (A)*, 1969, 1587.
[501] R. G. Hiskey, A. M. Thomas, R. L. Smith, and W. C. Jones, *J. Amer. Chem. Soc.*, 1969, **91**, 7525.
[502] C. Angst and F. Huegli, *Chimia (Switz.)*, 1969, **23**, 142.
[503] R. Rahman, S. Safe, and A. Taylor, *J. Chem. Soc. (C)*, 1969, 1665.

sulphide, but aryl thiols give mixtures of polysulphides, the proportions of di-, tri-, tetra-, and penta-sulphides obtained from $Ph \cdot CH_2 \cdot SH$ with $H_2S_2$ being 2 : 10 : 7 : 4.[503] Whereas hydrodisulphides give mixtures of thiol and trisulphide (with $H_2S + Et_3As \cdot S \cdot AsEt_3$) on treatment with $Et_3As$, polysulphides are not formed analogously with $Ph_3As$.[504] Thiol and polysulphides, together with $H_2S$, result from the reaction of hypotaurine, $NH_2 \cdot CH_2 \cdot CH_2 \cdot SO_2H$, with a hydrodisulphide;[505] thiotaurine, $NH_2 \cdot CH_2 \cdot CH_2 \cdot SO_2 \cdot SH$, is also formed. Tetrasulphides, $R \cdot S_4 \cdot R$, result from the oxidation of benzyl or diphenylmethyl hydrodisulphides in dioxan solution with $FeCl_3, 6H_2O$.[506]

Sporidesmin E (51; $R^1$ = Me, $R^2R^3$ = $\cdot S \cdot S \cdot S \cdot$) is a naturally-occurring trisulphide,[507] which may be obtained from the corresponding disulphide (sporidesmin) by treatment with $H_2S_2$; c.d. data for the di- and tri-sulphides may be interpreted [507] to show retention of configuration during the conversion of the disulphide into the trisulphide, suggesting a sulphur-insertion mechanism into the $\cdot S \cdot S \cdot$ grouping, rather than attack of nucleophilic sulphur at carbon. The studies with simple disulphides and $H_2S_2$ were initiated to give further support for the trisulphide structure assigned to sporidesmin E,[503, 507] since formation of trisulphides from disulphides appeared to be a novel reaction. Dehydrogliotoxin, derived from a related naturally-occurring disulphide-bridged dioxopiperazine, also yields tri- and tetra-sulphides on treatment with $H_2S_2$.[507] Acetylaranotin (107) carries a disulphide bridge in a similar environment and gives a

(107)

trisulphide with one equivalent of sulphur, or with $SCl_2$, and a tetrasulphide with excess $S_8$.[508]

**Reactions of Di- and Poly-sulphides.**—Reductive cleavage of disulphides into corresponding thiols is brought about usually by thiol–disulphide interchange, dithiothreitol ('Cleland's reagent'), 2-mercaptoethanol, and thioglycollic acid being typical reagents in protein chemistry; the former gives quantitative reduction of lysozyme, insulin, and prolactin, but leaves

---

[504] J. Tsurugi, T. Horii, T. Nakabayashi, and S. Kawamura, *J. Org. Chem.*, 1968, **33**, 4133.
[505] T. Nakabayashi, T. Horii, S. Kawamura, and J. Tsurugi, *Ann. Reports Radiat. Centre, Osaka Prefecture*, 1968, **9**, 64 (*Chem. Abs.*, 1969, **71**, 70,031).
[506] S. Kawamura, Y. Abe, and J. Tsurugi, *J. Org. Chem.*, 1969, **34**, 3633.
[507] S. Safe and A. Taylor, *J. Chem. Soc.* (C), 1970, 432.
[508] K. C. Murdock and R. B. Angier, *Chem. Comm.*, 1970, 55.

the disulphide bonds of lysozyme unaffected under standard conditions, pH 8·1, room temperature, 1 hr in 0·1M-KCl.[509] The extent of reduction of disulphides is assessed using iodoacetamide or 5,5′-dithiobis(2-nitrobenzoic acid),[509] or by *p*-chloromercuribenzoate titration [510] as exemplified in an assessment of the thiol–disulphide interchange between serum albumins and L-cystine or lysine vasopressin. Electrolytic reduction of disulphides, in the presence of minimum amounts of 2-mercaptoethanol, has been studied,[511] disulphide bridges in lysozyme are reduced by pulse radiolysis of aqueous solutions, HO· produced by a pulse giving a short-lived intermediate from simple disulphides ($\lambda_{max}$ 420 nm).[512]

Cystine is degraded by u.v.-irradiation ($\lambda < 300$ nm) in air-saturated aqueous solution;[513] at pH 10, pyruvic acid, cysteine, alanine-3-sulphinic acid, cysteic acid, cysteine-*S*-sulphonic acid, alanine, serine, glycine, $SO_4^{2-}$, $NH_3$, and hydroxylamine are formed, whereas in acid solution some differences in product composition are observed, but are due only to secondary reactions of some of these compounds. Ease of cleavage of the disulphide bond in 2-pyridyl aryl disulphides by benzoyl peroxide depends on the aryl substituent, mesityl compounds resisting cleavage but giving py·$SO_2$·S·mesityl, while phenyl analogues are easily cleaved.[389] The formation of thiobenzophenone by treatment of diphenylmethyl aryl disulphides with sodium isopropoxide involves a hybrid *E*1–*E*1cb-like transition state (108).[514] Nucleophiles such as $CN^-$, $SCN^-$, or $RS^-$ cleave

$$R\text{---}O\cdots\overset{\delta+}{H}\cdots\overset{\delta-}{\underset{\underset{Ph}{|}}{\overset{\overset{Ph}{|}}{C}}}\cdots\overset{\delta+}{S}\cdots\overset{\delta-}{S}\text{---}Ar$$

(108)

disulphides through an $S_N2$-like reaction pathway following attack on sulphur.[514]

Disulphide interchange involved in the equilibration of *p*-dimethylaminophenyl *p*-trimethylammoniophenyl disulphide iodide is complete within 30 min in daylight in the presence of 0·01 equivalents of NaOH in aqueous solution.[515] An ion-radical scheme is suggested since interchange is inhibited by added quinone. Acetylaranotin (107) gives a bis-disulphide with $Me_2S_2$ (R·S·S·R → 2R·S·S·Me) but a dithiol by reaction with Me·SH in pyridine at 25 °C (presumably R·S·S·R → R·S·S·Me + R·SH →

[509] T. A. Bewley and C. H. Li, *Internat. J. Protein Res.*, 1969, **1**, 117.
[510] F. B. Edwards, R. B. Rombauer, and B. J. Campbell, *Biochim. Biophys. Acta*, 1969, **194**, 234.
[511] G. M. Bhatnagar and W. G. Crewther, *Internat. J. Protein Res.*, 1969, **1**, 213.
[512] G. E. Adams, J. E. Aldrich, R. B. Cundall, and R. L. Willson, *Internat. J. Radiation Biol.*, 1969, **16**, 333.
[513] R. S. Asquith and L. Hirst, *Biochim. Biophys. Acta*, 1969, **184**, 345.
[514] U. Miotti, U. Tonellato, and A. Ceccon, *J. Chem. Soc.* (*B*), 1970, 325.
[515] N. V. Khromov-Borisov and V. E. Gmiro, *Doklady Akad. Nauk S.S.S.R.*, 1969, **185**, 377.

$2R \cdot SH + Me_2S_2$, with no interchange of $Me_2S_2$ with starting material).[508]

Borohydride reduction of sporidesmin E gives the dithiol, which gives sporidesmin D on treatment with MeI;[507, cf. 178] the trisulphide gives sporidesmin (the corresponding disulphide) with $Ph_3P$, or by successive reduction with borohydride and oxidation with 5,5'-dithiobis(2-nitrobenzoic acid).[503, 507]

Dielectric relaxation and dipole moment measurements confirm the rigidity of the skew conformation of dialkyl and diaryl disulphides;[516] earlier temperature-dependent n.m.r. studies[517] of disulphides and diselenides point to a significant barrier to rotation.

**Oxy-sulphur Analogues of Di- and Tri-sulphides.**—$S$-Sulphonates are valuable intermediates in the synthesis of thiols,[35] and selenosulphonates, also easily prepared from an alkyl halide (using sodium selenosulphite), have similar applications.[518] Oxidation of disulphides with 2 equivalents of $H_2O_2$, catalysed by $V_2O_5$, gives thiolsulphonates, $R \cdot SO_2 \cdot S \cdot R$, in moderate yields;[203] with 1 equivalent of $H_2O_2$, thiolsulphinates, $R \cdot SO \cdot S \cdot R$, are minor products with the thiolsulphonates. New di- and tri-sulphide oxides have been prepared in simple ways; 2-naphthyl thiol on treatment with thionyl chloride gives 100% $S$-chlorosulphinyl derivative, $R^1 \cdot S \cdot SO \cdot Cl$, which reacts further with thiols giving $R^1 \cdot S \cdot SO \cdot S \cdot R^2$ (10% yield with $p$-chlorothiophenol).[519] Chlorinolysis of the trisulphide oxide gives back $S$-chlorosulphinyl-2-naphthyl thiol;[519] factors determining the preferred cleavage site in this reaction now appear to be open to study. Thiolsulphonates show absorption bands in the regions $1110-1133$ cm$^{-1}$ and $1310-1333$ cm$^{-1}$, the latter characteristic of the $\cdot SO_2 \cdot$ grouping, proving that the isomeric sulphinylsulphoxide structure, $R \cdot SO \cdot SO \cdot R$, is not taken.[520]

Thiolsulphonates are susceptible to nucleophilic cleavage, and a number of papers have appeared this year using these compounds and sulphinyl analogues as substrates and providing considerable advances in knowledge of nucleophilic substitution at sulphur. Aminolysis of a thiolsulphonate gives a sulphenamide and alkylammonium salt of a sulphinic acid $(R^1 \cdot SO_2 \cdot S \cdot R^2 + R^3 \cdot NH_2 \rightarrow R^2 \cdot S \cdot NH \cdot R^3 + R^1 \cdot SO_2^- R^3 \cdot NH_3^+)$, a reaction which may be reversed, providing a useful thiolsulphonate synthesis.[372] Cleavage with sodium alkoxides proceeds similarly, giving a sulphoxide and a sodium sulphinate,[521] and thiolysis (with glutathione) of the thiolsulphonate analogue of pantethine has been reported.[522] Alkaline

---

[516] M. J. Aroney, H. Chio, R. J. W. Le Fèvre, and D. V. Radford, *Austral. J. Chem.*, 1970, **23**, 199.
[517] H. Kessler and W. Rundel, *Chem. Ber.*, 1968, **101**, 3350.
[518] A. Geens, G. Swaelens, and M. Anteunis, *Chem. Comm.*, 1969, 439.
[519] R. Steudel and G. Scheller, *Z. Naturforsch.*, 1969, **24b**, 351.
[520] B. G. Boldyrev, L. P. Slesarchuk, and T. A. Trofimova, *Khim. seraorg. Soedinenii, Soderzh. Neftyakh Nefteprod.*, 1968, **8**, 108.
[521] B. G. Boldyrev and L. V. Vid, *Zhur. org. Khim.*, 1969, **5**, 1093.
[522] B. Mannervik and G. Nise, *Arch. Biochem. Biophys.*, 1969, **134**, 90.

*Aliphatic Organo-sulphur Compounds etc.* 107

hydrolysis of aryl benzenethiolsulphinates[523] and corresponding thiolsulphonates[524] is first order in $OH^-$ and substrate.

Disproportionation of arylthiolsulphinates in acetic acid gives a mixture of disulphide and thiolsulphonate, and is catalysed by added sulphides.[525] E.s.r.-monitoring shows the absence of free radical intermediates, and nucleophilic cleavage by the sulphide, $R_2S$, into $R_2S^+ \cdot S \cdot Ar$ is suggested; the kinetic data are the same as those for sulphide-catalysed reaction between a thiolsulphinate and a thiol or a sulphinic acid, but the fact that different effects of structure of sulphide on reaction rates are observed is not inconsistent with the proposed mechanism. N.m.r. studies of the equilibration of $Me_2S$ with $Me \cdot S \cdot \overset{+}{S}Me_2$ in $CD_3CN$ reveal a rapid exchange, faster than the exchange of $MeS^-$ with $Me \cdot S \cdot S \cdot Me$ since $Me_2\overset{+}{S} \cdot$ is a better leaving group than $Me \cdot S \cdot$;[526] related studies of nucleophilic substitution at sulphenyl sulphur have been reported.[527]

The hydrolysis of aryl α-disulphones, $Ar \cdot SO_2 \cdot SO_2 \cdot Ar$, in acidic aqueous dioxan has been studied for comparison with earlier kinetic data relating to corresponding sulphinyl sulphones, $Ar \cdot SO \cdot SO_2 \cdot Ar$.[528] Substitution at sulphinyl sulphur is some $10^4$ times faster than at $\cdot SO_2 \cdot$ in these compounds, but this is due only to a difference in activation energy of 6 kcal $mol^{-1}$, and otherwise the two series show similar dependence of rate on the nature of substituents present in the aryl residues.[529] Both series show similarities also in rate increases with increased water content of solvent, in solvent isotope effects, and in $\Delta S\ddagger$, but there are significant differences in mechanism in comparison with the hydrolysis of sulphonic anhydrides, $Ar \cdot SO_2 \cdot O \cdot SO_2 \cdot Ar$.[529] The transition state suggested is (109).

Aryl α-disulphones show the relative reactivity $F^- \gg AcO^- \gg Cl^- > Br^- > H_2O$ towards nucleophiles, in displacement reactions.[530] The series for comparable displacement reactions with sulphinyl sulphones, $Ar \cdot SO \cdot SO_2 \cdot Ar$, is $Br^- > Cl^- \simeq AcO^- > F^- \gg H_2O$, showing that sulphonyl sulphur is a harder electrophilic centre than sulphinyl sulphur (adopting the terminology of the theory of hard and soft acids and bases);[530] sulphonyl sulphur is very much like carbonyl carbon in its reactivity towards nucleophiles.

Homolytic dissociation of aryl α-disulphones at 145—165 °C follows good first-order kinetics, but gives complex product mixtures.[531] Surprisingly,[531] these compounds are more stable than corresponding aryl

[523] Y. Yoshikawa and W. Tagaki, *Bull. Chem. Soc. Japan*, 1969, **42**, 2899.
[524] S. Oae, R. Nomura, Y. Yoshikawa, and W. Tagaki, *Bull. Chem. Soc. Japan*, 1969, **42**, 2903.
[525] J. L. Kice, C. G. Venier, G. B. Large, and L. Heasley, *J. Amer. Chem. Soc.*, 1969, **91**, 2028.
[526] J. L. Kice and N. A. Favstritsky, *J. Amer. Chem. Soc.*, 1969, **91**, 1751.
[527] J. H. Krueger, *J. Amer. Chem. Soc.*, 1969, **91**, 4974.
[528] J. L. Kice and G. Guaraldi, *J. Amer. Chem. Soc.*, 1967, **89**, 4113.
[529] J. L. Kice and G. J. Kasperek, *J. Amer. Chem. Soc.*, 1969, **91**, 5510.
[530] J. L. Kice, G. J. Kasperek, and D. Petterson, *J. Amer. Chem. Soc.*, 1969, **91**, 5516.
[531] J. L. Kice and N. A. Favstritsky, *J. Org. Chem.*, 1970, **35**, 114.

sulphinyl sulphones, which dissociate into $Ar \cdot SO^{\bullet} + Ar \cdot SO_2^{\bullet}$ *en route* to aryl sulphenyl sulphonates, $Ar \cdot S \cdot O \cdot SO_2 \cdot Ar$; the homolysis of $Ar \cdot SO \cdot SO_2 \cdot Ar$ is more readily brought about than the dissociation of disulphides, too, setting up the order $Ar \cdot SO^{\bullet} > Ar \cdot S^{\bullet} \cong Ar \cdot SO_2^{\bullet}$ for relative stabilities of these sulphur radicals; evidently, repulsive destabilisation thought to be built into these compounds (*e.g.* 110) is a less important

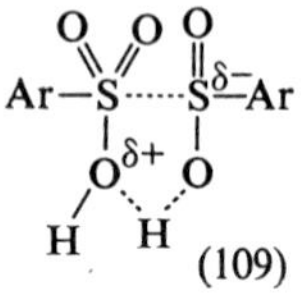

$$Ar-\overset{O}{\underset{\underset{H}{\overset{|}{O^{\delta+}}}}{\overset{\|}{S}}}\cdots\overset{O}{\underset{\underset{H}{\overset{|}{O}}}{\overset{\|}{S^{\delta-}}}}-Ar \qquad Ar-\overset{\overset{\delta-}{O}}{\underset{\underset{\delta-}{O}}{\overset{\|}{S^{\delta+}}}}\overset{\delta+}{\underset{\underset{\delta-}{O}}{\overset{\|}{S}}}-Ar$$

(109) 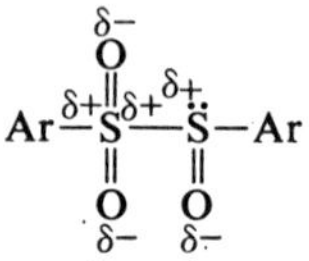 (110)

factor than stability of radicals formed on homolysis, when comparing their relative stabilities to thermolysis.

Thiolsulphonates suffer insertion by active methylene compounds, with concomitant reduction;[532] $Ph \cdot SO_2 \cdot S \cdot Ph$ reacts with $CH_2(CONH_2)_2$ at room temperature, giving $(Ph \cdot S)_2C(CONH_2)_2$, which gives $Ph \cdot S \cdot CH_2 \cdot CONH_2$ on gentle warming, through a mechanism which cannot be at all similar to that of thermal disproportionation reactions of (alkylthio)malonates.[175] The insertion reaction, however, is similar in its outcome to the reaction between diaryl diazomethanes and dialkylxanthogen disulphides, and other disulphides.[533]

[532] S. Hayashi, M. Furukawa, Y. Fujino, and H. Matsukura, *Chem. and Pharm. Bull. (Japan)*, 1969, **17**, 419.
[533] A. Schoenberg, E. Frese, W. Knöfel, and K. Praefcke, *Chem. Ber.*, 1970, **103**, 938.

# 3
# Small-ring Compounds of Sulphur and Selenium

## 1 Three-membered Rings

Interest in the chemistry of three-membered rings containing sulphur is increasing with the development of new methods for their construction and for their oxidation, and it has been enhanced by a greater understanding of the mechanisms of the reactions involved. Thiiran-1-oxides are now readily available, and the mechanism of extrusion of sulphur dioxide from thiiran-1,1-dioxides has been clarified. Almost sacrilegious doubt has been cast on the intermediacy of episulphonium ions in reactions where they would have been accepted hitherto, but evidence has been presented for thiirenium ion intermediates in at least one case. Theoretical aspects of the structure of thiiran have been considered,[1, 2] and orbital symmetries were conserved in an interpretation of the stereospecific addition of triplet sulphur atoms to *cis*- and *trans*-but-2-ene using delocalised $\sigma$-orbitals to describe the structure of thiiran.[2] Zeeman parameters and molecular quadrupole moments have been measured for thiiran.[3]

**Formation of Thiirans and Oxathiirans.**—The acid-catalysed cyclodehydration of $\beta$-hydroxythiols to give thiirans has not held preparative significance because of the ease of acid-catalysed polymerisation of the products, but under controlled conditions $\beta$-hydroxythiols may be converted into thiirans in acceptable yield.[4] A new method of preparing thiirans involved treating an olefin with sulphur monochloride to give mixtures of $\beta$-chloroalkylmono-, di-, and tri-sulphides, among which the di- and tri-sulphides were converted directly into thiirans by treatment with sodium sulphide or aluminium amalgam.[5] An excess of olefin was required to suppress formation of $\beta$-chloroalkylmonosulphides, which were useless for conversion to thiirans. The reaction proceeded stereospecifically with retention of olefin configuration, suggesting episulphonium ion intermediates, and accordingly the ratio of initial adducts of propylene with

---

[1] P. Lindner and O. Martensson, *Acta Chem. Scand.*, 1969, **23**, 429.
[2] E. Leppin and K. Gollnick, *Tetrahedron Letters*, 1969, 3819.
[3] D. H. Sutter and W. H. Flygare, *J. Amer. Chem. Soc.*, 1969, **91**, 4063.
[4] A. D. B. Sloan, *J. Chem. Soc. (C)*, 1969, 1252.
[5] F. Lautenschlaeger and N. V. Schwartz, *J. Org. Chem.*, 1969, **34**, 3991.

sulphur monochloride was similar to that obtained with alkylsulphenyl chlorides, and similar to that obtained on treatment of methylthiiran with chlorine, reactions for which episulphonium ion intermediates are accepted. Buta-1,3-diene was converted into 2-vinylthiiran by the new procedure, indicating almost exclusive 1,2-addition.

The assumption that dihydrothiadiazoles are intermediates in the formation of thiirans from diazo-compounds and thioketones has been verified by the reaction of bis(trifluoromethyl)diazomethane with hexafluorothioacetone to give the dihydrothiadiazole (1), which lost nitrogen on boiling to give tetrakis(trifluoromethyl)thiiran.[6] Bis(trifluoromethyl)thioketen reacted with bis(trifluoromethyl)diazomethane to give (2), which on heating

$$(CF_3)_2C\underset{S}{\overset{N=N}{\diagup\diagdown}}C(CF_3)_2$$
(1)

$$(CF_3)_2C=C\underset{S}{\overset{N=N}{\diagup\diagdown}}C(CF_3)_2$$
(2)

$$(CF_3)_2C=C\underset{S}{-}C(CF_3)_2$$
(3)

$$\underset{Ar}{\overset{Ar}{\diagdown}}C\underset{S}{-}C\underset{OR}{\overset{S-\overset{\|}{C}-OR}{\diagup}}$$
(4)

gave (3), the first example of an allene episulphide. It has also been pointed out that diazoalkanes react with many thioketones to give thiirans under conditions ($-78\,°C$, no catalyst, no irradiation) where carbenes are not generated, further suggesting that it is the diazoalkanes and not the derived carbenes which are the reactive species.[7] However, irradiation of a solution of thiobenzophenone and diethyl diazomalonate in cyclohexane gave diethyl-2,2-diphenylethylene-1,1-dicarboxylate (among other products), indicating the likely formation of diethyl-2,2-diphenyl-thiiran-3,3-dicarboxylate as an intermediate.[8] Unfortunately the temperature at which this reaction was performed was not recorded, and since thiobenzophenone is known to undergo thermally induced cycloaddition to diethyl diazomalonate, it would be imprudent to assume that photocatalysed cycloaddition necessarily occurred in this case. There are numerous examples of the addition of aromatic thioketones to diazoalkanes to give thiiran derivatives, and another case has been reported,[9] but it has been shown that aliphatic thioketones also react with diazomethane to give 2,2-dialkylthiirans.[10] Diazoalkanes also react with xanthogenic acid anhydrides to give thiirans typified by (4).[11] Thiobenzophenone reacted with tetracyanoethylene to

[6] W. J. Middleton, *J. Org. Chem.*, 1969, **34**, 3201.
[7] D. Seyferth and W. Tronich, *J. Amer. Chem. Soc.*, 1969, **91**, 2138.
[8] J. A. Kaufman and S. J. Weininger, *Chem. Comm.*, 1969, 593.
[9] A. Schonberg, E. Frese, W. Knofel, and K. Praefcke, *Ber.*, 1970, **103**, 938.
[10] D. Paquer and J. Vialle, *Bull. Soc. chim. France*, 1969, 3327.
[11] A. Schonberg, W. Knofel, E. Frese, and K. Praefcke, *Ber.*, 1970, **103**, 949.

## Small-ring Compounds of Sulphur and Selenium

give 1,1-dicyano-2,2-diphenylethylene, most probably *via* 2,2-dicyano-3,3-diphenylthiiran as intermediate,[12] whilst with phenyl(bromodichloromethyl)mercury it gave the known 2,2-chloro-3,3-diphenylthiiran.[7] Phenyl(bromodichloromethyl)mercury also reacted with thiophosgene or sulphur to give 2,2,3,3-tetrachlorothiiran, thiophosgene acting as an intermediate when sulphur was used. It was considered more likely that the thiocarbonyl compounds reacted directly with the organomercurial, rather than with the dichlorocarbene that it might generate.

The conventional preparation of the thiiran derivative (5) from bicyclo-[2,2,1]heptadiene *via* the corresponding oxiran derivative is not a viable proposition because of the relative inaccessibility of the latter. However, the readily available thietan derivative (6), has been converted into (5) in 30% yield by reaction with potassium cyanide,[13] the nonclassical ion (7)

being postulated as an intermediate. Insignificant amounts of 2-vinylthiiran were obtained on reaction of 3,7-dichloro-5-thianona-1,8-diene with potassium cyanide, confirming that the reaction was not of general utility. The Shionogi group has prepared more steroidal thiirans,[14] the influence of substituents on the formation of the thiiran ring by base-catalysed cyclisation of vicinal acetoxy thiocyanates being illustrated, for example, by the predominant formation of a thiiran from (8) and an oxiran from (9). Three stereoisomeric forms of the 2,2′-bisthiiran (10) have been prepared from the appropriate 2,2′-bisoxirans by treatment with potassium thiocyanate,[15] and 1,2:5,6-diepithio-L-iditol and some of its derivatives have been synthesised.[16] 3-Chloro-3-methylthietan-2-one has been cleaved by ammonia and dialkylamines to give 2-methyl-2-carbonamidothiiran derivatives,[17] and thiirans were obtained on treatment of oxirans with the alkali-metal salts of di-*O*-alkyl phosphorothiolic acid or di-*O*-alkylphosphorothiolthionic acid.[18]

[12] W. J. Linn and E. Ciganek, *J. Org. Chem.*, 1969, **34**, 2146.
[13] F. Lautenschlaeger, *J. Org. Chem.*, 1969, **34**, 3998.
[14] T. Komeno, S. Ishihara, H. Itani, H. Iwakura, and T. Takeda, *Chem. and Pharm. Bull. (Tokyo)*, 1969, **17**, 2110.
[15] P. W. Feit, *J. Medicin. Chem.*, 1969, **12**, 556.
[16] J. Kuszmann and L. Vargha, *Carbohydrate Res.*, 1969, **11**, 165.
[17] M. G. Lin'kova, A. M. Orlov, O. V. Kil'desheva, and I. L. Knunyants, *Izvest. Akad. Nauk S.S.S.R., Ser. khim.*, 1969, 1148.
[18] O. N. Nuretdinova, *Zhur. obshchei Khim.*, 1969, **39**, 2141.

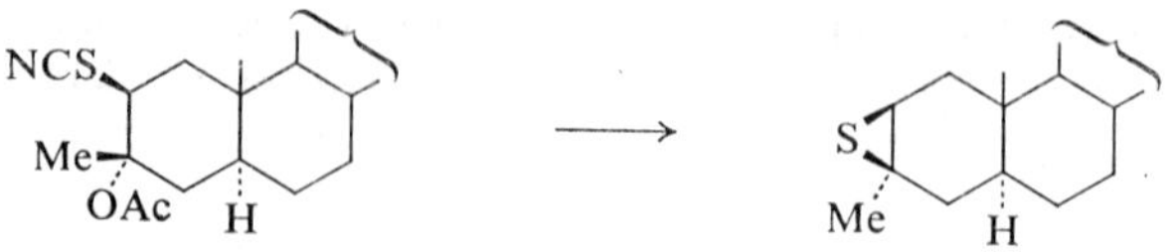

(8) 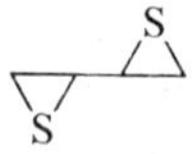 → (10)

(9) →

An unusual example of the formation of a thiiran ring was provided by the conversion of the spiroheterocycle (11) on treatment with ammonia into the tricyclic thiiran derivative (12).[19] The proposed mechanism

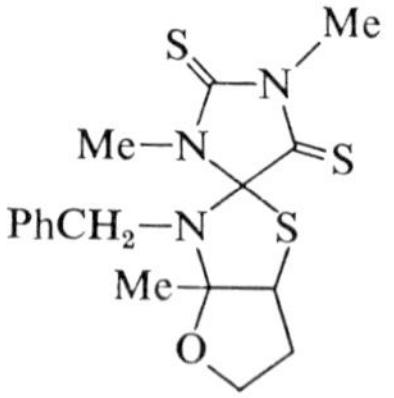

(11) 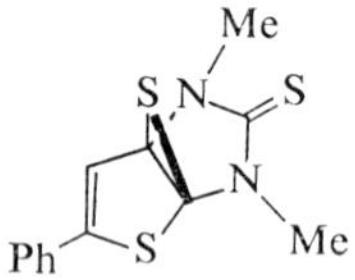     (12)

involved the formation of a thiiran ring by intramolecular nucleophilic attack of a $\beta$-thiolate anion on a thioketone. Two papers have described the probable intermediate formation of thiirans from 1,3-dipoles.[12,20] The dipole (13), generated by reaction of thiobenzophenone with tetracyanoethylene oxide, supposedly cyclised to 2,2-dicyano-3,3-diphenylthiiran which rapidly extruded sulphur or reacted further with thiobenzophenone,[12] whilst the dipole (14), generated from (15) by treatment with sodium

$Ph_2\overset{+}{C}-S-\bar{C}(CN)_2$     $(Me_2N)_2\overset{+}{C}-S-\bar{C}H-CO_2R$     $(Me_2N)_2C=\overset{+}{S}-CH_2-CO_2R$
(13)                          (14)                                  (15)

hydride, probably collapsed to 2,2-bis(dimethylamino)-3-alkoxycarbonyl-thiiran, which rapidly rearranged (see p. 113).[20] The dipole (14) cyclises so rapidly that it could not be trapped by diethyl acetylene dicarboxylate. Oxathiirans have not been isolated, but dichloro-oxathiiran has been proposed as an intermediate in the decomposition of the thermally unstable

[19] A. Takamizawa and S. Matsumoto, *Tetrahedron Letters*, 1969, 2875.
[20] M. Takaku, S. Mitamura, and H. Nozaki, *Tetrahedron Letters*, 1969, 3651.

thiophosgene-$S$-oxide,[21] and an oxathiiran has been implicated as an intermediate in the photodecomposition of aromatic thioketone-$S$-oxides.[22]

**Reactions of Thiirans.**—Two improved methods have been described for the oxidation of thiirans to their oxides.[23, 24] Thiiran and α-hydroxymethylthiiran were oxidised to their 1-oxides in 55—60% yield by an anhydrous solution of hydrogen peroxide in t-butyl alcohol containing vanadium pentoxide as catalyst.[23] Thiiran-1-oxides were formed at 15—20 °C, and the corresponding 1,1-dioxides at *ca.* 45 °C, and the conditions were readily adjusted to allow selective oxidation to the 1-oxides. Vanadium pentoxide acted as an indicator as well as a catalyst, and the reaction could be performed as a titration. Peroxybenzoic acid also oxidised thiirans to their 1-oxides efficiently at −20 to −30 °C in methylene chloride,[24] and n.m.r. data suggested that oxidation of 2-methylthiiran gave *trans*-2-methylthiiran-1-oxide stereospecifically (see p. 116). Oxidation of 1-vinylthiiran with hydrogen peroxide proceeded with rearrangement to give 2,5-dihydrothiophen-1-oxide,[13] whilst a thietan-1-oxide derivative was formed by oxidative rearrangement of the thiiran derivative (5).[13]

The controlled polymerisation of low-molecular-weight thiirans was efficiently initiated by thiolates of the general formula $(XCH_2 \cdot CH_2 \cdot S)_yM$, where M = Zn or Cd, $y$ = 1 or 2, and X is a group capable of co-ordinating with M, the cadmium derivatives being most effective.[25] Further examples of reactions of thiirans with amines comprised the reactions of primary and secondary amines with 2-methylthiiran to give derivatives of 1-amino-2-mercaptopropane [25a] (solvent has an effect upon product composition in such reactions) [26] and the reactions of 7-thiabicyclo[4,1,0]heptane with ammonia, methylamine, ethyl glycinate, and ethyl sarcosinate to give 2-aminocyclohexanethiol or its appropriate amino-derivatives.[27] In all cases further transformations occurred in which the thiol functions in the initially formed β-aminothiols reacted with the thiirans to give β-mercapto-β'-amino (or alkylamino) dialkyl sulphides. 2-Methyl-2-methoxycarbonylthiiran reacted with chlorine in carbon tetrachloride to give the products derived by both modes of cleavage of the thiiran ring, 1-chloro-2-methoxycarbonylpropane-2-sulphenyl chloride and 2-chloro-methoxycarbonylpropane-1-sulphenyl chloride.[17]

According to a mechanistic scheme devised to rationalise the rearrangement of the ylide (14), the proposed intermediate 2,2-bis(dimethylamino)-3-alkoxycarbonylthiiran rearranged to the enethiol (16a), which was characterised by its reaction with methyl iodide to give the enolthioether

---

[21] J. Silhanek and M. Zbirovsky, *Chem. Comm.*, 1969, 878.
[22] R. H. Schlessinger and A. G. Schutz, *Tetrahedron Letters*, 1969, 4513.
[23] F. E. Hardy, P. R. H. Speakman, and P. Robson, *J. Chem. Soc.* (C), 1969, 2334.
[24] K. Kondo, A Negishi, and M. Fukuyama, *Tetrahedron Letters*, 1969, 2461.
[25] (*a*) R. T. Wragg, *J. Chem. Soc.* (C), 1969, 2087; (*b*) *ibid.* p. 2582.
[26] E. Tobler, *Ind. and Eng. Chem. (Product Res. and Development)*, 1969, **8**, 415.
[27] K. Jankowski, R. Gauthier, and C. Berse, *Canad. J. Chem.*, 1969, **47**, 3179.

(16b).[20] Another interesting rearrangement involved the conversion on heating of the thiiran (4) into the compound (17).[11] This may involve a sigmatropic rearrangement as indicated in (18).

$$\underset{Me_2N}{\overset{Me_2N}{\Large >}}C=C\underset{CO_2R^1}{\overset{SR^2}{\Large <}}$$

(16a) $R^1$ = Me or Et, $R^2$ = H
(16b) $R^1$ = Me or Et, $R^2$ = Me

$$\underset{Ar}{\overset{Ar}{\Large >}}C\underset{\underset{S}{\overset{\|}{C}}-OR}{\overset{S-\overset{S}{\overset{\|}{C}}-OR}{\Large <}}$$

(17)

(18)

In an examination of the efficiency of the thiiran moiety as a neighbouring group in solvolysis, 2-chloromethylthiiran was found to undergo acetolysis 3000 times faster than allyl chloride, and 100 times faster than chloromethylcyclopropane.[28] Since the magnitude of the anchimeric assistance was comparable with that observed for 2-chloroethyl alkyl sulphides it appeared that participation involved only the lone pair on sulphur, and not the carbon–sulphur bond in a manner similar to that proposed for cyclopropyl participation in cyclopropyl carbinyl halide solvolysis. The formation of 3-acetoxythietane in the acetolysis of 2-chloromethylthiiran was established,[28] and further examples of this now well-recognised rearrangement of thiiran to thietan derivatives during the reaction of 2-chloromethylthiiran and related compounds with nucleophiles have been reported.[23, 29]

The stereochemistry of the desulphurisation of thiirans by organolithium compounds has been reinvestigated.[30] *trans*-2,3-Dimethylthiiran and butyl-lithium reacted to give *trans*-but-2-ene stereospecifically, whilst *cis*-2,3-dimethylthiiran gave *cis*-but-2-ene almost stereospecifically (0·6% *trans*-isomer). Two mechanisms compatible with these results were considered. The first (A, Scheme 1) (the *trans*-isomer is exemplified) involved the formation of a thiiran anion which collapsed in a concerted and hence stereospecific manner to the olefin. In the second proposed mechanism (B, Scheme 1) the lithium carbanion served as an intermediate which decomposed to products at a rate faster than bond rotation. In order to distinguish between these mechanisms the stereochemistry of the olefin obtained by decomposition of the *erythro*- and *threo*-carbanions generated (with retention of configuration) from the corresponding *erythro*- and *threo*-2-bromo-3-ethylthiobutanes was examined. Both carbanions gave mixtures of olefins suggesting that mechanism A operated for the desulphurisation. Although the Woodward–Hoffmann rules predict a non-concerted reaction for the decomposition of thiirans to olefins, it appears that they predict a concerted disrotatory mode of decomposition for the carbanionic intermediate of mechanism A, the change arising from *d*-orbital participation in the latter case.[30]

[28] H. Morita and S. Oae, *Tetrahedron Letters*, 1969, 1347.
[29] O. N. Nuretdinova, F. F. Guseva, and B. A. Arbusov, *Izvest. Akad. Nauk S.S.S.R., Ser. khim.*, 1969, 2851.
[30] B. M. Trost and S. Ziman, *Chem. Comm.*, 1969, 181.

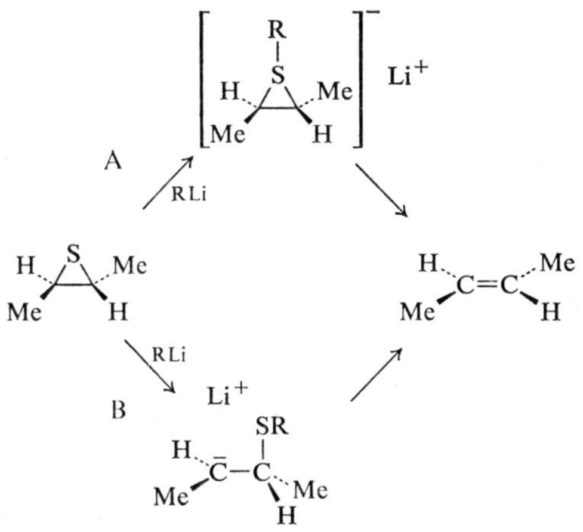

**Scheme 1**

Tetra-arylthiirans were reduced by lithium aluminium hydride to tetra-arylethylenes,[31] therefore behaving like steroidal thiirans and unlike simple thiirans, which give mercaptans. Further examples of the very rapid loss of sulphur from thiirans bearing electronegative substituents have been provided by the failure to isolate them from reactions in which thiirans were expected as products.[8, 12, 32] Phenyl, cyano, and ester groups, particularly if there are four present in any combination,[8, 12] are particularly effective in promoting extrusion, although it appears that the presence of only one cyano or ester group may be sufficient to cause spontaneous extrusion of sulphur at room temperature.[32] However, tetrachlorothiiran decomposed only after heating to 150 °C,[7] and 2,3-diphenyl-2,3-dibenzoyl-thiiran appears to be relatively stable.[33] Tetrachlorothiiran reacted with phosphorus trichloride to give tetrachloroethylene, but with triphenylphosphine, a common reagent for effecting this transformation, only traces of the olefin were produced.[7] Triphenylphosphine did react with the thiiran (4) to give the corresponding olefin.[11]

The photodecomposition of the thiiran (19) in methanol was dependent upon the wavelength of the incident light.[34] With 253·7 nm light desulphurisation to the tetramethylmethylene cyclobutanone (20) was the predominant process, but with light of wavelength longer than 280 nm the only reactions were loss of carbon monoxide to give (21), and rearrangement to the isomeric cyclic acetals (22). The products formed by photolysis with >280 nm light were consistent with Norrish type 1 cleavage of the $n \to \pi^*$

[31] N. Latif and N. Mishriky, *Chem. and Ind.*, 1969, 491.
[32] W. H. Mueller, *J. Org. Chem.*, 1969, **34**, 2955.
[33] D. C. Dittmar, G. C. Levy, and G. E. Kuhlmann, *J. Amer. Chem. Soc.*, 1969, **91**, 2097.
[34] J. G. Pacifici and C. Diebert, *J. Amer. Chem. Soc.*, 1969, **91**, 4595.

(19)   (20)   (21)   (22)

state of the carbonyl group, whilst loss of sulphur upon excitation with 253·7 nm light was associated with the 265 nm band in the u.v. spectrum of (19), interpreted as being due to a $n \to \pi^*$ sulphur transition. The geometry of the keto-thiiran (19) is such that significant orbital overlap between the carbonyl group and the lone pair on sulphur cannot occur, and accordingly no charge-transfer band was observed in its u.v. spectrum.

**Thiiran Oxides.**—These compounds are now more readily available by oxidation of thiirans.[23,24] Thiiran-1-oxide exhibited an $A_2B_2$ pattern in its n.m.r. spectrum, indicating a stable pyramidal configuration at sulphur,[24] and the presence of only one doublet signal attributable to the methyl group in 2-methylthiiran-1-oxide and cis-2,3-dimethylthiiran-1-oxide, obtained by oxidation of the corresponding thiirans with peroxybenzoic acid, suggested strongly that the oxidation proceeded stereospecifically, since the known magnetic anisotropy of the sulphinyl bond is such that the methyl groups in 2-methylthiiran-1-oxides isomeric at sulphur should resonate at different fields. Furthermore, the oxidation of 2-methylthiiran to its oxide was attended by an up-field shift in the methyl signal, from which it was tentatively deduced that the oxidation proceeded stereospecifically trans to the methyl group.[24] This followed by analogy with the n.m.r. characteristics of other cyclic sulphoxide systems, and the validity of this analogy was strengthened by an examination of the n.m.r. characteristics of 2,2-dimethylthiiran and trans-2,3-dimethylthiiran and their oxides (Scheme 2). The latter oxides were too unstable to isolate in a pure state.

Acid-catalysed ring opening of thiiran-1-oxide in methanol gave β-methoxyethanethiolsulphinate (23, R = H) (Scheme 3),[35] in contrast to a previous report that a mixture of the corresponding disulphide and thiolsulphonate was formed. Examination of the products obtained similarly from 2-substituted thiiran-1-oxides indicated that nucleophilic attack by methanol occurred preferentially at C-2, and a carbonium-ion mechanism was proposed for the reaction (Scheme 3).[35] It is pertinent, however, that previous evidence had indicated that thiiran-1-oxide underwent acid-catalysed methanolysis by a bimolecular mechanism. Whereas acyclic sulphoxides form complexes with various metal salts, thiiran-1-oxide reacted with copper(II) chloride in benzene at room temperature to give β-chloroethyl β-chloroethanethiolsulphonate (24) (Scheme 4).[36] In alcoholic solution below 0 °C the major products were alkyl β-chloroethyl-

[35] K. Kondo, A. Negishi, and G. Tsuchihashi, *Tetrahedron Letters*, 1969, 3173.
[36] K. Kondo, A. Negishi, and G. Tsuchihashi, *Tetrahedron Letters*, 1969, 2743.

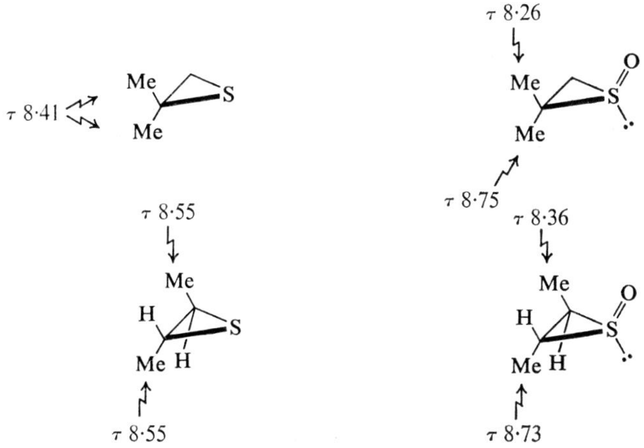

## Scheme 2

sulphinates (25). Copper(II) bromide reacted similarly. Since the presence of various olefins had no effect, an ionic rather than free-radical mechanism was proposed. In accord with this proposed mechanism thiiran-1-oxide reacted with ethanesulphinyl chloride to give a mixture of the thiolsulphinates (26) and (27) (Scheme 4).[36] Sulphur monoxide produced by pyrolysis of thiiran-1-oxide has been detected by microwave spectroscopy,[37] and the production of 9-dichloromethylene fluorene by addition of thiophosgene-$S$-oxide to 9-diazofluorene possibly involved the extrusion of sulphur monoxide from an intermediate thiiran-1-oxide derivative.[38]

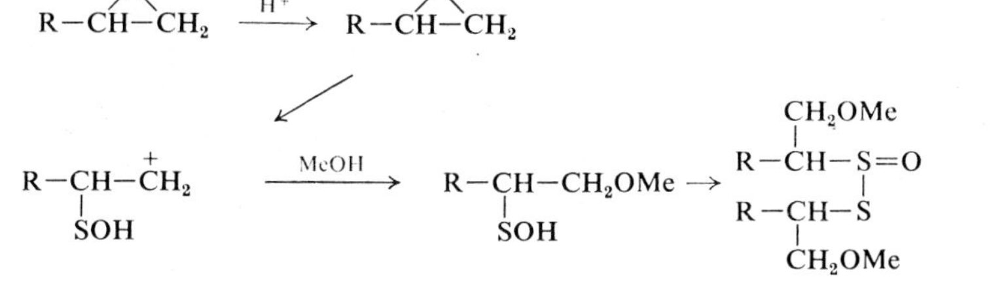

## Scheme 3

[37] S. Saito, *Bull. Chem. Soc. Japan*, 1969, **42**, 667.
[38] B. Zwanenburg, L. Thijs, and J. Strating, *Tetrahedron Letters*, 1969, 4461.

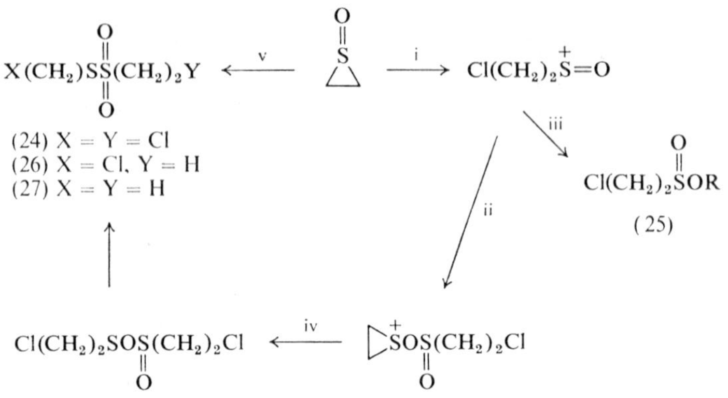

$$X(CH_2)S\overset{O}{\underset{O}{\overset{\|}{S}}}(CH_2)_2Y \xleftarrow{v} \underset{\triangle}{\overset{O}{\overset{\|}{S}}} \xrightarrow{i} Cl(CH_2)_2\overset{+}{S}=O$$

(24) X = Y = Cl
(26) X = Cl, Y = H
(27) X = Y = H

$$\downarrow iii$$

$$Cl(CH_2)_2 \overset{O}{\overset{\|}{S}}OR$$

(25)

$$\uparrow$$

$$\downarrow ii$$

$$Cl(CH_2)_2SO\underset{O}{\overset{\|}{S}}(CH_2)_2Cl \xleftarrow{iv} \triangleright \overset{+}{S}O\underset{O}{\overset{\|}{S}}(CH_2)_2Cl$$

Reagents: i, $CuCl_2$; ii, thiiran-1-oxide; iii, ROH; iv, $Cl^+$; v, EtSOCl

**Scheme 4**

**Thiiran-1,1-dioxides and Thiiren-1,1-dioxides.**—The thermal extrusion of sulphur dioxide from thiiran-1,1-dioxide is a synthetically useful reaction, but its mechanism has not been clearly understood despite extensive investigations. The reaction proceeds stereospecifically, although the concerted elimination of sulphur dioxide is forbidden according to the Woodward–Hoffmann rules. New light on this problem has been provided by the observation that pyrolysis at 300 °C of the unusually stable cis-2,3-diphenyl-2,3-dibenzoylthiiran-1,1-dioxide (28) gave benzil, diphenylacetylene, and the lactone (29) (Scheme 5).[33] These results were rationalised in terms of the sequential expansion in a stereospecific manner of the thiiran-1,1-dioxide (28) to the 1,2-oxathietan derivative (30) and the 1,3,2-dioxathiolan derivative (31), either of which could decompose to the observed products as shown in Scheme 5. It is significant that orbital correlation diagrams indicated that the concerted loss of sulphur dioxide from the 1,3,2-dioxathiolan derivative (31) is an allowed process, and this pathway was therefore considered to be the more likely in this case, and possibly in all cases of thermal extrusion of sulphur dioxide from thiiran-1,1-dioxides. Although the concerted loss of sulphur dioxide from the 1,2-oxathietan-2-oxide (30) was predicted to be a high energy (2+2) process, it was not entirely ruled out because the orbital energies and amplitude of orbital overlap in this sulphur-containing system might render the relevant transition state more easily accessible than in the corresponding all-carbon system. Photochemical extrusion of sulphur dioxide from the thiiran-1,1-dioxide (28) did not occur,[33] in contrast to the behaviour of some other sulphones.

Diphenylthiiren-1,1-dioxide is thermally more stable than its saturated analogues, and in an investigation of the role of conjugation in its relative stability,[39] the rates at which it reacted with alkoxide ions were compared

[39] F. G. Bordwell and S. C. Crooks, *J. Amer. Chem. Soc.*, 1969, **91**, 2084.

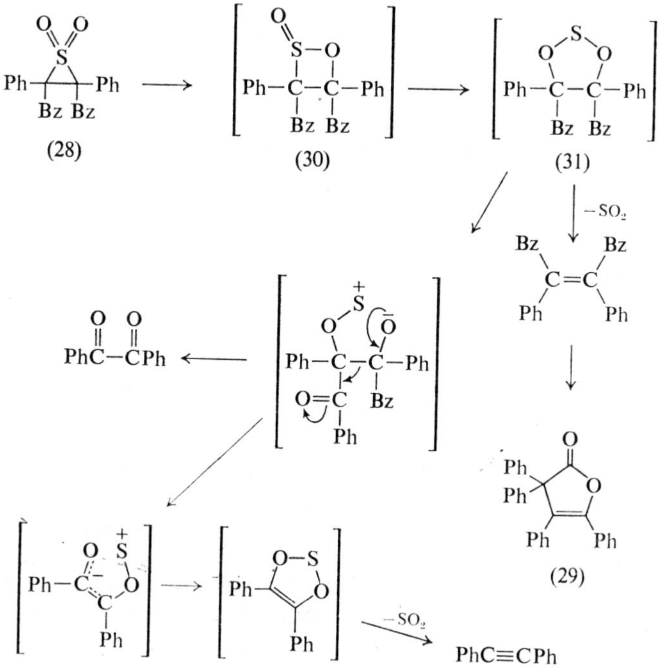

# Small-ring Compounds of Sulphur and Selenium

**Scheme 5**

with the corresponding rates for diphenylcyclopropenone, in which conjugative stabilisation has been established. With methoxide ion diphenylthiiren-1,1-dioxide gave predominantly the sulphonate (32), whilst diphenylcyclopropenone gave only methyl *cis*-2,3-diphenylpropenoic acid. Furthermore, incremental differences in rates of reaction with a series of alkoxide nucleophiles were similar for the two cyclic systems, suggesting a similarity of mechanism. However, the diphenylthiiren-1,1-dioxide reacted *ca.* 5000 times faster than diphenylcyclopropenone with a given alkoxide, a remarkable reversal of the situation in acyclic analogues, where carbonyl groups are attacked much more readily than sulphonyl groups. This was interpreted in terms of very little conjugative stabilisation of the thiiran dioxide ring, since possible mechanisms, one of which is indicated in Scheme 6, suggested that there should be a relationship between the

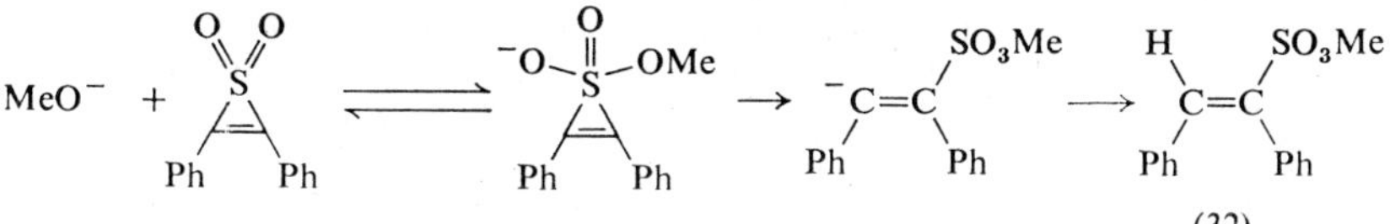

**Scheme 6**

aromaticity of the ring and rate of cleavage. Diphenylthiiren-1,1-dioxide reacted much faster than its saturated analogue with methoxide ion, reflecting the increased ring strain associated with the presence of the double bond,[39] and it readily underwent cycloaddition to acyclic and cyclic enamines to give unsaturated sulphones (Scheme 7),[40] and with ynamines to give stable 3-dialkylaminothiophen-1,1-dioxides.[40]

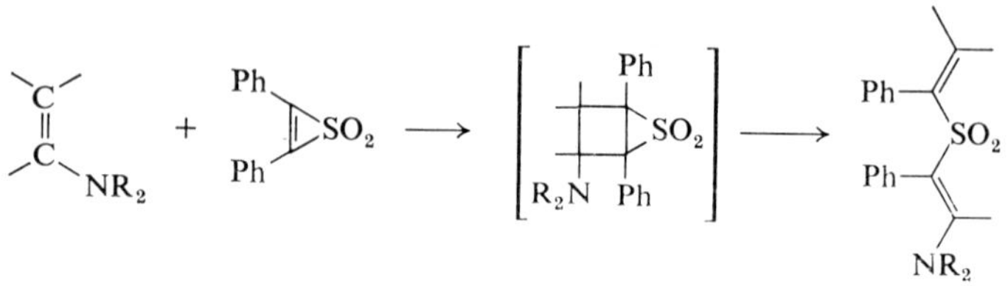

Scheme 7

Keten in liquid sulphur dioxide apparently gave the dipole (33), which may be in equilibrium with 2-oxo-thiiran-1,1-dioxide.[41] A spectroscopic search for this intermediate was fruitless, but it was trapped by Schiff's bases; for example with benzylidene aniline it gave 4-oxo-2,3-diphenyl-1,3-thiazolidine-1,1-dioxide. The relative stability of cis-2,3-diphenyl-2,3-dibenzoylthiiran-1,1-dioxide was attributed to the instability of a dipolar form analogous to (33).[33]

**Thiiranium, Thiirenium, and Seleniranium ions.**—Thiiranium ions continue to be invoked as reaction intermediates, and the subject has been reviewed.[42] They have been postulated as intermediates in the reaction of 2,4-dinitrobenzenesulphenyl chloride with ketones,[43] the reaction of olefins with thiocyanogen[44] or with sulphur monochloride,[5] the acid-catalysed cyclodehydration of β-hydroxythiols,[4] the reaction of 3,7-dichloro-5-thianona-1,8-diene with potassium cyanide,[13] the reaction of 2-chloropropyl sulphide with sodium thiosulphate,[25b] the cleavage of thiiran-1-oxides,[35, 36] and in the deoxydative β-substitution of pyridine-N-oxide and related compounds by mercaptans.[45] A kinetic study of the addition of 4-substituted-2-nitrobenzenesulphenyl halides to cyclohexene in acetic acid indicated that the formation of the thiiranium intermediate was the rate-limiting step in the reaction.[46] A seleniranium ion was invoked to account for the mode of reaction of sulphur monochloride with dienes.[47] According to extended

[40] M. H. Rosen and G. B. Bonet, Abstracts of the Papers of the 159th A.C.S. National Meeting, Houston, Texas, February 1970.
[41] A. de S. Gomes and M. M. Joullie, J. Heterocyclic Chem., 1969, 6, 729.
[42] W. H. Mueller, Angew. Chem. Internat. Edn., 1969, 8, 482; D. R. Christ and N. J. Leonard, ibid., p. 962.
[43] R. Gustafsson, C. Rappe, and J.-O. Levin, Acta Chem. Scand., 1969, 23, 1843.
[44] R. G. Guy, R. Bonnett, and D. Lanigan, Chem. and Ind., 1969, 1702.
[45] F. H. Hershenson and L. Bauer, J. Org. Chem., 1969, 34, 655.
[46] C. Brown and D. R. Hogg, J. Chem. Soc. (B), 1969, 1054.
[47] F. Lautenschlager, J. Org. Chem., 1969, 34, 4002.

Hückel calculations the angle between the S—H bond and the plane of the thiiran ring is 20° in the most stable conformation of protonated thiiran.[48] The calculations also indicated that the $\pi$-bond population in the C—C bond is smaller than that in cyclopropane, and that the highest occupied orbital in thiiran is the lone pair orbital on sulphur.

It has been argued that a thiiranium ion is not necessarily an intermediate in the hydrolysis of 2-chloroethyl methyl sulphide, although it hydrolysed some $2 \times 10^5$ times faster than ethyl bromide.[49] These arguments were based partly upon the finding that 2-hydroxyethyl methyl sulphide was the product, and no methanol or ethylene sulphide were formed. Although the possible influence of ring strain in directing the reaction was recognised, it was pointed out that the rates of hydrolysis of sulphonium salts are very slow compared to that of 2-chloroethyl methyl sulphide. Additionally, the temperature coefficient of the enthalpy of hydrolysis of 2-chloroethyl methyl sulphide was consistent with an $S_N 1$ type mechanism, so the hydrolysis did not involve an internal $S_N 2$ displacement.

Episulphonium salts are widely accepted as intermediates in the ionic addition of sulphenyl halides to olefins, but doubt has been cast upon the validity of this assumption.[50, 51] The stable (*i.e.* isolable) *cis*-cyclo-octene-*S*-methylepisulphonium ion (34) underwent nucleophilic attack predominantly or exclusively (depending upon the nucleophile) at sulphur to give mostly *cis*-cyclo-octene and methanesulphenyl compounds, and not at carbon, as analogy with the accepted mechanism for olefin-sulphenyl halide additions would suggest.[50] Some attack at carbon occurred only with acetate ion or pyridine as nucleophile. Increasing polarisability, and hence 'softness', in a series of nucleophiles was reflected in increased yields of *cis*-cyclo-octene, which was rationalised in terms of Pearson's theory of hard and soft bases, sulphur being a softer electrophile than carbon. The *trans*-isomer of (34) also gave *trans*-cyclo-octene with iodide or methanethiolate ion.[50] In a further investigation, the salt (34) and tetraphenylarsonium chloride in [$^2H_3$]tetranitromethane gave only 1-chloro-2-methylthiocyclo-octane as final product, but at short reaction times or at $-5$ °C n.m.r. spectroscopy revealed the presence of a quadrivalent sulphur derivative (35) which reacted immediately with chloride ion at $-5$ °C to give 1-chloro-2-methylthiocyclo-octane.[51] The structure of the compound (35) was allocated on the basis of its n.m.r. spectrum, and the square-pyramidal arrangement about sulphur was assigned after extended Hückel calculations on model compounds.

---

[48] T. Yonezawa, K. Shimizu, H. Morimoto, and H. Koto, *Nippon Kagaku Zasshi*, 1969, **90**, 1196.

[49] M. J. Blandamer, H. S. Golinkin, and R. E. Robertson, *J. Amer. Chem. Soc.*, 1969, **91**, 2678.

[50] D. C. Owsley, G. K. Helmkamp, and S. N. Spurlock, *J. Amer. Chem. Soc.*, 1969, **91**, 3606.

[51] D. C. Owsley, G. K. Helmkamp, and M. F. Rettig, *J. Amer. Chem. Soc.*, 1969, **91**, 5239.

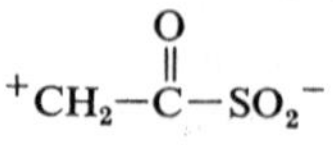

$^+CH_2-\overset{\overset{O}{\|}}{C}-SO_2^-$
(33)

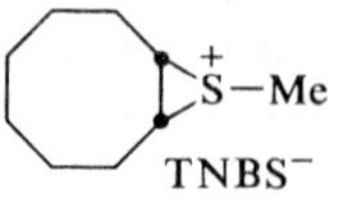

TNBS⁻
(34)

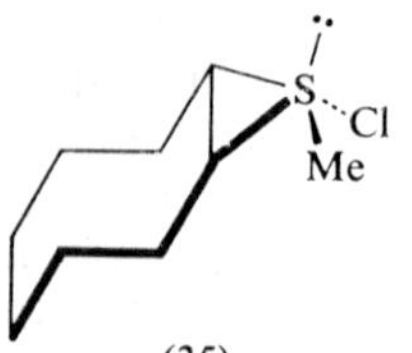

(35)

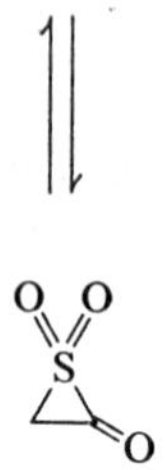

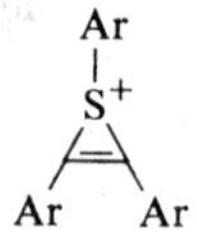

(36)

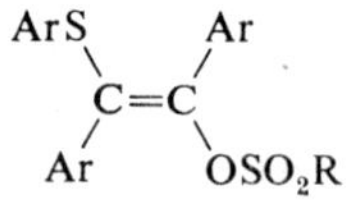

(37)

Convincing evidence has been presented supporting the intermediacy of thiirenium ions (36) in the unimolecular reactions of β-arylthiovinyl sulphonates (37),[52] and the mode of addition of toluene-*p*-sulphenyl chloride to alkylacetylenes has been rationalised in terms of the intermediacy of a thiirenium ion or its quadrivalent sulphur equivalent.[53]

## 2 Four-membered Rings

Interest in the formation of thietan derivatives by cycloaddition of sulphenes to enamines is flourishing, and optically active thietans have been prepared by this method. The photocatalysed addition of thioketones to olefins may also be of potential preparative utility. Configurational and conformational aspects of thietan-1-oxides, and their n.m.r. spectral characteristics, are now more clearly understood, and the stereochemical consequences of displacement reactions at sulphur and of ring cleavage have been examined.

**Formation of Thietan Derivatives.**—Addition of sulphenes to enamines is developing into a favoured method of entry into the thietan system. Thiet and thietan derivatives are rendered more accessible by an improved procedure for the preparation of 3-(*NN*-dimethylamino)thietan-1,1-dioxide by cycloaddition of sulphene and *NN*-dimethylvinylamine.[54] An efficient preparation of *NN*-dimethylvinylamine and its diethyl analogue was described. The presence of functional groups with a tendency to stabilise the incipient negative charge on carbon in sulphenes did not inhibit their cycloaddition to enamines, since ethoxycarbonyl sulphenes reacted with enamines to give 3-dialkylamino-2-ethoxycarbonylthietan-

---

[52] G. Cappozzi, G. Melloni, G. Modena, and V. Tonellato, *Chem. Comm.*, 1969, 1520.
[53] V. Cato, G. Scorrano, and G. Modena, *J. Org. Chem.*, 1969, **34**, 2020.
[54] P. L.-F. Chang and D. C. Dittmer, *J. Org. Chem.*, 1969, **34**, 2791.

1,1-dioxide derivatives, such as (38).[55] Similarly, 9-fluorenesulphonyl chloride with triethylamine and $NN$-dimethyl-1-isobutylamine gave (39).[55]

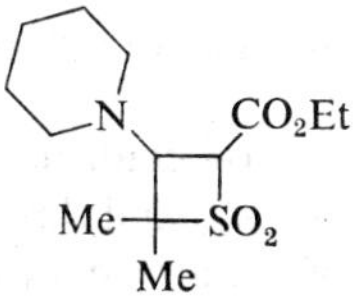

(38)

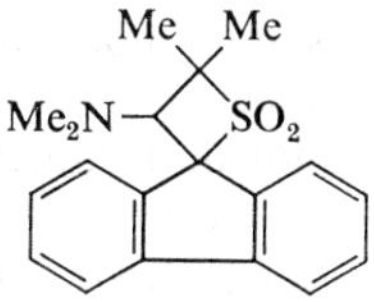

(39)

Assymetric induction has been observed in the formation of the 3-dialkylaminothietan-1,1-dioxide (40) by cycloaddition of sulphene to the enamine (41), derived from $(R)$-(+)-methyl-α-phenylethylamine.[56] This was revealed by conversion of the cycloadduct (40) to optically active $(S)$-(+)-methylthiet-1,1-dioxide (42) in 6% optical yield (Scheme 8). The absolute

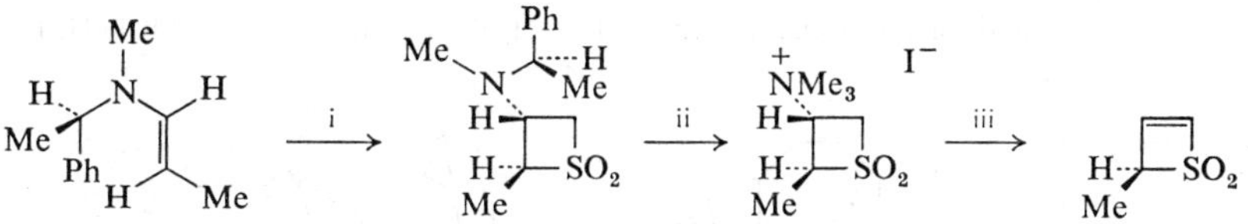

(41)　(40)　(42)

Reagents: i, $MeSO_2Cl/Et_3N$; ii, MeI; iii, $Ag_2O/CaSO_4$
One diastereoisomer of (40) is depicted

**Scheme 8**

configuration of (+)-4-methylthiet-1,1-dioxide (42) was established by catalytic reduction of its pure (−)-enantiomer (obtained by resolution of *trans*-2-methyl-3-piperidinothietan-1,1-dioxide *via* its *d*-camphor-10-sulphonate and subsequent Hofmann elimination) to $(R)$-(+)-2-methylthietan-1,1-dioxide, the absolute configuration of which was allocated by synthesis from $(R)$-(−)-1,3-dibromobutane.[56] Addition of phenylsulphene to 1-diethylamino-2-phenylethylene (of undefined configuration) gave 3-diethylamino-2,4-diphenylthietan-1,1-dioxide (43), the depicted stereochemistry of which was revealed by n.m.r. spectroscopy.[57] Oxidative deamination of the compound (43) gave 2,4-diphenylthiet-1,1-dioxide, which was reduced by sodium borohydride to the known *cis*-2,4-diphenylthietan-1,1-dioxide. Attempted reduction of the compound (43) with lithium aluminium hydride to 3-diethylamino-2,4-diphenylthietan apparently failed, although this reduction occurs smoothly in analogous compounds. 1-Diethylaminopropyne reacted with sulphene to give a tautomeric mixture of the thiet-1,1-dioxide derivatives (44 and 45; R = H),

[55] L. A. Paquette, J. P. Freeman, and R. W. Houser, *J. Org. Chem.*, 1969, **34**, 2901.
[56] L. A. Paquette and J. P. Freeman, *J. Amer. Chem. Soc.*, 1969, **91**, 7548.
[57] D. R. Eckroth and G. M. Love, *J. Org. Chem.*, 1969, **34**, 1136.

from which only (44, R = H) was obtained pure.[57] Similar results were recorded for phenylsulphene, whilst methylsulphene gave the expected sole isomer. These 3-diethylaminothiet-1,1-dioxide derivatives (44 and 45) were not reduced by sodium borohydride, apparently due to the influence of the diethylamino-group.

The adduct (46) of 1-dimethylamino-1,3-cyclo-octadiene and sulphene was very unstable,[58] decomposing into the thiet (47) and three other larger-ring sulphones (see p. 131). The formation of the adduct (46) provided an exception to the general rule that (4+2) cycloaddition competes favourably with (2+2) cycloaddition in the reaction of dienamines with sulphene,

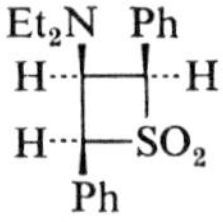

(43)
(one diastereoisomer shown)

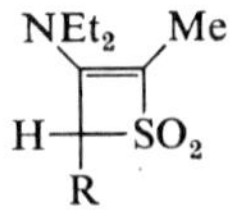

(44)
R = H, Ph, or Me

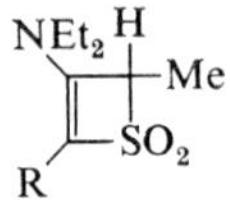

(45)
R = H, Ph

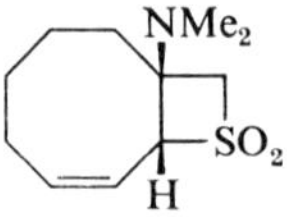

(46)

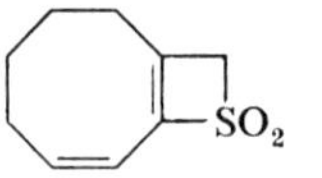

(47)

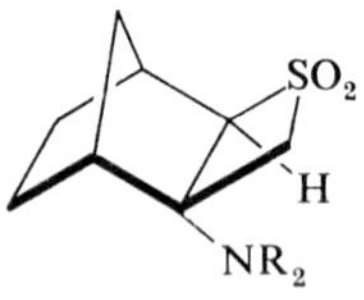

(48)

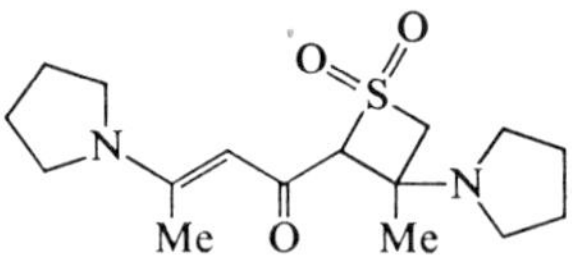

(49)

probably because of the nonplanarity of the $\pi$ system in 1-dimethylaminocyclo-octene associated with its tub conformation. Sulphene underwent cycloaddition to 2-alkylaminobicyclo[2,2,1]hept-2-ene and related systems to give only one type of adduct.[59] These were assigned the *exo*-configuration, for example (48), although confirmation from n.m.r. data was not possible. The apparent stereospecificity was treated with some reservation because of the low yields of products. 3-Dialkylaminothietan-1,1-dioxides derived by addition of sulphene to piperizinyl enamines of other cyclic ketones have been described.[60] Addition of sulphene to 2,6-bis(*N*-pyrrolidino)hepta-2,5-dien-4-one gave the adduct (49) which immediately rearranged to give an acyclic sulphone.[61]

[58] L. A. Paquette and R. W. Begland, *J. Org. Chem.*, 1969, **34**, 2896.
[59] J. F. Stephen and F. Marcus, *J. Org. Chem.*, 1969, **34**, 2535.
[60] H. Mazarguil and A. Lattes, *Bull. Soc. chim. France*, 1969, 3713.
[61] J. F. Stephen and F. Marcus, *Chem. and Ind.*, 1969, 416.

Cycloaddition of thiobenzophenone with olefins bearing electronegative substituents to give substituted thietans was catalysed by 366 nm light.[62] With other olefins 1,4-dithians were formed (see p. 171). The addition was usually regiospecific and always stereospecific, as illustrated by the conversion of acrylonitrile, *trans*-dichloroethylene, and *cis*-dichloroethylene respectively into the thietan derivatives (50), (51), and (52), the configurations of which were determined from dipole moments. No reaction occurred on irradiation with 589 nm light, and since two bands in the u.v. spectrum of thiobenzophenone at 609 nm ($\varepsilon$ 180) and 315 nm ($\varepsilon$ 15,500) were associated with $n \to \pi^*$ and $\pi \to \pi^*$ transitions respectively, it was proposed that the addition involved nucleophilic attack by the $\pi,\pi^*$ singlet

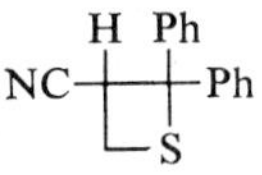

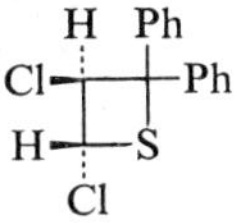

state of thiobenzophenone on the olefins. Photocycloaddition of thiobenzophenone and electron-rich olefins to give thietans occurred only if the olefin was sterically hindered. Irradiation of thiobenzophenone with either 366 or 589 nm light catalysed its addition to 1,3-cyclo-octadiene, and to either *cis*- or *trans*-propenyl benzene to give mixtures of *cis*- and *trans*-2,2,3-triphenyl-4-methylthietan. The mechanism proposed for these non-stereospecific additions involved initial attack of the $n,\pi^*$ triplet state of thiobenzophenone on the olefin to give a diradical which collapsed to give a thietan derivative if the olefin-derived portion was sterically hindered, but which underwent further attack by thiobenzophenone to give, ultimately, substituted 1,4-dithians if it was not (Scheme 9). An example of photocatalysed addition of a thiourea derivative to 3-methylbut-2-ene to give a substituted thietan has been reported,[63] and full details have now appeared on the reaction of benzylsulphonyl fluoride with keten diethylacetal in the presence of phenyl-lithium to give (in low yield) 2,2-diethoxy-3-phenyl-thietan-1,1-dioxide and 2-phenyl-3-ethoxythiet-1,1-dioxide.[64]

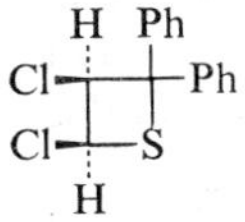

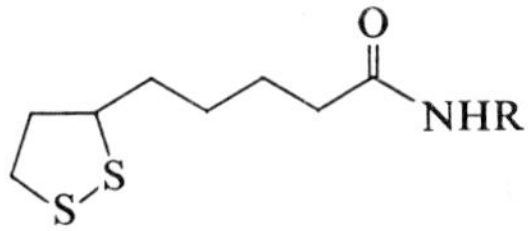

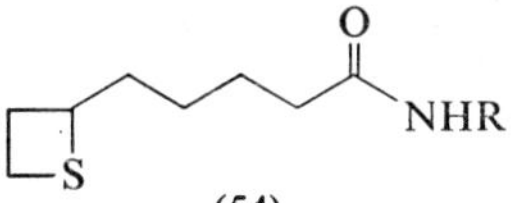

The reaction of 2-chloromethylthiiran with nucleophiles to give thietan derivatives has been further exemplified (refs. 23, 28, 29), and a novel ring

---

[62] A. Ohno, Y. Ohnishi, and G. Tsuchihashi, *J. Amer. Chem. Soc.*, 1969, **91**, 5038.
[63] T. Yonewawa, M. Matsumoto, Y. Matsumura, and H. Kato, *Bull. Chem. Soc. Japan*, 1969, **42**, 2323.
[64] Y. Shirota, T. Nagai, and N. Tokura, *Tetrahedron*, 1969, **25**, 3193.

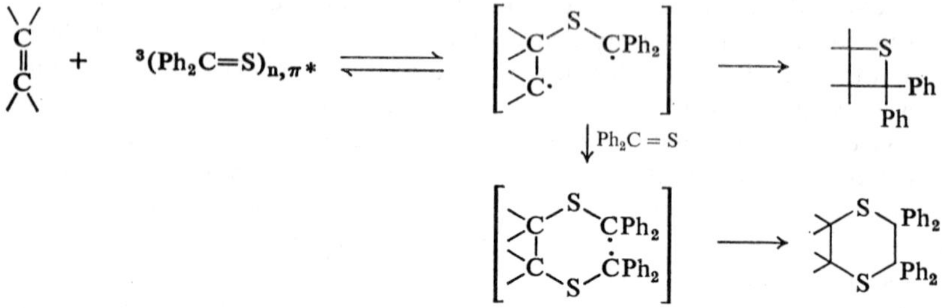

**Scheme 9**

contraction of a 1,2-dithiolan derivative (53) on treatment with tris(diethylamino)phosphine to give a thietan derivative (54) has been described.[65] Four dispirothietans have been prepared as outlined in Scheme 10,[66] and

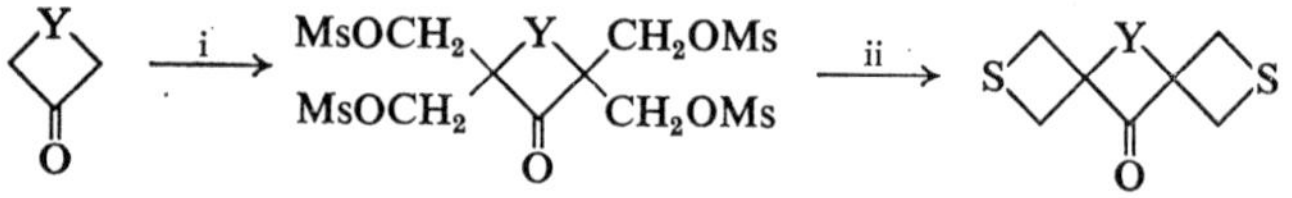

Y = $(CH_2)_n$, $n = 1, 2, 3$; and Y = $CH_2-O-CH_2$
Ms = $SO_2Me$
Reagents: i, $HCHO-H^+$, then MsCl–pyridine; ii, $Na_2S$

**Scheme 10**

pyrolysis of a co-polymer of tetrafluoroethylene and thiocarbonyl fluoride gave perfluorothietan as a major product.[67] Photolysis of 5-methyl-2,3-dihydro-2H,6H-thiapyran-2-one gave 2-isopropenylthietan-4-one.[68] This result, a brief report of which had appeared previously, suggested that thietanone intermediates were involved in the photolytic rearrangement of isothiochroman-4-one to thiochroman-3-one. Acetylenic thiols with the thiol function β to the acetylenic bond have been cyclised to 2-alkylidene derivatives of thietan.[69]

**Formation of 1,3-Dithietan and 1,3-Diselenetan Derivatives.**—Methylene di-iodide was converted by reaction with potassium trithiocarbonate into 1,3-dithietan-2-thione, which on treatment with nitric acid gave 1,3-dithietan-2-one.[70] The production of the latter compound by reaction of methylene halides and potassium ethyl xanthate has been confirmed.[71] One of the manifold synthetic uses of β-ketosulphoxides involved the

[65] J. G. Gleason and D. N. Harpp, *Abstracts of the Papers of the 158th A.C.S. National Meeting*, New York, September 1969.
[66] G. Seitz and W. D. Mikulla, *Tetrahedron Letters*, 1970, 615.
[67] R. James and D. G. Rowsell, *Chem. Comm.*, 1969, 1274.
[68] W. C. Lumma and G. A. Berchtold, *J. Org. Chem.*, 1969, **34**, 1566.
[69] J. M. Surzur, C. Dupuy, M. P. Crozet, and N. Aimar, *Compt. rend.*, 1969, **269**, C, 849.
[70] J. Wortmann and G. Gattow, *Z. Naturforsch.*, 1969, **24b**, 1194.
[71] T. L. Pickering, *Abstracts of the Papers of the 157th A.C.S. Meeting*, Minneapolis, April 1969.

synthesis of 2-acyl-1,3-dithietans according to Scheme 11.[72] Pyrolysis of olefins such as (55) [derived from thiirans such as (4) by treatment with triphenylphosphine] gave 1,3-dithietans typified by (56a) probably by dimerization of an intermediate thioketen.[11] Thioketen intermediates also

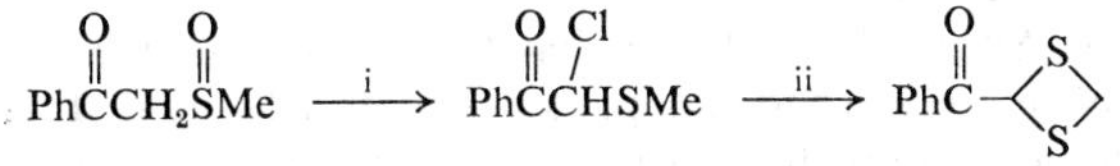

$$\underset{PhCCH_2SMe}{\overset{O\ \ \ \ O}{\underset{\|\ \ \ \ \|}{}}} \xrightarrow{i} \underset{PhCCHSMe}{\overset{O\ \ \ Cl}{\underset{\|\ \ \ /}{}}} \xrightarrow{ii} PhC\overset{O}{\underset{\|}{}}\!\!\!\begin{array}{c}S\\ \diagup\!\!\!\diagdown\\ S\end{array}$$

Reagents: i, SOCl$_2$; ii, HS(CH$_2$)$_2$SH

**Scheme 11**

probably were involved in the production of the 1,3-dithietan (56b) by treatment of di(sodiomercapto)methylene malonamide with 2,4-dinitrochlorobenzene,[73] although another less satisfactory mechanism was proposed. The diselenetan derivative (57) was formed from hydrogen

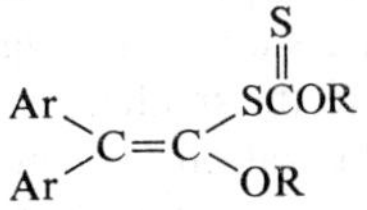

$$\underset{Ar}{\overset{Ar}{\diagdown}}C=C\underset{OR}{\overset{\overset{\overset{S}{\|}}{SCOR}}{\diagup}}$$

(55)

$$\begin{array}{c} R^1\ \ \ S\ \ \ R^2\\ \diagdown\!\!/\!\!\!\diagdown\!\!/\\ \diagup\!\!\!\diagdown\!\!\!\diagup\!\!\!\diagdown\\ R^2\ \ \ S\ \ \ R^1 \end{array}$$

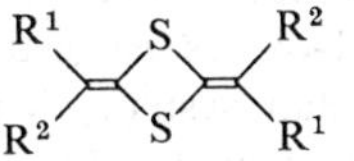

(56a) R$^1$ = R$^2$ = Ar
(56b) R$^1$ = CN; R$^2$ = CONH$_2$

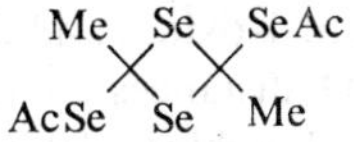

$$\begin{array}{c} Me\ \ \ Se\ \ \ SeAc\\ \diagdown\!\!/\!\!\!\diagdown\!\!/\\ \diagup\!\!\!\diagdown\!\!\!\diagup\!\!\!\diagdown\\ AcSe\ \ \ Se\ \ \ Me \end{array}$$

(57)

selenide in acetyl chloride in the presence of aluminium chloride as catalyst.[74] Apparently only one diastereoisomer was formed, but its configuration was not established.

**Formation of 1,2-Oxathietan and 1,3-Thietidine Derivatives.**—Sulphenes underwent cycloaddition to ketones to give 1,2-oxathietan-2,2-dioxide derivatives if the carbonyl group was activated and sterically unhindered.[75] This condition was satisfied in chloral and perhalogenated acetone derivatives containing CF$_2$Cl and CCl$_2$F groups, but not by, for example, acetone, hexachloroacetone, and trifluoroacetophenone. Under special conditions butene reacted with sulphuric acid to give 3,4-dimethyl-1,2-oxathietan-2,2-dioxide.[76] 4,4-Diaryl-1,3-thietidine-2-thiones were obtained as unstable intermediates in the reaction of carbon disulphide with 4,4'-diarylmethanimines, decomposing to a diaryl thione and diarylmethaniminium thiocyanate.[77] Derivatives of 1,3-thietidine were also postulated as intermedi-

---

[72] G. A. Russell and L. A. Ochryimowycz, *J. Org. Chem.*, 1969, **34**, 3618.
[73] M. Yokoyama, *J. Org. Chem.*, 1970, **35**, 281.
[74] K. Olsson and S.-O. Almqvist, *Acta Chem. Scand.*, 1969, **23**, 3271.
[75] W. E. Truce and L. K. Liu, *Chem. and Ind.*, 1969, 457; J. R. Norell, *Chem. Comm.*, 1969, 1291.
[76] M. Ionescu and M. Vagaonescu, *Rev. Roumaine Chim.*, 1969, **14**, 517.
[77] H. B. Williams, K. N. Yarbrough, K. L. Crochet, and D. V. Welsh, *Tetrahedron*, 1970, **26**, 817.

ates in the pyrolytic conversion of the adducts of $NN'$-dimethylbenzamidine with phenyl or methyl isocyanate into methyl thiobenzamide.[78]

**Configurational and Conformational Aspects of Thietan and Selenetan Derivatives.**—The i.r. and Raman spectra of selenetan [79] indicated a $C_s$ equilibrium conformation of the ring with a dihedral angle of 32·5°. Johnson and Siegl [80] have shown that the oxygen atom in 3-substituted thietan-1-oxides prefer to adopt an equatorial orientation (58), in contrast to the case for thiane-1-oxides, where oxygen prefers to be axial. Both thermal (175 °C) and chemical equilibration (HCl–dioxan at 25 °C or nitrogen tetroxide at 0 °C) of cis- and trans-3-t-butylthietan-1-oxide (58 and 59, R = Bu$^t$) and cis- and trans-3-p-chlorophenylthietan-1-oxide (58 and 59, R = $p$-ClC$_6$H$_4$) gave mixtures containing between 82 and 87% of the cis-isomers (58), according to g.l.c. and n.m.r. data. The near constancy of the equilibrium ratio at widely differing temperatures was surprising. The oxygen atom in the cis-isomers (58) was pseudoequatorially orientated, in view of the propensity of the bulky 3-substituents to be pseudoequatorial, and the difference in orientational preference of the sulphinyl oxygen in thietan-1-oxides and thiane-1-oxides was attributed to the smaller syn-axial H—O distances in the thietan system. The configurations of cis- and trans-3-p-chlorophenylthietan-1-oxide (58 and 59, R = $p$-ClC$_6$H$_4$) were allocated from dipole moment data, and these configurations were related to those of the cis- and trans-3-t-butylthietan-1-oxides by comparison of their n.m.r. spectra, the 3-proton resonating 65–68 Hz to lower field in the trans-isomers (59) than cis-isomers (58), due to the 'syn-axial' effect of the sulphinyl bond. In this respect the n.m.r. characteristics of 4-substituted thietan-1-oxides were analogous to those of 5- and 6-membered ring sulphoxides, and a further examination showed that another analogy existed,[81] in that the signals for the α-methylene protons were much broader for cis-isomers (58) than for trans-isomers (59). This phenomenon was comparable in general terms with the larger chemical shift difference between axial and equatorial protons adjacent to the sulphinyl group in thiane-1-oxides and oxathiane-1-oxides when the sulphinyl oxygen was equatorial. In this work [81] five pairs of isomeric 3-substituted thietan-1-oxides were separated chromatographically, and their configurational assignments were anchored by the known configuration of trans-3-chlorothietan-1-oxide and cis- and trans-2-p-chlorophenyl thietan-1-oxide as determined from dipole moment data. These results led to a reassignment of the configuration at sulphur in the thietan-1-oxide (60) and its diastereoisomer at sulphur.

A thorough examination of the n.m.r. spectra of two 3,3-disubstituted-thietans, their 1-oxides, and their 1,1-dioxides provided interesting informa-

[78] G. Schwenker and R. Kolb, *Tetrahedron*, 1969, **25**, 5549.
[79] A. B. Harvey, J. R. Durig, and A. C. Morrissey, *J. Chem. Phys.*, 1969, **50**, 4949.
[80] C. R. Johnson and W. O. Siegl, *J. Amer. Chem. Soc.*, 1969, **91**, 2796.
[81] C. R. Johnson and W. O. Siegl, *Tetrahedron Letters*, 1969, 1879.

tion about their configuration and conformation.[82] The chemical equivalence of the ring protons and of the methyl groups in 3,3-dimethylthietan (61a) and in its dioxide (62a) indicated either a planar ring conformation, or a rapid equilibrium of folded conformers. 3-Hydroxy-3-methylthietan-1,1-dioxide (62b) also displayed a sharp methyl resonance and chemically

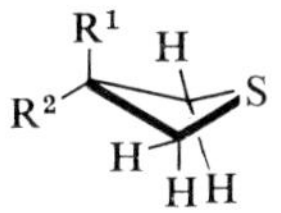

|       | $R^1$ | $R^2$ |
|-------|-------|-------|
| (61a) | Me    | Me    |
| (61b) | Me    | OH    |

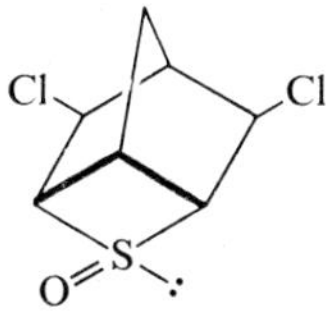

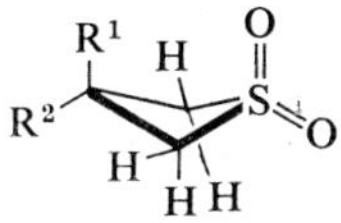

|       | $R^1$ | $R^2$ |
|-------|-------|-------|
| (62a) | Me    | Me    |
| (62b) | OH    | Me    |

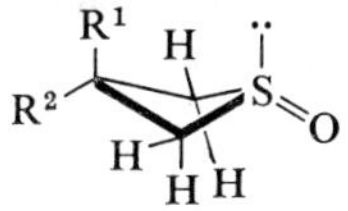

|       | $R^1$ | $R^2$ |
|-------|-------|-------|
| (63a) | Me    | Me    |
| (63b) | OH    | Me    |
| (63c) | Me    | OH    |

equivalent ring protons, providing no definitive evidence for its preferred conformation, although the conformation with a *syn*-axial methyl group and sulphonyl oxygen should be disfavoured for steric reasons. However, 3,3-dimethylthietan-1-oxide (63a), which displayed chemical nonequivalence of the methyl groups and of the ring protons, adopted a puckered conformation with the sulphinyl oxygen equatorial. The puckering was revealed by the long-range coupling ($\pm 0.8$ Hz) between an axial methyl group and the axial ring protons, and by the strong long-range coupling ($+5.7$ Hz) between the two equatorial ring protons; in both these cases the coupled protons could adopt the favoured planar W arrangement. The equatorial orientation of the sulphinyl group was indicated by the absence of strong deshielding of the methyl groups relative to those in 3,3-dimethylthietan, and by the positions of the equatorial and axial proton signals in the sulphoxide (63a) relative to those in the thietan (61a). It was similarly shown that 3-hydroxy-3-methylthietan (61b) adopted a slightly puckered conformation with the methyl group axial, whilst the ring in the isomeric 3-hydroxy-3-methylthietan-1-oxides (63b) and (63c) were strongly puckered, with the sulphinyl oxygen equatorial in both.

As with other cyclic and acyclic sulphonium salts, the base-catalysed hydrolysis of ethoxythietanium salts proceeded with inversion of configuration.[83] *cis*- and *trans*-1-Ethoxy-3-methylthietanium hexachloroantimonate

---

[82] W. Wucherpfennig, *Tetrahedron Letters*, 1970, 765.
[83] R. Tang and K. Mislow, *J. Amer. Chem. Soc.*, 1969, **91**, 5644.

(64) and (65) were prepared from cis- and trans-3-methylthietan-1-oxide (66) and (67) respectively (Scheme 12), the configurations of which were

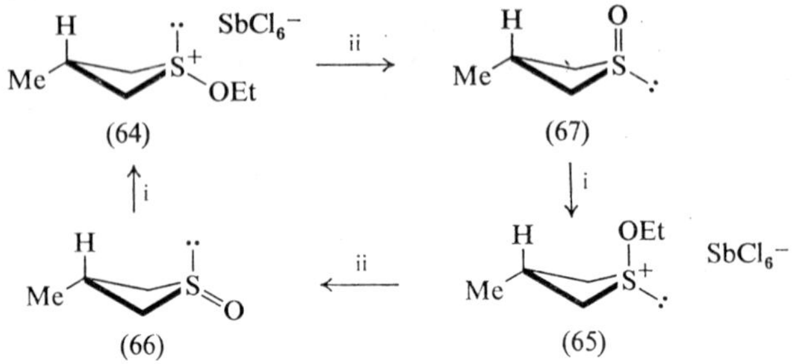

Reagents: i, $Et_3O^+SbCl_6^-$; ii, NaOH in aqueous dioxan
The conformations depicted may not be the most favoured

**Scheme 12**

determined by n.m.r. (cf. ref. 81). Hydrolysis of the cis-salt (64) gave the trans-sulphoxide (67) whilst the trans-salt (65) gave the cis-sulphoxide (66), each reaction proceeding with at least 95% inversion at sulphur. Direct nucleophilic displacement at sulphur was postulated, without the intervention of an intermediate of finite lifetime. The transient formation of a five-co-ordinate intermediate of approximate trigonal bipyramidal geometry was considered less likely, since it was not possible to explain the observed inversion of configuration according to the accepted rules governing the relative stability of the various geometrical arrangements of the trigonal bipyramidal intermediates and their interconversion by pseudorotation. In contrast, five-co-ordinate intermediates were invoked to explain the butyl-lithium-induced stereospecific fragmentation of cis- and trans-2,4-dimethylthietanium salts to trans- and cis-1,2-dimethylcyclopropane respectively.[84] Three mechanisms of ring cleavage were considered, all commencing with the formation of a five-co-ordinate intermediate (Scheme 13, illustrating the 2,4-trans-isomer). Mechanism C was eliminated because the butyl-lithium-induced dehalogenation of meso- and dl-2,4-dibromopentanes to 1,2-dimethylcyclopropanes (which proceeded by a mechanism analogous to C) proceeded in much poorer yield and was much less stereoselective than the fragmentation of the thietanium salt. However, it was not possible to distinguish between mechanisms A and B. It is pertinent that the direct fragmentation B of the five-co-ordinate sulphur intermediate to cis-1,2-dimethylcyclopropane is a thermally allowed conrotatory process, whilst extended Hückel calculations indicated that the trimethylene intermediate in pathway A should close in a conrotatory fashion to give cis-1,2-dimethylcyclopropane. The stereochemistry of cis- and trans-2,4-dimethylthietan was established by n.m.r. spectroscopy.

[84] B. M. Trost, W. L. Schinski, and I. B. Mantz, J. Amer. Chem. Soc., 1969, **91**, 4320.

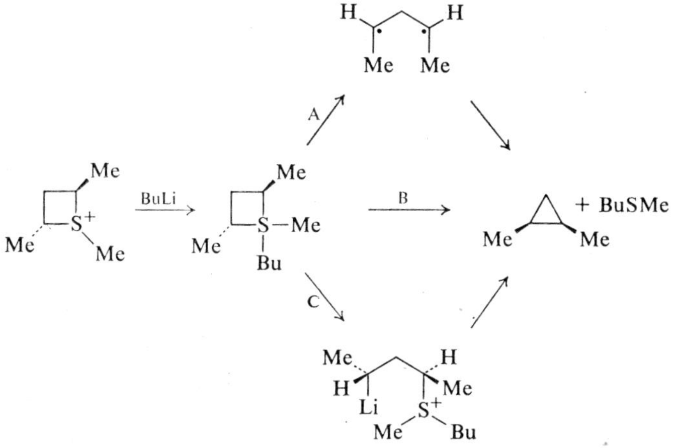

**Scheme 13**

**Oxidation and Cleavage of Thietan Derivatives.**—Thietan was readily oxidised to thietan-1-oxide by 1-chlorobenzotriazole,[85] and treatment of the thietan derivative (6) with aqueous sodium carbonate gave the sulphoxide (68), in which sulphinyl oxygen was *exo*-orientated according to n.m.r. data.[13] It appears that chlorine liberated in this reaction was the oxidising agent. The rearrangement of the thietan derivative (6) into the thiiran (5) by reaction with potassium cyanide probably involved the nonclassical ion (7).[13]

The adduct (46) of 1-dimethylamino-1,3-cyclo-octadiene and sulphene decomposed rapidly into the thiet (47) and the cyclic sulphones (69), (70), and (71).[58] The thiet (47) was formed by β-elimination of the elements of dimethylamine from the thietan (46), in which the bridgehead hydrogen is particularly strongly activated. In the mechanistic rationalisation of the formation of the sulphones (69) and (71) (Scheme 14) the driving force for the ring cleavage was attributed to allylic stabilisation of an α-sulphonyl carbanion in the zwitterionic intermediate and to relief of steric strain. Hydrolysis of (69) gave (70), whilst the final isomerisation leading to (71) finds analogy in other systems. The formation of an acyclic sulphone from the unstable thietan derivative (49) probably involved a similar mechanism for the initial cleavage.[61]

Chlorination of 3,3-dimethylthietan gave the expected 2,2-dimethyl-3-chloropropan-1-sulphonyl chloride [86] but treatment of 3-chloro-3-methylthietan-2-one with ammonia gave 2-methyl-2-carbonamidothiiran.[17] The thietan ring in 'thiamine anhydride' (72) was cleaved by hydrogen sulphide or carbon oxysulphide to give the 1,2-dithiolan derivative (73),

---

[85] W. D. Kingsbury and C. R. Johnson, *Chem. Comm.*, 1969, 365.
[86] E. J. Goethals, J. Huylebroeck, and W. Smolders, *Bull. Soc. chim. belges*, 1969, **78**, 191.

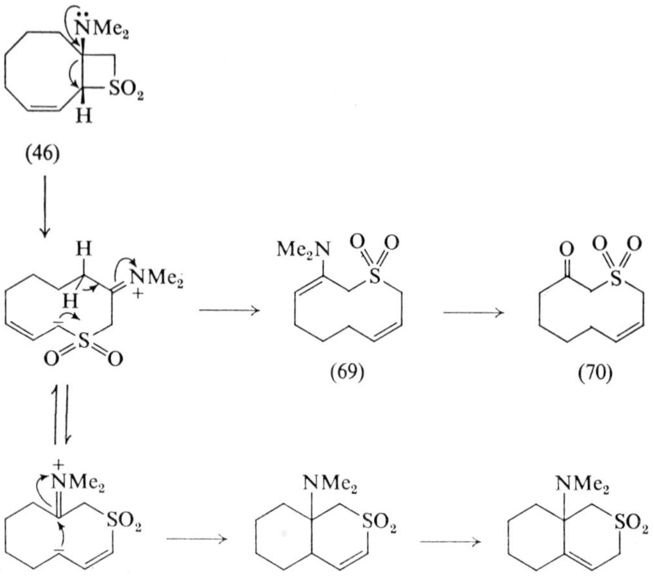


(71)

**Scheme 14**

but with various thiophenols other products were obtained, including acyclic disulphides, pyrimidodiazepin derivatives, and compounds related to thiamine containing a 2[*H*]-2-thiono-1,3-thiazole ring. The mechanisms of these reactions were rationalised in terms of initial cleavage of the thietan ring to give vinyl thioenol derivatives (74).[87] Perfluoro-1,2-oxathietan-

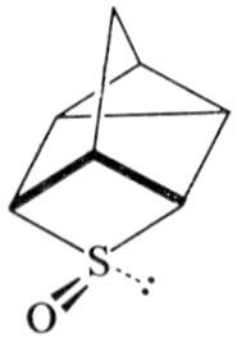


(68)

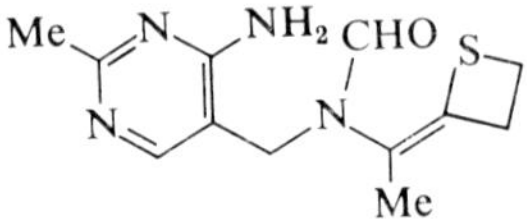


(72)

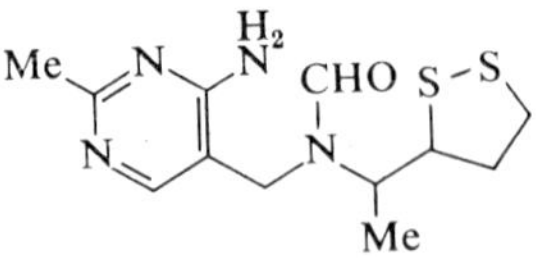


(73)

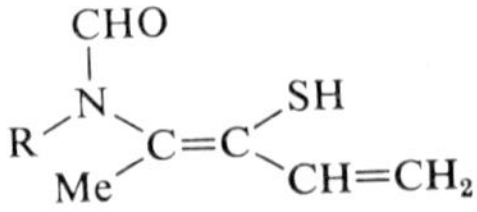


(74)

[87] A. Takamizawa, K. Hirai, and T. Ishiba, *Tetrahedron Letters*, 1970, 437, 441.

1,1-dioxide decomposed to give fluorosulphonic acid, perfluoroethylene and carbon monoxide on treatment with concentrated sulphuric acid.[88] Trapping experiments indicated the initial formation of an unstable trifluorovinylsulphate, which subsequently decomposed to give difluoroketen.

**Pyrolysis of Thiet-1,1-dioxides.**—Pyrolysis of the naphthothiet-1,1-dioxide (75) did not give naphthocyclopropene by extrusion of sulphur dioxide, but the thiophen derivatives 14$H$-benzo[$b$]benzo[3,4]fluoreno[2,1-$d$]thiophen and its 14-keto-derivative were formed.[89] However, pyrolysis in the presence of 9,10-dihydroanthracene gave the cyclic sulphinate ester (76), which on heating in the absence of 9,10-dihydroanthracene gave the above thiophen derivatives, suggesting that it was an initial transformation product of the thiet-1,1-dioxide (75). The role of 9,10-dihydroanthracene was unknown. A homolytic mechanism was proposed for the transformation of the thiet-1,1-dioxide (75) into the cyclic sulphinate (76) (Scheme 15),

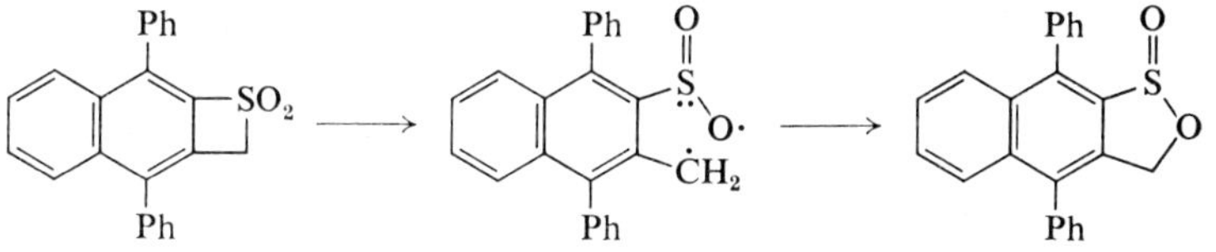

**Scheme 15**

but it is pertinent that evidence has been presented which suggested that certain thiet-1,1-dioxides underwent electrocyclic ring opening to vinyl sulphenes, which subsequently ring-closed to cyclic sulphinates.[90]

[88] L. I. Ragulin, M. A. Belaventsev, G. A. Sokol'ski, and I. L. Knunyants, *Izvest. Akad. Nauk S.S.S.R., Ser. khim.*, 1969, 2220.
[89] D. C. Dittmer, R. S. Henian, and N. Takashina, *J. Org. Chem.*, 1969, **34**, 1310.
[90] J. F. King, K. Piers, D. J. H. Smith, C. L. McIntosh, and P. de Mayo, *Chem. Comm.*, 1969, 31; C. L. McIntosh and P. de Mayo, *ibid.*, p. 32.

# 4
# Saturated Cyclic Compounds of Sulphur, Selenium, and Tellurium

The geometry and conformational properties of five- and six-membered heterocyclic compounds containing oxygen and sulphur have been reviewed.[1] Two reviews dealt in particular with conformational aspects of dithians,[2] and another contained material relating to the n.m.r. spectral characteristics, configuration, and conformation of dithians, oxathians, and oxides of the latter.[3] Information about the photochemistry of cyclic sulphur compounds,[4] and about aspects of cyclic sulphoxides[5] was contained in two other reviews.

## 1 Five-membered Rings

**Thiolan Derivatives.**—A variety of 2-thiabicyclo[2,2,1]heptane derivatives have been prepared from the adduct (1) of thiophosgene and cyclopentadiene.[6] Thiophosgene-$S$-oxide also underwent cycloaddition to cyclopentadiene to give the $S$-oxide of compound (1), but its stereochemistry was not investigated.[7] Both the adduct (1) and its $S$-oxide were very unstable, but oxidation gave a stable sulphone,[6,7] in which the double bond was relatively inert to electrophilic attack, although the corresponding epoxide was formed with $m$-chloroperoxybenzoic acid under forcing conditions.[6] The sulphone was reduced under various conditions to 2-thiabicyclo[2,2,1]hept-5-ene, 2-thiabicyclo[2,2,1]heptane and its dioxide,

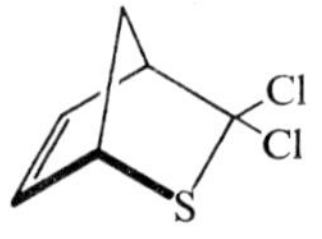

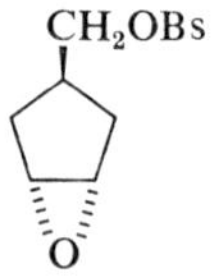

[1] C. Romers, C. Altona, H. R. Buys, and E. Havinga, *Topics Stereochem.*, 1969, **4**, 39.
[2] E. L. Eliel, *Bull. Soc. chim. France*, 1970, 517; *Accounts Chem. Res.*, 1970, **3**, 1.
[3] T. D. Inch, *Ann. Rev. N.M.R. Spectroscopy*, 1969, **2**, 35.
[4] E. Block, *Quart. Rep. Sulphur Chem.*, 1969, **4**, No. 4.
[5] C. R. Johnson and J. C. Sharp, *Quart. Rep. Sulphur Chem.*, 1969, **4**, No. 1.
[6] C. R. Johnson, J. E. Keiser, and J. C. Sharp, *J. Org. Chem.*, 1969, **34**, 860.
[7] B. Zwanenburg, L. Thijs, and J. Strating, *Tetrahedron Letters*, 1969, 4461.

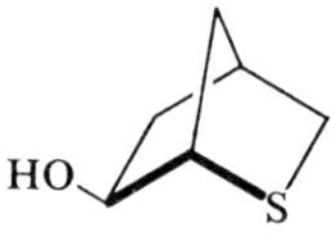

and *endo*-3-chloro-2-thiabicyclo[2,2,1]hept-5-ene and its saturated analogue.[6] A synthesis of *exo*-6-hydroxy-2-thiabicyclo[2,2,1]heptane (2) was also described,[6] involving in its last step the reaction of the epoxy *p*-bromobenzenesulphonate (3) with sodium sulphide.[6] Ginsburg and his co-workers have described the preparation of the 'propellanes' (4a—d) by conventional methods from the appropriate tetrahydroxymethyl compounds (5) (for 4a—c) or from the compound (6) (for 4d).[8] The absolute configurations of biotin and of (+)-lipoic acid have been verified by chemical means.[9] Degradation of biotin gave (*S*)-(−)-2-(5′-hydroxypentyl)-thiolan (7), the absolute configuration of which was established by syn-

(4)    (5)    (6)    (7)

a X = Y = Z = S        X = O or S,
b X = O, Y = Z = S     R = CH$_2$OH
c X = S, Y = Z = O
d X = CH$_2$, Y = S, Z = O

thesis from (*S*)-(−)-thiolan-2-carboxylic acid. However, (+)-biotin and (*S*)-(−)-2-(5′-hydroxypentyl)thiolan, although of equivalent configuration, gave Cotton effects of opposite sign centred near 240 nm in their o.r.d. curves, a phenomenon which was attributed to the adoption of different conformations by the thiolan ring in the two compounds. The (*S*)-configuration of (+)-lipoic acid was related to that of (*S*)-(−)-5-(2′-thiolano)pentanoic acid (which was conveniently prepared by hydrogenation of its thiophen analogue followed by resolution of its cinchonidine salt) by the formation of a quasi-racemate.

2-Thia-A-nor-5α-androstan-17β-ol (8), which was about as androgenic as testosterone, has been prepared from 17β-hydroxy-2,3-seco-5α-androstane-2,3-dioic acid by conventional methods.[10] Sulphur-containing sugars continue to attract attention, and the Pummerer rearrangement was used effectively to convert (±)-*trans*-3,4-diacetoxythiolan-1-oxide (9) (on treatment with acetic anhydride) into a mixture of the anomeric triacetates (10), from which one isomer crystallised.[11] *cis*-3,4-Diacetoxy-thiolan-1-oxide (of unknown configuration at sulphur) was similarly transformed into the 3,4-*cis*-diacetoxy analogue of (10). More conventional procedures were used to synthesise methyl-4-thio-α-D-arabino-furanoside (11) and its anomer,[12] a mixture of methyl-4-thio-α-D-lyxo-

---

[8] T. Altman, E. Cohen, T. Mayman, J. B. Petersen, N. Reshef, and D. Ginsburg, *Tetrahedron*, 1969, **25**, 5115.
[9] G. Claeson and H. G. Jonsson, *Arkiv Kemi*, 1969, **31**, 83.
[10] M. E. Wolff and G. Zanati, *J. Medicin. Chem.*, 1969, **12**, 629.
[11] J. E. McCormick and R. S. McElhinney, *Chem. Comm.*, 1969, 171.
[12] R. L. Whistler, U. G. Nayak, and A. W. Perkins, *J. Org. Chem.*, 1970, **35**, 519.

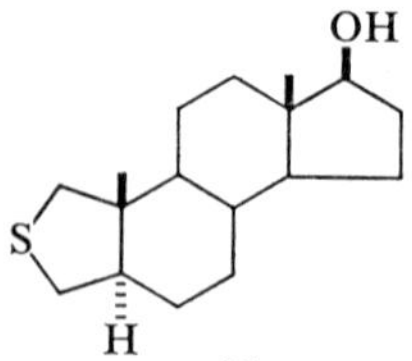

furanoside (12) with its anomer,[13] and 3,6-dideoxy-3,6-epithio-1,2-*O*-isopropylidene-α-D-glucofuranose (13).[14]

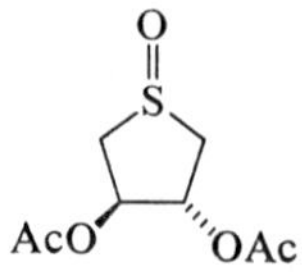

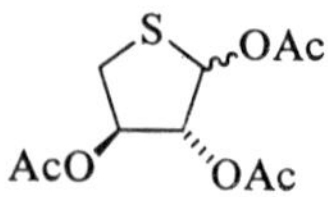

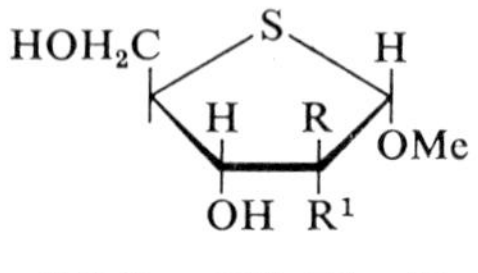

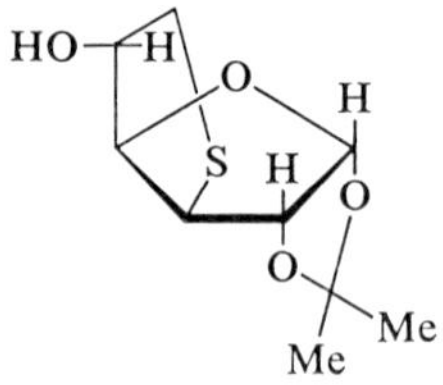

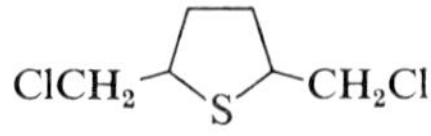

Irradiation of 1-phenyl-5-mercaptopentyne gave 2-benzylidenethiolan in 75% yield,[15] whilst pyrolysis of copolymers of tetrafluoroethylene and tetrafluorothiiran led to tetrafluorothiolan as a major product by a free-radical mechanism.[16] Selenium monochloride added to linear and cyclic olefins in a manner similar to that of sulphur dichloride. With hexa-1,5-diene it gave 2,5-di(chloromethyl)selenolan (14) stereospecifically, and a *cis*-configuration was assumed.[17] It was considered likely that the mechanism involved prior dissociation of selenium monochloride into selenium and selenium dichloride, which added in an anti-Markovnikov manner (as shown previously for sulphur dichloride) to one of the olefinic bonds *via* an episelenonium intermediate.

Electron diffraction data obtained for thiolan were in agreement with the angles and preferred conformation ($C_2$ symmetry) calculated by the Westheimer–Hendrickson method,[18] whilst its *X*-ray emission spectrum, and those of its 1-oxide and 1,1-dioxide, was characteristic of the sulphur oxidation state.[19] The effect of various catalysts upon the catalytic decomposition of thiolan into thiophen, hydrogen sulphide, and olefins has been investigated,[20] and it has been shown that base-catalysed elimination of hydrogen chloride from 2-allyl and 2-vinyl derivatives of 3-chlorothiolan proceeded with prototropic shifts to give 2-alkylthiophen derivatives.[21]

[13] J. P. H. Verheyden and J. G. Moffatt, *J. Org. Chem.*, 1969, **34**, 2643.
[14] J. M. Heap and L. N. Owen, *J. Chem. Soc.* (C), 1969, 707.
[15] J. M. Surzur, C. Dupuy, M. P. Crozet, and N. Aimar, *Compt. rend.*, 1969, 269C, 849.
[16] R. James and D. A. Rowsell, *Chem. Comm.*, 1969, 1274.
[17] F. Lautenschlaeger, *J. Org. Chem.*, 1969, **34**, 4002.
[18] Z. Nahlovska, B. Nahlovsky, and H. M. Seip, *Acta Chem. Scand.*, 1969, **23**, 3534.
[19] Y. Takahishi and K. Yabe, *Bull. Chem. Soc. Japan*, 1969, **42**, 3064.
[20] N. V. Shelemina, T. A. Danilova, and B. V. Romanovski, *Neftekhimiya*, 1969, **9**, 295.
[21] J. P. Gouesnard and G. J. Martin, *Bull. Soc. chim. France*, 1969, 4452.

The conversion of thiolan into 2-chlorothiolan by $N$-chlorosuccinimide, and its subsequent transformation into 2-alkyl and 2-aryl thiolan derivatives by Grignard reagents has been reported,[22] whilst 3-amino-, 3-isocyanato-, and 3-carbamido-derivatives of thiolan have been described.[23] Carbonyl stabilized ylides derived from thiolan sulphonium salts have been prepared and their reactions with dimethylvinylsulphonium bromide were described.[24]

Johnson and his co-workers have extended their investigations of the stereochemistry of oxidation of cyclic sulphur compounds to thiolan derivatives, and have investigated the stereochemistry of the oxidation of 2-thiabicyclo[2,2,1]heptane (15) by a variety of oxidising agents.[25] With ozone the ratio of *endo* (16) to *exo* (17) sulphoxides was 8:92, whereas with t-butyl hypochlorite it was 65:35, ten other oxidising agents giving ratios intermediate between these extremes. The stereochemistry of the sulphoxides (16) and (17) was assigned on the basis of their chromatographic behaviour (*endo* isomer eluted before *exo* isomer) and according to the expected stereoselectivity of some of the oxidising agents, since their n.m.r. spectra, although dissimilar for the isomers, appeared to be too complex to yield useful stereochemical information. Equilibration of the sulphoxides, either with hydrogen chloride in dioxan, or with dinitrogen tetroxide, or thermally, indicated that the *exo* isomer (17) was thermodynamically the more stable, but the values quoted for the differences in free energy between the isomers should be regarded with caution in view of the meagre experimental information. The greater stability of the *exo* isomer (17) was compared with the greater stability of axial than equatorial thian-1-oxides, the interatomic non-bonded distances between sulphinyl oxygen and proximate atoms being very similar in the *exo* isomer (17) and axial thian-1-oxides. Thiolan was conveniently oxidised to its 1-oxide by 1-chlorobenzotriazole without overoxidation to the sulphone,[26] and a kinetic investigation of the air-oxidation of thiolan, catalysed by

(15) X = Y = electron pair
(16) Y = electron pair, X = O
(17) Y = O, X = electron pair

(18)

[22] D. L. Tuleen and R. H. Bennett, *J. Heterocyclic Chem.*, 1969, **6**, 115.
[23] T. E. Bezmenova and P. I. Parkhomenko, *Deopovidi Akad. Nauk Ukrain. R.S.R. Ser. B*, 1969, **31**, 726.
[24] S. Gerhard and J. Gosselk, *Tetrahedron Letters*, 1969, 3445.
[25] C. R. Johnson, H. Diefenbach, J. E. Keiser, and J. C. Sharp, *Tetrahedron*, 1969, **25**, 5649.
[26] W. D. Kingsbury and C. R. Johnson, *Chem. Comm.*, 1969, 365.

vanadium pentoxide, to give the 1-oxide and 1,1-dioxide has been reported.[27] A carbon–sulphur bond in thiolan-1-oxide was cleaved by hydroxyl radicals, the products being trapped as nitroxyl radicals and detected by e.s.r. spectroscopy.[28]

The bond lengths in the compound (18), measured by $X$-ray crystallographic methods, indicated an appreciable mesomeric interaction involving sulphur and the ethylenic double bond, and, although there was no interannular sulphur–carbonyl interaction through the double bond, it appeared that sulphur to sulphur interaction made an important mesomeric contribution to the ground state of the compound.[29] The carbonyl group was probably necessary for the manifestation of this phenomenon, since the sulphur–oxygen bond distance was appreciably shorter than the sum of the van der Waals' radii for sulphur and oxygen. If compound (18) is taken as a model for the thioindigo chromophore, these results are incompatible with a previous suggestion that an interannular mesomeric interaction of sulphur with a carbonyl group was primarily responsible for the basic thioindigo chromophore. According to dipole moment data, $d$-orbital conjugation was important in thiolan-2-one and selenolan-2-one.[30] In accord with this, the basicity of the carbonyl oxygen progressively decreased in the series $\gamma$-butyrolactone, thiolan-2-one, and selenolan-2-one. Thiolan-2-thione contained none of the enethiol tautomer at equilibrium, according to n.m.r. and titrimetric data,[31] but n.m.r. data indicated that 3-thioacetylthiolan-2-one (19) existed entirely in the enethiol forms (20) and (21) in carbon tetrachloride.[32] The hydrogen-bonded form (20) was isolated virtually pure by crystallization.

A further example of the remarkable usefulness of the Ramburg–Backlund rearrangement for the introduction of double bonds has been provided by the conversion of the α-chlorosulphone (22) into tricyclo-[4,4,2,0]dodeca-3,11-diene (23),[33] whilst an ingenious new synthesis of *cis*-1,2-dialkylcyclohexanecarboxylic acids and angularly substituted poly-

[27] A. V. Mashkina, P. S. Makoveev, and N. I. Polovinkina, *Reakts. spos. org. Soedinenii*, 1969, **6**, 24.
[28] C. Lagercrantz and S. Forshult, *Acta Chem. Scand.*, 1969, **23**, 811.
[29] H. J. A. Hermann, H. L. Ammon, and R. E. Gibson, *Tetrahedron Letters*, 1969, 2559.
[30] I. Wallmark, M. H. Krakow, and H. G. Mautner, *Abs. 158th Nat. Meeting of the A.C.S.*, New York, Sept. 1969.
[31] R. Mayer, S. Scheithauer, S. Bleich, D. Kunz, G. Bahr, and R. Radeglia, *J. prakt. Chem.*, 1969, **311**, 472.
[32] F. Duus, E. B. Pedersen, and S.-O. Lawesson, *Tetrahedron*, 1969, **25**, 5703.
[33] L. A. Paquette and J. C. Philips, *Chem. Comm.*, 1969, 680.

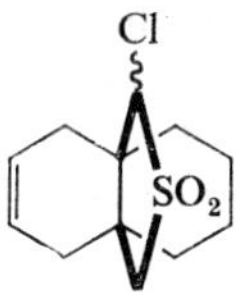

cyclic systems made use of Diels–Alder adducts of 2,5-dihydro-4-carbomethoxy-2-thiophen acetate;[34] for example, with 2-ethoxybutadiene it gave the adduct (24), which after dioxolan formation and desulphuration with Raney nickel gave the diester (25), a precursor in the synthesis of the 9-methyl-2-hydrindanone system. Cleavage of 3-aminothiolan-2-one with primary alkyl halides provided $S$-alkyl-DL-homocysteines of potential biological interest,[35] whilst pyrolysis of 1,2,3,6-tetrahydrophthalic thioanhydride brought about a retro-Diels–Alder reaction to give thiomaleic anhydride, a powerful dienophile.[36] The structure (26) previously suggested for the trimer derived by treatment of thiophen with phosphoric acid has been confirmed by spectral data and by dehydrogenation with chloranil to 2,2′;4′,2″-terthienyl.[37]

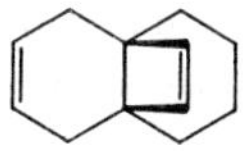

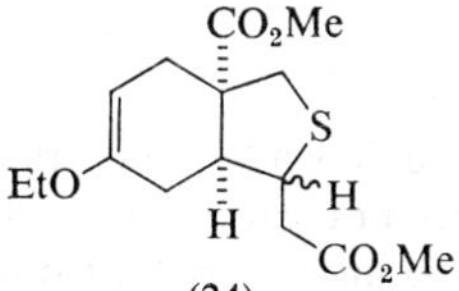

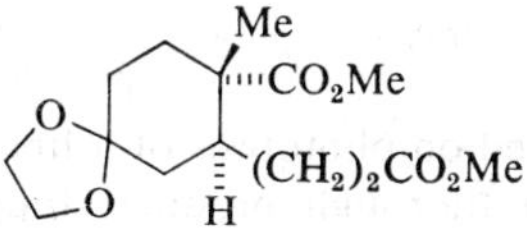

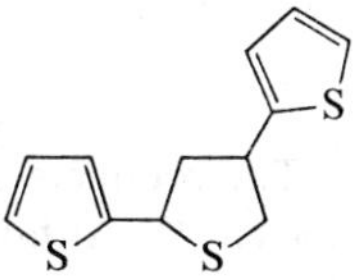

**1,2-Dithiolan Derivatives.**—In a study related to model systems for the catalytic reactions of riboflavin-requiring proteins it was found that flavins oxidised 1,3-propane dithiol to 1,2-dithiolan.[38] The reaction occurred only under basic conditions, and was first order in both mercaptan and flavin. Dithiols reacted about twice as fast as monofunctional mercaptans, and this was interpreted mechanistically in terms of initial attack of a mercaptide ion upon the flavin to form an intermediate in which sulphur was more susceptible to nucleophilic attack than in a thiol. The data would be explained by fast attack on this intermediate by a thiol or thiolate ion to produce disulphide. The 1,2-dithiolan derivative (27) was formed by spontaneous dimerization in solution of the thioacrylic acid derivative (28).[39] Among the interesting reactions described was the

---

[34] G. Stork and P. L. Stotter, *J. Amer. Chem. Soc.*, 1969, **91**, 7780.
[35] H. M. Kolenbrander, *Canad. J. Chem.*, 1969, 3271.
[36] M. Verbeek, H.-D. Scharf, and F. Korte, *Chem. Ber.*, 1969, **102**, 2471.
[37] R. F. Curtis, D. M. Jones, G. Ferguson, D. M. Hawley, J. G. Sime, K. K. Cheung, and G. Germain, *Chem. Comm.*, 1969, 165.
[38] M. J. Gibian and D. V. Winkelman, *Tetrahedron Letters*, 1969, 3901.
[39] L. Henriksen, *Chem. Comm.*, 1969, 1408; 1970, 132.

conversion of the compound (27) into the 1,3-dithiolan derivative (29) on treatment with aniline, for which a mechanistic rationalization was provided. The conversion of 'thiamine anhydride' into a 1,2-dithiolan derivative on treatment with hydrogen sulphide has been described (ref. 87, Chapter 3).

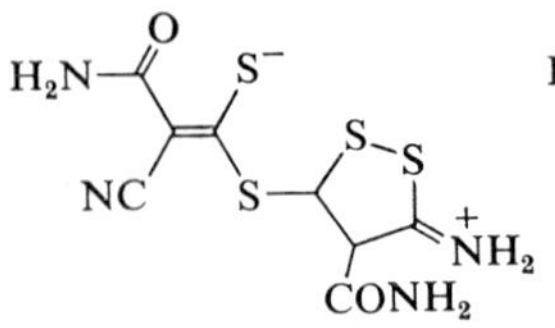

The products obtained on photolysis of α-lipoic acid depended upon the solvent.[40] Although a thorough product study was not carried out, it appeared that they were dihydrolipoic acid, linear oligomers derived from dihydrolipoic acid, and hydrogen sulphide, the relative proportions of which depended upon the ease of hydrogen abstraction from the solvent and upon the presence of water. The presence of oxygen had no apparent effect. In the same way the presence of oxygen had no effect upon product composition in the radiolysis of aqueous solutions of α-lipoic acid, but in this case the major product was tentatively regarded as a cyclic dimer.[41] Only small amounts of thiol were produced, and in contrast to the behaviour of acyclic disulphides, very little of any sulphonic or sulphinic acid derivatives. A proposed mechanism involving the initial formation of a radical cation of α-lipoic acid was supported by a kinetic investigation. Lipoic acid derivatives have been efficiently desulphurised by tris(diethylamino)phosphine to give the corresponding thietan derivatives (ref. 65, Chapter 3).

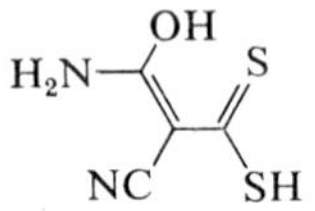

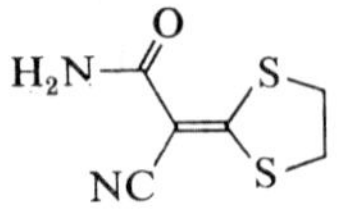

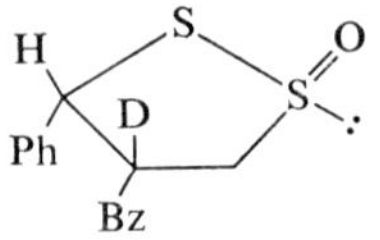

In contrast to the behaviour of acyclic thiolsulphinates which racemize (at sulphur) at 50 °C, 4-deuterio-*cis*-4-benzoyl-3-phenyl-1,2-dithiolan-1-oxide (30) did not racemize *via* pyramidal inversion at sulphur on heating to 116 °C, according to the invariance of its n.m.r. spectrum in the temperature range −50 to 116 °C.[42] The oxidation of *cis*-4-benzoyl-3-phenyl-1,2-dithiolan to its 1-oxide as (30) by *m*-chloroperoxybenzoic acid was apparently both position-specific and stereospecific, according to n.m.r.

[40] P. O. Brown and J. O. Edwards, *J. Org. Chem.*, 1969, **34**, 3131.
[41] T. C. Owen and A. C. Wilbraham, *J. Amer. Chem. Soc.*, 1969, **91**, 3365.
[42] F. Wudl, R. Gruber, and A. Padwa, *Tetrahedron Letters*, 1969, 2133.

evidence, the deshielding of the proton at C-3 which accompanied oxidation to the compound (30) suggesting a *syn*-relationship with the newly introduced sulphinyl oxygen. Oxidation of 4,4-dimethyl-1,2-dithiolan with hydrogen peroxide gave the corresponding 1,1-dioxide,[43] but attempts to oxidise 1,2-dithiolan to its 1,1-dioxide were unsuccessful, although this reaction was successfully performed by others.[43a] However, 1,2-dithiolan-1,1-dioxide was synthesised in low yield by treating 4-chloropropanesulphonyl chloride with sodium sulphide.[43]

The dihedral angles about the sulphur bonds in 1,2-dithiolan and 2,3-dithiaspiro[4,5]decane were altered on formation of their iodine charge-transfer complexes, according to u.v. data obtained at 77 K and at room temperature.[44] Full details have appeared of the isolation and desulphuration of the alkaloid brugine, (+)-tropine 1,2-dithiolan-3-carboxylate, the absolute configuration and synthesis of which were described previously.[45]

**1,3-Dithiolan Derivatives.**—Thermal degradation of polyethylene sulphide gave 2-methyl-1,3-dithiolan as a major product, whilst 2-ethyl-4-methyl-1,3-dithiolan and 2,2,4-trimethyl-1,3-dithiolan were important products from the pyrolysis of polypropylene sulphide.[46] It was suggested that these compounds were formed by a selective co-ordination-assisted zipping mechanism initiated at the metal site incorporated during the formation of the polymers. 2-Acyl-1,3-dithiolans have conveniently been prepared in two steps from β-ketosulphoxides (ref. 72, Chapter 3) (*cf.* Scheme 13, Chapter 3), whilst acetylacetone or malondiamide condensed with 1,2-bis(benzenesulphonylthio)ethane to give respectively 2,2-diacetyl-1,3-dithiolan (31a) and the corresponding diamide (31b).[47] This interesting reaction worked neither for other (unspecified) active methylene compounds, nor, surprisingly, for 1,2-bis(toluene-*p*-sulphonylthio)ethane. Arylaroyldiazomethanes (32; R = aryl) reacted with carbon disulphide to give 1,3-dithiolan derivatives (33; R = aryl), and not 1,3-diethetan derivatives as found previously for methylacetyldiazomethanes (32; R = Me)[48] the n.m.r. spectrum of the adduct from the compound (32; R = *p*-F-$C_6H_4$) showing three fluorine resonances in the ratio 1:1:2, consistent only with the structure (33). Treatment of tetramethylorthothiocarbonate with ethane dithiol gave 1,4,6,9-tetrathiaspiro[4,4]nonane (34).[49] The absolute configurations of the two enantiomeric forms of 1,3-dithiolan-*trans*-4,5-dicarboxylic acid (35), prepared from racemic 2,3-dimercapto-

---

[43] E. J. Goethals, J. Huylebroeck, and W. Smolders, *Bull. Soc. chim. belges*, 1969, **78**, 191.
[43a] L. Field and R. B. Barbee, *J. Org. Chem.*, 1969, **34**, 36.
[44] B. Nelander, *Acta Chem. Scand.*, 1969, **23**, 2127, 2136.
[45] J. W. Loder and G. B. Russell, *Austral. J. Chem.*, 1969, **22**, 1271.
[46] R. T. Wragg, *J. Chem. Soc. (B)*, 1970, 404.
[47] S. Hayashi, M. Furukawa, Y. Fujino, and H. Matsakura, *Chem. and Pharm. Bull. (Japan)*, 1969, **17**, 419.
[48] J. E. Baldwin and J. A. Kapecki, *J. Org. Chem.*, 1969, **34**, 724.
[49] D. L. Coffen, *J. Heterocyclic Chem.*, 1970, **7**, 201.

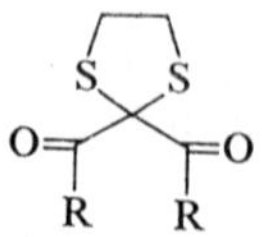

(31a) R = Me
(31b) R = NH$_2$

(32) RCCR with N$_2$ and C=O

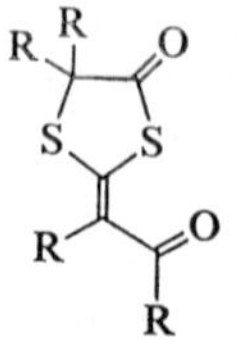

(33)

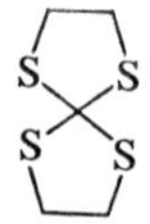

(34)

succinic acid followed by resolution, have been determined by the quasi-racemate method,[50] (−)-(35) having the 2(S),3(S) configuration.

4-Methylene-5-ethyl-5-methyl-1,3-dithiolan-2-thione (36) was obtained from 4-methylene-5-ethyl-5-methyl-1,3-oxathiolan-2-thione on treatment with potassium O-ethyl xanthate, the reaction being general for 4-alkylidene-1,3-oxathiolan-2-thiones.[51] The mechanism was not elucidated, but it was shown that carbon disulphide present in excess during the

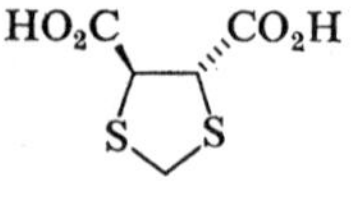

(35) 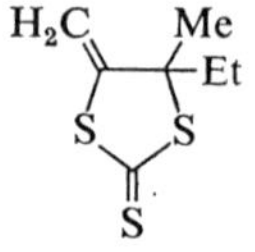 (36) 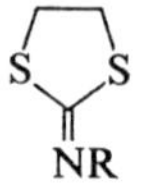 (37) 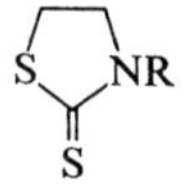 (38) 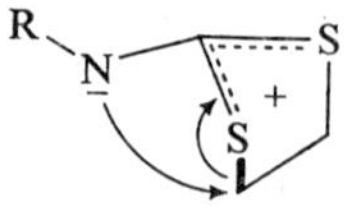 (39)

formation of 4-alkylidene-1,3-oxathiolan-2-thiones from acetylenic carbinolates was not responsible for the concomitant formation of 1,3-dithiolan-2-thiones such as (36) as byproducts. 2-Alkylimino-1,3-dithiolans (37) isomerized at 200 °C to 3-alkyl-1,3-thiazolidine-2-thiones (38),[52]

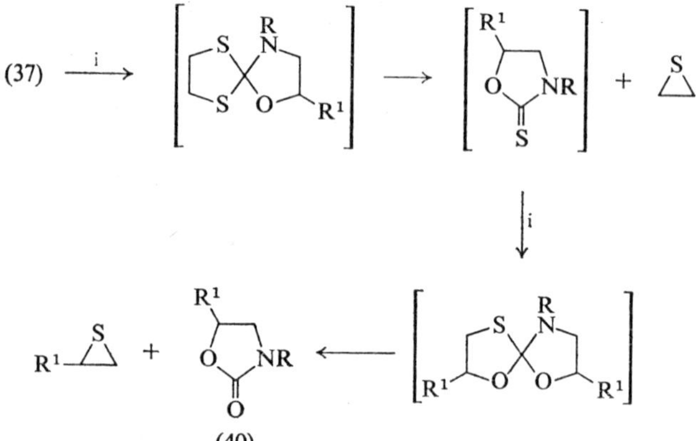

(40)

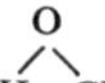

Reagent: i, R$^1$—CH—CH$_2$ (epoxide)

**Scheme 1**

[50] M. O. Hedblom, *Arkiv Kemi*, 1969, **31**, 489.
[51] K. Tomita and M. Nagano, *Chem. and Pharm. Bull. (Japan)*, 1969, **17**, 2442.
[52] Y. Ueno, T. Nakai, and M. Okawara, *Bull. Chem. Soc. Japan*, 1970, **43**, 162.

# Saturated Cyclic Compounds of Sulphur, Selenium, and Tellurium

and it was suggested that this involved a concerted rearrangement of a dipolar form (39), since control experiments indicated that fragmentation into an alkylisothiocyanate and thiiran had not occurred. Reaction of the 2-alkylimino-1,3-dithiolans (37) with oxirans gave 1,3-oxazolidone derivatives (40), for which the mechanistic rationalization depicted in Scheme 1 was offered. In support of this mechanism, 2-methyloxiran was found to undergo triethylamine-catalysed addition to 1,3-dithiolan-2-thione to give the spirocyclic compound (41), which decomposed into 1,3-dithiolan-2-one and 2-methylthiiran.[53] This was developed into a general method for the conversion of thiones into the corresponding oxo compounds. Chlorine in wet solvents also converted derivatives of 1,3-dithiolan-2-thione into the corresponding 1,3-dithiolan-2-one derivatives.[54]

The presence of an *ortho*-2-(1,3-dithiolano)-group, as in (42), was considerably more effective than an *ortho*-methoxy- or *ortho*-methylthio-group in accelerating the first-order rate of acetolysis of benzyl chloride.[55] Probably part of the explanation lay in the easier $R_2S$-5 participation by the dithiolan moiety than the $R_2S$-4 participation by the methylthio-group in the solvolysis. The significance of neighbouring group participation by the dithiolan ring was shown by the faster solvolysis of the *ortho* compound (42) than its *para* isomer (by a factor of 220), whereas *ortho*-methylthiobenzyl chloride reacted about 50 times slower than its *para* isomer. Certain 1,3-dithiolan derivatives were oxidised by 2,3-dichloro-5,6-dicyanoquinone, for example the dithiolan (43) gave the quinonoid

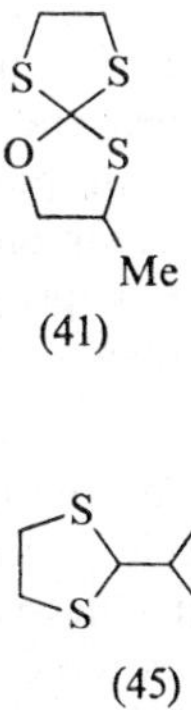

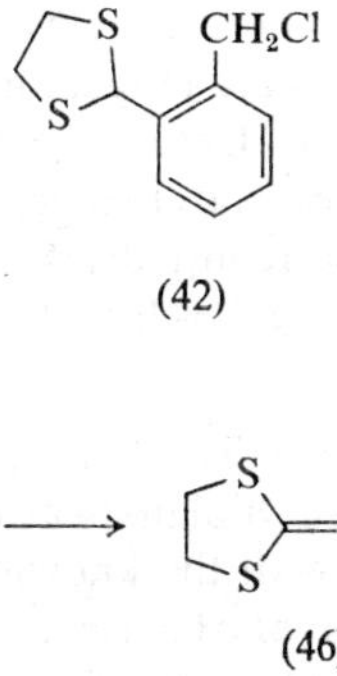

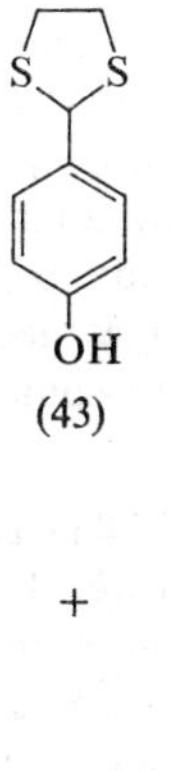

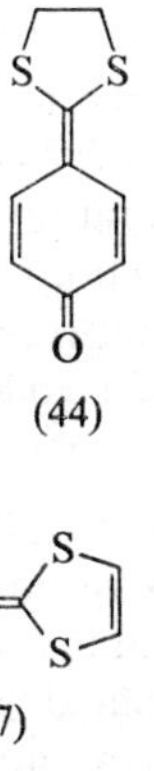

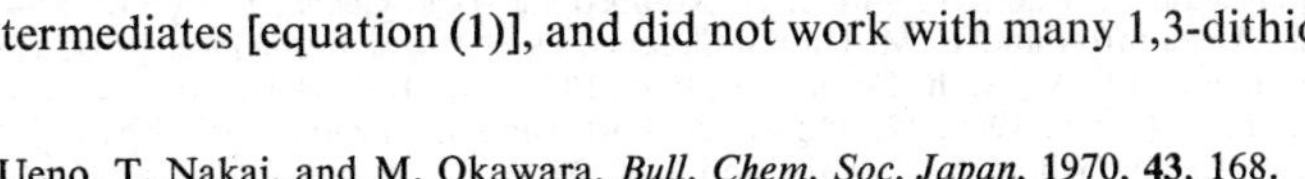

derivative (44), whilst 2,2'-bis(1,3-dithiolan) (45) gave a mixture of (46) and (47).[56] The reaction was considered to involve the formation of carbonium ion intermediates [equation (1)], and did not work with many 1,3-dithiolans.

[53] Y. Ueno, T. Nakai, and M. Okawara, *Bull. Chem. Soc. Japan*, 1970, **43**, 168.
[54] B. S. Sasha, W. M. Doane, C. R. Russell, and C. E. Rist, *J. Org. Chem.*, 1969, **34**, 1642.
[55] M. Hojo, T. Ichi, Y. Tamaru, and Z. Yoshida, *J. Amer. Chem. Soc.*, 1969, **91**, 5170.
[56] D. L. Coffen and P. E. Garrett, *Tetrahedron Letters*, 1969, 2043.

$$\text{(1)}$$

Fragmentation patterns in the mass spectra of 2-aryl-1,3-dithiolans have been elucidated by deuterium-labelling studies.[57] The most important mode of fission was as indicated in equation (2). A comparison of the mass

$$\text{(2)}$$

spectra of the 1,3-dioxolan, 1,3-oxathiolan, and 1,3-dithiolan compounds derived from 5α-cholestan-3-one has confirmed previous observations that dioxolan derivatives are the most suitable for mass spectral characterization of steroidal ketones.[58] Primary fragmentation occurred in the 1,3-oxathiolan ring and 1,3-dithiolan ring, the loss of a $C_2H_4S$ fragment again being an important process. For 4-keto-derivatives of the above compounds, however, primary fission of the sulphur-containing rings was less important, and diagnostic ions were obtained analogous to those produced by the dioxolan derivatives. For another example of mass spectral fragmentation of a 1,3-dithiolan derivative see p. 167. Solvent effects on the n.m.r. spectra of 2,2-dimethyl-1,3-dithiolan have been reported,[59] the signals due to the methylene groups undergoing a pronounced diamagnetic shift of 21 Hz, and the methyl signals a diamagnetic shift of 2 Hz on changing from carbon tetrachloride to benzene. The preparation of 1,3-dithiolan derivatives from 1,3-dithiols, for purposes of protection or reduction of aldehydes and ketones, is so commonplace that all recent examples need not be reported here, but it is appropriate to refer collectively to some interesting 1,3-dithiolan derivatives prepared in this manner.[60]

**1,2,3- and 1,2,4-Trithiolan Derivatives.**—The barrier to pseudorotation in 1,2,4-trithiolan was lower than in 1,2-dithiolan-4-carboxylic acid, according to n.m.r. investigations,[61] in accord with the known low torsional barrier associated with the C–S–C portion of the ring. When ammonia was used to catalyse condensation of hydrogen sulphide with dibenzylidene

---

[57] J. H. Bowie and P. Y. White, *Org. Mass Spectrometry*, 1969, **2**, 611.
[58] C. Fenselau, L. Milewich, and C. H. Robinson, *J. Org. Chem.*, 1969, **34**, 1374.
[59] N. E. Alexandrou and P. M. Hadjimihalakis, *Org. Magn. Resonance*, 1969, **1**, 401.
[60] G. Zinner and R. Vollrath, *Chem. Ber.*, 1970, **103**, 766; M. P. Mertes and A. A. Ramsay, *J. Medicin. Chem.*, 1969, **12**, 342; V. A. Portnyagina, *Ukrain. khim. Zhur.*, 1969, **35**, 734; G. E. McCasland, S. R. Naik, and A. K. M. Anisuzzman, Abs. of the 159th A.C.S. National Meeting, Houston, Texas, Feb. 1970.
[61] R. M. Moriarty, M. Ishibe, M. Kayser, K. C. Ramsey, and H. J. Gisler, *Tetrahedron Letters*, 1969, 4883.

## Saturated Cyclic Compounds of Sulphur, Selenium, and Tellurium

acetone,[62] the product was the 1,2,4-trithiolan derivative (48) and not 2,6-diphenylthian-4-one, the product previously obtained with sodium acetate as catalyst; hydrogen sulphide reacted with 2,6-diphenylthian-4-one to give (48). A minor product of the autoxidation of thiobenzophenone, 3,3,5,5-tetraphenyl-1,2,4-trithiolan (49), was obtained from thiobenzophenone and tetracyanoethylene oxide in ethanol at room temperature.[63] It was proposed that it was formed by reaction of thiobenzophenone with

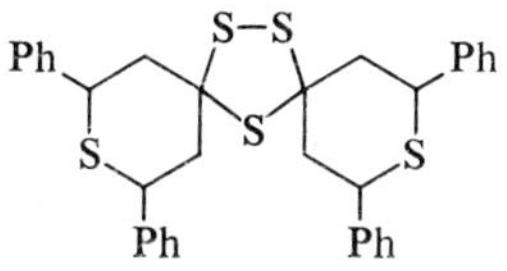

(48) 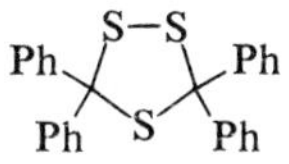  (49) 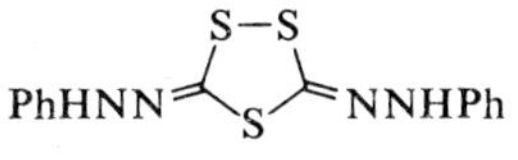  (50)

2,2-dicyano-3,3-diphenylthiiran present as an intermediate, but details of the mechanism were not given. The trithiolan (49) decomposed readily on heating into thiobenzophenone and sulphur. Oxidation of arylthiocarbazic acids, either with iodine or hydrogen peroxide, produced 1,2,4-trithiolan derivatives such as (50), together with tetrathian and pentathiepan derivatives.[64] Sulphuration of bicyclo[2,2,1]hept-2-ene by sulphur in the presence of ammonia gave the 1,2,3-trithiolan derivative (51) in high yield.[65] This new reaction was stereospecific, giving only the *exo* adduct (51), and selective, since addition took place only at the endocyclic double bond of 5-ethylidenebicyclo[2,2,1]hept-2-ene. Reduction of the trithiolan (51) by

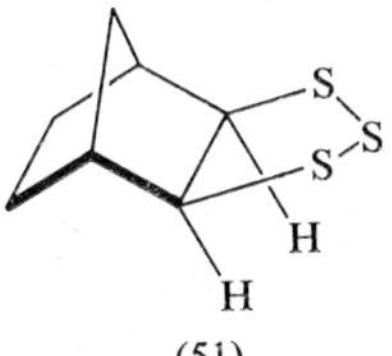

(51)

sodium in ammonia gave the corresponding 2,3-dithiol of *exo* configuration.

**1,2-Oxathiolan Derivatives.**—A new synthesis of cyclic sulphinates involved the desulphuration of cyclic thiolsulphinates, treatment of 1,2-dithiolan-1,1-dioxide with tris(diethylamino)phosphine giving 1,2-oxathiolan-2-oxide (52) in 92% yield.[66] The reaction possibly proceeded through an intermediate (53) containing the ambident sulphinate anion, which cyclised preferentially by an intramolecular nucleophilic attack involving sulphinyl

---

[62] G. C. Forward and M. C. Whiting, *J. Chem. Soc.* (*C*), 1969, 1647.
[63] W. J. Linn and E. Ciganek, *J. Org. Chem.*, 1969, **34**, 2146.
[64] L. Cambi, G. Bargigia, L. Colombo, and E. P. Dubini, *Gazzetta*, 1969, **99**, 780.
[65] T. C. Shields and A. N. Kurtz, *J. Amer. Chem. Soc.*, 1969, **91**, 5415.
[66] D. N. Harpp and J. G. Gleason, *Tetrahedron Letters*, 1969, 1447.

$$O=S\underset{(52)}{\overset{O}{\diagdown\diagup}}\underset{(53)}{\overset{O}{\underset{\diagdown\diagup}{\overset{\parallel}{S}}}-\overset{+}{S}P(NEt_2)_3 \leftrightarrow O=\underset{\diagdown\diagup}{\overset{O^-}{S}}-\overset{+}{S}P(NEt_2)_3}$$

oxygen. Reaction of the disulphide derived from 3-mercaptopropionic acid with chlorine gave 5-oxo-1,2-oxathiolan-2-oxide (54), presumably by loss of hydrogen chloride from 3-chlorosulphinylpropanoic acid formed as an intermediate.[67] Asymmetry at sulphur in (54) was revealed by the observation that the protons adjacent to sulphur were anisochronous, and the reactivity of the compound (a cyclic mixed anhydride of a sulphinic and carboxylic acid) was illustrated by its conversion into sulphonyldipropionanilides on treatment with aromatic amines. 1,2-Oxathiolan-2,2-dioxide was converted by n-butyl-lithium at $-78\,^\circ$C into its 3-lithio-derivative, which, on reaction with alkyl halides, aldehydes, or ketones,

(54) (55)

provided 3-substituted 1,2-oxathiolan-2,2-dioxides.[68] The temperature at which lithiation was performed is important, since 1,2-oxathiolan dioxides are known to react with metal alkyls at room temperature to give derivatives of sulphonic acids. Hydrolysis of 6-oxa-5-thiaspiro[2,4]heptane-5,5-dioxide (55), a cyclopropyl carbinyl sulphonate in which the leaving group has a fixed orientation with respect to the cyclopropane ring, proceeded only about twice as fast as for 1,2-oxathiolan-2,2-dioxide, indicating that participation by the cyclopropane ring (and in particular by the 1,2-bond) was insignificant.[69] The nature of the products (1′-hydroxymethylcyclopropylmethanesulphonic acid and 1′-hydroxycyclobutylmethanesulphonic acid) and the isotopic distributions with deuterium-labelled compounds were rationalized in terms of a bicyclobutonium ion intermediate.

**1,3-Oxathiolan Derivatives.**—Reactions of chlorosulphenylated carbonic acid derivatives which lead to derivatives of 2-alkylimino-1,3-oxathiolan have been reviewed.[70] Despite the lower ionization potential of sulphur, the predominant mode of mass spectral fragmentation of 1,3-oxathiolan and simple 2-, 4-, and 5-alkylated derivatives involved electron expulsion from oxygen, and the production of the predominant fragments was rationalized as indicated in Scheme 2.[71] A second analogous mode of

[67] Y. H. Chiang, J. S. Lufoff, and E. Schipper, *J. Org. Chem.*, 1969, **34**, 2397.
[68] T. Durst and J. du Manoir, *Canad. J. Chem.*, 1969, **47**, 1230.
[69] R. M. Coates and A. W. W. Ho, *J. Amer. Chem. Soc.*, 1969, **91**, 7544.
[70] G. Zumach and E. Kuhle, *Angew. Chem. Internat. Edn.*, 1970, **9**, 54.
[71] D. J. Pasto, *J. Heterocyclic Chem.*, 1969, **6**, 175.

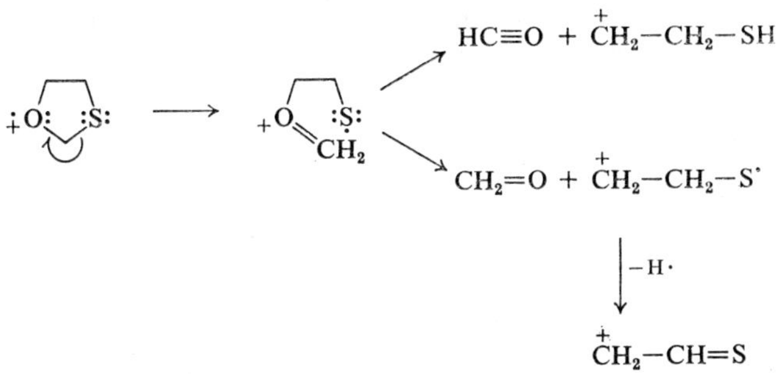

## Scheme 2

cleavage involved ionization at sulphur. Kinetic investigation of the interconversion of the epimeric ethylene hemithioketals of 3,3,5-trimethylcyclohexanone (56a) and (56b) showed that complexation of the oxothiolan ring with the equilibration catalyst (boron trifluoride etherate) was not a factor in the well-known preference of the oxygen in such compounds for the equatorial orientation.[72] The rates of acid-catalysed hydrolysis of 3′- and 4′-substituted-2-phenyl-1,3-oxathiolans (57; 4′-substituted isomer depicted) were much slower than those of the analogous 2-aryl-1,3-dioxolans, for example by a factor of $5 \times 10^4$ for the 4′-methoxy derivative.[73] The logarithms of the rate constants were linearly related to $\sigma^+$ constants in aqueous dioxan, but not to the Hammett substituent constant $\sigma$. The most likely mechanism was considered to involve unimolecular decomposition of an intermediate protonated on sulphur, with a transition state resembling a carbonium ion.

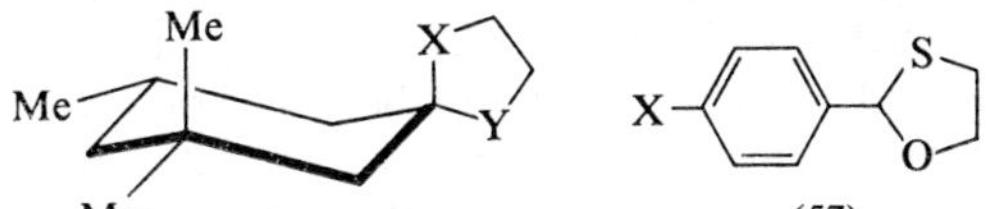

Oxidation of 2-aryl-4,4-diphenyl-1,3-oxathiolan-5-ones with hydrogen peroxide or polysebacic acid gave a diastereoisomeric mixture of the corresponding sulphoxides (58), which, remarkably, could not be further oxidised to the sulphone.[74] The presence of two isomers at sulphur was revealed by t.l.c. and n.m.r., two singlets attributed to the C-2 proton appearing in many cases. Analysis of n.m.r. spectra indicated that the behaviour of the C-2 proton (but not that of alkyl groups on the phenyl ring at C-2) was rational on the basis of acetylenic-type anisotropy of the sulphinyl bond. E.s.r. examination suggested that the sulphoxides (58), but

[72] M. P. Mertes, H.-K. Lee, and R. L. Schowen, *J. Org. Chem.*, 1969, **34**, 2080.
[73] T. U. Fife and L. K. Jao, *J. Amer. Chem. Soc.*, 1969, **91**, 4217.
[74] C. T. Pedersen, *Acta Chem. Scand.*, 1969, **23**, 489.

not the parent oxathiolans, formed cation radicals formulated as (59) in sulphuric acid. Authentic 1,3-oxathiolan-2-thione (60) has now been prepared,[75] earlier reports of its synthesis proving to be erroneous. The

(58) (59) (60)

compound was sensitive to heat and moisture, and was isomerized by halide ion into 1,3-dithiolan-2-one, whilst reaction with triphenylphosphine gave ethylene. Treatment with acid induced polymerization to polydithiol carbonates, and with alkyl halides β-halogenoethyl dithiolcarbonates were formed. The n.m.r. and i.r. characteristics of 1,3-oxathiolan-2-thione were compared with those of the five possible analogues containing variously oxygen or sulphur. 2-Phenylimino-1,3-oxathiolan was produced on reaction of 2,2-dibutyl-1,3,2-oxathiastannacyclopentane with phenyl isothiocyanate.[76]

**Derivatives of 1,3,2-Dioxathiolan, 1,2,4-Dioxathiolan, and 1,2,5-Oxadithiolan.**—The temperature dependence of the chemical shifts and coupling parameters associated with the AA'BB' n.m.r. spectral pattern of 1,3,2-dioxathiolan-2-oxide was considered to be due more probably to vibrational effects rather than to variations in conformational equilibria,[77] and with tertiary amines or dialkyl sulphides the compound reacted to give the zwitterionic 2-trialkylammoniumethanesulphonate or 2-dialkylsulphoniumethanesulphonate respectively.[78] Sulphur dioxide readily underwent 1,3-cycloaddition to 2,2-dimethylcyclopropanone to give 5,5-dimethyl-4-methylene-1,3,2-dioxathiolan-2-oxide (61).[79] The stereochemistry at sulphur of the isomeric sulphites (62) derived from (±)-10β-pinane-2,3-diol has been established by X-ray analysis of one of the isomers.[80] In the n.m.r. spectra of the isomers, the 10-methyl and 3-hydrogen were more deshielded in the isomer with a *syn*-relationship between the 10-methyl group and sulphinyl oxygen, in accord with previous observations on the anisotropy of the sulphinyl bond. Pyrolysis of the mixture of isomeric sulphites (62) gave pinocamphone (63) almost exclusively,[81] and pyrolysis of *cis*- and *trans*-4,5-dimethyl-1,3,2-dioxathiolan-1-oxide gave predominantly butan-2-

[75] F. N. Jones and S. Andreades, *J. Org. Chem.*, 1969, **34**, 3011.
[76] S. Sakai, Y. Kobayashi, and Y. Ishii, *Chem. Comm.*, 1970, 235.
[77] H. Finegold, *J. Phys. Chem.*, 1969, **73**, 4021.
[78] H. Distler, Abs. 158th A.C.S. Nat. Meeting, New York, Sept. 1969.
[79] N. J. Turro, S. S. Edelson, J. R. Williams, T. R. Darling, and W. B. Hammond, *J. Amer. Chem. Soc.*, 1969, **91**, 2283.
[80] M. D. Brice, J. M. Coxon, E. Danstead, M. P. Hartshorn, and W. T. Robinson, *Chem. Comm.* 1969, 356.
[81] J. M. Coxon, E. Danstead, M. P. Hartshorn, and K. E. Richards, *Tetrahedron*, 1969, **25**, 3307.

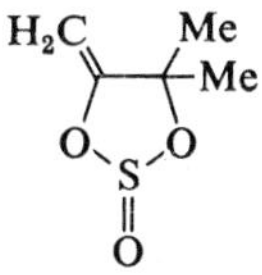

(61) 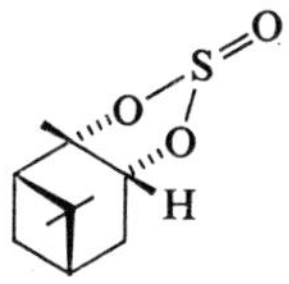 (62) 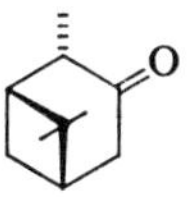 (63)

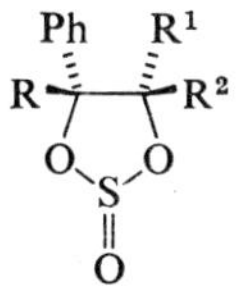

(64) R = R¹ = R² = Ph
(65) R = R² = H, R¹ = Ph
(66) R = R¹ = H, R² = Ph

one.[82] Photolysis of 4,4,5,5-tetraphenyl-1,3,2-dioxathiolan-1-oxide (64)[83] and cis- and trans-4,5-diphenyl-1,3,2-dioxathiolan-1-oxide, (65) and (66),[84] gave diphenylcarbene and phenylcarbene respectively, which were detected by typical insertion reactions. The mechanism of fragmentation was not established, but it did not appear to involve stilbene oxides as intermediates.

Photocatalysed addition of oxygen to the compound (67) gave the 1,2,4-dioxathiolan derivative (68), which was isolated chromatographically despite its extreme sensitivity to heat and light.[85] This was the first thio-ozonide isolated, although compounds of this type had been postulated as intermediates in the photosensitized autoxidation of dialkylthiophens. The thio-ozonide (68) readily decomposed on warming into a diketone and a keto-thiol ester, whilst photolytic decomposition gave a keto-thiono-ester. With triphenylphosphine, sulphur and oxygen were removed to give the hydrocarbon (69).

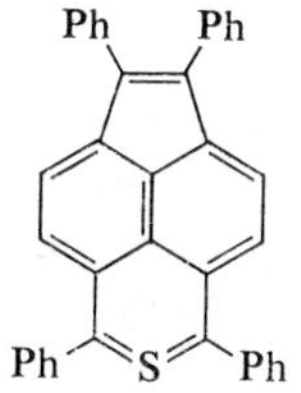

(67) 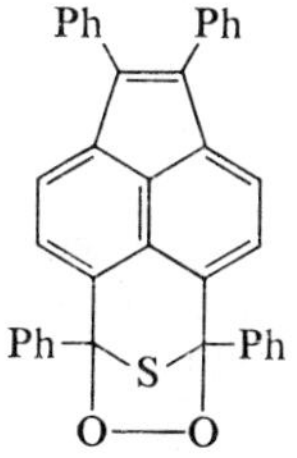 (68) 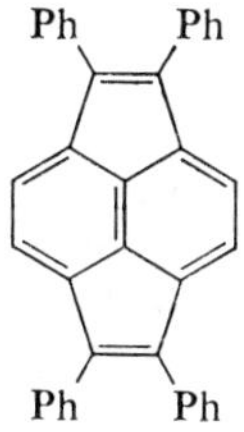 (69) 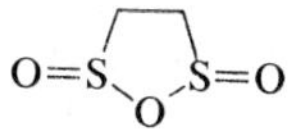 (70)

Following closely the first description of an acyclic sulphinic anhydride, the preparation of 1,2,5-oxadithiolan-2,5-dioxide (70) by controlled hydrolysis of ethane-1,2-disulphinyl chloride has been reported.[86] Hydrolysis of the anhydride (70) with boiling water gave the elusive ethane-1,2-disulphinic acid.

---

[82] J. M. Coxon, M. P. Hartshorn, A. J. Lewis, K. E. Richards, and W. H. Swallow, *Tetrahedron*, 1969, **25**, 4445.
[83] R. L. Smith, A. Manmade, and G. W. Griffin, *J. Heterocyclic Chem.*, 1969, **6**, 443.
[84] R. L. Smith, A. Manmade, and G. W. Griffin, *Tetrahedron Letters*, 1970, 663.
[85] J. M. Hoffman and R. H. Schlessinger, *Tetrahedron Letters*, 1970, 797.
[86] W. H. Mueller and M. B. Dines, *Chem. Comm.*, 1969, 1205.

**1,2- and 1,3-Thiazolidine Derivatives.**—2-t-Butyl-3-methyl-1,2-thiazolidine 1,1-dioxide was prepared from $N$-t-butyl-$N$-chlorobutanesulphonamide by photolytic conversion to $N$-t-butyl-3-chlorobutanesulphonamide followed by base-catalysed ring closure,[87] whilst 3-oxo-2-chloroalkyl-1,2-thiazolidine-1,1-dioxides were formed when chlorosulphonyl isocyanate reacted with olefins in the presence of a radical initiator.[88] Thiadiazine derivatives typified by (71) underwent acid-catalysed rearrangement to give 3-oxo-1,2-thiazolidine-1-oxide derivatives (72), probably by the mechanism indicated in Scheme 3.[89]

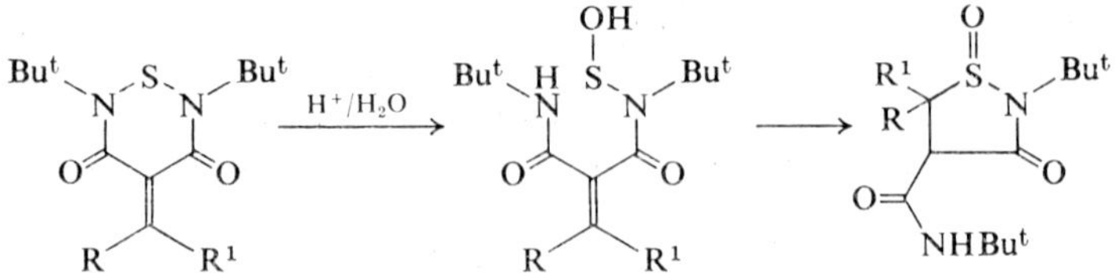

R = H, R¹ = Ph
or R = R¹ = Me

**Scheme 3**

The reaction of L-cysteine and 4-oxopentanoic acid in water to give 1,3-thiazolidine derivatives has been studied by pH determinations, n.m.r. spectroscopy, and polarimetry,[90] and some new 2-substituted 1,3-thiazolidines have been prepared for biological evaluation.[91] The usefulness has been exemplified of 2-cyclohexylaminoethanethiol in the selective preparation of mono-1,3-thiazolidine derivatives of dicarbonyl compounds.[92] Thiamine and related thiazolium ylides underwent novel cycloaddition to phenylisocyanate to give spiro-1,3-thiazolidine derivatives [93] [as compound (11), Chapter 3]. Dialkyl trithiocarbonates reacted with 2-aminoethanol to give a mixture of 1,3-thiazolidine-2-thione, 1,3-thiazolidine-2-one, and 1,3-oxazolidine-2-one,[94] whilst the aziridine ring in 1-$N$-aziridino-2-cyanoethylene was cleaved by carbon disulphide in the presence of sodium iodide to give the compound (73).[95] Keten reacted with benzylidene aniline in liquid sulphur dioxide to give 4-oxo-2,3-diphenyl-1,3-thiazolidine-1,1-dioxide.[96] The reaction was rationalized in terms of the formation of

[87] R. S. Neale, and N. L. Marcus, *J. Org. Chem.*, 1969, **34**, 1808.
[88] D. Gunther and F. Soldan, *Chem. Ber.*, 1970, **103**, 663.
[89] H. Wittman, E. Ziegler, H. Sterk, and G. Dworak, *Monatsh.*, 1969, **100**, 959.
[90] N. Hellstrom, S.-O. Almqvist, and M. Aamisepp, *J. Chem. Soc. (B)*, 1969, 1103.
[91] B. J. Sweetman, M. Bellas, and L. Field, *J. Medicin. Chem.*, 1969, **12**, 888.
[92] K. Ueno, F. Ishikawa, and T. Naito, *Tetrahedron Letters*, 1969, 1283.
[93] A. Takamizawa, K. Hirai, S. Matsumoto, and T. Ishiba, *Chem. and Pharm. Bull. (Japan)*, 1969, **17**, 462.
[94] T. Taguchi, Y. Kiyoshima, O. Komori, and M. Mori, *Tetrahedron Letters*, 1969, 3631.
[95] T. Sasaki, T. Yoshioka, and K. Shoji, *J. Chem. Soc. (C)*, 1969, 1086.
[96] A. de S. Gomes and M. M. Joullie, *J. Heterocyclic Chem.*, 1969, **6**, 729.

# Saturated Cyclic Compounds of Sulphur, Selenium, and Tellurium 151

an intermediate 1,3-dipole [(33), Chapter 3], and its postulated presence was substantiated by the reactions of keten and sulphur dioxide with 2-phenyl-4$H$,5$H$-1,3-thiazole to give the adduct (74), and with cinnamylidene aniline to give a 1,3-thiazolidine derivative, no addition occurring at the ethylenic bond. Methyl acetylene dicarboxylate formed 2:1 adducts typified by (75) with 2-alkyl-2-thiazolines,[97] whilst acid-catalysed cyclization of 2-aldehydo-4,5-methylenedioxyphenyl-$N$-allylthiourea (derived from 6-aminopiperonal and allyl isocyanate) gave the 1,3-thiazolidine derivative (76).[98] A much higher barrier to rotation (measured by n.m.r.) about the

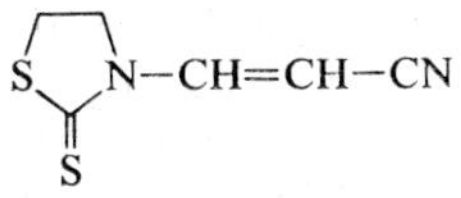

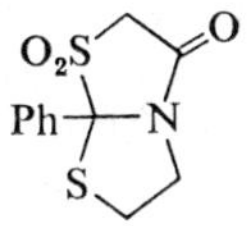

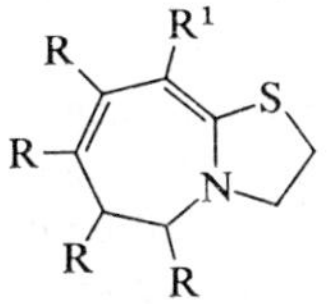

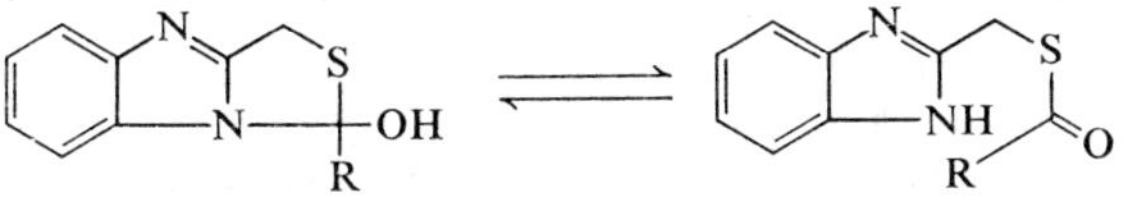

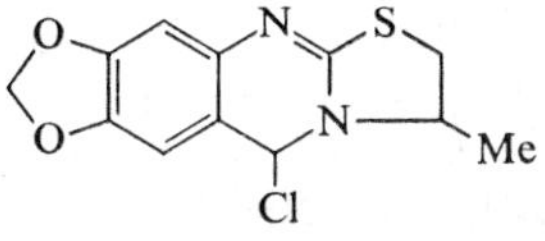

$R = CO_2Me$
$R^1 = H$ or alkyl

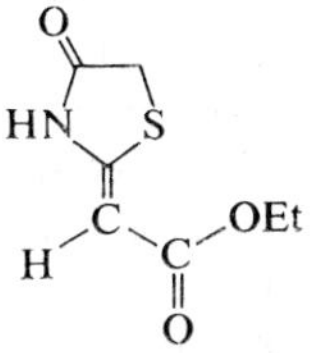

double bond in 2-(bismethoxycarbonyl)methylene-1,3-thiazolidine than in the $N$-methyl analogue was attributed to intramolecular hydrogen bonding between the secondary amino-group and an ester carbonyl oxygen in the former compound.[99] The position of equilibrium in the ring–chain tautomeric system (77a ⇌ 77b) depended upon the inductive effect of the group R, the equilibrium being exclusively in favour of the cyclic tautomer (77a) for R = H or $CF_3$, and entirely in favour of the open-chain tautomer (77b) for R = $Bu^t$.[100] A spectroscopic examination (i.r. and n.m.r.) of a series of 2-alkylidene-1,3-thiazolidine-4-ones typified by (78) revealed that for electronegative substituents at C-2 the compounds existed as the

[97] C. Divome and J. Roggero, *Compt. rend.*, 1969, **269**, C, 615.
[98] H. Singh, I. Singh, K. S. Bhandhari, K. S. Dhami, S. S. Aroia, and N. S. Narang, *J. Indian Chem. Soc.*, 1969, **46**, 367.
[99] Y. Shvo and I. Belsky, *Tetrahedron*, 1969, **25**, 4649.
[100] H. Alper, *Chem. Comm.*, 1970, 383.

depicted enamine tautomer. The *cis* and *trans* forms of these tautomers could be isolated, the *trans* isomer being the more stable except when internal hydrogen bonding involving the secondary amino-hydrogen stabilized the *cis* form.[101] The preparations of other interesting derivatives of 1,3-thiazolidine derivatives are worthy of reference.[102]

There has been considerable interest recently in the chemistry of the penicillins, but this report is confined to those aspects directly related to the chemistry of the 1,3-thiazolidine ring in these compounds. These are concerned with cleavage reactions and with oxidation at sulphur. Treatment of methyl 6-phthalimidopenicillanate (79a) with triethylamine in methylene chloride at room temperature resulted not only in epimerization at C-6 as reported previously, but also in the formation of the 7-oxo-2,3,4,7-tetrahydro-1,4-thiazepin derivative (80a).[103] In a related case, the penicillin derivative (79b) rearranged to (80b) to the exclusion of epimerization at C-6.[104] The formation of the compounds (80a) and (80b) supported, but did not prove, a hypothesis that epimerization at C-6 proceeded *via* the intermediate thiolate (81), since this was a rational intermediate both for the epimerization at C-6, and for the formation of the compounds (80a) and (80b), in the latter case by nucleophilic attack of the thiolate anion upon the β-lactam carbonyl group and subsequent fission of the

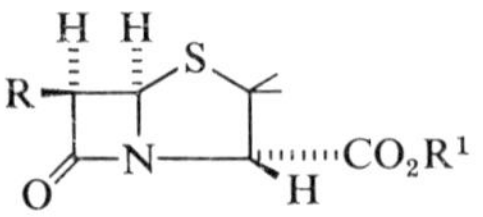

(79) 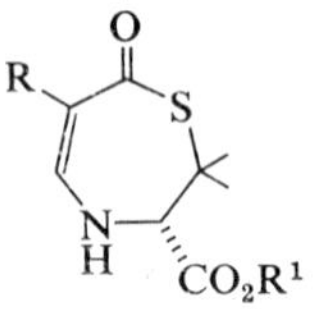    (80) 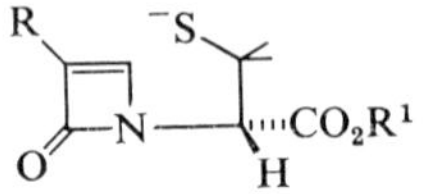    (81)

(a) R = phthalimido, $R^1$ = Me
(b) R = $p$-$NO_2$–$C_6H_4$–CH=N–, $R^1$ = $CH_2OMe$

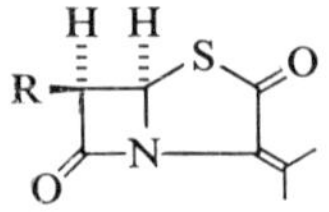

(84) 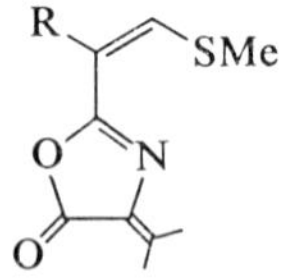    (82) 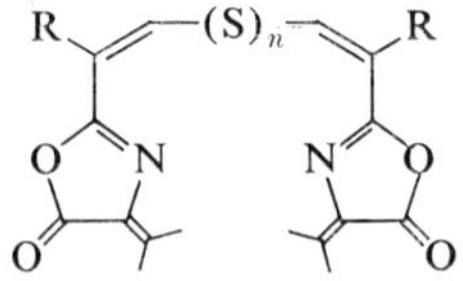    (83) $n = 1$
                (85) $n = 2$

[101] P. J. Taylor, *Spectrochim. Acta*, 1970, **26A**, 153, 165.
[102] V. Hahnkamm and G. Gattow, *Naturwiss.*, 1969, **56**, 515; M. R. Chandramohan, M. S. Sardessai, S. R. Shah, and S. Seshadri, *Indian J. Chem.*, 1969, **7**, 1006; A. U. Rahman and H. S. E. Gatica, *Rec. Trav. chim.*, 1969, **88**, 905; M. Z. Peretyazhko and P. S. Pel'kis, *Ukrain. khim. Zhur.*, 1969, **35**, 532; V. K. Chadha, H. S. Chaudhary, and H. K. Pujari, *Austral. J. Chem.*, 1969, **22**, 2697.
[103] O. K. J. Kovacs, B. Ekstrom, and B. Sjoberg, *Tetrahedron Letters*, 1969, 1863.
[104] J. R. Jackson and R. J. Stoodley, *Chem. Comm.*, 1970, 14.

amide link. Penicillin V (79, R = PhOCH$_2$.CONH, R$^1$ = H) rearranged on treatment with triethylamine and methyl chloroformate to give the oxazolones (82) and (83),[105] and not the expected anhydropenicillin (84). The compound (83) and the disulphide (85) were obtained, together with the expected phenoxymethyl anhydropenicillin (84), when the acid chloride of penicillin V was treated with triethylamine.[105] Related rearrangements were observed when penicillin V was treated with acetic anhydride. These results may perhaps be rationalized mechanistically as indicated in Scheme 4, although a mechanistic explanation for the formation of the sulphide (83) is not obvious. The proposed mechanism of the anhydropenicillin rearrangement has been substantiated by the isolation of

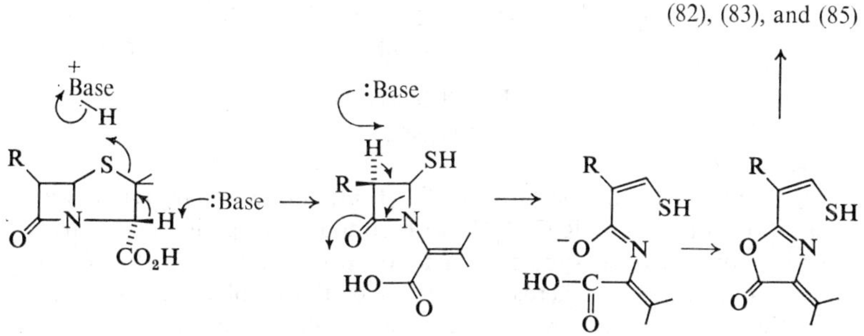

(82), (83), and (85)

Scheme 4

compounds derived from postulated ring-opened intermediates,[106] and a reversal of the anhydropenicillin rearrangement has been reported [107] involving a thiazolidine-forming cyclization, which is significant in relation to one of the postulated pathways for the biosynthesis of the penicillins.

In a procedure for the conversion of the penicillin into the cephalosporin ring system, methyl 6$\beta$-phenoxyacetamidopenicillanate-($S$)-1-oxide (86), on treatment with hot acetic anhydride, gave a mixture of the penicillin derivative (87) and the cephalosporin derivative (88).[108] The mechanism probably involved the formation of an intermediate sulphenic anhydride (Scheme 5) which subsequently underwent intramolecular addition to a double bond. It is interesting that the compounds (87) and (88) were obtained diastereoisomerically pure, indicating that the rearrangements occurred stereospecifically.

Although penicillin sulphoxides have long been known, stereochemical aspects of these compounds have not been investigated until very re-

[105] S. Kukolja, R. D. G. Cooper, and R. B. Morin, *Tetrahedron Letters*, 1969, 3381.
[106] J. P. Clayton, *J. Chem. Soc.* (*C*), 1969, 2123.
[107] S. Wolfe, R. N. Bassett, S. M. Caldwell, and F. I. Wasson, *J. Amer. Chem. Soc.*, 1969, **91**, 7205.
[108] R. B. Morin, B. G. Jackson, R. A. Mueller, E. R. Lavagnino, W. B. Scanlon, and S. L. Andrews, *J. Amer. Chem. Soc.*, 1969, **91**, 1401.

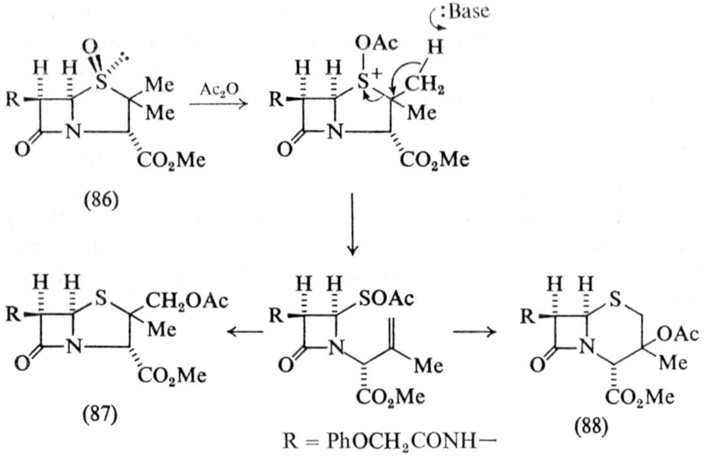

## Scheme 5

(86) → (with Ac$_2$O, :Base) → (87) ← SOAc intermediate → (88)

R = PhOCH$_2$CONH—

cently.[106, 108−112] Oxidation of penicillins bearing a secondary amide group at position 6 (79, R = R$^2$CONH) with sodium metaperiodate, hydrogen peroxide, *m*-chloroperoxybenzoic acid or ozone gave only the (S)-sulphoxide (as 86),[108, 109] whereas oxidation of the 6β-phthalimidopenicillin derivatives with the same reagents gave only sulphoxides of (R)-configuration.[110, 111] The β-face of the molecule is more sterically compressed than the α-face, so that the (R)-sulphoxide is the product of steric approach control. The stereospecific formation of the (S)-sulphoxide in penicillins with 6β-secondary amide functions was attributed to hydrogen bonding of the amide proton with the oxidant, so directing attack onto the more hindered β-face,[109−111] and this interpretation was substantiated by the observation that oxidation of methyl 6,6-dibromopenicillanate with sodium metaperiodate gave a mixture (3:7) of the (R)- and (S)-sulphoxides.[106] This directing influence did not extend to iodobenzene dichloride, which oxidised methyl 6β-phenylacetamidopenicillanate to a *ca.* 1:1 mixture of the (R)- and (S)-sulphoxides.[111] This may possibly be due to the intermediacy of a tetracovalent dichlorosulphide intermediate, analogous to that found for the oxidation of sulphides by t-butyl hypochlorite (see p. 157). Remarkably, methyl 6β-phenylacetamidopenicillanate-(R)-1-oxide isomerized to the (S)-isomer on heating in benzene.[111] If this proceeds by pyramidal inversion at sulphur as seems very likely, it represents by far the easiest such inversion of configuration of a dialkyl sulphoxide. Attempted inversion of configuration at sulphur of penicillin (S)-sulphoxides using

[109] R. D. G. Cooper, P. V. DeMarco, J. C. Cheng, and N. D. Jones, *J. Amer. Chem. Soc.*, 1969, **91**, 1408.
[110] R. D. G. Cooper, P. V. DeMarco, and D. O. Spry, *J. Amer. Chem. Soc.*, 1969, **91**, 1528.
[111] D. H. R. Barton, F. Comer, and P. G. Sammes, *J. Amer. Chem. Soc.*, 1969, **91**, 1529.

trimethyloxonium fluoroborate failed, as did attempted racemization using acetyl chloride,[109] but photochemical inversion of methyl 6β-acetamidopenicillanate-(S)-1-oxide and its phenylacetamido analogue proceeded to give solely the (R)-isomers.[112] In all these investigations the stereochemistry of the sulphoxides was assigned by n.m.r. techniques, and the configuration of 6β-phenoxyacetamidopenicillin-(S)-1-oxide was established by X-ray analysis.[109] Pertinent evidence included the deshielding of the hydrogens at C-3 and C-6 in the (S)-sulphoxides (relative to the parent penicillins) due to the operation of the 'syn-axial' effect of the sulphinyl bond, and the relative deshielding of the C-5 proton which is in a trans relationship with the sulphur lone pair. However, in the (R)-sulphoxides the C-5 hydrogen was relatively shielded, despite the cis relationship with the sulphoxide oxygen,[110-112] illustrating further the complicated nature of the anisotropy of the sulphinyl bond. N.m.r. spectroscopy also revealed strong intramolecular hydrogen bonding between the N–H of a 6β-amide side-chain and the sulphinyl oxygen in the (S)-sulphoxides, but not in the (R)-sulphoxides, suggesting that this may be a factor contributing towards the greater thermodynamic stability of the (S)-isomers.[109, 112] An investigation of nuclear Overhauser effects in 6β-phenoxyacetamidopenicillin, its (S)-sulphoxide, and the corresponding sulphone, coupled with the X-ray analysis of the sulphoxide, indicated that in solution and in the solid state the thiazolidine ring in the sulphoxide and sulphone adopted the conformation depicted in (89),[109] in contrast to the conformation (90)

(89) (X = O or electron pair)    (90)

previously established by X-ray analysis for the thiazolidine ring in potassium benzylpenicillin.

**Miscellaneous Five-membered Rings containing Sulphur.**—Cycloaddition of NN'-bis(p-ethoxycarbonylphenyl)sulphur di-imide and diphenylketen gave the 1,2,5-thiadiazolidine derivative (91),[113] and the preparation of the related 3,4-dioxo-1,2,5-thiadiazolidine-1-oxide from 3,4-diethoxy-1,2,5-thiadiazole has been described.[114] 1,3-Dimethylurea reacted with benzoyl peroxide in a mass spectrometer to give 2,4-dimethyl-3,5-bis(methylimino)-1,2,4-thiadiazolidine.[115] 2,2-Dimethyl-1,3,4-thiadiazolidine-2-thione (92)

[112] R. A. Archer and P. V. DeMarco, *J. Amer. Chem. Soc.*, 1969, **91**, 1530.
[113] H. H. Hoerhold and H. Eibisch, *Tetrahedron*, 1969, **25**, 4277.
[114] M. Carmack, I. W. Stephen, and R. Y. Wen, *Org. Prep. Proced.*, 1969, **1**, 255.
[115] T. Kinoshita and C. Tamura, *Tetrahedron Letters*, 1969, 4963.

was obtained by reaction of acetone with hydrazinium dithiocarbazate,[115] in a general reaction which applied to aliphatic and aromatic ketones and alkyl (and dialkyl) hydrazinium dithiocarbazates. Such 1,3,4-thiadiazolidine-2-thiones as (92) existed in tautomeric equilibrium with their 'hydrazone form' (93) in solution, according to n.m.r. and chemical investigations,[116, 117] the equilibrium shifting entirely in favour of the salt of the acyclic tautomer in basic solution.[117] Methylamine reacted with ethane-1,2-disulphenyl chloride to give 2-methyl-1,3,2-dithiazolidine, but in poor yield.[118]

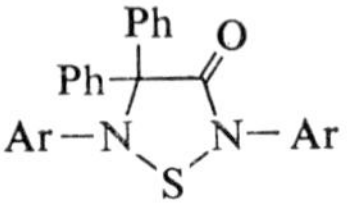

(91) Ar = $p$-EtCO$_2$—C$_6$H$_4$—

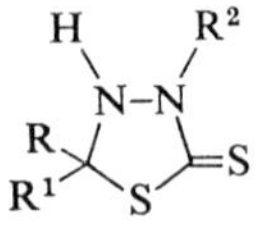

(92)

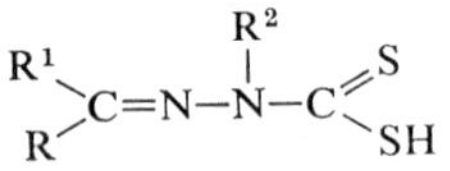

(93)

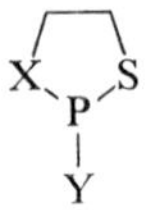

(94) X = S, Y = OMe
(95) X = O, Y = OMe
(98) X = O, Y = Cl

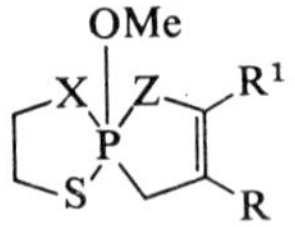

(97)

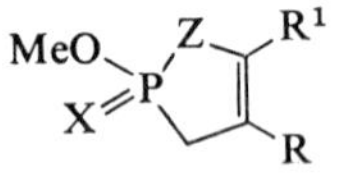

(96)

(99)

X = S or O, Z = O or CH$_2$

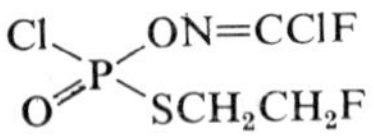

Sulphur was eliminated from 3-methoxy-1,3,2-dithiaphospholan (94) and 3-methoxy-1,3,2-oxathiaphospholan (95) on treatment with conjugated dienes or $\alpha\beta$-unsaturated ketones, the final products being (96). The reaction most probably involved the formation of an intermediate (97) which underwent Arbuzov rearrangement eliminating a C$_2$H$_4$S fragment.[119] Fission of only the carbon–oxygen bond occurred in 2-chloro-1,3,2-oxathiaphospholan (98) and related compounds on treatment with fluorodichloronitrosomethane, the product being (99) or related compounds.[120] 1-Ethylthio- and 1-phenyl-1,3,2-oxathiaborolan have been prepared from 2-mercaptoethanol on treatment respectively with triethylthioborane and bis(ethylthio)phenylborane.[121]

[116] U. Anthoni, C. Larsen, and P. H. Nielsen, *Acta Chem. Scand.*, 1970, **24**, 179.
[117] K. H. Mayer and D. Lauerer, *Annalen*, 1970, **731**, 142.
[118] W. H. Mueller and M. Dines, *J. Heterocyclic Chem.*, 1969, **6**, 627.
[119] L. S. Kovalev, N. A. Razumova, and A. A. Petrov, *Zhur. obshchei Khim.*, 1969, **39**, 869.
[120] I. V. Martynov, Yu. L. Kruglyak, G. A. Leibovskaya, Z. I. Khromova, and O. G. Strukov, *Zhur. obshchei Khim.* 1969, **39**, 996; I. V. Martynov, Yu. L. Kruglyak, G. A. Leibovskaya, and O. G. Strukov, *ibid.*, p. 999; Yu. L. Kruglyak, S. I. Malekin, I. V. Martynov, and O. G. Strukov, *ibid.*, p. 1265.
[121] R. H. Cragg, *Chem. Comm.*, 1969, 832.

## 2 Six-membered Rings

**Thian Derivatives.**—The differential rates of exchange of methylene protons adjacent to a sulphinyl group have attracted much interest recently, and an investigation of the rate of exchange of the α-methylene protons in *cis*- and *trans*-4-phenylthian-1-oxide (100) and (101) has contributed much to

(100)    (101)

the understanding of this phenomenon.[122] Since these compounds are relatively fixed in chair conformations [at least 96% even in (100)], the spatial relationships of the methylene protons and sulphinyl bond are fixed, in contrast to the acyclic sulphoxides investigated previously, and axial and equatorial α-methylene protons in both (100) and (101) may be identified unambiguously by n.m.r. spectroscopy. Base-catalysed hydrogen–deuterium exchange was stereoselective in water and in MeOD, but non-stereoselective in Bu$^t$OD and dimethyl sulphoxide–methanol mixtures. This solvent effect was attributed to the stability of carbanion configuration in water and methanol, and rapid inversion of carbanion configuration in the other solvents. In all solvents the *cis* isomer (100) exchanged overall more rapidly than the *trans* isomer (101), and the order of decreasing ease of proton exchange was a > b > c. This conflicted partly with previous molecular orbital calculations by Wolfe and his co-workers on the relative stabilities of carbanions derived from methyl sulphenic acid, which suggested the order b > a > c, and with previous experimental results from acyclic systems. Convincing evidence has been presented that tetracovalent sulphur intermediates were formed during the oxidation of thiane to its 1-oxide by t-butyl hypochlorite,[123] (*cf.* ref. 51, Chapter 3). The mechanistic proposals were that an initially formed chlorosulphonium salt was converted rapidly into a tetracovalent alkoxysulphonium chloride, which subsequently collapsed thermally, or on treatment with base, to the sulphoxide. Thians were also oxidised to the sulphoxides by 1-chlorobenzotriazole, without concomitant sulphone formation.[26] Thian-1-oxide derivatives (102) were obtained by reaction of dimsyl sodium with αβ-unsaturated ketones.[124] A mechanistic rationalization involved initial 1,4-addition followed by cyclization (Scheme 6). The mass spectrum of the

[122] B. J. Hutchinson, K. K. Andersen, and A. R. Katritzky, *J. Amer. Chem. Soc.*, 1969, **91**, 3839.
[123] C. R. Johnson and J. J. Rigau, *J. Amer. Chem. Soc.*, 1969, **91**, 5398.
[124] J. A. Gautier, M. Miocque, M. Plat, H. Moskowitz, and J. Blanc-Guenee, *Tetrahedron Letters*, 1970, 895.

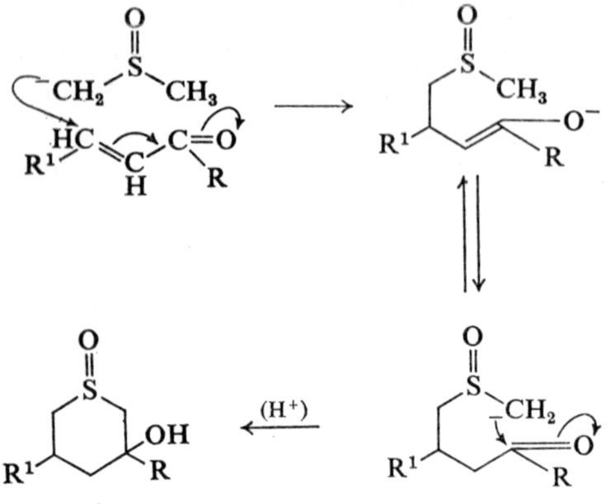

**Scheme 6**

*endo*-sulphoxide (103) showed a higher ratio of hydroxyl to oxygen loss from its molecular ion than its *exo* isomer (104), which was consistent with the ability of the molecular ion from (103) to undergo both 1,4 and 1,2 loss of hydroxyl, whereas (104) could lose hydroxyl only in a 1,2 manner.[125] The utility of mass spectral information for allocation of sulphoxide configuration has been demonstrated previously. Dodson and his co-workers [126] have confirmed that treatment of sulphonate esters of 1,2,2-trimethyl-1,3-bis(hydroxymethyl)cyclopentane with sodium sulphide proceeded without rearrangement to give 1,8,8-trimethyl-3-thiabicyclo-[3,2,1] octane (105), and not alternative thiabicyclo-octanes as claimed by others.

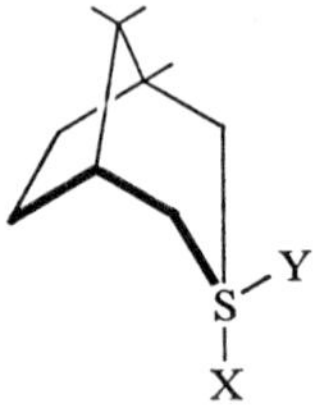

(103) X = O, Y = electron pair
(104) X = electron pair, Y = O
(105) X = Y = electron pair

There is continued interest in the photochemistry of ketosulphoxides. Irradiation of (106) in pentane with either 257 or 366 nm light gave (107).[127] It was suggested that a singlet excited state of (106) was involved

[125] P. J. Cahill, *Diss. Abs.* (*B*), 1970, **30**, 3090.
[126] R. M. Dodson, P. J. Cahill, and B. H. Chollar, *Chem. Comm.*, 1969, 310.
[127] A. Padwa, A. Battisti, and E. Shefter, *J. Amer. Chem. Soc.*, 1969, **91**, 4000.

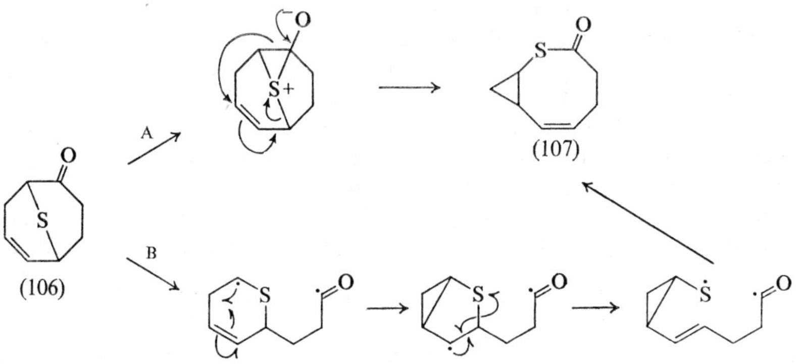

**Scheme 7**

in the rearrangement, which was rationalized either in terms of a charge-transfer intermediate (pathway A) or a Norrish Type 1 homolysis (pathway B) (Scheme 7). The saturated analogue of (106) is known to behave differently on irradiation, so the double bond played an important role in the rearrangement of (106). Photolysis of (108) in methanol gave mixtures of (109) and (110), and the δ-lactone corresponding to (111), whilst in benzene the ester derived from (108) and (111) was also obtained.[128] The product composition depended upon the excited transition. Charge-transfer excitation led almost exclusively to (109), whilst $n \rightarrow \pi^*$ excitation gave mixtures of the compounds mentioned. Acid-catalysed intramolecular cyclization of (108) afforded the diheterotwistane (112).

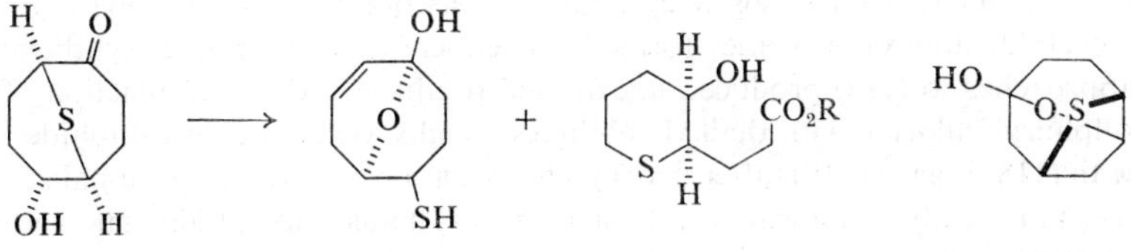

The Ramburg–Backlund reaction affords a useful method for introducing a double bond into bicyclic systems, as exemplified by the conversion of (113) and (114) into (115).[129, 130] However, the reaction failed when applied to 1-chloro-8-thiabicyclo[3,2,1]octane-8,8-dioxide [129] and 1-bromo-7-thiabicyclo[2,2,1]heptane-7,7-dioxide.[130] The sulphone (116) was converted directly into a mixture of (115) and (117) on treatment with potas-

---

[128] C. Ganter and J.-F. Moser, *Helv. Chim. Acta*, 1969, **52**, 725, 967.
[129] L. A. Paquette and R. W. Houser, *J. Amer. Chem. Soc.*, 1969, **91**, 3870.
[130] E. J. Corey and E. Block, *J. Org. Chem.*, 1969, **34**, 1233.

sium hydroxide in carbon tetrachloride,[131] presumably by way of Ramburg–Backlund rearrangement of the α-chlorosulphone (113) formed *in situ*, followed by some addition of dichlorocarbene to (115) to give (117). The ease of the transformation of (113) or (114) into (115), which must involve inversion of configuration at both reacting centres, confirmed previous conclusions regarding the stereochemical consequences of the Ramburg–Backlund reaction in acyclic systems, and it was considered probable that the relevant bond fissions and formations were synchronous. Thermal expulsion of sulphur dioxide from (116) to give *cis*-bicyclo[3,3,0]octane and

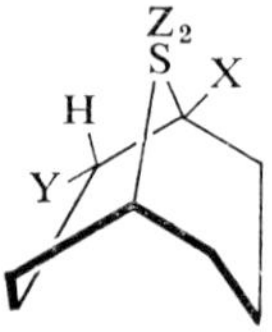

(113) X = Cl, Y = H, Z = O  (115)  (117)
(114) X = Br, Y = H, Z = O
(116) X = H, Y = H, Z = O
(118) X = H, Y = H, Z = electron pair
(118a) X = H, Y = Cl, Z =   ,,   ,,
(119) X = H, Y = Cl, Z = O
(120) X = H, Y = OBu$^t$, Z = O

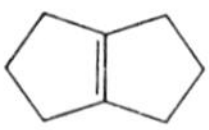

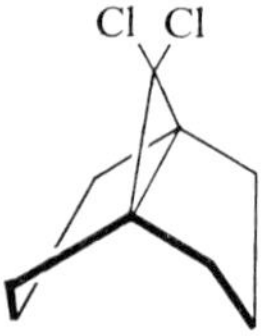

cyclo-octene occurred at 710 °C, probably *via* diradical intermediates, whilst irradiation of (116) in trivalent organophosphorus solvents (chosen to desulphurize intermediate thiyl radicals to carbon free radicals) gave the same products together with cyclo-octane.[130] Pyrolytic expulsion of sulphur dioxide from 7-thiabicyclo[2,2,1]heptane-7,7-dioxide gave mainly hexa-1,5-diene, whilst mostly cyclohexene was obtained on photolysis, no bicyclo[2,2,0]hexane being formed in either case.[130] Some synthetic approaches to (113) produced unexpected results.[129] Although reaction of sulphuryl chloride with dialkyl sulphides usually yields α-chlorosulphides, with (118) it gave (119) (after peroxyacid oxidation), which was rationalized mechanistically as indicated in Scheme 8. 8-Thiabicyclo[3,2,1]octane with sulphuryl chloride gave mainly 1,4-dichlorocycloheptane with some *endo*-2-chloro-8-thiabicyclo[3,2,1]octane-8,8-dioxide (after subsequent oxidation). The latter compound and (119) reacted with potassium t-butoxide to give the corresponding *endo*-2-t-butoxy-ethers [*e.g.* (120)] possibly *via* Michael addition of t-butoxide ion to an intermediate αβ-unsaturated sulphone (formed by dehydrohalogenation) in which the double bond occupied a bridgehead position. If so, this represented the most flagrant known violation of Bredt's rule, and explanation for this was sought in the reduction of torsional forces on the olefinic π-orbital because of the large van der Waals radius of sulphur.

[131] C. Y. Meyers, A. M. Malte, and W. S. Matthews, *J. Amer. Chem. Soc.*, 1969, **91**, 7510.

# Saturated Cyclic Compounds of Sulphur, Selenium, and Tellurium

**Scheme 8**

The ability of divalent sulphur to participate in the solvolysis of alkyl tosylates in an $R_2S$-4 manner depends critically upon the distance between sulphur and the incipient carbonium ion, according to Paquette et al.[132] The exo-tosylate (121) underwent acetolysis $3 \times 10^3$ times faster than its endo isomer (122) and $5 \times 10^3$ times faster than a reference pentacyclic alkyl tosylate [as (121), with sulphur expelled]. However, a similar rate enhancement was not observed for the exo-tosylate (123), in which the distance between sulphur and the incipient carbonium ion is slightly larger than in (121). Neighbouring group participation of the $R_2S$-4 variety was invoked to explain the enhanced reactivity of (121), which gave only the acetate of retained configuration (124), consistent with the intermediacy of the sulphonium ion (125). Extensive rearrangement occurred in the acetolysis of the other tosylates (122), (123), and (126). In agreement with these conclusions, 4-tosyloxythian underwent solvolysis more slowly than cyclohexyl tosylate (but slightly faster than its oxygen analogue),

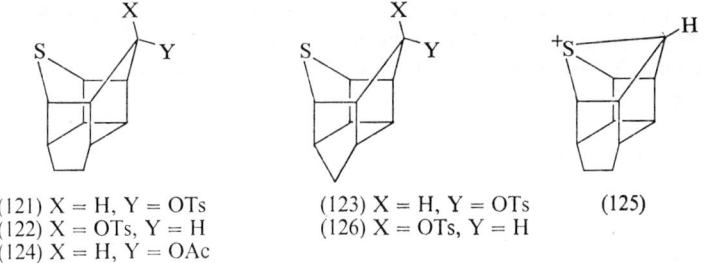

(121) X = H, Y = OTs
(122) X = OTs, Y = H
(124) X = H, Y = OAc

(123) X = H, Y = OTs
(126) X = OTs, Y = H

(125)

[132] L. A. Paquette, G. V. Meehan, and L. D. Wise, *J. Amer. Chem. Soc.*, 1969, **91**, 3231.

indicating that neighbouring group participation of sulphur involving boat conformations of the thian ring was not important.[133] The rate retardation was attributed to a transannular dipolar field effect exerted by the sulphur atom.

The crystal structure of 4-oxothian has been determined,[134] the C–S bond distance being 1·796 Å and the C–S–C angle 97°. According to dipole moment data the *cis* isomers of 2,6-diphenylthian-4-one, its 1-oxide, and its 1,1-dioxide have chair conformations with equatorial phenyl groups, the same applying to their *p*-chlorophenyl analogues,[135] whilst similar conformations were demonstrated for the isomeric carbinols obtained on treatment of *cis*-2,6-diphenylthian-4-one with methyl magnesium halide or *p*-fluorophenylmagnesium halide.[136] 2,6-Diarylthian-4-ones and their 1,1-dioxides have been converted into their oximes, hydrazones, and azines,[137] and 2,6-diphenylthian-4-one with hydrogen sulphide gave (48).[62] Keto–enol equilibria in 3-alkoxycarbonylthian-4-ones[138] and thione–enethiol equilibria in 3-thioacetylthian-2-one[32] have been investigated. The latter compound was prepared from 3-acetylthian-2-one on treatment with hydrogen sulphide and hydrogen chloride in acetonitrile, but with these reagents in ethanol 3-benzoylthian-2-one gave only 5-ethoxycarbonyl-6-phenyl-3,4-dihydrothiapyran, presumably by rapid rearrangement of the initially formed 3-thiobenzoylthian-2-one.[32] Benzylidene acetone, on treatment with alcoholic sodium sulphide, gave the thian derivative (127), a minor product of the ammonia-catalysed reaction of benzylidene acetone with hydrogen sulphide.[62] With triethylamine as catalyst at $-15\,°C$ the product was (128), which on base-catalysed cyclization gave a mixture of (127) and its C-4 epimer. The depicted stereochemistry and conformation of (127) was established by n.m.r. spectroscopy, except for configuration at C-4, which was tentatively assigned on the assumption that stronger intramolecular hydrogen bonding (detected by i.r. spectroscopy) occurred in the epimer with an equatorial C-4 hydroxy-group. Perfluorothian was a major product of thermal decomposition of a copolymer of tetrafluoroethylene and thiocarbonyl fluoride,[16] and *N*-chlorosuccinimide converted thian into 2-chlorothian, which reacted with Grignard reagents to give 2-alkyl or 2-aryl derivatives.[22] 3,6-Di-*O*-benzyl-5-thio-D-glucopyranose[12] and 6-amino-6-deoxy-5-thio-D-glucopyranose[139] have been prepared.

Selenium monochloride added to *cis*,*cis*-cyclo-octa-1,5-diene at $-20\,°C$ to give (129), which had the twin-chair conformation according to i.r.

[133] D. S. Tarbell and J. R. Hazen, *J. Amer. Chem. Soc.*, 1969, **91**, 7657.
[134] M. S. Wood, *Diss. Abs. (B)*, 1970, **30**, 3131.
[135] B. A. Arbuzov, L. K. Yuldasheva, and R. P. Arshinova, *Izvest. Akad. Nauk S.S.S.R., Ser. khim.*, 1969, 2385.
[136] R. Haller and J. Ebersberg, *Arch. Pharm.*, 1969, **302**, 677.
[137] E-S. El-Kholy and F. K. Rafla, *J. Chem. Soc. (C)*, 1969, 315.
[138] O. A. Erastov and S. N. Ignat'eva, *Izvest. Akad. Nauk S.S.S.R. Ser. khim.*, 1969, 1309.
[139] R. H. Whistler and R. E. Pyler, *Carbohydrate Res.*, 1970, **12**, 201.

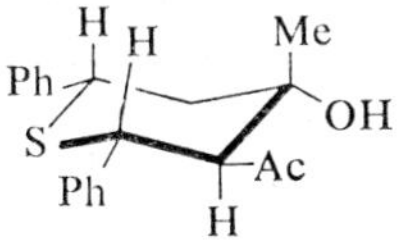


(127)

$(AcCH_2CHPh)_2S$

(128)

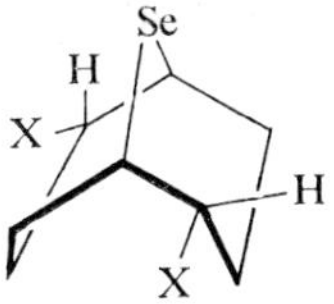


(129) X = Cl
(130) X = OH

data.[17] The compound (129) was as reactive as its sulphur analogue towards aqueous sodium bicarbonate, giving (130) with retention of configuration. With potassium cyanide, (129) gave *cis,cis*-cyclo-octa-1,5-diene, possibly by the mechanism indicated in Scheme 9. In this work, 4,8,9,10-tetrachloro-2-selena-6-thia-adamantane was obtained by addition of selenium

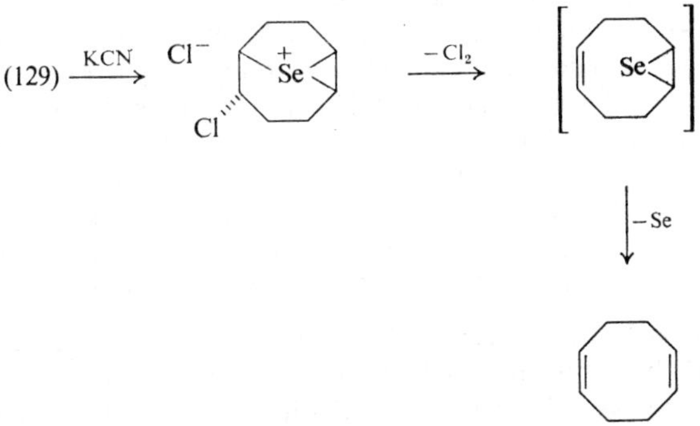


**Scheme 9**

monochloride to 4,8-dichloro-9-thiabicyclo[3,3,1]nona-2,6-diene, and others have prepared 6-thia-1,3-diazadamantane,[140] and 4,8-dialkyl-, dibromo-, and dihydroxy-2-thia-adamantan-6,6-dioxide.[141] The extent of ring puckering in thian, and its selenium and tellurium analogues, has been determined using the expression (i) relating the dihedral angle $\psi$ in the

$$\cos^2 \Psi = \frac{3}{(2+4R)} \quad \ldots\ldots(i)$$

$CH_2$–$CH_2$ fragment of the rings with the ratio $R = J_{trans}/J_{cis}$ of the vicinal coupling constants.[142] In agreement with expectation, the calculated values of ring puckering increased along the series thian, selenan, and telluran, and they were in excellent accord with those derived from *X*-ray or electron diffraction results.

[140] H. Stetter and J. Schoeps, *Chem. Ber.*, 1970, **103**, 205.
[141] S. Landa and J. Janku, *Coll. Czech. Chem. Comm.*, 1969, **34**, 2014.
[142] H. R. Buys, *Rec. Trav. chim.*, 1969, **88**, 1003.

**1,2-Dithian Derivatives.**—The reactions of nucleophiles with 1,2-dithian, its 1,1-dioxide and its tetroxide have been investigated, and compared with the corresponding reactions of 1,2-dithiolan, 1,2-dithiepan, and their 1,1-dioxides and tetroxides.[143] The tetroxides were hydrolysed more readily than the dioxides, whilst the disulphides were inert. The order of reactivity of 1,2-dithian tetroxide with nucleophiles was $PhS^- > OH^- > CN^- > SCN^- > I^-$, the reaction with sodium thiophenoxide leading quantitatively to diphenyl disulphide and disodium 1,4-butanedisulphinate. This was suggested as a potentially useful general method for the oxidation of thiols to disulphides. In general, the dithian derivatives were less readily cleaved than the dithiolan derivatives, with the dithiepan derivatives occupying a variable position. The more highly oxidised derivatives were the more readily reduced, according to polarography, but they were less susceptible to attack by thiophenol. At 280 °C, sulphur dioxide was expelled from 1,2-dithian tetroxide giving thiolan-1,1-dioxide, but in poor yield; 1,2-dithian-1,1-dioxide appeared to be inert on heating at 132 °C. 1,2-Dithian-1,1-dioxide had been obtained [43a] by oxidation of 1,2-dithian by hydrogen peroxide, and others have reported the same reaction;[43] treatment of 5-chloropentanesulphonyl chloride with sodium sulphide also gave 1,2-dithian-1,1-dioxide.[43] Pyrolysis of copolymers of tetrafluoroethylene and tetrafluorothiiran gave perfluoro-1,2-dithian as a minor product.[16] The dihedral angle about the sulphur bonds in 1,2-dithian was altered on formation of its iodine charge-transfer complex, according to u.v. data.[44] Flavins oxidised butane-1,4-dithiol to 1,2-dithian [38] (see p. 139).

**1,3-Dithian and 1,3-Diselenan Derivatives.**—Eliel and Hutchins have investigated the conformational preferences of alkyl substituents and the chair–boat energy differences in 1,3-dithians.[144] Suitable compounds (131—136) were prepared by condensations of an appropriate aldehyde with suitably substituted 1,3-propanethiols, followed by chromatographic separation of isomers. Equilibration was achieved using boron trifluoride etherate at room temperature. Conformational preferences for the methyl, ethyl, and isopropyl groups at the 2-position, determined by equilibrating (131) and (132), were similar to those observed for the same substituents in cyclohexane, but the value for the 2-t-butyl group was appreciably lower than that in cyclohexane (see Table). N.m.r. data were consistent with chair conformations (A) and (B) for compounds (131a—d) and (132a—c) respectively, but suggested a rigid boat conformation (C) for the *trans*-2-t-butyl compound (132d). Models confirmed that the boat conformation of (132d) should be rigid because pseudorotation would be inhibited by strong alkyl–ring interactions. The observed $-\Delta G^\circ$ value for the equilibrium (131d ⇌ 132d) therefore most probably represented the free-energy difference between boat and chair forms, which implied that the actual conformational preference

---

[143] L. Field and R. B. Barbee, *J. Org. Chem.*, 1969, **34**, 1792.
[144] E. L. Eliel and R. O. Hutchins, *J. Amer. Chem. Soc.*, 1969, **91**, 2703.

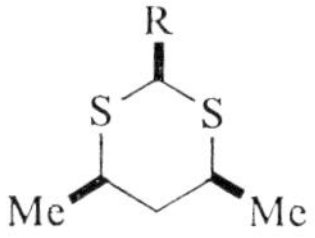

(131) 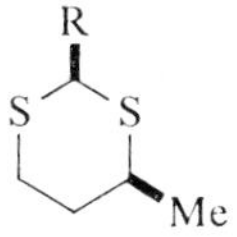         (133) 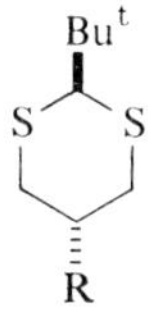         (135)

                                           R
                                           (a) Me
                                           (b) Et
                                           (c) Pr$^i$
                                           (d) Bu$^t$

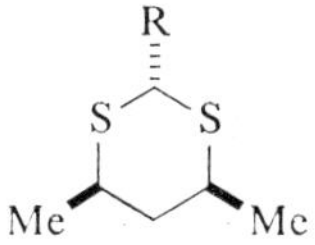

            (132) 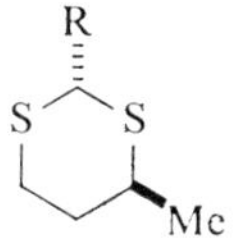         (134) 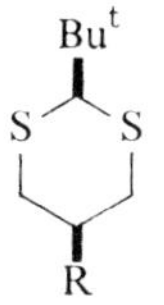         (136)

of a 2-t-butyl group was much larger than the observed 2·15 kcal mol$^{-1}$. Accordingly, the 4-methyl group in (134d) adopted predominantly an axial position in a chair conformation (D), according to n.m.r. spectroscopy, indicating that the conformational preference of the 2-t-butyl group was much larger than that of the 4-methyl group. N.m.r. data also showed that the *cis* isomers (133a—d) adopted chair conformations (as A), so that the equilibration (133d ⇌ 134d) provided a value for the conformational preference of the 4-methyl group. For the other *trans*-2-alkyl-4-methyl-1,3-dithians (134a—c), both chair conformations (as B and D) were populated, and it was shown that the $-\Delta G°$ values in the 2-alkyl and

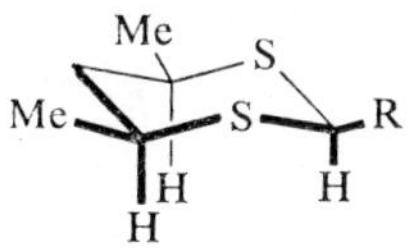

    (A) 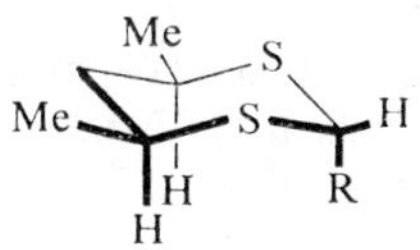           (B) 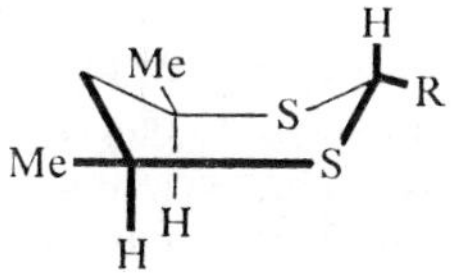           (C)

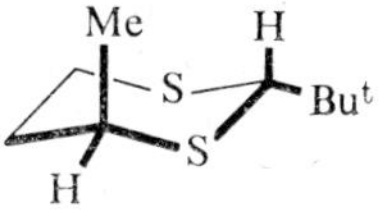

    (D) 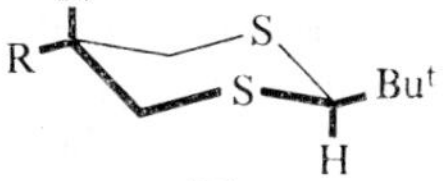           (E) 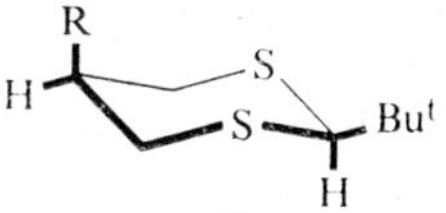           (F)

    (G)

**Table** *Conformational preferences* ($-\Delta G^{\circ}$, *in kcal mol*$^{-1}$) *for alkyl groups in 1,3-dithian and in cyclohexane*

|  | 1,3-dithian Position of substitution | | | Cyclohexane |
|---|---|---|---|---|
|  | 2 | 4 | 5 |  |
| Me | 1·77 | 1·69 | 1·04 | 1·7 |
| Et | 1·54 |  | 0·77 | 1·75 |
| Pr$^i$ | 1·95 |  | 0·85 | 2·15 |
| Bu$^t$ | 2·15 |  |  | >4·9 |

4-alkyl series were additive and consistent. For the determination of conformational preferences at C-5, the equilibration (135 ⇌ 136) was investigated. N.m.r. data suggested that the *trans* isomers (135a—d) adopted the chair conformation (E) whereas the *cis* isomers (136a—c) existed predominantly in a chair conformation (F) with an axial 5-alkyl substituent, and partly in a flexible boat conformation (G); the *cis*-2,5-di-t-butyl derivative existed entirely in the flexible boat form (G). Therefore the free-energy differences for the equilibration of the *cis*- and *trans*-5-methyl, 5-ethyl-, and 5-isopropyl-1,3-dithians represented approximately the conformational preferences of the 5-alkyl groups, but for the 5-t-butyl derivatives it represented the chair–boat energy difference. The values obtained for the conformational preferences of the 5-methyl, 5-ethyl, and 5-isopropyl groups were substantially lower than at C-2 and C-4 in 1,3-dithians, and lower than for the corresponding groups in cyclohexane (see Table). This was rationalized in terms of the smaller steric requirements of sulphur than methylene in the six-membered ring. With regard to this feature, the authors preferred to avoid terms such as the 'steric requirement' of electron lone pairs. Equilibration of *cis*- and *trans*-2,5-di-t-butyl-1,3-dithian at different temperatures furnished values for the chair–boat enthalpy difference (3·42 kcal mol$^{-1}$) appreciably lower than those for 1,3-dioxan and cyclohexane, indicating that 1,3-dithian adopted the twist-boat conformation much more readily than its oxygen and carbon analogues. This was explained in terms of the lesser non-bonded interactions in the larger 1,3-dithian ring, and the smaller torsional barriers about the C–S than C–O and C–C bonds. The entropy difference ($-\Delta S^{\circ}$) of 5·3 cal deg$^{-1}$ mol$^{-1}$ confirmed the flexibility of the twist-boat conformation. Variable temperature n.m.r. studies on (136c) and (136d) also furnished chair–boat enthalpy and entropy differences, and *cis*-2-phenyl-5-t-butyl-1,3-dithian apparently existed appreciably in the flexible boat conformation at room temperature in chloroform.

The rates of chair to chair inversion of 1,3-dithian, 1,3-oxathian, and various *gem*-dimethyl derivatives of these compounds have been determined by n.m.r. spectroscopy and compared with those of the corresponding 1,3-dioxan derivatives.[145] The energy barrier to inversion was some-

---

[145] H. Friebolin, H. G. Schmid, S. Kabuss, and W. Faisst, *Org. Magn. Resonance*, 1969, **1**, 67.

what higher for 1,3-dithian (10.4 kcal mol$^{-1}$) than for 1,3-oxathian (9.3 kcal mol$^{-1}$) and 1,3-dioxan (9.9 kcal mol$^{-1}$). 5,5-Dimethyl-1,3-dithian inverted twice as slowly as 1,3-dithian, whilst 2,2-dimethyl-1,3-dithian inverted twice as fast as the unsubstituted compound at $-70$ °C. These effects were qualitatively similar to those observed for the corresponding 1,3-dioxan derivatives, but they were far more pronounced in the latter case; the differences in magnitude were associated with the longer C–S than C–O bond distance. For the 1,3-oxathian derivatives the effects of geminal methyl groups at C-2 or C-5 were intermediate between those observed for 1,3-dithian and 1,3-dioxan. Gelan and Anteunis [146] have also found that the barrier to ring inversion in various *trans*-4-alkyl-6-methyl-1,3-dithians was *ca.* 9.5 kcal mol$^{-1}$, according to n.m.r. data. For the 6-ethyl-, 6-propyl-, 6-butyl, and 6-isobutyl derivatives the chair conformation with methyl equatorial was preferred, but flexible ring conformations of *trans*-4-isopropyl-6-methyl-1,3-dithian were populated to a significant extent. 1,3-Diselenan has been prepared, and the barrier to ring inversion found to be 8.2 kcal mol$^{-1}$ by n.m.r. spectroscopy.[147]

Fragmentation patterns in the mass spectra of 2-aryl-1,3-dithians have been elucidated by deuterium labelling studies.[57] Two important modes of fragmentation of 2-phenyl-1,3-dithian are outlined in Scheme 10, and

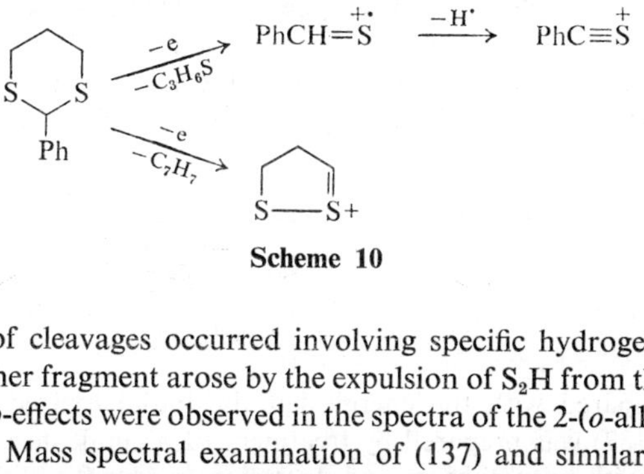

**Scheme 10**

other series of cleavages occurred involving specific hydrogen rearrangements. Another fragment arose by the expulsion of $S_2H$ from the molecular ion, and *ortho*-effects were observed in the spectra of the 2-(*o*-alkoxyphenyl)-1,3-dithians. Mass spectral examination of (137) and similar compounds showed that two important fragmentation pathways leading to ions of *m/e* 177 and 145 involved migration of hydrogen,[148] which was rationalized as indicated in Scheme 11. Abundant ions at *m/e* 164 and 132 indicated that similar cleavages occurred in the 1,3-dithian ring, whilst ions at *m/e* 105 and 120 revealed the expected α-cleavage to ions (138) and (139). Type A cleavage (Scheme 11) formally resembles a McClafferty rearrangement, the first reported example involving rearrangement to a saturated carbon atom. The mass spectrum of the compound derived from glyoxal and ethane

---

[146] J. Gelan and M. Anteunis, *Bull. Soc. chim. belges*, 1969, **78**, 599.
[147] A. Geens, G. Swaelens, and M. Anteunis, *Chem. Comm.*, 1969, 439.
[148] D. L. Coffen, K. C. Bank, and P. F. Garrett, *J. Org. Chem.*, 1969, **34**, 605.

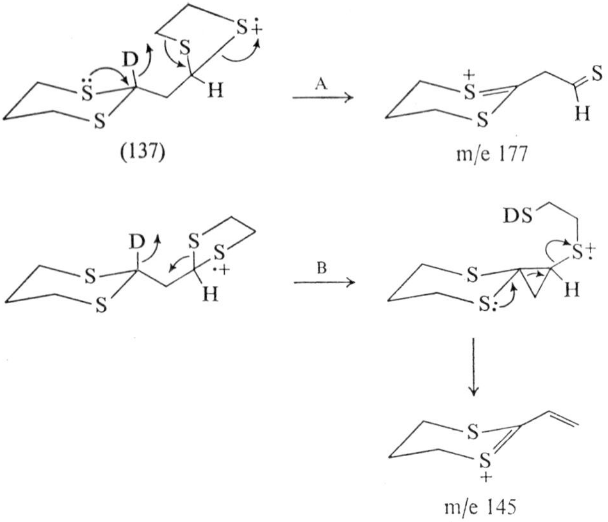

**Scheme 11**

dithiol indicated that it had structure (140) and not (141). The preponderance of Type A cleavage over Type B cleavage was attributed to the equatorial orientation of the surviving ring to the cleaved ring in the former case, and its axial orientation in the latter, but this conclusion appears tenuous in view of the very small magnitude of conformational energies compared with the energies of the initial ionized species. The compound (137) was prepared by treatment of its undeuteriated analogue sequentially with butyl-lithium and deuterium oxide. The lithio-derivative of the 1,3-dithian ring was therefore formed selectively, despite previous reports of the greater kinetic acidity of 1,3-dithiolans than 1,3-dithians. Saturated 19-methyl- and 19-nor-steroids containing a 3-keto-group have readily been differentiated by examination of the mass spectra of their 2-spiro-2′-(1′,3′-dithian) derivatives.[149] The ratio of peaks at $m/e$ 145 and 159 due respectively to $C_6H_9S_2^+$ and $C_7H_{11}S_2^+$ was greater than 20 : 1 for the 19-nor-steroids, and the reverse for the 19-methyl compounds, but in derivatives of unsaturated steroids the difference was less sharply defined.

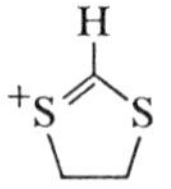

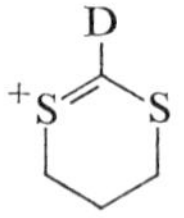

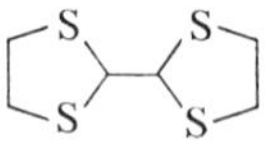

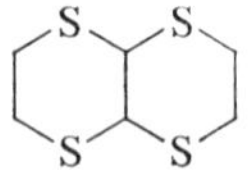

[149] J. M. Midgley, B. J. Millard, W. B. Whalley, and C. J. Smith, *Chem. Comm.*, 1969, 84.

The chemical shift difference between the intrinsically diastereotopic methyl groups in (142) has been measured,[150] this being the first measurement of time-averaged chemical shift difference due exclusively to intrinsic diastereotopism. The value was solvent dependent, having a maximal value (2·3 Hz at 60 MHz) in pyridine. Chemical shift differences between diastereotopic nuclei or groups are generally due partly to differences in conformer population, and partly to an intrinsic effect. The ingenious choice of (142) for study ensured that contributions due to conformer

(142)

population are zero, since the molecule, although chiral, has a threefold degenerate axis of rotation, along which lies the bond joining it to the prochiral centre.

The synthetic utility of 1,3-dithian continues to be exploited. 5α-Cholestan-2α,3α- and 2β,3β-oxide reacted with 2-lithio-1,3-dithian to give, after desulphuration and oxidation, both epimers of 2-methyl-3-oxo- and 3-methyl-2-oxo-5α-cholestane.[151] In contrast to the reactions of these oxiran derivatives with Grignard reagents, contraction of ring A did not occur. 2-Lithio-1,3-dithian also underwent 1,4-addition to nitrostyrenes,[152] in contrast to the exclusive 1,2-addition to αβ-unsaturated ketones; this result is surprising in view of the particular propensity of organolithium compounds to undergo 1,2-addition to αβ-unsaturated electrophilic functions. The synthetic use of 1,3-dithian in other cases may be noted.[153] In the conversion of (143) into (144) a new non-oxidative method for ketone transposition was used, involving the 1,3-dithian derivative (145) (Scheme 12).[154] In this work, attempted methanesulphonylation of the alcohol (146) gave (147) presumably by the indicated pathway (Scheme 13). Treatment of (148) with *m*-xylylene dibromide and butyl-lithium, in tetrahydrofuran, unexpectedly gave a mixture of (149) and (150), possibly by cleavage of a 1,3-dithian ring to give propane-1,3-dithiol, which reacted with the *m*-xylylene dibromide.[155]

[150] G. Binsch and G. R. Franzen, *J. Amer. Chem. Soc.*, 1969, **91**, 3999.
[151] J. B. Jones and R. Grayshan, *Chem. Comm.*, 1970, 141.
[152] D. Seebach and H. F. Leitz, *Angew. Chem.*, 1969, **81**, 1047.
[153] A. G. Brook and P. J. Dillon, *Canad. J. Chem.*, 1969, **47**, 4347; J. P. O'Brien, A. I. Rachlin, and S. Teitel, *J. Medicin. Chem.*, 1969, **12**, 1112.
[154] J. A. Marshall and H. Roebke, *J. Org. Chem.*, 1969, **34**, 4188.
[155] R. H. Mitchell and U. Boekelheide, *J. Heterocyclic Chem.*, 1969, **6**, 981; *Tetrahedron Letters*, 1969, 2013.

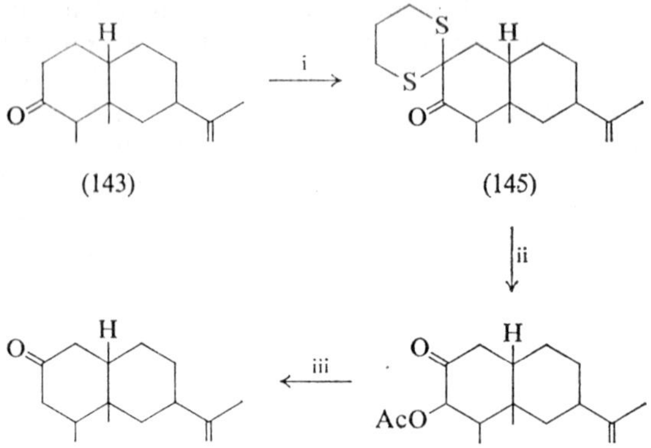

(143) → (145)

↓ ii

(144) ← iii ← (AcO derivative)

Reagents: i, (a) H·CO₂Et, NaH; (b) TsS(CH₂)₃STs, KOAc; ii, (a) LiAlH₄; (b) Ac₂O; (c) H₂O/HgCl₂; iii, Ca/NH₃

**Scheme 12**

Keto–enol equilibria in 4-methoxycarbonyl-1,3-dithian-5-one, prepared by Dieckmann condensation of dimethyl 2,4-dithiapentane-1,5-dicarboxylate, have been measured by absorption spectroscopy.[156] The spiroheterocycle (151) was obtained from tetramethylorthothiocarbonate and butane-1,4-dithiol,[49] and other spirocyclic 1,3-dithians have been prepared.[60] I.r. data indicated that the mercury–sulphur sigma bonds in mercuric halide adducts of 1,3-dithian and 1,4-dithian were formed by donation of an equatorial sulphur electron pair, but in the silver nitrate adducts both axial and equatorial electron pairs were involved.[157] There

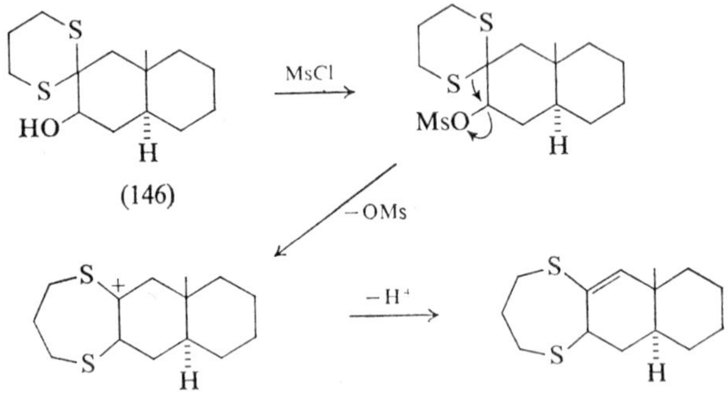

(146) —MsCl→ 

—OMs↙   —H⁺→ (147)

**Scheme 13**

[156] O. A. Erastov, A. B. Remirov, and S. N. Ignat'eva, *Izvest. Akad. Nauk S.S.S.R., Ser. khim.*, 1969, 2438.
[157] J. A. W. Dalziel, M. J. Hitch, and S. D. Ross, *Spectrochim. Acta*, 1969, **25A**, 1055, 1061.

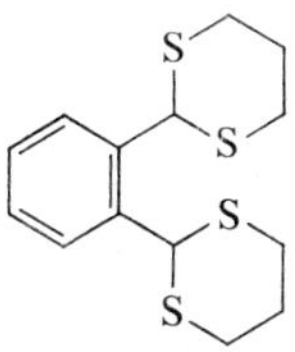

(148)

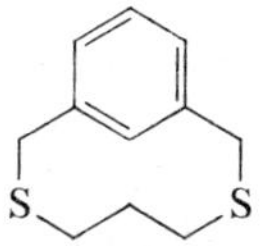

(149)

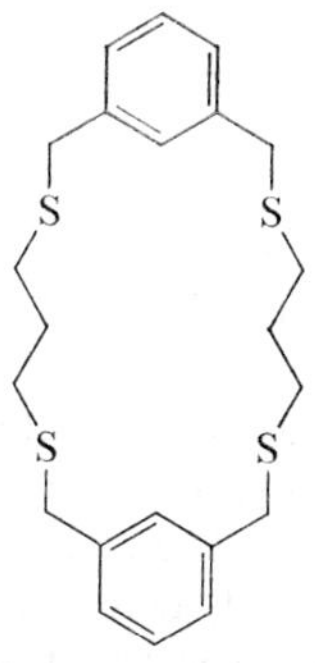

(150)

was no evidence for boat conformations of the rings in the adducts. 1,3,5,7-Tetramethyl-2,4,6,8-tetraselenadamantane has been prepared (ref. 74, Chapter 3) by the method used previously for its sulphur analogue.

**1,4-Dithian and 1,4-Diselenan Derivatives.**—Thiobenzophenone underwent photocatalysed cycloaddition to electron deficient olefins to give thietan derivatives,[158] but with electron-rich olefins either thietans or 1,4-dithian derivatives were formed, depending upon the structure of the olefin (see p. 125, Chapter 3). With olefins such as cyclohexene, *trans*-addition occurred to give (152). It is interesting that the products (153) and (154) obtained from butyl vinyl ether and butyl vinyl sulphide, respectively, adopted different preferred conformations, according to n.m.r. data. Allyl disulphide and methallyl disulphide reacted with sulphuryl chloride to give a mixture of *cis* and *trans* isomers of (155) and (156) respectively.[159] The mixtures were not separated into their components, but n.m.r. spectroscopy suggested that the *trans* isomer of (155) predominated in the mixture from allyl disulphide, and that it adopted the chair conformation with the chloromethyl groups equatorial. However,

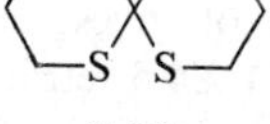

(151)

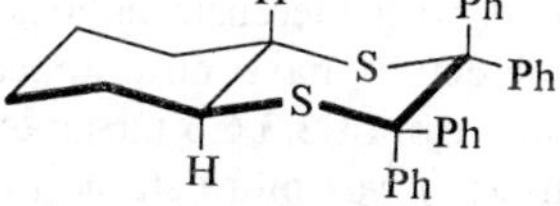

(152)

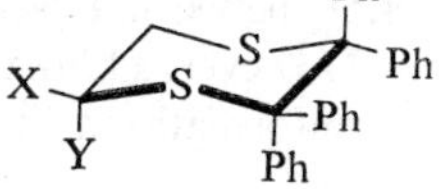

(153) X = H, Y = Bu$^t$O
(154) X = Bu$^t$S, Y = H

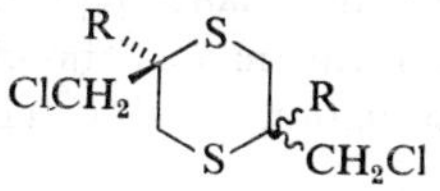

(155) R = H
(156) R = Me

[158] A. Ohno, Y. Ohnishi, and G. Tsuchihashi, *J. Amer. Chem. Soc.*, 1969, **91**, 5038.
[159] W. A. Thaler and P. E. Butler, *J. Org. Chem.*, 1969, **34**, 3389.

the *trans* isomer of (156) had the chloromethyl groups axial to a chair ring, according to n.m.r. data. Addition of acetylene and but-2-yne to ethane-1,2-disulphenyl chloride gave *trans*-2,3-dichloro-1,4-dithian and *trans*-2,3-dichloro-2,3-dimethyl-1,4-dithian, respectively, in good yield.[118] With propyne the product was 2-chloro-3-methyl-5,6-dihydro-1,4-dithiin, the formation of which was taken as an indication that the addition involved carbonium ion rather than episulphonium ion intermediates. Thermal degradation of polyethylene sulphide gave 1,4-dithian as a major product, whilst polypropylene sulphide gave 2,5-dimethyl-1,4-dithian among other products.[46] The absolute configuration of (−)-*trans*-1,4-dithian-2,3-dicarboxylic acid, determined by the quasi-racemate method,[50] was 2(*S*),3(*S*). The mass spectral fragmentation of 1,4-dithian and 1,4-oxathian were qualitatively very similar, but they differed considerably from those of 1,4-dioxan.[160] The similarity of the ionization potentials of 1,4-dithian and 1,4-oxathian suggested that primary ionization occurred at sulphur in both cases (in contrast to the behaviour of 1,3-oxathiolans, see p. 146). Furthermore, 1,4-oxathian did not give any prominent oxygen-containing ions, other than the molecular ion. Absorption spectral data for 1,4-dithian, its 1-oxide, and various of their metal complexes have been investigated;[157, 161] they were consistent with a chair conformation for 1,4-dithian, and indicated that 1,4-dithian-1-oxides co-ordinated with various metals through the electron pairs on oxygen (see 1,2-dithian section). The ring puckering in 1,4-dithian and 1,4-diselenan has been determined by n.m.r. spectroscopy, using the '*R*-value' method,[142] (see thian section).

**1,3,5-Trithian and 1,2,4,5-Tetrathian Derivatives.**—Hydrogen sulphide reacted with aromatic aldehydes to give predominantly *cis-trans*-2,4,6-triaryl-1,3,5-trithians, *e.g.* (157), according to n.m.r. examination of the crude reaction product,[162] and not predominantly the *cis,cis* isomer as thought previously. 1,3,5-Trithian exhibited rapid ring-inversion at temperatures down to −20 °C, according to n.m.r. data,[163] and the i.r. spectra of its adducts with mercuric halides and silver nitrate have been investigated.[157] Full details have now appeared of investigations of the conformational isomerism of 3,3,6,6-tetramethyl-1,2,4,5-tetrathian.[164] The twist conformation (158) was more stable than the chair by 0·5—0·6 kcal mol⁻¹, which was attributed to the absence of Pitzer strain in the twist form, and to the smaller repulsive interaction between 1,3-electron pairs in the twist form than in the chair form. However, (159) preferred the chair conformation at 0 °C (20% twist conformation), indicating the

[160] G. Conde-Caprace and J. E. Colin, *Org. Mass Spectrometry*, 1969, **2**, 1277.
[161] P. Klaboe, *Spectrochim. Acta*, 1969, **25A**, 1437; M. J. Hitch and S. D. Ross, *ibid.*, p. 1041; J. Reedijk, A. H. M. Fleur, and W. L. Groeneveld, *Rec. Trav. chim.*, 1969, **88**, 1115; J. D. Donaldson and D. G. Nicholson, *J. Chem. Soc.* (*A*), 1970, 145.
[162] E. Campaigne and M. P. Georgadis, *J. Heterocyclic Chem.*, 1969, **6**, 339.
[163] B. J. Christ and A. N. Hambly, *Austral. J. Chem.*, 1969, **22**, 2471.
[164] C. H. Bushweller, *J. Amer. Chem. Soc.*, 1969, **91**, 6019.

importance of the value of the C–S–C bond angle in influencing the conformational equilibrium, since the 'gem-dialkyl' effect could operate in the tetramethyl compound but not in (159). The high energy barrier to chair–boat interconversion (16 kcal mol$^{-1}$), in which electron pair repulsions were considered to be important, enabled the pure twist form to be obtained

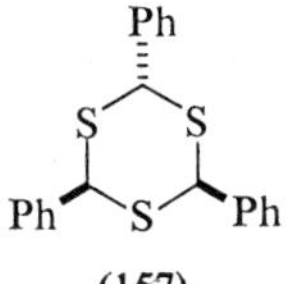

(157) 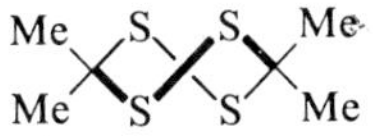    (158)

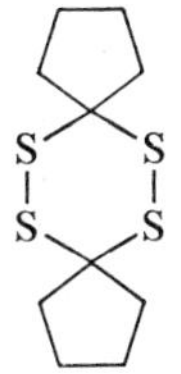

(159) 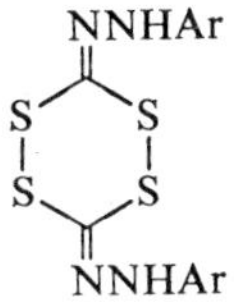    (160) 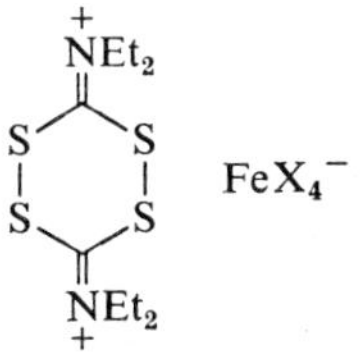    (161) X = Cl, Br

by crystallization from carbon disulphide at $-80$ °C. Measurement of the rate of conversion of this pure twist conformation to the chair by n.m.r. spectroscopy provided precise values for the activation parameters,[165] the low positive entropy of activation (2·8 cal deg$^{-1}$ mol$^{-1}$) contrasting strikingly with the high positive value (21·0 cal deg$^{-1}$ mol$^{-1}$) obtained in a parallel investigation.[166] Oxidation of arylthiocarbazic acids gave (160) as one of the products,[64] whilst halogenation of tris-(NN-diethyldithiocarbamato)-iron gave the salt (161).[167]

**Six-membered Rings containing Sulphur and Other Heteroatoms.**—Desulphuration of 1,2-dithian-1,1-dioxide with tris(diethylamino)phosphine gave 1,2-oxathian-2-oxide in excellent yield,[66] probably *via* an intermediate analogous to (53). 1,2-Oxathian-2,2-dioxide has been converted into its 2-lithio-derivative (which reacted with alkyl halides, aldehydes, and ketones) without cleavage to sulphonic acids,[68] and 3,4,6,6-tetramethyl-1,2-oxathian-2,2-dioxide was obtained from the reaction of butylene with sulphuric acid (ref. 76, Chapter 3). Because of the appreciable difference in C–O and C–S bond lengths, chair conformations of 1,3-oxathian are distorted, and it is possible to construct two chair forms which differ in the degree of puckering of the S and O sides of the ring. In the conformation where puckering is greater on the O side of the ring, models reveal a strong *syn*-axial interaction between hydrogens at

---

[165] C. H. Bushweller, J. Golini, G. V. Rao, and J. W. O'Neil, *Chem. Comm.*, 1970, 51.
[166] B. Magnusson, B. Rodmar, and S. Rodmar, *Arkiv Kemi*, 1969, **31**, 65.
[167] J. Willemse and J. J. Steggerda, *Chem. Comm.*, 1969, 1123.

C-2 and C-6, and n.m.r. investigations using the '$R$-value' method (see p. 163) confirmed that this was the less stable conformation.[168] The dihedral angles about the 4,5 and 5,6 bonds in 1,3-oxathian, and its 2,2-, 4,4-, and 6,6-dimethyl derivatives were intermediate between those for the corresponding 1,3-dioxan and 1,3-dithian derivatives. The puckering in 2,2,3,3-tetrachloro-1,4-oxathian has been determined by the same method.[142] Benzene formed a solute–solvent complex with 1,4-oxathian and 1,4-oxaselenan, in which it was in closer proximity to sulphur or selenium than oxygen, according to n.m.r. data, the signals due to the methylene groups adjacent to sulphur or selenium undergoing a larger diamagnetic shift than those adjacent to oxygen on changing solvent from carbon tetrachloride to benzene.[59] 1,4-Oxathian was cleaved by hydrobromic acid to give di(2-bromoethylsulphide), which subsequently alkylated the 1,4-oxathian *in situ* to give the corresponding disulphonium salt.[169] Conformational aspects [145] of 1,3-oxathian (see p. 166) and the mass spectral behaviour [160] of 1,4-oxathian (see p. 172) have been mentioned previously. Addition of selenium monochloride to allyl ether gave 3,5-di(chloromethyl)-1,4-oxaselenan which was stereochemically homogeneous, a *cis* configuration being assumed.[17]

1,3,2-Dioxathian-2-oxide derivatives with an equatorially orientated sulphinyl oxygen have been isolated for the first time.[170, 171] Dipole moment studies of the three diastereoisomeric forms of 4,6-dimethyl-1,3,2-dioxathian-2-oxide indicated that the two *meso* forms had the conformations (162) and (163), whilst the racemic isomer existed either in a twist conformation (164) or as a distorted chair,[170] in which the barrier to pseudorotation was substantial. It was also considered that 5-t-butyl-1,3,2-dioxathian-*cis*-2-oxide adopted a twist conformation, and did not exist in two equilibrating chair conformations as suggested previously by others. The isomer (162) rapidly epimerized to (163) under acid catalysis, in accord with the greater thermodynamic stability of the axial than equatorial orientation of sulphinyl oxygen in these systems. The three

(162) R = Me, R¹ = H, X = O, Y = electron pair
(163) R = Me, R¹ = H, Y = O, X = ,, ,,
(165) R = Pr$^i$, R¹ = Me, X = O, Y = ,, ,,
(166) R = Pr$^i$, R¹ = Me, Y = O, X = ,, ,,

(164) R = Me, R¹ = H
(167) R = Pr$^i$, R¹ = Me

---

[168] N. de Wolf and H. R. Buys, *Tetrahedron Letters*, 1970, 551.
[169] E. A. Allen, N. P. Johnson, D. T. Rosevear, and W. Wilkinson, *J. Heterocyclic Chem.*, 1969, **6**, 393.
[170] G. Wood and K. Miskow, *Tetrahedron Letters*, 1969, 1109.

diastereoisomeric forms of 5,5-dimethyl-4,6-di-isopropyl-1,3,2-dioxathian-1-oxide were separated by distillation, and their configurations and conformations were unambiguously established as (165), (166), and (167) from n.m.r. and i.r. data.[171] It is noteworthy that the i.r. band at 1223 cm$^{-1}$ (CCl$_4$) in (167) was intermediate between that (1191 cm$^{-1}$) in (165) and that (1243 cm$^{-1}$) in (166). Small proportions of the chair conformation of *trans*-4-methyl-1,3,2-oxathian-2-oxide with methyl axial and sulphinyl oxygen equatorial exist in equilibrium with the chair conformation with oxygen axial and methyl equatorial, according to the very sensitive ultrasonic relaxation technique,[172] which also provided activation parameters for ring inversion in 4-methyl-1,3,2-dioxathian-2,2-dioxide;[173] this compound adopted predominantly (95%) the chair conformation with methyl equatorial at 30 °C in xylene. Pyrolysis of (168) gave only *p*-cymene and 4-propenyltoluene,[80] in contrast to the behaviour of (62) on pyrolysis, whilst with hot pyridine (168) gave a mixture of (169) and (170).

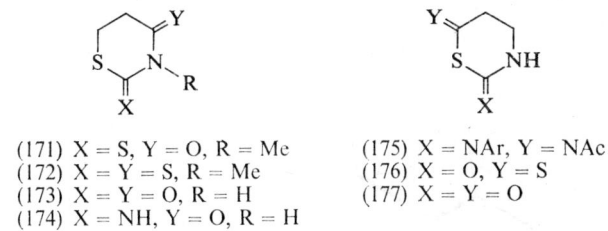

(168)

(169) R = CHO, R$^1$ = H
(170) R = H, R$^1$ = CHO

Sequential reaction of acrylic acid with methyldithiocarbamic acid and acetic anhydride–sulphuric acid gave (171), which gave (172) with phosphorus pentasulphide;[174] crotonic acid, methacryllic acid, and allyldithiocarbamic acid were used to prepare analogous compounds. The preparation and u.v. spectral characteristics of (173) and (174) have also been described.[175] Cyclization of 1-aryl-3-(β-cyanoethyl)thiocarbamides

(171) X = S, Y = O, R = Me
(172) X = Y = S, R = Me
(173) X = Y = O, R = H
(174) X = NH, Y = O, R = H

(175) X = NAr, Y = NAc
(176) X = O, Y = S
(177) X = Y = O

---

[171] M. Casaux and P. Maroni, *Tetrahedron Letters*, 1969, 3667.
[172] D. C. Hamblin, R. F. M. White, G. Eccleston, and E. Wyn-Jones, *Canad. J. Chem.*, 1969, **47**, 2731.
[173] R. A. Pethrick, E. Wyn-Jones, P. C. Hamblin, and R. F. M. White, *J. Chem. Soc.* (A), 1969, 2552.
[174] P. S. Jogdeo, V. R. Mamdapur, and M. S. Chadha, *J. Indian Chem. Soc.*, 1969, **46**, 919.
[175] N. M. Turkevich, E. V. Vladzimirskaya, and L. M. Vengrinovich, *Khim. geterotsikl. Soedinenii*, 1969, 504.

gave (175), which under basic conditions rearranged to 4-aryliminohexahydropyrimidine-2-thiones.[176] Other perhydrothiazine derivatives have been prepared.[177] Thione–enethiol tautomerism in (176) and some of its derivatives has been investigated,[178] and hydrolysis of the enethiol benzoate of (176) gave (177). The ring–chain tautomeric equilibria (178a ⇌ 178b) and (179a ⇌ 179b) lay entirely in favour of the open chain tautomer, in contrast to the case for (180a ⇌ 180b), in which the ring tautomer was

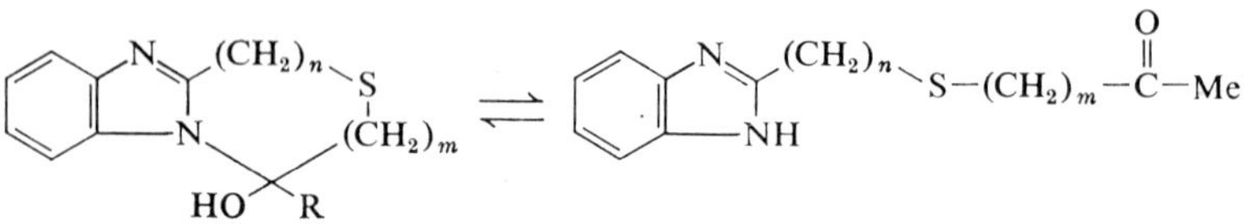

(178a) $n = 0, m = 2$
(179a) $n = m = 1$
(180a) $n = 0, m = 1$

(178b) $n = 0, m = 2$
(179b) $n = m = 1$
(180b) $n = 0, m = 1$

solely present at equilibrium.[179] Di-t-butylthiodi-imine reacted with alkyl malonyl chlorides to give the compounds (71), which were oxidised by chromium trioxide to the corresponding sulphoxides.[89] The acid-catalysed rearrangement of (71) is summarized in Scheme 3. The intense bands at ca. 250 and 290 nm in the u.v. spectra of 3,5-disubstituted tetrahydro-1,3,5-thiadiazine-2-thiones were attributed to electron transfer between the thione group and neighbouring nitrogen, and the thione group

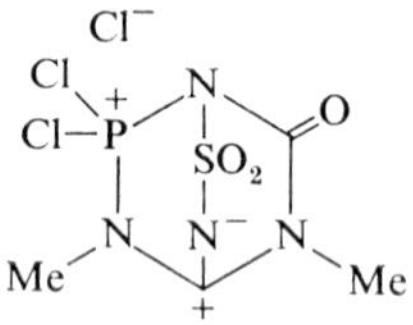

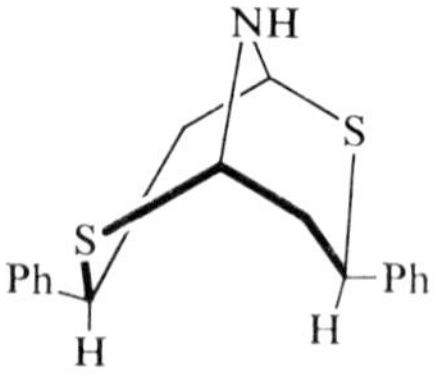

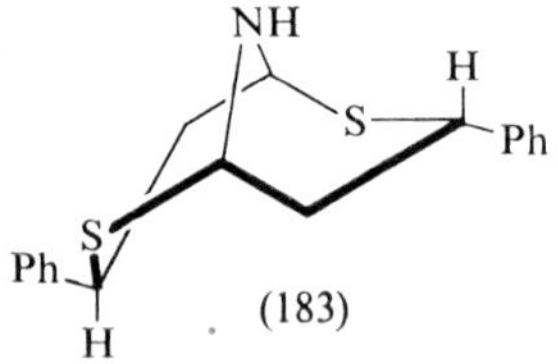

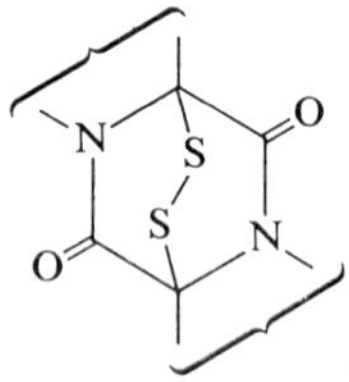

---

[176] Y. R. Rao, *Indian J. Chem.*, 1969, **7**, 772.
[177] S. Rachlin and J. Enemark, *J. Medicin. Chem.*, 1969, **12**, 1089; B. Helferich and I. Zeid, *J. prakt. Chem.*, 1969, **311**, 172.
[178] B. E. Zhitar and S. N. Baranov, *Zhur. org. Khim.*, 1969, 1893.
[179] H. Singh and S. Singh, *Tetrahedron Letters*, 1970, 585.

and neighbouring sulphur, respectively.[180] The preparation and reactions with water of (181) have been described.[181] The major products of reaction of benzylidene acetone with hydrogen sulphide and ammonia were (182) and (183), the conformations of which were established by n.m.r. spectroscopy.[62] The strained disulphide bridge in acetylaranotin, which contains the moiety (184), was cleaved readily by sulphur, sulphur dichloride, methanethiol, dimethyl disulphide, and hydrogen cyanide.[182]

### 3 Seven- and Higher-membered Rings

Although 4-hydroxythiepan readily forms a transannular hydrogen bond between oxygen and sulphur, such bonding was inhibited in (185) and considerably reduced in importance in (186) and (187), according to i.r. data.[183] However, the 4-t-butyl derivative (188) exhibited strong transannular hydrogen bonding, which was attributed to the forcing of the hydroxy-group, by the strong equatorial preference of the t-butyl group, into an axial orientation suitable for hydrogen-bond formation. In the other derivatives steric constraints apparently inhibited the adoption of favourable conformations. Transannular reactions normally require at least eight atoms in a ring, but the diol (189) readily underwent acid-catalysed rearrangement to (190), perhaps by the mechanism indicated in Scheme 14. With triphenylphosphine dibromide, 4-oxo-5-hydroxy-3,3,6,6-tetramethylthiepan gave (191), a transformation also rational in

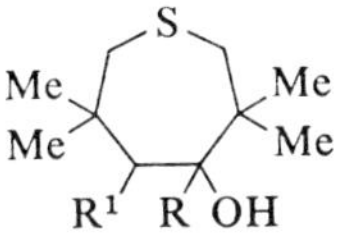

| | R | R¹ |
|---|---|---|
| (185) | H | H |
| (186) | Me | H |
| (187) | Pr$^i$ | H |
| (188) | Bu$^t$ | H |
| (189) | H | OH |

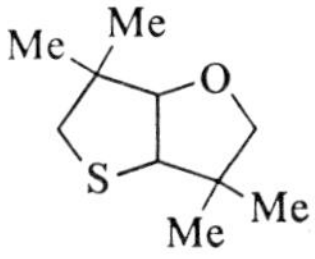

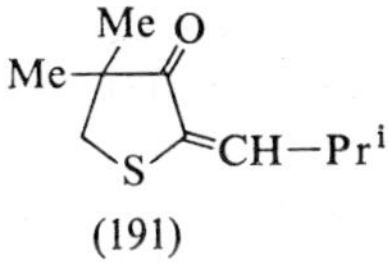

terms of a transannular reaction forming a bicyclic sulphonium ion. 3,3,6,6-Tetramethylthiepan-4,5-dione has been converted in two steps to 3,3,6,6-tetramethyl-1-thiacycloheptyne, the first seven-membered ring acetylene,[184] whilst α-chlorothiepan was formed from thiepan and N-chlorosuccinimide.[22] 5-Amino-1,2-dithiepan and (−)-(S)-4-amino-1,2-diethiepan

---

[180] P. Kristian and J. Bernat, *Coll. Czech. Chem. Comm.*, 1969, **34**, 2952.
[181] H. P. Latscha and W. Klein, *Angew. Chem.*, 1969, **81**, 291.
[182] K. C. Murdock and R. B. Angier, *Chem. Comm.*, 1970, 55.
[183] A. E. DeGroot, J. A. Boerma, and H. Wynberg, *Rec. Trav. Chim.*, 1969, **88**, 995.
[184] A. Krebs and H. Kimling, *Tetrahedron Letters*, 1970, 761.

have been synthesised,[185] whilst perfluoro-1,2-dithiepan [16] and 1,6,8,13-tetrathiaspiro[6,6]tridecane [49] have been prepared. Variable-temperature n.m.r. studies indicated that the barrier to ring inversion in 1,2,3,5,6-pentathiepan was 12·9 kcal mol$^{-1}$, but it was too low to measure in 1,2,4,6-tetrathiepan;[61] both compounds exhibited rapid pseudorotation.

Scheme 14

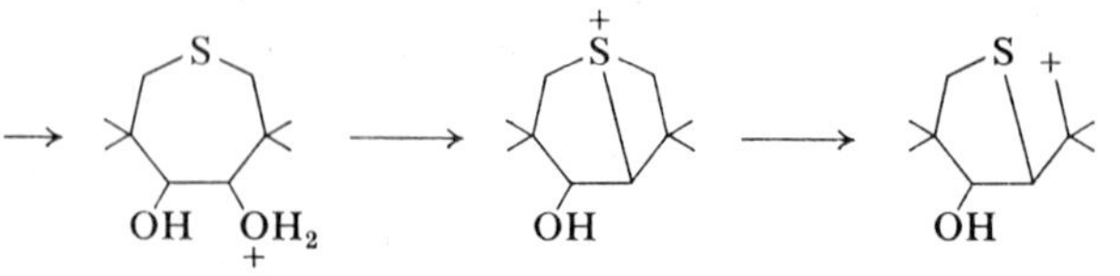

The increase in the barrier to conformational inversion reflected similar effects in six-membered rings on sequential replacement of carbon by sulphur. The mechanism of desulphuration of di- and tri-sulphides by triphenylphosphine, which is not well understood, has been illuminated by the observation that thiodehydrogliotoxin (192) was converted by triphenylphosphine into (193) with retention of configuration of the sulphur bridge, but (193) was converted into (194) with inversion at the bridge.

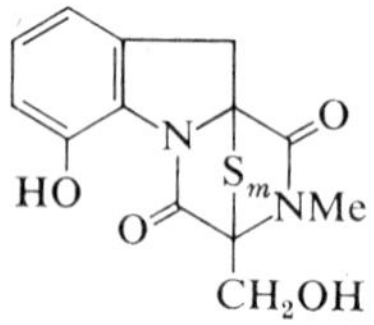

(192) $m = 3$
(193) $m = 2$
(194) $m = 1$

according to circular dichroic data.[186] Isotopic labelling experiments showed that the central sulphur atom was removed in the transformation of (192) into (193). Sulphuration of (193) by dihydrogen disulphide gave (192) with retention of configuration, suggesting that C–S bond fission did not occur, [187] and (192) was similarly converted into the corresponding tetrasulphide. Irradiation of (192) induced disproportionation into the tetrasulphide and (193). Another metabolite of *Pithomyces chartarum*, sporidesmin E, with a trisulphide bridge (as 192) has been isolated.[188]

1,3,5,7-Tetrathiocan, formed from methylene chloride and sodium sulphide, gave a tetra-anionic species on treatment with butyl-lithium, which was detected by its reaction with deuterium oxide to give its 2,4,6,8-tetradeuterio-derivative, and by its reaction with methyl iodide to give

[185] H. F. Herbrandson and R. H. Wood, *J. Medicin. Chem.*, 1969, **12**, 617, 620.
[186] S. Safe and A. Taylor, *Chem. Comm.*, 1969, 1466.
[187] S. Safe and A. Taylor, *J. Chem. Soc.* (*C*), 1970, 432.
[188] R. Rahman, S. Safe, and A. Taylor, *J. Chem. Soc.* (*C*), 1969, 1665.

mainly two isomeric 2,4,6,8-tetramethyl-1,3,5,7-tetrathiocans.[189] N.m.r. data and models suggested that the major tetramethyl derivative had the all-*cis* configuration with the methyl groups equatorial to a crown ring; the minor isomer had one axial methyl group. It was suggested that sulphur *d*-orbitals were involved in the stabilization of the tetra-anion, which in this case would involve quite extensive π-molecular orbitals. However, it is pertinent that calculations by Wolfe *et al.* on model systems suggested that sulphur *d*-orbitals are not necessarily involved in the stabilization of carbanions in thioacetal systems. In the presence of ammonia, acetophenone or its derivatives reacted with sulphur to give hexathiocan derivatives (195), the reaction proceeding *via* intermediate imines.[190] Reference may be made to large rings containing sulphur prepared incidentally to other studies.[191]

Metal complexes of macrocyclic polythioethers have received further attention. 1-Oxa-4,10-dithia-7-aza-cyclododecane readily formed octahedral complexes with nickel(II) chloride, cobalt(II) chloride, and cobalt(II) bromide, whilst 1,4,7,10,13,16-hexathiaoctadecane gave octahedral complexes [M (ligand)](picrate)$_2$ where M = Ni or Co, for which the configuration (196) was proposed.[192] The complexes derived from 1,2-dioxa-4,7,13,16-tetrathiaoctadecane were allocated configurations like (196) in which the oxygen atoms occupied apical positions. In the nickel(II) complexes of 1,4,8,11-tetrathiacyclotetradecane and related quadridentate macrocycles, the metal ion was surrounded in its equatorial plane by the sulphur atoms.[193] The *trans-anti-trans* configuration was tentatively assigned to 2,2,7,7-tetramethyl-4,5:9,10-biscyclopentano-1,3,6,8-tetrathia-

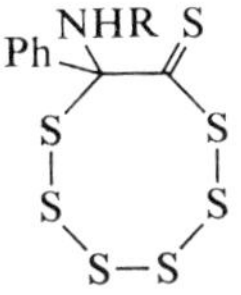

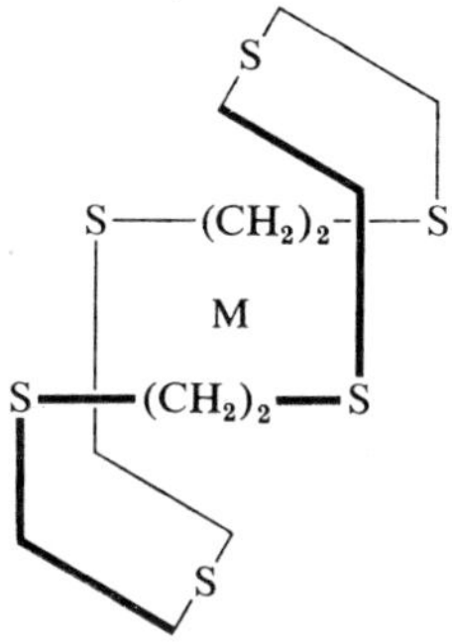

---

[189] R. T. Wragg, *Tetrahedron Letters*, 1969, 4959.
[190] F. Asinger, A. Sans, H. Offermanns, and F. A. Dagga, *Annalen*, 1969, **723**, 119.
[191] J. D. Bower and R. H. Schlessinger, *J. Amer. Chem. Soc.*, 1969, **91**, 6891; A. Miyake and M. Tomoeda, *Chem. Comm.*, 1970, 240; L. A. Paquette and R. W. Begland, *J. Org. Chem.*, 1969, **34**, 2896.
[192] D. St. C. Black, and I. A. McLean, *Tetrahedron Letters*, 1969, 3961.
[193] W. A. Rosen and D. H. Busch, *Chem. Comm.*, 1969, 148; *J. Amer. Chem. Soc.*, 1969, **91**, 4694.

cyclodecane on the basis of n.m.r. data,[194] and a re-examination of the reaction between pentane-2,4-diol and thionyl chloride indicated that a minor product was 4,6,10,12-tetramethyl-1,3,7,9-tetraoxa-2,8-dithiacyclododecane-2,8-dioxide, which on heating gave *cis*-4,6-dimethyl-1,3,2-dioxathian-2-oxide.[195]

[194] T. B. Grindley, J. F. Stoddart, and W. A. Szarek, *J. Amer. Chem. Soc.*, 1969, **91**, 4722.
[195] R. E. Lack and L. Tarasoff, *J. Chem. Soc.* (*B*), 1969, 1095.

# 5
# Thiocarbonyls, Selenocarbonyls, and Tellurocarbonyls

## 1 Introduction

This chapter reports comprehensively on the progress during January 1969—March 1970 in all aspects of the chemistry of organic compounds formally containing the C=S or C=Se linkage. This field has been the subject of an increasing volume of research in recent years, as the study of such compounds provides data concerning an important aspect of multiple bonding of sulphur and selenium in organic compounds.

In the thiocarbonyl function, bivalent sulphur is multiply-bonded to carbon, a situation comparatively rare in organic compounds because of its inherent instability. The necessary overlap of the carbon $2p$-orbitals and sulphur $3p$-orbitals is less efficient than the $2p$–$2p$ overlap of the C=O link, owing to the different spatial symmetry of $2p$- and $3p$-orbitals. The situation is even more extreme in the case of selenocarbonyl and tellurocarbonyl compounds in which the C=Se and C=Te bonds, which involve $2p$–$4p$ and $2p$–$5p$ overlap to form the $\pi$-bond, are several magnitudes less stable even than C=S. The most stable compounds containing the C=X linkage (X = S, Se, Te) tend to be those in which the carbon atom of the C=X group is directly bonded to an atom possessing available lone-pair electrons, *e.g.* a nitrogen atom in the case of thioamides. Interaction of the lone pair with the C=X group results in a polarisation towards a more stable singly-bonded form, *e.g.* $\overset{+}{C}-\overset{-}{X}$, the dipole being stabilised by delocalisation of the positive charge within the remainder of the molecule. Simple compounds containing the $\underset{H}{\overset{C}{>}}C=X$ and $\underset{C}{\overset{C}{>}}C=X$ units possess no such stabilising influence. In these cases the instability of the system is alleviated either by enolisation, where $\alpha$-CH groups are available, or by oligomerisation. Unfortunately, the very compounds which would be of most value in the determination of the physical and chemical properties of the CX group, *i.e.* those in which the CX bond order approaches 2 or a 'true' double bond, are thus the least readily available for study. Only with the recent application of spectroscopic techniques to structure elucidation and the detection of transient species has reliable data on such simple thiocarbonyl compounds been obtained.

In view of the foregoing remarks, greater emphasis will be placed in this chapter on the discussion of progress reported in the chemistry of thials, thiones, and related compounds. However, the more stable compounds containing hetero-atoms directly bonded to the thiocarbonyl carbon atom, e.g. thioamides, thioureas, thiocarboxylic acids, have considerable utility in synthesis, and the much greater volume of work in this field will also be covered in some detail, with particular emphasis on the cases where the thiocarbonyl carbon atom is not part of a heterocyclic ring.

In the case of the related seleno- and telluro-carbonyl compounds, such is the relatively small output of papers that no particular emphasis will be placed on any one class of compounds to the exclusion of others.

The chemistry of organic isothiocyanates, 6a-thiathiophthenes, 1,2(3)-dithiole-3(2)-thiones, 4(2)$H$-thiapyran- and 4(2)$H$-pyran-4(2)thiones are discussed in Chapters 2, 8, 9, and 10 respectively. Metal complexes of thiocarbonyl compounds in general fall outside the scope of this report and a full coverage has not been attempted. Some papers of particular interest will be discussed, e.g., in the context of trapping of unstable ligands, or the use of complexes as a means of elucidating the i.r. spectra of the thiocarbonyl ligands.

**Reviews.**—Here is a list of the more important reviews of thiocarbonyl and selenocarbonyl compounds published since 1945.

*Thiocarbonyl Compounds.* (*i*) Synthesis from the Reaction of Sulphur with Organic Compounds.[1] (*ii*) Chemistry,[2] Reactions,[3] Polymerisation,[4] Photochemistry,[5] As Dienophiles in 1,4-Cycloaddition Reactions.[6] (*iii*) Physical Properties,[7] Charge Distribution,[8] I.R. Spectra,[9,10] U.V. Spectra.[11] (*iv*) As Ligands in Metal Complexes.[12]

---

[1] S. Pryor, 'Mechanisms of Sulphur Reactions', McGraw-Hill Co., New York, 1962, p. 172.
[2] N. Lozac'h, *Rec. Chem. Progr.*, 1959, **20**, 23.
[3] E. E. Reid, 'Organic Chemistry of Bivalent Sulphur', Chemical Publishing Co., New York, 1965, vol. VI, p. 352.
[4] E. E. Reid, ref. 3, p. 350.
[5] A. Schönberg, 'Preparative Organic Photochemistry', Springer Verlag, Berlin, 1968, pp. 106, 444, 448.
[6] J. Hamer and J. A. Turner, '1,4-Cycloaddition Reactions', ed. J. Hamer, Academic Press, New York, 1967, p. 211.
[7] E. Campaigne, 'The Chemistry of the Carbonyl Group', ed. S. Patai, Interscience, New York, 1966, p. 934.
[8] M. J. Janssen and J. Sandström, *Tetrahedron*, 1964, **20**, 2339.
[9a] L. J. Bellamy, 'Organic Sulphur Compounds', ed. N. Kharasch, Pergamon, Oxford/New York, 1961, vol. I, p. 47; [b] 'The Infra-red Spectra of Complex Molecules', Methuen, London, 2nd ed., 1958, p. 355.
[10] F. A. Billig and N. Kharasch, *Quart. Reports Sulfur Chem.*, 1966, **1**, 227.
[11] S. F. Mason, *Quart. Rev.*, 1961, **15**, 314; R. Mayer, J. Morgenstern, and J. Fabian, ref. 18, p. 284.
[12] L. F. Lindsay, *Co-ordination Chem. Rev.*, 1969, **4**, 41.

# Thiocarbonyls, Selenocarbonyls, and Tellurocarbonyls

*Thioaldehydes.* Synthesis and Properties.[13-15]
*Thioketones.* (*i*) Synthesis and Properties.[14, 16-18] (*ii*) Some Reactions.[5, 19]
*Sulphines.* Refs. 20 and 21.
*Sulphenes.* Refs. 20—23.
*Thioamides.* (*i*) Synthesis and Properties.[24-34] (*ii*) Reactions and Mechanism.[35]
*Thioureas.* (*i*) Synthesis and Properties.[36-40] (*ii*) Reactions and Uses.[41-43]
*Thiosemicarbazides.* (*i*) Synthesis and Properties.[30, 44, 45] (*ii*) Reactions.[46, 47]
*Thiocarboxylic Acids and Derivatives.* (*i*) Synthesis and Properties.[48-51] (*ii*) Reactions.[52]

[13] A. Wagner, 'Methoden der organische Chemie', ed. E. Müller, Houben-Weyl, Berlin, 1955, vol. 9, p. 699.
[14] E. Campaigne, *Chem. Rev.*, 1946, **39**, 1; ref. 7, p. 917; *Mechanisms Reactions Sulfur Compounds*, 1967, **2**, 111.
[15] E. E. Reid, ref. 3, 1960, vol. III, p. 148; ref. 3, 1965, vol. VI, p. 368.
[16] A. Schönberg, ref. 13, p. 704.
[17] E. E. Reid, ref. 3, 1960, vol. III, p. 159; 1965, vol. VI, p. 359.
[18] R. Mayer, J. Morgenstern, and J. Fabian, *Angew. Chem. Internat. Edn.*, 1964, **3**, 277.
[19] *a* A. K. Wiersema, *Chem. Weekblad.*, 1967, **63**, 87; *b* D. S. Tarbell and D. P. Harnish, *Chem. Rev.*, 1951, **49**, 64.
[20] G. Opitz, *Angew. Chem. Internat. Edn.*, 1967, **6**, 109.
[21] N. Kharasch, B. S. Thyagarajan, and A. I. Khodair, *Mechanisms Reactions Sulfur Compounds*, 1966, **1**, 97.
[22] T. J. Wallace, *Quart. Rev.*, 1966, **20**, 67.
[23] R. B. Scott jun., ref. 9a, vol. IV.
[24] P. Chabrier and S. H. Renard, *Bull. Soc. chim. France*, 1949, D272.
[25] R. N. Hurd and G. de la Mater, *Chem. Rev.*, 1961, **61**, 45.
[26] F. Kurzer, *Chem. and Ind.*, 1961, 1333.
[27] W. Walter and K.-D. Bode, *Angew. Chem. Internat. Edn.*, 1966, **5**, 447.
[28] W. Walter, 'Organosulphur Chemistry', ed. M. J. Janssen, Interscience, New York, 1967, Chapter 13.
[29] K. A. Petrov and L. N. Andreev, *Russ. Chem. Rev.*, 1969, **5**, 21.
[30] R. Wegler, E. Kühle, and W. Schäfer, 'Newer Methods of Preparative Organic Chemistry', ed. W. Foerst, Academic Press, New York/London, vol. III, p. 1.
[31] R. Mayer and K. Gewald, *Angew. Chem. Internat. Edn.*, 1967, **6**, 294.
[32] A. Schöberl and A. Wagner, ref. 13, p. 762.
[33] E. E. Reid, ref. 3, 1962, vol. IV, p. 45; 1965, vol. VI, p. 367.
[34] J. F. Willems, *Fortschr. Chem. Forsch.*, 1963, **4**, 554.
[35] R. N. Hurd, *Mechanisms Reactions Sulfur Compounds*, 1968, **3**, 79.
[36] D. C. Schroeder, *Chem. Rev.*, 1955, **55**, 181.
[37] F. Kurzer, *Chem. Rev.*, 1952, **50**, 1.
[38] E. E. Reid, ref. 3, 1963, vol. V, p. 11.
[39] M. Bogemann *et al.*, ref. 13, pp. 799, 884.
[40] J. T. Edward, ref. 9a, 1966, vol. II, p. 287.
[41] P. C. Gupta, *Z. analyt. Chem.*, 1963, **196**, 412.
[42] H. A. Staab and W. Rohr, ref. 30, 1968, vol. V, p. 91.
[43] D. S. Tarbell and D. P. Harnish, ref. 19b, p. 74.
[44] E. E. Reid, ref. 3, 1963, vol. V, p. 194.
[45] M. Bogemann *et al.*, ref. 13, p. 905.
[46] J. F. Willems, *Fortschr. Chem. Forsch.*, 1965, **5**, 147.
[47] F. Kurzer and M. Wilkinson, *Chem. Rev.*, 1970, **70**, 111.
[48] E. E. Reid, ref. 3, 1962, vol. IV, p. 11; 1965, vol. VI, p. 364.
[49] K. A. Jensen, *Acta Chem. Scand.*, 1961, **15**, 1067.
[50] A. Schöberl, ref. 13, pp. 745, 759.
[51] M. G. Linkova, N. D. Kuleshova, and I. L. Knunyants, *Russ. Chem. Rev.*, 1964, **33**, 493.
[52] D. S. Tarbell and D. P. Harnish, ref. 19b, p. 61.

Thionocarbonates, Xanthates, and Derivatives. Refs. 53 and 54.
Thiocarbamic Acids and Derivatives. Refs. 55—59.
Seleno- and Telluro-carbonyl Compounds. Refs. 60 and 61.

## 2 Thioaldehydes

**Synthesis.**—Two series of heterocyclic thioaldehydes have been prepared by means of novel variations of the Vilsmeier–Haack aldehyde synthesis. The first series is obtained [62] by treatment of the Vilsmeier salts (1a) of various alkylindolizines with sodium hydrogen sulphide, affording the stable thioformyl derivatives (2a). Such compounds can also be obtained less readily from the related ethoxymethyleneindolizinium salts (1b), and from the corresponding formylindolizines in the latter case by treatment with phosphorus pentasulphide.

The unusual stability of the thioformyl group in compounds of type (2) undoubtedly arises from a considerable ground-state contribution of the polarised form (2b). Significantly, the authors report that attempts to obtain the dithioformyl derivative (3) afforded only polymeric material,

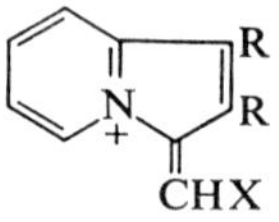

(1)
a X = NMe$_2$
b X = OEt

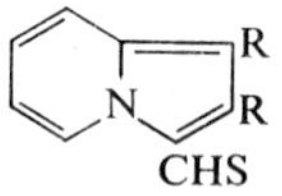

(2a)

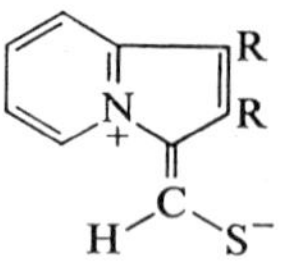

(2b)

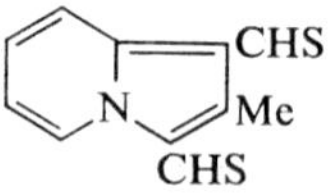

(3)

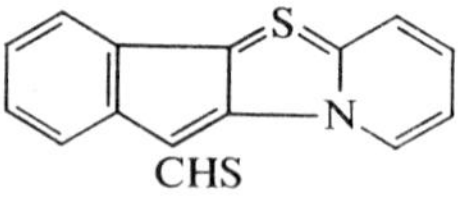

(4a)

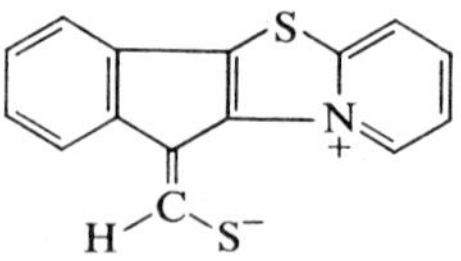

(4b)

---

[53] M. Bogemann et al., ref. 13, p. 804.
[54] D. S. Tarbell and D. P. Harnish, ref. 19b, p. 56.
[55] M. Bogemann et al., ref. 13, p. 823.
[56] E. E. Reid, ref. 3, 1962, vol. IV, p. 196.
[57] W. Walter and K.-D. Bode, *Angew. Chem. Internat. Edn.*, 1967, **6**, 281.
[58] W. Haas and K. Irgolic, *Z. analyt. Chem.*, 1963, **193**, 248.
[59] D. J. Halls, *Mikrochim. Acta*, 1969, 62.
[60] T. W. Campbell, H. G. Walker, and G. M. Coppinger, *Chem. Rev.*, 1952, **50**, 326.
[61] H. Rheinbolt, ref. 13, p. 1187.
[62] D. H. Reid and S. McKenzie, *J. Chem. Soc. (C)*, 1970, 145.

presumably because efficient stabilisation through conjugation with the electron-rich indolizine nucleus is not possible for both thioformyl groups.

The same group of workers have also reported the synthesis of stable thioformyl derivatives (4) of the 10*H*-indeno[2,1-*d*]pyrido[2,1-*b*]thiazolium and related systems *via* similar procedures based on the Vilsmeier–Haack reaction.[63] As a further interesting variation on the general procedure, dimethylthioformamide is employed as a more reactive reagent in the formation of Vilsmeier salts. The neutral structure (4a) of these thioaldehydes formally contains a tetracovalently bonded sulphur atom in the heterocyclic ring, but their stability undoubtedly arises from polarisation towards the dipolar form (4b).

**Transient Species.**—The simplest member of the thioaldehyde series, thioformaldehyde, has been detected by means of its electronic spectrum during the flash photolysis of dimethyl disulphide,[64] and has been proposed as an important intermediate in the reaction of methanesulphenyl halides with secondary amines.[65] In the latter hypothesis, transient thioformaldehyde arises from an overall elimination proceeding by two possible routes, and further reacts with secondary amines giving the diamine (5) together with hydrogen sulphide, which is the source of other by-products.

$$[HCHS] \xrightarrow{2R_2NH} (R_2N)_2CH_2 + H_2S \xrightarrow{MeSCl} Me_2S_2 + Me_2S_3$$
(5)

Two groups of workers have proposed that cyclopropene thioaldehydes, such as (6) and (7), are intermediates in photochemical reactions of thiophen derivatives, including phenyl migration in phenylthiophens,[66] and the

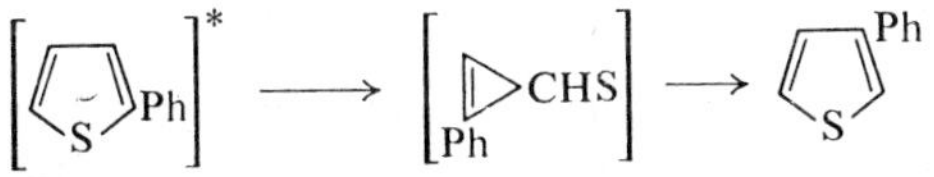

(6)

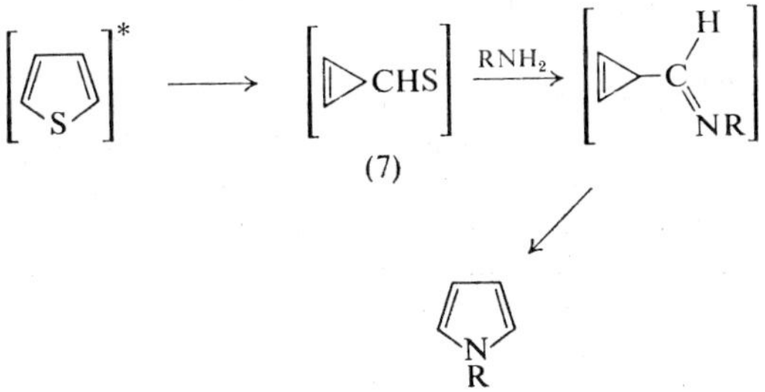

(7)

[63] J. G. Dingwall, D. H. Reid, and K. Wade, *J. Chem. Soc.* (*C*), 1969, 913.
[64] A. B. Callear, J. Connor, and D. R. Dickson, *Nature*, 1969, **221**, 1238.
[65] D. A. Armitage and M. J. Clark, *Chem. Comm.*, 1970, 104.
[66] H. Hiraoka, *J. Phys. Chem.*, 1970, **74**, 574.

transformation of thiophen to *N*-alkylpyrroles in the presence of amines.[67] Aliphatic thioaldehydes are also believed to be important intermediates in the synthesis of thiapyrans from propargyl vinyl sulphide,[68] and in postulated electron-impact-induced fragmentation pathways of malonaldehyde bisthioacetals.[69]

The participation of transient monomeric thiobenzaldehyde is thought to be an important factor in two recently reported reactions. Kuhlemann and Dittmer[70] have elucidated the mechanism of the formation of the oxathiole (9) from desyl thiocyanate (8) on the basis of the greater reactivity of thiobenzaldehyde as compared with benzaldehyde—also a likely intermediate—towards the enolate anion of (8). Transient thiobenzaldehyde also plays a part in the mechanism envisaged[71] as accounting for the

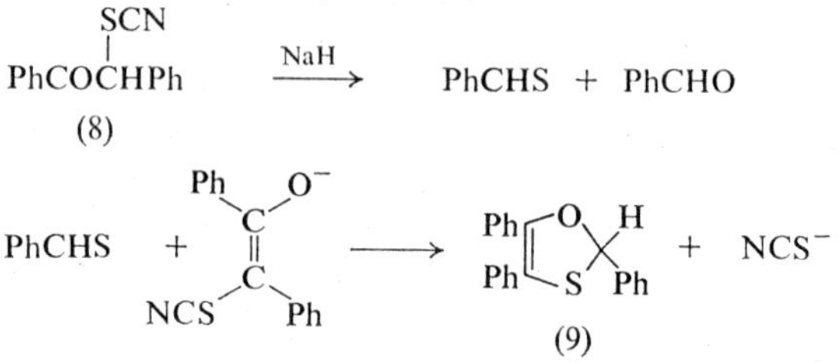

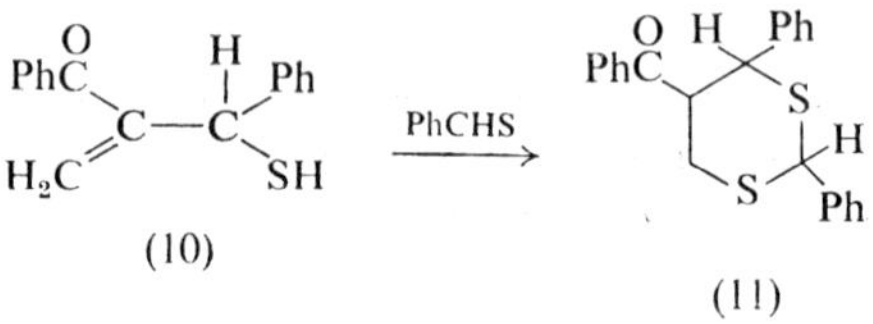

$$\text{PhCOCHPh} \xrightarrow{\text{NaH}} \text{PhCHS} + \text{PhCHO}$$
$$\overset{|}{\text{SCN}}$$
$$(8)$$

products obtained in the reaction of α-bromomethylchalcone with sodium hydrogen sulphide. In this scheme, the initially formed thiol (10) may be cleaved by hydrogen sulphide to thiobenzaldehyde, which then reacts with unchanged (10) to give the dithian derivative (11).

### 3 Thioketones

**Synthesis.**—A new general synthesis of thiobenzophenone derivatives has evolved from the study[72] of an interesting class of reactions analogous to the Wittig reaction. The new procedure involves the reaction of ketimine anions with thiocarbonyl compounds such as dimethylthioformamide or carbon disulphide, which proceeds through the 4-membered cyclic inter-

---

[67] A. Couture and A. Lablache-Combier, *Chem. Comm.*, 1969, 524.
[68] L. Brandsma and P. J. W. Schuijl, *Rec. Trav. chim.*, 1969, **88**, 30.
[69] D. L. Coffen, K. C. Bank, and P. E. Garrett, *J. Org. Chem.*, 1969, **34**, 605.
[70] G. E. Kuhlemann and D. C. Dittmer, *J. Org. Chem.*, 1969, **34**, 2006.
[71] A. Padwa and R. Gruber, *Chem. Comm.*, 1969, 5.
[72] R. Ahmed and W. Lwowski, *Tetrahedron Letters*, 1969, 3611.

mediates (12), the latter decomposing to the relevant thione. Similar derivatives have been prepared [73] by a second procedure, also involving the participation of a postulated 4-membered cyclic intermediate (14), in

$$RR^1C-N^- \;\; Li^+ \longrightarrow RR^1C\cdots N^- \;\; Li^+ \longrightarrow R^1 \!\!\!\diagdown \!\!\! C=S$$
$$\;\;\;\; | \;\;\;\; | \qquad\qquad\qquad\; \vdots \;\;\;\; \vdots \qquad\qquad\qquad R \!\!\! \diagup$$
$$\;\; S-CHNR_2^2 \qquad\quad S\cdots CHNR_2^2$$

(12) $\qquad\qquad\qquad\qquad\qquad\qquad\qquad\qquad$ + $R_2^2NH$ + $Li^+CN^-$

$$\begin{array}{c} Ar \\ \diagdown \\ Ar \diagup \end{array} C=NH \;\; + \;\; CS_2 \;\; \longrightarrow \;\; \begin{array}{c} S-C \!\!\!\diagup\!\!\!^S \\ Ar \!\!\!\diagdown | \;\;\; | \\ \;\;\;\; \diagup C-NH \\ Ar \end{array} \;\; \longrightarrow \;\; \begin{array}{c} R^1 \\ \diagdown \\ R \diagup \end{array} C=S \;\; + \;\; HNCS$$

(13) $\qquad\qquad\qquad\qquad\qquad$ (14)

which the sealed-tube reaction of 4,4′-dimethoxydiphenylmethanimine with carbon disulphide affords diaryl thiones and isothiocyanic acid, which is further transformed. In addition, the latter authors report that the presence of metallic copper in the reaction media provides a distinct rate enhancement which may be catalytic.

The interesting dithione (16), a member of the little-studied class of 1,3-dithio-1,2,3-triketones, has been prepared in 40% yield *via* the decomposition of the tetraxanthate (15) derived from tetrabromodibenzyl ketone.[74] The product (16) shows unusual properties for a thioketone, in

$$\begin{array}{c} \;\;\;\;\;\; \diagup SCSOEt \\ PhC \!\!\!\diagdown \\ \;\;\; | \;\;\; \diagdown SCSOEt \\ \;\; CO \\ \;\;\; | \;\;\; \diagup SCSOEt \\ PhC \!\!\!\diagdown \\ \;\;\;\;\;\; \diagdown SCSOEt \end{array} \qquad\qquad \begin{array}{c} PhCS \\ | \\ CO \\ | \\ PhCS \end{array}$$

(15) $\qquad\qquad\qquad\qquad$ (16)

$$RR^1C \!\!\!\diagup\!\!\!^{SCSOR^2}_{\diagdown SCSOR^2} \;\; \longrightarrow \;\; \begin{array}{c} R \\ \diagdown \\ R^1 \diagup \end{array} C=S \;\; + \;\; S \!\!\!\diagup\!\!\!^{CSOR^2}_{\diagdown CSOR^2}$$

(17)

that it is relatively unreactive towards both oxygen and diaryldiazomethanes. The same authors have also described [75] a related but more general synthesis of thiones from xanthate derivatives such as methylene bisxanthates (17), which decompose readily to thiones at room temperature. Substituted thiobenzophenones and thiofluorenone have been prepared by

---

[73] H. B. Williams, K. N. Yarborough, K. L. Crochet, and D. V. Wells, *Tetrahedron.*, 1970, **26**, 817.
[74] A. Schönberg and E. Frese, *Tetrahedron Letters*, 1969, 4063.
[75] A. Schönberg, E. Frese, W. Knöfel, and K. Praefcke, *Chem. Ber.*, 1970, **103**, 938.

this method, and the former have also been shown to occur as by-products of the decomposition of thiosulphates derived from diarylmethanes.[76]

Several of the traditional syntheses have been the subject of renewed study and application. Paquer and Vialle [77] have shown that the synthesis of simple aliphatic thiones, *via* the classical method involving the thionation of ketone analogues with hydrogen sulphide in the presence of hydrogen chloride, is best carried out by performing the reaction at low temperature ($-55\ °C$) with subsequent thermal decomposition of the resulting *gem*-dithiol. The product is normally a mixture of the required thione and its enethiol tautomer, which is separable by g.l.c. Adamantanethione (19) [78]

(18)   (19)

and a series of thioacyl-lactones and thiolactones [79] have been obtained by similar methods. The former reaction, which is carried out in ethanolic solution, is regarded by the authors as the first example of the preparation of a thione by elimination of alcohol from a *gem*-alkoxy-thiol—for example the intermediate product (18).

Interesting examples of the numerous reports of syntheses of thiones by the well-established technique of thionation of ketones using phosphorus pentasulphide are the preparation of a new series of thioacyl thiophens [80] and of 3-amino-1-aryl-prop-2-ene-1-thiones.[81] In the latter case the phosphorus pentasulphide method appears to be particularly suitable for derivatives of arylamines, the hydrogen sulphide–hydrogen chloride procedure being equally useful in the case of aliphatic members.

In a general comment [82] on the thionation of ketones using inorganic sulphides, Dean *et al.* have proposed that the synthesis would be much improved by use of silicon disulphide and boron sulphide in place of phosphorus pentasulphide. These workers believe that the greater utility of the former reagents lies both in their reactivity under less forcing conditions and also in the production of less acidic co-products during their reactions, thus minimising secondary condensations and isomerisations of the thione product.

Perhaps the most significant of various studies concerned with the synthesis of thioketones by other previously published methods or those

[76] U. Tonellato, O. Rossetto, and A. Fava, *J. Org. Chem.*, 1969, **34**, 4032.
[77] D. Paquer and J. Vialle, *Bull. Soc. chim. France*, 1969, 3595.
[78] J. W. Greidamus and W. J. Schwalm, *Canad. J. Chem.*, 1969, **47**, 3715.
[79] F. Duus, E. B. Pedersen, and S.-O. Lawesson, *Tetrahedron*, 1969, **25**, 5703, 5720.
[80] C. Andrieu, Y. Mollier, and N. Lozac'h, *Bull. Soc. chim. France*, 1969, 827.
[81] F. Clesse and H. Quiniou, *Bull. Soc. chim. France*, 1969, 1940.
[82] F. M. Dean, J. Goodchild, and A. W. Hill, *J. Chem. Soc.* (*C*), 1969, 2192.

utilising variants of known reactions—the latter group including, *inter alia*, the syntheses of previously unknown dipyrryl thiones by reaction of thiophosgene with alkylpyrroles,[83] and of enaminothiones by thiolysis of the corresponding ethoxymethylene fluoroborates [84]—is that dealing with thiocarbonyl elimination from sulphenic systems.[85] Miotti, Tonellato, and their co-workers [85] have recorded a Hammett $\rho$ value of $+3\cdot5$—one of the highest ever reported for a 1,2-elimination—for the reaction of diarylmethyl thiocyanates (20) with alkoxides giving diaryl thiones. This result

$$\begin{array}{c} Ar \\ \phantom{Ar}\diagdown \\ \phantom{Ar}C \\ \phantom{Ar}\diagup\phantom{Ar}\diagdown \\ Ar \phantom{CCCC} SCN \end{array} \longrightarrow \begin{array}{c} Ar \\ \phantom{Ar}\diagdown \\ \phantom{AA}C=S \\ \phantom{Ar}\diagup \\ Ar \end{array} + HCN$$

(20)

$$\begin{array}{c} Ar \\ \phantom{Ar}\diagdown \\ \phantom{Ar}C \\ \phantom{Ar}\diagup\phantom{Ar}\diagdown \\ Ar \phantom{CCCC} SSAr^1 \end{array} \longrightarrow \begin{array}{c} Ar \\ \phantom{Ar}\diagdown \\ \phantom{AA}C=S \\ \phantom{Ar}\diagup \\ Ar \end{array} + Ar^1SH$$

(21)

leads the authors to the conclusion that the transition-state has considerable anionic character, and further that the favoured mechanism, involving concerted rupture of C–H and S–CN bonds simultaneously with C=S formation, is not fully synchronous, *i.e.* it is $E$1cb-like. Comparison of these results with the conclusions from a study of the related synthesis of thiones from aryldiphenylmethyl disulphides (21)—also a diphenylmethyl sulphenic system—indicates that the mechanistic responses of thiocarbonyl elimination in the two systems are in accord with postulated transition-states involving balanced concerted geometry in the case of the disulphides, and $E$1cb-like structures for the more acidic diphenylmethyl thiocyanates.

**Transient Species.**—The thioketocarbene (22), previously suggested as an intermediate in various reactions, such as the thermolysis and photolysis of 1,2,3-thiadiazoles and the pyrolysis of metal dithienes, has been trapped for the first time as the metal complex (23a).[86] The structure (23b) of this complex, as elucidated by $X$-ray methods, shows a remarkable bonding bifunctionality, with the $Fe^2-C^1$ bond essentially $\sigma$, and the $Fe^1-C^1$ and $Fe^2-C^2$ bonds essentially $\pi$ in nature. The structure is regarded as being consistent with the intermediate thioketocarbene 1,3-dipolar structure (22b).

Australian workers [87] have obtained the complex bis(monothioacetylacetonato)nickel(II) from the reaction of acetylacetone with hydrogen sulphide–hydrogen chloride in the presence of nickel(II) ions. From this

---
[83] P. S. Clezy and G. A. Smythe, *Austral. J. Chem.*, 1969, **22**, 239.
[84] D. H. Gerlach and R. H. Holm, *J. Amer. Chem. Soc.*, 1969, **91**, 3457.
[85] A. Ceccon, U. Miotti, U. Tonellato, and M. Padovan, *J. Chem. Soc.* (*B*), 1969, 1084; U. Miotti, U. Tonellato, and A. Ceccon, *ibid.*, 1970, p. 325.
[86] G. N. Schrauzer, H. N. Rabinowitz, J. A. K. Frank, and I. C. Paul, *J. Amer. Chem. Soc.*, 1970, **92**, 212.
[87] G. Barraclough, R. L. Martin, and I. M. Stewart, *Austral. J. Chem.*, 1969, **22**, 891.

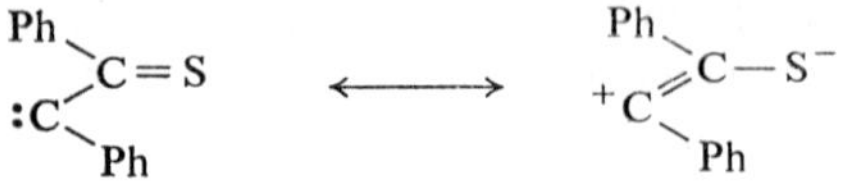

$$Ph_2C=S \quad \longleftrightarrow \quad {}^+C(Ph)_2-C(Ph)=S^-$$
(22a) (22b)

$Ph_2C_2SFe_2(CO)_6$
(23a)

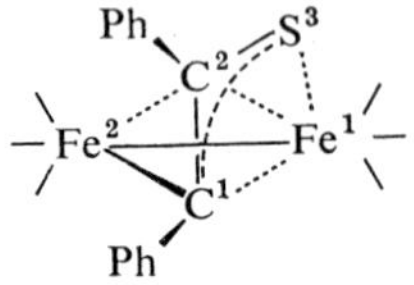

(23b) [Fe$^2$–Fe$^1$ cluster with Ph, C$^2$, S$^3$, C$^1$, Ph labels]

they conclude that transient monomeric dithioacetylacetone is not necessarily an intermediate in the formation of bis(dithioacetylacetonato) metal complexes, prepared by a similar procedure. The mechanism now proposed involves the initial formation of monothioacetylacetone in its stable enethiol form, the formation of a bis(monothioacetylacetonato) complex, and the attack of hydrogen sulphide on this complex, affording a bis(dithioacetylacetonato) complex, as successive stages.

Also of interest is the complex (24), derived from the relatively stable hexafluorothioacetone [88] and prepared directly from the thione dimer at

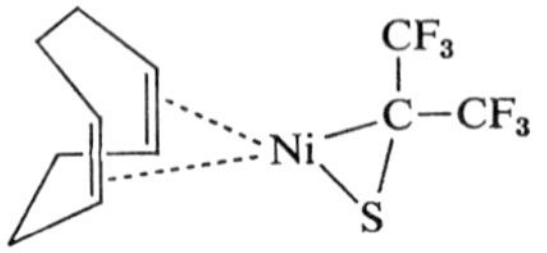

(24) [Ni complex with cyclo-octa-1,5-diene and S–C(CF$_3$)$_2$ ligand]

0 °C by reaction with bis(cyclo-octa-1,5-diene)nickel(0). Since the thermal cleavage of the dimer requires a reaction temperature of 600 °C, as expected for a thermally forbidden concerted reaction, the synthesis of (24) at 0 °C illustrates the ability of transition metals to alter the orbital-symmetry requirements of reactions.

Thione intermediates have been postulated in several recently reported reactions, one of which is the photolysis of lipoic acid (25).[89] The proposed

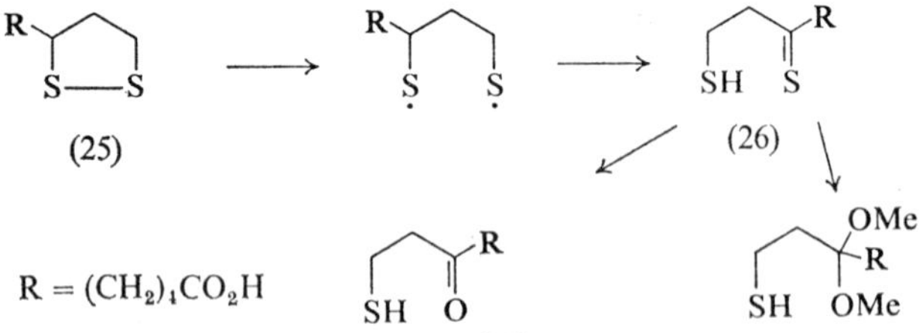

(25) R–[S–S ring] → (26) R–[SH S] → R–[SH S]

$R = (CH_2)_4CO_2H$

(27) [SH, O, R] 

(28) [SH, OMe, OMe, R]

[88] J. Browning, C. S. Cundy, M. Green, and F. G. A. Stone, *J. Chem. Soc.* (*A*), 1969, 20.
[89] P. R. Brown and J. O. Edwards, *J. Org. Chem.*, 1969, **34**, 3131.

mechanism, incorporating the transient species (26), is regarded as generally applicable to this reaction in solvents from which a hydrogen radical is not abstracted, e.g., water and methanol. In the latter solvents the respective products of the photolysis, viz. (27) and (28), are explicable on the basis of solvent attack on the thione (26).

Further examples of recently proposed thione intermediates are monothio-p-benzoquinone (29) in the formation of polythiobisphenols by condensation of phenols with sulphur,[90] the dithione (30) in the preparation of the heterocycle (31) by thionation of tetrabenzoylnaphthalene,[91] the enaminothione (32) in the thionation of the corresponding ketone leading to isothiazole derivatives,[92] and radical cations such as (33) in the electron-impact-induced fragmentation pathways of phenyl vinyl sulphides.[93]

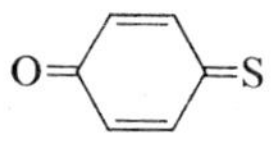

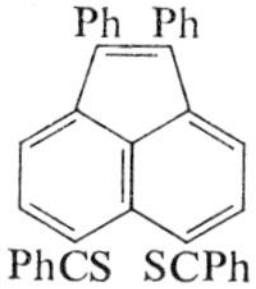

**Reactions.**—The reactivity of thioketones as nucleophiles, electrophiles, and dienophiles is well known, and a great deal of the interest shown in these compounds springs from their greater reactivity, and hence potentially greater utility for synthesis, in many such reactions, as compared with their ketone isosteres. The increased nucleophilic reactivity of thiones is well illustrated by results from a study of the nucleophilic ring-opening of tetracyanoethylene oxide (34).[94] Treatment of the latter compound with thiobenzophenone in boiling benzene, or at room temperature, affords the dicyanoethylene (36) or the trithiolan (37) respectively. It is likely that both products are derived from a common precursor, the unstable dicyanoepisulphide (35). In the case of 4,4'-bis(dimethylamino)-thiobenzophenone, the formation of the bis(dimethylamino)-derivative of (36) occurs readily at −50 °C. In contrast, a reaction of (34) with benzaldehyde at elevated temperatures gave rise to a poor yield (10%) of a 1 : 1 adduct.

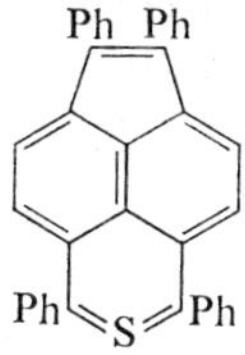

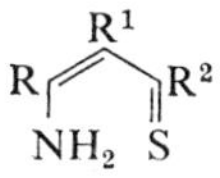

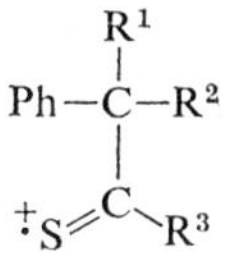

---

[90] A. J. Neale, P. J. S. Bain, and T. J. Rawlings, *Tetrahedron*, 1969, **25**, 4583, 4593.
[91] J. M. Hoffmann jun. and R. H. Schlessinger, *J. Amer. Chem. Soc.*, 1969, **91**, 3953.
[92] D. N. McGregor, U. Corbin, J. E. Swigor, and L. C. Cheney, *Tetrahedron*, 1969, **25**, 389.
[93] W. D. Weringa, *Tetrahedron Letters*, 1969, 273.
[94] W. J. Linn and E. Ciganek, *J. Org. Chem.*, 1969, **34**, 2146.

$$\underset{(34)}{\overset{NC}{\underset{NC}{>}}\!\!\!\!\overset{O}{\triangle}\!\!\!\!\overset{CN}{\underset{CN}{<}}} \xrightarrow{Ph_2CS} \left[\underset{(35)}{\overset{Ph}{\underset{Ph}{>}}\!\!\!\!\overset{S}{\triangle}\!\!\!\!\overset{CN}{\underset{CN}{<}}}\right] \xrightarrow{Ph_2CS}$$

$$\underset{(37)}{\overset{S-S}{\underset{Ph}{Ph}\overset{}{\underset{S}{\searrow}}\!\!\!\!\overset{}{\underset{Ph}{Ph}}}}$$

$$\underset{(36)}{\overset{Ph}{\underset{Ph}{>}}C=C\overset{CN}{\underset{CN}{<}}}$$

The high nucleophilic reactivity of thiocarbonyl compounds is also demonstrated [95] in the cycle of reactions (38) ⇌ (41), representing the interconversion of a series of novel azathiopyrilium salts with their oxygen analogues. The crucial factor in determining the mode of cyclisation of the acyclic intermediates (39) and (41) is the higher nucleophilic reactivity of thiocarbonyl sulphur.

$$\underset{(38)}{\overset{R^1}{R^2\underset{R^3}{\diagdown}\!\!\overset{N}{\underset{O^+}{\diagup}}\!\!\overset{}{\diagdown}R}} \xrightarrow{SH^-} \underset{(39)}{\overset{R^3CS}{R^2}\!\!\overset{R^1}{\underset{NHCOR}{>}}C=C\overset{R^1}{\underset{NHCOR}{<}}}$$

↑ HClO$_4$   ↓ HClO$_4$

$$\underset{(41)}{\overset{R^3CO}{R^2}\!\!\overset{R^1}{\underset{NHCSR}{>}}C=C\overset{R^1}{\underset{NHCSR}{<}}} \xleftarrow[H^+]{OH^-} \underset{(40)}{\overset{R^1}{R^2\underset{R^3}{\diagdown}\!\!\overset{N}{\underset{S^+}{\diagup}}\!\!\overset{}{\diagdown}R}}$$

The reactivity of thioketones towards electron-deficient species such as carbenes has been much studied in the past, and several recent papers in this field are worthy of mention. Seyferth and Tronich,[96a] in attempting to generate thiophosgene by the reaction of dichlorocarbene with elemental sulphur, found that the main product is the perchlorothiiran (42), which may arise from further reaction of the initially formed thiophosgene with dichlorocarbene, and may be an example of a general reaction involving thiones and dichlorocarbene. The authors have found support for this hypothesis in the isolation of (42), and its dichlorodiphenyl derivative (43), from the reaction of authentic thiophosgene, and of thiobenzophenone, respectively, with the dichlorocarbene precursor phenyl(bromodichloromethyl)mercury, but regard this as insufficient corroboration, in the absence of mechanistic evidence for the participation of the carbene. The fact that the same product (43) has previously been obtained from thiophosgene and

---

[95] R. R. Schmidt and D. Schwille, *Chem. Ber.*, 1969, **102**, 269.

diphenyldiazomethane does not resolve the uncertainty for, in the view of the authors, the latter reaction probably does not involve diphenylcarbene.

Kaufmann and Weininger,[96b] however, are less cautious in invoking the participation of a carbene intermediate in the photolysis of diethyl diazomalonate in the presence of thiobenzophenone. They consider that the isolation of γ-butyrolactone as a major product points to a mechanism involving the second recorded instance of an intramolecular insertion of an alkoxycarbonyl carbene. As the reaction is so strongly promoted by the presence of thiobenzophenone, the authors regard the reaction of the thione with diethyl diazomalonate as the first stage, and are investigating the possibility that the ylide (44) or the thiiran (45) are intermediates.

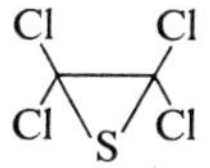

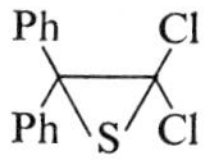

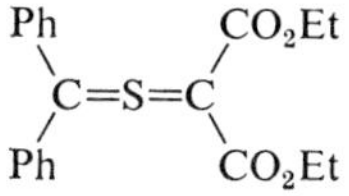

One example of the reaction between a thione and a diazo-compound has been shown not to involve the initial addition of a carbene across the CS double bond.[97a] The reaction of hexafluorothioacetone with bis-(trifluoromethyl)diazomethane leads to the episulphide (46) at high temperatures, but under milder conditions the fluorine-stabilised 1,3,4-thiadiazoline (47) is isolated. The initial stage of the reaction is then akin to a 1,3-dipolar cycloaddition of the thiocarbonyl group, rather than carbene insertion. An overall insertion of carbene into the thiocarbonyl bond of aliphatic thioketones using diazomethane in ether at 0 °C has been described,[97b] though no mechanistic details are given. This behaviour towards diazomethane contrasts with that of the corresponding enethiol tautomers, which are S-methylated under similar conditions.

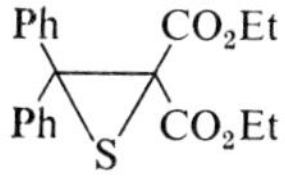

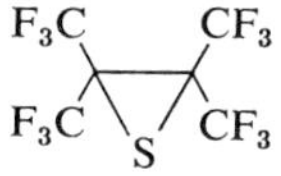

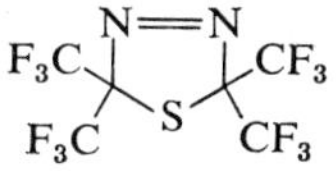

A significant clarification of the mechanism and scope of another much-studied reaction of thiones has been presented by Japanese workers in an excellent series of papers [98] on the photocycloaddition of thiones to olefins. Thiobenzophenone is employed as the thione participant, and its reactions with a series of olefins are classified into three different groups, based on a correlation of the type of product obtained with the nature of the olefin.

[96] *a* D. Seyferth and W. Tronich, *J. Amer. Chem. Soc.*, 1969, **91**, 2138; *b* J. A. Kaufmann and S. J. Weininger, *Chem. Comm.* 1969, 593.
[97] *a* W. J. Middleton, *J. Org. Chem.*, 1969, **34**, 3201; *b* D. Paquer and J. Vialle, *Bull. Soc. chim. France*, 1969, 3327.
[98] A. Ohno, Y. Ohnishi, and G. Tsuchihashi, *Tetrahedron Letters*, 1969, 161, 283; *J. Amer. Chem. Soc.*, 1969, **91**, 5038.

Case I olefins contain electron-releasing groups and produce 1,4-dithians, *e.g.* (49) on reaction with thiobenzophenone under 5890 Å or 3660 Å light; Case II olefins contain electron-withdrawing groups and afford thietans, *e.g.* (51) stereospecifically, under 3660 Å light only; Case III contain electron-releasing groups but, in contrast to Case I, give rise to thietans,

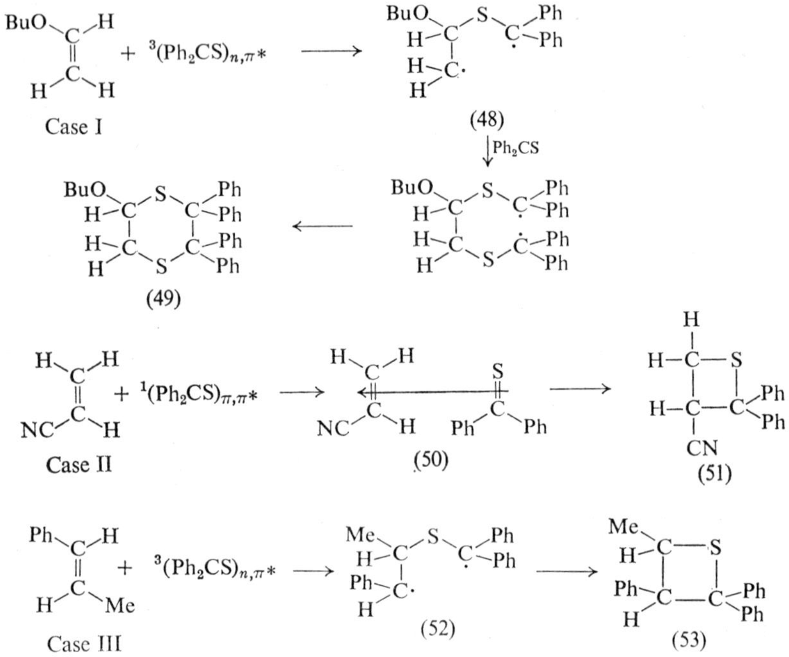

*e.g.* (53), under 5890 Å light. The scheme is then rationalised in a most elegant manner by consideration of the likely modes of reaction of the various excited states of thiobenzophenone with each olefin-type. Thus, the reaction of Case I and Case III olefins, proceeding under excitation by 5890 Å light, must involve the triplet ($n, \pi^*$) state of the thione, giving rise to a diradical intermediate such as (48) or (52). The distinction between Case I and Case III arises out of steric factors, which inhibit attack by a further molecule of thione on the diradical (Case III), thereby promoting non-stereospecific cyclisation yielding thietans such as (53). The reaction of Case II olefins is possible only under 3660 Å light, hence involves the singlet ($\pi, \pi^*$) excited state of thiobenzophenone, and proceeds *via* nucleophilic attack of the latter species on the electron-deficient double bond, leading to thietans such as (51) with complete retention of the olefin configuration.

The special bonding characteristics of diphenylcyclopropenethione, which to a large extent exists as the dipolar form (54), is illustrated in a report [99] of its reaction with electron-rich olefins. Electrophilic attack of the olefin on the negative cyclopropene ring, leading to the dipolar adduct (55), is the preferred reaction pathway, rather than cycloaddition reaction of the CS bond.

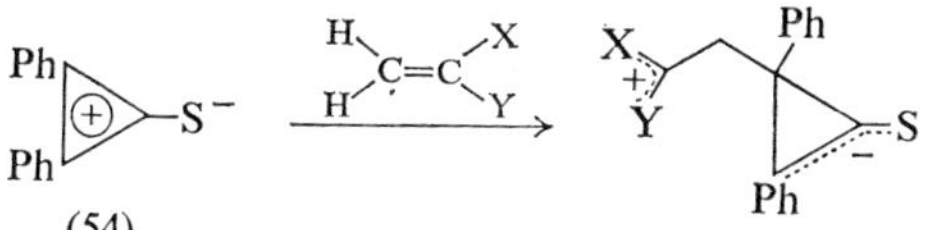

X, Y = NMe$_2$, OEt, SMe

Ohno and his co-workers have investigated [100] the thermal 1,4-cycloaddition reactions of thiobenzophenone with open-chain dienes, and have elucidated for the first time the isomer ratios of the dihydrothiopyran products (56) and (57). The results of this study confirm earlier comments

X = H, Me, Cl
R = Me, Ph

of other workers on the greatly increased reactivity of thiocarbonyl relative to carbonyl compounds in such reactions, and also show that the reactivity of the thiocarbonyl compound depends upon its structure, e.g., thiobenzophenone reacts twice as fast as thioacetophenone. As the reaction is not affected by radical scavengers, the authors conclude that it does not proceed by an out-of-cage free-radical mechanism, but find no evidence to distinguish between in-cage free-radical and ionic mechanisms.

A successful synthesis of 2-thiabicyclo[2,2,1]heptanes using the known 1,4-cycloaddition reaction between thiophosgene and cyclopentadiene,[101] and an unsuccessful attempt to synthesise derivatives of benzothiete via the 1,2-cycloaddition of benzyne, generated from benzenediazonium-2-carboxylate, and thiobenzophenone [102] have been reported. Other recent studies on the reactivity of thiones include the preparation of 2,3;6,7;-2′,3′;6′,7′-tetrabenzoheptafulvalene (59) [103] from the thione (58) by the copper-metal-promoted desulphuration–coupling procedure, and the reactions of thiobenzophenone [104] and thiobenzoyl chloride [105] with organometallic compounds.

[99] J. Sauer and H. Krapf, *Tetrahedron Letters*, 1969, 4279.
[100] A. Ohno, Y. Ohnishi, and G. Tsuchihashi, *Tetrahedron*, 1969, **25**, 871.
[101] C. R. Johnson, J. E. Keiser, and J. C. Sharp, *J. Org. Chem.*, 1969, **34**, 860.
[102] D. C. Dittmer and E. S. Whitman, *J. Org. Chem.*, 1969, **34**, 2004.
[103] A. Schönberg, U. Sodtke, and K. Praefcke, *Chem. Ber.*, 1969, **102**, 1453.
[104] H. Tada, K. Yasuda, and R. Okawara, *J. Organometallic Chem.*, 1969, **16**, 215.
[105] H. Lindner, H. Weber, and H.-G. Karmann, *J. Organometallic Chem.*, 1969. **17**, 303.

## 4 Thioketens

Thioketen and derivatives thereof are very unstable species, but have been postulated as intermediates in various reactions, including the formation of dithioesters from alkynethiolates and the formation of thiatriazoles from 1-acetylthioalkynes (see ref. 108 for some leading references). A rare relatively stable derivative, bis(trifluoromethyl)thioketen (60) has recently been prepared, and a report [97a] of its reaction with bis(trifluoromethyl)diazomethane has appeared during 1969. The reaction proceeds in an analogous manner to that of hexafluorothioacetone (see above), leading to the thiadiazoline (61) and, by thermal decomposition, the thiiran (62)—regarded as the first example of an allene episulphide. The product (61) appears to be more stable than the analogous product (47) from the hexafluorothioacetone reaction, being stabilised by conjugation with the olefinic double bond. Two instances of the trapping of thioketenic species as organometallic complexes have been lately recorded. The first examples of complexes of (60), such as (63), have been prepared [106] by reaction of various arylphosphine-containing low-valent metal complexes with mixtures of fluorinated tetrathian and trithiolan. The latter compounds are precursors of the recently prepared dimer (64) of the thioketen (60), but reaction of (64) with the samẽ metal complexes did not afford a complex

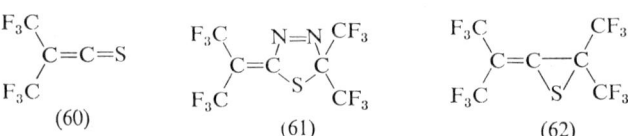

[106] M. Green, R. B. L. Osborn, and F. G. A. Stone, *J. Chem. Soc. (A)*, 1970, 944.

similar to (63), although this procedure had been successful in the case of hexafluorothioacetone dimer (see above ref. 88). As an interesting corollary to the above results, other workers [107] have reported that carbon sub-sulphide ($C_3S_2$) reacts with tetrakis(triphenylphosphine)platinum to afford a complex of structure (65).

There have been reports that thioketen derivatives may be concerned as unstable intermediates in the base-catalysed dimerisations of $N$-alkyl- and the polymerisation of $N$-acyl-3-isothiazolones,[108] in the formation of cycloadducts from aryldiazomethanes and carbon disulphide,[109] and in the preparation of thioamides from the reaction of arylacetylenic sulphides and amines.[110]

### 5 Sulphines

Sulphines, *i.e.* sulphoxides derived from thiones, may be regarded as one of the simpler classes of thione derivatives and are as such of considerable theoretical interest. They are also of some importance as reaction intermediates, though not to the same extent as the related sulphenes (see below, section 6). There have been several recent reports of their synthesis, including the claim [111] of a novel route to the sulphine (66) *via* interaction of *m*-methoxyphenol with various compounds containing the trichloromethanesulphonyl moiety. The authors also believe that sulphines analogous to (66) are responsible for the yellow colour obtained in the standard analytical test for captan, by fusion with resorcinol. This conclusion has been subsequently refuted, however, other workers [112] elucidating the structure of the captan product as (67) by $X$-ray analysis.

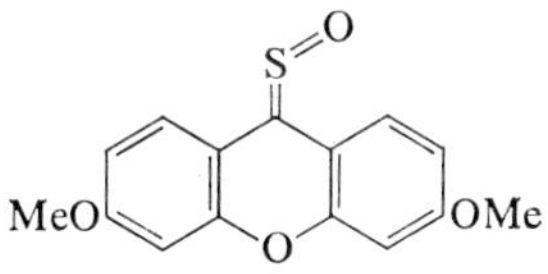

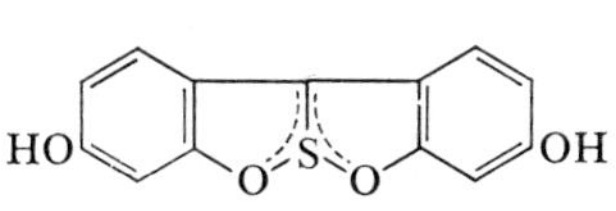

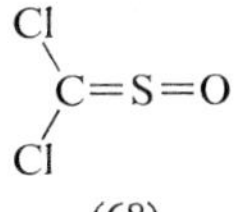

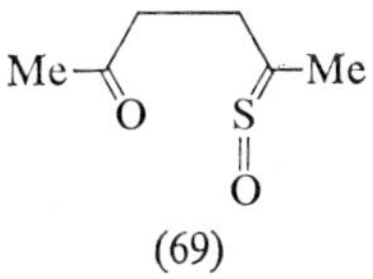

[107] A. P. Ginsberg and W. E. Silverthorn, *Chem. Comm.*, 1969, 823.
[108] A. W. K. Chan, W. D. Crow, and I. Gosney, *Tetrahedron*, 1970, **26**, 1493.
[109] J. E. Baldwin and J. A. Kapecki, *J. Org. Chem.*, 1969, **34**, 725.
[110] M. L. Petrov, B. S. Kunin, and A. A. Petrov, *Zhur. org. Khim.*, 1969, **5**, 1759.
[111] G. D. Thorn, *Canad. J. Chem.*, 1969, **27**, 2898.
[112] I. Pomerantz, L. Miller, E. Lustig, D. Mastbrook, E. Hansen, R. Barron, N. Oates, and J.-Y. Chen, *Tetrahedron Letters*, 1969, 5307.

Dichlorosulphine (68) has been synthesised by two different routes, one of which [113] utilises the most common general synthesis of sulphines, *i.e.* peracid oxidation of the relevant thione—in this case thiophosgene. The sulphine structure of the product is verified through its reaction with diazofluorene, yielding 9-dichloromethylenefluorene and the sulphine derived from thiofluorenone. This (to date) simplest stable sulphine (68) has also been isolated as the product of the aqueous hydrolysis of trichloromethanesulphenyl chloride.[114] Zwanenberg, Thijs, and Strating [113] note the great reactivity of (68) with amines, thiophenol, and with cyclopentadiene in a cycloaddition reaction, while both groups record a sluggish decomposition reaction in aqueous solution and a strong lachrymatory property. The acyclic sulphine derivative (69) has also been obtained independently by two groups of workers [115] from the photoreaction of 2,5-dimethylthiophen with singlet oxygen. This result contrasts with the previously supposed inertness of thiophen to photosensitised autoxidation.

By performing the peracid oxidation of an unsymmetrical diaryl thione, Mangini and his co-workers [116] have shown the possibility of obtaining 1:1 mixtures of sulphine geometric isomers, from which the *syn*-isomer (70) can be isolated and purified. It is evident, therefore, that the C=S=O system is a rigid, non-linear group in aromatic sulphines, as in thioacyl chloride *S*-oxides and in thiono-oxides of dithiocarboxylic acids. Both

(70)      (71)      (72)

*syn*- (71) and *anti*- (72) isomers of 1-napthyl phenyl sulphine have subsequently been obtained from the peracid oxidation of 1-naphthyl phenyl thione.[117] The latter paper also comprises a study of the photodesulphuration of (71) and (72), in which the mechanism of sulphine desulphuration is elucidated for the first time.

The participation of sulphine intermediates has been reported in the synthesis of thiazolo[4,5-*d*]pyrimidine derivatives from uracils,[118] and in the formation of benzothiazoles from *o*-toluidine.[119] The parent sulphine (74) has been shown to be a fragment in the mass spectral fragmentation of the sulphoxide (73).[120]

[113] B. Zwanenberg, L. Thijs, and J. Strating, *Tetrahedron Letters*, 1969, 4461.
[114] J. Silhanek and M. Zbirovsky, *Chem. Comm.*, 1969, 878.
[115] C. N. Skold and R. H. Schlessinger, *Tetrahedron Letters*, 1970, 791; H. H. Wassermann and W. Strehlow, *ibid.*, 1970, 795.
[116] S. Ghersetti, L. Lunazzi, G. Maccagnani, and A. Mangini, *Chem. Comm.*, 1969, 834.
[117] R. H. Schlessinger and A. G. Schultz, *Tetrahedron Letters*, 1969, 4513.
[118] I. M. Goldman, *J. Org. Chem.*, 1969, **34**, 3285.
[119] M. Davis and A. W. White, *J. Org. Chem.*, 1969, **34**, 2986.
[120] W. R. Oliver and T. H. Kinstle, *Tetrahedron Letters*, 1969, 4317.

PhCOCH₂SOMe   ⟶   Ph—C(⁺OH)(=CH₂) + CH₂=SO
(73)                                      (74)

## 6 Sulphenes

Despite the importance of sulphenes, otherwise thione $S$-dioxides, as reactive intermediates in mechanistic studies, scant success has attended attempts at their synthesis. Nevertheless, the direction of the reaction of sulphenes with activated olefins, and results of molecular orbital calculations on the charge distribution within the sulphene molecule, have led Paquette and his co-workers [121] to conclude that stabilisation might be achieved by delocalisation of the predicted negative charge on the sulphene carbon atom. Thus far, however, attempts by this group to synthesise stable sulphenes possessing electron-withdrawing substituents, such as carbalkoxy- or cyclopentadienyl groups, have failed. In the belief that the deficiency of the charge stabilisation approach lies in the simultaneous necessity of minimising polymerisation, the authors suggest that a study of sterically-hindered sulphenes may be more fruitful.

De Mayo and his co-workers have investigated the possibility of generating and detecting sulphene in the gaseous phase. The results of this study [122] indicate that flash thermolysis of thietan and thiete $S$-dioxides probably produces parent sulphene (75) and the olefinic derivative (76) respectively,

H₂C=SO₂          R—C(=SO₂)—CH=    R = H or Ph
(75)                  (76)

but that the initial products undergo further transformations in the hot-zone, thus eluding trapping procedures. In comparable experiments with chlorosulphonylacetic acid, however, the most significant product of methanol-trapping is methyl methanesulphonate. The incorporation of deuterium into the latter compound by use of [²H]methanol as trapping agent requires that some species of formula $CH_2SO_2$ is generated in the flash-zone, travels to the cold trap, and there reacts with [²H]methanol at 77 K. In the opinion of Paquette and his co-workers, this entity must be sulphene (75), which is thus shown to be a discrete volatile compound capable of a finite existence in the gaseous phase. The most common method of generating sulphene in solution, however, remains the base-catalysed elimination of hydrogen halide from alkanesulphonyl halides. This approach

---

[121] L. A. Paquette, J. P. Freeman, and R. W. Hauser, *J. Org. Chem.*, 1969, **34**, 2901.
[122] J. F. King, K. Piers, D. J. H. Smith, C. L. McIntosh, and P. deMayo, *Chem. Comm.*, 1969, 31; C. L. McIntosh and P. deMayo, *ibid.*, 32; J. F. King, P. deMayo, and D. L. Verdun, *Canad. J. Chem.*, 1969, **47**, 4509.

has now been used [123] to indicate the transient existence of disulphene (77), generated from the dehydrochlorination of methanedisulphonyl chloride and trapped with keten diethyl acetal to afford a spiro-bithietan tetroxide. The mechanism of the formation of (77) is still unclear, however, the

$$O_2S=C=SO_2$$
(77)

authors being unable to distinguish between a one- or two-stage elimination on present evidence.

Evidence to corroborate the generally accepted participation of sulphenes in the dehydrohalogenation of alkanesulphonyl halides possessing at least one α-hydrogen atom has been presented [124a] in a mechanistic study of the reaction of methanesulphonyl chloride with triethylamine in the presence of aniline. The composite rate equation thus obtained shows that part of the reaction is zero order in respect of the trapping agent, requiring the formulation of an intermediate which reacts with aniline but whose influence on the rate is not detectable. A sulphene intermediate produced by an E2 elimination is most closely in accord with the experimental evidence. German workers [124b] have drawn a similar conclusion from results of a study of deuterium isotope effects in the reaction of alkanesulphonyl chlorides with sodium phenolate.

Results of several studies of the reactions of sulphene generated from methanesulphonyl chloride have been recorded. Most significant is the report [125] of the first isolation of the dimer of sulphene (78), which is

(78)

obtained if the generation is carried out at low temperature. Other papers record the isolation of novel vinyl methanesulphonates from the reaction of sulphene with highly halogenated ketones;[126] an improved synthesis of thietan and thiete derivatives *via* the reaction of sulphene and NN-dimethylvinylamine;[127] studies of the stereoselectivity of sulphene attack on an asymmetric enamine [128] and on bicyclic enamines;[129] the isolation of

[123] K. Hirai and N. Tokura, *Bull. Chem. Soc. Japan*, 1970, **43**, 488.
[124] *a* J. F. King and T. W. S. Lee, *J. Amer. Chem. Soc.*, 1969, **91**, 6524; *b* K. Gunther, M. Hampel, W. Höbold, H. Hübner, G. Just, J. Müller-Hagen, W. Pritzkow, W. Rolle, M. Wahren, and H. Winter, *J. prakt. Chem.*, 1969, **311**, 596.
[125] G. Opitz, and H. R. Mohl, *Angew. Chem. Internat. Edn.*, 1969, **8**, 73.
[126] J. R. Norell, *Chem. Comm.*, 1969, 1291; W. E. Truce and L. K. Liu, *Tetrahedron Letters*, 1970, 515.
[127] P. L. F. Chang and D. C. Dittmer, *J. Org. Chem.*, 1969, **34**, 2791.
[128] L. A. Paquette and J. P. Freeman, *J. Amer. Chem. Soc.*, 1969, **91**, 7550.
[129] J. F. Stephen and E. Marcus, *J. Org. Chem.*, 1969, **34**, 2535.

cyclic adducts from the reaction of sulphene with a mesocyclic dienamine,[130] pyridine,[131] alkynylamines,[132] benzoyl imines,[133] aryl nitrones, and N-phenylhydroxylamines;[134] and the isolation of an acyclic adduct from the reaction of sulphene with an enaminoketone.[135] Derivatives of sulphene have also been quoted as possible intermediates in the hydrolysis of substituted α-toluenesulphonic acid sulphones,[136] in the reaction of amide and ester derivatives of sulphoacetic acid with nucleophiles,[137] and in studies of the synthesis and chlorinolysis of 2H-1,2,3-benzothiadiazine-1,1-dioxide.[138]

## 7 Thioamides

**Synthesis.**—Two novel syntheses of derivatives of thioformamide, the simplest member of the thioamide series, have been achieved during 1969. In one case,[139] a variant of a known reaction of thiocarbamoyl chlorides (79) with metal alcoholates, normally affording the thiourethane (80), is

$$\underset{(79)}{\underset{R^2}{\overset{R^1}{\diagdown}}\underset{S}{\overset{Cl}{\diagup}}\,\,\,\underset{\overline{OP_{r^-}}}{\overset{OR^-}{\longrightarrow}}\,\,\underset{(80)}{\underset{R^2}{\overset{R^1}{\diagdown}}\underset{S}{\overset{OR}{\diagup}}\,\,\,\underset{(81)}{\underset{R^2}{\overset{R^1}{\diagdown}}\underset{S}{\overset{H}{\diagup}}}}$$

employed. It appears that in the sole case of sodium isopropoxide NN-disubstituted thioformamides such as (81) are obtained, their formation being favoured by the presence of bulky N-alkyl groups in the starting material, and of excess isopropanol to provide greater solvation of the isopropoxide ions in the reaction media. Thioformamide itself has now been synthesised from hydrogen cyanide by reaction with hydrogen sulphide in the presence of ammonia or tertiary amines.[140] The future utility of this relatively simple procedure, which can be carried out either in protic

---

[130] L. A. Paquette and R. W. Begland, *J. Org. Chem.*, 1969, **34**, 2897.
[131] J. S. Grossert, *Chem. Comm.*, 1970, 305.
[132] M. H. Rosen, *Tetrahedron Letters*, 1969, 647; D. R. Eckroth and G. M. Love, *J. Org. Chem.*, 1969, 1136.
[133] N. P. Gambaryan and Y. V. Zefman, *Izvest. Akad. Nauk S.S.S.R., Ser. khim.*, 1969, 2059.
[134] W. E. Truce, J. W. Fieldhouse, D. J. Vrenair, J. R. Norell, R. W. Campbell, and D. G. Brady, *J. Org. Chem.*, 1969, **34**, 3097.
[135] J. F. Stephen and E. Marcus, *Chem. and Ind.*, 1969, 416.
[136] P. Muller, D. F. Mayers, O. R. Zaborsky, and E. T. Kaiser, *J. Amer. Chem. Soc.*, 1969, **91**, 6732; O. R. Zaborsky and E. T. Kaiser, *ibid.*, 1970, **92**, 860.
[137] B. E. Hogenboom, M. S. El-Faghi, S. C. Fink, E. D. Hoganson, S. E. Lindberg, T. J. Lindell, C. J. Linn, D. J. Nelson, J. O. Olsen, L. Rennerfeldt, and K. A. Wellington, *J. Org. Chem.*, 1969, **34**, 3414.
[138] J. F. King, A. Hawson, D. M. Deaken, and J. Komery, *Chem. Comm.*, 1969, 33.
[139] W. Walter and R. F. Becker, *Annalen*, 1969, **725**, 234.
[140] R. Tull and H. Weinstock, *Angew. Chem. Internat. Edn.*, 1969, **8**, 279.

or aprotic solvents, will probably be in the synthesis of thiazoles without the necessity of isolating the thioformamide. The base-catalysed reaction of nitriles with hydrogen sulphide is a well-known procedure, previously successful only in the higher thioamide homologues, and has also lately been applied to the synthesis of malonodithioamide from dimethylaminomalonodinitrile.[141]

The reaction of isothiocyanates with compounds containing nucleophilic carbon atoms, notably with enamines, is a common source of thioamides, and has been used in the synthesis of 2-arylaminocyclohexene-1-thioamides (82),[142] a tetrahydroisoquinoline derivative,[143] and thioamides derived from pyrrolo[1,2-a]benzimidazole,[144] 4-hydroxycoumarin,[145] and arylsulphonylacetonitrile.[146] Two groups of workers[147] have reported the synthesis of thioamides such as (83) via the reaction between isothiocyanates and methylenephosphoranes. The subsequent reactions of (83) offer new

preparative possibilities,[147a] exemplified in a Wittig reaction with aldehydes producing N-substituted thioamides (84) of α,β-unsaturated carboxylic acids.

Among several reports of the synthesis of thioamides utilising variations of the Willergrodt—Kindler reaction, is an interesting paper[148] describing the reaction of aldehydes with sulphur in the presence of morpholine. The reaction apparently does not proceed well with aliphatic aldehydes but succeeds with formaldehyde, affording thioformamides. NN-Dimethyl derivatives (85) of arylthioacetamides and arylthiopropionamides have been prepared[149] in a related procedure, involving the reaction of styrene and 3-phenylpropene with sulphur in dimethylformamide solution. Asinger and his co-workers, in continuation of their studies of Willergrodt–Kindler-type reactions, report[150a] that the reaction of acetone with sulphur

[141] H. Eilingsfeld and M. Patsch, *Angew. Chem. Internat. Edn.*, 1969, **8**, 750.
[142] J. Schoen and K. Bogdanowicz-Szwed, *Roczniki. Chem.*, 1969, **43**, 65.
[143] M. D. Nair and S. R. Mehta, *Indian J. Chem.*, 1969, **7**, 684.
[144] F. S. Babichev and Nguyen Tkhy Nguyet Fyong, *Ukrain. khim. Zhur.*, 1969, **35**, 932.
[145] A. Metallidis, H. Junek, and E. Zeigler, *Monatsh.*, 1970, **101**, 88.
[146] V. M. Penlynev, Yu. N. Usenko, R. G. Dubenko, and P. S. Pel'kis, *Zhur. org. Khim.*, 1970, **6**, 164.
[147] [a] H. J. Bestmann and S. Pfohl, *Angew. Chem. Internat. Edn.*, 1969, **8**, 762; [b] M. I. Shevchuk, I. S. Zabrodskaya, and A. V. Dombrovskii, *Zhur. obschchei Khim.*, 1969, **39**, 1282.
[148] L. Maier, *Angew. Chem. Internat. Edn.*, 1969, **8**, 141.
[149] J. Chauvin and Y. Mollier, *Compt. rend.*, 1969, **268**, C, 294.
[150] [a] F. Asinger, H. Offermanns, D. Neuray, and F. A. Dagga, *Monatsh.*, 1970, **101**, 500; [b] F. Asinger, H. Offermanns, and A. Saus, *ibid.*, 1969, **100**, 724; [c] F. Asinger, A. Saus, H. Offermanns, and F. A. Dagga, *Annalen*, 1969, **723**, 119.

Ar(CH$_2$)$_n$CSNMe$_2$
(85)

Ar—C(=NR)—NHMe
(86)

(NR)C(Ar)—C(=S)(NHR)
(87)

(S)C(Ar)—C(=S)(NHR)
(88)

(89)

in the presence of ammonia affords an imidazoline-5-thione, and have shown [150b, c] that the α-iminophenylthioglyoxylamide (87) and the α-thioxophenylthioglyoxylamide (88), formed via thionation of their imine precursor (86), are important intermediates in a proposed mechanism of the Willgerodt–Kindler synthesis of arylthioacetamides from acetophenone derivatives in the presence of primary amines. If great excess of amine is employed, the hexathiocan derivatives (89) are formed rather than thioamides.[150c]

The thiocarbonyl sulphur atom has been introduced into 1-hydroxypyridine-2-thiones [151] and thioamide derivatives [152] by means of reactions of elemental sulphur, and into thioamides similar to (87) by means of the reaction of arylacetylenic sulphides with amines.[110]

Thionation of amides with phosphorus pentasulphide is perhaps the most commonly used thioamide synthesis. Demands of space preclude the discussion of most recent examples, mainly in the field of heterocyclic chemistry. Worthy of mention, however, are studies of the thionation of hypoxanthines [153] and of oxindoles.[154]

The cleavage of sulphur heterocycles appears to have been a fruitful source of thioamides during 1969. Products obtained by this route include dithiomalonamide from 3,5-diamino-1,2-dithiolium perchlorate,[155] S-aryl esters of β-thioamidothiocarboxylic acids from 5-amino-1,2-dithiacyclopentenone,[156] and thiobenzamide and thioacetamide derivatives from the photochemical fragmentation of 1,3,4-thiadiazoles [157] and the photosensitised oxygenation of 2,4,5-triphenylthiazole.[158] Thioamides have also been obtained by the ring-contraction of benzodiazepin thiones [159] and

---

[151] R. A. Abramovitch and E. E. Knaus, *J. Heterocyclic Chem.*, 1969, **6**, 989.
[152] F. Becke and H. Hagen, *Chem.-Ztg., Chem. Appl.*, 1969, **93**, 474.
[153] F. Bergmann and M. Rashi, *J. Chem. Soc. (C)*, 1969, 1831.
[154] T. Hino, K. Tsuneoka, M. Nakagawa, and S. Akaboshi, *Chem. and Pharm. Bull. (Japan)*, 1969, **17**, 550.
[155] M. B. Kolesova, L. I. Maksimova, and A. V. El'tsov, *Zhur. org. Khim.*, 1970, **6**, 610.
[156] F. Boberg and R. Schardt, *Annalen*, 1969, **728**, 44.
[157] R. M. Moriarty and R. Mukherjee, *Tetrahedron Letters*, 1969, 4627.
[158] T. Matsuura and I. Saito, *Bull. Chem. Soc. Japan*, 1969, **42**, 2973.
[159] R. I. Fryer, J. V. Earley, and L. H. Sternbach, *J. Org. Chem.*, 1969, **34**, 649.

the alkaline cleavage of disulphides.[160, 161] The latter procedure, using bis(2-pyridyl)disulphide, is a useful method for quantitative analysis of thiols,[160] the 2-thiopyridone product being determined photometrically.

**Reactions.**—As a further aspect of their wide-ranging studies on the properties of thioamides, the group led by Professor Walter have now reported[162] attempts to find an alkylation procedure non-specific for *N*- and *S*-substitution. They find that, whereas reaction of thioacetamide and thiobenzamide with triphenylmethyl chloride affords no alkylated product in the absence of base, bis(*p*-methoxyphenyl)methyl chloride leads to thiolimidate ester hydrochlorides (90) under these conditions, and to *N*-alkylated products (91) in the presence of potassium carbonate.

$$\underset{\underset{NH, HCl}{\|}}{RC-S-CH(p\text{-}OMeC_6H_4)_2} \qquad \underset{\underset{S}{\|}}{RC-NHCH(p\text{-}OMeC_6H_4)_2}$$
$$\qquad\qquad (90) \qquad\qquad\qquad\qquad\qquad (91)$$

The authors also report the versatility of alkylation using substituted benzhydryl carbinols in acidic media, and the thermal rearrangement of thiolimidate esters into the thermodynamically more favoured *N*-alkylated thioamide isomer.

The reaction of thioamides with amine derivatives has been studied by several groups of workers. Larsen and his co-workers[163] record the unusual direct substitution of a thioamide group by hydrazine, the reaction of thioacetamide in the presence of nickel(II) ions leading to the complex *trans*-bis(thioacethydrazato)nickel(II). In the absence of nickel(II) ions, the primary adduct (92) decomposes by liberation of hydrogen sulphide rather than to thioacethydrazide (93) by loss of ammonia, suggesting that

$$\underset{\underset{NH_2}{|}}{\overset{\overset{SH}{|}}{Me-C-NHNH_2}} \qquad\qquad Me\overset{\overset{S}{\|}}{C}\diagdown_{NHNH_2}$$
$$\qquad (92) \qquad\qquad\qquad\qquad (93)$$

aliphatic thioacylhydrazides possessing α-hydrogen atoms may prove too unstable to be isolated. The well-known reaction of dithio-oxamide with ethylenediamine, generally regarded as leading to 2,2′-bis(2-imidazolinyl) (94), has been reinvestigated.[164] The previous assignment of structure (94) to this product is supported by the *X*-ray crystallographic confirmation[164b] of analogous bis-thiazolinyl structures for products of the

[160] D. R. Grasetti and J. F. Murray, *Analyt. Chim. Acta*, 1969, **46**, 139.
[161] B. C. Pal, M. Uziel, D. G. Doherty, and W. E. Cohn, *J. Amer. Chem. Soc.*, 1969, **91**, 3634.
[162] W. Walter and J. Krohn, *Chem. Ber.*, 1969, **102**, 3786.
[163] E. Larsen, P. Trinderup, B. Olsen and K. J. Watson, *Acta Chem. Scand.*, 1970, **24**, 261.
[164] *a* G. Barnikow and G. Saeling, *Z. Chem.*, 1969, **9**, 107; *b* S. C. Mutha, and R. Ketcham, *J. Org. Chem.*, 1969, **74**, 2053.

# Thiocarbonyls, Selenocarbonyls, and Tellurocarbonyls

$$\underset{(94)}{\begin{array}{c}\text{HN}\diagup\diagdown\text{N}\\ \|\quad\|\\ \text{N}\diagdown\diagup\text{NH}\end{array}}$$

mechanistically and structurally similar reaction between cyanogen and mercaptoacetic acids. The recent investigators point out, however, that equally-likely fused six-membered ring structures have been disregarded in earlier studies of the reaction of dithio-oxamide with diamines and with α-halogenoketones. The use of dimethylthioformamide in the synthesis of thiobenzophenones *via* ketimines has been reported.[72]

Russian workers, in a series of papers [165] concerning the interaction of various thioamide derivatives with aryldiazonium salts, report that *S*-arylation takes place in the presence of sodium acetate or acetate buffer— but not in acidic solution—in the case of *N*-substituted simple thioamides [165a, e] and thioamidonitriles.[165b, e] Reaction with derivatives (95) of 1-benzoyl-1-cyanothioacetanilide, however, proceeds *via* displacement of the thioamide group to the arylhydrazone (96),[165c] while the acetyl derivative of thioacetanilide affords the arylazohydrazone (97).[165d]

$$\underset{(95)}{\text{ArNHC}\overset{S}{\underset{\underset{\text{CN}}{|}}{\overset{\|}{\diagup}}}\underset{\text{CH}}{\diagdown}\text{COPh}} \quad \xrightarrow{\text{Ar}^1\text{N}_2{}^+} \quad \underset{(96)}{\text{Ar}^1\text{NHN=C}\diagup\diagdown\overset{\text{CN}}{\text{COPh}}} \qquad \underset{(97)}{\text{ArN=N}\diagdown\underset{\text{ArNHN}}{\overset{}{\diagup}}\text{C}-\overset{S}{\underset{}{\overset{\|}{\text{C}}}}\diagdown\text{NHPh}}$$

Halogenations, affording thioimidoyl halides, have been reported for substituted thioformanilides[166] and for α-cyanoacetamide,[167] perhalogenoderivatives being formed in the latter case. α-Polychloroamines have been obtained from the photochlorination of *NN*-dialkylthioformamides.[168] The reaction of arylsulphenyl chlorides with thiobenzamide in the absence of base has been shown to give iminomethyl disulphide hydrochlorides (98).[169] The latter compounds are unstable as the free base, but their *N*-aryl derivatives may be isolated and are oxidised by peracids to the corresponding thioamide *S*-oxides (99).[169b] The reaction of *N*-methylthiobenzamide with sulphur dichloride, however, is reported to give the dithiazetine derivative (100).[170] Di-iminomethyl disulphide dihydro-iodides are

---

[165] a E. P. Nesynov, M. M. Besprozvannaya, and P. S. Pel'kis, *Zhur. org. Khim.*, 1970, **6**, 540; b 1969, **5**, 58; c R. G. Dubenko, E. F. Gorbenko, and P. S. Pel'kis, *ibid.*, 1969, **5**, 529; d A. N. Borisevich and P. S. Pel'kis, *ibid.*, 1969, **5**, 180; e E. P. Nesynov, M. M. Besprozvannaya, and P. S. Pel'kis, *Dopovidi Akad. Nauk Ukrain. R.S.R. Ser. B*, 1969, 27.
[166] W. Walter and R. F. Becker, *Annalen*, 1970, **733**, 195.
[167] J. V. Burakevich, A. M. Love, and S. P. Volpp, *Chem. and Ind.*, 1970, 288.
[168] K. Grohe, E. Klauke, H. Hottschmidt, and H. Heitzer, *Annalen*, 1969, **730**, 140.
[169] W. Walter and P.-M. Hell, *Annalen*, 1969, **727**, a 22; b 50.
[170] A. V. El'tsov, V. E. Lonatin, and M. G. Mikhel'son, *Zhur. org. Khim.*, 1970, **6**, 402.

Ph—C—S—S—Ar
||
$\overset{+}{N}H_2$  Cl$^-$
(98)

Ph
\
C=SO
/
PhHN
(99)

Ph—C$\overset{\overset{Me}{N_+}}{\underset{S}{\diagdown}}$S  Cl$^-$

(100)

formed from the reaction of thiobenzamide with iodine in organic solvents.[171]

Several examples of the Hantzsch synthesis of thiazoles *via* the reaction of thioamides with α-halogenoketones have been reported.[172–177] Thiazoles and thiazolines have also been obtained in related procedures, involving the reaction of thioamides with α-halogenoacids [178] and α-nitroepoxides,[179] the spontaneous cyclisation of thioamides derived from *N*-(β-halogenoalkyl)imide halides by treatment with hydrogen sulphide,[180] the cyclisation of amino-acids and their *N*-acyl and *N*-thioacyl derivatives with thioacetic acid and with trifluoroacetic acid,[181] the bromocyclisation of thioanilides derived from α-cyanomalonate esters,[182] and the cyclisation of α-carbethoxyacetamides using oxalyl chloride.[183] In addition, the use of thioamides in the synthesis of derivatives of pyrrole,[184] thiophen,[185,186] pyrazole,[187] imidazole,[164a] pyrimidine,[188–191] and 1,2,4-triazine [192] has been reported.

[171] P. H. Deshpande, *J. Indian Chem. Soc.*, 1969, **46**, 1096.
[172] K. Brown and R. A. Newbery, *Tetrahedron Letters*, 1969, 2796.
[173] M. Lozinskii, S. Kukota, P. S. Pel'kis, and T. Kudrya, *Khim. geterotsikl. Soedinenii*, 1969, 757.
[174] Y. I. Smushkevich, T. M. Babuyeva, and N. N. Suvorov, *Khim. geterotsikl. Soedinenii*, 1969, 91.
[175] M. Robba and Y. LeGuen, *Bull. Soc. chim. France*, 1969, 1762.
[176] R. Vivaldi, H. J. M. Dou, G. Vernin, and J. Metzger, *Bull. Soc. chim. France*, 1969, 4014.
[177] R. Bognar, I. Farkas, L. Szilagyi, M. Menyhart, E. N. Nemes, and I. F. Szabo, *Acta Chim. Acad. Sci. Hung.*, 1969, **62**, 179.
[178] A. Shaikh, A. Chinone, and M. Ohta, *Bull. Chem. Soc. Japan*, 1970, **43**, 453.
[179] H. Newman and R. B. Angier, *Chem. Comm.*, 1969, 369.
[180] J. Beger, D. Schöde, and J. Vogel, *J. prakt. Chem.*, 1969, **311**, 408.
[181] G. C. Barrett, A. R. Khokhar, and J. R. Chapman, *Chem. Comm.*, 1969, 818; G. C. Barrett and A. R. Khokhar, *J. Chem. Soc.* (*C*), 1969, 1117.
[182] H. Kunzek, and G. Barnikow *Chem. Ber.*, 1969, **102**, 351.
[183] G. Barnikow and G. Saeling, *Z. Chem.*, 1969, **9**, 145.
[184] K. T. Potts and U. P. Singh, *Chem. Comm.*, 1969, 66.
[185] E. J. Smutny, *J. Amer. Chem. Soc.*, 1969, **91**, 208.
[186] S. Rajappa and B. G. Advani, *Tetrahedron Letters*, 1969, 5067.
[187] A. N. Borisevich and P. S. Pel'kis, *Khim. geterotsikl. Soedinenii*, 1969, 312.
[188] R. W. Lamon, *J. Heterocyclic Chem.*, 1969, **6**, 37, 261.
[189] G. Wagner and L. Rothe, *Z. Chem.*, 1969, **9**, 446.
[190] M. S. Manhas, V. V. Rao, P. A. Seetharaman, D. Succardi, and J. Padzera, *J. Chem. Soc.* (*C*), 1969, 1937.
[191] E. Ziegler, W. Steiger, and T. Kappe, *Monatsh.*, 1969, **100**, 150, 948.
[192] S. Kwee and M. Lund, *Acta Chem. Scand.*, 1969, **23**, 2711.

Further aspects of the chemical reactivity of acyclic thioamide derivatives are described in recently published papers, dealing with the reaction of Grignard reagents at the methylene group of substituted thioacetamides,[149] and the attack of ethyl chloroformate and mercury bis(phenylacetylide) at the sulphur atom of thioamides leading to imidate hydrochlorides [193] and symmetrical mercury mercaptides [194] respectively. Limitations of space preclude even a cursory examination of the many papers comprising studies of the chemical reactivity of heterocyclic thiones containing the thiolactam grouping. However, a selection of papers dealing with the chemistry of thiones derived from pyrrolidine,[195] pyrazole,[196, 197] thiazoline,[198] imidazoline,[199] pyridine,[200, 201] pyridazine,[202] pyrimidine,[203-209] oxazine,[210] 1,3- and 1,4-thiazine,[211-213] 1,2,3-triazine,[214] and 1,4-thiazepine [215] may be of interest to workers in this field.

**The Thio-Claisen Rearrangement.**—Several recent papers describe examples of the thio-Claisen rearrangement—a 3,3-sigmatropic process typified by the reaction (101) → (103)—of heterocyclic derivatives, in which thiolactams, their vinylogues, or their cyclisation products are isolated. Makisumi and his co-workers [216] have shown that whereas in the thermal rearrangement of allyl 3- and allyl and propargyl 4-quinolyl sulphides, the initially formed thiones such as (104) undergo a prototropic cyclisation leading to thiophenic systems (105), the rearrangement of allyl 2-quinolyl sulphides

---

[193] F. Suydam, W. E. Greth, and N. R. Langermann, *J. Org. Chem.*, 1969, **34**, 292.
[194] W. Ried, W. Merkel, and R. Oxenius, *Chem. Ber.*, 1970, **103**, 32.
[195] A. Mondon and P. R. Seidel, *Chem. Ber.*, 1970, **103**, 1298.
[196] I. I. Grandberg, V. G. Vinokurov, V. S. Troitskaya, T. A. Ivanova, and V. A. Mosalenko, *Khim. geterotsikl. Soedinenii*, 1970, 202.
[197] C. Dittli, J. Elguero and R. Jacquier, *Bull. Soc. chim. France*, 1969, 4466, 4474.
[198] N. Y. Plevachuk and I. D. Komaritsa, *Khim. geterotsikl. Soedinenii*, 1970, 195.
[199] J. Nyitrai and K. Lempert, *Tetrahedron*, 1969, **25**, 4265.
[200] P. Beak and J. T. Lee jun., *J. Org. Chem.*, 1969, **34**, 2125.
[201] K. Undheim, V. Nordal, and K. Tjønneland, *Acta Chem. Scand.*, 1969, **23**, 1704
[202] B. Stanovnik, M. Tišler, and A. Vrbanic, *J. Org. Chem.*, 1969, **34**, 996.
[203] H. Hayatsu, *J. Amer. Chem. Soc.*, 1969, **91**, 5693; H. Hayatsu and M. Yano, *Tetrahedron Letters*, 1969, 755.
[204] E. Sato and Y. Kanaoka, *Tetrahedron Letters*, 1969, 3547.
[205] V. M. Vedenskii and N. M. Turkevich, *Dopovidi Akad. Nauk Ukrain. R.S.R., Ser B.*, 1970, 60.
[206] J. Mirza, W. Pfleiderer, A. D. Brewer, A. Stuart, and H. C. S. Wood, *J. Chem. Soc.*, (C), 1970, 437.
[207] K. Undheim and J. Røe, *Acta Chem. Scand.*, 1969, **23**, 2437.
[208] E. Dyer, R. E. Harris jun., C. E. Minnier, and M. Tokizawa, *J. Org. Chem.*, 1969, **34**, 973.
[209] A. Albert, *J. Chem. Soc.* (C), 1969, 152.
[210] M. Mazharuddin and G. Thyagarajan, *Indian J. Chem.*, 1969, **7**, 658.
[211] M. Turkevich, Y. V. Vladzimirskaya, and L. M. Vengrmovich, *Khim. geterotsikl. Soedinenii*, 1969, 504; N. M. Turkevich and B. S. Zimenovskii, *ibid.*, 1969, 59.
[212] B. E. Zhitar and S. N. Baranov, *Zhur. org. Khim.*, 1969, **5**, 1893.
[213] C. R. Johnson and C. B. Thanawalla, *J. Heterocyclic Chem.*, 1969, **6**, 247.
[214] E. E. Gilbert and B. Veldhuis, *J. Heterocyclic Chem.*, 1969, **6**, 779.
[215] K. H. Wunsh and A. Ehlers, *Chem. Ber.*, 1969, **102**, 1869.
[216] Y. Makisumi and A. Murabayshi, *Tetrahedron Letters*, 1969, *a* p. 1970; *b* p. 2449; *c* p. 2453; *d* Y. Makisumi and T. Sasatani, *ibid.*, 1969, 1975.

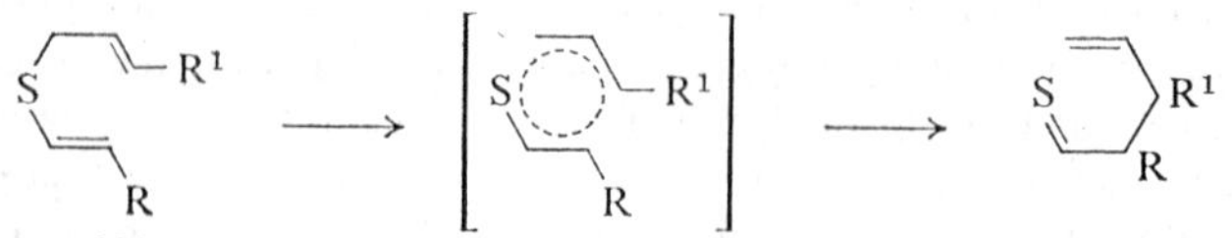

(106) leads to the corresponding propenyl sulphides (107). They also observe [216d] that the expected thio-Claisen product of the latter reaction, the N-allylthiocarbostyril (108), is thermodynamically less stable than the allyl sulphide (106), thus explaining the reported reverse thio-Claisen rearrangement (108) → (106). The isolation of two cyclised products from the thermal rearrangement of allyl 3-quinolyl sulphides, and the observed effect of acid and free-radical initiator catalysts on the product ratio, lead the authors to conclude that competing ionic and free-radical processes occur.[216c, d] The thio-Claisen rearrangement of prop-2-enyl and prop-2-ynyl 2-indolyl sulphides to indole-2-thione derivatives has been described.[217]

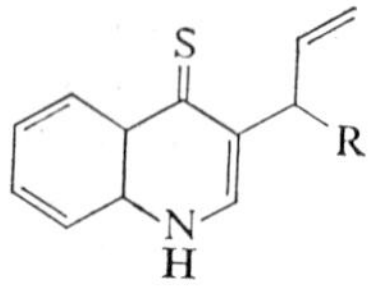

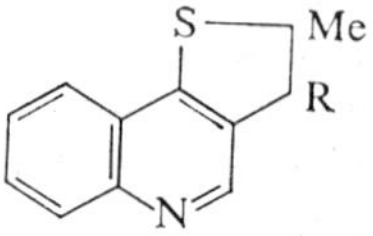

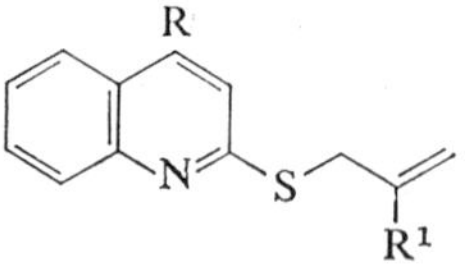

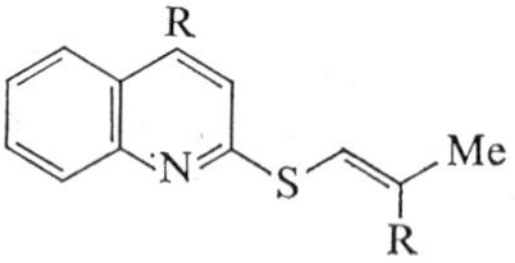

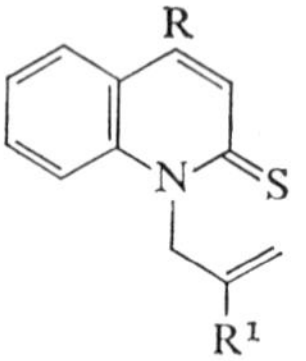

Several recent papers [68, 218-220] dealing with more general aspects of the thio-Claisen rearrangement and not necessarily involving nitrogen heterocycles may be conveniently cited at this juncture. Of particular interest is the paper of Kwart and George,[218] in which the successful

[217] B. W. Bycroft and W. Landon, *Chem. Comm.*, 1970, 168.
[218] H. Kwart and T. J. George, *Chem. Comm.*, 1970, 433.
[219] [a] P. J. W. Schuijl and L. Brandsma, *Rec. Trav. chim.*, 1969, **88**, 1201; [b] P. J. W. Schuijl, H. J. T. Bos, and L. Brandsma, *ibid.*, p. 597.
[220] L. Brandsma and D. Schuijl-Laros, *Rec. Trav. chim.*, 1970, **89**, 110.

thio-Claisen rearrangement of propargyl phenyl sulphides is reported. The rearrangement is regarded as proceeding both by a simple thio-Claisen mechanism, and by a process involving an initial isomerisation of the acetylenic moiety into an allene. The authors suggest that this latter step is also important in the aforementioned rearrangement of propargyl 4-quinolyl sulphides,[216a] although the possibility is not discussed by the Japanese workers.

## 8 Thioureas

**Synthesis.**—A great majority of recent thiourea syntheses have employed the reaction between alkyl or aryl isothiocyanates and amines. Among the $N$-substituent groups thus introduced into thiourea are perfluoroalkenyl,[221] 2-hydroxy-2-$p$-nitrophenylethyl,[222] 1-hydroxymethyl-2-hydroxy-2-$p$-nitrophenylethyl,[223] norbornenyl,[224] 1-adamantyl,[225] and variously substituted aryl.[226-229] Thiourea derivatives of 3,6-diamino-1,4-benzoquinone,[230] 6,7-benzo-1,4-thiazepin,[231] 2-aminopyridine,[232] and of various 2-mercapto-1,3-diazacycloalkenes [233] have also been prepared by similar methods. Other isothiocyanato-derivatives which have been used in the synthesis of thioureas include vinyl isothiocyanate,[234] acyl isothiocyanates,[234, 235] carbamoyl isothiocyanates,[236] and ammonium isothiocyanate.[237] Russian workers [238] have investigated the dual reactivity of benzyl isothiocyanate with primary and secondary amines, and have shown that reaction can take place at the benzyl carbon atom as well as at the isothiocyanate sulphur atom, affording benzylammonium isothiocyanates in addition to the more usual thiourea product. In this respect, benzyl isothiocyanate resembles the behaviour of allyl isothiocyanates.

[221] S. R. Sterlin, B. L. Dyatkin, L. G. Zhuravkova, and I. L. Knunyants, *Izvest. Akad. Nauk S.S.S.R., Ser. khim.*, 1969, 1176.
[222] M. O. Lozinskii, T. N. Soroka, D. F. Yarovskii, S. S. Kirienko, Y. Y. Yakovleva, and P. S. Pel'kis, *Khim. geterotsikl. Soedinenii*, 1969, 501.
[223] M. O. Lozinskii, Yu. V. Karabanov, T. N. Kudrya, P. S. Pel'kis, and T. M. Chevrechenko, *Dopovidi Akad. Nauk Ukrain. R.S.R., Ser.B*, 1969, 125.
[224] W. R. Dively, G. A. Buntin, and A. D. Lohr, *J. Org. Chem.*, 1969, **34**, 616.
[225] A. Kreutzberger and H. H. Schröders, *Tetrahedron Letters*, 1969, 5101.
[226] A. S. Narang, A. N. Kaushal, Harjit Singh, and K. S. Narang, *Indian J. Chem.*, 1969, **7**, 1191.
[227] D. M. Wiles and T. Suprunchuk, *J. Medicin. Chem.*, 1969, **12**, 526.
[228] A. Sausins and V. Y. Grin'stein, *Izvest. Akad. Nauk Latv. S.S.R., Ser. khim.*, 1970, 85.
[229] E. V. Vanag, A. E. Sausins, and V. Y. Grin'stein, *Zhur. org. Khim.* 1969, **5**, 889.
[230] E. Winkelmann, *Tetrahedron*, 1969, **25**, 2447.
[231] M. D. Nair and S. M. Kalbag, *Indian J. Chem.*, 1969, **7**, 862.
[232] N. Krishnaswami and H. D. Bhargava, *Indian J. Chem.*, 1969, **7**, 710.
[233] C. DiBello, A. C. Veronese, F. Filira, and F. D'Angeli, *Gazzetta*, 1970, **100**, 86.
[234] K. Nagarajan, V. P. Arya, A. Venkateswarlu, and S. P. Ghate, *Indian J. Chem.*, 1969, **7**, 1195.
[235] V. S. Etlis, L. N. Grobov, and A. Sineokov, *Khim. geterotsikl. Soedinenii*, 1969, 799.
[236] J. Goerdeler and D. Wobig, *Annalen*, 1970, **731**, 120.
[237] N. B. Galstukhova, N. N. Shchukina, Z. Budeshinski, and A. Shrab, *Zhur. org. Khim.*, 1969, **5**, 1142.
[238] I. M. Fedin and S. P. Miskidzh'yan, *Zhur. obshchei Khim.*, 1969, 1447.

As further variations on the general procedure, thioureas have been synthesised from isothiocyanates in their reactions with hydroxylamines,[239] benzamidines,[240] and ketimines.[241, 242] Phosphorus analogues of thioureas (109) have been obtained [243] from the reaction of isothiocyanates with

$$\underset{(109)}{R_2PCSNHR^1}\overset{X}{\underset{\|}{}} \quad \underset{(110)}{R_2PNHCSNHR^1}\overset{X}{\underset{\|}{}} \quad X = O \text{ or } S$$

phosphine oxides and phosphine sulphides. Phosphinyl derivatives of thioureas such as (110) have been prepared *via* the reaction of isothiocyanates, derived from diphenylphosphinic [244] and diphenylphosphinothioic [245] acids, with amines.

Two reports of the synthesis of thioureas from the reaction of amines with carbon disulphide have appeared in the recent literature. Haugwitz [246] has shown that the reaction of carbon disulphide with cyclohexyl and aromatic hydroxylamines affords the corresponding *N*-hydroxythiourea, presumably formed from the *NN'*-dihydroxy-derivative by reduction with hydrogen sulphide. Under prolonged reaction conditions the monohydroxythiourea is further reduced to the thiourea. Schlatmann and his co-workers [247] have reported the unusual reaction of the homo-adamantane derivative (111), which affords the thiourea (112) on reaction with carbon disulphide, rather than the expected thiazoline.

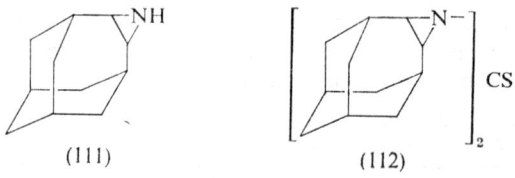

Reactions.—Much of the utility of thioureas in synthesis is founded on the nucleophilic reactivity of the thiocarbonyl sulphur atom, exemplified by the reaction with alkyl halides, leading to *S*-alkylisothiouronium salts. Several recent papers [248-251] report examples of the latter reaction, the

---

[239] L. Capuano, W. Ebner, and J. Schrepfer, *Chem. Ber.*, 1970, **103**, 82.
[240] R. Schwenker and R. Kolb, *Tetrahedron*, 1969, **25**, 5437.
[241] N. S. Koslev, B. D. Pak, I. A. Ivanov, and G. N. Kaluzin, *Zhur. org. Khim.*, 1969, **5**, 2195.
[242] A. de Savignac and A. Lattes, *Compt. rend.*, 1969, **268**, *C*, 2325.
[243] L. Ojima, K. Akiba, and N. Inamoto, *Bull. Chem. Soc. Japan*, 1969, **42**, 2975.
[244] G. Tomaschewski and D. Zanke, *Z. Chem.*, 1970, **10**, 117.
[245] R. A. Spence, J. M. Swan, and S. H. B. Wright, *Austral. J. Chem.*, 1969, **22**, 2359.
[246] R. D. Haugwitz, *Annalen*, 1970, **731**, 171.
[247] J. L. M. A. Schlatmann, J. G. Korsloot, and J. Schut, *Tetrahedron*, 1970, **26**, 949.
[248] G. Tóth, I. Tóth, and L. Toldy, *Tetrahedron Letters*, 1969, 5299.
[249] K. Hartke and G. Salamon, *Chem. Ber.* 1970, **103**, 133.
[250] B. Paul and W. Korytnyk, *Tetrahedron*, 1969, **25**, 1071.
[251] C. Toniolo and A. Fontana, *Gazzetta*, 1969, **99**, 1017.

most significant in terms of theoretical interest being the study by Toth and his colleagues [248] of the structure and *cis–trans* isomerisation of the *S*-alkylisothioureas (113) and (114), obtained by methylation of the

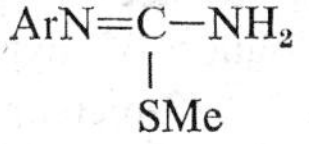

$$\text{ArN}=\text{C}-\text{NH}_2 \qquad\qquad \text{ArNH}-\text{C}=\text{NH}$$
$$\qquad\quad | \qquad\qquad\qquad\qquad\quad |$$
$$\qquad\;\text{SMe} \qquad\qquad\qquad\qquad\;\;\text{SMe}$$
$$\qquad\;(113) \qquad\qquad\qquad\qquad\quad (114)$$

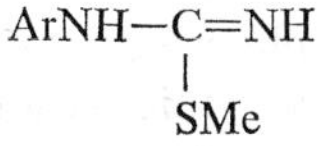

corresponding *N*-arylthiourea. *S*-Arylation of thioureas can be achieved using aryl halides possessing activating ring substituents,[252] but a more versatile procedure, employing an aryldiazonium salt as the arylating reagent, has been used to obtain *S*-aryl derivatives of thiourea[253] and *NN'*-tetramethylthiourea.[252] *S*-Arylisothiouronium salts may also be synthesised *via* 'nascent' quinones,[254] whereas synthesis of an *S*-(3-indolyl)-isothiouronium salt (115) has been achieved by the reaction between indole

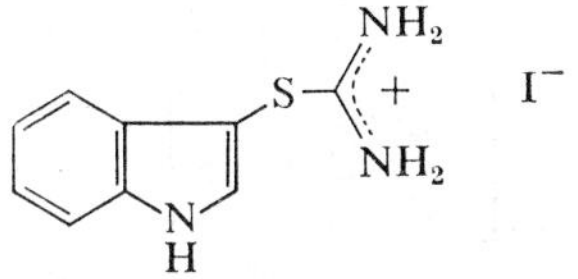

(115)

and thiourea, in the presence of iodine and potassium iodide.[255] A novel and versatile synthesis of *S*-alkylthioisothioureas (117), in which alkyl thiols are the source of the alkylthio-moiety, has lately been reported by Japanese workers.[256] The mechanism of the reaction, which takes place in ethanolic solution containing aqueous hydrochloric acid and in the presence of hydrogen peroxide, is regarded as involving nucleophilic

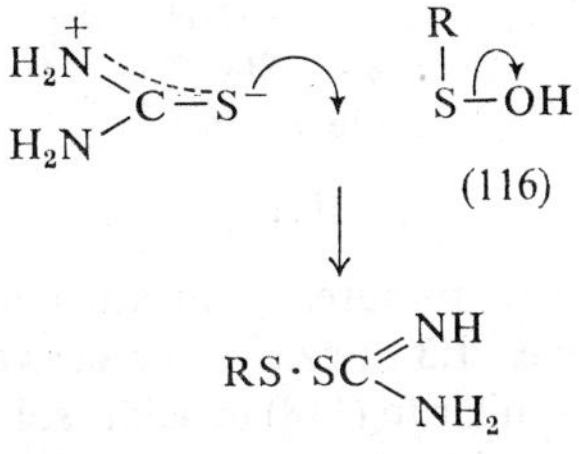

(116)

(117)

---

[252] H. Kessler, H. O. Kalinowski, and C. von Chamier, *Annalen*, 1969, **727**, 228.
[253] B. V. Kopglova, M. N. Khasanova, and R. Kh. Freidlina, *Izvest. Akad. Nauk S.S.S.R., Ser. khim.*, 1970, 633; B. V. Kopglova and M. N. Khasanova, *ibid.*, 1969, 2619.
[254] J. Daneke, U. Jahnke, B. Pankow, and H.-W. Wanzlick, *Tetrahedron Letters*, 1970, 1271.
[255] R. L. N. Harris, *Tetrahedron Letters*, 1969, 4465.
[256] K. Sirakawa, O. Aki, T. Tsujikawa, and T. Tsuda, *Chem. and Pharm. Bull. (Japan)*, 1970, **18**, 235.

attack of the thiourea sulphur atom on the sulphenic acid (116), formed by oxidation of the thiol.

Several of the aforementioned syntheses have been undertaken in order to utilise the well-known route to thiols, involving the action of base on isothiouronium salts. By this means, thio-analogues of pyridoxol,[250] halogenated thiophenols,[253] mercaptocatechols,[254] 3-mercaptophenothiazine,[254] and 3-indolylthiol [255] have been prepared. In addition, isothiouronium salts have been employed in the synthesis of 6,6'-diaminofulvenes,[249] and in the removal of α-halogenoacetyl blocking groups in amino-acid chemistry.[251]

The reaction of thiourea and various $N$-alkyl and $N$-aryl derivatives with chlorocyanoacetylene is reported [257] to proceed via attack of the thiourea sulphur atom on the triple bond, affording the isothiourea-intermediate (118). This is transformed by elimination of nitrile and hydrogen chloride to cyanothioketen (119), which dimerises, yielding the dithiafulvene (120).

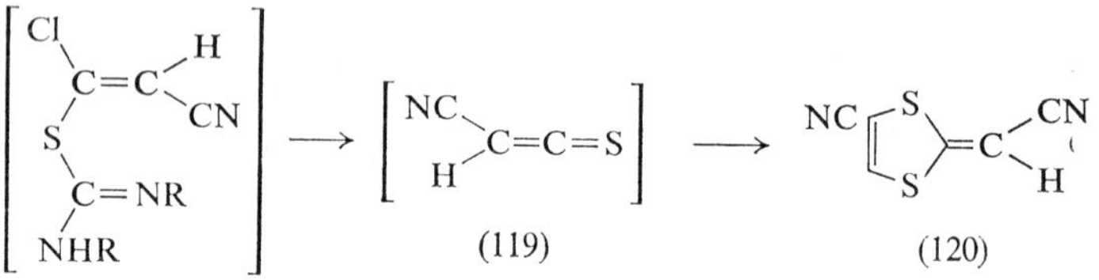

A similar reaction with cyanoacetylene also involves the formation of a primary intermediate analogous to (118), but in this case thiadiacrylonitrile is isolated in good yields. Exceptionally, $NN'$-dimethylthiourea affords the 2,3-dihydro-1,3-thiazine derivative (121). Russian workers,[258] in similar

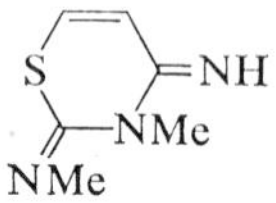

studies on the reactivity of thioureas and selenoureas with electrophilic triple bonds, have obtained 1,3-thiazin-4-one derivatives in neutral media and open-chain adducts similar to (118) in acidic solution. The nucleophilic reactivity of thiourea is further demonstrated [94] by its reaction with tetracyanoethylene oxide, leading to cleavage of the oxiran ring and formation of the thiouronium ylide (122). In the case of $NN'$-diphenylthiourea, the corresponding ylide cyclises to a thiazoline. A similar oxiran-cleavage reaction of thiourea is reported in the cyclopenin series,[259] but a rather

---

[257] T. Sasaki, K. Kanematsu, and K. Shoji, *Tetrahedron Letters*, 1969, 2371.
[258] E. G. Kataev, L. K. Konovalova, and E. G. Yarkova, *Zhur. org. Khim.*, 1969, **5**, 621.
[259] P. K. Martin, H. Rapoport, H. W. Smith, and J. L. Wong, *J. Org. Chem.*, 1969, **34**, 1359.

## Thiocarbonyls, Selenocarbonyls, and Tellurocarbonyls

$$\underset{H_2N}{\overset{H_2N}{>}}\overset{+}{C}-S-\bar{C}\underset{CN}{\overset{CN}{<}} \qquad PhSCH_2-HC\underset{O}{\overset{\diagdown\diagup}{-}}CH_2 \qquad PhSCH_2-HC\underset{S}{\overset{\diagdown\diagup}{-}}CH_2$$

(122) \qquad\qquad (123) \qquad\qquad (124)

different mode of action is found in the case of the epoxide (123), which is converted into the thiiran (124) by treatment with thiourea.[260] The catalytic effect of thiourea on the iodide ion reduction of $SS$-dimethylsulphiminium perchlorate is thought to depend largely on its particular nucleophilic character.[261]

The peroxide and peracid oxidation of thioureas has been studied by Walter and his co-workers,[262] who find that the $S$-monoxides, *e.g.* (125),

$$\underset{SO}{\overset{RNHCNHR^1}{\underset{\parallel}{}}}$$

(125)

of thiourea and $N$-monosubstituted thioureas can be detected only in solution, while those of $NN$-disubstituted thioureas can be detected chromatographically. $NN'$-Disubstituted thiourea $S$-monoxides, however, can be isolated in cases where the substituents carry bulky groups. $S$-Dioxides can be isolated for $N$-monoarylthioureas containing 2,6-disubstituted phenyl groups, and for $NN'$-dialkylthioureas, while the reaction of peracetic acid on $N$-mono-, $NN$-di-, and $NN'$-di-substituted thioureas, or their $S$-oxides and -dioxides, affords $S$-trioxides of considerably higher thermal stability. The authors note that structural conditions favouring the formation of $S$-monoxides, *i.e.* bulky substituents, tend to prejudice the formation of higher oxides. The hydrogen peroxide oxidation of trisubstituted thioureas tends to give the corresponding urea *via* an unstable $S$-monoxide, and often a formamidine derivative such as (126) through

$$RN=CHN\underset{R^2}{\overset{R^1}{<}}$$

(126)

decomposition of an $S$-dioxide. The yield of formamidine is found to increase with the increasing bulkiness of substituent groups. The oxidation of certain less sterically hindered trisubstituted thioureas with peracetic acid can give rise to $S$-trioxides.

Japanese workers [263] have conceived a novel method of speeding up certain synthetic investigations, by carrying out small-scale reactions in

---

[260] B. A. Arbuzov, O. N. Nuretdinova, and L. Z. Nikonova, *Izvest. Akad. Nauk S.S.S.R., Ser. khim.*, 1969, 167.
[261] J. H. Kreuger, *J. Amer. Chem. Soc.*, 1969, **91**, 4974.
[262] W. Walter and G. Randau, *Annalen*, 1969, **722**, 52, 80, 98; W. Walter and K. P. Ruess, *Chem. Ber.* 1969, **102**, 2640.
[263] T. Kinoshita and C. Tamura, *Tetrahedron Letters*, 1969, 4963.

the ionisation chamber of a mass spectrometer, thereby gaining a rapid insight into possible reaction courses and possible reaction intermediates. As the first test reaction for this procedure they have chosen the oxidation of various thioureas with benzoyl peroxide, and have been able to show the presence of the intermediate (127) in the oxidative cyclisation of $NN'$-diethylthiourea to a 1,2,4-thiadiazolidine derivative. Two groups of workers have reported the oxidation of thioureas to carbodi-imides, the oxidant being mercuric oxide in the preparation of phosphinylcarbodi-imides,[264] and ethyl diazodicarboxylate in the case of dialkylcarbodi-imides.[265] Triphenylphosphine, whose function is to remove sulphur from the initially-formed disulphide, is required as co-reagent in the latter procedure.

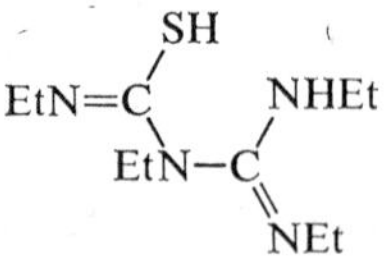

(127)

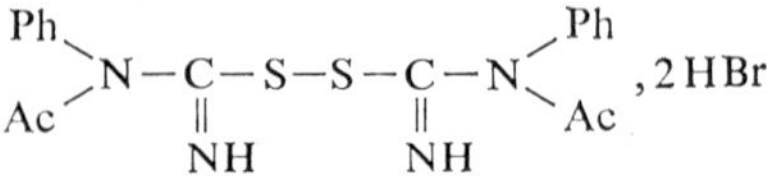

(128)

Srivastava,[266] in a mechanistic study of the oxidation of $N$-acetylthioureas, has shown that the formation of bis($N$-phenyl-$N$-acetylformamidine)-disulphides (128), and related monosulphides, are intermediate stages in the bromocyclisation of $N$-acetyl-$N$-phenylthiourea to phenylimino-1,2,4-thiazolidines. Di-iminomethyl disulphides similar to (128) have recently been isolated in the oxidation of thiobenzamide with iodine [171] (see section 7). Indian workers [267] have also studied the mechanism of the different modes of action of the halogen–thiourea redox polymerisation initiator system in aqueous and non-aqueous polar media. The major initiating species are thought to be the isothioureide radical (129), and the thioureide radical cation (130) respectively, thus accounting for the different polymer

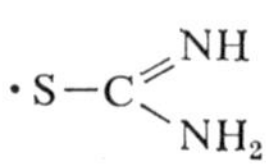

(129) 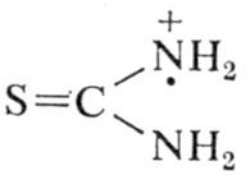   (130)

end-groups detected in the two media. Potentiometric and mechanistic studies on the oxidation of thiourea to formamidine disulphide by means of $Cu^{II}$ [268] and $Co^{III}$ [269] ions have been reported. The oxidation and reduction

---

[264] G. Tomaschewski, B. Breitfeld, and D. Zanke, *Tetrahedron Letters*, 1969, 3191.
[265] O. Mitsunobu, K. Kato, and F. Kakese, *Tetrahedron Letters*, 1969, 2473.
[266] P. K. Srivastava, *Indian J. Chem.*, 1969, 7, 323.
[267] T. K. Sengupta, D. Pramanick, and S. R. Palit, *Indian J. Chem.*, 1969, 7, 908.
[268] [a] H. C. Mruthyunjaya and A. R. Vasudeva Murthy, *Indian J. Chem.*, 1969, 7, 403; [b] *Analyt. Chem.*, 1969, 41, 186.
[269] A. McAuley and U. D. Gomwalk, *J. Chem. Soc. (A)*, 1969, 977.

of thiourea by ionising radiations have been investigated,[270] using e.s.r. and endor techniques at temperatures low enough (liquid helium) to stabilise the free-radical oxidation and reduction products.

The reaction of thioureas with nucleophiles has not received the same attention during the past year as that with electrophiles. However, a mechanistic study of the ethanolysis of 1,1'-thiocarbonylbispyrazoles leading to 1-ethoxythiocarbonyl pyrazoles (131),[271] the synthesis of 1-(2,2-dialkylthiocarbazoyl)imidazoles (132), via the hydrazinolysis of 1,1'-thiocarbonylbisimidazoles,[272] and the use of the hydrazinolysis of thioureas,

leading to 1,2,4-triazoles, as an analytical technique for the detection of thioureide groups,[273] have been described.

The utility of thioureas in heterocyclic synthesis is illustrated in a number of papers dealing with the synthesis of thiazoles by means of intramolecular cyclisation of thiourea-derivatives,[273, 274] and by Hantzsch-type procedures.[174, 222, 275–282] Other heterocyclic systems synthesised via thiourea-derivatives include oxazoles,[283] imidazoles,[284] 1,2,4-triazoles,[273, 285] pyrimidines and quinazolines,[279, 286, 287] 1,3-oxazines,[288] 1,3-thiazines,[289, 290] and 1,3,5-triazines.[285, 291]

[270] H. C. Box, H. G. Freund, K. T. Lilga, and E. E. Budzinski, *J. Phys. Chem.*, 1970, **74**, 40.
[271] L.-O. Carlsson and J. Sandström, *Acta Chem. Scand.*, 1970, **24**, 299.
[272] U. Anthoni, C. Larsen, and P. H. Nielsen, *Acta Chem. Scand.*, 1969, **23**, 1231.
[273] R. G. Dickinson and N. W. Jacobsen, *Analyt. Chem.*, 1969, **41**, 1324.
[274] J. Rabinowitz, *Helv. Chim. Acta*, 1969, **52**, 255.
[275] L. Pentimali, *Gazzetta*, 1969, **99**, 362.
[276] B. C. Das and G. N. Mahapatra, *J. Indian Chem. Soc.*, 1970, **47**, 98.
[277] J. Bartoszewski, *Roczniki Chem.*, 1969, 324.
[278] W. Hampel and I. Müller, *J. prakt. Chem.*, 1969, **311**, 684.
[279] Amrik Singh, A. S. Uppal, and K. S. Narang, *Indian J. Chem.*, 1969, **7**, 881.
[280] A. Y. Perkone, N. O. Saldabol, and S. A. Hiller, *Khim. geterotsikl. Soedinenii*, 1969, 498.
[281] A. Friedman and J. Metzger, *Compt. rend.*, 1969, **269C**, 1000.
[282] U. R. Rahman and H. S. E. Gatica, *Rec. Trav. chim.*, 1969, **9**, 905.
[283] W. L. Budde and O. Le Roy Salerni, *J. Heterocyclic Chem.*, 1969, **6**, 419.
[284] D. V. Bite and A. K. Aren, *Khim. geterotsikl. Soedinenii*, 1969, 719, 1083.
[285] J. S. Davidson, *J. Chem. Soc. (C)*, 1969, 194.
[286] W. Steiger, T. Kappe, and E. Ziegler, *Monatsh.*, 1969, **100**, 529.
[287] Z. F. Solomko, M. S. Malmovskii, L. N. Polovina, and V. I. Gorbalenko, *Khim. geterotsikl. Soedinenii*, 1969, 536.
[288] E. Ziegler, G. Kollenz, and T. Kappe, *Monatsh.*, 1969, **100**, 540.
[289] Yu. V. Migalina, V. I. Staninets, and I. V. Smolanka, *Ukrain. khim. Zhur.*, 1969, **35**, 526.
[290] Y. Ramachandra Rao, *Indian J. Chem.*, 1969, **7**, 772.
[291] D. L. Klayman and G. W. A. Milne, *Tetrahedron*, 1969, **25**, 191.

Among miscellaneous studies of thiourea reactivity, perhaps the most interesting is the attempt by Anthoni and his co-workers [292] to obtain alkoxyisothiocyanates (134) *via* thermal fragmentation of alkoxythioureas (133). Two fragmentation paths are possible, but in fact the aniline moiety is eliminated, affording the unstable required product (134). In a related study of the pyrolysis of $N$-(aryloxythiocarbonyl)thioureas (135),[293]

$$\text{EtONHCNHPh} \longrightarrow \text{EtON=C=S} + \text{PhNH}_2$$
$$\text{(133)} \qquad\qquad\qquad \text{(134)}$$

$$\text{EtONH}_2 + \text{PhN=C=S}$$

$$\underset{\underset{S}{\overset{\|}{\text{Me}_2\text{N}-\text{C}-\text{N}-\text{C}-\text{OAr}}}}{\overset{\text{Ar}^1}{|}}\qquad\qquad \underset{S}{\overset{\text{C}-\text{N}-\text{Ar}^1}{\text{Me}_2\text{N}\cdots\text{C}-\text{OAr}}}$$
$$\text{(135)} \qquad\qquad\qquad \text{(136)}$$

thionocarbamates and aryl isothiocyanates have been obtained. The latter reaction involves the four-membered cyclic intermediate (136). The synthesis of cyclic thionocarbonates from 1,2-diols, using 1,1′-thiocarbonylbisimidazole,[294] and the preparation of mercury mercaptides from the reaction of thioureas with mercury bis(phenylacetylide) [194] have been described.

Recent papers concerning the reactivity of heterocyclic thiones containing the thioureide group include studies on derivatives of imidazole,[295-300] pyrimidine,[301-304] and triazine.[303, 305, 306]

[292] C. Larsen, U. Anthoni, C. Cristophersen, and P. H. Nielsen, *Acta Chem. Scand.*, 1969, **23**, 324.
[293] K. Miyasaki, *Bull. Chem. Soc. Japan*, 1969, **42**, 1697.
[294] T. L. Nagabhushan, *Canad. J. Chem.*, 1970, **48**, 383.
[295] T. Yonezawa, M. Matsumoto, Y. Matsumura, and H. Kato, *Bull. Chem. Soc. Japan*, 1969, **42**, 2323.
[296] P. Neelakantan and G. Thyagarajan, *Indian J. Chem.*, 1969, **7**, 189.
[297] E. Dyer and C. E. Minnier, *J. Heterocyclic Chem.*, 1969, **6**, 23.
[298] O. H. Hankovsky and K. Hideg, *Acta Chim. Acad. Sci. Hung.*, 1969, **61**, 69.
[299] J. Nyitrai, R. Markovits-Korms, and K. Lempert, *Acta Chim. Acad. Sci. Hung.*, 1969, **60**, 141.
[300] E. N. Prilezhaeva and L. I. Shimonin, *Izvest. Akad. Nauk S.S.S.R., Ser. khim.*, 1969, 670.
[301] V. M. Vedenskii, M. P. Makucha, and L. Ya. Makarina-Kibak, *Khim. geterotsikl. Soedinenii*, 1969, 1096.
[302] V. Škaric and B. Gašpert, *J. Chem. Soc. (C)*, 1969, 2631.
[303] H. Vorbrüggen, P. Strehlke, and G. Schulz, *Angew. Chem. Internat. Edn.*, 1969, **8**, 976; H. Vorbrüggen and P. Strehlke, *ibid.*, p. 977.
[304] J. Moravek and J. J. Kopecky, *Coll. Czech. Chem. Comm.*, 1969, **34**, 4013.
[305] G. Hornyak, L. Lang, K. Lempert, and G. Menczel, *Acta Chim. Acad. Sci. Hung.*, 1969, **61**, 93.
[306] J. Sletten, *J. Amer. Chem. Soc.*, 1969, **91**, 4545.

## 9 Thiosemicarbazides and Derivatives

**Synthesis.**—Thiosemicarbazides are commonly prepared from the reaction of isothiocyanates with hydrazines, recent examples of this general procedure being the syntheses of aryl-,[227] styryl-,[307] and β-hydroxyethylthiosemicarbazides,[308] and thiosemicarbazones (137) derived from amidra-

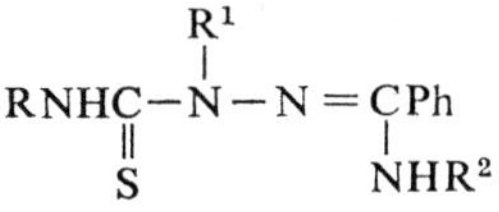

$$RNHC(=S)-N(R^1)-N=CPh(NHR^2)$$

(137)

zones.[309] In a modification of this procedure, Japanese workers[310] have obtained thiosemicarbazide from the reaction of hydrazine and isothiocyanic acid in ethanol, by means of hydrolysis of the intermediate 1,2,4-triazolidine-3-thione derivative (138). Anthoni and his co-workers[311] have

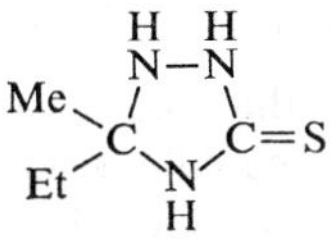

(138)

attempted to prepare phosphorus analogues of thiosemicarbazide such as (139) from the reaction of N-isothiocyanatodi-isopropylamine with secondary phosphines, but find that the initially-formed product (139) is cleaved by further reaction with the phosphine to di-isopropylammonium isothiocyanate and the diphosphine (140). Stable adducts (141), however, can be obtained from tertiary phosphines.

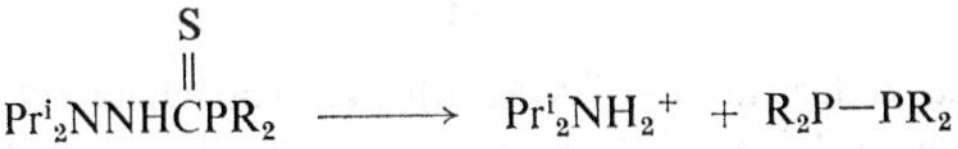

$$Pr^i_2NNHCPR_2(=S) \longrightarrow Pr^i_2NH_2^+ + R_2P-PR_2$$

(139)      (140)

$$Pr^i_2N-\bar{N}-C(=S)(PR_3^+) \longleftrightarrow Pr^i_2N-N=C(S^-)(PR_3^+)$$

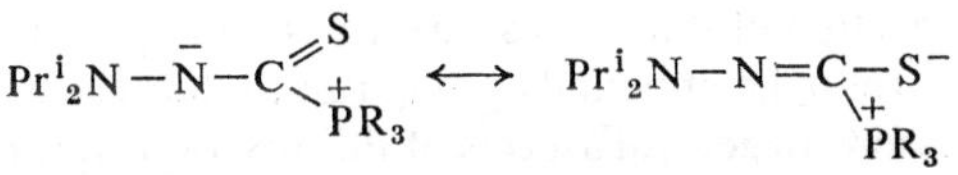

(141)

Other recent syntheses of thiosemicarbazides involve the aminolysis of 1,1′-thiocarbonylbisimidazole,[272] affording (132), and the reaction of

---

[307] N. A. Barba, A. M. Shur, and Nguyen Kong-Zinh, *Zhur. Vsesoyuz. Khim. obshch. im. D.I. Mendeleeva*, 1969, **14**, 464.
[308] J. Rabinowitz, S. Chang, J. M. Majes, and F. Woeller, *J. Org. Chem.*, 1969, **34**, 372.
[309] G. Barnikow and W. Abraham, *Z. Chem.*, 1969, **9**, 183.
[310] I. Arai, Y. Satoh, I. Muramatsu, and A. Hagitani, *Bull. Chem. Soc. Japan*, 1969, **42**, 2739.
[311] U. Anthoni, O. Dahl, C. Larsen, and P. H. Nielsen, *Acta Chem. Scand.*, 1969, **23**, 943.

*N*-aryldithiocarbamates with hydrazine and phenylhydrazine.[312] Kurzer and Wilkinson[313] have corrected a previous report by other workers, claiming the synthesis of 1-benzoylthiocarbohydrazides (143) from the hydrazinolysis of esters of 3-benzoyldithiocarbazic acid (142). The present

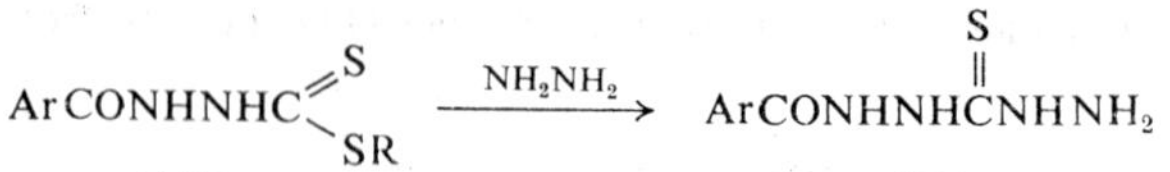

$$ArCONHNHC\begin{smallmatrix}\nearrow S \\ \searrow SR\end{smallmatrix} \xrightarrow{NH_2NH_2} ArCONHNHC(=S)NHNH_2$$

(142)   (143)

$$Ar\underset{\underset{NH_2}{|}}{\overset{N-NH}{\underset{N}{\diagdown}}}S$$

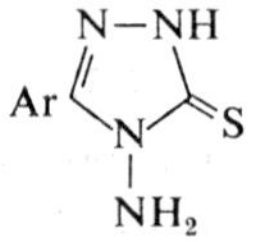

(144)

authors[313] find that the products of this reaction are in fact 4-amino-1,2,4-triazoline-5-thiones (144).

**Reactions.**—The reactivity of thiosemicarbazides and related compounds tends to be dominated by nucleophilic reactions of the hydrazine moiety, although the nucleophilic reactivity of the thiocarbonyl sulphur atom is important in many of their cyclisation reactions. However, thiosemicarbazides can be alkylated to alkyl halides at sulphur, recent examples including the *S*-methylation of *N*-acylthiosemicarbazides[314, 315] and of the dianilide (145),[316] which spontaneously cyclises to a 1,2,4-triazole deriva-

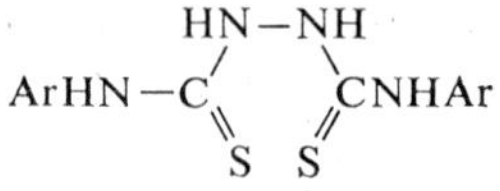

$$ArHN-C\underset{\underset{S}{\diagdown\diagdown}}{\overset{HN-NH}{\diagup}}\underset{S}{\diagdown}CNHAr$$

(145)

tive. The *S*-methyl derivatives of *N*-acylthiosemicarbazides can be thermally cyclised to amino-1,3,4-oxadiazoles[314, 315] and to derivatives of mercapto-1,2,4-triazole.[314]

The classic reaction of thiosemicarbazides is, of course, the condensation reaction with carbonyl compounds giving rise to thiosemicarbazones. Two recent papers deal with general aspects of this reaction. Sayer and Jencks[317] report that the formation of the thiosemicarbazone from *p*-chlorobenzaldehyde exhibits a change of rate-determining step at *ca.* pH 4, with nucleophilic attack on the carbonyl group rate-determining above, and dehydration rate-determining below this pH. Lamon has shown[318] that mono-

---

[312] M. G. Paranjpe and P. H. Deshpande, *Indian J. Chem.*, 1969, **7**, 186.
[313] F. Kurzer and M. Wilkinson, *J. Chem. Soc. (C)*, 1969, 1218.
[314] J. E. Giudicelli, J. Menin, and H. Najer, *Bull. Soc. chim. France*, 1969, 870, 874.
[315] Y. Kitamura, M. Yamashita, M. Kashihara, and I. Hirao, *J. Chem. Soc. Japan*, 1969, **90**, 713.
[316] I. Simiti and I. Proinov, *J. prakt. Chem.*, 1969, **311**, 687.
[317] J. M. Sayer and W. P. Jencks, *J. Amer. Chem. Soc.*, 1969, **91**, 6353.
[318] R. W. Lamon, *J. Org. Chem.*, 1969, **24**, 756.

hydrazones obtained from thiocarbohydrazide and simple aldehydes and ketones often have the tetrazinethione structure (147), rather than the open-chain structure (146). Straight-chain aliphatic aldehydes, in fact,

$$\underset{R^1}{\overset{R}{\diagdown}}C=NNHCNHNH_2 \quad \quad \underset{R \quad R^1}{\overset{S}{\underset{HN \diagdown NH}{\overset{HN \diagup NH}{\|}}}}$$
$$\overset{\|}{S}$$

(146)         (147)

produce entirely the cyclic isomer. Thiosemicarbazones prepared by the usual method have been cyclised by various procedures to derivatives of thiazoline,[319] amino-1,3,4-thiadiazole,[320-322] and 1,2,4-triazolinethione.[321-323] In addition, thiosemicarbazones derived from α-dicarbonyl compounds have been cyclised to 1,2,4-triazinethiones,[324, 325] and the reaction of thiosemicarbazides with polyfunctional carbonyl compounds such as α-dichloroacetyl chloride [326] and diazoketones [327] has been observed to yield derivatives of amino-1,3,4-thiadiazole [326] and amino-1,3,4-thiadiazine,[327] presumably *via* thiosemicarbazone intermediates.

As further examples of the nucleophilicity of the hydrazine moiety of thiosemicarbazides, Howe and Bolluyt [328] have shown that reaction with ethoxymethylenemalonates or 2-cyano-3-ethoxyacrylates affords open-chain substitution products such as (148). The cyclisation of (148a) gives an amino-1,3,4-thiadiazole while (148b) is cyclised to a 5-amino-1-thiocarbamoylpyrazole derivative. 1-Phenylthiocarbohydrazide has been

$$\overset{S}{\underset{\|}{RNHCNHNHC}}=C\overset{X}{\underset{CO_2Et}{\diagdown}} \quad \quad \begin{array}{l}(a)\ X = CO_2Et \\ (b)\ X = CN\end{array}$$

(148)

$$\overset{S}{\underset{\|}{PhNHNHC}}\overset{NR}{\underset{\|}{NHNHCNHR}}$$

(149)

[319] I. V. Smolanka and N. P. Manyo, *Ukrain. khim. Zhur.*, 1969, **35**, 509.
[320] I. S. Ioffe, and A. B. Tomchin, *Zhur. obshchei Khim.*, 1969, **39**, 1150.
[321] J. K. Landquist, *J. Chem. Soc. (C.)* 1970, 63, 323.
[322] T. Radha Vakula, V. Ranga Rao, and V. R. Srinivasan, *Indian J. Chem.*, 1969, **7**, 577.
[323] R. Grashey and M. Baumann, *Tetrahedron Letters*, 1969, 2173.
[324] I. S. Ioffe, A. B. Tomchin, and E. A. Rusakov, *Zhur. obshchei Khim.*, 1969, **39**, 2345.
[325] J. Daunis, R. Jacquier, and P. Viallefont, *Bull. Soc. chim. France*, 1969, 3675.
[326] W. A. Remers, G. J. Gibbs, and M. J. Weiss, *J. Heterocyclic Chem.*, 1969, **6**, 835.
[327] W. Hampel, *Z. Chem.*, 1969, **9**, 61; W. Hampel, M. Kapp, and G. Proksch, *ibid.*, 1970, **10**, 142.
[328] R. K. Howe and S. C. Bolluyt, *J. Org. Chem.*, 1969, **34**, 1713.

shown [329] to react with carbodi-imides to yield open-chain adducts (149), which can be cyclised to derivatives of 1,2,4-triazoline-5-thione or 2-amino-1,3,4-thiadiazole, under different conditions.

Anthoni and his co-workers,[330] in continuation of their studies of N-isothiocyanato-compounds (see above section 8, ref. 292), have shown that pyrolysis of the thiosemicarbazide (150), which exists in the dipolar form (151), affords the relatively stable N-di-isopropylaminoisothiocyanate (152), whereas thiosemicarbazides, which exist in the non-polar form

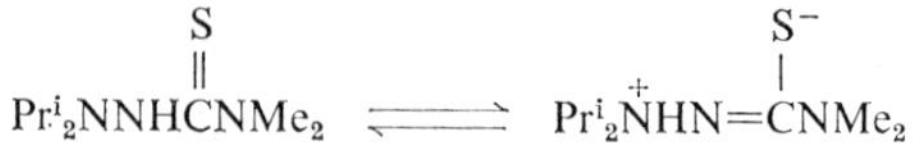

$$Pr^i_2NNHCNMe_2 \text{ (S)} \rightleftharpoons Pr^i_2\overset{+}{N}HN=CNMe_2 \text{ (S}^-\text{)}$$

(150)     (151)

$$Pr^i_2NN=C=S$$

(152)

similar to (150), afford unstable isothiocyanates which dimerise to 1,2,4-triazoline-5-thione derivatives.

Other recent papers dealing with thiosemicarbazide chemistry include reports on their cyclisation to 2-amino-5-mercapto-1,3,4-thiadiazoles using carbon disulphide,[331] and on the supposed thermal cyclisation of thiocarbohydrazide to the tetrazolidinethione (153).[332] Results quoted

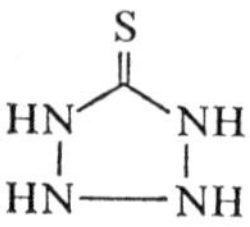

$$\begin{array}{c} S \\ \| \\ HN \diagup \diagdown NH \\ | \quad | \\ HN \text{———} NH \end{array}$$

(153)

in the latter paper show clearly that this unlikely 'cyclisation' leads rather to polymeric products.

## 10 Thionocarboxylic and Dithiocarboxylic Acids and Derivatives

**Synthesis.**—Several groups of workers have employed the base-catalysed reaction of carbon disulphide with reactive methylene groups in the synthesis of dithiocarboxylic acids and their esters derived from dimethylmalonate,[333] 1,3-dibenzoylpropane,[334] cyclopentanone,[335] and 3-alkyl- or

---

[329] F. Kurzer and M. Wilkinson, *J. Chem. Soc.* (*C*), 1970, 26.
[330] C. Larsen, U. Anthoni, and P. H. Nielsen, *Acta Chem. Scand.*, 1969, **23**, 320; C. Larsen, U. Anthoni, C. Cristophersen, and P. H. Nielsen, *ibid.*, p. 322.
[331] V. S. Misra and N. S. Agarwal, *J. prakt. Chem.*, 1969, **311**, 697.
[332] K.-H. Linke and J. Dahm, *Z. Naturforsch.*, 1969, **24b**, 260.
[333] Y. Shvo and I. Belsky, *Tetrahedron*, 1969, **25**, 4649.
[334] F. Clesse and H. Quiniou, *Compt. rend.*, 1969, **268** *C*, 637.
[335] T. Takeshima, M. Yokoyama, T. Imamoto, M. Akano, and H. Asaba, *J. Org. Chem.*, 1969, **34**, 730.

# Thiocarbonyls, Selenocarbonyls, and Tellurocarbonyls

3-aryl-camphor.[336] Most significantly, Treibs[337] has shown that the reaction of carbon disulphide with highly reactive alkylpyrroles proceeds readily to pyrrole-2-dithiocarboxylic acids in the presence of bases, without requiring the formation of an intermediate 2-pyrrolyl Grignard reagent. Other recent examples of the use of well-known procedures include the syntheses of trifluorodithioacetic acid[338] and esters of aminodithioacids[339] from the reaction of the corresponding nitriles and hydrogen sulphide, and the phosphorus pentasulphide thionation of derivatives of anthranilic acid leading to benzoisothiazolinethiones, presumably *via* cyclisation of an intermediate dithiocarboxylic acid.[340] The use of anthranilate esters in the latter procedure, however, affords derivatives of the interesting phosphorus heterocycle (154), which can be cleaved by alkoxide

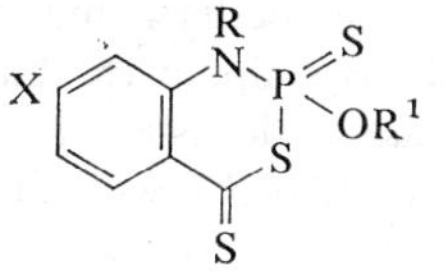

(154) 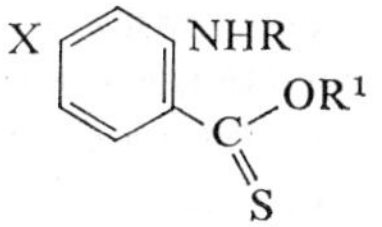     (155)

to thionoesters such as (155).[340b] A thionoester (157) has also been obtained from the facile photoconversion of the thio-ozonide (156), the first example of a new class of potentially useful synthetic intermediates.[341]

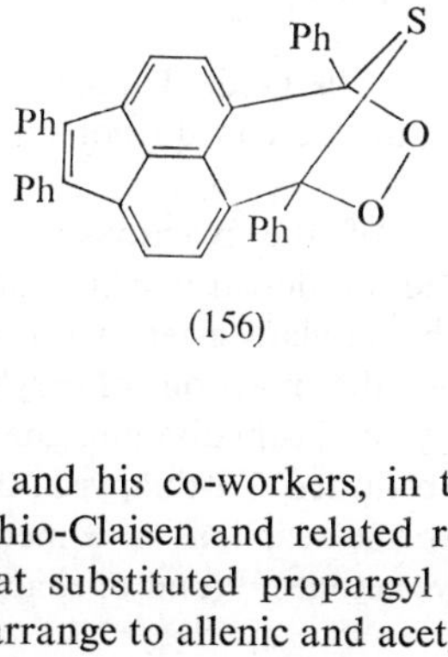

(156) 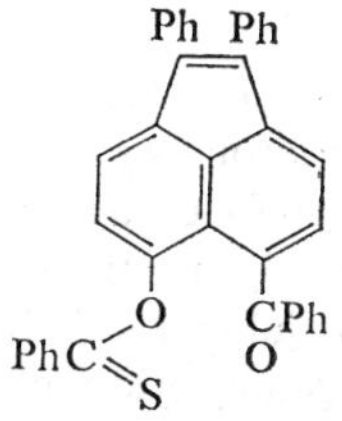     (157)

Brandsma and his co-workers, in the course of their investigations in the field of thio-Claisen and related rearrangements (see section 7), have shown[219] that substituted propargyl and allenyl vinyl sulphides can be caused to rearrange to allenic and acetylenic dithioesters, such as (158) and (159), respectively. Keten dithioacetals which possess *S*-alkyl and *S*-acyl groups have also been converted into dithioesters by means of their respective reaction with lithium in liquid ammonia[342] and with amines.[333]

[336] A.-M. Lamazouère, J. Sotiropoulos, and P. Bedos, *Compt. rend.*, 1970, **270**, C, 828.
[337] A. Treibs, *Annalen*, 1969, **723**, 129.
[338] E. Lindner and U. Kunze, *Chem. Ber.* 1969, **102**, 3347.
[339] H. Hartmann, W. Stapf, and J. Heidberg, *Annalen*, 1969, **728**, 237.
[340] L. Legrand and N. Lozac'h, *Bull. Soc. chim. France*, 1969, [a] p. 1170; [b] p. 1173.
[341] J. M. Hoffmann and R. H. Schlessinger, *Tetrahedron Letters*, 1970, 797.
[342] L. Brandsma and P. J. W. Schuijl, *Rec. Trav. chim.*, 1969, **88**, 513.

$$R^1CH=C\underset{SEt}{\overset{SCH_2C\equiv CR^2}{<}} \xrightarrow{\Delta} H_2C=C=\underset{}{\overset{R^2\ R^1}{C}}-\underset{}{C}-C\underset{SEt}{\overset{S}{<}}$$
(158)

$$\underset{H_2C=C}{\overset{H_2C=C=C}{<}}\underset{SEt}{\overset{R}{<}}\ \xrightarrow{\Delta}\ RC\equiv CCH_2CH_2C\underset{SEt}{\overset{S}{<}}$$
(159)

**Reactions.**—Olah and his colleagues [343] have now extended their expansive n.m.r. study of the protonation of organic compounds in 'super-acidic' media to include dithiocarboxylic acids, thionoesters, and dithioesters. Not unexpectedly, protonation at thiocarbonyl sulphur leading to cations of general type (160) is observed. In the case of the esters, an equilibrium mixture of the protonated isomers (160a) and (160b), with the latter predominant, is detected. With dithiocarboxylic acids, however, only signals

$$RC\overset{S}{\underset{R^1}{\overset{+}{\underset{X}{<}}}}H \qquad RC\overset{H}{\underset{X}{\overset{S}{\underset{+}{<}}}}R^1 \qquad X = O\ or\ S$$

(160a) \qquad\qquad (160b)

due to the cation (160a) ($R^1 = H$) are observed. Unlike their thiolester isomers,[343b] protonated thiono- and dithio-esters do not readily undergo alkyl-$S$ or acyl-$S$ cleavage.[343a]

Other recently recorded examples of the reactions at nucleophilic thiocarbonyl sulphur of thiono- and dithio-carboxylic acid derivatives include the $S$-alkylation of aminothioacrylate esters with alkyl iodides and with α-halogenoketones,[185] and the reaction of aryldithioacetate esters with dithiolium salts, leading to 6a-thiathiophthenes or to 2H-thiapyran-2-thiones.[334] Variously substituted 2H-thiapyran-2-thione derivatives have also been prepared from the reactions of α-methylenedithiocarboxylic acids with β-halogenovinyl ketones [344] and of α-acyldithioacetic acids with α-acetylenic ketones.[345]

The high reactivity of thionoesters with nucleophiles, and particularly with amines, is much exploited in heterocyclic synthesis. Reynaud et al.[346] have now employed the reaction of thionoesters with diamines, which are present as their monoammonium tosylate salts, in the synthesis

[343] *a* G. A. Olah and A. T. Ku, *J. Org. Chem.*, 1970, **35**, 331; *b* G. A. Olah, A. T. Ku, and A. M. White, *ibid.*, 1969, **34**, 1827.
[344] S. Scheithauer and R. Mayer, *Z. Chem.*, 1969, **9**, 59.
[345] J. P. Pradère, A. Guenec, G. Deguay, J. P. Guemas, and H. Quiniou, *Compt. rend.*, 1969, **269**, *C*, 929.
[346] P. Reynaud, Phan Chi Dao, and I. Ismaili, *Compt. rend.*, 1969, **268**, *C*, 432.

of cyclic amidines, including imidazolines and tetrahydropyrimidines. In contrast, Lozac'h and his co-workers [347] report that the nucleophilic attack of amines at the thiocarbonyl carbon atom of the cyclic dithioester, benzo-1,3-thiazine-4-thione, results in a ring-cleavage leading to thioanilide derivatives. The hydrazide (161) has been obtained from the corresponding dithiocarboxylic acid on hydrazinolysis.[348]

$$\underset{(161)}{\overset{RN-NR^1}{\underset{S}{\overset{O}{\diagup}}\overset{O}{\underset{C}{\diagdown}}\overset{}{\underset{NHNH_2}{}}}}$$

Brandsma and his co-workers [219, 349] have studied the reaction of α-methylenedithioesters with strong bases in liquid ammonia, and have shown that abstraction of a proton from the α-carbon atom leads to an enolate anion such as (162). Similar anions can also be produced by the action of base on keten dithioacetals.[342] α-Allenyl derivatives of (162)

$$\underset{R^1}{\overset{R}{\diagdown}}C=C\underset{SR^2}{\overset{S^-}{\diagup}}$$

(162)

have been cyclised to 2H-thiapyrans [219b] and to thiophens.[349] The chemistry of dithiolate dianions (163), derived from cyanodithioacrylic acid, has been studied by three groups of workers. Klemm and Geiger report [350] the synthesis of benzo-1,3-dithioles from the reaction of (163) with *p*-quinones, while Henriksen [351] has shown that various reactions of (163) (R = $CONH_2$) in acidic media probably pass through the free dithiocarboxylic acid dimer (164) as an intermediate. Benzo-1,3-dithioles have also been

$$\underset{NC}{\overset{R}{\diagdown}}C=C\underset{S^-}{\overset{S^-}{\diagup}} \qquad \underset{NC}{\overset{H_2NOC}{\diagdown}}C=C\underset{S}{\overset{S^-}{\diagup}}\underset{H}{\overset{S}{\diagdown}}\overset{S}{\underset{}{\diagdown}}C=\overset{+}{N}H_2 \\ \qquad\qquad\qquad\qquad\qquad\qquad C-CHCONH_2$$

(163)          (164)

obtained from the reaction of (163) with di- and tri-nitrochlorobenzenes.[352]

In a paper which may be of significance to workers in the field of carbohydrate chemistry, Hedgley and Leon [353] describe various transformations of glyceryl esters of thionobenzoic acid. These include thermal isomerisation to *S*-benzoyl esters, Raney nickel desulphuration to alcohols, and

---

[347] C. Denis-Garez, L. Legrand, and N. Lozac'h, *Bull. Soc. chim. France*, 1969, 3727.
[348] C. Papini and G. Auzzi, *Gazzetta*, 1969, **99**, 97.
[349] D. Schuijl-Laros, P. J. W. Schuijl, and L. Brandsma, *Rec. Trav. chim.*, 1969, **88**, 1343.
[350] K. Klemm and B. Geiger, *Annalen*, 1969, **726**, 103.
[351] L. Henriksen, *Chem. Comm.*, 1969, 1408.
[352] M. Yokoyama, *J. Org. Chem.*, 1970, **35**, 283.
[353] E. J. Hedgley and N. H. Leon, *J. Chem. Soc. (C)*, 1970, 467.

specific replacement of thiocarbonyl sulphur by carbonyl oxygen, through silver nitrate treatment, in both thiono- and dithio-benzoates. The acetic anhydride–boron trifluoride cyclisation of thiobenzoylthioglycollic acid, previously regarded as leading to the mesoionic system (165), has now been shown to give the cyclised dimer (166).[354] The monomer (165) can be

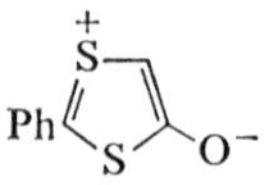

(165)

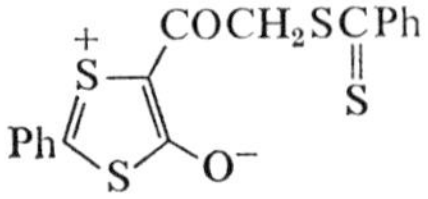

(166)

obtained by use of an acetic anhydride–triethylamine cyclisation agent, under rigorously moisture-free conditions.

## 11 Thionocarbonates, Thionodithiocarbonates, and Trithiocarbonates

**Synthesis.**—Jones and Andreades [355] have described the first synthesis of the thermally unstable ethylene thionocarbonate (167) by means of the reaction between ethylene glycol and thiophosgene in the presence of

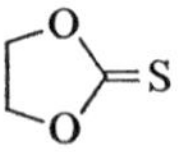

(167)

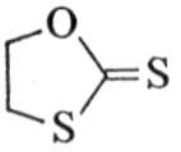

(168)

potassium carbonate. Similar treatment of the lead salt of 2-mercaptoethanol provides the first satisfactory synthesis of 1,3-oxathiolan-2-thione (168). Cyclic thionocarbonates have also been obtained [356] by room-temperature treatment of organo-tin derivatives, such as (169), with carbon disulphide, and by the reaction of 1,2-diols with $NN'$-thiocarbonyl-bis-heterocycles.[294, 357]

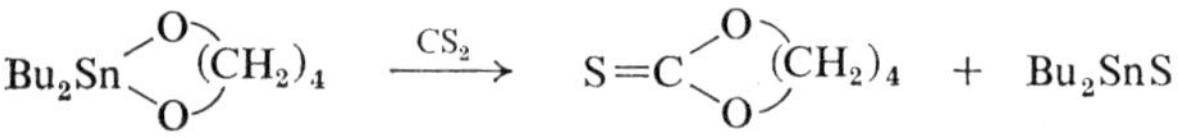

(169)

The low-temperature treatment of trimethyl-lead alkoxides with carbon disulphide is reported to yield the trimethyl-lead xanthate (170).[358] Sodium xanthate salts (171) have been prepared [359] from the reaction of carbon disulphide with the corresponding acetylenic alkoxide, and are found to be

[354] K. T. Potts and U. P. Singh, *Chem. Comm.*, 1969, 569.
[355] F. N. Jones and S. Andreades, *J. Org. Chem.*, 1969, **34**, 3011.
[356] S. Sakai, Y. Kobayashi, and Y. Ishii, *Chem. Comm.*, 1970, 235.
[357] E. Fujita, T. Fujita, H. Katayama, and Y. Nagao, *Tetrahedron*, 1969, **25**, 1335.
[358] R. Hönigschmid-Grossich and E. Amberger, *Chem. Ber.*, 1969, **102**, 3589.
[359] K. Tomita and M. Nagano, *Chem. and Pharm. Bull. (Japan)*, 1969, **17**, 2442.

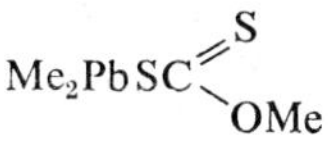

(170) Me₂PbSC(=S)OMe

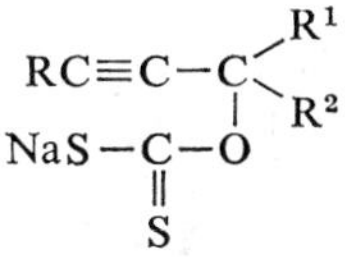

(171) RC≡C−C(R¹)(R²)−C(=S)−SNa... 

Structures (170) and (171) shown.

readily converted into 4-alkylidene derivatives of 1,3-oxathiolan-2-thione and 1,3-dithiolan-2-thione. German workers[360] have reported the isolation of thermally unstable methylenetrithiocarbonate (172) from the reaction

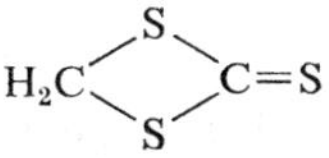

$$H_2C\underset{S}{\overset{S}{\diagup\!\!\!\diagdown}}C=S$$

(172)

of potassium trithiocarbonate with di-iodomethane. An example of the synthesis of derivatives of ethylene trithiocarbonate by means of the base-catalysed addition of carbon disulphide to episulphides has been described by Russian workers.[361]

**Reactions.**—Recent investigations of the chemistry of thiocarbonate derivatives have dealt predominantly with xanthates. This is not unexpected in view of the synthetic utility of these compounds, which is exemplified in the well-known Tschugaev synthesis of olefins. An important modification of the latter procedure has been reported by Turbak and his co-workers,[362] in which the sodium xanthate corresponding to the required olefin product is treated with a compound which will provide hetero-atom unsaturation one, two, or three methylene groups distant from the xanthate alkylthio sulphur atom in the adduct thus formed. Such an adduct, e.g. (173), may be converted in weakly acidic media to the required olefin at a much lower temperature than required for the thermal decomposition of simple xanthates. This mild procedure greatly improves the viability of the Tschugaev reaction in industrial processes, and also affords a method of steroid dehydration incorporating specific positioning of the double bond produced. The thermal fragmentation of S-p-bromophenacyl xanthates (174), which are in fact examples of the general class (173), has been reported[363] to yield non-Tschugaev products, particularly alkyl bromides, in significant amounts. The authors conclude that these products arise from homolytic cleavage either of bond A or of bond B in structure (174),

---

[360] J. Wortmann and G. Gattow, Z. Naturforsch., 1969, **24b**, 1194.
[361] M. G. Linkova, A. M. Orlov, O. V. Kil'disheva, and I. L. Knunyants, Izvest. Akad. Nauk. S.S.S.R., Ser. khim., 1969, 1148.
[362] A. Turbak, N. Burke, and D. Bridgeford, Chem. Eng. News, 1969, **47**, No. 40, 41.
[363] R. E. Gilman, J. D. Henion, S. Shakshooki, J. I. H. Patterson, M. J. Bogdanowicz, R. J. Griffith, D. E. Harrington, R. K. Crandall, and K. T. Finley, Canad. J. Chem., 1970, **48**, 971.

$$\underset{(173)}{\overset{R^2\ R^3}{\underset{R^1-CH}{\overset{|}{\underset{R}{C}}}\overset{O}{\underset{|}{\overset{|}{C}}}\overset{S}{\underset{S}{\overset{|}{\underset{C}{C}}}\overset{(CH_2)_n}{\underset{Y\ Z}{|}}}} \longrightarrow \underset{R^1}{\overset{R}{>}}C=C\underset{R^3}{\overset{R^2}{<}} + COS + RSH$$

$$\underset{(174)}{RCH_2O\overset{S}{\overset{\|}{C}}\overset{A}{-}S\overset{B}{-}CH_2\overset{O}{\overset{\|}{C}}-\left\langle\!\!\!\begin{array}{c}\phantom{x}\\\phantom{x}\end{array}\!\!\!\right\rangle-Br}$$

the free-radical reactions competing with the normal cis-elimination Tschugaev processes. In related studies, the isolation of isobutene from the metal-halide-promoted decomposition of t-butyl xanthates,[364] the thermal rearrangement of O,S-diaryldithiocarbonates, which do not possess the Tschugaev requirement of β-hydrogens in the alkoxy-moiety, to SS-diaryldithiocarbonates,[365] and the spectral detection of carbon disulphide in the thermal decomposition of xanthic acid salts [366] have been reported.

A further important synthetic use of xanthates is in the formation of thiols via aminolysis of the alkylthio-moiety. Using this procedure, alkanethiols have recently been obtained from the reaction of trithiocarbonates, xanthates, and dithiocarbonates with 2-aminoethanol,[367] alkanethiols and dialkyl sulphides from xanthates and ethylenediamine,[368] and N-(mercaptoacyl)amino-acids from the corresponding xanthate and ammonia.[368a] The alkylthio-group of xanthates is also subject to hydrazinolysis, as has been shown by Anthoni and his co-workers,[369] who have obtained alkoxythiocarbonylhydrazides such as (175) by this means.

$$\underset{(175)}{RO\overset{S}{\overset{\|}{C}}NR^1NH_2}$$

Yoshida [370] has shown that disulphide bonds in bisxanthates and in xanthic thioanhydrides can be cleaved by triethylamine and trimethylamine, affording the corresponding S-ethyl xanthates and quaternary ammonium xanthates respectively. The latter salts are also obtained from the reaction of trimethylamine with S-ethoxycarbonyl xanthates.

---

[364] J. P. Fackler jun., W. C. Seidel, and M. Myron, *Chem. Comm.*, 1969, 1133.
[365] Y. Araki, *Bull. Chem. Soc. Japan*, 1970, **43**, 252.
[366] J. F. Devlin, *J. Polymer Sci.*, Part B, Polymer Letters, 1969, **7**, 515.
[367] T. Taguchi, Y. Kiyoshima, O. Komori, and M. Mori, *Tetrahedron Letters*, 1969, 3631.
[368] K. Mori and Y. Nakamura, *J. Org. Chem.*, 1969, **34**, 4170.
[368a] V. L. Vasilevskii, K. Georgiadis, and I. M. Abdel-Mageb, *Zhur. org. Khim.*, 1970, **6**, 244.
[369] K. A. Jensen, U. Anthoni, and A. Holm, *Acta Chem. Scand.*, 1969, **23**, 1916.
[370] H. Yoshida, *J. Chem. Soc. Japan*, 1969, **90**, 1036.

Schönberg and his colleagues [75, 371] have studied the reactions of diaryl-diazomethanes with bisxanthate derivatives. From the reaction of dixanthogens, they have obtained [75] methylene bisxanthates (17) which can be thermally converted into thiones (see section 3) and xanthic anhydrides (176). The latter products also react with diaryldiazomethanes [371] to give thiirans such as (177), which can be further converted to derivatives of α,α'-diarylthionoacetic acid and to 2,4-bis(diarylmethylene)-1,3-thietans (178).

$$\begin{array}{ccc} \overset{S}{\underset{\underset{S}{\overset{|}{\underset{COR^1}{\overset{COR}{|}}}}{\parallel}}{}} & R^3\underset{R^2}{\overset{}{>}}C\overset{S}{\underset{}{\diagdown}}\underset{}{\overset{}{-}}C\overset{SCOR^1}{\underset{OR}{\diagup}} & R^3\underset{R^2}{\overset{}{>}}C=C\overset{S}{\underset{S}{\diagdown}}C=C\overset{R^3}{\underset{R^2}{\diagdown}} \\ (176) & (177) & (178) \end{array}$$

Japanese workers [372] have studied the effect of metal ions on the acid decomposition of alkali-metal xanthates. They find that lead(II), cadmium(II), and zinc(II) ions retard the rate of decomposition of $O$-ethyl xanthates to ethanol and carbon disulphide through the formation of M(xanthate)$^+$ complex ions. Finkelstein [373] has shown that trialkylammonium halides catalyse the decomposition of alkali-metal xanthates at high pH, and regards this as proceeding through an oxidation process involving atmospheric oxygen, rather than the normal uncatalysed mechanism of thiocarbonyl bond hydration. Although the known oxidation products of xanthates, dixanthogens, have not been detected in the foregoing oxidative decomposition process, their formation from xanthates using copper(II) salts as oxidants has now been studied potentiometrically.[268a] The mechanism of the reaction is thus shown to be a two-stage process involving initial formation of dixanthogen together with copper xanthate, which is in turn oxidised by copper(II) ions to dixanthogen in the second stage. Certain cyclic thiono- and trithio-carbonate sugar derivatives have been found to undergo oxidation at the thiocarbonyl group, leading to the corresponding carbonyl compounds, on chlorination.[374] With similar xanthate derivatives, however, the expected addition of chlorine across the thiocarbonyl double bond was observed. The authors suggest that the sulphenyl halides isolated in the latter case are less sensitive to moisture, which was not rigorously excluded from the reaction media, than their cyclic analogues which are hydrolysed to the observed carbonyl

[371] A. Schönberg, W. Knöfel, E. Frese, and K. Praefcke, *Chem. Ber.*, 1970, **103**, 949.
[372] M. Nanjo and T. Yamasaki, *Bull. Chem. Soc. Japan*, 1969, **42**, 972.
[373] N. P. Finkelstein, *J. Appl. Chem.*, 1969, **19**, 73.
[374] B. S. Shasha, W. M. Doane, C. R. Russell, and C. E. Rist, *J. Org. Chem.*, 1969, **34**, 1642.

products. The catalytic hydrogenolysis of cyclic and acyclic trithio-carbonates, leading to 1,3-dithiolans and the corresponding acyclic sulphides, respectively, has been described.[375]

Various reports on miscellaneous reactions of thionocarbonate derivatives include descriptions of the preparation of $S$-triphenylgermanium, $S$-triphenyltin, and $S$-triphenyl-lead xanthates,[376] the reaction of sodium xanthates with arylsulphenyl chlorides leading to alkoxythiocarbonyl aryl disulphides,[377] the phosphorus pentasulphide cyclisation of $S$-phenacyl xanthates to 1,3-dithiole-2-thiones,[378] the formation of the tris(phenylthio)-methyl radical from diphenyl trithiocarbonate,[379] and the photodethiocarbonylation of diaryl trithiocarbonates.[380]

## 12 Thionocarbamic and Dithiocarbamic Acids and Derivatives

**Synthesis.**—The parent dithiocarbamic acid has been liberated from its ammonium salt by the reaction of the latter in ethereal suspension with gaseous hydrogen chloride at 0 °C.[381] Under these conditions dithiocarbamic acid does not form an addition compound with the medium. The decomposition of the product to carbon disulphide and ammonium dithiocarbamate occurs on raising the temperature to 20 °C, though it is still evidently more stable than its analogue, carbamic acid. Martin and Rackow[382] have reported a new general synthesis of $N$-substituted $O$-aryl thionocarbamates (180) which involves the thiolysis of the chloro-iminoether (179) with hydrogen sulphide in the presence of triethylamine.

$$RN=C\begin{matrix}OAr\\Cl\end{matrix} \qquad\qquad RNHC\begin{matrix}S\\OAr\end{matrix}$$
$$(179) \qquad\qquad\qquad (180)$$

Since the ether, which is prepared from a dichloromethylene imine precursor, need not be isolated, the procedure is readily carried out. A new synthesis of dithiocarbamates *via* the Willgerodt–Kindler-type reaction of $NN$-disubstituted thioformamides with sulphur in the presence of a secondary amine has been reported.[148]

The well-known general synthesis of dithiocarbamates *via* the addition of primary and secondary amines to carbon disulphide has now been studied in respect of tertiary benzylamines.[383] The investigators have shown

---

[375] D. J. Martin, *J. Org. Chem.*, 1969, **34**, 473.
[376] M. Schmidt, H. Schumann, F. Gliniecki, and J. F. Jaggard, *J. Organometallic Chem.*, 1969, **17**, 277.
[377] F. Klivenyi, E. Vinkler, and A. E. Szabó, *Acta Chim. Acad. Sci. Hung.*, 1970, **63**, 437.
[378] K. M. Pazdro, *Roczniki Chem.*, 1969, **43**, 1089.
[379] K. Uneyama, T. Sadokage, and S. Oae, *Tetrahedron Letters*, 1969, 5193.
[380] H. G. Heine and W. Metzner, *Annalen*, 1969, **724**, 223.
[381] J. L. Fourquet, *Bull. Soc. chim. France*, 1969, 3001.
[382] D. Martin and S. Rackow, *Z. Chem.*, 1969, **9**, 21.
[383] A. O. Fitton, A. Rigby, and R. J. Hurlock, *J. Chem. Soc.* (*C*), 1969, 230.

that the mechanism involves an overall insertion of carbon disulphide into a C–N bond, the initial product being the primary adduct (181). Fragmentation of this adduct leads to a dithiocarbamate anion (182) and a quinone methide (183), which recombine by means of nucleophilic attack of anionic sulphur on the reactive methylene group to afford the observed product (184). The formation of the quinone methide intermediate is the

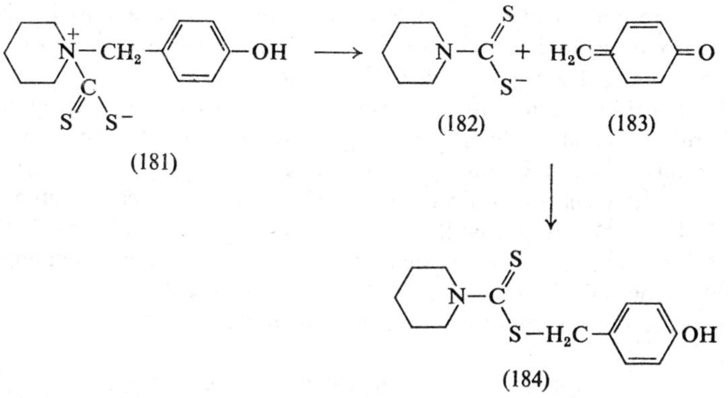

important step, making the presence of a phenolic hydroxy-group at position 4 of the benzyl group the crucial structural factor. Several recent examples of the general synthesis of dithiocarbamates from amines and carbon disulphide can be found in refs. 233 and 384—386.

As a variation on the procedure, Anthoni and his colleagues have described the synthesis of dithiocarbazic acid from the base-catalysed reaction of hydrazine with carbon disulphide.[387] The reaction between carbon disulphide and isothioureas has been investigated,[388] and support for the participation of the trithioallophanic acid derivative (185) obtained. Under non-oxidising conditions the S-benzyl dithiocarbamate (186) is

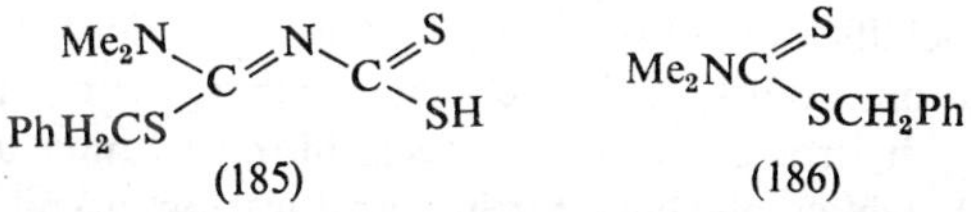

isolated, rather than the previously obtained 1,2,4-thiazolidine-3-thione. The insertion of carbon disulphide into N–S [389] and N–Si [390] bonds has been described, leading to S-alkylthio and S-silyl dithiocarbamates re-

---

[384] G. Kjellin and J. Sandström, *Acta Chem. Scand.*, 1969, **23**, 2879.
[385] A. Takimazawa and K. Hirai, *Chem. and Pharm. Bull. (Japan)*, 1969, **17**, 1924.
[386] E. J. Rukhadze, E. V. Pavlova, V. V. Dunina, and A. P. Terent'ov, *Zhur. obshchei. Khim.*, 1969, **39**, 2545.
[387] U. Anthoni, B. M. Dahl, C. Larsen, and P. H. Nielsen, *Acta Chem. Scand.*, 1969, **23**, 1061.
[388] M. S. Chande, *Indian J. Chem.*, 1969, **7**, 941.
[389] J. E. Dunbar and J. H. Rogers, *J. Org. Chem.*, 1970, **35**, 279.
[390] C. Glidewell and D. W. H. Rankin, *J. Chem. Soc. (A)*, 1970, 279.

spectively. Two groups of workers [391, 392] report the synthesis of alkali-metal dialkyl- or diaryl-phosphinodithioformates (187) from the reaction

$$R_2PC\begin{matrix}S\\S\end{matrix} \quad M^+$$
(187)

of the corresponding metal phosphides, or of the phosphine in basic media, with carbon disulphide. In the presence of the corresponding quaternary phosphonium salts, a bisdithioformate is obtained.[392]

$O$-Alkyl thionocarbamates and $S$-alkyl dithiocarbamates are commonly prepared by the alcoholysis and thiolysis, respectively, of isothiocyanates. Using this procedure thionocarbamates derived from perfluoroisobutylene [221] and $N$-benzoyldithiocarbamates [393] have been recently prepared. Similarly, $NN$-di-isopropylthionocarbazate and $NN$-di-isopropyldithiocarbazate esters (189) have been obtained from the corresponding $N$-isothiocyanatodi-isopropylamine (188).[394] Thionocarbazate esters have also been obtained from the hydrazinolysis of xanthates.[370]

$$R_2NN=C=S \qquad R_2NNC\begin{matrix}S\\XR^1\end{matrix} \quad X = O \text{ or } S$$
(188) (189)

Thionocarbamate esters have been isolated from the thermal decomposition of $N$-(aryloxythiocarbonyl)thioureas,[293] and dithiocarbamate esters in Reissert compound studies.[395]

**Reactions.**—Chakrabarti and his co-workers, in an important series of papers,[396] have reinterpreted the unusual acid–base titration curves of dithiocarbamates, wherein the requirement of two equivalents of strong acid for neutralisation has previously been taken to indicate the dibasic character of dithiocarbamic acids. The present investigators have now shown that on acidification of solutions containing dithiocarbamate anions, (190), protonation occurs exclusively at sulphur, with the immediate formation of a hydrogen bridge between nitrogen and sulphur, as in structure (191), which prevents further protonation at nitrogen. Since dithiocarbamic acids readily decompose in solution to amines and carbon disulphide—the ease being attributed to the proximity of fractional positive

---

[391] R. Kramolowsky, *Angew. Chem. Internat. Edn.*, 1969, **8**, 202.

[392] O. Dahl, N. C. Gelting, and O. Larsen, *Acta Chem. Scand.*, 1969, **23**, 3369; O. Dahl and O. Larsen, *ibid.*, 1969, **23**, 3613.

[393] B. W. Nash, R. A. Newbery, R. Pickles, and W. K. Warburton, *J. Chem. Soc. (C)*, 1969, 2794.

[394] U. Anthoni, C. Larsen, and P. H. Nielsen, *Acta Chem. Scand.*, 1969, **23**, 1439, 3385.

[395] F. D. Popp and C. W. Klinowski, *J. Chem. Soc. (C)*, 1969, 741.

[396] S. J. Joris, K. I. Aspila, and C. L. Chakrabarti, *Analyt. Chem.*, 1969, **41**, 1441; *J. Phys. Chem.*, 1970, **74**, 860; K. I. Aspila, V. S. Sastri, and C. L. Chakrabarti, *Talanta*, 1969, **16**, 1099.

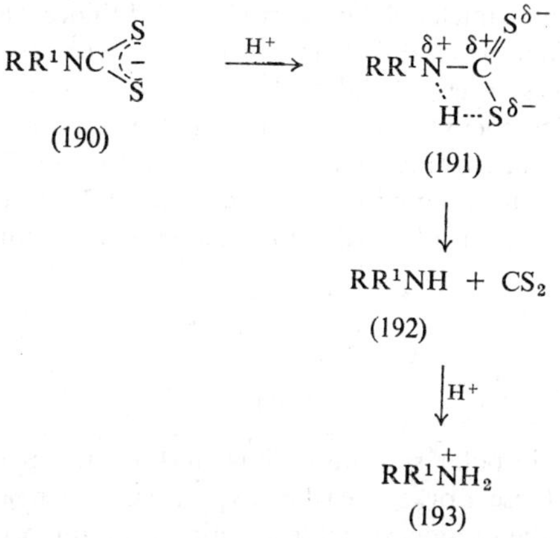

$$RR^1NC\overset{S}{\underset{S}{\lessgtr}}_H \xrightarrow{H^+} RR^1\overset{\delta+}{N}-\overset{\delta+}{C}\overset{S^{\delta-}}{\underset{H\cdots S^{\delta-}}{\diagdown}}$$

(190) (191)

$$\downarrow$$

$$RR^1NH + CS_2$$

(192)

$$\downarrow H^+$$

$$RR^1\overset{+}{N}H_2$$

(193)

charges on nitrogen and carbon in structure (191)—the requirement of a second equivalent of acid for neutralisation of dithiocarbamate solutions is explained in the overall reaction sequence (190) → (193).

The pyrolysis of O-alkyl NN-dimethylthiocarbamates, which contain β-hydrogen atoms in the alkoxy-group, is reported [397] to afford olefins in high yield and at lower temperatures than in the Tschugaev reaction of the corresponding xanthates (see section 11). Exceptionally, O-neopentyl dimethyldithiocarbamate is recovered unchanged on heating to 300 °C. Other recent reports on the chemistry of dithiocarbamic acid derivatives describe the reaction of dithiocarbamic acids with α-halogenoketones, leading to thiazoline-2-thiones,[398] the oxidation of dithiocarbamate ligands in metal complexes to dimeric dications such as (194),[399] and the hydrazino-

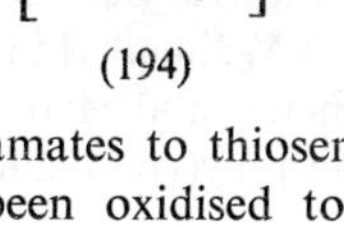

$$\left[\begin{array}{c} NR_2 \\ \| \\ S^{\diagup C}\diagdown S \\ | \quad | \\ S\diagdown_C\diagup S \\ \| \\ NR_2 \end{array}\right]^{2+}$$

(194)

lysis of S-benzyl dithiocarbamates to thiosemicarbazides.[312] In addition, dithiocarbazic acids have been oxidised to thiourams, trithiolans, and tetrathians [400] and cyclised with cyanogen bromide to amino-1,3,4-thiadiazoles.[401]

---

[397] M. S. Newman and F. H. Hetzel, *J. Org. Chem.*, 1969, **34**, 3604.
[398] G. Westphal, *J. prakt. Chem.*, 1969, **311**, 527.
[399] J. Willemse and J. J. Steggarda, *Chem. Comm.*, 1969, 1123.
[400] L. Cambia, G. Bargigia, L. Colombo, and E. Dubini Paglia, *Gazzetta*, 1969, **99**, 680.
[401] K.-H. Uteg, *Z. Chem.*, 1969, **9**, 450.

Examples of the nucleophilic reactivity of dithiocarbamate anions in reactions of alkali-metal dithiocarbamate salts with alkyl halides,[402, 403] α-halogenoketones,[385] acyl halides,[404] acetylenic halides,[405] 1,3-dithiolium cations,[406] alkylsulphenyl halides,[407] and halogenoboranes[408] have been described. Most interesting of these studies is that of Okawara and his colleagues[402] who have now isolated salts of the 1,3-dithiolanylium cation (195) from the reaction of 1,2-dichloroethane with sodium $NN$-dimethyl-

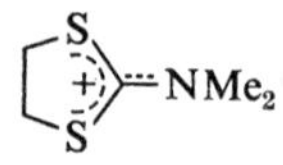

(195)

dithiocarbamate in polar solvents. Thus, additional experimental support is obtained for these workers' earlier hypothesis that neighbouring-group participation of the dithiocarbamate function is an important factor in the rate-enhancement in polar solvents of the related reaction between dithiocarbamate salts and 1,3-dichloropropane, the proposed mechanism of which is shown in the scheme (196) → (198). The ion (195) may be regarded as the anchimeric species—equivalent to (197)—in the related process starting from 1,2-dichloroethane. This view is strengthened by the

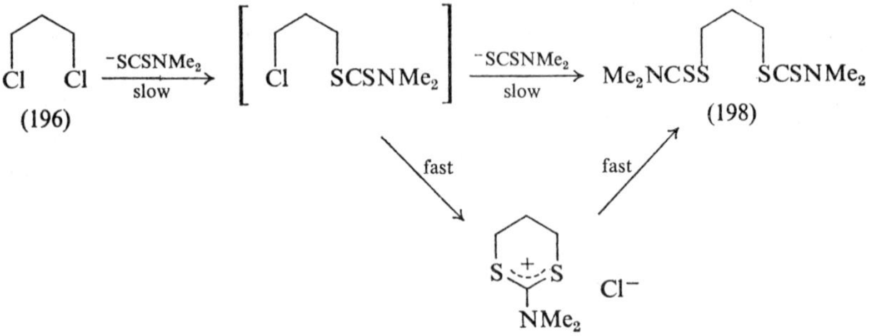

isolation of ethylene bis($NN$-dimethyldithiocarbamate)—equivalent to the final product (198)—in quantitative yield from the reaction of the perchlorate salt of (195) with $NN$-dimethyldithiocarbamate. Similar conclusions are drawn from kinetic studies on the rate of substitution of chlorine by dithiocarbamate in poly(vinyl chloride)[402a]—the first time that

---

[402] *a* T. Nakai, H. Kawaoka, and M. Okawara, *Bull. Chem. Soc. Japan*, 1969, **42**, 508; *b* T. Nakai, Y. Ueno, and M. Okawara, *ibid.*, 1970, **43**, 156.
[403] L. I. Aristov and A. A. Shamshurin, *Khim. geterotsikl. Soedinenii*, 1969, 228.
[404] P. M. Weintraub and F. E. Highman, *J. Org. Chem.*, 1969, **34**, 254.
[405] I. N. Azerbaev, M. Zh. Aytkhodzhaeva, and L. A. Tsoy, *Izvest. Akad. Nauk Kazakh. S.S.R., Ser. khim.*, 1969, No. 3, 77.
[406] A. Takamizawa and K. Hirai, *Chem. and Pharm. Bull. (Japan)*, 1969, **17**, 1930.
[407] W. F. Gilmore and R. N. Clark, *J. Chem. and Eng. Data*, 1969, **14**, 119.
[408] H. Noth and P. Schweizer, *Chem. Ber.*, 1969, **102**, 161.

neighbouring-group participation has been shown to occur in polymer main chains.

Several studies of the reactivity of heterocyclic thiones, possessing a cyclic thionocarbamate or dithiocarbamate structure, have been described.[409-411]

### 13 Selenocarbonyl and Tellurocarbonyl Compounds

**Synthesis.**—Not unexpectedly, the syntheses of seleno- and tellurocarbonyl compounds reported during the past year deal with the less unstable members, such as selenoureas, all of which possess nitrogen atoms directly bonded to the seleno- or telluro-carbonyl carbon atom. As far as the reviewer is aware, no simple aldehyde or ketone analogue has been recently isolated. Indeed, results obtained in a study[87] of the formation of bis(diselenoacetylacetonato)nickel(II) complexes and their sulphur analogues cast some doubt on the previously assumed transient presence of uncomplexed monomeric diselenoacetylacetone in the reaction medium.

Most interesting of the reported successful syntheses is that of the cyclic tellurourea-derivative (199), obtained in low yield from the reaction of

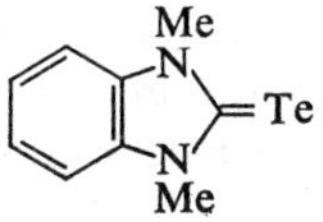

(199)

the corresponding benzimidazoline with tellurium.[412] The u.v. spectrum of (199) corresponds well with the spectra of less rare sulphur and selenium analogues, which are prepared in a similar manner, but in much better yields. $NN$-Dimethylselenoformamide, one of the simplest known selenoamides, has now been obtained from the base-catalysed reaction of the bisamidrazone of $NN$-dimethylformamide with hydrogen selenide.[413] $N$-Alkyl and $N$-aryl selenoureas have been obtained [414] by a new procedure involving the reaction of hydroselenide with $S$-methylisothiouronium salts in weakly basic media. At lower pH, however, examples of the previously unreported series of $S$-alkyl selenothiocarbamates (200) are isolated.

$$\text{R R}^1\text{NCSMe}$$
with Se double bond
(200)

---

[409] W. J. Humphlett, *J. Heterocyclic Chem.*, 1969, **6**, 397.
[410] T. Radha Vakula and V. R. Srinivasan, *Indian J. Chem.*, 1969, **7**, 583.
[411] P. Sohar, G. H. Denny jun., and R. D. Babson, *J. Heterocyclic Chem.*, 1969, **6**, 163.
[412] M. Z. Girshovich and A. V. El'tsov, *Zhur. obshchei Khim.*, 1969, 941.
[413] R. V. Kendall and R. A. Olofson, *J. Org. Chem.*, 1970, **35**, 806.
[414] D. L. Klayman and R. J. Shine, *J. Org. Chem.*, 1969, **34**, 3549.

Selenoureas and selenosemicarbazides may be obtained from isoselenocyanates through their reaction with amines and hydrazines, recently reported examples being the synthesis of $N$-($\alpha$-iminoalkyl)selenoureas [415] and the phosphinothioyl derivatives (201).[416] Using procedures which are

$$\text{ArPNHCNHAr}^1 \quad \overset{\text{S Se}}{\underset{\| \|}{}}$$

(201)

also analogous to common synthetic routes to thiocarbonyl compounds, diselenocarbazic acid,[387] and diselenocarbamates derived from morpholine and pyrrolidine [417] have been obtained from the reaction of carbon diselenide with hydrazine and the corresponding amines, respectively.

**Reactions.**—Russian workers [196] have shown that protonation of thio-, seleno-, and imino-derivatives of 3- and 5-pyrazolone occurs at the exocyclic hetero-atom in all cases. The nucleophilic reactivity of selenium in selenocarbonyl compounds has also been demonstrated in the reported reaction of selenourea with electrophilic triple bonds,[258] which affords both open-chain adducts structurally similar to (118) and 1,3-selenazin-4-one derivatives, under different conditions. Thioureas give analogous products under these circumstances (see section 8). The similar reactivity of selenourea and thiourea is further shown in their respective oxidation by metal ions to formamidine diselenide (202) [418, 419] and formamidine disulphide [268, 269] (see section 8).

$$\underset{\text{HN}}{\overset{\text{H}_2\text{N}}{\diagdown}}\text{C}-\text{Se}-\text{Se}-\text{C}\underset{\text{NH}_2}{\overset{\text{NH}}{\diagup}}$$

(202)

Selenourea has recently been employed in the synthesis of 2-seleno-6-methyluracil [420] and derivatives of 6-selenopurine,[208] and its greater selectivity, relative to urea and thiourea, in the formation of inclusion compounds with geometric isomers of substituted cyclohexanes, has been noted.[421] Some physical and chemical properties of selenocarbohydrazide have been described.[422]

## 14 Physical Properties

Sections 2—13 have dealt largely with the chemical reactivity of the Group VIA element isosteres of carbonyl compounds. This reactivity is determined

[415] N. A. Kirsanova and G. I. Derkach, *Ukrain. khim. Zhur.*, 1970, **36**, 372.
[416] T. Gabrio and G. Barnikow, *Z. Chem.*, 1969, **9**, 183.
[417] M. L. Shankaranarayana, *Acta Chem. Scand.*, 1970, **24**, 351.
[418] A. Chiesi, G. Grossoni, M. Nardelli, and M. E. Vidoni, *Chem. Comm.*, 1969, 404.
[419] V. L. Varand, V. M. Shul'man, and E. V. Khlystunova, *Izvest. Akad. Nauk S.S.S.R., Ser. khim.*, 1970, 450.
[420] M. Sy and J. Oiry, *Bull. Soc. chim. France*, 1969, 331.
[421] H. Van Bekkum, J. D. Remijnse, and B. M. Wepster, *Chem. Comm.*, 1969, 67.
[422] K.-H. Linke and D. Skupin, *Z. Naturforsch*, 1970, **25b**, 1, 3.

by the fundamental property of compounds containing C=X linkages (X = S, Se, Te), *viz.* the tendency to 'escape' to a more favourable bonding environment. Such considerations are also dominant in determining the physical properties of the various classes of C=X compounds, such as structure, spectra, dipole moments, and basicity. In attempting to collate the results of many recent physical investigations of thiocarbonyl compounds and their isosteres, an effort has been made to distinguish between studies which aim to evaluate the effect of the C=X group on the structure of the entire molecule, and those which set out to measure a fundamental property, *e.g.* C=X stretching frequency, and take account of the effect of total molecular environment on this.

**Structure.**—*Enethiol Tautomerism.* The practicability of the common iodine titration method for estimation of the enethiol content of various thiocarbonyl compounds has been tested by Mayer and his co-workers.[423] Their conclusions are that reliable results are obtained only in non-aqueous media, and that the end-points can be determined optically, or better potentiometrically. The latter technique, however, is not suitable for certain thiones and thionoesters which possess acidic C–H groups, and which can thus produce enethiolates in the presence of basic light-earth salts. Paquer and Vialle [77] have suggested a general structural rule for the formation of enethiols in aliphatic thiones, *viz.* α-CH groups not forming part of a methyl group lead to enethiol formation.

Two groups of workers have studied enethiol tautomerism in monothio-β-dicarbonyl compounds and have come to conflicting conclusions. On the basis of n.m.r. spectral data, Arnold, Klose, and their colleagues [424] suggest that β-thioxoketones of general type (203a) exist almost exclusively in the chelated thione-enol form (203b, where $R^3$ = methyl or aryl), and have obtained a value of 2·0 kcal mol$^{-1}$ for the heat of formation of the O–H···S hydrogen bond. In contrast, results of i.r. and n.m.r. spectroscopic study of similar compounds [425] suggest to Duus and Lawesson that the predominant form is the enethiolketone chelate (203c). The latter group has obtained similar results in studies of β-thioxoesters (203a, $R^1$ = alkoxy),[425] α-thioacyl-lactones,[79] and α-thioacylthiolactones.[79]

Thioamides do not normally show thiolimide tautomerism because they can be stabilised through conjugation with the amidic nitrogen atom leading to a polarised C=S bond. Walter and his co-workers [426] have recently shown that previous reports on thiolimide tautomerism in thionicotinamides are incorrect. α-Thiocarbamidomethylenephosphoranes (83, R = acyl),[147b] S-phenyl α-thiocarbamidothiolacetate derivatives,[156] and

---

[423] R. Mayer, S. Scheithauer, S. Bleisch, D. Kunz, G. Bahr, and R. Radeglia, *J. prakt. Chem.*, 1969, **311**, 472.
[424] K. Arnold, G. Klose, P. H. Thomas, and E. Uhlemann, *Tetrahedron*, 1969, **25**, 2957; K. Arnold and G. Klose, *ibid.*, 1969, **25**, 3775.
[425] F. Duus and S.-O. Lawesson, *Arkiv. Kemi*, 1969, **29**, 127.
[426] W. Walter, H. P. Kubersky, and D. Ahlquist, *Annalen*, 1970, **733**, 170.

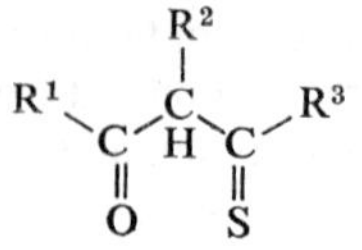

(203a)

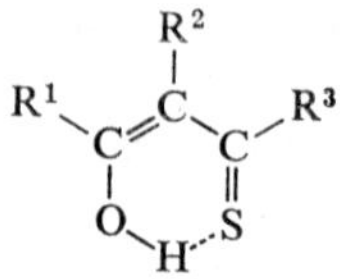

(203b)

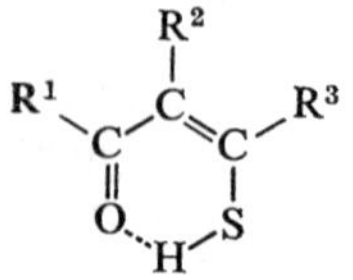

(203c)

$N$-thiobenzoyl-L-α-amino-acids [427] have also been shown to exist in the thione form, whereas thioanilides derived from α-cyanomalonate esters are regarded as being fixed in the enethiol form (204).[182] Various spectroscopic

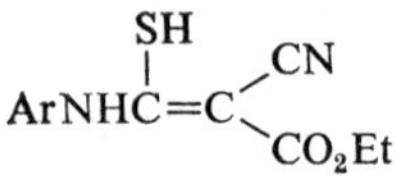

(204)

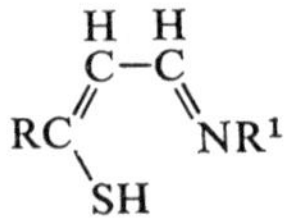

(205)

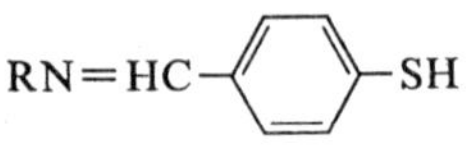

(206a)

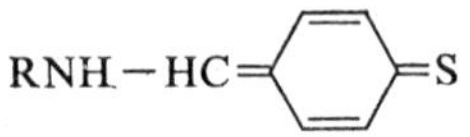

(206b)

studies [84, 428, 429] of enaminothiones have demonstrated that these vinylogous thioamides do not form imine-enethiol tautomers (205). The thioquinoid system (206b), however, does exhibit benzoid–quinoid tautomerism,[430] the mercaptoaldimine form (206a) predominating in non-polar solvents and in the solid state. Similar mercaptoaldimines in the furan series, however, are found to exist predominantly in a chelated thiol form at ambient temperature.[431] The thione form also predominates in dipyrryl thiones,[83] and in 1-aminoindene-3-thiones.[432]

[427] G. C. Barrett and A. R. Khokhar, *J. Chem. Soc. (C)*, 1969, 1120.
[428] M. L. Filleux-Blanchard, H. Durand, M. T. Bergein, F. Clesse, H. Quiniou, and G. J. Martin, *J. Mol. Structure*, 1969, 3, 351.
[429] J. Bignebat, H. Quiniou, and N. Lozac'h, *Bull. Soc. chim. France*, 1969, 127.
[430] L. P. Olekhnovich, V. I. Minkin, V. T. Panyushkin, and V. I. Kriul'kev, *Zhur. org. Khim.*, 1969, 5, 123; V. I. Minkin, L.P. Olekhnovich, L. E. Nivopozhkin, Yu. A. Zhdanov, and M. I. Knyazhanskii, *ibid.*, 1970, 6, 348; V. I. Minkin, L. P. Olekhnovich, Yu. A. Zhdanov, and Yu. A. Ostroumov, *ibid.*, 1970, 6, 549; Yu. A. Zhdanov, V. I. Minkin, L. P. Olekhnovich, and E. N. Malisheva, *ibid.*, 1970, 6, 554.
[431] V. S. Bogdanov, Ya. L. Danyushevskii, and Ya. A. Goldfart, *Izvest. Akad. Nauk S.S.S.R., Ser. khim.*, 1970, 675.
[432] D. Y. Murnietze and Ya. A. Freymanis, *Zhur. org. Khim.*, 1969, 5, 1482.

Walter and Reubke [433] have investigated the effect of the introduction of an N-amino-group on the tautomerism of thioamides. They have found that in 1,1-diethylthioformylhydrazide tautomerism occurs between the thione form and the betaine (207) rather than the expected thiolimide. Various

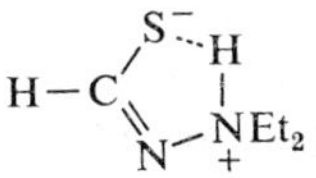

$$H-C\begin{smallmatrix}S^-\cdots H\\ \\N-NEt_2\\+\end{smallmatrix}$$

(207)

physical investigations have shown that the thione form predominates both in the solid state and in solution for N-hydroxythioureas,[434] 2-indolinethione,[154] benzimidazole-2-thiones,[435,436] various azoline-2-thiones,[437] pyrimidine-4-thiones,[438] thiopurinol,[439] 2H-1,4-benzoxazine-3-thiones,[440] and 3H-1,2,4-triazino[5,6-b]indole-3-thione.[441] Thioacridone [442] and thiosemicarbazones derived from 9-formylacridine,[320] however, exist as thiones in the solid state and exhibit enethione tautomerism in solution. 5-Phenylimidazolidine-2,4-dithione has been shown to have the thione-enethiol structure (208).[443]

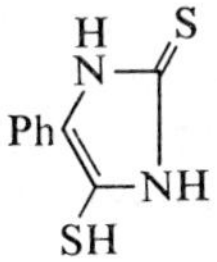

(208)

*Ring–Chain Isomerism.* Two groups of workers have reported conflicting results from studies of the ring–chain isomerism of 1,3,4-thiadiazoline-2-thiones. Mayer and Lauerer,[444] on the basis of n.m.r. spectral data, have detected the tautomeric equilibrium (209) ⇌ (210), whereas Anthoni and his co-workers [445] cite i.r. and n.m.r. evidence in support of the 1,3,4-

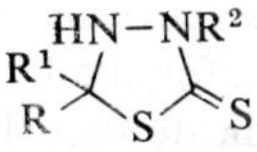

(209) 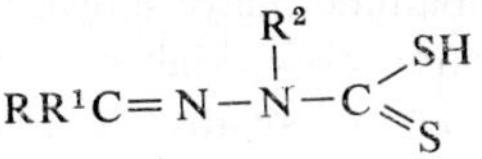   (210)

[433] W. Walter and K.-J. Reubke, *Chem. Ber.*, 1969, **102**, 2117.
[434] F. Grambal, J. Mollin, and M. Hejsek, *Monatsh.*, 1970, **101**, 120.
[435] P. Nuhn, G. Wagner, and S. Leistner, *Z. Chem.*, 1969, **9**, 152.
[436] Y. Takahashi and K. Yabe, *Bull. Chem. Soc. Japan*, 1969, **42**, 3064.
[437] G. Kjellin and J. Sandström, *Acta Chem. Scand.*, 1969, **23**, 2889.
[438] M. Kuchar, J. Strof, and J. Vachek, *Coll. Czech. Chem. Comm.*, 1969, **34**, 2278.
[439] K. Vang Thang and J. L. Olivier, *Compt. rend.*, 1969, **268**, C, 1798.
[440] M. Mazharuddin and G. Thyagarajan, *Tetrahedron*, 1969, **25**, 517.
[441] I. S. Ioffe, A. B. Tomchin, and E. N. Zhukova, *Zhur. obshchei Khim.*, 1969, **39**, 78, 640.
[442] V. P. Maksimets and O. N. Popilin, *Khim. geterotsikl. Soedinenii*, 1970, 191.
[443] J. T. Edward and J. K. Liu, *Canad. J. Chem.*, 1969, **47**, 1123.
[444] K. H. Mayer and D. Lauerer, *Annalen*, 1970, **731**, 142.
[445] U. Anthoni, C. Larsen, and P. H. Nielsen, *Acta Chem. Scand.*, 1970, **24**, 179.

thiadiazolidine-2-thione structure (209), rather than the alkylidene dithiocarbazic acid (210). The former group also discuss [444] the possibility of ring–chain isomerism between 1,3,4-thiadiazolines and 1-thiocarbonylhydrazone derivatives.

*Polarisation Effects and Restricted Rotation.* The phenomenon of conjugation between the thiocarbonyl group and the amide nitrogen in thioamides has been established through the n.m.r. spectral detection of effects attributed to restricted rotation around the thioamide C–N bond, and has been alluded to in interpretations of thioamide i.r. spectra. Restricted rotation is also observed in the n.m.r. spectra of amides and it has been shown recently, that not only are the free-energy and entropy barriers to rotation in thioanilides higher than in corresponding anilides, but they are also less subject to solvent effects.[446] Walter and his co-workers [447] have evaluated the dipole moments of various thioamides and have been able to assign the predominant configuration of groups around the C–N partial double bond on this basis. They have also noted that the restriction on rotation in *NN*-dimethylthiopivalimide is much reduced by steric factors.[447, 448] The barriers to rotation in selenoamides have been found to be slightly higher (1 kcal mol⁻¹) than in corresponding thioamides.[449] Restricted-rotation effects have been noted in the n.m.r. spectrum of 2,3-dimethyl-1-thioformylindolizine, indicating a large ground-state contribution of the polarised form (211).[62]

(211)

Studies of the restricted C–N bond rotation in enaminothiones, *i.e.* vinylogous thioamides, have shown [450] that the barriers to rotation in these compounds are also higher than in the corresponding ketones ($E_a = 27\cdot2$ kcal mol⁻¹ relative to 15·6 kcal mol⁻¹ for the ketones), though less than in some thioamides. These results indicate that the substitution of sulphur for oxygen strongly enhances the electron delocalisation within the molecule.

Two groups of workers have demonstrated the occurrence of restricted C–N rotation in thioureas.[451, 452] The calculated free energies of activation

---

[446] T. H. Siddall, E. L. Pye, and W. E. Stewart, *J. Phys. Chem.*, 1970, **74**, 594; T. H. Siddall and W. E. Stewart, *J. Org. Chem.*, 1969, **34**, 2927.
[447] W. Walter and H. Hühnerfüss, *J. Mol. Structure*, 1969, **4**, 435.
[448] W. Walter, E. Schaumann, and H. Paulsen, *Annalen*, 1969, **727**, 61.
[449] K. A. Jensen and J. Sandström, *Acta Chem. Scand.*, 1969, **23**, 1911.
[450] M. L. Filleux-Blanchard, F. Clesse, J. Bignebat, and G. J. Martin, *Tetrahedron Letters*, 1969, 981.
[451] A. S. Tompa, R. D. Barefoot, and E. Price, *J. Phys. Chem.*, 1969, **73**, 435.
[452] B. T. Brown and G. F. Katekar, *Tetrahedron Letters*, 1969, 2343.

(*ca.* 12 kcal mol$^{-1}$) are considerably lower than for thioamides. In tetramethylselenourea, restricted rotation could not be observed even at low temperatures,[449] probably owing to steric factors inhibiting coplanarity.

Thioformyl- and thioacyl-hydrazides have been shown to undergo tautomerism leading to betaines such as (207) rather than to enethiols.[433, 453] However, in the non-polar form (212), rotational isomerism about the

$$R^1_2NNHC \underset{R}{\overset{S}{\diagup}}$$

(212)

C–N$^1$ bond is observed,[453, 454] as in thioamides. Similar thione–betaine tautomerism is exhibited by thiosemicarbazones,[272, 330] dithiocarbazate esters,[394] and thionocarbazate esters.[394] The latter compounds, however, are also found to exhibit C–N$^1$ rotational isomerism.[394]

Lemire and Thomson[455] have investigated the hindered rotation about C–N bonds in dithio- and thiono-carbamates and have evaluated the various thermodynamic activation parameters for rotation. These are of the order of the values obtained for thioureas,[451, 452] probably because of conjugation effects arising from interaction of the alkylthio- and alkoxy-groups with the thiocarbonyl function, which partially compensate for the loss in C–N $\pi$-bonding energy on rotation. The fact that the barriers are higher for thionocarbamates than for dithiocarbamates indicates that alkylthio-groups conjugate more efficiently. These workers also make some general comments on the reliability of certain published activation parameters and suggest that total n.m.r. line-shape fitting methods, and possibly also spin-echo techniques, are the only reliable methods for obtaining such data. Restricted-rotation effects have also been observed in *NN:NN'*-tetraalkylthiuram disulphides and their metal complexes,[456] and in methyl dithioacetate and dimethyl trithioacetate.[457] The latter result is significant in that it supports the aforementioned observations[455] as to the efficiency of conjugation in the system S—C=S.

Also pertinent to the subject of thiocarbonyl group polarisation are recent papers describing the evaluation, by means of *X*-ray emission spectroscopy, of the number of valence electrons of the thiocarbonyl sulphur atom in various organic compounds,[458] and with spectral and *X*-ray crystallographic studies of potentially mesionic heterocyclic thiones. The latter group include dithiocoumarins,[459] 3*H*-imidazoline-2-thiones,[460] dehydrodithizone—a

[453] U. Anthoni, C. Larsen, and P. H. Nielsen, *Acta Chem. Scand.*, 1969, **23**, 3513; U. Anthoni, P. Jakobsen, C. Larsen, and P. H. Nielsen, *ibid.*, p.1820.
[454] W. Walter and H. Weiss, *Angew. Chem. Internat. Edn.*, 1969, **8**, 989.
[455] A. I. Lemire and J. C. Thomson, *Canad. J. Chem.*, 1970, **48**, 824.
[456] H. C. Brinkoff, A. M. Grotens, and J. J. Steggarda, *Rec. Trav. chim.*, 1970, **89**, 11.
[457] K. Herzog, E. Steger, P. Rosmus, S. Scheithauer, and R. Mayer, *J. Mol. Structure*, 1969, **3**, 339.
[458] Y. Takahashi, K. Yabe, and T. Sato, *Bull. Chem. Soc. Japan*, 1969, **42**, 2707.
[459] A. Ruvet and M. Renson, *Bull. Soc. chim. belges*, 1970, **79**, 89.
[460] G. B. Ansell, D. M. Forkey, and D. W. Moore, *Chem. Comm.*, 1970, 56.

derivative of tetrazole,[461] and 4,5-dioxo-1,3-dithiolan-2-thione,[462] all of which are found to possess a lengthened exocyclic C=S bond. The crystal structure of $N$-acetyl-$N'$-phenylselenourea shows a C–Se distance approaching that of a single bond.[463]

Other recently determined crystal structures include those of 1-phenyl-3-(2-thiazolinyl)-2-thiourea,[464] the thiourea–iodine adduct $[(NH_2)_2CS]_2I^+I^-$,[465] thioacetanilide $S$-oxide,[466] thiosemicarbazide,[467, 468] the thiosemicarbazones of 2-keto-3-ethoxybutyraldehyde [469] and of 2-formylthiophen,[470] thiocarbohydrazide,[471] hydrazinium dithiocarbazate,[472] and t-butyl $NN$-dimethyltrithiopercarbamate.[473] In addition to those quoted in refs. 460—462, the following heterocyclic thione structures have been elucidated by $X$-ray crystallographic methods, viz. 2-thiohydantoin,[474] thiouridine,[475] 6-methylpurine-2-thione,[476] and purine-6-thione monohydrate.[477]

**Spectra.**—*U.v. Spectra.* Results from PPP molecular orbital calculations [478] indicate that monomeric thioformaldehyde should exhibit an $(n,\pi^*)$ band in its electronic spectrum at 531 nm, and should therefore be red in colour. Owing to the extreme instability of thioformaldehyde, experimental verification of this prediction is not likely to be obtained. The calculated value, however, is consistent with that obtained from extrapolation of experimental data on other simple thioaldehydes. A band at 211·7 nm observed in the flash-photolysis of dimethyl disulphide is attributed to an electronic transition of transient thioformaldehyde.[64] Thioformylindolizines (2a) exhibit weak $(n,\pi^*)$ bands of the thiocarbonyl group at 520 and 550 nm.[62]

The energies of $(n,\pi^*)$ states of thiobenzophenone, xanthione, and $N$-methylacridine-9-thione are shown to be reduced with respect to their ketone analogues,[479] and the corresponding bands are successively blue-

[461] Y. Kushi and Q. Fernando, *Chem. Comm.*, 1969, 1240.
[462] B. Krebs and P. F. Koenig, *Acta Cryst.*, 1969, **25B**, 1022.
[463] M. Perez Rodriguez and A. Lopez Castro, *Acta Cryst.*, 1969, **25B**, 532.
[464] J. L. Flippen and I. L. Karle, *J. Phys. Chem.*, 1970, **74**, 769.
[465] H. Hope and G. Hung-Yin Lin, *Chem. Comm.*, 1970, 169.
[466] O. H. Jarchow, *Acta Cryst.*, 1969, **25B**, 267.
[467] P. Dominiano, G. Fava Gasparri, M. Nardelli, and P. S. Sgarabotto, *Acta Cryst.*, 1969, **25B**, 343.
[468] F. Hansen and R. G. Hazell, *Acta Chem., Scand.* 1969, **23**, 1359.
[469] E. J. Gabe, M. R. Taylor, J. Pickworth Glusker, J. A. Minkin, and A. L. Patterson, *Acta Cryst.*, 1969, **25B**, 1620.
[470] M. Mathew and G. J. Palenik, *Chem. Comm.*, 1969, 1086.
[471] A. Braibanti, A. Tiripicchio, and M. Tiripicchio Camellini, *Acta Cryst.*, 1969, **25B**, 2286; *Ricerca sci.*, 1969, **39**, 269.
[472] A. Braibanti, A. M. Manotti Lanfredi, A. Tiripicchio, and F. Logiudice, *Acta Cryst.*, 1969, **25B**, 93.
[473] D. J. Mitchell, *Acta Cryst.*, 1969, **25B**, 998.
[474] L. A. Walker, K. Folting, and L. L. Merritt, *Acta Cryst.*, 1969, **25B**, 88.
[475] W. Saenger and K. H. Scheit, *Angew. Chem. Internat. Edn.*, 1969, **8**, 139.
[476] J. Donohue, *Acta Cryst.*, 1969, **25B**, 2418.
[477] E. Sletten, J. Sletten, and L. H. Jensen, *Acta Cryst.*, 1969, **25B**, 1330.
[478] J. Fabian and L. Mehlhorn, *Z. Chem.*, 1969, **9**, 271.
[479] R. N. Nurmukhametov, L. A. Mileshina, D. N. Shigorin, and G. T. Khachaturova, *Zhur. fiz. Khim.*, 1969, **43**, 51.

shifted within the thione series, showing that conjugation of the thiocarbonyl group is more efficient with nitrogen atoms than with sulphur atoms in heterocycles, and much more so than with carbocycles. In the same study [479] $(n,\pi^*)$ fluorescence in thiocarbonyl compounds is established for the first time. The conjugative effects of various exocyclic groups in fluorene derivatives (213, X = $CH_2$, $CMe_2$, NOH, NH, O, S) on the

(213)

$(\pi,\pi^*)$ band of these molecules has been studied.[480] The anomalous position of the thiofluorenone band, which does not fit into the expected order based on electronegativity criteria for the X group, is attributed to the weak $\pi$-overlap in the C=S bond. Weak $(n,\pi^*)$ bands have been observed in the u.v. spectrum of adamantane-2-thione.[78] The $(n,\pi^*)$ band-splitting observed in tetramethylcyclobutane-1,3-dione, and attributed to 1,3-transannular $\pi$-interactions, has not been detected in the corresponding dithione.[481]

In a study of solvent shifts in the electronic spectra of thioamides, amides, and nitriles, Gramstad and Sandström [482] have shown that the Badger–Brauer relationship between the strength of hydrogen bonds and the stretching frequency of the X–H bond in the proton donor is not upheld in general. The hydrogen-bonding ability of thioamides is found to be markedly lower than for amides and nitriles. A comparison of u.v. solvent shifts and hydrogen-bond energies supports previous views on the changes in $\pi$-electron distribution on excitation of thioamides. The $(n,\pi^*)$ bands of thioamide–iodine complexes have been observed.[483]

Ritchie and Spedding [484] have assigned the observed blue-shift effects of N-alkyl groups on the $(n,\pi^*)$ and $(\pi,\pi^*)$ bands of thioureas to $\sigma$–$\pi$ interactions which raise the energy of the $\pi^*$-orbital. The smaller shifts in $(\pi,\pi^*)$ transitions caused by N-aryl substitution in thiourea indicate a weaker interaction between phenyl and thiourea-groups than between carbonyl and olefinic groups. Ring-substituent effects in the u.v. spectrum of N-arylthioureas have been noted.[484, 485]

The long-wavelength bands in the u.v. spectrum of aromatic dithiocarboxylate esters have been discussed in relation to M.O. calculated transition energies.[486] A similar theoretical study of organic sulphur compounds,

---

[480] J. Dehler and K. Fritz, *Tetrahedron Letters*, 1969, 2157.
[481] R. E. Ballard and Chong Hoe Pak, *Spectrochim. Acta*, 1970, **26A**, 43.
[482] T. Gramstad and J. Sandström, *Spectrochim. Acta*, 1969, **25A**, 31.
[483] A. F. Grand and M. Tamres, *J. Phys. Chem.*, 1970, **74**, 208.
[484] R. K. Ritchie and H. Spedding, *Spectrochim. Acta*, 1970, **26A**, 9.
[485] F. H. Jackson and A. T. Peters, *J. Chem. Soc.* (C), 1969, 268.
[486] J. Fabian, S. Scheithauer, and R. Mayer, *J. prakt. Chem.*, 1969, **311**, 45.

including thioacids, thiocarbamates, and xanthates, has been described.[487] Ohno and his co-workers have discussed the effect of ring-substituents on the $(n,\pi^*)$ transition of thionobenzoate esters in terms of the Hammett relationship.[488] The $(n,\pi^*)$ thiocarbonyl band of dithiocarboxylate esters derived from 3-alkyl- and 3-aryl-camphor has been detected at 464—475 nm.[336]

U.v. spectroscopic studies of the structure and protonation of hydantoin-2,4-dithiones[489] and of thiones and selenoketones derived from pyrazolines[196] have been described. Results from SCF–PPP calculations on the protonation of thiourea are now shown to be in agreement with u.v. spectroscopic data.[490] The electronic spectra of thio-, seleno-, and telluro-analogues of 1,3-dimethylbenzimidazolin-2-one,[412] thionaphthostyryl derivatives,[491] perhydro-1,3-thiazine-2-thiones,[211] and tetrahydro-1,3,5-thiadiazine-2-thiones[492] have been reported.

Barrett and his co-workers[427, 493] have studied the circular dichroism of N-thiobenzoyl derivatives of amino-acids as an aid to structural assignment. Reliable assignments of absolute configuration to N-terminal amino-acid residues in oligopeptides are obtained only for measurements in aqueous media.[493] Rippberger[494] has devised a simple quadrant rule which predicts the sign of $(n,\pi^*)$ Cotton effects of dithiocarbamates and dithiourethanes. O.r.d. and c.d. studies of thiocarbamate salts,[386] cyclic thionocarbonates,[495] xanthate derivatives of cyclodextrin,[496] and nucleosides derived from pyrimidine-2-thiones[497] and purine-6-thione[498] have been reported.

*I.r. Spectra.* Several groups of workers have continued their efforts to corroborate and to rationalise many conflicting published data concerning the C=S stretching frequency [$\nu$(C=S)] in various classes of thiocarbonyl compounds. The band at 1170—1200 cm$^{-1}$ has been attributed to $\nu$(C=S) for a series of dithienyl thiones and thienyl aryl thiones.[499] Comparison of these results with the carbonyl stretching [$\nu$(C=O)] frequencies of the corresponding ketones indicate that the value of the ratio $\nu$(C=O) : $\nu$(C=S) for these compounds is 1·37, *i.e.* reasonably close to the value of 1·5 previously proposed as general for ketone–thione analogues. Moule and

---

[487] M. Bossa, *J. Chem. Soc. (B)*, 1969, 1182.
[488] A. Ohno, T. Koizumi, Y. Ohnishi, and G. Tsuchihashi, *Bull. Chem. Soc. Japan*, 1969, **42**, 3556.
[489] J. T. Edward and J. K. Liu, *Canad. J. Chem.*, 1969, **47**, 1117.
[490] A. Azman, B. Lukman, and D. Hadzi, *J. Mol. Structure*, 1969, **4**, 468.
[491] I. L. Belaits, R. N. Nurmukhametov, D. N. Shigorin, and G. I. Bystritskii, *Zhur. fiz. Khim.*, 1969, **43**, 1673.
[492] P. Kristian and J. Bernat, *Coll. Czech. Chem. Comm.*, 1969, **34**, 2952.
[493] G. C. Barrett, *J. Chem. Soc. (C)*, 1969, 1123.
[494] H. Rippberger, *Tetrahedron*, 1969, **25**, 725.
[495] A. H. Haines and C. S. P. Jenkins, *Chem. Comm.*, 1969, 350.
[496] F. Cramer, G. Mackensen, and K. Sensse, *Chem. Ber.*, 1969, **102**, 494.
[497] T. Ueda and H. Nishino, *Chem. and Pharm. Bull. (Japan)*, 1969, **17**, 920.
[498] K.-K. Cheong, Y.-C. Fu, R. K. Robins, and H. Eyring, *J. Phys. Chem.*, 1969, **73**, 4219.
[499] C. Andrieu and Y. Mollier, *Bull. Soc. chim. France*, 1969, 831.

Subramaniam[500] have assigned bands at 1242 and 1257 $cm^{-1}$ in the gas-phase i.r. and Raman spectra, respectively, of thiocarbonyl chlorofluoride to $v(C=S)$. Frequencies calculated from normal-co-ordinate analysis support this assignment. A vibrational frequency of 1130 $cm^{-1}$ active in the $(n,\pi^*)$ fluorescence spectrum of xanthione is attributed to $v(C=S)$ for this compound.[479]

Contrary to previous experience in respect of primary thioamides, results of normal-co-ordinate treatment of dimethylthioformamide and dimethylthioacetamide indicate that i.r. bands at 915 $cm^{-1}$ and 1010 $cm^{-1}$ respectively for these compounds are essentially $v(C=S)$ in character.[501] A similar result is reported for other $NN$-dimethylthioamides.[138] In the case of trifluorothioacetamide, however, mixed $C=N$ and $C=S$ vibrations are found to occur near 1400 $cm^{-1}$.[338] The i.r. spectral elucidation of the configuration of thio- and seleno-lactams has been reported.[502] Results from this study[503] confirm that the difficulties in assigning $v(C=S)$ and $v(C=Se)$ frequencies in thioamides and selenoamides are caused by vibrational coupling. For five- and six-membered thio- and seleno-lactams, however, the vibrational frequency appears to be essentially localised at 1115 $cm^{-1}$ for $v(C=S)$ and 1085 $cm^{-1}$ for $v(C=Se)$.[503] An i.r. spectral study of intramolecular hydrogen-bonding in thiobenzamidopyridines has been described.[504] The technique of partial deuteriation at nitrogen has been used to distinguish between the i.r. spectra of $NN'$- and $NN$-disubstituted urea and thiourea isomers.[505]

Volka and Holzbacher[506] have employed normal-co-ordinate analysis methods in the localisation of $v(C=S)$ and $v(C=Se)$ vibrations of thio- and seleno-semicarbazides.[506] A considerable contribution of $v(C=S)$ is found for a band at 802 $cm^{-1}$. The corresponding band for selenosemicarbazide is at somewhat lower frequency and appears to be less characteristic of $v(C=Se)$. The various $v(N-H)$ vibrations of thiosemicarbazide have also been located.[507] Assignments of $C=X$ stretching vibrations have been made in the i.r. spectra of 2-formyladamantane thiosemicarbazone,[508] thio- and seleno-carbohydrazide,[509] and cyclic thionocarbonates.[356]

The i.r. spectra of metal $NN$-dialkyldithio- and $NN$-dialkyldiselenocarbamates have been investigated by two groups of workers,[510, 511] and

---

[500] D. C. Moule and C. R. Subramaniam, *Canad. J. Chem.*, 1969, **47**, 1011.
[501] C. A. I. Chary and K. V. Ramiah, *Proc. Indian Acad. Sci.*, 1969, **69**, 18.
[502] H. E. Hallam and C. M. Jones, *J. Chem. Soc. (A)*, 1969, 1033.
[503] H. E. Hallam and C. M. Jones, *Spectrochim. Acta*, 1969, **25A**, 1791.
[504] J. Mirek and B. Kawalek, *Tetrahedron*, 1970, **26**, 1261.
[505] A. R. Katritzky, H. J. Keogh, K. Lempert, and J. Puskas, *Tetrahedron*, 1969, **25**, 2265.
[506] K. Volka and Z. Holzbacher, *Coll. Czech. Chem. Comm.*, 1969, **34**, 1353.
[507] D. M. Wiles and T. Suprunchuk, *Canad. J. Chem.*, 1969, **47**, 1087.
[508] J. Scharp, H. Wynberg, and J. Strating, *Rec. Trav. chim.*, 1970, **89**, 18.
[509] K.-H. Linke and R. Turley, *Z. Naturforsch.*, 1969, **24b**, 821.
[510] G. Durgaprasad, D. N. Sathyanarayana, and C. C. Patel, *Canad. J. Chem.*, 1969, **47**, 631.
[511] A. T. Pilipenko and N. V. Mel'nikova, *Zhur. neorg. Khim.*, 1969, 462, 1843.

bands characteristic of the $CS_2$ and $CSe_2$ groups are assigned. The i.r. spectrum of dithiocarbamic acid has also been published.[381] Intense bands at 1065 cm$^{-1}$ and 986 cm$^{-1}$ in the spectrum of dithiocarbazic acid and at 1020 cm$^{-1}$ and 896 cm$^{-1}$ for diselenocarbazic acid have been attributed to the stretching vibrations of the $CS_2$ and $CSe_2$ groups respectively.[387]

The i.r. spectra of various heterocyclic thiocarbonyl compounds derived from pyrazole,[196] imidazole,[296, 512] 1,3,4-oxadiazole,[410] 1,2,4-thiadiazole,[513] thio- and seleno-coumarin,[459] 1,3-thiazine,[290] and 1,3,5-triazine[291] have been recorded. I.r. spectral studies of metal complexes of monothio-acetylacetone,[87, 514] dithio- and diseleno-acetylacetone,[87] 2-thioamido-pyridine,[515] dithio-oxamides,[516] thiourea,[517] 2-amidinothiourea,[518] xanthates,[376] dithiocarbamates,[510, 511, 519] thionocarbamates,[520] purine-6-thione,[521] and tetrazoline-5-thione[522] have been reported.

*N.m.r. Spectra.* The considerable interest currently being shown in the n.m.r. spectral properties of thiocarbonyl compounds stems from evidence of the greater magnetic anisotropy of the C=S bond relative to C=O. In an important paper, Demarco and his co-workers[523] have examined the causes of increased deshielding of the thiocarbonyl carbon atom in the $^{13}$C spectrum of thiocamphor. The conclusion is that the important factor is the higher paramagnetic screening of the thiocarbonyl group relative to the carbonyl group. These authors have also shown that a shielding cone similar to that for the C=O bond can be drawn for C=S. The same conclusion has been drawn from an n.m.r. study[62] of the thioformylindol-izines (2a) which has shown that the signal attributed to the thioformyl proton is *ca.* 1·0 p.p.m. deshielded with respect to the formyl proton signal in the corresponding aldehyde. Increased deshielding of neighbouring protons has also been detected in aliphatic thiones.[77]

Walter and his co-workers[448] have developed models of the magnetic anisotropy of the thioformamide and thioacetamide systems with the aid of sterically fixed 1-thioacylpiperidines. In this way the different assignments of the signals of *cis-* and *trans-N-*methyl groups are explained. In a systematic study of the n.m.r. spectra of *NN'-*dialkyl- and *NN'-*diaryl-thioureas, Ritchie and Spedding[524] report that anisotropic effects of the

[512] K. Lempert, P. Sohar, and K. Zauer, *Acta Chim. Acad. Sci. Hung.*, 1970, **63**, 87.
[513] U. Anthoni, C. Larsen, and P. H. Nielsen, *Acta Chem. Scand.*, 1969, **23**, 537.
[514] U. Agarwala and P. B. Rao, *Appl. Spectroscopy*, 1969, **23**, 224.
[515] G. J. Sutton, *Austral. J. Chem.*, 1969, **22**, 2475.
[516] H. O. Desseyn, W. A. Jacob, and M. A. Herman, *Spectrochim. Acta*, 1969, **25A**, 1685.
[517] L. D. Dremetskaya, N. B. Lyubimova, and S. D. Beskov, *Zhur. fiz. Khim.*, 1969, **43**, 850.
[518] A. Syamal, *Z. Naturforsch.*, 1969, **24b**, 1192.
[519] D. C. Bradley and M. H. Gitlitz, *J. Chem. Soc. (A)*, 1969, 1152.
[520] K. Kawai, F. Mizukami, and Y. Kodama, *Bull. Chem. Soc. Japan*, 1969, **42**, 2396.
[521] J. Brigardo and D. Colaitis, *Bull. Soc. chim. France*, 1969, 3440.
[522] U. Agarwala and B. Singh, *Indian J. Chem.*, 1969, **7**, 727.
[523] P. V. Demarco, D. Doddrell, and E. Wenkert, *Chem. Comm.*, 1969, 1418.
[524] R. K. Ritchie and H. Spedding, *Spectrochim. Acta*, 1970, **26A**, 1.

# Thiocarbonyls, Selenocarbonyls, and Tellurocarbonyls 245

thiocarbonyl group on α-methylene groups may account for the deshielding observed relative to the α-methylene groups in corresponding amine derivatives. This effect may also operate in the observed deshielding of the N–H protons, but appears too small in respect of $o$-protons in $N$-aryl groups. Anisotropic deshielding of neighbouring protons by thiocarbonyl groups has also been observed in thiocarbamoyl sugar derivatives,[525] cyclic thionocarbonates,[355] thiono- and dithio-carboxylate esters,[526] glucopyranosides possessing thiopyridone substituents,[527] and in 2$H$-1,4-benzoxazine-3-thione derivatives.[440]

*Mass Spectra.* Schönberg and his co-workers[528] report that the mass spectra of thiobenzophenone derivatives are differentiated from those of the corresponding ketones by a new type of fragmentation involving a single-stage radical cleavage of a thiol group, leading to a fluorenyl cation. Non-enolisable thiones, in which the thiocarbonyl group is part of a ring, show no $M-$ SH fragmentation, though some exhibit an intensive $M-1$ fragmentation. Japanese workers,[529] however, report that the main fragments in the electron-impact fragmentation of indoline-2-thiones are $M-1$, $M-$S, and $M-$SH.

The mass spectra of β-thioketothiolesters,[530] 1-thiocarbamoyl-2-pyrazolines,[531] and phenylthiohydantoin derivatives of amino-acids[532] have been recently studied.

**Other Physical Properties.**—*Polarography.* Two groups of workers have studied the polarographic reduction of diaryl thiones in organic solvents.[533, 534] Polish workers[533] report that, with thiobenzophenones in DMF, two cathodic waves corresponding to mono-electron transfers are obtained. Elofson and his co-workers,[534] however, observe two-electron reduction waves for diaryl thiones, including thiobenzophenone, in acetonitrile. This phenomenon is interpreted as indicating one-electron reduction followed by rapid protonation and further reduction at the same potential. The pH-dependence of the half-wave potential suggests that, at low pH, reduction occurs simultaneously with, or immediately following, protonation. Only in the case of xanthione at low pH are two one-electron waves observed. The facile polarographic reduction of thiones relative to ketones is regarded as being related to the differences in respective $(n,\pi^*)$

[525] J. B. Lee and J. F. Scanlon, *Tetrahedron*, 1969, **25**, 3413.
[526] R. Radeglia, S. Scheithauer, and R. Mayer, *Z. Naturforsch.*, 1969, **24b**, 283.
[527] P. Nuhn, A. Zschunk, and G. Wagner, *Z. Chem.*, 1969, **9**, 335.
[528] D. Schumann, E. Frese, and A. Schönberg, *Chem. Ber.*, 1969, **102**, 3192.
[529] T. Hino, M. Nakagawa, K. Tsuneoka, S. Misawa, and S. Akaboshi, *Chem. and Pharm. Bull. (Japan)*, 1969, **17**, 1651.
[530] F. Duus, S.-O. Lawesson, J. H. Bowie, and R. G. Cooks, *Arkiv. Kemi*, 1969, **29**, 191; F. Duus, G. Schroll, S.-O. Lawesson, J. H. Bowie, and R. G. Cooks, *ibid.*, 1969, **30**, 347.
[531] S. W. Tam, *Org. Mass Spectrometry*, 1969, **2**, 729.
[532] B. W. Melvas, *Acta Chem. Scand.*, 1969, **23**, 1679.
[533] W. Kemula, H. Kryszczynska, and M. K. Kalinowski, *Roczniki Chem.*, 1969, **43**, 1071.
[534] R. M. Elofson, F. F. Gadallah, and L. A. Gadallah, *Canad. J. Chem.*, 1969, **47**, 3979.

transition energies rather than to the previously assumed greater polarity of thiones.

Other recently reported polarographic studies deal with dithiocarbamate esters,[535] metal xanthates,[535] and 1-hydroxypyridine-2-thione.[536]

*Dipole Moments.* Walter and Hühnerfüss [447] have obtained experimental electric moments for various *N*-alkyl-, *N*-aryl-, and *NN*-dialkyl-thioamides, and have shown how these relate to the rotational isomerism of these compounds. Dipole–dipole interactions are found to be the main cause of the preference of *N*-alkylthioamides for the configuration (214), while in

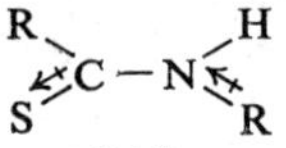

$$\underset{S}{\overset{R}{>}}C-N\underset{R}{\overset{H}{<}}$$
(214)

the case of thioanilides, steric factors are also important. Polish workers [537] have studied the association of thiocaprolactam on the basis of dipole moment measurements. The structure of the chelated complex of 1,2,4-triazoline-5-thione and secondary amines has been evaluated by similar methods.[538]

Dipole moment measurements have confirmed the mesionic character of 1,3,4-thiadiazole derivatives (215) and have shown that an exocyclic

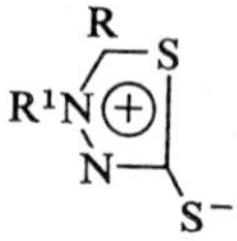

$$R^1N\overset{R}{\underset{N}{\oplus}}\overset{S}{\underset{S^-}{|}}$$
(215)

sulphur atom carries a greater charge than an exocyclic oxygen atom in this system.[539] Dipole moments have been calculated for thionoformic, thionoacetic, and thionopropionic acids.[540]

*Acidity and Basicity Measurements.* The acidities of several aromatic thioamides have been determined by potentiometric titration in dimethylsulphoxide–ethanol solution.[541] The thioamides are shown to be stronger N–H acids than the corresponding amides. Picolinamide has also been shown to be a stronger base than thiopicolinamide in dioxan–water solution.[542]

---

[535] V. I. Gorokhovskaya and L. I. Klimova, *Zhur. obshchei Khim.*, 1969, **39**, 31.
[536] A. F. Krivis and E. S. Gazda, *Analyt. Chem.*, 1969, **41**, 212.
[537] J. Hurwic, W. Waclawek, and M. Wejroch, *Roczniki Chem.*, 1969, **43**, 1063.
[538] L. A. Vlasova, V. I. Minkin, and I. Ya. Postavskii, *Zhur. obshchei Khim.*, 1970, **40**, 372.
[539] C. W. Atkin, A. N. M. Barnes, P. G. Edgerley, and L. E. Sutton, *J. Chem. Soc. (B)*, 1969, 1194.
[540] N. Saraswathi and S. Soundararajan, *J. Mol. Structure*, 1969, **4**, 419.
[541] W. Walter and R. F. Becker, *Annalen*, 1969, **727**, 71.
[542] J. C. Pariaud, J. Dumas, and R. Mauger, *Bull. Soc. chim. France*, 1969, 743.

The observed reduction of the basicity of semicarbazide on substitution of sulphur and selenium for carbonyl oxygen has been attributed to the increasing mesomeric ($-M$) effects in the series $O < S < Se$.[543] Selenourea, however, does not fit into this picture since it is found to be more basic than either thiourea or urea.

[543] D. R. Goddard, B. D. Lodam, S. O. Ajayi, and M. J. Campbell, *J. Chem. Soc. (A)*, 1969, 506.

# 6
# Ylides of Sulphur, Selenium, and Tellurium and Related Structures

Ylides have been defined [1] as 'substances in which a carbanion is attached directly to a heteroatom carrying a high degree of positive charge.... The definition also includes those molecular systems whose heteroatoms carry less than a formal full positive charge.' Therefore, an ylide may be represented by the general structure (1). However, structures such as (2) also

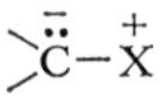

$$\overset{..}{\underset{}{>}}\overset{-}{C}-\overset{+}{X} \qquad \overset{..}{\underset{}{>}}\overset{-}{C}-\overset{+}{X}-\overset{-}{O}$$

(1) 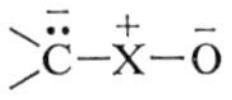                   (2)

may be classified as ylides and often exhibit ylide characteristics. The spelling 'ylid' predominated in the earlier literature but the new spelling 'ylide' seems to be preferred by editors at present.

## 1 Introduction

Since it is the objective of the Specialist Periodical Reports to present accounts of progress in specialized areas of chemistry on a regular basis, it would seem advantageous if this first volume containing such a report on sulphur ylide chemistry could review all of the pertinent work which has been reported in the field to date. Such a task clearly is beyond the scope of this Report. Therefore, as a second-best alternative, this first Report will cover the field in its entirety since the last (and only) complete review, that which appeared in the book 'Ylid Chemistry'.[1] The organization of this report will be virtually identical to that previously used,[1] thereby permitting an easier comparison of developments.

Only two reviews in the past five years have covered any significant portion of the chemistry of sulphur ylides. Bloch[2] presented a rather extensive but non-critical review of the chemistry of sulphonium ylides, dimethyloxysulphonium ylides, and methylsulphinyl carbanion. Hochrainer[3] reviewed only the chemistry of sulphonium ylides.

One of the features of ylide chemistry which adds interest to their study is the unique molecular structure, involving as it does the question of

---

[1] A. Wm. Johnson, 'Ylid Chemistry,' Academic Press, New York and London, 1966.
[2] J. C. Bloch, *Ann. Chim. (France)*, 1965, **10**, 419.
[3] A. Hochrainer, *Österr. Chem.-Ztg.*, 1966, **67**, 297.

Ylides of Sulphur, Selenium, and Tellurium and Related Structures    249

multiple bonding between the carbanion and the 'onium atom. Such bonding seems possible through overlap of the orbital containing the carbanion electrons with a vacant 3d-orbital of sulphur. While it is difficult to prove absolutely the existence of such bonding and its contribution to the uniquely charged molecules, a substantial accumulation of evidence clearly points toward such bonding.[1,4] While there have been two recent reviews concerning the use of outer d-orbitals in sulphur bonding,[5,6] both have focused only on divalent sulphur. In the past five years there has been no significant contribution to a theoretical understanding of the carbon–sulphur bonding in ylides of sulphur. Such study is sorely needed at this time as it might contribute to an understanding of the electronic characteristics of such bonds, the stability and stabilization of such systems, and the molecular geometry of ylides.

To the knowledge of the author, there has been no work reported in the literature directed toward the chemistry of tellurium ylides. Therefore, this report will not discuss such substances. There have been but a few reports concerning the chemistry of selenium ylides and these will be discussed separately in the last section of this chapter.

This chapter will discuss sulphur ylide chemistry in the following sequence: Sulphonium Ylides, Oxysulphonium Ylides, Sulphinyl Ylides, Sulphonyl Ylides, Sulphenes and Sulphines, Sulphur Imines, Sulphonyl and Sulphinyl Amines, Selenonium Ylides.

## 2 Sulphonium Ylides

Sulphonium ylides, represented by general structure (3), were the earliest known class of sulphur ylides. Their chemistry has been reviewed by Bloch[2] in 1965, Johnson[7] in 1966, and Hochrainer[3] in 1966.

$$R^1_{\phantom{1}}\!\!\diagdown\!\!\overset{+}{S}\!-\!\overset{\;\!\!..}{C}\!\diagdown\!\!R^3$$
$$R^2\!\diagup\phantom{\overset{+}{S}}\phantom{\overset{..}{C}}\diagup R^4$$

(3)

**Methods of Synthesis.**—Prior to 1965 the most widely employed method for the preparation of sulphonium ylides was the so-called 'salt method', whereby a sulphide effected a nucleophilic substitution on an alkyl halide (primary or secondary) to form a sulphonium salt. A proton alpha to the sulphur atom subsequently was removed utilizing the appropriate base to afford the ylide. This method still is the most widely used and still appears to offer the most flexibility. Robson et al.[8] reported a synthetic scheme employing this method to afford a series of ylides:

---

[4] C. C. Price and S. Oae, 'Sulfur Bonding', Ronald Press, New York, 1962.
[5] D. L. Coffen, Rec. Chem. Prog., 1969, **30**, 275.
[6] K. A. R. Mitchell, Chem. Rev., 1969, **69**, 157.
[7] Ref. 1, pp. 310–344.
[8] R. Robson, P. R. H. Speakman, and D. G. Stewart, J. Chem. Soc. (C), 1968, 2180.

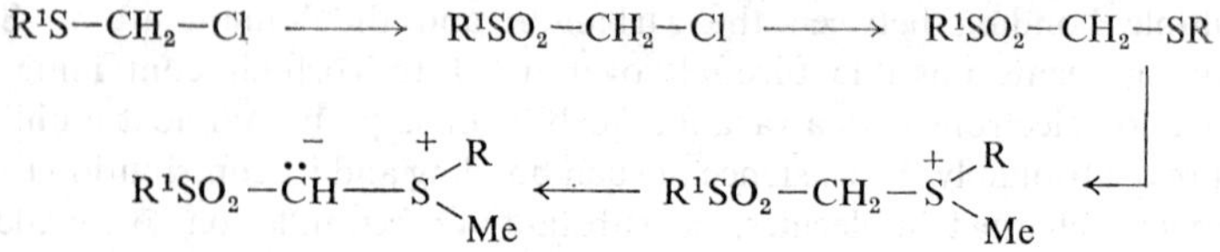

$$R^1S-CH_2-Cl \longrightarrow R^1SO_2-CH_2-Cl \longrightarrow R^1SO_2-CH_2-SR$$

$$R^1SO_2-\overset{-}{\underset{}{C}}H-\overset{+}{S}\diagdown_{Me}^{R} \longleftarrow R^1SO_2-CH_2-\overset{+}{S}\diagdown_{Me}^{R} \longleftarrow$$

Ratts and Yao[9] reported the preparation of a series of $\beta$-carbonyl sulphonium ylides, $Me_2S=CH-CO-R$, using the 'salt method'. Corey et al.[10] have utilized the 'salt method' for the preparation of a simple ylide, ethylidenediphenylsulphurane (4), which subsequently was converted by alkylation

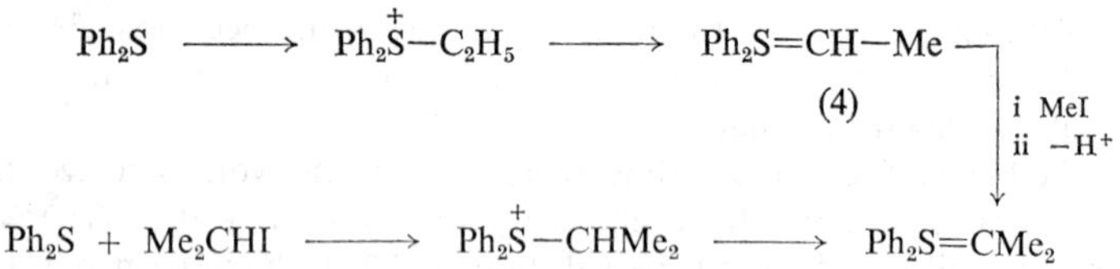

$$Ph_2S \longrightarrow Ph_2\overset{+}{S}-C_2H_5 \longrightarrow Ph_2S=CH-Me$$
$$(4) \quad \begin{array}{l} \text{i MeI} \\ \text{ii } -H^+ \end{array}$$

$$Ph_2S + Me_2CHI \longrightarrow Ph_2\overset{+}{S}-CHMe_2 \longrightarrow Ph_2S=CMe_2$$

and proton abstraction into more complex ylides. This seemed to be a better route to alkyl-substituted ylides than the direct 'salt method' route from the secondary alkyl halide.

In 1965 Diekmann[11] reported that, under either photochemical or thermal conditions, bis(benzenesulphonyl)diazomethane reacted with alkyl sulphides to expel molecular nitrogen and afford sulphonium ylides in high yield. Shortly thereafter Lloyd and Singer[12] applied this method to the preparation of the previously elusive cyclopentadienylidenesulphurane system (5). In 1969, Ando and his co-workers[13–15] explored much further

$$(PhSO_2)_2CN_2 + R_2S \xrightarrow{h\nu \text{ or } \Delta} (PhSO_2)_2C=SR_2 \quad (R = Me \text{ or } Bu^n)$$

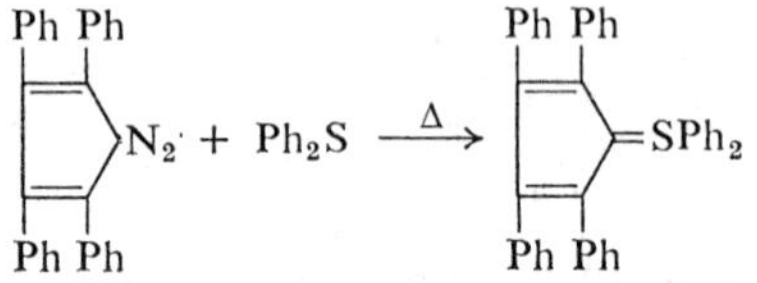

the scope of this kind of reaction by utilizing a series of diazo-compounds and a series of sulphides. In all cases the diazo-compounds carried strongly electronegative groups (e.g. carbethoxy), and the yields of the ylides

[9] K. W. Ratts and A. N. Yao, J. Org. Chem., 1966, 31, 1185.
[10] E. J. Corey, M. Jautelat, and W. Oppolzer, Tetrahedron Letters, 1967, 2325.
[11] J. Diekmann, J. Org. Chem., 1965, 30, 2272.
[12] D. Lloyd and M. I. C. Singer, Chem. and Ind. (London), 1967, 118.
[13] W. Ando, T. Yagihara, S. Tozune, S. Nakaido, and T. Migita, Tetrahedron Letters, 1969, 1979.
[14] W. Ando, T. Yagihara, S. Tozune, and T. Migita, J. Amer. Chem. Soc., 1969, 91, 2786.
[15] W. Ando, K. Nakayama, K. Ichibari, and T. Migita, J. Amer. Chem. Soc., 1969, 91, 5164.

decreased considerably as less nucleophilic and more sterically hindered sulphides were employed. The addition of benzophenone as a sensitizer in the photochemical synthesis prevented ylide formation, indicating that the ylide synthesis probably involved singlet carbene attack on the nucleophilic sulphide.

In 1966 four separate groups reported a new method for the preparation of sulphonium ylides from active methylene compounds and sulphoxides. Gomper and Euchner [16] utilized acetic anhydride as a catalyst to condense bis(benzenesulphonyl)methane with dimethylsulphoxide to afford the stabilized ylide (6). Diefenbach et al.[17,18] also condensed sulphonyl-activated methylenes with DMSO. Hochrainer [19,20] reported the use of a

$$(PhSO_2)_2CH_2 + Me_2SO \longrightarrow (PhSO_2)_2C=SMe_2$$
(6)

$$A\begin{bmatrix}C(=O)\\ \\C(=O)\end{bmatrix}CH_2 + R^1R^2SO \longrightarrow A\begin{bmatrix}C(=O)\\ \\C(=O)\end{bmatrix}C=SR^1R^2$$
(7)

series of cyclic β-dicarbonyl compounds and a series of alkyl sulphoxides for the synthesis of the corresponding sulphonium ylides (7), utilizing either acetic anhydride or dicyclohexylcarbodi-imide as the dehydrating agent. Nozaki et al.[21,22] reported that dimethylsulphoxide or phenylmethylsulphoxide, but not diphenylsulphoxide, would condense with a series of active methylene compounds in the presence of either acetic anhydride or triethylamine–phosphorus pentoxide to afford the corresponding ylide. These workers claimed that dicyclohexylcarbodi-imide was ineffective in this reaction.[21] However, more recently Cook and Moffatt [23] have reported numerous additional examples of this type of synthesis using dicyclohexylcarbodi-imide and a trace of phosphoric acid as the catalyst. Seitz [24] has reported the preparation of a series of cyclopentadienylidene sulphonium ylides from the appropriate cyclopentadiene and a series of acyclic and cyclic sulphoxides.

[16] R. Gompper and H. Euchner, *Chem. Ber.*, 1966, **99**, 527.
[17] H. Diefenbach and H. Ringsdorf, *Angew. Chem. Internat. Edn.*, 1966, **5**, 971.
[18] H. Diefenbach, H. Ringsdorf, and R. E. Wilhelms, *Chem. Ber.*, 1970, **103**, 183.
[19] A. Hochrainer and F. Wessely, *Monatsh. fur Chem.*, 1966, **97**, 1.
[20] A. Hochrainer, *Monatsh. fur Chem.*, 1966, **97**, 823.
[21] H. Nozaki, Z. Morita, and K. Kondo, *Tetrahedron Letters*, 1966, 2913.
[22] H. Nozaki, D. Tunemoto, Z. Morita, K. Nakamura, K. Wanatabe, M. Takaku, and K. Kondo, *Tetrahedron*, 1967, **23**, 4279.
[23] A. F. Cook and J. G. Moffatt, *J. Amer. Chem. Soc.*, 1968, **90**, 740.
[24] G. Seitz, *Chem. Ber.*, 1968, **101**, 585.

The mechanism of the reaction between active methylene compounds and sulphoxides probably involves first the conversion of the sulphoxide to an alkoxysulphonium salt (8). Attack by the carbanion of the active methylene compound on the sulphur atom could afford a sulphonium salt (9). This salt would be extremely acidic, due to the presence of three

$$R_2SO + (MeCO)_2O \longrightarrow R_2\overset{+}{S}-OCOMe$$
(8a)

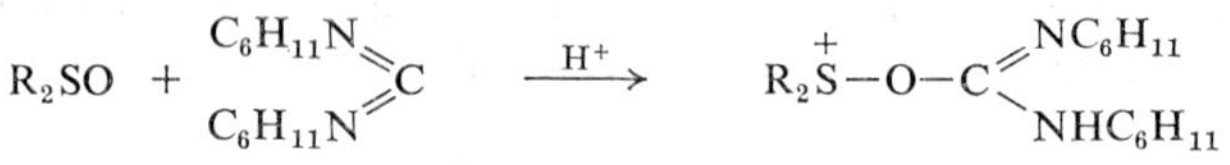

$$R_2SO + \begin{matrix} C_6H_{11}N \\ C_6H_{11}N \end{matrix}\!\!>\!\!C \xrightarrow{H^+} R_2\overset{+}{S}-O-C\!\!<\!\!\begin{matrix} NC_6H_{11} \\ NHC_6H_{11} \end{matrix}$$
(8b)

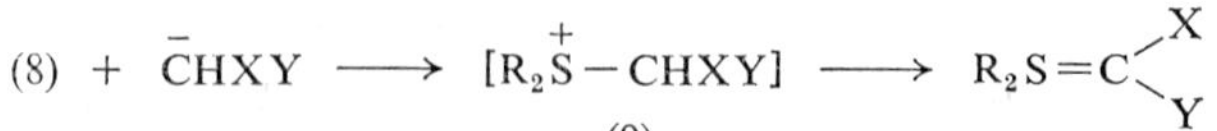

$$(8) + \bar{C}HXY \longrightarrow [R_2\overset{+}{S}-CHXY] \longrightarrow R_2S=C\!\!<\!\!\begin{matrix} X \\ Y \end{matrix}$$
(9)

powerful electronegative groups on the methine carbon, and by loss of a proton would afford the ylide. It has been shown in a separate experiment [22] that the sodium salt of active methylene compounds will react with phenyl-methylethoxysulphonium tetrafluoroborate to afford the expected ylide. To date, this synthesis has been restricted to those methylene derivatives which carry strong activating groups. However, based on the proposed mechanism almost any carbanion should work equally well. Similarly, the only requirement for the sulphoxide component is that it be sufficiently nucleophilic to be converted into an alkoxysulphonium salt.

In summary, during the past five years no new method which is completely general for the synthesis of sulphonium ylides has been discovered. Those new methods reported in this section seem quite general and very efficient for the preparation of sulphonium ylides which carry highly electronegative groups on the carbanion.

**Physical Properties.**—In 1968 and 1969 two separate reports on an $X$-ray crystallographic structure determination were published for dicyano-methylenedimethylsulphurane (10),[23, 25] the first for any sulphur ylides.

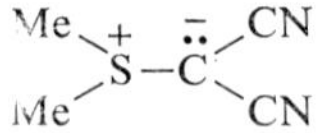

$$\begin{matrix} Me \\ Me \end{matrix}\!\!>\!\!\overset{+}{S}-\bar{C}\!\!<\!\!\begin{matrix} CN \\ CN \end{matrix}$$
(10)

The significant structural characteristics reported were as follows: the sulphur was pyramidal; the carbanion was planar; the entire grouping S–C(CN)$_2$ was planar; the methyl–sulphur distance was normal for a single bond (1·81—1·84 Å); the sulphur–carbanion distance (1·73 Å) was inter-

[25] A. T. Christensen and W. G. Whitmore, *Acta Cryst.*, 1969, **25B**, 73.

mediate between the values known for C–S single bonds (1·81 Å) and those known for C–S double bonds (1·55 Å); the carbanion–cyanocarbon distance (1·40 Å) was slightly shorter than that expected for a carbon–carbon bond of $sp$–$sp^2$ hybridization (1·42—1·46 Å); the C–N bonds (1·16 Å) seemed to be normal length; the carbanion bond angles (118—122°) were within the range known for trigonally-hybridized carbon. These geometric properties seem to confirm earlier speculation both on the geometry of such molecules and the general nature of the electronic distribution in sulphonium ylides. It seems apparent that the trigonal carbanion is stabilized by delocalization of the negative charge through all of the carbanion substituents, probably via $\pi$-bonding. The shortening of the carbanion–carbon distances and the carbanion–sulphur distance provides evidence to this effect. The rather striking shortening of the carbanion–sulphur bond is consistent with the proposal of considerable $p\pi$–$d\pi$ overlap between the 2$p$-orbital of the carbanion and a 3$d$-orbital of the sulphur atom. The actual C–S bond distance is remarkably close to the 1·7 Å distance suggested five years ago for such bonds,[26] and may indicate transfer of about one-third of the electron density of the carbanion to the sulphur atom.

Darwish and Tomlinson [26a] reported the conversion of an optically-active sulphonium salt into optically-active phenacylidenemethylethylsulphurane. The ylide racemized readily, probably via pyramidal inversion facilitated by $p\pi$–$d\pi$ overlap.

Additional evidence regarding the role the carbanion substituents play in stabilizing sulphonium ylides is available from an examination of the i.r. spectra of $\beta$-carbonyl sulphonium ylides. Dimethylsulphoniumphenacylide (11) exhibited carbonyl absorption in the i.r. at 1508 cm$^{-1}$ while the precursor sulphonium salt absorbed at 1665 cm$^{-1}$,[27] implying that a considerable contribution is made by the enolate form (12). $\beta$-Sulphonyl sulphonium

$$\underset{(11)}{\overset{+}{Me_2S}-\overset{-}{CH}-\overset{\overset{O}{\|}}{C}-Ph} \quad \rightleftharpoons \quad \underset{(12)}{\overset{+}{Me_2S}-CH=\overset{\overset{O^-}{|}}{C}-Ph}$$

ylides also show the S–O stretching frequency shifted to a lower value than that of the precursor salt or sulphone.[18] The fact that sulphonium ylides which do not carry strongly electronegative groups on the carbanion usually are chemically rather unstable also indicates the importance of delocalization of the negative charge from the carbanion by its substituents.

Study of the n.m.r. spectra of sulphonium ylides has led to an increased understanding of the geometry and electronic interactions of these substances. Several groups have shown that conversion of a dimethylsulphonium salt, $Me_2\overset{+}{S}-CH_2-R$, to the corresponding ylide, $Me_2\overset{+}{S}-\overset{-}{CH}-R$,

---

[26] Ref. 1, p. 318.
[26a] D. Darwish and R. L. Tomlinson, *J. Amer. Chem. Soc.*, 1968, **90**, 5938.

results in a significant *increase* in the shielding of the *S*-methyl groups.[9, 27, 28] This effect is attributed to the transfer of some electron density from the carbanion to sulphur, resulting in lowered positive charge character for sulphur in the ylide than in the sulphonium salt, and therefore more shielding by sulphur.

The shielding of the methine proton in sulphonium ylides also provides an indication of the kind and direction of electronic effects in the ylides. It appears that attachment to sulphur of groups which are more electron withdrawing has the effect of increasing the delocalization of negative charge from the carbanion to sulphur, and *vice versa*. Thus, in a series of bis-ylides (13), stepwise replacement of the two methyl groups on sulphur with the *less* electronegative ethyl groups has the effect of increasing the shielding of the methine proton (*i.e.* shifts it from $\delta 2 \cdot 50$ to $2 \cdot 33$ then to $2 \cdot 23$).[29] Similarly, in a series of phenacylides (14) stepwise replacement of

$$Ph_3 \overset{+}{P} - \overset{-}{C}H - \overset{+}{S} \overset{R^1}{\underset{R^2}{\diagdown}} \qquad PhCO\overset{-}{C}H - \overset{+}{S} \overset{R^1}{\underset{R^2}{\diagdown}}$$

(13) (14)

(a) $R^1 = R^2 = Me$
(b) $R^1 = Me; R^2 = Et$
(c) $R^1 = R^2 = Et$

the *S*-methyl groups with the *more* electronegative phenyl groups has the effect of decreasing the shielding of the methine proton (*i.e.* shifts it from $\delta 4 \cdot 31$ to $4 \cdot 55$ then to $4 \cdot 75$).[27] These observations are consistent with the view that as the positive character of the sulphur atom is increased by the attachment of appropriate substituents, it provides more effective delocalization for the carbanion electrons through $p\pi$–$d\pi$ overlap, probably by contraction of the 3$d$-orbitals.

In the first report on the preparation of dimethylsulphoniumphenacylide (11) Ratts and Yao[9] indicated that the methine absorption in the n.m.r. spectrum was quite broad. It was suggested that this broadening was due to an equilibrium between the two geometric isomers of the enolate form of the ylide (12). Trost[30] studied this system using variable temperature n.m.r. spectroscopy and reported changes in methine peak position and bandwidth with temperature as follows: 30 °C at $\delta 4 \cdot 30$ (bandwidth at half-height 9·0 c.p.s.); $-20$ °C at $\delta 4 \cdot 42$ (1·5 c.p.s.); 118 °C at $\delta 4 \cdot 24$ (2·0 c.p.s.). It was concluded that the single most stable isomer, that with the sulphonium group and the oxyanion *cis*, was present at $-20$ °C and that as the temperature was raised rotation about the carbon–carbon bond

---

[27] A. Wm. Johnson and R. T. Amel, *Tetrahedron Letters*, 1966, 819; *J. Org. Chem.*, 1969, **34**, 1240.
[28] J. Casanova and D. A. Rutolo, *Chem. Comm.*, 1967, 1224.
[29] J. Gosselck, H. Schenk, and H. Ahlbrecht, *Angew. Chem., Internat. Edn.*, 1967, **6**, 249.
[30] B. Trost, *J. Amer. Chem. Soc.*, 1966, **88**, 1587; *ibid.*, 1967, **89**, 138.

became rapid on the n.m.r. time scale. More recently, Caserio et al.[31] have found that with an anhydrous sample of the phenacylide (11) the n.m.r. spectrum was invariant from $-30$ °C to $+100$ °C, exhibiting a sharp singlet for the methine proton. A double resonance experiment indicated that the broadening reported by others [9, 30] was the result of a protolytic exchange reaction, analogous to that reported for the phosphoniumphenacylide.[32] Methylation of the phenacylide (11) with trimethyloxonium tetrafluoroborate afforded the two isomeric vinyl ethers (15) and (16) in a 94 : 6 ratio,

$$\begin{matrix}(11)\\(12)\end{matrix} \xrightarrow{Me_3O^+BF_4^-} \underset{H}{\overset{Me_2S^+}{>}}C=C\underset{Ph}{\overset{OMe}{<}} \quad + \quad \underset{H}{\overset{Me_2S^+}{>}}C=C\underset{OMe}{\overset{Ph}{<}}$$
$$\qquad\qquad\qquad\qquad (15) \qquad\qquad\qquad (16)$$

indicating clearly that the ylide (11) does exist in two geometric forms. Interestingly, the ether (15) formed in highest yield is that derived from the enolate isomer which is expected to be the most stable. The fact that the two enolate isomers of the ylide cannot be detected in the n.m.r. spectrum is as yet unexplained, but may be due to rapid (rotational) interconversion, even at low temperatures. No geometric isomers have been detectable in the n.m.r. spectrum of the analogous phosphonium phenacylide.[32]

Geometric isomers have been detected by n.m.r. spectroscopy of carbomethoxymethylenedimethylsulphurane (17). Casanova and Rutolo [28]

$$Me_2\overset{+}{S}-\overset{-}{C}HCOOMe \rightleftharpoons \underset{H}{\overset{Me_2S^+}{>}}C=C\underset{OMe}{\overset{O^-}{<}} \rightleftharpoons \underset{H}{\overset{Me_2S^+}{>}}C=C\underset{O^-}{\overset{OMe}{<}}$$
$$(17)$$

found that the methine proton of this ylide was a sharp singlet at $+45$ °C, a very broad peak at 0 °C, and had split into two peaks at $-45$ °C. In fact, the three singlets present in the spectrum of the ylide at $+45$ °C were split into six singlets at $-45$ °C, with the ratio between the two sets being 83 : 17. The dominant isomer exhibited methine absorption at lowest field and therefore was assigned a structure in which the sulphonium group and the oxyanion were *cis*. These observations are similar to those made for the related phosphonium ylide, carbomethoxymethylenetriphenylphosphorane.[32] Therefore, in both the sulphonium and phosphonium ylides there appears to be a barrier to rotation for the carbomethoxy-ylides, but such a barrier has not been detected for the benzoyl ylides (phenacylides).

Nozaki et al.[22] have examined the n.m.r. spectra of a series of bis-β-carbonyl ylides. Bisacetylmethylenedimethylsulphurane showed a singlet for both the *S*-methyls and the *C*-methyls at room temperature. However, the *C*-methyl peak gradually broadened and split as the temperature was

---

[31] S. H. Smallcombe, R. J. Holland, R. H. Fish, and M. C. Caserio, *Tetrahedron Letters*, 1968, 5987.

[32] F. J. Randall and A. Wm. Johnson, *Tetrahedron Letters*, 1968, 2841.

lowered, and at −60 °C exhibited two sharp singlets. This was attributed to either inversion of the sulphur atom or geometric isomerism about the enolate system. The former explanation seems unlikely since the geometry of the bisacetylmethylene portion of the molecule is expected to be symmetrical.

Further information regarding the geometry of sulphonium ylides has been obtained as a result of the observation of magnetic nonequivalence in some ylides (18) where R ≠ H. Ratts[33] reported that in a series of

$$R^1-S-R^2$$
$$H_A-C-H_B$$
$$R$$

(18)

phenacylides (18; $R^2$ = $CHCOC_6H_5$, $R^1$ = Me, Et, Pr, or Ph), when R = Me, Et, or Ph, but not H, the methylene protons ($H_A$ and $H_B$) were nonequivalent. One of the protons was shifted to higher field by about 0·9 p.p.m. and the coupling constant between the two protons was about 12 c.p.s. In structure (18) $H_B$ would be expected to be shielded by the phenacylide group ($R^2$). Hochrainer and Silhan,[34] at about the same time, demonstrated the presence of nonequivalent methylene protons in a series of cyclic bis-β-carbonyl sulphonium ylides (7). The geminal coupling constant was also 12 c.p.s. in these ylides. Similar reports of magnetic nonequivalence of S-methylene protons appeared in 1967[22] and in 1968,[23] both dealing with β-carbonyl sulphonium ylides. In all of these reports, the nonequivalence has been attributed to the unique ylide structure because the corresponding salts were claimed not to show nonequivalence. However, Kondo and Mislow[35] reported that a number of sulphonium salts containing α-methylene groups also showed magnetic nonequivalence. In addition, it is apparent that the extent of nonequivalence is a function of solvent. Accordingly, it is not possible at this time to elaborate a consistent explanation for such nonequivalence in sulphonium systems.

A study of the basicity of sulphonium ylides, or of the acidity of the conjugate acids, has shed some light on the kinds of electronic interactions taking place between the various structural components of sulphonium ylides. Relatively little comparative work has been reported to date on the effect of the carbanion substituents on the $pK_a$ of sulphonium salts. However, Ratts and Yao[9] did show that in a series of dimethylsulphoniumphenacylides, $Me_2S=CH-CO-C_6H_4-X$, the basicity of the ylide decreased as X was made more electronegative. This trend, analogous to that reported for the triphenylphosphoniumphenacylides,[36] was attributed

---

[33] K. W. Ratts, *Tetrahedron Letters*, 1966, 4707.
[34] A. Hochrainer and W. Silhan, *Monatsh.* 1966, **97**, 1477.
[35] K. Kondo and K. Mislow, *Tetrahedron Letters*, 1967, 1325.
[36] S. Fliszar, R. F. Hudson, and G. Salvadori, *Helv. Chim. Acta*, 1963, **46**, 1580.

to increased delocalization of the electron density of the carbanion to the phenacyl group.

Substituents on the sulphur atom also affect the basicity of sulphonium ylides. Aksnes and Songstad,[37] and later Johnson and Amel,[38] determined the $pK_a$'s of a variety of phenacyl sulphonium salts. They found that replacement of methyl by higher alkyl and then by phenyl led to an increase in the acidity of the sulphonium salts. It was argued that a diphenyl-sulphonio-group provided more stabilization for the ylide carbanion than did the alkyl groups, as also was indicated by the n.m.r. evidence cited earlier in this section. The acidifying effect of ethyl over methyl was attributed to a decreased hyperconjugative interaction with the sulphur atom. This latter point was evident again in the report by Lillya and Miller[39] concerning the effect of sulphur substituents on the acidity of a series of bis-sulphoniomethanes (19). For the system where R = methyl,

$$\begin{matrix} R^1 \\ \diagdown \\ R \diagup \end{matrix} \overset{+}{S} - CH_2 - \overset{+}{S} \begin{matrix} \diagup R^1 \\ \diagdown R \end{matrix} \quad 2BF_4^-$$
(19)

the $pK_a$ of the salts increased from 9·19 to 7·76 as $R^1$ was changed from methyl to ethyl to isopropyl. A similar change was evident when R = ethyl and $R^1$ was changed through the same series. In summary, it is evident that the same structural changes which increase the stabilization of a sulphonium ylide also lead to a decrease in the basicity of the ylide, and both presumably result from increased delocalization of the negative charge from the carbanion to the various substituents attached thereto.

Among the stabilized ylides which are known the sulphonium ylides seem to be the least basic. In other words, sulphur seems able to provide the most stabilization for an adjacent carbanion, presumably via $p\pi-d\pi$ overlap. In the phenacylide series, the stabilizing effect has been shown by two groups,[37,38] to decrease in the order S > P > As as long as the heteroatom substituents are identical (i.e. all methyl or all phenyl, etc.). In a series of cyclopentadienylides, Lloyd and Singer[40] found the stabilizing effect to be S ~ Se > P > As > Sb based on the basicity of the ylides carrying all phenyl substituents on the heteroatoms.

**Chemical Stability.**—As was discussed in an earlier review,[7] sulphonium ylides not carrying electronegative substituents stabilizing the carbanion function generally are chemically rather reactive, often decomposing in the presence of moisture or oxygen. On the other hand, numerous stabilized sulphonium ylides have been prepared, isolated, and purified in order that the pure ylide in a free state might be studied. In the earlier review, the spontaneous decomposition of simple ylides was rather thoroughly

---

[37] G. Aksnes and J. Songstad, *Acta Chem. Scand.*, 1964, **18**, 655.
[38] A. Wm. Johnson and R. T. Amel, *Canad. J. Chem.*, 1968, **46**, 461.
[39] C. P. Lillya and E. F. Miller, *Tetrahedron Letters*, 1968, 1281.
[40] D. Lloyd and M. I. C. Singer, *Chem. and Ind. (London)*, 1968, 1277.

discussed. Since no additional information has become available in the intervening five years, this aspect of sulphonium ylide chemistry will not be reviewed. Instead, attention will be focused on the decomposition, rearrangement, photochemical, and hydrolytic chemistry of the more complex sulphonium ylides.

In 1950 Krollpfeiffer and Hartmann [41] reported obtaining 1,2,3-*trans*-tribenzoylcyclopropane (20) upon treatment of phenacylphenylmethylsulphonium bromide with aqueous base, and proposed that it might have been formed through the intermediacy of an as yet unknown zwitterion, phenacylidenephenylmethylsulphurane. In 1964 Horak and Kohout [42] reported that treatment of phenacyldi-n-butylsulphonium bromide with diazomethane afforded methyl bromide and the same cyclopropane (20). It was proposed that the sulphonium salt was deprotonated to the ylide, which rapidly decomposed to the carbene, benzoylmethylene. The latter was thought to trimerize to (20). In 1966, Johnson and Amel [27] reported the isolation and purification of phenacylidenedimethylsulphurane (11) and its thermal conversion to the cyclopropane (20). Trost [30] reported the same conversion to occur in even higher yield when the decomposition was effected in the presence of cupric sulphate. Ratts and Yao [43] also reported the thermal conversion of *p*-nitrophenacylidenedimethylsulphurane into the corresponding tri-*p*-nitrobenzoylcyclopropane. Johnson and Amel [27] and Casanova and Rutolo [28] reported a similar decomposition for carbomethoxymethylenedimethylsulphurane. Two distinct mechanisms have been proposed to account for all of these observations.

The first mechanism [27] involves the intermediate formation of a carbene fragment (21) as a result of a unimolecular dissociation of the ylide. This fragment would be likely to be trapped by the nucleophilic ylide to form a zwitterion which would undergo elimination to afford the olefin (22).

$$\text{PhCO}-\overset{-}{\text{CH}}-\overset{+}{\text{S}}\text{Me}_2 \longrightarrow [\text{PhCOCH:}] \xrightarrow{(11)} \left[ \text{PhCO}\overset{-}{\text{CH}}-\underset{\overset{+}{\text{SMe}_2}}{\text{CH}}-\text{COPh} \right]$$
(11) (21)

$$\text{PhCO}-\text{HC}-\text{CH}-\text{COPh}$$
$$\underset{\underset{\text{COPh}}{|}}{\overset{\diagdown\diagup}{\text{CH}}} \xleftarrow{(11)} \text{PhCO}-\text{CH}=\text{CH}-\text{COPh}$$
(20) (22)

Michael addition and displacement of methyl sulphide would give the observed cyclopropane (20). The major evidence for this mechanism is that it is a thermal reaction occurring in inert solvents and that the carbene

---

[41] F. Krollpfeiffer and H. Hartmann, *Chem. Ber.*, 1950, **83**, 90.
[42] V. Horak and L. Kohout, *Chem. and Ind. (London)*, 1964, 978.
[43] K. W. Ratts and N. A. Yao, *J. Org. Chem.*, 1966, **31**, 1689.

fragment has been trapped with dimethylbenzylamine,[27] triphenylphosphine,[43] and cyclohexene,[30] although in low yield in each case. The low yields may be due to the fact that remaining ylide would be a more effective trapping agent. The Michael addition of ylide (11) to olefin (22) separately has been shown to be a rapid reaction occurring in high yield.

In no instance with $\beta$-carbonyl ylides has the olefin (22) been isolated. However, numerous examples of ylide decomposition are known in which olefins are the major products. Earlier, Johnson et al.[44] obtained stilbenes from diphenyl- and dimethyl-sulphoniumbenzylide, Swain and Thornton[45] obtained p,p'-dinitrostilbene from dimethyl-p-nitrobenzylsulphonium tosylate in basic media, and Hayasi et al.[46] obtained 1,2-diphenylthioethene from phenylthiomethylenedimethylsulphurane. Johnson and Amel[27] proposed that in such instances olefins were obtained because they were not susceptible to Michael addition by the ylide remaining. (Interestingly, the latter ylide did effect a Michael addition with 1,2-dibenzoylethene to afford a cyclopropane[46]). Furthermore, failure to obtain cyclopropane indicates that carbene addition to olefin is an unlikely mechanism for cyclopropane formation in these reactions. The observation by Serratosa and Quintana[47] that an acyl diazo-compound (23) would decompose in an

RCOCHN$_2$ (23)
- Cu-bronze → RCOCH=CHCOR
- Cu-bronze, Me$_2$C=CMe$_2$ → ▷—COR
- Cu-bronze, Ph—S—Me → RCO—C(COR)—COR

inert solvent in the presence of copper-bronze to the diacyl olefin, whereas repetition of the experiment but with the addition of thioanisole afforded the cyclopropane, clearly indicates that the carbene will not add to the olefin to form cyclopropane but that a sulphonium ylide, formed from the carbene and sulphide, would do so. The carbene would add to a more nucleophilic olefin, tetramethylethylene, when it was added to the reaction. It seems clear, therefore, that the last step in the formation of cyclopropanes from sulphonium ylides probably involves a Michael addition of ylide to the olefin.

The second overall mechanism[30, 48] includes an ionic chain reaction, again resulting in the formation of an olefin which is converted to the cyclopropane via a Michael addition of the ylide. It was suggested that any

---

[44] A. Wm. Johnson, V. J. Hruby, and J. L. Williams, *J. Amer. Chem. Soc.*, 1964, **86**, 918.
[45] C. G. Swain and E. R. Thornton, *J. Amer. Chem. Soc.*, 1961, **83**, 4033.
[46] Y. Hayasi, M. Takaku, and H. Nozaki, *Tetrahedron Letters*, 1969, 3179.
[47] F. Serratosa and J. Quintana, *Tetrahedron Letters*, 1967, 2245.
[48] H. Nozaki, K. Kondo, and M. Takaku, *Tetrahedron Letters*, 1965, 251.

proton source, moisture, or traces of sulphonium salt, *etc*. would serve to afford phenacyldimethylsulphonium salt. Nucleophilic attack by ylide (11)

$$PhCO\overset{-}{C}H-\overset{+}{S}Me_2 \xrightarrow{H^+} PhCOCH_2-\overset{+}{S}Me_2$$
(11)

$$PhCOCH_2-\overset{+}{S}Me_2 + (11) \longrightarrow PhCOCH-CH_2COPh$$
$$\qquad\qquad\qquad\qquad\qquad\qquad |$$
$$\qquad\qquad\qquad\qquad\qquad\quad ^+SMe_2$$
(24)

$$\downarrow (11)$$

$$PhCO-HC-CH-COPh \xleftarrow{(11)} PhCOCH=CHCOPh$$
$$\qquad\quad \backslash\;/$$
$$\qquad\quad\; CH \qquad\qquad\qquad\qquad (22)$$
$$\qquad\qquad |$$
$$\qquad\quad COPh$$
(20)

on this salt would afford a new salt (24) which could undergo 1,2-elimination to afford the olefin (22). The basic reagent to effect this elimination could be ylide, thereby generating new starting sulphonium salt with which to repeat the chain reaction. The major evidence for this alkylation–elimination mechanism is the separate observation that reaction of ylide (11) with added conjugate acid or phenacyl bromide leads to a high yield of cyclopropane in a fast reaction. In summary, it seems agreed that cyclopropane formation proceeds *via* Michael-type addition to an initially formed olefin, but the mechanism of olefin formation has not been definitively elaborated.

Some sulphonium ylides undergo intramolecular eliminations when warmed, especially those carrying higher alkyl groups on sulphur. For example, (25) underwent a 1,2- or 1′,2-elimination to afford dodecene.[18] Similarly, (26) afforded ethylene.[49] However, replacement of the ethyl group (with its β-protons) with a group which cannot undergo such an elimination may lead to another product (27), the result of a Stevens rearrangement.[49]

The first Stevens rearrangement of a sulphonium salt or ylide was reported by Stevens in 1932[50] to involve the conversion of phenacylmethylbenzylsulphonium bromide in methanolic sodium methoxide to α-thiomethyl-β-phenylpropiophenone (28), presumably *via* an ylide intermediate. This result was verified in 1949[51] and in 1950.[41] Recently, Schollkopf *et al*.[52] effected a series of analogous rearrangements, only starting with the ylide rather than the sulphonium salt and using substituted-benzyl derivatives. On the basis of kinetic analysis and CIDNP-n.m.r.

---

[49] W. Ando, T. Yagihara, and T. Migita, *Tetrahedron Letters*, 1969, 1983.
[50] T. Thomson and T. S. Stevens, *J. Chem. Soc.*, 1932, 69.
[51] H. Bohme and W. Krause, *Chem. Ber.*, 1949, **82**, 426.
[52] U. Schollkopf, G. Ostermann, and G. Schossig, *Tetrahedron Letters*, 1969, 2619.

$$MeSO_2CH=S\overset{Me}{\underset{C_{12}H_{25}}{\diagup}} \longrightarrow MeSO_2CH_2SMe + C_{10}H_{21}CH=CH_2$$
(25)

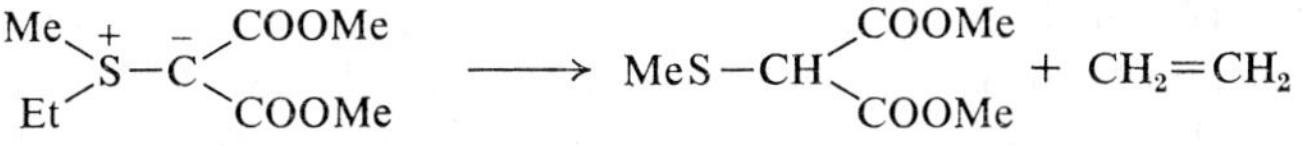

$$\overset{Me}{\underset{Et}{\diagup}}\overset{+}{S}-\overset{-}{C}\overset{COOMe}{\underset{COOMe}{\diagup}} \longrightarrow MeS-CH\overset{COOMe}{\underset{COOMe}{\diagup}} + CH_2=CH_2$$
(26)

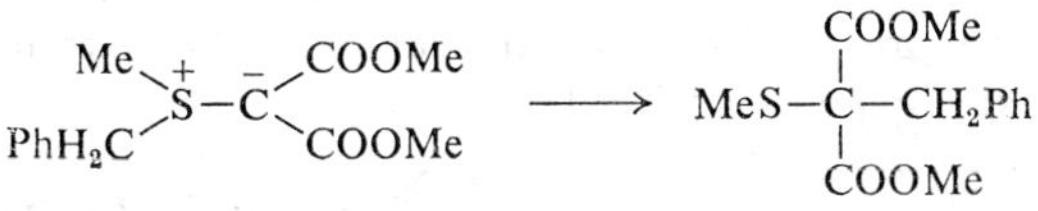

$$\overset{Me}{\underset{PhH_2C}{\diagup}}\overset{+}{S}-\overset{-}{C}\overset{COOMe}{\underset{COOMe}{\diagup}} \longrightarrow MeS-\underset{COOMe}{\overset{COOMe}{\underset{|}{C}}}-CH_2Ph$$
(27)

spectroscopy it was claimed that the reaction proceeded through a homolytic dissociation and 1,3-migration of the benzyl radical. Some question arose in 1958, however, regarding the actual structure of the Stevens rearrangement products since Ruiz [53] reported that repetition of the original Stevens reaction afforded two products, (28) and an olefin (29), in 6% and 79%

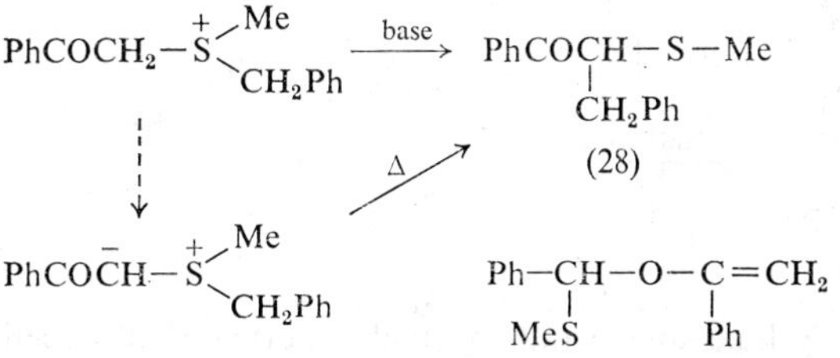

$$PhCOCH_2-\overset{+}{S}\overset{Me}{\underset{CH_2Ph}{\diagup}} \xrightarrow{base} PhCOCH-S-Me$$
$$\underset{CH_2Ph}{|}$$
(28)

$$PhCO\overset{-}{C}H-\overset{+}{S}\overset{Me}{\underset{CH_2Ph}{\diagup}} \qquad Ph-\underset{MeS}{\overset{|}{CH}}-O-\underset{Ph}{\overset{|}{C}}=CH_2$$
(29)

yields respectively. More recently, Ratts and Yao [54] have clarified the situation with the report that in aqueous solution the methylbenzylsulphoniumphenacylide does undergo the Stevens 1,3-rearrangement to the thioether (28) but that in methanolic sodium methoxide a 1,5-rearrangement to the olefin (29) occurs. Hayashi and Oda [55] have subsequently reported that in aprotic solvents three rearrangement processes occur: the Stevens 1,3-rearrangement, the Sommelet rearrangement, and the olefin 1,5-rearrangement. The proportion of 1,3-rearrangement product decreases as the temperature is lowered.

Dual rearrangements of sulphonium ylides to a mixture of products resulting from both 1,3- and 1,5-rearrangements have been reported for rather different systems than those discussed above. Parham and Groen [56]

---

[53] E. B. Ruiz, *Acta Salmanticensa Ser. Cienc.*, 1958, **2**, 64; *Chem. Abs.*, 1960, **54**, 7623.
[54] K. W. Ratts and A. N. Yao, *Chem. and Ind.* (*London*), 1966, 1963; *J. Org. Chem.*, 1968, **33**, 70.
[55] Y. Hayashi and R. Oda, *Tetrahedron Letters*, 1968, 5381.
[56] W. E. Parham and S. H. Groen, *J. Org. Chem.*, 1965, **30**, 728; *ibid.*, 1966, **31**, 1694.

found that treatment of an allylic sulphide (30) with sodium trichloroacetate, a source of dichlorocarbene, resulted in the formation of two unsaturated halogenosulphides. The diene was probably formed *via* initial ylide formation, then a Stevens 1,3-rearrangement, and finally elimination of HCl. The dichloro-olefin was probably formed *via* initial ylide formation followed by a 1,5-rearrangement. Utilizing hydrocarbon systems, Blackburn and Ollis [57] and Baldwin and Heckler [58] also have reported a mixture of 1,3- and 1,5-rearrangements occurring, with the proportion of 1,3-product decreasing as the temperature was lowered. An example is the rearrangement of the symmetrical sulphonium salt (31).[58]

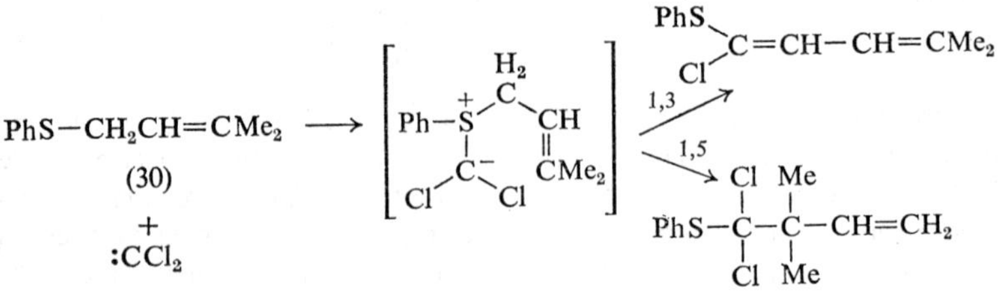

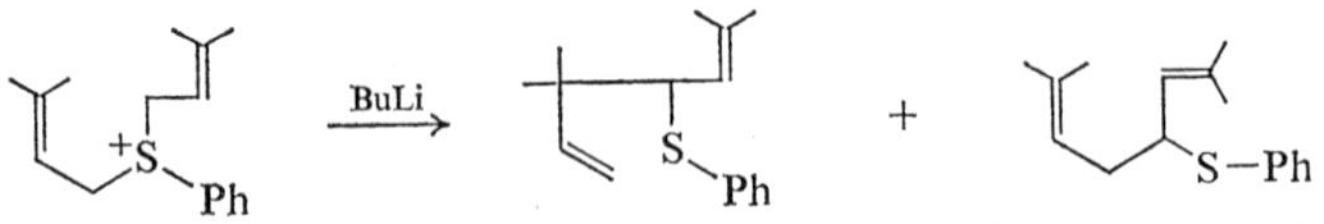

In 1949 Bohme and Krause [51] had reported that dimethylphenacylsulphonium bromide rearranged under basic conditions to afford the 1,3-rearrangement product, a methyl group supposedly migrating. However, in 1966 Ratts and Yao [54] found that the structure of the product was not as earlier proposed, but instead was the olefin (32), the result of a 1,5-rearrangement. Such a structure seemed to demand a mechanism involving the formation of a bond between the two negative dipoles of the ylide intermediate, the ylide (11) having been shown to undergo the rearrangement simply upon warming in aqueous solution for 5 hours. Nonetheless, this mechanistic proposal seems to have been the forbear of a wide variety of 1,5-rearrangements of sulphonium ylides discovered in the past few years and represented by the general mechanism indicated for (33),[59] the conversion of sulphur(IV) to sulphur(II) being a major driving force for the reaction.

The year 1968 saw at least ten separate communications appear regarding the 1,5-rearrangement of sulphonium ylides. Kirmse and Kapps [60] found

[57] G. M. Blackburn and W. D. Ollis, *Chem. Comm.*, 1968, 1261.
[58] J. E. Baldwin and R. E. Heckler, *J. Amer. Chem. Soc.*, 1969, **91**, 3646.
[59] J. E. Baldwin, R. E. Heckler, and D. P. Kelly, *Chem. Comm.*, 1968, 538.
[60] W. Kirmse and M. Kapps, *Chem. Ber.*, 1968, **101**, 994.

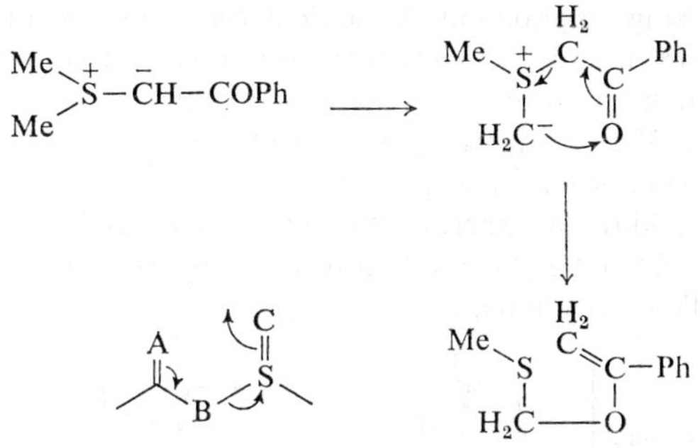

that treatment of 1,1,1′,1′-tetradeuteriodiallylsulphide (34) with diazomethane in the presence of cuprous chloride afforded a 35% yield of a 17 : 83 mixture of addition product (35) and insertion product (36). There was no rearrangement of the deuterium labels in the addition product, but in the insertion product one of the dideuteriomethylene groups had migrated. These workers proposed a 1,5-rearrangement of a sulphonium ylide, formed by attack of the nucleophilic sulphide on the electrophilic carbene precursor, diazomethane. Ando et al.[15] reported similar results for the photochemical reaction of dimethyldiazomalonate with butylallylsulphide. A similar mechanism was proposed by Kirmse[60] for the rearrangement, under aqueous base conditions, of a methyldiallylsulphonium salt to 3-methylthio-1,5-hexadiene.[61] In Kirmse's experiments, isotopic labelling was required to elaborate the mechanism since the products could be accounted for by a 1,3-rearrangement mechanism as well.

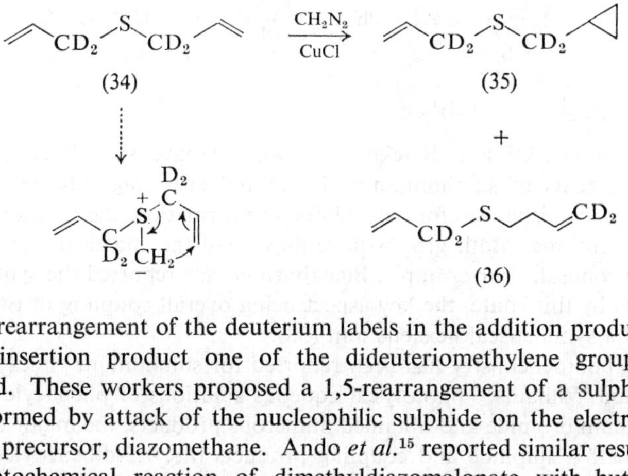

Trost and LaRochelle[62] and Bates and Feld[63] both found that similar rearrangements occurred in the presence of a much stronger base, butyllithium in THF. They reported the rearrangement of dimethylallyldimethylsulphonium bromide to 4-methylthio-3,3-dimethyl-1-butene under such

---

[61] W. Kirmse and M. Kapps, Chem. Ber., 1968, **101**, 1004.
[62] B. M. Trost and R. LaRochelle, Tetrahedron Letters, 1968, 3327.
[63] R. D. Bates and D. Feld, Tetrahedron Letters, 1968, 417.

conditions. Using sulphonium salt dideuteriated in the allylic position, Bates and Feld [63] found that ylide formation (*i.e.* proton abstraction) was faster at the allylic position. However, the methylide rearranged to olefin faster than it underwent a transylidation to the more stable allylic ylide. These experiments provided structural evidence for the 1,5-rearrangement. An interesting additional experiment under similar conditions was reported by Baldwin *et al.*,[64] interesting in that it forced the conversion of an alkyne (37) into an allenic grouping.

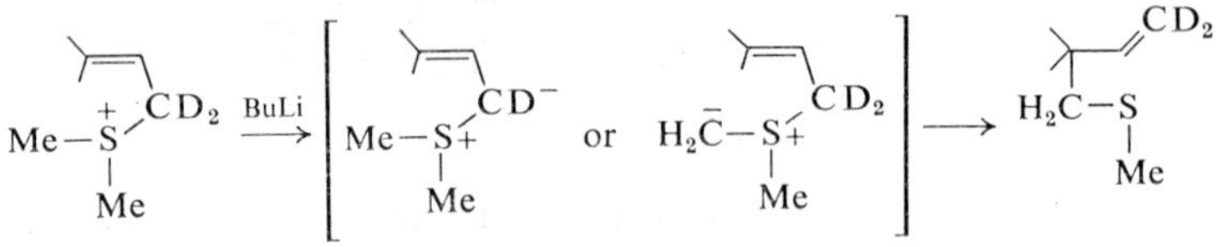

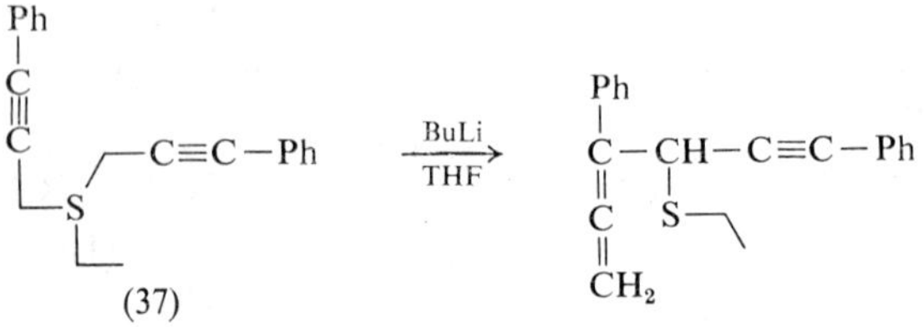

Baldwin *et al.*[65] and Blackburn *et al.*[57, 66] have speculated that 1,5-rearrangements of sulphonium ylides of polyenes might be part of the biosynthetic mechanism for natural polyene formation, the sulphur coming from an enzyme. Both groups have illustrated the chemical feasibility of such a proposal. For example, Blackburn *et al.*[66] reported the synthesis of squalene by this route, the key aspect being overall coupling of two triene units to a symmetrical hexaene unit (38).

Little photochemistry has been reported for sulphonium ylides or salts. Laird and Williams [67] photolysed aqueous solutions of phenacyldimethyl-sulphonium bromide and obtained numerous products, the major products being methyl sulphide and coupling products (*i.e.* 1,2-dibenzoylethane in 25% yield). Homolytic cleavage of the phenacyl–sulphur bond was proposed to account for the products. Nozaki *et al.*[68] reported briefly that phenylmethylsulphoniumphenacylide afforded a 30% yield of *trans*-1,2,3-tribenzoylcyclopropane (20) upon photolysis in methanol solution. In a more extensive investigation, Trost [30] found that dimethylsulphonium-phenacylide (11) afforded a 92% yield of (20) upon photolysis in benzene solution. In ethanolic solution, (20) was accompanied by acetophenone and ethyl phenylacetate. In cyclohexene solution, (11) was photolysed to

[64] J. E. Baldwin, R. E. Heckler, and D. P. Kelly, *Chem. Comm.*, 1968, 1083.
[65] J. E. Baldwin and D. P. Kelly, *Chem. Comm.*, 1968, 899.
[66] G. M. Blackburn, W. D. Ollis, C. Smith, and I. O. Sutherland, *Chem. Comm.*, 1969, 99.
[67] T. Laird and H. Williams, *Chem. Comm.*, 1969, 561.
[68] H. Nozaki, M. Takaku, and K. Kondo, *Tetrahedron*, 1966, **22**, 2145.

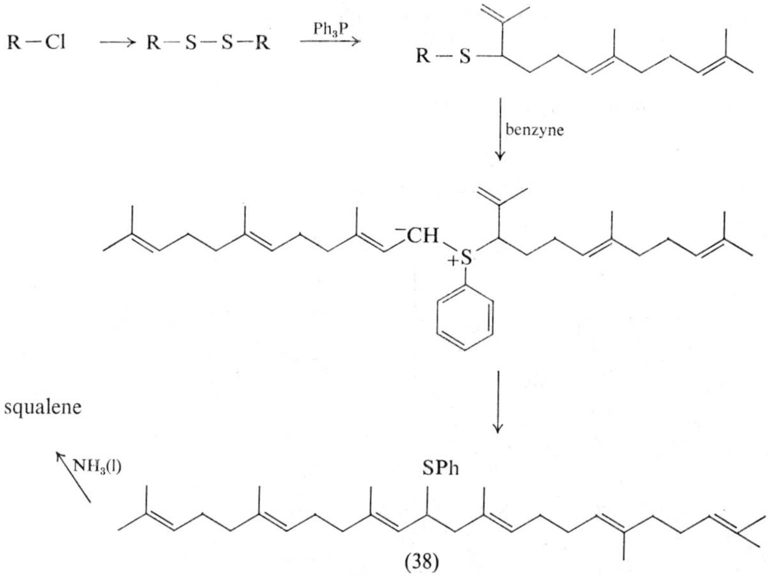

$$R-Cl \longrightarrow R-S-S-R \xrightarrow{Ph_3P}$$

[Reaction scheme showing R-S- adding to squalene-like structure, then benzyne addition to form a sulfonium ylide with -CH and +S-Ph, followed by reaction with NH$_3$(l) to give product (38) bearing an SPh group on the squalene chain.]

squalene

(38)

acetophenone (37%), (20) (40%), and a 5% yield of benzoylnorcarane. On the basis of these results it was proposed that the ylide was cleaved to methyl sulphide and benzoylmethylene by photolysis. The methylene could abstract protons to form acetophenone, rearrange to phenyl keten (which would be converted to ethyl phenylacetate in ethanolic solution), or react with the original ylide to form sym-dibenzoylethylene and eventually (20). This proposal is analogous to the carbene mechanism mentioned earlier in this section for the thermal decomposition of the same ylides. More recently, Caserio et al.[69] reported the photolysis of a cyclic β-carbonyl ylide (39) to indanone in 45% yield. This product can be accounted for by methine–sulphur cleavage to a carbene, followed by carbene insertion into the methylene–sulphur bond, forming 2-methylthioindanone. This substance is known to cleave photochemically to indanone. Further evidence for the insertion process in such cases was provided by the observation that the ortho-methyl derivative of the phenacylide (11) could be photochemically converted to a variety of products, but including indanone in 7% yield.[69]

New information has become available on the hydrolytic stability of sulphonium ylides. Most commonly, sulphonium ylides are cleaved, through the formation of a sulphonium hydroxide, to an alcohol and the sulphide.[70] At approximately the same time four groups were studying the chemistry of carboalkoxymethylenedimethylsulphuranes, the methyl [27, 71]

---

[69] R. H. Fish, L. C. Chow, and M. C. Caserio, *Tetrahedron Letters*, 1969, 1259.
[70] Ref. 1, p. 320.
[71] J. Casanova and D. A. Rutolo, *J. Amer. Chem. Soc.*, 1969, **91**, 2347.

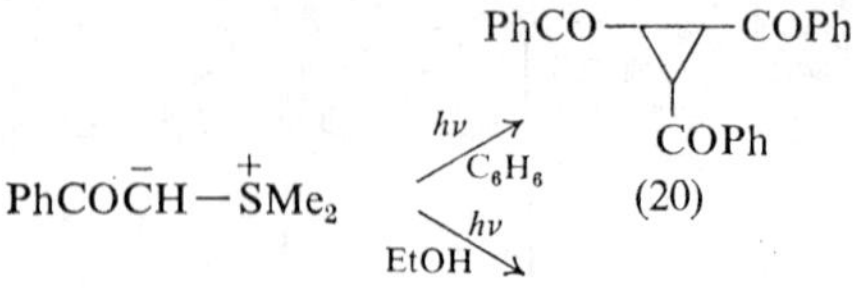

$$\text{PhCO}\bar{\text{C}}\text{H}-\overset{+}{\text{S}}\text{Me}_2 \xrightarrow[\text{EtOH}]{\overset{hv}{C_6H_6}} \begin{array}{c}\text{PhCO}-\bigtriangledown-\text{COPh}\\ \text{COPh} \\ (20)\end{array}$$

$$(20) + \text{PhCOMe} + \text{PhCH}_2\text{COOEt}$$

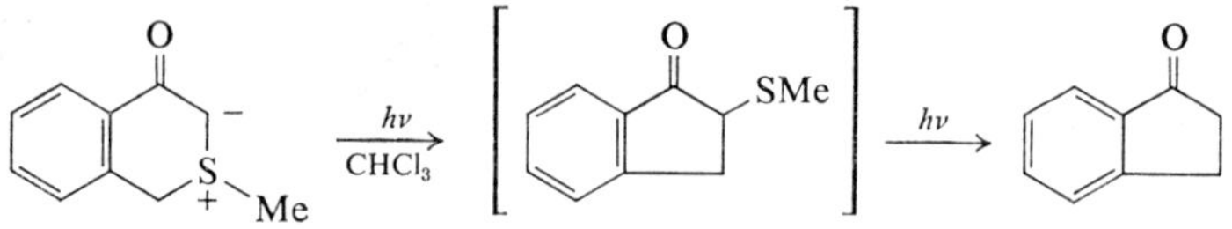

[structure (39)] $\xrightarrow[\text{CHCl}_3]{hv}$ [indanone-SMe intermediate] $\xrightarrow{hv}$ [indanone]

(39)

[o-methylphenyl-CO–$\bar{\text{C}}$H–$\overset{+}{\text{S}}$Me$_2$] $\xrightarrow[C_6H_6]{hv}$ [indanone] + others

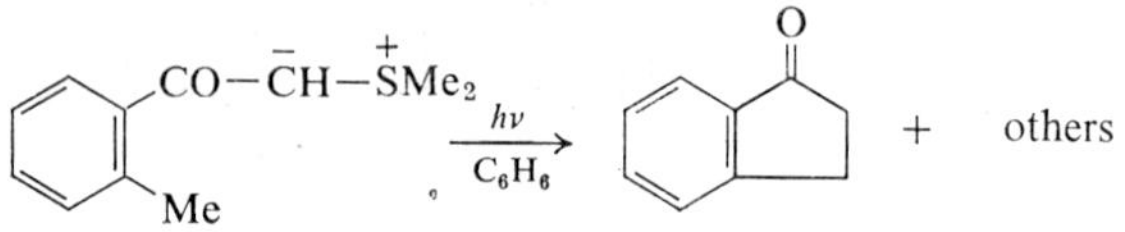

and ethyl [54, 72] esters, and their conjugate acids, the sulphonium salts. In all of these cases it was found that the methine–sulphur bond did not break but, instead, the ester function was rapidly hydrolysed to afford the zwitterionic dimethylthetin (40). Casanova and Rutolo [71] studied the kinetics

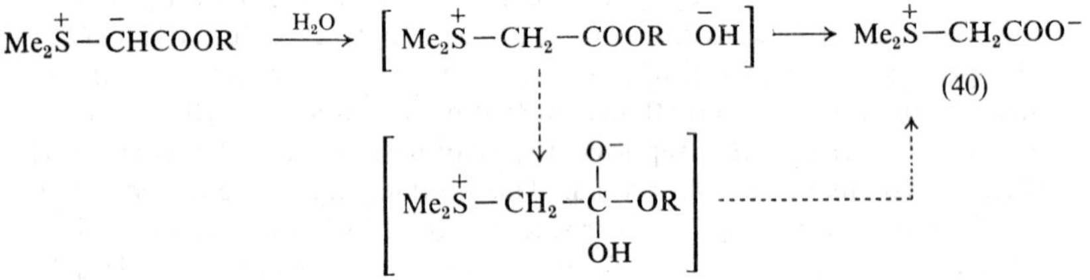

$$\text{Me}_2\overset{+}{\text{S}}-\bar{\text{C}}\text{HCOOR} \xrightarrow{\text{H}_2\text{O}} \left[ \text{Me}_2\overset{+}{\text{S}}-\text{CH}_2-\text{COOR} \quad \bar{\text{O}}\text{H} \right] \longrightarrow \text{Me}_2\overset{+}{\text{S}}-\text{CH}_2\text{COO}^-$$

(40)

$$\left[ \text{Me}_2\overset{+}{\text{S}}-\text{CH}_2-\underset{\text{OH}}{\overset{\text{O}^-}{\underset{|}{\overset{|}{\text{C}}}}}-\text{OR} \right]$$

of the alkaline hydrolysis of carbomethoxymethyldimethylsulphonium tosylate and found it to be first order in hydroxide and in sulphonium salt. They reported that conversion of the salt to ylide was approximately 120 times as fast as the hydrolysis and, therefore, proposed that the hydrolytic reaction proceeded *via* the sulphonium hydroxide and an oxysulphonium zwitterion, with the slow step being the formation of the latter. All of these reactions are quite rapid with the ylides hydrolysing simply upon exposure to the atmosphere.

There have been no studies on the oxidation of sulphonium ylides. One report [73] indicated that sulphonium ylides may be reduced, however,

[72] G. B. Payne, *J. Org. Chem.*, 1967, **32**, 3351.
[73] E. Winterfeldt, *Chem. Ber.*, 1965, **98**, 1581.

resulting in the cleavage of the ylidic bond to form a sulphide and a hydrocarbon residue.

**Reactions of Sulphonium Ylides.**—Five years ago there had been no reports of the alkylation of sulphonium ylides. In the interim there have been several reports but all with $\beta$-carbonyl ylides where there may be $C$-alkylation or $O$-alkylation. Using dimethylsulphoniumphenacylide (11) three

$$\text{Me}_2\text{SCHCOPh} \xrightarrow[\text{Me}_3\text{O}^+]{\overset{\text{PhCH}_2\text{Br}}{\underset{\text{PhSCl}}{\longrightarrow}}} \begin{array}{l} \text{MeS}-\text{CH}\!\!\begin{array}{c}\diagup\text{CH}_2\text{Ph}\\\diagdown\text{COPh}\end{array}\\ \text{Me}_2\text{S}=\text{C}\!\!\begin{array}{c}\diagup\text{SPh}\\\diagdown\text{COPh}\end{array}\\ \overset{+}{\text{Me}_2\text{S}}-\text{CH}=\text{C}\!\!\begin{array}{c}\diagup\text{OMe}\\\diagdown\text{Ph}\end{array}\end{array}$$

(11)

groups [27, 43, 46] have reported $C$-alkylation using methyl iodide, benzyl bromide, or phenylsulphenyl chloride. In the first two instances, the initial alkylation product, the sulphonium salt, was demethylated by the anion present, so the isolated product was a sulphide. Use of trimethyloxonium tetrafluoroborate as the methylating agent led to an $O$-alkylation, but the characteristics of the products have yet to appear in the literature.[31] Use of the very reactive oxonium salt may indicate the reaction is under kinetic control, thereby accounting for the change from the usual $C$-alkylation when under thermodynamic control.

Acylation of sulphonium ylides also was unknown five years ago. With simple sulphonium ylides the reaction proceeds as expected to afford an acylated sulphonium salt, but one which usually loses a proton to afford a more highly stabilized ylide. Thus, the net effect is a substitution for an ylidic proton, as for the benzoylation of phenylthiomethylenedimethylsulphuran.[46] Most acylation studies have been effected with $\beta$-carbonyl ylides where the competition between $O$-acylation and $C$-acylation could be studied. Johnson and Amel [27] and, later, Nozaki *et al.*[68] found that the course of the benzoylation of phenacylides (14) depended on the benzoylating agent used. Use of benzoic anhydride resulted in $C$-acylation and formation of a new, highly stabilized dibenzoylmethylenesulphurane (41).

$$\text{Me}_2\text{S}=\text{CHSPh} \xrightarrow{\text{PhCOCl}} \text{Me}_2\text{S}=\text{C}\!\!\begin{array}{c}\diagup\text{COPh}\\\diagdown\text{SPh}\end{array}$$

$$\text{R}_2\text{S}=\text{CHCOPh} \xrightarrow[\text{PhCOCl}]{(\text{PhCO})_2\text{O}} \begin{array}{l}\text{R}_2\text{S}=\text{C}\!\!\begin{array}{c}\diagup\text{COPh}\\\diagdown\text{COPh}\end{array}\\(41)\\ \text{RS}-\text{CH}=\text{C}\!\!\begin{array}{c}\diagup\text{OCOPh}\\\diagdown\text{Ph}\end{array}\\(42)\end{array}$$

(14)

However, use of the more reactive benzoyl chloride resulted in *O*-acylation followed by demethylation to form the enol ether (42). Based on the closely related work of Chopard *et al.*[74] with phosphoniumphenacylides, it appears that *C*-acylation results from thermodynamic control of the reaction while *O*-acylation results from kinetic control. In somewhat of an anomalous report, Ratts and Yao indicated that *p*-nitrophenacylidenedimethylsulphurane was *C*-acylated by ethyl oxalyl chloride.[43] Others[73,75] have reported *O*-acylations with acyl halides, while use of tosyl chloride[76,77] also results in *O*-acylation to afford a sulphonate ester rather than a sulphonated ylide. However, use of sulphonyl halides having an α-methylene group leads to *C*-sulphonation, probably through the intermediacy of a sulphene (see section 6).

It is interesting to note that carboalkoxymethylenesulphonium ylides do not undergo *O*-acylation, only *C*-acylation, whereas the benzoylmethylene ylides (phenacylides) undergo both *C*-acylation and *O*-acylation, depending on the reagent. Benzoylation of carbomethoxymethylenedimethylsulphurane (17) with either benzoic anhydride or benzoyl chloride afforded the benzoylated ylide (43).[27] Similar results have been reported for

$$Me_2\overset{+}{S}-\overset{-}{C}HCOOMe \xrightarrow[\text{OR (PhCO)}_2\text{O}]{\text{PhCOCl}} Me_2S=C\begin{smallmatrix}COOMe\\COPh\end{smallmatrix}$$
(17) (43)

other sulphonium ester ylides.[78,79] The fact that replacement of the phenyl group of the phenacylide with an alkoxy-group would preclude *O*-acylation may be attributed to less enolate character in the ester ylides (17). This conclusion is consistent with that proposed for a series of phenacylides in which the competition between *O*-acylation and *C*-acylation seemed to depend on the necessity for enolate stabilization of the ylides.[27]

One of the more important reactions of sulphonium ylides is that with carbonyl compounds to form oxirans. During the past five years this reaction has become widely used for synthetic purposes and no attempt will be made to document all of the cases. Several aspects of the reaction are of continuing interest, however. In a most interesting report Hatch[80] described a procedure for generating a sulphonium ylide from its salt in aqueous media containing a high concentration of base, and then trapping the ylide with a reactive aldehyde. For example, trimethylsulphonium salts in 50% aqueous sodium hydroxide reacted with benzaldehyde to afford styrene oxide in 68% yield. Numerous other examples were reported, but the procedure seems limited by the reactivity of the carbonyl component

[74] P. O. Chopard, R. J. G. Searle, and F. H. Devitt, *J. Org. Chem.*, 1965, **30**, 1015.
[75] A. Winkler and J. Gosselck, *Tetrahedron Letters*, 1969, 4229.
[76] H. Nozaki, M. Takaku, and Y. Hayasi, *Tetrahedron Letters*, 1967, 2303.
[77] H. Nozaki, M. Takaku, Y. Hayasi, and K. Kondo, *Tetrahedron*, 1968, **24**, 6563.
[78] H. Nozaki, D. Tunemoto, S. Matubara, and K. Kondo, *Tetrahedron*, 1967, **23**, 545.
[79] G. B. Payne, *J. Org. Chem.*, 1968, **33**, 3517.
[80] M. J. Hatch, *J. Org. Chem.*, 1969, **34**, 2133.

and the stability of both the ylide and the carbonyl compound to the reaction conditions. Corey et al.[10, 81] have reported experimental conditions which lead to very high yields of oxirans from alkylides in anhydrous media. They have reported that in contrast with diphenylsulphonium methylide, which reacts with α,β-unsaturated carbonyl compounds to afford oxirans and not cyclopropanes, diphenylsulphoniumisopropylide seems exclusively to attack the olefin group and afford cyclopropanes. Therefore, the structure of the ylide must be considered in predicting the focus of attack in molecules with more than one electrophilic site. Drefahl et al.[82] found that 17-keto-5,6-unsaturated steroids were converted exclusively to oxirans by dimethylsulphonium methylide.

Some interest has been directed to the reaction of stabilized sulphonium ylides with carbonyl compounds. Three groups reported that sulphoniumphenacylides would not react with cyclohexanone, benzaldehyde, or p-nitrobenzaldehyde.[30, 43, 68] However, another group did report such a reaction, in low yield, with p-nitrobenzaldehyde to afford 2-p-nitrophenyl-3-benzoyloxiran, the expected product.[27] Later, Payne[79] reported the successful reaction of the ester ylide, carbethoxymethylenedimethylsulphurane, with chloral or ethyl pyruvate to afford the expected oxirans, but in rather low yield. It seems apparent, and not at all surprising, that the stabilized sulphonium ylides are considerably less reactive to carbonyl compounds than are the non-stabilized ylides. This is simply a reflection of the lowered electron density on the ylide carbanion, and thereby a lowered nucleophilicity, as a result of delocalization of the charge throughout the molecule. Such behaviour is well known for phosphonium ylides.

Only a little work has been reported concerning the mechanism of the reaction of sulphonium ylides with carbonyl compounds. It has been suggested[83] that the reaction proceeds via dissociation of the ylide to sulphide and carbene, with the latter then adding to the carbonyl group. This suggestion can be dismissed by the observation that ylides normally will not transfer their carbon fragments to isolated olefins unless the ylide is photochemically or thermally dissociated. Furthermore, olefins and cyclopropanes would be expected frequently to be by-products of ylide reactions with carbonyl compounds if carbenes were involved as intermediates. Yoshimine and Hatch[84] have studied the mechanism of the reaction of dimethylbenzylsulphonium chloride with formaldehyde in aqueous sodium hydroxide. Based on deuterium exchange studies, ylide formation was found to be rapid and reversible. The reaction clearly involved nucleophilic attack of the ylide on the carbonyl carbon since competition reactions showed the most electrophilic carbonyl compound to be attacked the most

[81] E. J. Corey and M. Jautelat, *J. Amer. Chem. Soc.*, 1967, **89**, 3912.
[82] G. Drefahl, K. Ponsold, and H. Schick, *Chem. Ber.*, 1964, **97**, 3529.
[83] R. Sivaramakrishnan and D. S. Radhakrishnamurti, *Current Sci.*, 1965, **34**, 404; *Chem. Abs.*, 1965, **63**, 6816.
[84] M. Yoshimine and M. Hatch, *J. Amer. Chem. Soc.*, 1967, **89**, 5831.

$$\text{PhCH}_2-\overset{+}{\text{S}}\text{Me}_2 + \overset{-}{\text{OH}} \underset{\text{fast}}{\overset{\text{fast}}{\rightleftarrows}} \text{Ph}-\overset{-}{\text{CH}}-\overset{+}{\text{S}}\text{Me}_2$$

$$\text{Ph}-\overset{-}{\text{CH}}-\overset{+}{\text{S}}\text{Me}_2 + \text{CH}_2\text{O} \xrightarrow{\text{slow}} \begin{array}{c} \text{Ph}-\text{CH}-\overset{+}{\text{S}}\text{Me}_2 \\ | \\ \text{CH}_2-\text{O}^- \end{array}$$

$$\downarrow \text{fast}$$

$$\text{Ph}-\text{HC}-\text{CH}_2 + \text{Me}_2\text{S}$$
$$\diagdown \diagup$$
$$\text{O}$$

readily. Study of the betaine intermediate obtained from another source indicated that betaine formation was not reversible and that betaine decomposition was faster than betaine formation. Therefore, the reaction of this ylide with carbonyl compounds seems to involve slow nucleophilic attack on the carbonyl compound followed by rapid betaine decomposition to oxiran. Earlier Johnson et al.[44] had suggested, based on stereochemical arguments, that the first step of the reaction (betaine formation) was slow but reversible. However, no proof was available regarding the proposed reversibility.

Some new work on the stereochemistry of the addition of sulphonium ylides to cyclic ketones has been reported, but only with one ylide, dimethylsulphonium methylide. Carlson and Behn[85] reported addition to *trans*-2-decalone afforded two oxirans, in a ratio of 2:3 with the axial adduct predominant. Similar implications can be drawn from the reaction of the same ylide with 17-ketosteroids in which the oxiran ring is β, implying a pseudo-axial attack by the ylide.[82] Therefore, in these and earlier reports, the sulphonium ylide seems to approach from the most hindered side. Use of the sulphoxonium ylides, which are bulkier, invariably results in equatorial attack. In the bicyclic systems Bly et al.[86] reported the ylide to attack a carbonyl ring from the *exo*-side. Such is the case, for example, with norcamphor and dehydronorcamphor and such behaviour is analogous to that exhibited by other nucleophiles, such as methylmagnesium iodide.

Sulphonium ylide addition to isolated double bonds is not known and this was verified in additional examples by Trost.[87] However, there are many examples known of the addition of reactive sulphonium ylides to olefins when the functional group is conjugated with an electronegative group. In such cases a Michael-type addition occurs with the intermediate carbanion collapsing by displacement of the sulphide and formation of a cyclopropane (*e.g.* refs. 46, 81). In such instances ylide attack might be at the olefin or at the carbonyl group. The determining factors seem to be ylide nucleophilicity, carbonyl electrophilicity, steric bulk of the ylide, and

[85] R. G. Carlson and N. S. Behn, *J. Org. Chem.*, 1967, **32**, 1363.
[86] R. S. Bly, C. M. DuBose, and G. B. Konizer, *J. Org. Chem.*, 1968, **33**, 2188.
[87] B. M. Trost, *Tetrahedron Letters*, 1966, 5761.

steric hindrance to attack at the olefinic group. A clear example of the latter effect was reported by Corey and Jautelat.[81] Diphenylsulphonium-isopropylide effected a conjugate addition with 2-methylcyclohex-2-enone but attacked the carbonyl group of 3-methylcyclohex-2-enone. Goralski[88] reported that $\alpha,\beta$-unsaturated sulphones also undergo conjugate addition with dimethylsulphonium methylide.

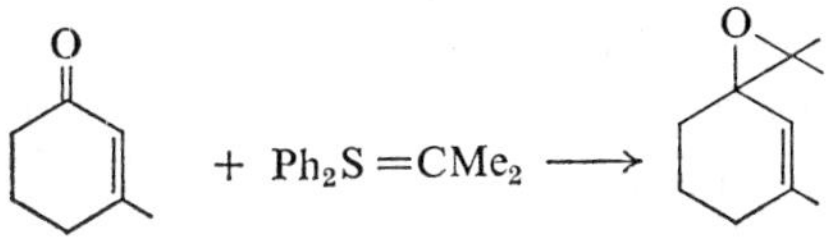

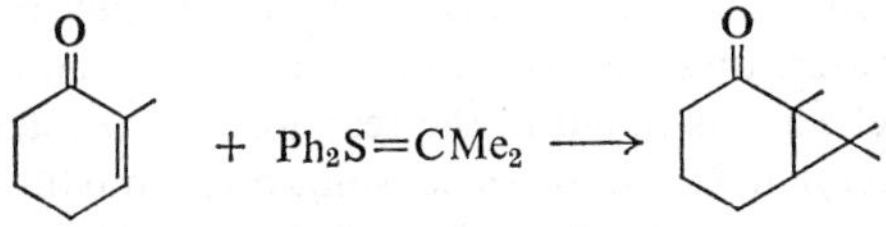

Several reports have appeared of the reaction of stabilized sulphonium ylides with conjugated carbonyl compounds to afford cyclopropanes, such reactions often proceeding in high yield even though the ylide may have been quite unreactive towards carbonyl compounds. Nozaki et al.[68] and Trost[30] reported the addition of phenacylides (14) to sym-dibenzoyl-

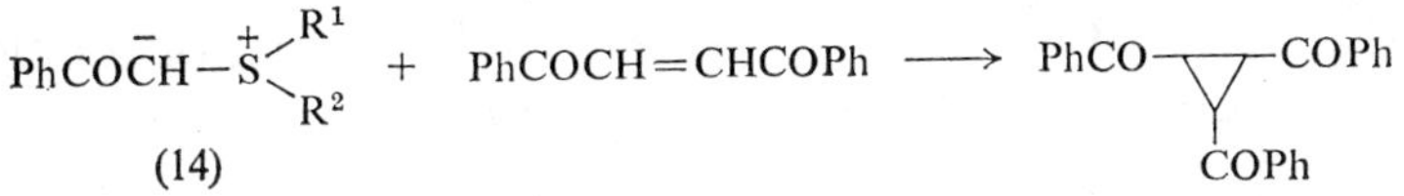

$$\text{PhCOCH}^- - \overset{+}{\text{S}} \overset{R^1}{\underset{R^2}{\diagdown}} + \text{PhCOCH}=\text{CHCOPh} \longrightarrow \text{PhCO}-\triangledown-\text{COPh}$$
(14)                                                                COPh
                                                                                    (20)

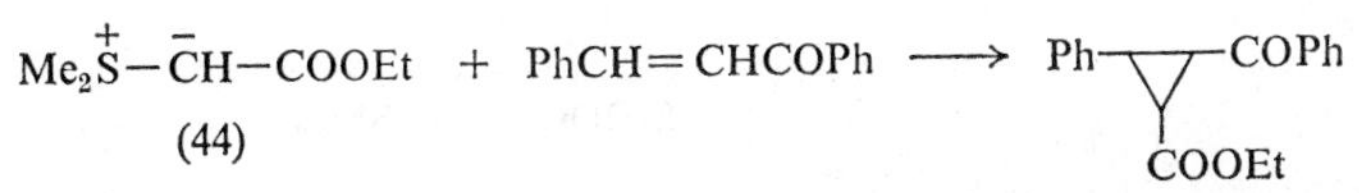

$$\text{Me}_2\overset{+}{\text{S}}-\overset{-}{\text{CH}}-\text{COOEt} + \text{PhCH}=\text{CHCOPh} \longrightarrow \text{Ph}-\triangledown-\text{COPh}$$
(44)                                                                  COOEt

ethylene to afford trans-1,2,3-tribenzoylcyclopropane (20). Additions were also reported to benzalacetophenone[30] and diethyl-maleate and -fumarate.[68] Similar additions were reported for carboalkoxymethylenesulphuranes with dimethyl-maleate [27, 28, 72] and -fumarate,[28, 72] dibenzoylethylene[78] and benzalacetophenone,[78] in all cases forming the trans-isomers. All of these reactions were carried out in aprotic solvents, and presumably involved sulphonium group displacement by the intermediate carbanion. Carrying out the same reactions in protic solvents led to the formation of quite different products. For example, Payne[89] found that carbethoxymethylene-dimethylsulphurane (44) reacted with diethyl maleate to afford an olefin 'coupling' product, triethyl aconitate, in 87% yield. It was suggested that

---

[88] C. T. Goralski, Diss. Abs., 1969, 30B, 1591.
[89] G. B. Payne, J. Org. Chem., 1968, 33, 1284.

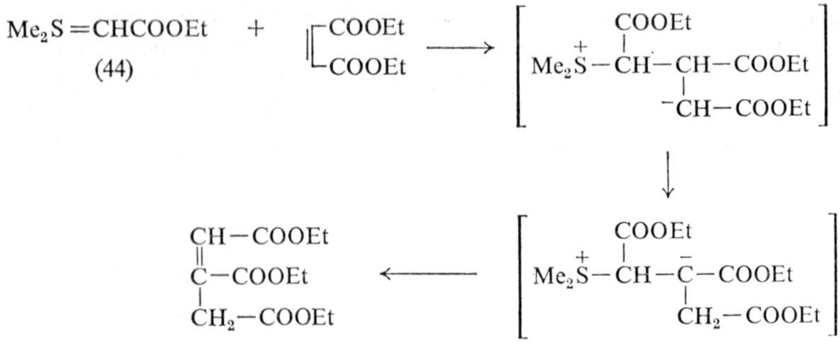

$Me_2S=CHCOOEt$ + ⌈−COOEt  ⟶  [ $Me_2\overset{+}{S}-\overset{|}{CH}-\overset{COOEt}{\underset{\overset{|}{-CH-COOEt}}{CH-COOEt}}$ ]
(44)            ⌊−COOEt

↓

$\overset{CH-COOEt}{\underset{CH_2-COOEt}{\overset{\|}{C}-COOEt}}$  ⟵  [ $Me_2\overset{+}{S}-\overset{COOEt}{\underset{CH_2-COOEt}{\overset{|}{CH}-\overset{|}{\overset{-}{C}}-COOEt}}$ ]

the intermediate carbanion underwent a proton migration and then sulphonium group elimination. Similar migrations appeared to afford new ylides in other cases, these new ylides then adding to additional olefin. A similar solvent effect was noted in the reaction of the ester ylide (44) with ethyl α-bromoacrylate.[90] In benzene solution a mixture of two cyclopropanes was formed, the bromocyclopropane presumably from carbanion displacement of the sulphonium group and the thioether cyclopropane presumably *via* a proton migration then bromide displacement. In a more polar solvent (acetone), only the thioether cyclopropane was formed, presumably the result of the facilitation of proton migration. Reaction of (44) with vinyldimethylsulphonium salts also presented an opportunity to

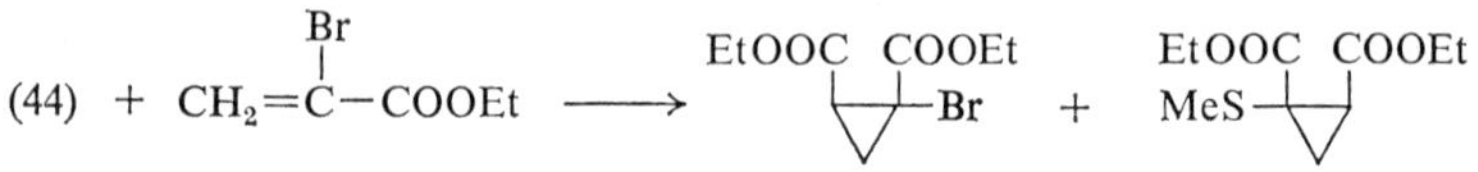

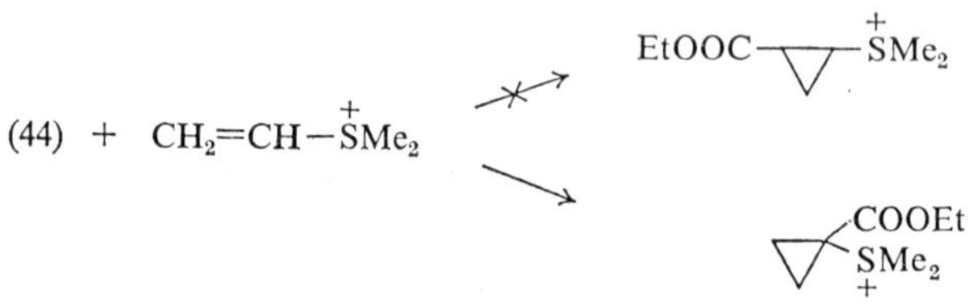

(44) + $CH_2=\overset{\overset{Br}{|}}{C}-COOEt$ ⟶ $\underset{\triangledown}{EtOOC\ \ COOEt}$–Br + $MeS\underset{\triangledown}{\overset{EtOOC\ \ COOEt}{\biggl\vdash\!\!\!\!\dashv}}$

(44) + $CH_2=CH-\overset{+}{S}Me_2$ ⟶̸ $\underset{\triangledown}{EtOOC-\overset{+}{S}Me_2}$

⟶ $\underset{\triangledown}{\overset{COOEt}{\underset{+}{SMe_2}}}$

obtain two cyclopropanes. However, only the 1,1-di-substituted isomer was obtained, indicating that proton migration to form the more stable ylidic intermediate took place before a sulphonium group was displaced for ring formation.[91] The competition between direct cyclopropane formation and prototropic shifts to form olefins, new cyclopropanes, or new ylides is newly discovered for sulphonium ylides, but the findings parallel exactly the considerable work already reported for similar phos-

[90] G. B. Payne, *J. Org. Chem.*, 1968, **33**, 1285.
[91] G. Schmidt and J. Gosselck, *Tetrahedron Letters*, 1969, 3445.

phonium ylides.[92] The route followed seems to depend on the solvent system and the relative stabilities of the various potential carbanions. Sulphonium ylides have been found to undergo a variety of additional reactions with unsaturated systems. Stabilized ylides effect a nucleophilic addition to conjugated alkynes which is followed by a proton transfer, resulting in the formation of new stabilized ylides (45).[30, 79] Using dibenzoylmethylenedimethylsulphurane, Takaku et al.[93] reported the formation of the furan (46), presumably via attack of the enolate form of the ylide on the alkyne, followed by carbanion displacement of the sulphonium group.

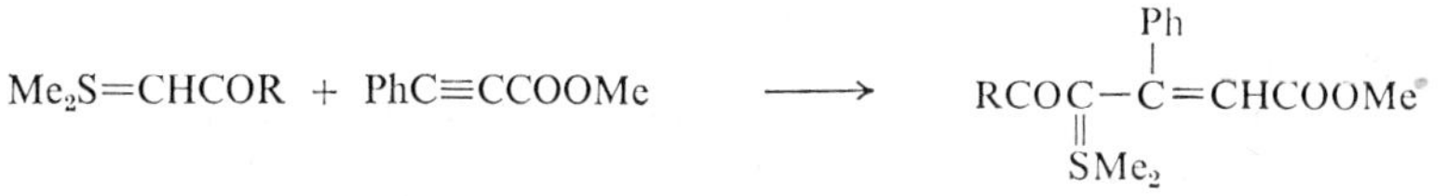

$$Me_2S=CHCOR + PhC\equiv CCOOMe \longrightarrow RCOC(SMe_2)=C(Ph)CHCOOMe$$

(45)

$$Me_2S=C(COPh)_2 + EtOOC-C\equiv C-COOEt \longrightarrow$$ PhCO, Ph, O, COOEt, COOEt (furan 46)

(46)

$$Ph_2S=CMe_2 + MeOOC-C\equiv C-COOMe \longrightarrow$$ MeOOC-[bicyclobutane]-COOMe

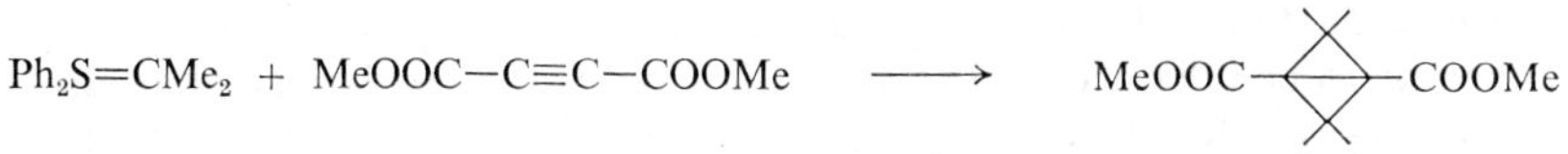

(47)

Using a highly reactive alkylide, Corey and Jautelat[81] found a bicyclobutane (47) to result. In this case, the disubstituted ylide precluded any proton migration. The initial product should have been a conjugated cyclopropene which would be expected to be susceptible to additional nucleophilic attack by a second ylide molecule.

Several stabilized sulphonium ylides have been shown to undergo addition to phenylisocyanate. All cases reported to date have been monosubstituted ylides (i.e. carry a proton on the ylide carbon) so the initial attack on the isocyanate carbon has been followed by a tautomeric proton transfer to nitrogen, resulting in the formation of an amido-ylide.[43, 68, 78, 79] The only report to date of reaction of a sulphonium ylide with a keten, in this case diphenylketen, indicates that an analogous reaction was not followed, but instead two equivalents of the keten were consumed to afford a new ylide (48).[79]

Stabilized sulphonium ylides react with nitrosobenzene to afford nitrones[27, 43] as previously reported[94] for other sulphonium ylides. DeAngelis and Hess[95] reported addition to the pyrimidinium ring system

[92] Ref. 1, pp. 116–120.
[93] M. Takaku, Y. Hayasi, and H. Nozaki, *Tetrahedron Letters*, 1969, 2053.
[94] Ref. 1, p. 338.
[95] G. C. DeAngelis and H. J. Hess, *Tetrahedron Letters*, 1969, 1451.

$$Me_2S=CHCOR + PhNCO \longrightarrow Me_2S=C{\overset{COR}{\underset{CONHPh}{}}}$$

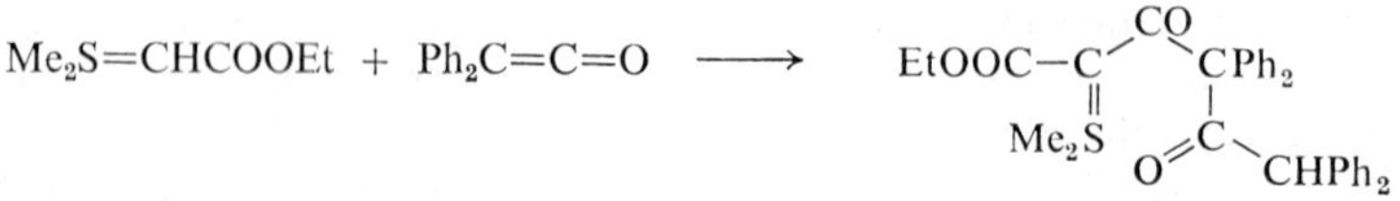

$$Me_2S=CHCOOEt + Ph_2C=C=O \longrightarrow EtOOC-\underset{\underset{Me_2S}{\|}}{C}{\overset{CO}{\diagdown}}\underset{\overset{C}{\diagup}\diagdown CHPh_2}{\underset{O}{|}}CPh_2$$

(48)

and Hayashi et al.[96, 97, 98] reported several examples of reaction with 1,3-dipolarophiles.

Reaction of stabilized sulphonium ylides with simple electrophilic reagents also has been reported. For example, treatment of a monosubstituted ylide with bromine [43, 90] or iodine [43] affords the halogenosulphonium salt which readily gives up a proton to form a new bromo- or iodo-ylide. Whereas such phosphonium ylides are known to be quite stable, the halogeno-sulphonium ylides seem to be quite unstable. An interesting homologative-type reaction has been discovered from the reaction of stabilized sulphonium ylides with trialkylboranes. Tufariello et al.[99, 100] showed that a variety of sulphonium ylides, both stabilized and non-stabilized, reacted with tri-n-hexylborane or triphenylborane to afford, after oxidation, a homologated substance. They proposed a new trialkylborane to be intermediate in the reaction and that secondary alkyl groups migrated to the ylide carbanion faster than did primary alkyl groups. However, Thelman [101] later claimed that the effectiveness of migration was

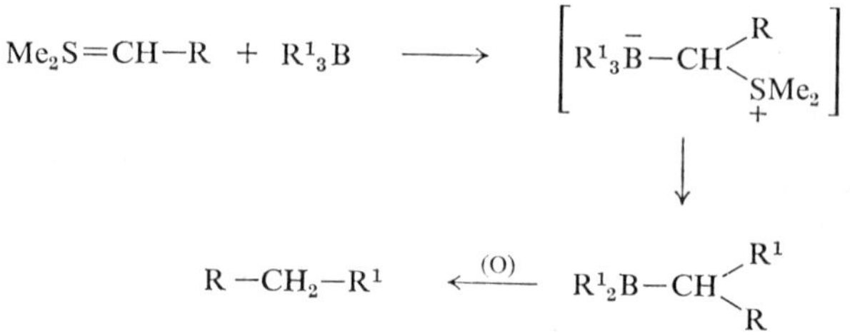

$$Me_2S=CH-R + R^1_3B \longrightarrow \left[ R^1_3\bar{B}-CH{\overset{R}{\underset{\overset{+}{SMe_2}}{\diagdown}}} \right]$$

$$\downarrow$$

$$R-CH_2-R^1 \xleftarrow{(O)} R^1_2B-CH{\overset{R^1}{\underset{R}{\diagdown}}}$$

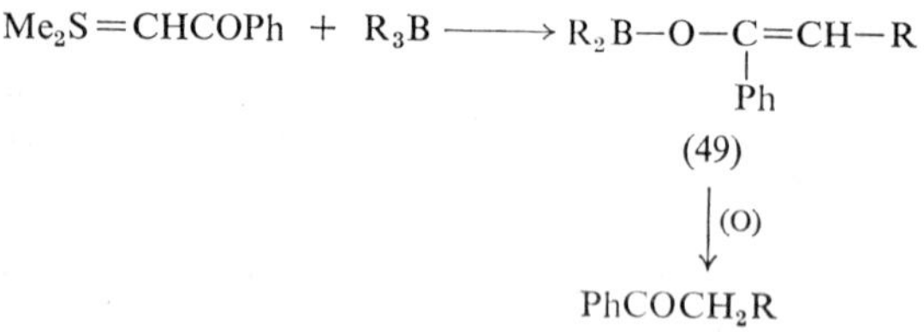

$$Me_2S=CHCOPh + R_3B \longrightarrow R_2B-O-\underset{\underset{Ph}{|}}{C}=CH-R$$

(49)

$$\downarrow (O)$$

$$PhCOCH_2R$$

[96] Y. Hayashi and R. Oda, *Tetrahedron Letters*, 1969, 853.
[97] Y. Hayashi, T. Watanabe, and R. Oda, *Tetrahedron Letters*, 1970, 605.
[98] S. Ueda, Y. Hayashi, and R. Oda, *Tetrahedron Letters*, 1969, 4967.
[99] J. J. Tufariello, P. Wojtkowski, and L. T. C. Lee, *Chem. Comm.*, 1967, 505.
[100] J. J. Tufariello, L. T. C. Lee, and P. Wojtkowski, *J. Amer. Chem. Soc.*, 1967, **89**, 6804.
[101] J. P. Thelman, *Diss. Abs.*, 1969, **30B**, 2618.

vinyl > 1° > 2° > 3°, indicating migration as an anionic group. More recently, Pasto and Wojtkowski [102] have shown that in the case of β-carbonyl ylides, attack on the borane occurs through the enolate form of the ylide, an alkoxydialkylborane (49) being the intermediate subject to oxidation. In any event, the reaction effectively couples the carbon portion of the ylide molecule and the alkyl group of the borane and may be of some synthetic use. However, the same overall effect (joining of an alkyl group to the carbon portion of the ylide) usually can be accomplished by simply alkylating the ylide directly with an alkyl halide.

**Sulphonium Ylides as Reaction Intermediates.**—In any reaction involving a sulphonium system in the presence of basic conditions, the possibility of the involvement of a sulphonium ylide in the reaction must be considered. Recently, the well-known DMSO oxidation of primary alcohols or alkyl halides to aldehydes has been shown, through the efforts of three groups,[103−105] to involve a sulphonium ylide intermediate. The dicyclohexylcarbodi-imide–DMSO procedure has been shown to involve an adduct of those two compounds by an $^{18}$O transfer experiment and to involve an α′,β-elimination through the ylide by a deuteriation experiment.[103]

$$(MeCO)_2O \xrightarrow{DMSO} (8) \xrightarrow{RCH_2OH} \underset{Me}{\overset{Me}{>}}\overset{+}{S}-O-CH_2R$$

or

$$C_6H_{11}N=C=NC_6H_{11}$$

$$RCHO + Me_2S \longleftarrow \left[ \underset{H_2C}{\overset{Me}{>}}\overset{+}{\underset{\diagdown}{S}}\overset{O}{\underset{H}{\diagup}}CH-R \right]$$

DMSO oxidation of alkyl bromides has been shown to involve an alkoxysulphonium salt (by isolation) which has been found to undergo the same α′,β-elimination.[104] Under separate conditions, Johnson and Phillips [105] also proved the nature of this elimination for alkoxydialkylsulphonium salts. A similar mechanism prevails for alcohol oxidation with DMSO–acetic anhydride, again involving the alkoxydimethylsulphonium salt intermediate.[106]

In a structurally quite different but mechanistically similar reaction, alkylation of phenol with DMSO in the presence of dicyclohexylcarbodi-imide has also been proposed to involve an ylide intermediate. In 1965,

---

[102] D. J. Pasto and P. W. Wojtkowski, *Tetrahedron Letters*, 1970, 215.
[103] A. H. Fenselau and J. G. Moffatt, *J. Amer. Chem. Soc.*, 1966, **88**, 1762.
[104] K. Torsell, *Acta Chem. Scand.*, 1967, **21**, 1.
[105] C. R. Johnson and W. G. Phillips, *J. Org. Chem.*, 1967, **32**, 1926.
[106] J. D. Albright and L. Goldman, *J. Amer. Chem. Soc.*, 1967, **89**, 2416.

two groups [107, 108] found that the above reaction afforded 2-methylthiomethylphenol and both proposed the reaction to proceed by attack of phenol on the DMSO-carbodi-imide complex to afford an intermediate phenoxydimethylsulphonium salt (50). Removal of an acidic proton would

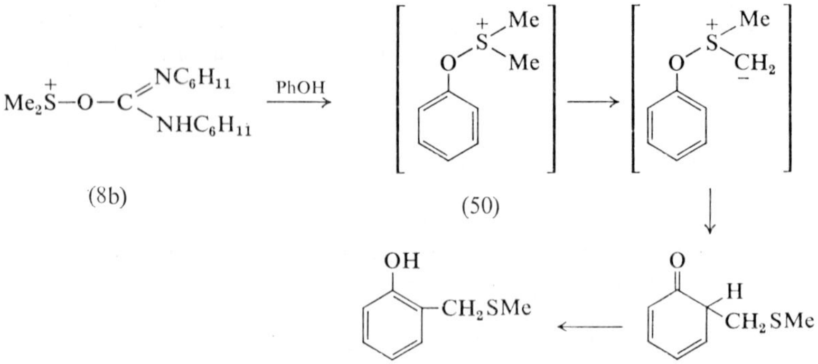

$Me_2\overset{+}{S}-O-C\overset{NC_6H_{11}}{\underset{NHC_6H_{11}}{\diagdown}}$ →PhOH→ $\left[\overset{+}{S}(Me)(Me)-O-C_6H_5\right]$ → $\left[\overset{+}{S}(Me)(CH_2^-)-O-C_6H_5\right]$

(8b)      (50)

↓

(2-hydroxyphenyl)-CH$_2$SMe ← (cyclohexadienone with H and CH$_2$SMe)

afford an ylide which would undergo a Sommelet rearrangement to the ortho position. This mechanistic proposal has not been proven to date, however.

### 3 Oxysulphonium Ylides

Oxysulphonium ylides are oxidized forms of sulphonium ylides which exhibit characteristics similar to those of the sulphonium ylides and have become of considerable use in synthetic organic chemistry. Their chemistry was reviewed in 1965 by Bloch[2] and was discussed by Johnson in the book 'Ylid Chemistry' in 1966.[1]

**Synthesis.**—Although oxysulphonium ylides appear to be an oxidized form of sulphonium ylides, they have not been obtainable directly by that route. Prior to 1965, virtually the only chemistry known of this class of compounds was that of the simplest member of the class, methylenedimethyloxysulphurane [$CH_2=S(O)Me_2$]. Even to this date, very few more complex oxysulphonium ylides are known, and most of those have been obtained by substitution reactions (*i.e.* acylation) on the simplest member. These kinds of reactions will be discussed later.

One of the methods of synthesis newly reported in the last five years holds some hope for general application. Diekmann[11] found that bisbenzenesulphonyldiazomethane reacted with DMSO under thermochemical conditions for long periods of time to afford a highly stabilized ylide. Later, Ando *et al.*[13] expanded the scope of this reaction such that diphenylsulphoxide could be used as well, a wide variety of stabilized diazo-compounds could be employed, and the reaction could be effected

---

[107] M. G. Burden and J. G. Moffatt, *J. Amer. Chem. Soc.*, 1965, **87**, 4656.
[108] K. E. Pfitzner, J. P. Marino, and R. A. Olofson, *J. Amer. Chem. Soc.*, 1965, **87**, 4658.

under either thermal or photochemical conditions. However, yields the in most cases were below 50%. This reaction presumably involves attack of a carbenic species on the nucleophilic sulphur, and is analogous to that employing a sulphide rather than a sulphoxide.

In a similar manner, C. R. Johnson et al.[109] have reported that reaction of hydrazoic acid with sulphoxides affords imino-oxysulphuranes which can be methylated to the azasulphoxonium salts. Removal of a proton from these salts afforded the oxysulphonium ylides (51).

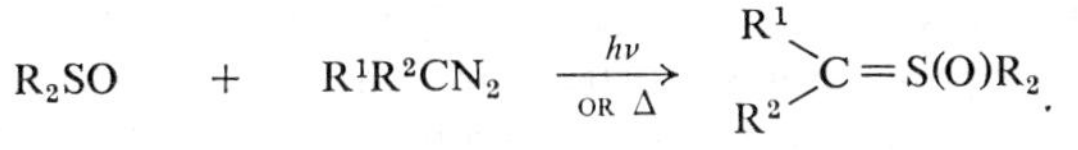

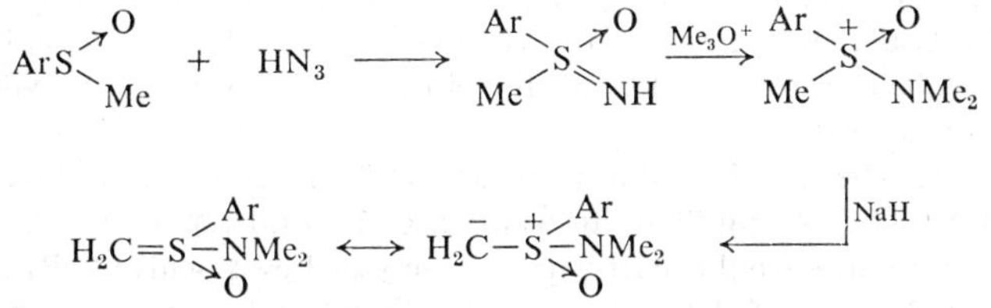

$$R_2SO + R^1R^2CN_2 \xrightarrow[\text{OR }\Delta]{hv} \underset{R^2}{\overset{R^1}{>}}C=S(O)R_2.$$

$$ArS\underset{Me}{\overset{O}{<}} + HN_3 \longrightarrow \underset{Me}{\overset{Ar}{>}}S\underset{NH}{\overset{O}{\nearrow}} \xrightarrow{Me_3O^+} \underset{Me}{\overset{Ar}{>}}\overset{+}{S}\underset{NMe_2}{\overset{O}{\nearrow}}$$

$$H_2C=S\underset{O}{\overset{Ar}{\underset{\nwarrow}{<}}}NMe_2 \longleftrightarrow H_2\bar{C}-\overset{+}{S}\underset{O}{\overset{Ar}{<}}NMe_2 \xleftarrow{\quad NaH \quad}$$

(51)

These two procedures are somewhat complementary in that in Ando's procedure, structural variation is effected by varying the nature of the hydrocarbon component of the reaction (i.e. the diazo-compound), while in Johnson's procedure, structural variation in the ylide mainly is obtained by choice of the sulphoxide to be used. The presence of the dimethylamine function in Johnson's procedure has no appreciable effect on the chemistry of the ylide, since in most of the reactions the sulphur-containing portion of the ylide is eliminated from the desired product.

In an observation that could hold considerable significance for oxysulphonium ylide chemistry, it was found that ethylenediaminedichloroplatinum formed a stable complex with DMSO. This complex underwent proton exchange in $D_2O$–NaOD, indicating the formation of a carbanion intermediate, perhaps of structure (52).[110] If such a structure could be verified and if it had appreciable lifetime, its ylide character could be

$$Pt(en)Cl_2 + DMSO \longrightarrow [Pt(en)(Cl)(DMSO)]Cl$$

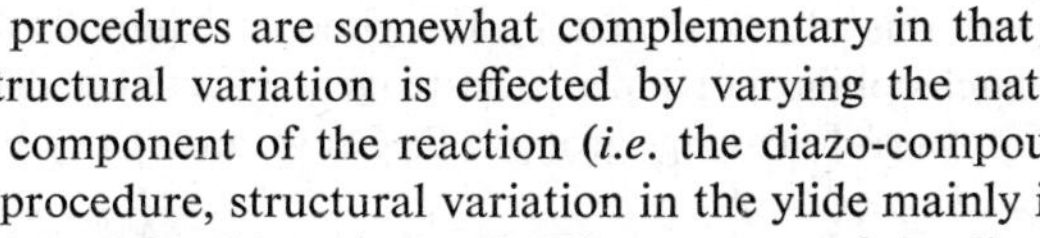

$$Pt\overset{Me}{\underset{\downarrow}{-\overset{+}{S}-Me}}\quad \rightleftharpoons \quad Pt\overset{Me}{\underset{\downarrow}{-\overset{+}{S}-CH_2^-}} + H^+$$
$$\phantom{Pt-}O\phantom{xxxxxxxxxxxxxx}O$$

(52)

[109] C. R. Johnson, E. R. Janiga, and M. Haake, *J. Amer. Chem. Soc.*, 1968, **90**, 3890.
[110] D. A. Johnson, *Inorg. Nuclear Chem. Letters*, 1969, **5**, 225.

utilized in organic synthesis. A wide variety of such ylides could be readily available, limited only by the availability of the appropriate sulphoxide.

In 1968 Schmidbaur and Tronich [111] isolated and purified, for the first time, methylenedimethyloxysulphurane. It was a low-melting (9—10 °C), highly-reactive substance which showed a singlet in the n.m.r. spectrum. The singlet could not be split by changing solvents or by lowering the temperature to $-30$ °C. This would imply that rapid proton exchange is occurring between the two methyl groups and the methylene carbanion.

**Chemical Stability.**—While the simplest oxysulphonium ylide is susceptible to decomposition upon exposure to the atmosphere and moisture, those ylides carrying strongly electronegative groups, such as carbonyl, are sufficiently stable to be isolated, purified, and handled in the open atmosphere. The stability of such ylides undoubtedly is due to delocalization of the negative charge through the carbonyl system, leading to a lowered nucleophilicity for the ylide and, therefore, much lowered reactivity with, for example, water.

Although the stabilized ylides fail to react with water, they will react with stronger acids. Konig,[112] in an abstract of an oral report, indicated that $\beta$-carbonyl ylides would react with a variety of Lewis acids (A–B) to effect eventual cleavage of the carbon–sulphur bond and result in overall A–B addition to the carbanion fragment. It is likely that the electrophilic portion of the reagent was attacked first by the ylide carbanion, after which the remaining nucleophilic portion of the reagent effected a displacement of DMSO, a powerful leaving group. DeGraw and Cory [113] found that

$$R-CO-CH=S(O)Me_2 \xrightarrow{A-B} R-CO-CH{<}^{A}_{B} + DMSO$$

$$A = H^+, Br^+, CH_3^+ \qquad B = Cl^-, Br^-, PhS^-, PhNH^-$$

$$RCOCH_2COR + CH_2=S(O)Me_2 \longrightarrow RCOCH(Me)-COR + DMSO$$

phenacylidenedimethyloxysulphurane was similarly cleaved by acetic acid or hydrochloric acid. Holt et al.[114] reported a similar reaction, that between methylenedimethyloxysulphurane and $\beta$-diketones, with the latter being the Lewis acid system through the acidic methylene protons. The net result of this reaction was methylation of the $\beta$-diketone. If this reaction were to be generalized, it would appear that any potential carbanion system, such as an active methylene system, could be alkylated with the carbanion fragment

[111] H. Schmidbaur and W. Tronich, Tetrahedron Letters, 1968, 5335.
[112] H. Konig, Angew. Chem. Internat. Edn., 1965, 4, 982.
[113] J. I. DeGraw and M. Cory, Tetrahedron Letters, 1968, 2501.
[114] B. Holt, J. Howard, and D. A. Lowe, Tetrahedron Letters, 1969, 4937.

# Ylides of Sulphur, Selenium, and Tellurium and Related Structures

of an oxysulph oniumylide. This reaction could be of use in the formation of new carbon–carbon bonds, essentially effecting the insertion of an alkyl group into a C–H bond.

Reduction of oxysulphonium ylides has not been generally investigated. However, Ide and Kishida [115] reported one such reduction using zinc–acetic acid, converting the ylide methine into a methyl group. Development of such procedures is important for if such ylides are to be utilized in synthetic procedures, effective means must be available for the final removal of the sulphur-containing group.

In 1964 Corey and Chaykovsky [116] had reported the photochemical cleavage of a stabilized oxysulphonium ylide into a carbene which rearranged to a keten. Truce and Madding [117] more recently have reported that a series of sulphonylmethylenedimethyloxysulphuranes were inert to both thermal and photochemical decomposition. Kishida et al.[118, 119] have reported the photochemistry of some complex, highly-substituted oxysulphonium ylides leading to cyclization reactions, the mechanisms or generality of which have yet to be established.

**Reactions of Oxysulphonium Ylides.**—Oxysulphonium ylides exhibit many reactions characteristic of nucleophiles, but few of these have been thoroughly explored. These ylides have been shown to undergo acylation under a variety of conditions. Corey [116] and Konig [112] earlier had reported the benzoylation of methylenedimethyloxysulphurane (53) with benzoyl chloride. More recently, Johnson et al.[109] found that the azaoxysulphonium ylide (51) underwent an analogous benzoylation, and both Ide [120] and Nozaki [78] reported similar benzoylations of more complex ylides. Van

$$\underset{(51)}{\underset{Me_2N}{Ar}\!\!\diagdown\!\!\underset{CH_2}{S}\!\!\diagup\!\!O} + PhCOCl \longrightarrow \underset{Me_2N}{\underset{Me_2N}{Ar}\!\!\diagdown\!\!\underset{CHCOPh}{S}\!\!\diagup\!\!O}$$

$$\underset{(53)}{Me_2S(O)\!=\!CH_2} + Cl\!-\!COOEt \longrightarrow \underset{(54)}{Me_2S(O)\!=\!C}\!\diagdown\!\!\underset{COOEt}{\overset{COOEt}{\diagup}}$$

Leusen and Taylor [121] reported that analogous results could be obtained using an anhydride as the acylating agent. Nozaki et al.[78] found that double acylation could occur using ethyl chloroformate to form the bis-ester ylide (54). The mono-ester ylide could be isolated as an intermediate and separately acylated a second time with additional chloroformate or with a

---

[115] J. Ide and Y. Kishida, *Tetrahedron Letters*, 1966, 1787.
[116] E. J. Corey and M. Chaykovsky, *J. Amer. Chem. Soc.*, 1964, **86**, 1640.
[117] W. E. Truce and G. D. Madding, *Tetrahedron Letters*, 1966, 3681.
[118] Y. Kishida, T. Hiraoka, and J. Ide, *Tetrahedron Letters*, 1968, 1139.
[119] Y. Kishida, T. Hiraoka, and J. Ide, *Chem. Pharm. Bull.*, 1969, **17**, 1591.
[120] J. Ide and Y. Kishida, *Chem. Pharm. Bull.*, 1968, **16**, 784.
[121] A. M. Van Leusen and E. C. Taylor, *J. Org. Chem.*, 1968, **33**, 66.

different acylating agent. To date there have been no reports of *O*-acylation of β-carbonyl oxysulphonium ylides, consistent with the conclusion that there is a high degree of $p\pi$–$d\pi$ overlap between the carbanion and sulphur, thereby necessitating little enolate character in such ylides.[27]

Speakman and Robson [122] reported a sulphination of methylenedimethyloxysulphurane (53) which occurred in a manner analogous to the acylations to form a β-sulphinyl ylide. In a similar vein, Truce and Madding,[117] and later Ide and Yura,[123] reported the sulphonation of the same ylide using a sulphonyl chloride. The latter group suggested that the reaction might have proceeded by initial conversion of the sulphonyl chloride to a sulphene which then was attacked by the nucleophilic ylide on sulphur. However, this proposal has not been proven.

Numerous examples of the reaction of methylenedimethyloxysulphurane (53) with aldehydes and ketones have been reported as means of preparing epoxides in high yield. The past five years have seen some of the fine points of this reaction elaborated, in many instances these being specific limitations on the reaction. Holt *et al.*[114] found that the reaction was often unsuccessful with β-dicarbonyl compounds because the methylene protons were too acidic, thereby protonating the ylide before it could react with the carbonyl group. In many instances where other functional groups were near the carbonyl group under attack, the intermediate betaine, formed initially between the ylide and the carbonyl group, reacted with that functional group affording products other than the expected epoxides.[114, 124] Several groups have found that oxysulphonium ylides will not react with the carbonyl function of an ester group, and House *et al.*[125] found that (53) would react with a ketone group in the presence of an ester group.

With cyclic ketones, (53) has been shown to afford the expected exocyclic epoxide.[125, 126] Carlson and Behn [85] have shown that reaction of (53) with *trans*-2-decalone afforded a mixture of two epoxides, that resulting from equatorial attack of the ylide present to the extent of 90% in the mixture.

[122] P. R. H. Speakman and P. Robson, *Tetrahedron Letters*, 1969, 1373.
[123] J. Ide and Y. Yura, *Tetrahedron Letters*, 1968, 3491.
[124] B. Holt and P. A. Lowe, *Tetrahedron Letters*, 1966, 683.
[125] H. O. House, S. G. Boots, and V. K. Jones, *J. Org. Chem.*, 1965, 30, 2519.
[126] W. N. Speckamp, R. Neeter, P. D. Rademaker, and H. O. Huisman, *Tetrahedron Letters*, 1968, 3795.

This is in contrast to the results with methylenedimethylsulphurane which gave the same products but with the other isomer dominating the 60:40 mixture. It appears that the oxysulphurane has larger steric requirements, thereby effecting more equatorial attack. Similar results were reported by King et al.[127] in the reaction of (53) with 3-ketopyranosides in which the dominant epoxide isomer formed was that resulting from equatorial attack of the ylide.

Bly and his co-workers [86, 128-130] have studied the reaction of (53) with several bicyclic ketones of the norbornane series. In all cases the major or only product was the expected epoxide but the stereochemistry varied somewhat. Norbornan-2-one afforded a 90:10 mixture of epoxides, *exo* attack dominating, but norborn-5-ene-2-one afforded a 70:30 mixture of epoxides, *endo* attack dominating. In a related experiment with norbor-2-ene-7-one, the ylide attacked from the *syn* side. Therefore, it appears that in spite of *exo* attack being sterically favoured, the presence of the double bond facilitates ylide attack on the carbonyl group from the side on which the olefinic group is located. Bly et al.[86] suggested that complexation of the oxysulphonium group with the olefinic group, probably through a $\pi$-overlap mechanism, facilitated the rate of ylide attack from the *syn* side.

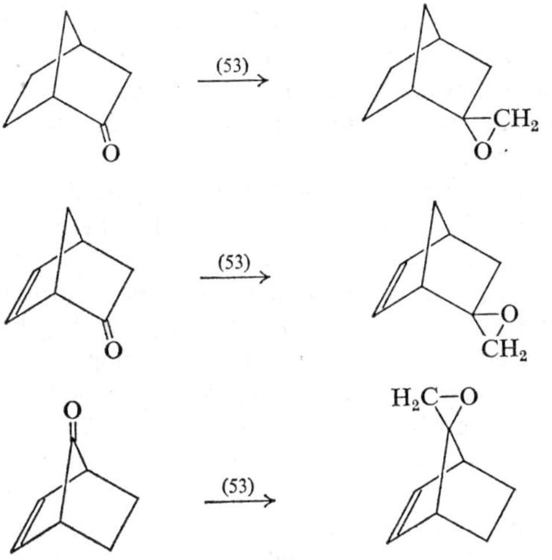

There have been very few reports of any attempts to react oxysulphonium ylides other than the simplest member (53) with carbonyl compounds to form substituted epoxides. Johnson et al.[109] found that the azaoxysulphonium ylide (51) would transfer its methylene group to an aldehyde

---

[127] R. D. King, W. G. Overend, J. Wells, and N. R. Williams, *Chem. Comm.*, 1967, 726.
[128] R. K. Bly and R. S. Bly, *J. Org. Chem.*, 1963, **28**, 3165.
[129] R. K. Bly and R. S. Bly, *J. Org. Chem.*, 1969, **34**, 304.
[130] G. B. Konizer, *Diss. Abs.*, 1969, **30B**, 2608.

in good yield to form the expected epoxide in 60% yield. However, Truce and Madding [117] reported that β-sulphonyl ylides, Me$_2$S(O)=CH—SO$_2$R, would not react with *p*-nitrobenzaldehyde to afford epoxides. Based on this meagre evidence, it may be tentatively concluded that the oxysulphonium ylides are more stabilized, and therefore less nucleophilic, than the corresponding sulphonium ylides. This would be expected if the positive character of the sulphur atom is a key determinant of the magnitude of $p\pi$–$d\pi$ overlap.[27] Attachment of an electronegative group to the carbanion of an oxysulphonium ylide therefore seems sufficient to reduce the nucleophilicity of the ylide to a point where it is of little use in organic synthesis. Accordingly, for conversion of carbonyl groups into epoxide functions, the oxysulphonium ylide method seems applicable only when one of the carbon atoms of the epoxide is to remain a methylene group. If the desired epoxide is to have substituents on both carbons, then a sulphonium ylide should be used.

Numerous examples have been reported in the past five years of the reaction of methylenedimethyloxysulphurane (53) with α,β-unsaturated aldehydes, ketones, and esters to afford cyclopropanes resulting from addition at the olefinic group rather than at the carbonyl group. Johnson *et al.*[109] reported similar behaviour with the aza ylide (51), benzalacetophenone being converted to 1-phenyl-2-benzoylcyclopropane in 100% yield. Lehmann *et al.*[131] found that an olefin precursor could be added to the reaction under conditions which would produce the olefin *in situ*, and the cyclopropane product could be obtained directly. Thus, a series of 2-ammoniomethyl-3-keto-steroids afforded the spirocyclopropane derivative when treated with trimethyloxysulphonium salt and two equivalents of base. In a related attempt, Breuer and Melumad [132] found that trimethyl-

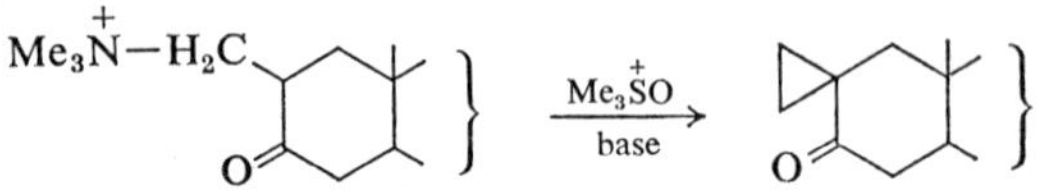

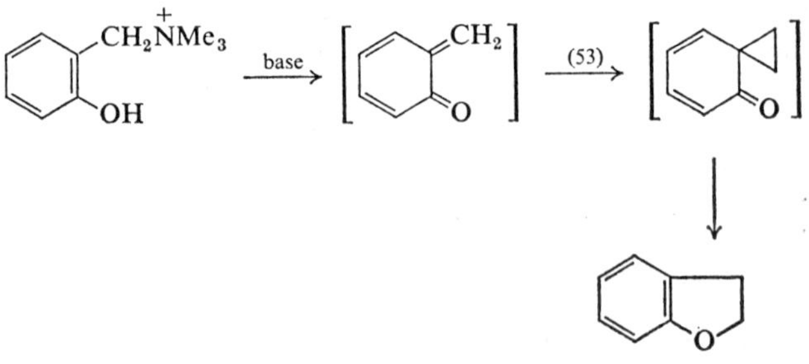

---

[131] H. G. Lehmann, H. Muller, and R. Wiechert, *Chem. Ber.*, 1965, **98**, 1470.
[132] E. Breuer and D. Melumad, *Tetrahedron Letters*, 1969, 1875.

(2-hydroxybenzyl)ammonium salts afforded dihydrobenzofuran rather than the spirocyclopropane ketone when treated with (53). However, they did suggest that the cyclopropane might have been an intermediate since cyclopropyl carbonyl compounds are known to rearrange to dihydrofurans.

Two reports have appeared of the addition of (53) to the $\gamma,\delta$-bond of dienone systems. Dyson et al.[133] found that addition of 1—2 equivalents of (53) to steroidal 4,6-dien-3-ones carrying no ring substituents afforded the 6,7-methano-4-en-3-ones, with nearly equal attack from the $\alpha$- and $\beta$-side of the rings. Addition of a $\beta$-substituent at C-11 led to formation of exclusively the $\alpha$-isomer. Addition of four equivalents of ylide (53) to ethyl hexa-2,4-dienoate resulted in cyclopropane formation at both of the double bonds, presumably via initial attack at the 4,5-bond, followed by attack at the still-conjugated 2,3-double bond.[134]

Relatively few complications have appeared in the reaction of (53) with conjugated carbonyl systems. Reaction under non-hydrolytic conditions with flavone afforded the expected cyclopropane,[135] but in the presence of moisture a diketone was obtained, apparently the result of ring-opening of the cyclopropane adduct.[136] However, reaction with isoflavone afforded a mixture of two products, one the expected cyclopropane, and the other the result of addition, ring-opening, and reclosure.[136] Nozaki et al.[78] have

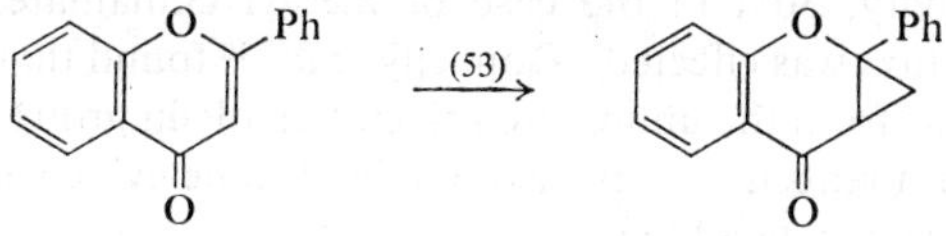

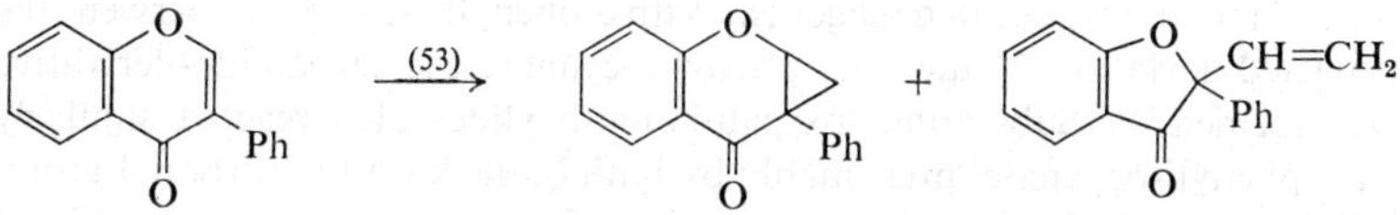

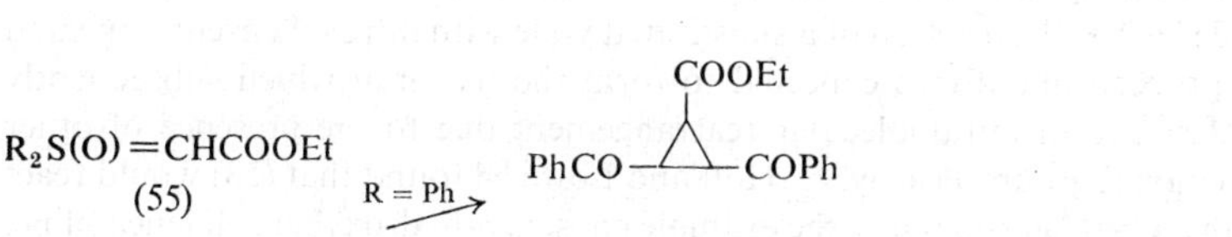

[133] N. H. Dyson, J. A. Edwards, and J. H. Fried, *Tetrahedron Letters*, 1966, 1841.
[134] S. R. Landor and N. Punja, *J. Chem. Soc. (C)*, 1967, 2495.
[135] P. Bennett and J. A. Donnelly, *Chem. and Ind. (London)*, 1969, 783.
[136] G. A. Caplin, W. D. Ollis, and J. O. Sutherland, *Chem. Comm.*, 1967, 575; *J. Chem. Soc. (C)*, 1968, 2302.

reported the only addition reactions involving substituted oxysulphonium ylides. They found that the ester ylide (55) effected normal addition to dibenzoylethylene when R = phenyl, but when R = methyl, a cyclic sulphoxide resulted, presumably from proton transfer in the betaine stage to afford a new ylide which attacked the carbonyl group.

Considerable interest has been focused on the stereochemistry of the reaction of methylenedimethyloxysulphurane (53) with α,β-unsaturated carbonyl compounds. Several groups have found that with a wide variety of α,β-unsaturated carbonyl compounds, the methylene group transferred by the ylide effects a *cis*-addition.[137-140] Thus, (53) with *trans*-benzalacetophenone affords the *trans*-1-phenyl-2-benzoylcyclopropane.[139] On this basis Agami and Prevost [140] suggested that the addition mechanism allows no time for a bond rotation after the initial addition (*i.e.* almost a concerted reaction). However, based on the observations of Caplin *et al.*[136] with the flavone series and the observations by Kaiser *et al.*[138] that mixtures of stereoisomers are formed in some instances, it cannot be said that such additions characteristically are concerted or stereospecific. In fact, with dimethyl maleate Johnson and Schroeck [141] reported exclusive formation of the *trans*-adduct when using the aza-ylide (51). Using optically active ylide (51), these workers also found that the resulting cyclopropanes possessed some optical activity, 30% in the case of methyl cinnamate. Therefore, asymmetric induction was effected. Donnelly *et al.*[142] found that methylene transfer by (53) was from the unhindered side of an olefin group, just as for the Simmons–Smith reagent. They also obtained some evidence for bond rotation in the intermediate adduct.

Oxysulphonium ylides have been reported to react with a variety of miscellaneous unsaturated reagents. With diphenylketen, (53) underwent the expected acylation [138] and with phenylisocyanate the carbamido-derivative was formed.[138] Substituted oxysulphonium ylides also reacted similarly with phenylisocyanate, presumably by initial attack on the carbonyl group followed by intramolecular proton transfer to form a new stabilized ylide.[78, 109, 120] Reaction of a substituted ylide with nitrosobenzene appeared to proceed initially as expected to form the oxaziran which subsequently underwent an intramolecular rearrangement due to the presence of other functional groups nearby.[121] Holt and Lowe [124] found that (53) would react with a C=N group but in the example chosen, *o*-hydroxybenzylideneaniline, the initial adduct underwent an intramolecular displacement of DMSO, resulting in the formation of 3-anilinodihydrobenzofuran. Truce and Madding [117] found that β-sulphonyl oxysulphuranes would not effect an

[137] M. Julia, S. Julia, and H. Brisson, *Compt. rend.*, 1964, **259**, 833.
[138] C. Kaiser, B. M. Trost, J. Beeson, and J. Weinstock, *J. Org. Chem.*, 1965, **30**, 3972.
[139] C. Agami, *Compt. rend.*, 1967, **264**, C, 1128; *Bull. Soc. chim. France*, 1967, 1391.
[140] C. Agami and C. Prevost, *Bull. Soc. chim. France*, 1967, 2299.
[141] C. R. Johnson and C. W. Schroeck, *J. Amer. Chem. Soc.*, 1968, **90**, 6852.
[142] J. A. Donnelly, D. D. Keane, K. G. Marathe, D. C. Meaney, and E. M. Philbin, *Chem. and Ind.* (*London*), 1967, 1402.

$Me_2S(O)=CH_2$ + $Ph_2C=C=O$ ⟶ $Me_2S(O)=CHCOCHPh_2$

(53) + PhNCO ⟶ $Me_2S(O)=CHCONHPh$

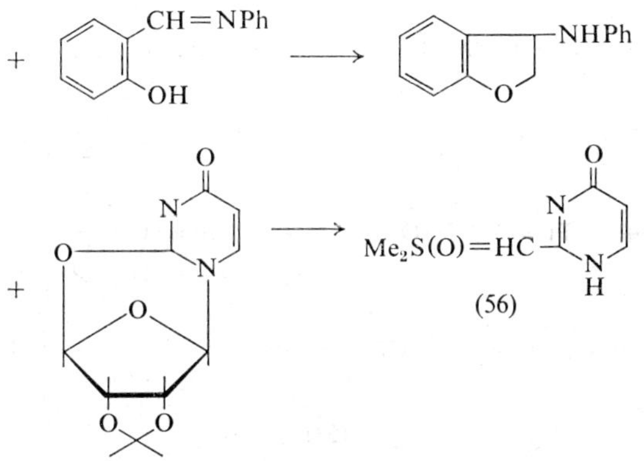

[structure: 2-(N-phenylimino)methylphenol + (53) → 2,3-dihydrobenzofuran-3-yl-NHPh]

[structure: uridine derivative + (53) → $Me_2S(O)=HC$–(pyrimidinone) (56)]

addition with benzylideneaniline. Kunieda and Witkop [143] reported that (53) would displace an alkoxy-group from the 2-position of a 1-alkyl-2-alkoxyuridine to form a new oxysulphonium ylide (56).

Oxysulphonium ylides have been found to react with conjugated alkynes. Kaiser et al.[138] and Ide and Kishida [115] both reported addition of (53) to ethyl phenylpropiolate to afford a new complex ylide (57), the result of initial addition of the nucleophilic ylide to the triple bond, followed by intramolecular proton transfer. Interestingly, (57) still was reactive enough to add to dimethyl acetylenedicarboxylate in the same manner.[120] In a rather interesting rearrangement, reminiscent of the 1,5-rearrangements

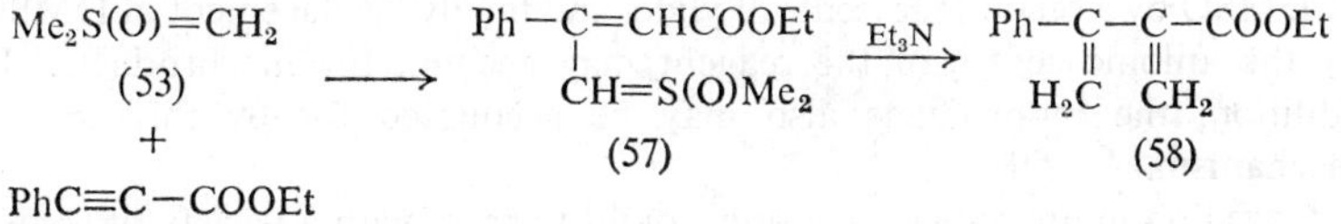

$Me_2S(O)=CH_2$   $Ph-C=CHCOOEt$   $\xrightarrow{Et_3N}$   $Ph-C-C-COOEt$
(53)   ⟶   $\overset{|}{CH}=S(O)Me_2$    $\overset{\|}{H_2C}\ \overset{\|}{CH_2}$
+      (57)       (58)
$PhC≡C-COOEt$

undergone by sulphonium ylides,[59] (57) was converted by triethylamine into the diene (58).[144]

Gaudiano and co-workers have reported extensively on the reaction of (53) with a series of 1,3-dipolarophiles. In nearly all cases, one equivalent of the dipolarophile appeared to react with two equivalents of the ylide, the structures of the major products being isoelectronic. Reaction with azides afforded the triazolines (59) or the triazenes (60), the latter forming only if the carbon portion of the azide was strongly electron-withdrawing.[145, 146] Diphenylnitrilimine [146] afforded a pyrazoline (61) with (53) and

---

[143] T. Kunieda and B. Witkop, *J. Amer. Chem. Soc.*, 1969, **91**, 7752.
[144] J. Ide and Y. Kishida, *Chem. Pharm. Bull.*, 1968, **16**, 793.
[145] G. Gaudiano, C. Ticozzi, A. Umani-Ronchi, and P. Bravo, *Gazzetta*, 1967, 1411.
[146] G. Gaudiano, A. Umani-Ronchi, P. Bravo, and M. Acampora, *Tetrahedron Letters*, 1967, 107.

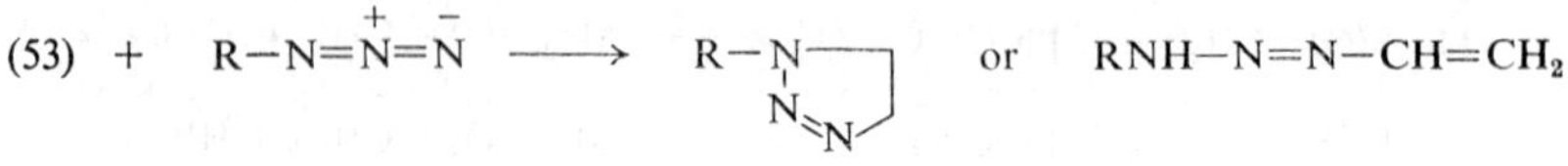

$$(53) + R-\overset{+}{N}=\overset{}{N}=\overset{-}{N} \longrightarrow R-N\underset{N\approx N}{\overset{}{\big|}} \quad \text{or} \quad RNH-N=N-CH=CH_2$$

(59)           (60)

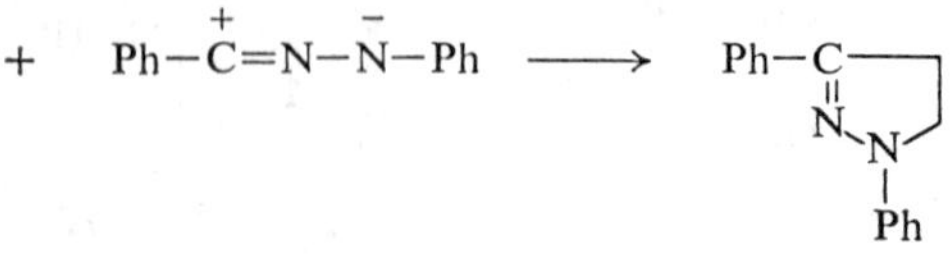

$$+ \quad Ph-\overset{+}{C}=N-\overset{-}{N}-Ph \longrightarrow Ph-C\underset{\underset{Ph}{N}}{\overset{\|}{\big|}}$$

(61)

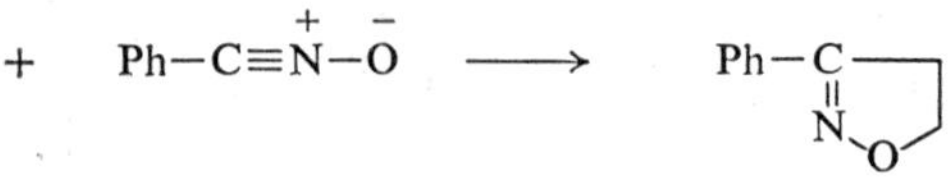

$$+ \quad Ph-C\equiv\overset{+}{N}-\overset{-}{O} \longrightarrow Ph-C\underset{\overset{}{N}\diagdown O}{\overset{\|}{\big|}}$$

(62)

$$R-\overset{+}{A}-B-\overset{-}{C} + CH_2=S(O)Me_2 \longrightarrow \left[ \begin{array}{c} R-A-B-C^{-} \\ | \\ CH_2-\underset{+}{S(O)Me_2} \end{array} \right]$$

(53)

↓ (53)

$$\begin{array}{c} R-A-B \\ | \quad\quad \diagdown C \\ C-C \diagup \\ H_2 \; H_2 \end{array} \longleftarrow \left[ \begin{array}{c} R-A-B-C^{-} \\ | \\ CH_2 \\ | \\ CH_2-\underset{+}{S(O)Me_2} \end{array} \right]$$

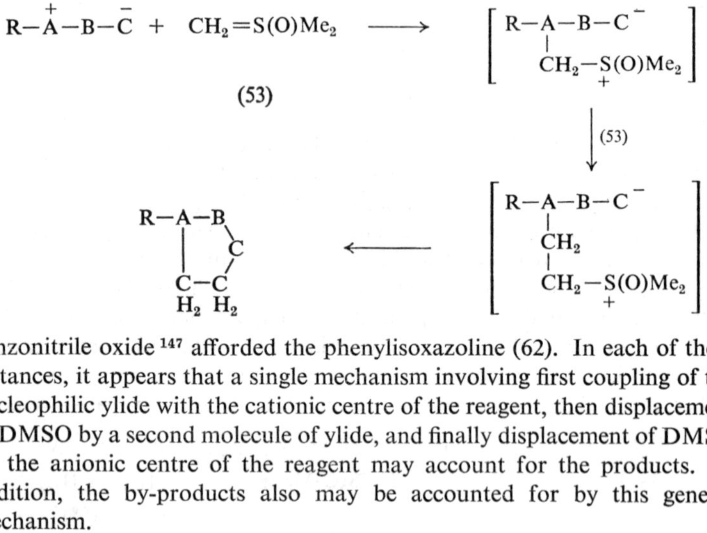

benzonitrile oxide [147] afforded the phenylisoxazoline (62). In each of these instances, it appears that a single mechanism involving first coupling of the nucleophilic ylide with the cationic centre of the reagent, then displacement of DMSO by a second molecule of ylide, and finally displacement of DMSO by the anionic centre of the reagent may account for the products. In addition, the by-products also may be accounted for by this general mechanism.

Oxysulphonium ylides have been found to react with a variety of simple electrophilic reagents. Reaction of (53) with trialkylboranes proceeded exactly as for sulphonium ylides, the overall effect being the homologation of the alkyl group.[148] For example, tri-n-hexylborane afforded a 26% yield of n-heptyl alcohol after reaction with (53) then oxidation with hydrogen peroxide. Reaction of (53) with α-halogenoketones has been found to afford cyclopropyl ketones. The reaction involves three equivalents of ylide per equivalent of halide and probably involves displacement of the halide, elimination of DMSO to afford an α,β-unsaturated ketone, then final addition of a third molecule of ylide to afford the cyclopropane.[149]

---

[147] A. Umani-Ronchi, P. Bravo, and G. Gaudiano, *Tetrahedron Letters*, 1966, 3477.
[148] J. J. Tufariello and L. T. C. Lee, *J. Amer. Chem. Soc.*, 1966, **88**, 4757.
[149] P. Bravo, G. Gaudiano, C. Ticozzi, and A. Umani-Ronchi, *Tetrahedron Letters*, 1968, 4481.

Through the use of methylenedimethyloxysulphurane (53), four different groups have reported the preparation of substituted 1-methylthiabenzene 1-oxides. Since the structure of these substances has not been proven unambiguously, and since the mechanism of the formation of the substances is not yet known, these reactions will not be discussed in detail. Suffice it to mention the compounds reportedly have been prepared from reaction of (53) with $\beta$-diketones,[114] benzonitrile,[112] and benzoylphenylacetylene,[150] and by intramolecular reaction of a complex multi-functional oxysulphonium ylide.[151]

## 4 Sulphinyl Ylides

The chemistry of sulphinyl carbanions [R—CH=S(O)—R¹] was reviewed by Bloch[2] in 1965 and by Johnson in 1966.[1] Nearly all such chemistry was concerned with methylsulphinyl (dimsyl) carbanion and such continues to be the case. Sjoberg[152] reported that dimsyl sodium, when prepared from sodium hydride in mineral oil with ultrasound, could be successfully stored for up to 10 weeks. This substance behaves as a normal powerful nucleophile, having been alkylated and acylated, and can be used to carry out nucleophilic additions.

Treatment of DMSO with excess sodium in liquid ammonia afforded a solution of the bis-carbanion which could be alkylated with an alkyl halide.[153] Similarly, phenacylmethylsulphoxide could be converted to its carbanion with sodium hydride and then alkylated with methyl iodide.[154,155] Interestingly, there was no evidence of O-alkylation in this case. In a similar vein, dimsyl sodium underwent acylation with esters of carboxylic acids to afford methyl(acylmethyl) sulphoxides.[154]

Dimsyl sodium has been found to react with p-tolualdehyde[156] and benzophenone[157] to afford the carbinol adduct. Heating the latter to 160 °C resulted in elimination of MeSOOH and the formation of diphenylethylene in 45% yield. Likewise, the carbanion of methylphenyl sulphoxide also reacted with benzophenone and, after treatment for 15 hours with 10% sulphuric acid, afforded a 65% yield of diphenylethylene.[157]

In a similar reaction, but one which has been adapted for effective use in organic synthesis, Corey and Durst[157] have found that methylsulphinamides of primary amines, when treated with two equivalents of butyl-lithium at −78 °C, were converted into the bis-anions (63) which would add to one equivalent of carbonyl compound in very high yield, the alcohol being

---

[150] A. G. Hartmann, *J. Amer. Chem. Soc.*, 1965, **87**, 4972.
[151] Y. Kishida and J. Ide, *Chem. Pharm. Bull.*, 1967, **15**, 360.
[152] K. Sjoberg, *Tetrahedron Letters*, 1966, 6383.
[153] H. Moskowitz, J. Blanc-Guenee, and M. Miocque, *Compt. rend.*, 1968, **267**, C, 898.
[154] G. A. Russell, E. Sabourin, and G. J. Mikol, *J. Org. Chem.*, 1966, **31**, 2854.
[155] P. G. Gassman and G. D. Richmond, *J. Org. Chem.*, 1966, **31**, 2355.
[156] G. A. Russell, E. G. Janzen, H. Becker, and F. Smentowski, *J. Amer. Chem. Soc.*, 1962, **84**, 2652.
[157] E. J. Corey and T. Durst, *J. Amer. Chem. Soc.*, 1966, **88**, 5656; *ibid.*, 1968, **90**, 5548.

Me$_2$SO + Ph$_2$CO $\xrightarrow{\text{NaH}}$ Ph$_2$C(OH)-CH$_2$S(O)Me $\xrightarrow{\Delta}$ Ph$_2$C=CH$_2$

MeS(O)NHAr $\xrightarrow{\text{BuLi}}$ [$\bar{\text{C}}$H$_2$-SO$\bar{\text{N}}$Ar] $\xrightarrow{\text{Ph}_2\text{CO}}$ Ph$_2$C(OH)-CH$_2$S(O)NHAr

(63)

Ph$_2$C=CH$_2$ + ArNH$_2$ + SO$_2$ $\xleftarrow{\Delta}$

capable of isolation. Heating the alcohol in benzene solution afforded the olefin, the amine, and sulphur dioxide in greater than 90% yield. This reaction has been shown to have rather general application, since both aldehydes and ketones can be used and ethyl sulphinamides have been used to transfer an ethylidene group.[157] The mechanism of the reaction has not been proven, but probably involves a cyclic *cis*-elimination[158] in a manner reminiscent of the Wittig reaction. Using benzaldehyde as the carbonyl component afforded the expected secondary alcohol. Oxidation of this alcohol with manganese dioxide should have afforded the β-ketosulphinamide, but instead afforded acetophenone.[157] The implication that the

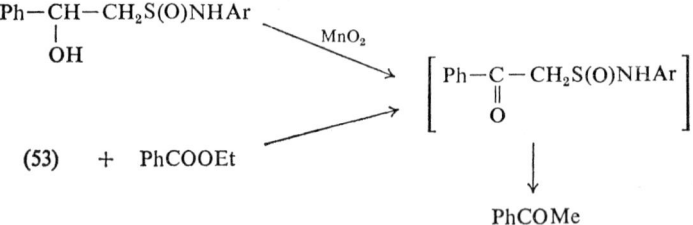

β-ketosulphinamide was unstable and spontaneously decomposed was confirmed by the observation that its attempted synthesis by benzoylation of (63) with ethyl benzoate also afforded acetophenone directly. The presence of the β-keto-group obviously rendered the carbon–sulphur bond extremely labile.

Finally, the nucleophilic strength of dimsyl sodium has been utilized to effect attack on aromatic rings, leading to methylation. For example, quinoline was converted to 4-methylquinoline[159] and phenanthrene was converted to 9-methylphenanthrene.[160] Direct attack of the carbanion on the ring was proposed, followed by either elimination to the *exo*-methylene derivative which tautomerized,[159] or followed by hydride migration and displacement of MeSO$^-$.[160] Neither mechanism has been proven.

[158] E. J. Corey and T. Durst, *J. Amer. Chem. Soc.*, 1968, **90**, 5553.
[159] G. A. Russell and S. A. Weiner, *J. Org. Chem.*, 1966, **31**, 248.
[160] H. Nozaki, Y. Yamamoto, and R. Noyori, *Tetrahedron Letters*, 1966, 1123.

## 5 Sulphonyl Ylides

Sulphonyl ylides first were reviewed by Johnson,[1] in 1966. In the interim, no special 'ylide-type' chemistry has been reported for these carbanions, only typical 'nucleophile-type' chemistry has been reported.

Truce et al.[161] have found that the carbanion generated by treatment of an α-methylenesulphone could be nitrated by alkyl nitrates and could be brominated by bromine to the α-nitro- and α-bromo-derivatives of the sulphone. Kaiser and Hauser [162] have reported the reaction of α-sulphonyl carbanions with alkyl halides to form the alkyl sulphone and with carbonyl compounds to form the carbinol adducts. Durst and Du Manoir [163] reported the formation of a carbanion (64) at low temperatures from a cyclic sulphonate ester. This carbanion would react normally with alkyl halides and with carbonyl compounds. At room temperature, however, the carbanion was not formed, the base instead opening the ester ring.

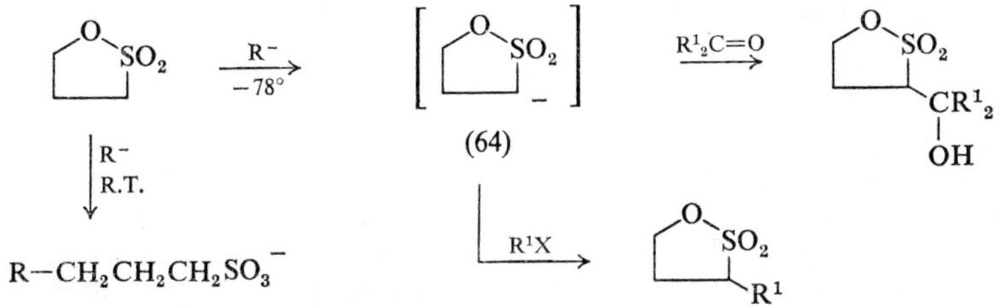

Carroll and O'Sullivan [164] reported the reactions of a substituted sulphonyl ylide (65), one stabilized by an attached benzoyl group and thereby capable of stabilization through the enolate form. Alkylation of this carbanion afforded the C-alkyl derivative, consistent with the observation that all β-ketosulphur ylides known to date have undergone C-alkylation with no evidence for O-alkylation. However, acylation of the carbanion (65) with benzoyl chloride afforded the O-acylation product, as was the

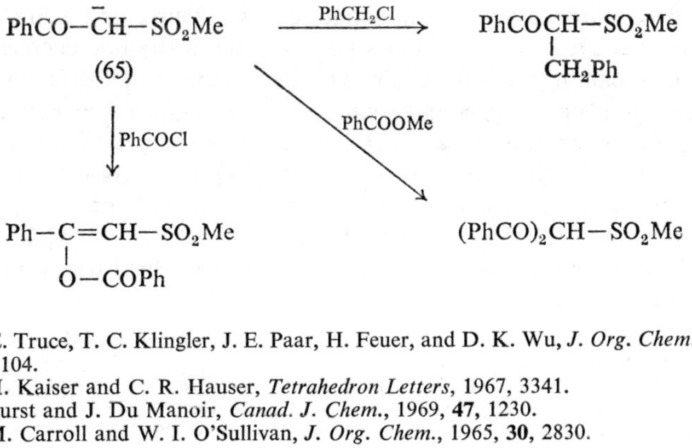

---

[161] W. E. Truce, T. C. Klingler, J. E. Paar, H. Feuer, and D. K. Wu, *J. Org. Chem.*, 1969, **34**, 3104.
[162] E. M. Kaiser and C. R. Hauser, *Tetrahedron Letters*, 1967, 3341.
[163] T. Durst and J. Du Manoir, *Canad. J. Chem.*, 1969, **47**, 1230.
[164] N. M. Carroll and W. I. O'Sullivan, *J. Org. Chem.*, 1965, **30**, 2830.

case for the analogous phenacylidenedimethylsulphurane.[27] Benzoylation with methyl benzoate, however, resulted in C-acylation. Therefore, it may be concluded that the behaviour of the carbanion (65) is similar to that of the analogous sulphonium ylide, in that kinetic control of the acylation yields the O-acylation product and thermodynamic control yields the more stable C-acylation product. Thus, the electronic characteristics of the sulphonyl carbanion (65) are more closely related to those of the sulphonium ylide than to those of the oxysulphonium ylide, the former two having more enolate stabilization of the carbanion.[27]

Finally, in a reaction similar to that of dimsyl sodium,[159, 160] the carbanion of methylphenyl sulphone was found to attack an aromatic ring to effect methylation with loss of phenylsulphinate ion.[165] .Thus, anthracene could be converted to 9-methylanthracene in 53% yield under mild conditions.

### 6 Sulphenes and Sulphines

Sulphenes, $RR^1C=SO_2$, and sulphines, $RR^1C=SO$, may be looked upon as being alkylidene sulphones and alkylidene sulphoxides, respectively, or as being the internal anhydrides of sulphonic acids and sulphinic acids, respectively. They will be discussed in this chapter since many of their reactions appear to involve an 'ylide-type' structure, the carbon end of the double bond carrying substantial negative charge.

The chemistry of sulphenes first was reviewed by Johnson [1] in 1966 but at that time relatively little work had been reported. More recently, Wallace [166] in 1966 and Opitz [167] in 1967 have reviewed the field. The chemistry of sulphines is all post-1964 and has been reviewed only by Opitz.[167]

**Preparation of Sulphenes.**—Three methods of preparation have commonly been used for sulphenes and these have been reviewed and summarized. Little has been added in the interim regarding the application of these methods: (a) the elimination of HX from a sulphonyl halide; (b) the reaction of sulphur dioxide with diazo-compounds; (c) photochemical ring-opening of sultones (cyclic sulphonate esters). Durst and King [168] have reported that cyclic sulphonamides also are subject to photochemical ring-opening, apparently to afford sulphenes in a manner analogous to that of sultones. Shirota et al.[169] have studied the ever-present competition in method (a) between elimination, to form sulphene, and substitution and have reported the competition to be affected by the solvent used and the metal that may be used in the organometallic base. For example, with benzylsulphonyl fluoride, phenyl-lithium lead to sulphene while phenyl-magnesium bromide lead to substitution. The same held true with phenyl benzylsulphonate.

[165] H. Nozaki, Y. Yamamoto, and T. Nisimura, *Tetrahedron Letters*, 1968, 4625.
[166] T. J. Wallace, *Quart. Rev.*, 1966, **20**, 67.
[167] G. Opitz, *Angew. Chem.*, 1967, **79**, 161; *Angew. Chem. Internat. Edn.*, 1967, **6**, 107.
[168] T. Durst and J. F. King, *Canad. J. Chem.*, 1966, **44**, 1869.
[169] Y. Shirota, T. Nagai, and N. Tokura, *Tetrahedron*, 1969, **25**, 3193.

Mulder et al.[170] have reported yet a fourth route to sulphenes, although one that does not promise to be of synthetic use as yet. Photolysis of arylsulphonyldiazomethanes presumably initially affords the sulphonyl carbene which, in methanolic solution, was trapped to form the ether.

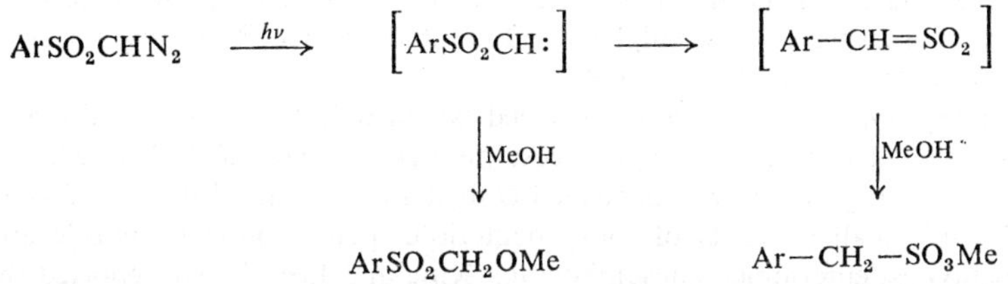

$$ArSO_2CHN_2 \xrightarrow{h\nu} [ArSO_2CH\colon] \longrightarrow [Ar-CH=SO_2]$$

$$\downarrow MeOH \qquad\qquad \downarrow MeOH$$

$$ArSO_2CH_2OMe \qquad\qquad Ar-CH_2-SO_3Me$$

However, up to 12% of the methyl benzylsulphonate ester was isolated. This observation implies that the sulphonyl carbene probably underwent a Wolff-type rearrangement to form a sulphene which was trapped by methanol.

The review by Opitz [167] included a detailed description of the conditions affecting method (a) to which readers are referred for further discussion.

**Evidence for the Existence of Sulphenes.**—No sulphene has been isolated and characterized to date. Therefore, all evidence regarding such species has been indirect. Most of the evidence has involved a trapping or interceptive experiment. The original such evidence, the reaction of a sulphene with deuterioalcohols [171, 172] resulting in the appearance of a single deuterium atom *alpha* to the sulphonyl group of the sulphonate ester product, has been followed by many other trapping experiments, some of which will be discussed in the next section.

Opitz and co-workers have reported several attempts to store or isolate sulphenes. They have found that sulphene itself, upon storage for 90 hours, was converted into head-to-tail dimer (66).[173] Methylsulphonylsulphene (67) also could be stored for about the same length of time. After 90 hours

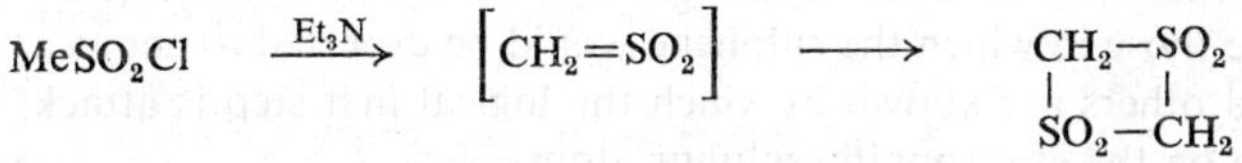

$$MeSO_2Cl \xrightarrow{Et_3N} [CH_2=SO_2] \longrightarrow \begin{array}{c} CH_2-SO_2 \\ | \quad\quad | \\ SO_2-CH_2 \end{array}$$

(66)

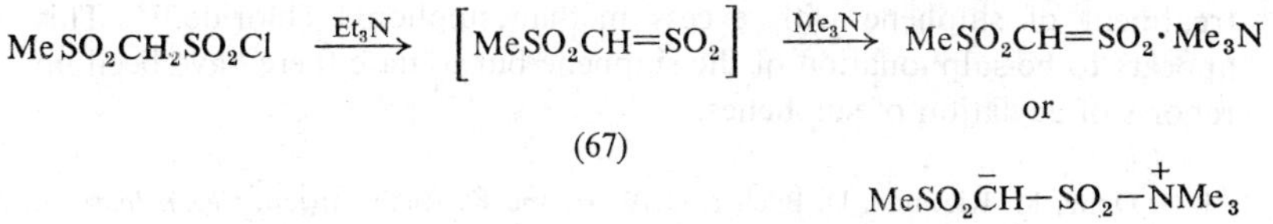

$$MeSO_2CH_2SO_2Cl \xrightarrow{Et_3N} [MeSO_2CH=SO_2] \xrightarrow{Me_3N} MeSO_2CH=SO_2 \cdot Me_3N$$

(67)

or

$$MeSO_2\overset{-}{C}H-SO_2-\overset{+}{N}Me_3$$

---

[170] R. J. Mulder, A. M. Van Leusen, and J. Strating, *Tetrahedron Letters*, 1967, 3057.
[171] J. F. King and T. Durst, *J. Amer. Chem. Soc.*, 1965, **87**, 5684.
[172] W. E. Truce and R. W. Campbell, *J. Amer. Chem. Soc.*, 1966, **88**, 3599.
[173] G. Opitz and H. R. Mohe, *Angew. Chem. Internat. Edn.*, 1969, **8**, 73.

at $-40$ °C, it still could be trapped in 52% yield.[174] The same sulphene, when prepared directly from the corresponding sulphonyl chloride with excess triethylamine, could be precipitated in the form of a 1:1 complex with trimethylamine. This yellow complex melted at 126—130 °C, showed three singlets in the n.m.r. spectrum in the ratio 1:3:9, and underwent reactions expected of a sulphene. Opitz and Bucher[175] proposed a zwitterionic structure for the complex.

King et al.[176] recently have provided evidence that sulphene is a discrete, volatile substance. Heating chlorosulphonylacetic acid to 640 °C under a cold finger on which was deposited $O$-deuterio-methanol led to the formation of small amounts of mono-deuteriomethanesulphonyl chloride and methyl $\alpha$-deuteriomethanesulphonate. King and Lee[177] have reported the first detailed kinetic study of the formation of sulphenes. The reaction between methanesulphonyl chloride and propan-1-ol or propan-2-ol in the presence of triethylamine was found to have a substantial second-order component, first order in amine and first order in sulphonyl chloride. The fact that it was zero order in alcohol indicates than an intermediate must have been formed which reacted with the alcohol in a separate step. The existence of a primary deuterium isotope effect and a $\rho$-value of $+2 \cdot 35$ for substituted benzylsulphonyl chlorides, the comparison of the rate of sulphene formation with the rate of $\alpha$-deuterium exchange in benzylsulphonyl derivatives, and the relative rates of sulphene formation for the two epimeric 2-decalysulphonyl chlorides all provided strong new evidence for the actual existence of sulphenes.

It is probably safe to assume that someone will discover a particular sulphene whose structure provides sufficient stabilization to permit isolation. There is no evidence that sulphenes are thermodynamically unstable, they seem simply to be extremely reactive.

**Reactions of Sulphenes.**—Most of the reactions of sulphenes seem to depend on a polarization of the sulphene, with the carbon atom negatively charged and the sulphur atom positively charged. Some of the reactions undergone by sulphenes are those in which the sulphene would be expected to act as a nucleophile, and others are known in which the logical first step is attack of a nucleophile on the electrophilic sulphur atom.

One of the simplest reactions involving nucleophilic attack by sulphene is the alternative preparation of methanesulphonyl sulphene (67) involving treatment of sulphene with excess methanesulphonyl chloride.[178] This appears to be sulphonation of the sulphene but to date there have been no reports of acylation of sulphenes.

[174] G. Opitz, M. Kleeman, D. Bucher, G. Walz, and K. Rieth, *Angew. Chem. Internat. Edn.*, 1966, **5**, 594.
[175] G. Opitz and D. Bucher, *Tetrahedron Letters*, 1966, 5263.
[176] J. F. King, P. deMayo, and D. L. Verdun, *Canad. J. Chem.*, 1969, **47**, 4509.
[177] J. F. King and T. W. S. Lee, *J. Amer. Chem. Soc.*, 1969, **91**, 6524.
[178] G. Opitz, K. Rieth, and G. Walz, *Tetrahedron Letters*, 1966, 5269.

$$CH_2=SO_2 + MeSO_2Cl \longrightarrow MeSO_2CH=SO_2$$
$$(67)$$

(67) + [tetrahydropyran with O] ⟶ [bicyclic product with $SO_2$, $SO_2Me$, O]

$$RCH=SO_2 + R^1CH=C(OR^2)_2 \longrightarrow \underset{\underset{SO_2-CH-R}{|}}{\overset{}{R^1-CH-C(OR^2)_2}}$$

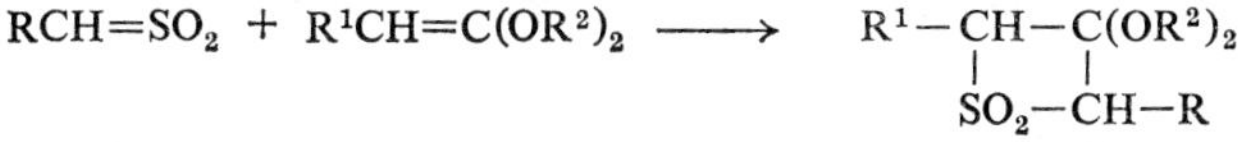

Evidence for the formation of (67) was provided by one of the usual trapping reactions characteristic of sulphenes, reaction with a vinyl ether, for example dihydropyran.[178] Reaction with vinyl acetals [179] proceeded analogously and both reactions probably involved electrophilic attack by the positive sulphur of the sulphene on the double bond. More complex products resulted when additional functional groups (*e.g.* carbonyl) were present.[179]

A similar reaction resulted between a sulphene and a vinyl amine to afford a four-membered sulphone.[180] The fact that the sulphone group was *gamma* to the nitrogen is consistent with the proposed polarization in the sulphene molecule. Use of vinyl diamines led to a mixture of two products which could be accounted for on the basis of a common zwitterionic intermediate [179, 181] and indicates that the reaction of sulphenes with olefins is not a concerted reaction. Use of more complex vinyl amine systems has

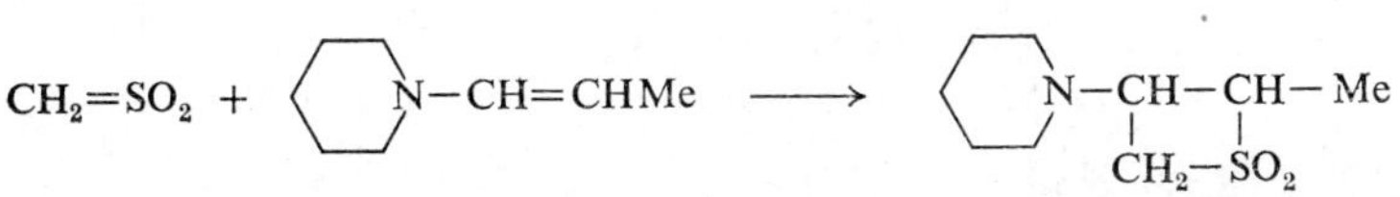

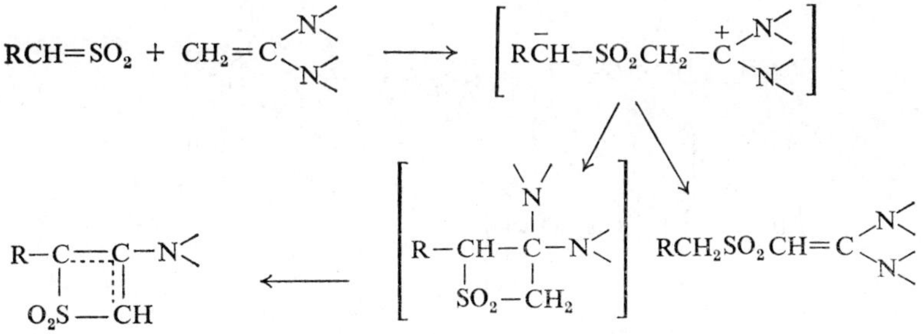

$$CH_2=SO_2 + \text{[piperidine]}N-CH=CHMe \longrightarrow \text{[piperidine]}N-CH-CH-Me$$
$$\underset{CH_2-SO_2}{|}$$

$$RCH=SO_2 + CH_2=C(N<)_2 \longrightarrow [RCH-SO_2CH_2-\overset{+}{C}(N<)_2]$$

$$\underset{O_2S-CH}{\overset{R-C\cdots C-N<}{|}} \longleftarrow [R-CH-\overset{|}{C}-N<,\; SO_2-CH_2] \longrightarrow RCH_2SO_2CH=C(N<)_2$$

---

[179] W. E. Truce, D. J. Abraham, and P. Son, *J. Org. Chem.*, 1967, **32**, 990.
[180] G. Opitz, H. Schempp, and H. Adolph, *Annalen*, 1965, **684**, 92.
[181] G. Opitz and H. Schempp, *Annalen*, 1965, **684**, 103.

afforded cyclic and acyclic products, all of which are accountable in terms of sulphene attack initially on the most nucleophilic site.[182, 183]

Opitz and Fischer[184] reported that sulphenes reacted with diazo-compounds initially to form episulphones which, as is known from other work, decompose slowly to eliminate sulphur dioxide and afford olefins. Somewhat later, Rossi and Maiorana[185] confirmed this observation using *p*-nitrobenzylsulphonyl bromide and diazomethane. They were able to isolate the episulphone and observe its decomposition, but also found that

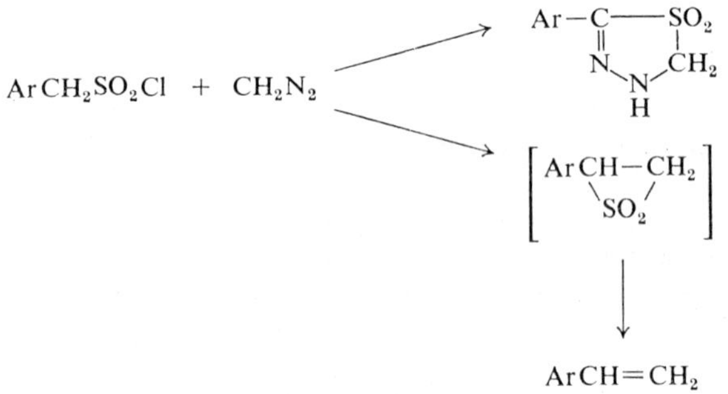

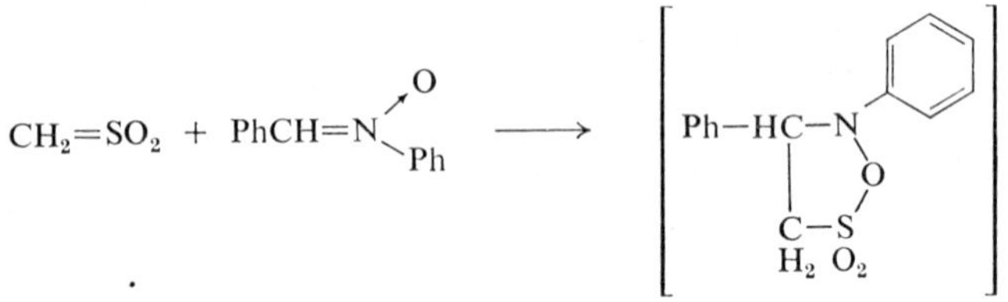

$$ArCH_2SO_2Cl + CH_2N_2 \longrightarrow \begin{matrix} Ar-C-SO_2 \\ \parallel \quad \quad | \\ N \quad \quad CH_2 \\ \diagdown N \diagup \\ H \end{matrix}$$

$$\longrightarrow \left[ \begin{matrix} ArCH-CH_2 \\ \diagdown SO_2 \diagup \end{matrix} \right] \longrightarrow ArCH=CH_2$$

$$CH_2=SO_2 + PhCH=N\begin{matrix}O \\ \diagdown Ph\end{matrix} \longrightarrow \left[ \begin{matrix} Ph-HC-N-C_6H_5 \\ | \quad \quad \diagdown O \\ C-S \diagup \\ H_2 \, O_2 \end{matrix} \right]$$

(68)

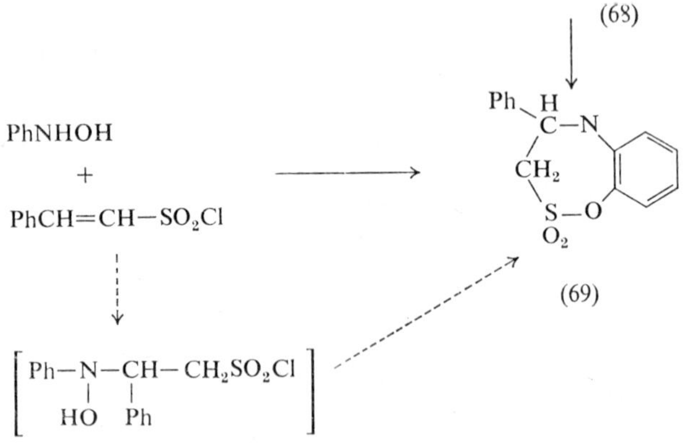

PhNHOH
+
PhCH=CH—SO$_2$Cl

$\longrightarrow$

Ph H
 \\C—N—C$_6$H$_4$
 CH$_2$
  \\S—O
  O$_2$

(69)

$$\left[ \begin{matrix} Ph-N-CH-CH_2SO_2Cl \\ | \quad | \\ HO \quad Ph \end{matrix} \right]$$

---

[182] G. Opitz and E. Tempel, *Annalen*, 1966, **699**, 68.
[183] L. A. Paquette and M. Rosen, *Tetrahedron Letters*, 1967, 703; *J. Amer. Chem. Soc.*, 1967, **89**, 4102.
[184] G. Opitz and K. Fischer, *Angew. Chem. Internat. Edn.*, 1965, **4**, 70.
[185] S. Rossi and S. Maiorana, *Tetrahedron Letters*, 1966, 263.

a thiadiazoline dioxide was formed, the result of a 1,3-addition by the sulphene. A similar 1,3-addition seems to have been involved in the reaction of sulphene with nitrones. Diphenylnitrone probably was converted initially to the oxathiazole (68) but the isolated product was an azasultone (69).[186-188] The structure of the azasultone was proven by two different groups through degradation reactions. In addition, it was found that use of an $^{18}$O-nitrone led to incorporation of 80% of the $^{18}$O in the *o*-hydroxyaniline formed by LiAlH$_4$ reduction [187] of (69). The azasultone was also produced by an alternative reaction which also must have involved a rearrangement including oxygen transfer to the aryl ring. Interestingly, reaction of phenylsulphene (from benzylsulphonyl chloride) with α-chlorobenzaldoxime in the presence of triethyl amine, conditions for the formation of benzonitrile oxide, a well-known 1,3-dipolarophile, did not result in a 1,3-addition.[189, 190] Instead, the reaction simply afforded the *O*-benzylsulphonate ester of α-chlorobenzaldoxime. The same reaction occurred using the unsubstituted benzaldoxime. These products probably were formed through a sulphene intermediate, since reaction of the aldoximes with tosyl chloride under the same conditions did not result in sulphonation. It is not clear why reaction with benzonitrile oxide did not proceed *via* 1,3-addition, since the product would be an oxathiazole structurally analogous to that probably formed from the nitrone reaction except for the presence of C=N function.

Finally, sulphenes are known to react with phosphonium, sulphonium, and oxysulphonium ylides. Nozaki *et al.*[76] have shown that reaction of a sulphonium ylide with tosyl chloride simply leads to sulphonation. In the case of β-carbonyl ylides, *O*-tosylation resulted with the loss of one of the *S*-alkyl groups. However, using ethanesulphonyl chloride and triethylamine as the sulphene source, reaction with phenacylidenephenylmethylsulphurane afforded three products: the sulphonated ylide, an olefin, and a cyclic product (70).[76, 77] The cyclic product only appeared when ethanesulphonyl chloride was the sulphene source, and the olefin was usually the dominant product. All three products can be accounted for by the mechanism shown from a common intermediate. The proposals for olefin and ylide formation are well known but the sulphonium-transfer step proposed for the formation of (70) has no precedent. Use of an oxysulphonium ylide in a reaction with a sulphene only afforded the olefin or sulphonated ylide products.[123] In an extensive series of reactions of phosphonium ylides with sulphenes, Ito *et al.*[191] obtained olefins, episulphones, sulphonated ylide,

---

[186] W. E. Truce, J. R. Norell, R. W. Campbell, D. G. Brady, and J. W. Fieldhouse, *Chem. and Ind.* (*London*), 1965, 1870.
[187] W. E. Truce, J. W. Fieldhouse, D. J. Vrencur, J. R. Norell, R. W. Campbell, and D. G. Brady, *J. Org. Chem.*, 1969, **34**, 3097.
[188] F. Eloy and A. Van Overstraeten, *Bull. Soc. chim. belges*, 1967, **76**, 63.
[189] W. E. Truce and A. R. Naik, *Canad. J. Chem.*, 1966, **44**, 297.
[190] J. F. King and T. Durst, *Canad. J. Chem.*, 1966, **44**, 409.
[191] Y. Ito, M. Okano, and R. Oda, *Tetrahedron*, 1967, **23**, 2137.

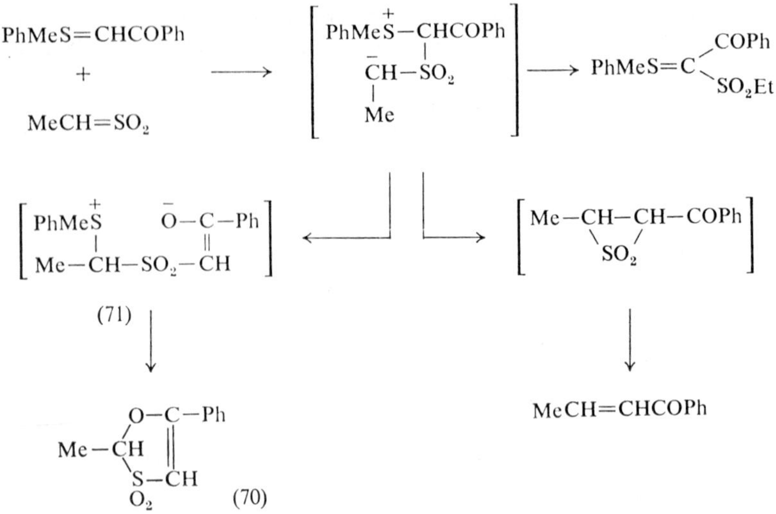

PhMeS=CHCOPh + MeCH=SO$_2$ ⟶ [PhMe$\overset{+}{\text{S}}$−CHCOPh | $\overset{-}{\text{CH}}$−SO$_2$ | Me] ⟶ PhMeS=C(COPh)(SO$_2$Et)

↓↑

[PhMe$\overset{+}{\text{S}}$   O−C−Ph
 |         ‖
 Me−CH−SO$_2$−$\overset{-}{\text{CH}}$]   ⟵⟶   [Me−CH−CH−COPh
                                                    \  /
                                                    SO$_2$]

(71) ↓                                          ↓

Me−CH(O−C−Ph / ‖ \ S−CH / O$_2$)   (70)        MeCH=CHCOPh

and a new rearranged ylide in which the sulphene group had been inserted between the phosphorus atom and its carbanion. All four of the products can be accounted for on the basis of the same mechanism as proposed for reaction with the sulphonium ylide, the rearranged ylide being derived by intramolecular proton transfer in (71) (Ph$_3$P in place of PhMeS) resulting in a sulphonyl-stabilized phosphonium ylide. It should be mentioned that the structure of the rearranged ylide was not unambiguously proven, the n.m.r. spectrum not being completely consistent with the proposed structure and the reaction products obtained from the rearranged ylide not being known.

**Sulphines.**—Sulphine chemistry all is post-1964. That year two groups independently reported the preparation of two crystalline sulphines by the same method. Sheppard and Diekmann [192] dehydrochlorinated 9-fluorenylsulphinyl chloride using triethylamine, as had been reported for sulphene preparation, and isolated the orange 9-fluorenylidenesulphine (72). Shortly after this report on the preparation of a thioketone-$S$-oxide, Strating *et al.*[193] reported the preparation of a thioaldehyde-$S$-oxide, 2-methoxy-1-naphthylsulphine (73), by the same method. Two years later, the latter group also reported the preparation of (72) from thiofluorenone by oxidation with monoperphthalic acid.[194] Several other sulphines have been prepared by each of these methods. Recently, Schultz and Schlessinger [195] reported the photochemical cleavage of a cyclic sulphoxide to afford a mixture of the two geometric isomers of phenyl-1-(8-benzylnaphthyl)

[192] W. A. Sheppard and J. Diekmann, *J. Amer. Chem. Soc.*, 1964, **86**, 1891.
[193] J. Strating, L. Thijs, and B. Zwanenburg, *Rec. Trav. chim.*, 1964, **83**, 631.
[194] J. Strating, L. Thijs, and B. Zwanenburg, *Tetrahedron Letters*, 1966, 65.
[195] A. G. Schultz and R. H. Schlessinger, *Chem. Comm.*, 1969, 1483.

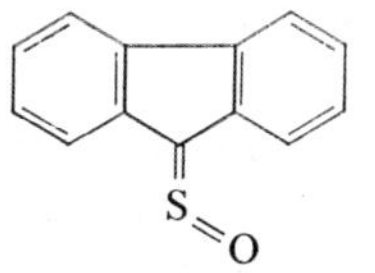

(72)

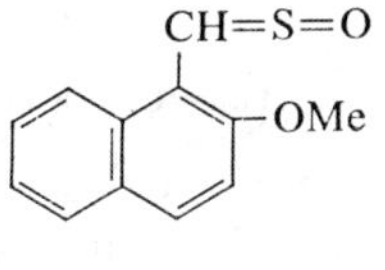

(73) CH=S=O, -OMe (on naphthalene)

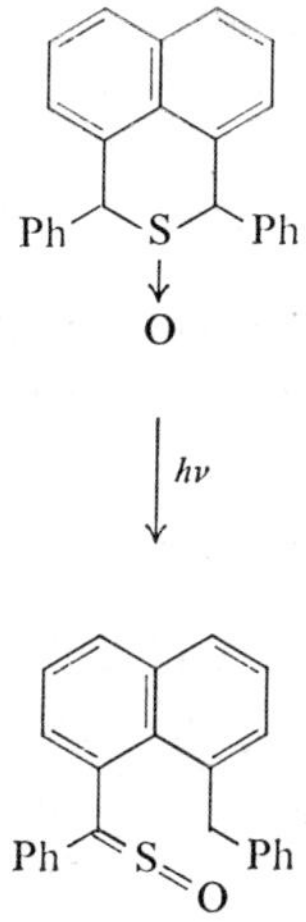

(74)

sulphine (74). All attempts to prepare very simple sulphines so far have been unsuccessful; sulphine ($CH_2=S=O$) and dimethylsulphine (thio-acetone-$S$-oxide) are unknown, although there is some evidence the latter may have a fleeting existence in solution.[192]

The larger sulphines, such as (72) and (73), are isolable, crystalline compounds, orange or yellow in colour. Evidence for the structure rests on analyses, i.r. absorption in the 1050 cm$^{-1}$ region characteristic of the S=O group, and for (72) the n.m.r. spectrum which showed a singlet at $-0.53\tau$, indicative of a highly deshielded proton. The course of some of the reactions of sulphines also served to verify the proposed structure. The stability of the sulphines is in marked contrast to the instability of the sulphenes.

At room temperature, the stable sulphines slowly decomposed. For example, (72) evolved sulphur dioxide to form difluorenylidene[192] and phenylsulphine afforded cis- and trans-stilbenes.[196] Oxidation of sulphines with peracid also resulted in the evolution of sulphur dioxide and the formation of the corresponding carbonyl compound. Thus, xanthione-$S$-oxide was converted to xanthone.[197] Photolysis of sulphines also resulted in formation of the carbonyl compound.[193, 195, 196] The photochemical synthesis of a sulphine from a sulphoxide appears to involve a triplet state while conversion of the sulphine into the carbonyl compound appears to involve a singlet state.[195] 2,4-Dinitrophenylhydrazine has been reported to react with sulphines[193, 196] to afford the hydrazone derivative of the corresponding carbonyl compound. In view of the mechanism of hydrazone formation and in view of the presumed polarization of the sulphine C=S

[196] A. M. Hamid and S. Trippett, *J. Chem. Soc.* (*C*), 1968, 1612.
[197] B. Zwanenburg, L. Thijs, and J. Strating, *Rec. Trav. chim.*, 1967, **86**, 577.

bond, it seems likely that the sulphine was first hydrolysed to the carbonyl compound which then reacted with the hydrazine derivative.

The polarization of the sulphine was indicated by the fact that oxidation occurred on carbon, not on sulphur.[197] Additional evidence for the positively-charged character of the sulphur atom was provided by the observation that phenylsulphine reacted with carbethoxymethylenetriphenylphosphorane to afford the carbethoxy-sulphinyl ylide (75; R = Ph).[196] This reaction is analogous to those reported for sulphenes.

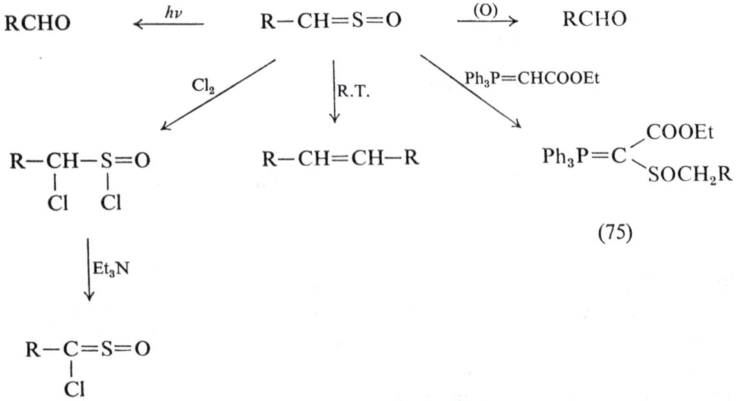

$$RCHO \xleftarrow{h\nu} R-CH=S=O \xrightarrow{(O)} RCHO$$

$$\swarrow Cl_2 \quad \downarrow R.T. \quad \searrow Ph_3P=CHCOOEt$$

$$R-CH-S=O \quad\quad R-CH=CH-R \quad\quad Ph_3P=C\begin{array}{c}COOEt\\SOCH_2R\end{array}$$
$$\ \ |\ \ \ \ |$$
$$\ \ Cl\ \ Cl$$

(75)

$$\downarrow Et_3N$$

$$R-C=S=O$$
$$\ \ |$$
$$\ \ Cl$$

Sulphines undergo addition reactions. Reaction with halogens (chlorine or bromide) resulted in straightforward addition to afford a halogenoalkylsulphinyl halide.[192, 198] Treatment of the adduct with triethylamine repeated the sulphine-forming reaction to afford a halogenoalkidene sulphoxide. Reaction of 9-fluorenylidenesulphine with chloranil resulted in addition across the dione system[198] whereas reaction of the thioketone (thiofluorenone) with chloranil resulted in 1,4-addition across the diene system. Sulphine addition to olefins did not result in the formation of four-membered rings as did sulphenes. For example, reaction of 9-fluorenylidenesulphine (72) with 1-morpholinocyclohex-1-ene afforded an adduct formulated as (76), mainly on the basis of i.r. absorption at 1710 cm$^{-1}$.[192] The full details of this reaction have not been published as yet, but it seems likely that the negative pole of the enamine reacted with the positive sulphur atom. In this instance failure to close the prospective four-membered ring might be attributed to the stability of the product, a sulphinyl carbanion. Repetition of the reaction, but using phenylsulphine,[196] indicates that this reason probably is valid since the latter compound afforded the sulphinated enamine (77), probably formed through an intermediate analogous to (76) which underwent a proton transfer. The same course of reaction was observed in the reaction between phenylsulphine and 1-pyrrolidinocyclohex-1-ene.[196] Phenylsulphine would not react with

[198] J. Strating, L. Thijs, and B. Zwanenburg, *Rec. Trav. chim.*, 1967, **86**, 641.

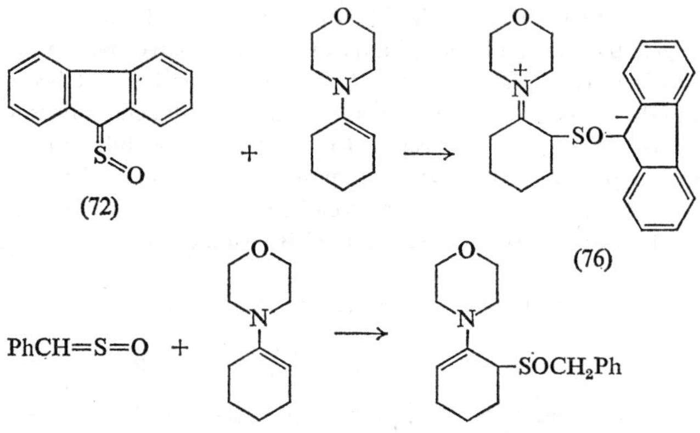

[Structures (72), (76), (77) shown]

PhCH=S=O +

aldehydes, acyl halides, isocyanates, or 1,3-dipolarophiles, all of which are known to react with many sulphur ylides.

## 7 Sulphur Imines

For each sulphur ylide, an isoelectronic sulphur-nitrogen compound can be visualized and many of these are known. The best known are the imino-sulphuranes (sulphidimines) and the imino-oxysulphuranes (sulphoximines). The chemistry of these substances was reviewed by Johnson[1] in 1966. Additional chemistry of these substances is reported in this section together with some chemistry of sulphonium di-imides, the nitrogen analogues of sulphones. Although their chemistry will not be discussed, mention should be made of the existence of a series of S-(dihalogeno)-N-substituted imino-sulphuranes. The N-substituents range from perfluoroalkyl to pentafluoro-sulphato[199-202] and the sulphur substituents are fluoro or chloro. Little of the chemistry of these compounds has been reported to date.

**Iminosulphuranes.**—No new methods have been reported for the preparation of iminosulphuranes (sulphidimines or sulphidimides), but there have been reported additional examples and slight modifications of the known methods. Saika et al.[203] reported reaction of N-chloroamines with sulphides. Okahara and Swern[204] reported the reaction of sulphides with N-sulphonato-sulphonamides, and Kucsman et al.[205] reported the reaction of amides with sulphides in the presence of sodium hypochlorite, all three of these procedures being slight modifications on one general route to iminosulphuranes, a

[199] M. Lustig, *Inorg. Chem.*, 1966, **5**, 1317.
[200] A. Muller, O. Glemser, and B. Krebs, *Z. Naturforsch.*, 1967, **22b**, 550.
[201] A. F. Clifford and R. G. Goel, *Inorg. Chem.*, 1969, **8**, 2004.
[202] A. F. Clifford and J. W. Thompson, *Inorg. Chem.*, 1966, **5**, 1424.
[203] D. Saika, H. S. Beilan, and D. Swern, Abstracts of 158th meeting of the American Chemical Society, 1969, Organic Division paper No. 155.
[204] M. Okahara and D. Swern, *Tetrahedron Letters*, 1969, 3301.
[205] A. Kucsman, F. Ruff, I. Kapovits, and J. G. Fischer, *Tetrahedron*, 1966, **22**, 1843.

displacement reaction by sulphide on a substituted amine carrying a leaving group. Many of these kinds of reactions have been proposed to proceed *via* nitrene intermediates.[204] However, this has not been proven to date and evidence from those reactions using *N*-halogenoamines indicates simple displacement first occurs to form amino-sulphonium salts.[203] Reaction of phenylsulphide with ammonium trichloride afforded the bis-imine (78) by the same route.[206] Neidlein and Heukelbach [207] reported additional examples of the reaction of isocyanates with sulphoxides to

$$Ph_2S + NCl_3 \longrightarrow [Ph_2S\!\!=\!\!N\!\!=\!\!SPh_2]^+ \; Cl_3^-$$
(78)

(79)

afford imines, and Abramovitch *et al.*[208] reported the intramolecular reaction between a sulphonyl azide and a sulphide to afford a cyclic imine (79).

Little has been reported regarding the physical characteristics of the iminosulphuranes. Kucsman *et al.*[205, 209] have studied the i.r. spectra of these compounds and have reported the S=N absorption appears at 780—820 cm$^{-1}$ for *N*-acyl derivatives and at 935—1010 cm$^{-1}$ for *N*-sulphonyl derivatives. This group has attempted to draw conclusions regarding details of the electronic interactions in the imines but on the basis of extremely small frequency shifts. Kalman [210] has reported an *X*-ray crystallographic analysis of *N*-(methylsulphonyl)iminodimethylsulphurane. He reported a sulphonium–nitrogen bond distance of 1·63 Å, somewhat longer than analogous distances reported later in this section. The other bond distances and angles seemed normal.

Little chemistry of the iminosulphuranes had been reported during 1965, and only very little has been reported since then. Cogliano and Braude [211] reported that iminodiethylsulphurane reacted with chloramine in an alkylation-type reaction proceeding by chloride displacement. However, later work by Laughlin and Yellin [212] and an *X*-ray analysis by Webb and Gloss [213] have shown that alkylation occurred on sulphur, not on the

[206] R. Appel and G. Buchler, *Annalen*, 1965, **684**, 112.
[207] R. Neidlein and E. Heukelbach, *Arch. Pharm.*, 1966, **299**, 64.
[208] R. A. Abramovitch, C. I. Azogu, and I. T. McMaster, *J. Amer. Chem. Soc.*, 1969, **91**, 1219.
[209] A. Kucsman, F. Ruff, and I. Kapovits, *Tetrahedron*, 1966, **22**, 1575.
[210] A. Kalman, *Acta Cryst.* 1967, **22**, 501.
[211] J. A. Cogliano and G. L. Braude, *J. Org. Chem.*, 1964, **29**, 1397.
[212] R. G. Laughlin and W. Yellin, *J. Amer. Chem. Soc.*, 1967, **89**, 2435.
[213] N. C. Webb and R. A. Gloss, *Tetrahedron Letters*, 1967, 1043.

nitrogen of the imine. Johnson et al.[214] effected the $N$-methylation and $N$-ethylation of iminosulphuranes utilizing trialkyloxonium salts. Other than for this alkylation, iminosulphuranes have not been found to undergo the typical reactions known for ylides, reactions such as those with aldehydes, epoxides, and $\alpha,\beta$-unsaturated olefins.[203]

**Imino-oxysulphuranes.**—Once again, no completely new methods have been reported for the preparation of imino-oxysulphuranes(sulphoximines). Okahara and Swern[204] found that in the presence of triethylamine, a sulphonatosulphonamide reacted with DMSO, supposedly *via* a nitrene intermediate, to afford the sulphoximine (80). At about the same time,

$$PhSO_2NH-OSO_2Ar \xrightarrow{Et_3N, DMSO}$$
$$PhSO_2N_3 \xrightarrow{Cu, DMSO} PhSO_2\bar{N}-\overset{+}{S}(O)Me_2 \quad (80)$$

Carr et al.[215] found that chloramine-T would react with DMSO in the presence of copper to afford the $p$-tolyl analogue. Earlier, Kwart and Kahn[216] had reported the reaction of sulphonyl azides with DMSO, also in the presence of copper, to be an effective synthetic method. The same reaction had been reported earlier under thermal or photochemical conditions but it proceeded in low yield. It seems likely that the azide reaction does involve a nitrene or nitrene–copper complex intermediate.

In 1965 Fusco and Tenconi[217] prepared phenylmethylsulphoximine (iminophenylmethyloxysulphurane) by known methods and reported its optical resolution using camphor-10-sulphonic acid. Oae and coworkers[218, 219] determined the $pK_a$ values of the conjugate acids of a series of iminomethyl(substituted phenyl)oxysulphuranes
[R—C$_6$H$_4$—S(O)(NH)Me] and found them to correlate with sigma, giving a $\rho$-value of $-1\cdot49$. Therefore, there appears to be both inductive and conjugative interaction through the nitrogen–sulphur–aryl ring system. These workers also reported the i.r., n.m.r., and mass spectra of these compounds.[218]

Methionine sulphoximine had earlier been identified as the toxic factor from agenized zein and wheat flour[220] and had been found to inhibit glutamine synthetase. This compound has been synthesized from L-methionine by oxidation to the sulphoxide then treatment with sodium

---

[214] C. R. Johnson, J. J. Rigau, M. Haake, D. McCants, J. E. Keiser, and A. Gertsema, *Tetrahedron Letters*, 1968, 3719.
[215] D. Carr, T. P. Seden, and R. W. Turner, *Tetrahedron Letters*, 1969, 477.
[216] H. Kwart and A. A. Kahn, *J. Amer. Chem. Soc.*, 1967, **89**, 1950.
[217] R. Fusco and F. Tenconi, *Chimica e Industria*, 1965, **47**, 61.
[218] N. Furukawa, K. Tsujihara, Y. Kawakatsu, and S. Oae, *Chem. and Ind. (London)*, 1969, 266.
[219] S. Oae, K. Tsujihara, and N. Furukawa, *Chem. and Ind. (London)*, 1968, 1569.
[220] H. R. Bentley, E. E. McDermott, T. Moran, J. Pace, and J. K. Whitehead, *Proc. Roy. Soc.*, 1950 B, **137**, 402.

azide in sulphuric acid.[221] The sulphoximine was separated into two diastereoisomers and the X-ray analysis of the 2(S),S(R) isomer proved the structure to be (81).

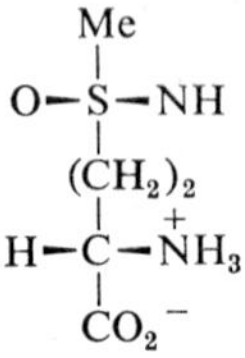

$$\begin{array}{c} \text{Me} \\ | \\ \text{O}-\text{S}-\text{NH} \\ | \\ (\text{CH}_2)_2 \\ | \quad + \\ \text{H}-\text{C}-\text{NH}_3 \\ | \\ \text{CO}_2^- \end{array}$$

(81)

Once again very little chemistry has been reported for the sulphoximines. Johnson et al.[214] reported the methylation of phenylmethylsulphoximine with trimethyloxonium tetrafluoroborate. Treatment of the product of this reaction with aqueous base afforded a new sulphoximine, N-methylimino-phenylmethyloxysulphurane. Similarly, Schmidbaur and Kammel [222] found dimethylsulphoximine to undergo silylation with trimethylsilyl chloride, and in the presence of triethylamine the product lost a proton to afford the new sulphoximine, N-(trimethylsilyl)iminodimethyloxysulphurane. Rave and Logan [223] removed the proton from dimethylsulphoximine and then reacted it with dodecyldimethylphosphine dichloride. In a sense a

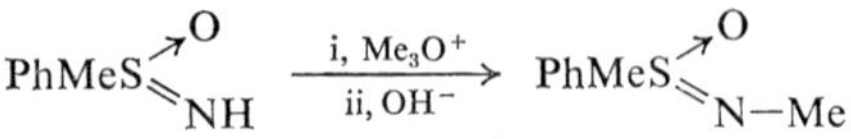

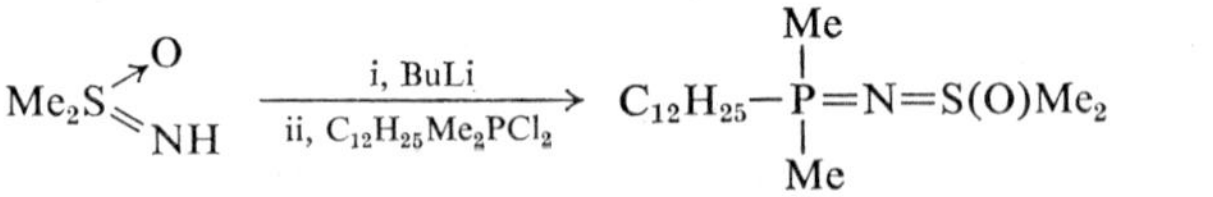

PhMeS(O)(NH) $\xrightarrow[\text{ii, OH}^-]{\text{i, Me}_3\text{O}^+}$ PhMeS(O)(N–Me)

Me$_2$S(O)(NH) $\xrightarrow[\text{ii, C}_{12}\text{H}_{25}\text{Me}_2\text{PCl}_2]{\text{i, BuLi}}$ C$_{12}$H$_{25}$–P(Me)(Me)=N=S(O)Me$_2$

(82)

$\downarrow$ H$_2$O

C$_{12}$H$_{25}$–P(Me)(Me)→O + Me$_2$S(O)(NH)

PhMeS(O)(NH) + CH$_2$=CHCOOMe $\xrightarrow{\text{NaH}}$ PhMeS(O)(N–CH$_2$CH$_2$COOMe)

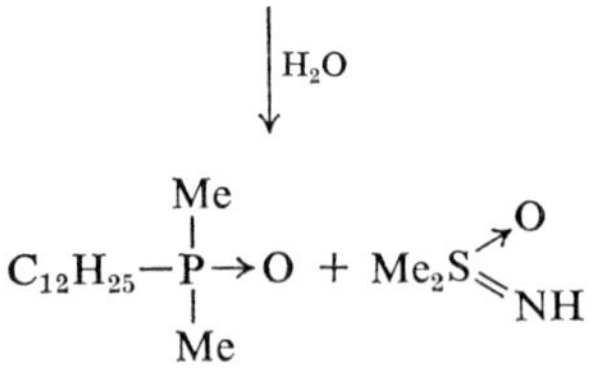

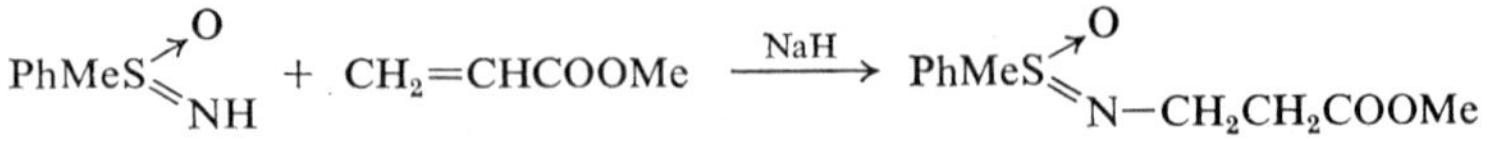

[221] B. W. Christensen, A. Kjaer, S. Neidle, and D. Rogers, Chem. Comm., 1969, 169.
[222] H. Schmidbaur and G. Kammel, Chem. Ber., 1969, **102**, 4128.
[223] T. W. Rave and T. J. Logan, J. Org. Chem., 1967, **32**, 1629.

double alkylation occurred to afford the mixed iminophosphorane-sulphoximine (82). The stability of the S=N bond in sulphoximines was clearly evidenced by the fact that (82), under basic or acidic conditions, underwent hydrolysis to the phosphine oxide and the unsubstituted sulphoximine. In the only reaction involving multiply-bonded compounds, phenylmethylsulphoximine was found to react with methyl acrylate in a normal Michael addition, the presence of base converting the initial adduct to a new sulphoximine and thereby effecting an overall alkylation of the sulphoximine.[214]

**Sulphurdi-imines.**—Sulphurdi-imines are the nitrogen analogues of sulphones, just as sulphidimines(iminosulphuranes) are the nitrogen analogues of sulphoxides. In an early synthesis attempt Cogliano and Braude [211] reported that reaction of diethyl sulphide with hydroxylamine-O-sulphonic acid afforded the known iminodiethylsulphurane which, upon reaction with chloroamine, was claimed to afford diethylhydrazinosulphonium chloride. A dimeric structure was proposed for the salt, and treatment with sodium methoxide was claimed to afford a dimer (83) of the free base. These workers claimed the same products resulted directly from the reaction of diethyl sulphide with chloroamine. Somewhat later, Appel

$$Et_2S\overset{+}{\underset{-}{\diagup}}\overset{NH-\overset{-}{N}H}{\underset{NH-NH}{\diagdown}}\overset{+}{\diagdown}SEt_2$$

(83)

$$\underset{Me}{\overset{Me}{\diagdown}}S\overset{\nearrow NH}{\underset{\searrow NH}{}}$$

(84)

et al.[224] reported that reaction of iminodimethylsulphurane with chloroamine afforded an amination product which, in the presence of sodium ethoxide, was converted to bis-iminodimethylsulphurane (84). This substance gave a mass spectrum similar to that of dimethylsulphoximine, and it was inert to oxidation (iminosulphuranes are known to be oxidizable to sulphoximines). Laughlin and Yellin [212] found that (84) could be prepared directly by reaction of methyl sulphide with chloroamine in acetonitrile–ammonia. These workers also supported the di-imine structure on the basis of thermal stability, hydrolytic stability comparable to that of sulphoximines rather than to sulphidimines, by comparison of $pK_a$ values and by Raman spectroscopy. A final determination of the structure of the di-imines was provided by the X-ray analysis of (84) reported by Webb and Glass.[213] They reported a distorted tetrahedron about sulphur (C–S–C angle of 103° and N–S–N angle of 122°) and a sulphur–nitrogen bond distance of 1·53 Å. The sulphur–carbon bond distance was 0·05 Å shorter than expected.

The di-imines are inert to basic hydrolysis, but one of the imine groups can be hydrolysed in acidic media, affording a sulphoximine.[225] The

[224] R. Appel, H. W. Felhaber, D. Hanssgen, and R. Schollhorn, *Chem. Ber.*, 1966, **99**, 3108.
[225] R. G. Laughlin, *J. Amer. Chem. Soc.*, 1968, **90**, 2651.

imine–sulphur bond is even more resistant to hydrolytic cleavage than an amine–sulphur bond. The di-imines are sufficiently basic to act as nucleophiles, since (84) would undergo double silylation with trimethylchlorosilane, and in triethylamine the alkylation product was deprotonated to

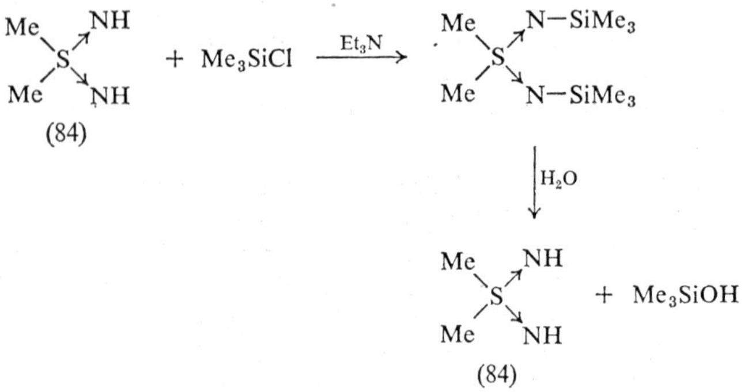

$$\begin{array}{c}\text{Me}\diagdown\text{S}\diagup^{\text{NH}}_{\text{NH}}\\ \text{Me}\diagup\text{S}\diagdown^{\text{NH}}_{\text{NH}}\\ (84)\end{array} + \text{Me}_3\text{SiCl} \xrightarrow{\text{Et}_3\text{N}} \begin{array}{c}\text{Me}\diagdown\text{S}\diagup^{\text{N}-\text{SiMe}_3}\\ \text{Me}\diagup\text{S}\diagdown^{\text{N}-\text{SiMe}_3}\end{array}$$

$$\downarrow \text{H}_2\text{O}$$

$$\begin{array}{c}\text{Me}\diagdown\text{S}\diagup^{\text{NH}}\\ \text{Me}\diagup\text{S}\diagdown^{\text{NH}}\\ (84)\end{array} + \text{Me}_3\text{SiOH}$$

afford a new bis(N-trimethylsilylimino)dimethylsulphurane.[226] Interestingly, this di-imine could be hydrolysed by water, but with the cleavage only of the silicon–nitrogen bonds, not the nitrogen–sulphur bonds. Appel and Ross[227] treated (84) with potassium amide in order to deprotonate the di-imine, then the latter was ethylated with ethyl bromide to N-ethyldi-iminodimethylsulphurane. Strenuous treatment of (84) with potassium amide led gradually to replacement of the methyl groups with imine groups, eventually forming $K_3SN_4H_3$ and $K_6S_2N_7H_3$.[228] The di-imine (84) was sufficiently nucleophilic to react with bromine, affording N-bromodi-iminodimethylsulphurane.[224] The bromide was a useful reactive intermediate since the bromine was readily displaced, for example by phosphines or sulphides.

**Miscellaneous Properties of Sulphur Imines.**—Two groups have reported $pK_a$ data on sulphur imines and their data agree remarkably well.[212, 225, 229] In the dimethylsulphur series, the basicity of the compounds decreases in the order $Me_2SNH > Me_2S(NH)_2 > Me_2S(O)(NH)$. However, in the diethylsulphur series basicity decreases in the order $Et_2S(NH)_2 > Et_2SNH$. Cross comparisons reveal that basicity decreases in the following series as shown: $Me_2SNH > Et_2SNH$, but $Et_2S(NH)_2 > Me_2S(NH)_2$. It would be expected that the order first shown above for the dimethylsulphur series is the 'normal' order in that oxygen is more electronegative than nitrogen, thereby making the sulphur more positive and better able to delocalize the negative charge on nitrogen. Furthermore, in the same sense that oxysulphonium ylides are less basic than sulphonium ylides, the di-imines

---

[226] R. Appel, L. Siekmann, and H. D. Hoppen, *Chem. Ber.*, 1968, **101**, 2861.
[227] R. Appel and B. Ross, *Chem. Ber.*, 1969, **102**, 1020.
[228] R. Appel and B. Ross, *Chem. Ber.*, 1969, **102**, 3769.
[229] F. Knoll, J. Gronebaum, and R. Appel, *Chem. Ber.*, 1969, **102**, 848.

would be expected to be less basic than the imines. A delicate balancing of steric, hyperconjugative, and inductive effects is probably responsible for the inconsistent position of the diethylsulphur derivatives in the basicity sequences.

The stereochemistry about sulphur in sulphoxides and sulphur imines has been of considerable interest in the past five years. Commencing with the report by Fusco and Tenconi [217] that phenylmethylsulphoximine could be resolved, and the early observations by Day and Cram on the stereochemistry of the interconversion of sulphoxides and iminosulphuranes,[230] the stereochemistry of the entire sulphide–sulphoxide–sulphidimine–sulphoximine cycle has been elaborated. The Scheme indicates the general-

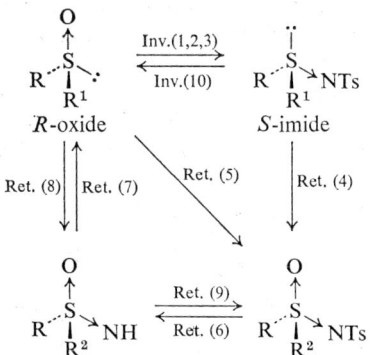

Ret. = with retention; Inv. = with inversion

1. Ref. 231; (TsN)$_2$S
2. Ref. 232; TsNH$_2$+P$_2$O$_5$
3. Ref. 230, 233; TsNCO or TsNSO
4. Ref. 231—234; KMnO$_4$
5. Ref. 231, 233; TsN$_3$+Cu
6. Ref. 231; H$_2$SO$_4$
7. Ref. 231; NOPF$_6$
8. Ref. 233; NaN$_3$+H$_2$SO$_4$
9. Ref. 231, 233; TsCl
10. Ref. 230, 231, 233; KOH

**Scheme**

ized structure of the compounds, the stereochemistry of each step, and the nature of the reagents employed together with a reference for each reagent. Four separate groups [231–234] have studied these interconversions using three different specific starting compounds, but the stereochemical results all were the same with but one exception. Christensen and Kjaer [234] reported that the conversion of methionine sulphoxide to the N-tosyl sulphidimide using N-sulphinyltoluenesulphonamide occurred with retention of configuration rather than with inversion as three other groups reported. It

[230] J. Day and D. J. Cram, *J. Amer. Chem. Soc.*, 1965, **87**, 4398.
[231] D. Rayner, D. M. Von Schriltz, J. Day, and D. J. Cram, *J. Amer. Chem. Soc.*, 1968, **90**, 2721.
[232] M. A. Sabol, R. W. Davenport, and K. K. Andersen, *Tetrahedron Letters*, 1968, 2159.
[233] C. R. Johnson and J. J. Rigau, *J. Org. Chem.*, 1968, **33**, 4340.
[234] B. W. Christensen and A. Kjaer, *Chem. Comm.*, 1969, 934.

is conceivable that the other asymmetric centre in methionine sulphoxide influenced the course of this reaction. The configurations in the methionine series cannot be questioned since the structure of one of the diastereoisomers of the sulphoximine has been determined by $X$-ray crystallographic analysis.[221]

## 8 Sulphonyl and Sulphinyl Amines

Sulphinyl amines ($R-N=S=O$) and sulphonyl amines ($R-N=SO_2$) are the nitrogen analogues of sulphines and sulphenes, respectively (see section 6). Since the latter two groups of compounds have been shown to exhibit nucleophilic and even ylidic characteristics, the same might be expected of sulphonyl and sulphinyl amines. The chemistry of sulphinyl amines recently has been reviewed by Kresze and Wucherpfennig.[235]

Only one report has appeared regarding the chemistry of sulphonyl amines. Atkins and Burgess [236] reported the dehydrohalogenation of the $N$-sulphonyl chloride of benzamide with triethylamine to afford a solution of $N$-sulphonylbenzamide (85). The latter reacted with ethylamine to

$$PhCONHSO_2Cl \xrightarrow{Et_3N} [PhCON=SO_2]$$
$$(85)$$

with EtOCH=CH$_2$ giving EtO−CH(CH$_2$−SO$_2$)−N−COPh

with EtNH$_2$ giving PhCONH−SO$_2$−NHEt

afford a sulphone diamide and reacted with ethoxyethylene to afford a cyclic sulphonamide. From the orientation of the cycloaddition reaction it appears that the nitrogen atom of (85) is the negative centre of the molecule. Similar chemistry was reported for the sulphonyl chloride of ethylamine.

No new syntheses of $N$-sulphinyl amines have been reported. As a result, most of the chemistry of this class of compounds is effected using either $N$-sulphinylaniline(thionyl aniline) or $N$-sulphinylsulphonamides. The conversion of $N$-sulphinyl amines to sulphodi-imides, $R-N=S=N-R^1$, has been reported to occur simply upon heating [237] or upon treatment with base.[238] In both instances, it appears likely that the nucleophilic centre of one molecule (nitrogen) attacked the electrophilic centre (sulphur) of a second molecule, to be followed by four-centre elimination of sulphur dioxide. This reaction presents a most useful method for the synthesis of sulphodi-imides.

---

[235] G. Kresze and W. Wucherpfennig, *Angew. Chem. Internat. Edn.*, 1967, **6**, 149.
[236] G. M. Atkins and E. M. Burgess, *J. Amer. Chem. Soc.*, 1967, **89**, 2502.
[237] W. Wucherpfennig and G. Kresze, *Tetrahedron Letters*, 1966, 1671.
[238] T. Minami, H. Miki, H. Matsumoto, Y. Ohshiro, and T. Agawa, *Tetrahedron Letters*, 1968, 3049.

*Ylides of Sulphur, Selenium, and Tellurium and Related Structures* 307

Numerous reactions of N-sulphinyl amines are known in which sulphur dioxide, or a closely related highly stable molecule, is eliminated through an apparent four-centre reaction. For example, reaction of thionyl aniline with nitrogen dioxide afforded benzenediazonium nitrate and sulphur dioxide, while reaction with nitrosyl chloride afforded benzenediazonium chloride.[239] Reaction with thioureas lead to the extrusion of $S_2O$ and the formation of guanidines,[240] while reaction with phosphine oxides or phosphine sulphides lead to extrusion of $SO_2$ or $S_2O$, respectively, and formation of iminophosphoranes.[241] Reaction with diphenylnitrone appeared to proceed *via* nucleophilic attack of the sulphinyl amine to

$$R-N=S=O \xrightarrow{\Delta} \begin{bmatrix} R-N=S-O^- \\ \phantom{R-N}|\phantom{S}\overset{+}{\phantom{S}} \\ R-N-S=O \end{bmatrix} \xrightarrow{-SO_2} R-N=S=N-R$$

$$Ph-N=S=O + N_2O_4 \text{ or } NOCl \xrightarrow{-SO_2} PhN_2^+$$

$$R-N=S=O + Ph_3PO \longrightarrow \begin{bmatrix} Ph_3P-O^- \\ \phantom{Ph_3P}|\phantom{O}\overset{+}{\phantom{O}} \\ R-N-SO \end{bmatrix} \xrightarrow{-SO_2} Ph_3P=N-R$$

$$\begin{array}{c} PhSO_2-N=S=O \\ + \\ PhCH=\overset{+}{N}\diagup^{O^-}_{\diagdown Ph} \end{array} \longrightarrow \begin{bmatrix} PhSO_2-N-\overset{+}{S}=O \\ \phantom{PhSO_2}|\phantom{N}\phantom{S} \\ Ph-CH-N-O^- \\ \phantom{Ph-CH-N}|\phantom{O} \\ \phantom{Ph-CH-N}Ph \end{bmatrix} \xrightarrow{-SO_2} PhSO_2N=CH-NPh_2$$

(86)

afford a Schiff base, probably *via* a zwitterionic intermediate (86) which extrudes sulphur dioxide and undergoes a phenyl migration.[242]

A few examples are known of reactions in which it appears that sulphinyl amines reacted, initially at least, as electrophiles. Pyridine-N-oxide attacked the sulphur atom of N-sulphinylbenzenesulphonamide to produce a zwitterionic intermediate, $Ph-SO_2-\bar{N}-S(O)-O-\overset{+}{N}C_5H_5$.[242] N-Sulphinyltosylamide underwent attack by N-phenyliminotriphenyl-phosphorane in benzene solution to afford N-sulphinylaniline and N-tosyliminotriphenylphosphorane.[241] The net effect of this reaction was exchange of nitrogen substituents and the reaction proceeded under those conditions to

---

[239] M. Kobayashi and K. Honda, *Bull. Chem. Soc. Japan*, 1966, **39**, 1778.
[240] A. Senning, *Acta Chem. Scand.*, 1967, **21**, 1293.
[241] A. Senning, *Angew. Chem. Internat. Edn.*, 1965, **4**, 357; *Acta Chem. Scand.*, 1965, **19**, 1755.
[242] R. Albrecht and G. Kresze, *Chem. Ber.*, 1965, **98**, 1205.

afford the most stable phosphonium imine. Finally, *cis*- and *trans*-1-ethoxy-1-butene underwent stereospecific *cis* addition by *N*-sulphinyl-*p*-chlorobenzenesulphonamide to afford the four-membered sulphinamides (87).[243]

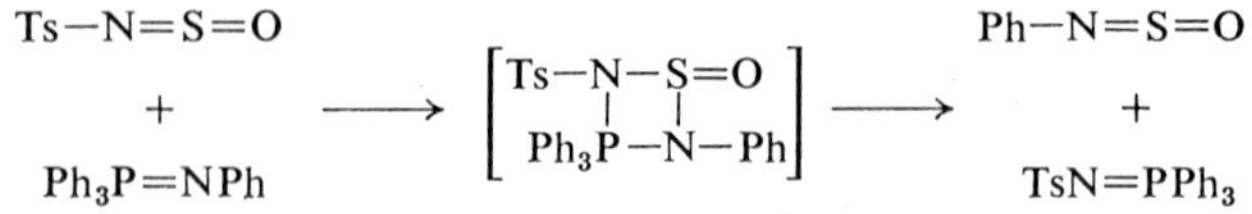

$$\begin{array}{c} \text{Ts}-\text{N}=\text{S}=\text{O} \\ + \\ \text{Ph}_3\text{P}=\text{NPh} \end{array} \longrightarrow \begin{bmatrix} \text{Ts}-\text{N}-\text{S}=\text{O} \\ | \quad | \\ \text{Ph}_3\text{P}-\text{N}-\text{Ph} \end{bmatrix} \longrightarrow \begin{array}{c} \text{Ph}-\text{N}=\text{S}=\text{O} \\ + \\ \text{TsN}=\text{PPh}_3 \end{array}$$

$$\begin{array}{c} \text{ArSO}_2-\text{N}=\text{S}=\text{O} \\ + \\ \text{EtOCH}=\text{CHEt} \end{array} \longrightarrow \begin{array}{c} \text{ArSO}_2\text{N}\!\!-\!\!\!-\!\!\text{S}=\text{O} \\ | \qquad | \\ \text{EtO}-\text{CH}-\text{CH}-\text{Et} \\ (87) \end{array}$$

Here again, the orientation of the addition seems to indicate that the sulphur atom is the positive end of the N=S dipole.

### 9 Selenonium Ylides

Although it would be expected that the chemistry of organoselenides and their derivatives would be very similar to that of organic sulphides, practically no work has been reported on such derivatives that might lead to selenonium ylides or related compounds. The only report in the literature to date regarding such ylides is the finding by Lloyd and Singer [244] that their synthesis of sulphonium and other ylides, achieved by heating a sulphide with tetraphenyldiazocyclopentadiene causing elimination of nitrogen, could be applied to diphenylselenide resulting in a 93% yield of diphenylselenoniumtetraphenylcyclopentadienylide (88). The ylide was stable in air

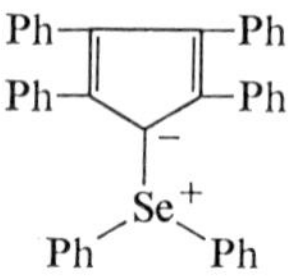

(88)

in the dark but decomposed upon exposure to light. It was very weakly basic, being insoluble in dilute hydrochloric acid. No chemical reactions were reported for the ylide.

[243] F. Effenberger and G. Kiefer, *Angew. Chem. Internat. Edn.*, 1967, **6**, 951.
[244] D. Lloyd and M. I. C. Singer, *Chem. and Ind. (London)*, 1967, 390.

# 7
# Heterocyclic Compounds of Quadricovalent Sulphur

## 1 Introduction

Heterocyclic compounds reviewed in this chapter are those for which the only possible conventional neutral covalent formulations contain quadricovalent sulphur. It is emphasised that the correctness of using such structures may be open to question, and the decision to use them in this review was made primarily for convenience.

Since the chemistry of the heterocyclic systems discussed in this chapter has hitherto been reviewed [1] only briefly and in part, it was decided to review the subject fully and include all older material as well as that which has appeared during the past year.

## 2 Thiabenzenes

Several thiabenzenes (1)—(5) have been known for some time.[2-5] All have been synthesised by the same method, which involves the reaction of a thiapyrylium, thianaphthalenium, or thiaxanthylium perchlorate with

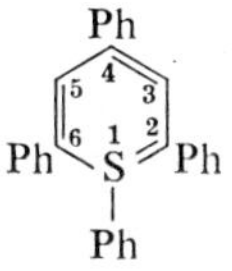

(1)

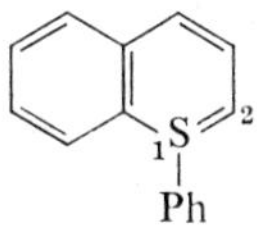

(2)

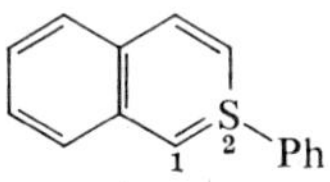

(3)

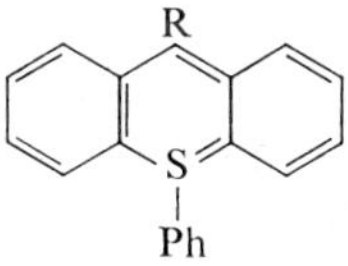

(4) R = H
(5) R = Ph

---

[1] W. G. Salmond, *Quart. Rev.*, 1968, **23**, 253.
[2] G. Suld and C. C. Price, *J. Amer. Chem. Soc.*, 1961, **83**, 1770.
[3] G. Suld and C. C. Price, *J. Amer. Chem. Soc.*, 1962, **84**, 2090.
[4] G. Suld and C. C. Price, *J. Amer. Chem. Soc.*, 1962, **84**, 2094.
[5] C. C. Price, M. Hori, T. Parasaran, and M. Polk, *J. Amer. Chem. Soc.*, 1963, **85**, 2278.

phenyl-lithium. The tetraphenylthiabenzene (1) is an unstable violet amorphous solid [2,4] which reacts readily with oxygen to give an unstable peroxide formulated as (6). Decomposition of this peroxide with hydrogen chloride liberated phenyl mercaptan together with the pyrylium oxide (7). The structure of the zwitterion (7) followed from an independent synthesis.[4] Formation of phenyl mercaptan establishes that a carbon–sulphur bond is formed in the reaction of the perchlorate with phenyl-lithium. 1,2,4,6-Tetraphenylthiabenzene (1) rearranges slowly and spontaneously to the isomer (8).

In contrast to compound (1) the ring homologues (2)—(5) are red-brown solids, much more stable to heat, light, and oxygen.[5] Thus compound (5) is stable to boiling acetic acid, and extended treatment of compound (3) with oxygen followed by hydrogen chloride gave a small amount only of phenyl mercaptan. In agreement with structures (3) and (5) Raney nickel desulphurisation of compound (3) gave a mixture of toluene and o-ethyltoluene, and similar treatment of 9,10-diphenyl-10-thia-anthracene (5) gave triphenylmethane.

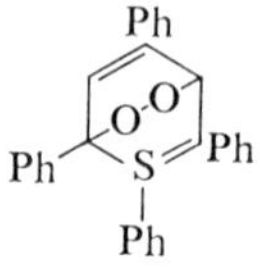

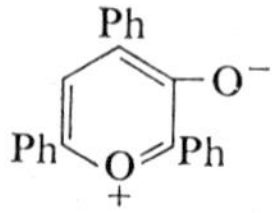

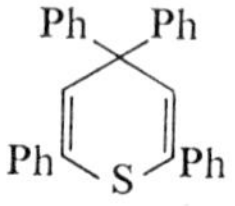

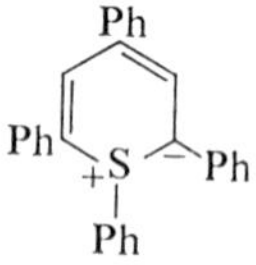

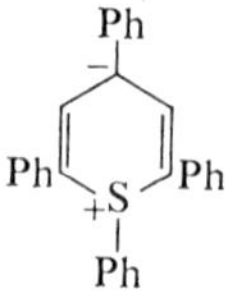

Dipole moment studies favour the quadricovalent sulphur formulations (1)—(5) and suggest that the contribution of dipolar structures, such as (9a) and (9b) in the case of 1,2,4,6-tetraphenylthiabenzene (1), is minimal. The moments of compounds (1)—(5) (in benzene) are small, and lie in the range 1·5—1·9 D. For comparison the dipole moment (6·2 D) of 9-dimethylsulphoniofluorenylide (10) may be cited.[6] Compounds (1)—(5) show $^1$H n.m.r. absorption in the region δ7—7·5 characteristic of aromatic-type hydrogen, whereas the hetero-ring proton signal of compound (8) appears at higher field (δ6·0). It is concluded that the thiabenzene ring is an aromatic-type π-electron system.

Acceptance of the neutral quadrivalent structures (1)—(5) raises the question of the nature of the bonding at the sulphur atom in these compounds. Price considers two possibilities.[5] (A) The sulphur atom uses its three 3p-orbitals for σ-bonding (to C-2, C-6, and the S-phenyl group), retains its unshared electron pair in the 3s-orbital, and utilises one (or

[6] G. M. Phillips, J. M. Hunter, and L. E. Sutton, *J. Chem. Soc.*, 1945, 146.

perhaps two) d-orbitals for cyclic conjugation. (B) Alternatively, three sulphur $sp^2$ hybrid orbitals are used in σ-bonding (to C-2, C-6, and the S-phenyl group), the unshared electrons are promoted to a d-orbital, and

Me₂S+      Ph P Ph      S Ph

(10)      (11)      (12)

the remaining $3p_z$ orbital is available for cyclic conjugation. The π-bonding at the sulphur atom in arrangement (A) recalls that suggested for the phosphorus atom in the phosphorins, for example (11),[7] in which cyclic conjugation through d-orbitals is unavoidable. The orbital arrangment for cyclic conjugation in (B) is similar to that in pyridine. Emphasising the marked difference in stability and colour between 1,2,4,6-tetraphenylthiabenzene (1) and the thiabenzenes (2)—(5), Price concludes that the bonding at sulphur in compound (1) differs from that in compounds (2)—(5), and is a consequence of steric hindrance to coplanarity of the phenyl groups at positions 2, 1, and 6 in compound (1). Bonding arrangement (A), involving mainly unhybridised p-orbitals, accommodates the distortion from coplanarity in compound (1). Cyclic conjugation is therefore of the $3d_\pi$–$2p_\pi$ type. The bonding arrangement (B), involving the more efficient $3p_\pi$–$2p_\pi$ overlap, is preferred, however, in the case of compounds (2)—(5) in which steric effects are unimportant.

More recently,[8] the synthesis and properties of 1-phenylthiabenzene (12) have been reported. 1-Phenylthiabenzene, prepared by treating thiapyrylium perchlorate or iodide with phenyl-lithium, is a red-brown amorphous solid, more similar in properties to compounds (2)—(5) than to compound (1). It is stable to light and oxygen, has a low dipole moment ($\mu$ = 0·8 D), and shows ¹H n.m.r. absorption at δ7·2. Its u.v. spectrum [$\lambda_{max}$ (log ε) 246 (3·63) and 202 (4·01) nm] is similar to that of biphenyl [$\lambda_{max}$ (log ε) 250 (4·26) and 205 (ca. 4·50) nm], and its characteristic colour is due to a weak absorption tailing into the visible region. The bonding at the sulphur atom and the nature of the cyclic π-electron system in 1-phenylthiabenzene are considered to be of the type in the derivatives (2)—(5).

In further studies on the effect of steric factors on the stability of thiabenzenes, the derivatives (13) and (14) were synthesised [9] from 2-thianaphthalenium perchlorate as illustrated in Scheme 1. 1,2-Diphenyl-2-thianaphthalene (13) could be isolated as a purple solid which, however, reacted rapidly with oxygen and could not be obtained analytically pure.

[7] G. Märkl, Angew. Chem. Internat. Edn., 1965, 4, 1023.
[8] M. Polk, M. Siskin, and C. C. Price, J. Amer. Chem. Soc., 1969, 91, 1206.
[9] C. C. Price and D. H. Follweiler, J. Org. Chem., 1969, 34, 3202.

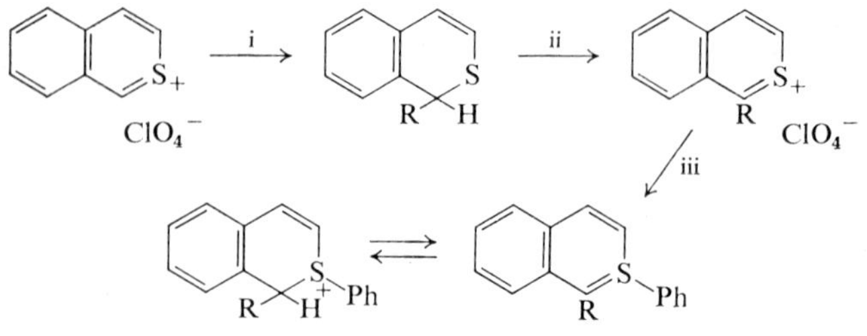

(15) R = Ph
(16) R = Bu$^t$
(13) R = Ph
(14) R = Bu$^t$

Reagents: i, RMgBr; ii, SO$_2$Cl$_2$–HClO$_4$; iii, PhLi

Scheme 1

Also, unlike compounds (2)—(5), it protonates reversibly in aqueous acid (p$K_a$ 2·9) to form the conjugate acid (15). 1-t-Butyl-2-phenyl-2-thianaphthalene (14) could not be isolated, but hydrolysis of the mixture from the reaction of 1-t-butyl-2-thianapthalenium perchlorate with phenyllithium in the presence of ammonium chloride gave the chloride of the cation (16). This salt could be deprotonated only by powerful bases such as potassium ethoxide in DMSO, indicating a p$K_a$ in the range 20—25. The resulting thiabenzene (14), moreover, could not be isolated, and decomposed rapidly in solution. The progressive destabilisation caused by steric effects of substituents in passing along the series (3) → (13) → (14) is ascribed to a diminished contribution from planar $sp^2$ geometry at sulphur which allows participation of the sulphur $3p_z$ orbital in the aromatic $\pi$-bonding. In the sterically favoured non-planar $p^3$ geometry only 3$d$-orbitals on sulphur would be free to participate in the aromatic $\pi$-bonding. It is significant that the phosphorins,[7] which must also use $d$-orbitals for aromatic $\pi$-bonding, are protonated reversibly even in the absence of steric effects.

## 3 Thiabenzene-1-oxides

Several thiabenzene-1-oxides have been synthesised, in all cases by routes involving at some stage the use of dimethyloxosulphonium methylide. 2-Benzoyl-5-hydroxy-1-methyl-3-phenylthiabenzene-1-oxide (17) and its methyl ether (18) were obtained [10] by the reactions outlined in Scheme 2. Dimethyloxosulphonium methylide in DMSO also reacted with benzoylphenylacetylene to give 1-methyl-3,5-diphenylthiabenzene-1-oxide (19),[11] with 2-acylcyclohexanones to form a series of tetrahydrobenzo[c]-1-thiabenzene-1-oxides (20),[12] and with benzonitrile to form 1-methyl-3,5-

---

[10] Y. Kishida and J. Ide, *Chem. and Pharm. Bull. (Japan)*, 1967, **15**, 360.
[11] A. G. Hortmann, *J. Amer. Chem. Soc.*, 1965, **87**, 4972.
[12] B. Holt, J. Howard, and P. A. Lowe, *Tetrahedron Letters*, 1969, 4937.

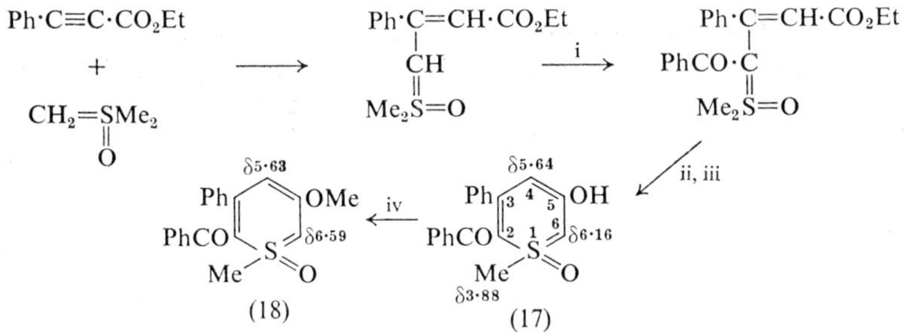

Ph·C≡C·CO₂Et + CH₂=SMe₂ → Ph·C=CH·CO₂Et, CH, Me₂S=O →i Ph·C=CH·CO₂Et, PhCO·C, Me₂S=O

(18) Ph, PhCO, Me, S=O, OMe  δ5·63, δ6·59  ←iv  (17) Ph, PhCO, Me, S=O, OH  δ5·64, δ6·16, δ3·88  ↙ii, iii

Reagents: i, PhCOCl–Et₃N; ii, NaOEt–THF–DMSO; iii, CH₃CO₂H; iv, Me₂SO₄

### Scheme 2

diphenyl-4-azathiabenzene-1-oxide (21).[13] Thiabenzene-1-oxides are stable yellow compounds, but information about their properties is limited. ¹H N.m.r. spectral studies showed that the signals from 6-H in the compounds (17) and (18) (in [²H₆]DMSO) and from 2-H(≡6-H) in the compound (19) (in CDCl₃) occur at unusually high field. Also, 4-H and 6-H

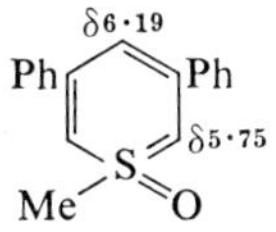

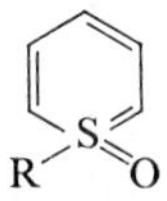

(19) Ph, Ph, Me, S=O  δ6·19, δ5·75

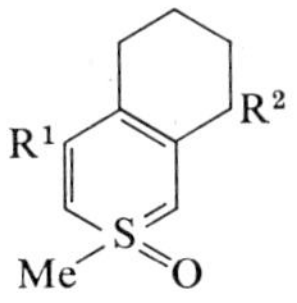

(20) R¹, R², Me, S=O   R¹, R² = H, alkyl

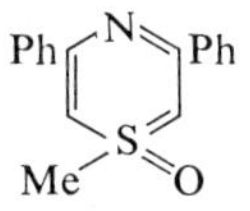

(21) Ph, N, Ph, Me, S=O

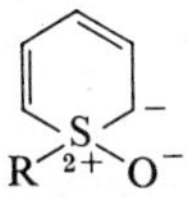

(22a) R, S, O⁻ (2+)
(22b) R, S, O⁻ (+)
(22c) R, S, O

in compounds (17) and (18), and 2-H in compound (19) were exchanged for deuterium in CD₃CO₂D solution. These properties were taken as indicative of ylide character for positions 2 and 6. It was also noted [10] that the *S*-methyl group in compound (17), but not that in the ether (18), underwent H–D exchange in CD₃CO₂D, whence it was concluded that the thiabenzene ring C—S bonds must have more ylene character in the ether than in the compound (17). The *S*-methyl group in compound (19) underwent base-catalysed H–D exchange slowly. Clearly valence-shell expansion has occurred in the thiabenzene-1-oxides, and structures such as (22a)—(22c) with quadri-, quinque-, and even sexi-valent sulphur may be written for these compounds.

[13] H. König, *Angew. Chem. Internat. Edn.*, 1967, **6**, 575.

## 4 2-Thiaphenalenes

Attempts to obtain 2-thiaphenalene (naphtho[1,8-cd]thiapyran) (24) were reported simultaneously by two groups of workers [14,15] who used almost identical synthetic approaches (Scheme 3). Boiling the sulphoxide (23)

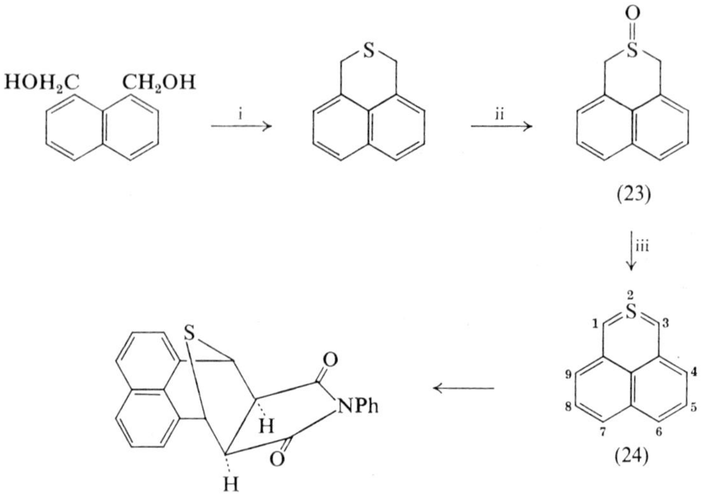

Reagents: i, $P_2S_5$–$CS_2$; ii, $NaIO_4$–aq. EtOH; iii, N-phenylmaleimide–boiling $Ac_2O$

**Scheme 3**

alone in acetic anhydride gave a complex mixture of unidentified products, but in the presence of N-phenylmaleimide led to the *exo*-adduct (25). It was inferred that 2-thiaphenalene (24) had been formed as an unstable intermediate. An analogous sequence of reactions led to the adduct (27) *via* the equally unstable derivative (26).[15] Later the *cis*-sulphoxide (28) was synthesised by a similar route, and reactions implicating 1,3-diphenyl-2-thiaphenalene (29) were described.[16] Thus, boiling the sulphoxide (28) in acetic anhydride, in air, containing N-phenylmaleimide, gave three products (30) (15%), (31) (7%), and (32) (11%). In an oxygen atmosphere all three compounds were again formed, but with the diketone (31) as the major product, whereas in oxygen-free solutions only compounds (30) and (32) were isolated. The intermediacy of a peroxide (33) was suggested. Of further interest was the result that heating the sulphoxide (28) alone in acetic anhydride at 120 °C gave the hydrocarbon (30) quantitatively, evidently *via* the episulphide (34) which could, in fact, be isolated when this reaction was carried out at a lower temperature. The episulphide (34) was

[14] M. P. Cava, N. M. Pollack, and D. A. Repella, *J. Amer. Chem. Soc.*, 1967, **89**, 3640.
[15] R. H. Schlessinger and I. S. Ponticello, *J. Amer. Chem. Soc.*, 1967, **89**, 3641.
[16] R. H. Schlessinger and A. G. Schultz, *J. Amer. Chem. Soc.*, 1968, **90**, 1676.

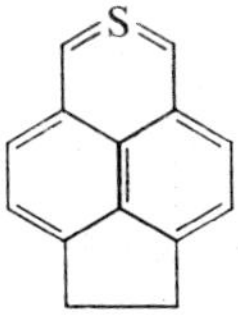

(26) 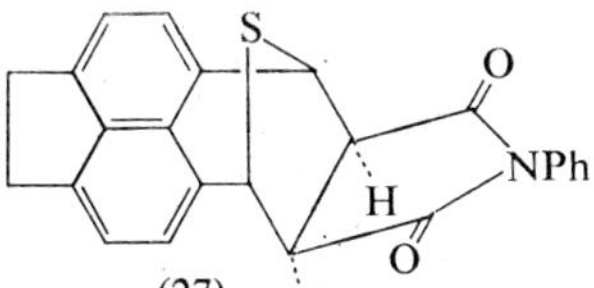 (27) 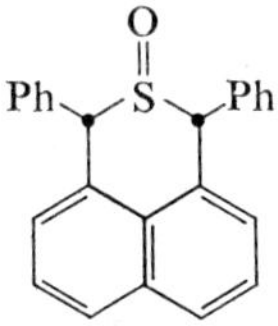 (28) 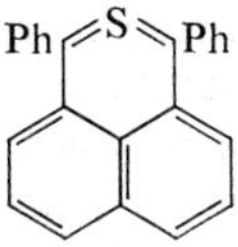 (29) 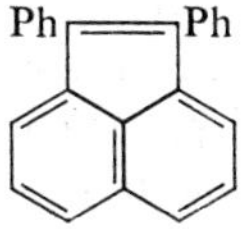 (30) 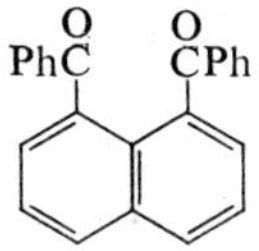 (31) 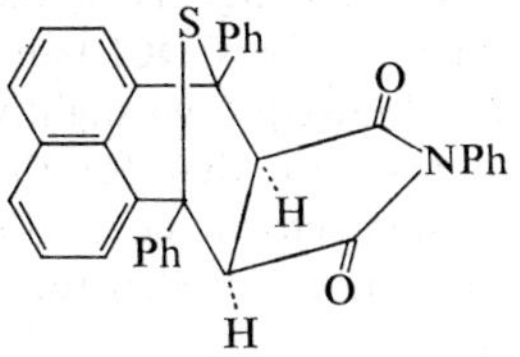 (32) 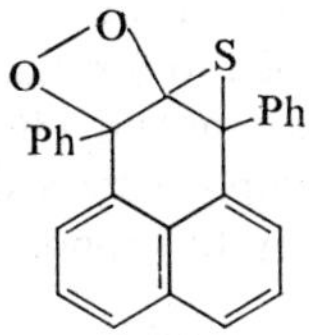 (33) 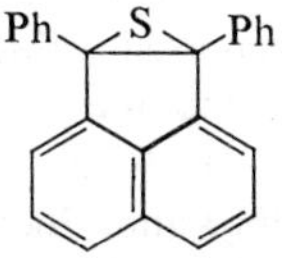 (34)

transformed thermally or photochemically into the hydrocarbon (30). These results are readily rationalised in terms of the transient thiaphenalene (29).

A synthetic approach essentially the same as those already outlined was used in an attempted synthesis of the 2-thiaphenalene (36).[17] This compound is additionally interesting in that it contains $(4n+2)$ peripheral

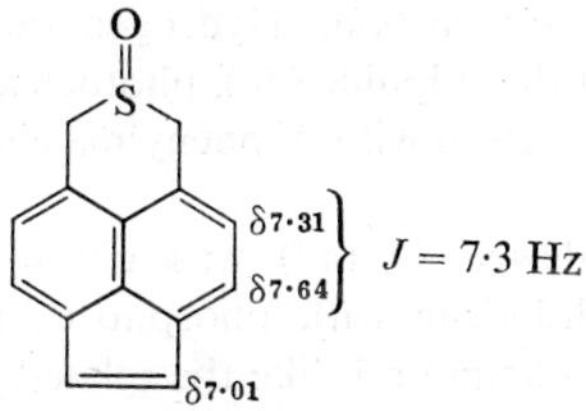

(35) $\delta 7.31$, $\delta 7.64$, $J = 7.3$ Hz, $\delta 7.01$

$\pi$-electrons, and may be described by the more conventional quinonoid form (36b) as well as by a dipolar form (36c). Despite these facts compound (36) could not be isolated from boiling acetic anhydride solutions of its precursor (35), but its transient existence was inferred as in previous cases

[17] R. H. Schlessinger and I. S. Ponticello, *Tetrahedron Letters*, 1967, 4057.

(36a)   (36b)   (36c)

from the isolation of an *exo*-adduct with *N*-phenylmaleimide. In a later study,[18] however, reaction of the sulphoxide (35) with methylmagnesium bromide in ether or benzene gave blue solutions containing the compound (36). Attempts to isolate the thiaphenalene gave a yellow polymer. From $^1$H n.m.r. studies of the solutions it was concluded that a diamagnetic peripheral $\pi$-electron ring-current is present in the thiaphenalene (36). This would account for the substantial down-field shift of the signals from 4-H, 5-H, and 6-H in compound (36) compared with the signals of the corresponding protons in the sulphoxide (35) [chemical shifts (CDCl$_3$) and coupling constants are appended to formulae (35) and (36a)]. The slightly increased value of $J_{4,5}$ for compound (36) compared with that for the sulphoxide (35) was taken to imply increased double-bond character for the (C-4)—(C-5) bond and thereby to support the contribution of structures (36b) and (36c).

Recently, two stable derivatives of the thiaphenalene (36) have been isolated. 6,7-Dibromo-1,3-diphenylacenaphtho[5,6-*cd*]thiapyran (40) was synthesised [19] by the sequence in Scheme 4. It formed dark blue crystals, $\lambda_{\text{max}}(\varepsilon)$, 245 (20,600), 250 (21,400), 322 (10,800), 403 (54,000), 548 (3040), 612 (2980), and 652 (2140) nm, with visible absorption tailing to 725 nm. The spectrum was not solvent dependent. The stability of compound (40) is reflected in its mode of formation, in its mass spectrum in which a strong peak due to the molecular ion was present, and by its resistance to hydrogenation over palladium-on-carbon. Hydrogenation over platinum, however, gave a low yield of the sulphide (38), photochemical oxygenation gave the diketone (37), and reaction with *N*-phenylmaleimide in boiling benzene gave the *exo*-adduct (41).

The second stable derivative (42) was obtained simply by heating 1,4,5,8-tetrabenzoylnaphthalene with phosphorus pentasulphide in pyridine.[20] This blue-black compound, like the related compound (40), shows a parent peak (100%) in its mass spectrum, and is only slowly hydrogenated over palladium-on-carbon to the sulphide (43). Oxidation with chromic anhydride in pyridine gives a mixture of the diketone (44) and the monothione (45). The thione (45) is smoothly converted back into the thiaphenalene (42) by hydrogen chloride in ether or by triphenylphosphine, in the absence of oxygen.

[18] I. S. Ponticello and R. H. Schlessinger, *Chem. Comm.*, 1969, 1214.
[19] I. S. Ponticello and R. H. Schlessinger, *J. Amer. Chem. Soc.*, 1969, **91**, 4190.
[20] J. M. Hoffmann and R. H. Schlessinger, *J. Amer. Chem. Soc.*, 1969, **91**, 3953.

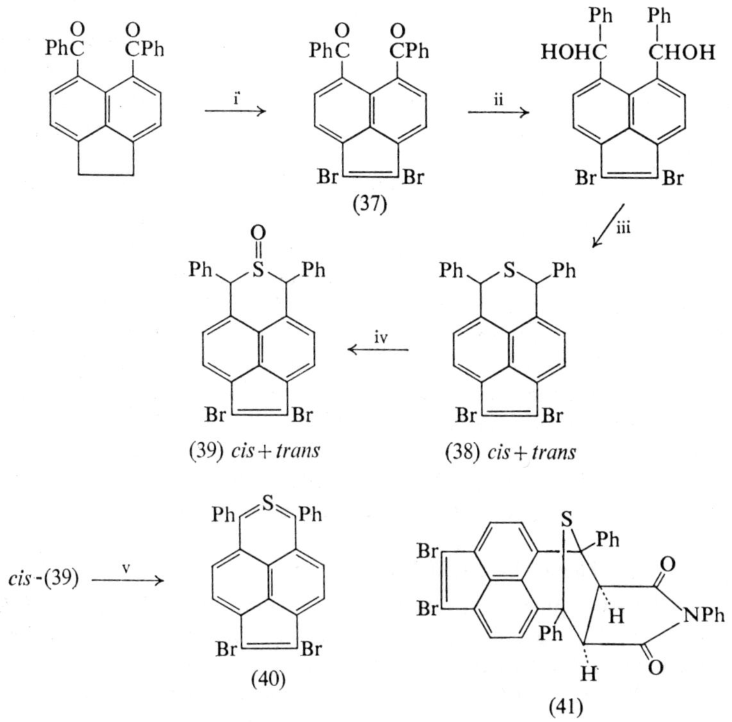

Reagents: i, NBS; ii, $NaBH_4$; iii, $P_2S_5$–pyridine; iv, *m*-chloroperbenzoic acid; v, Boil in 'degassed' $Ac_2O$

**Scheme 4**

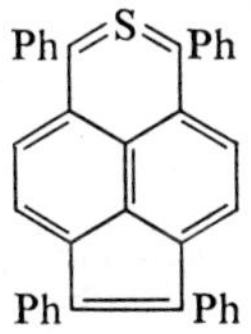

(42)

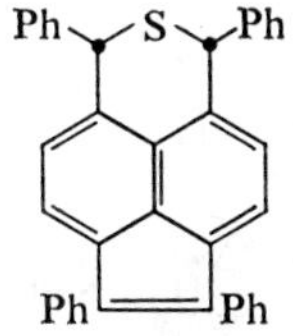

(43)

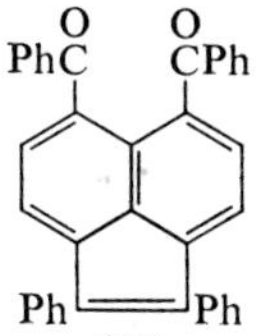

(44)

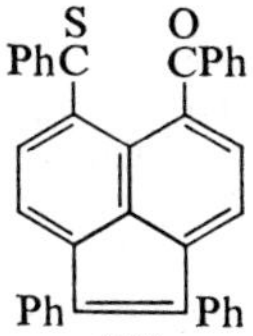

(45)

Since compounds (36), (40), and (42) may be described adequately by conventional structures, exemplified by formulae (36b) and (36c) in the case of compound (36), it is not yet possible to draw conclusions about the aromaticity of the 2-thiaphenalene system, no simple members of which have yet proved to be sufficiently stable to be isolated or examined in solution.

Results of calculations of the $\pi$-electronic structure of the thiaphenalenes (24) and (36) have appeared.[21] The SCF–LCI procedure was used without inclusion of $d$-orbitals. The lowest energy state of the thiaphenalene (36) is a singlet which corresponds in terms of resonance structures to an ionic formulation. Satisfactory correlation was found between the predicted (604 and 503 nm) and experimentally found long wavelength transitions of the stable derivatives (40) (625, 612, and 548 nm) and (42) (665, 610, and 565 nm). It was concluded that it is premature to invoke unusual structures involving sulphur $d$-orbitals to describe the thiaphenalene (36). Ready addition of dienophiles at the 1,3-positions of 2-thiaphenalene (24) is predicted on the basis of the calculated localisation energy for these positions. This had previously been established experimentally.[14, 15]

## 5 Thieno[3,4-c]thiophens

Thieno[3,4-c]thiophens and their aza-analogues, the thieno[3,4-c]-1,2,5-thiadiazoles discussed in section 6, have the same potentiality as the 2-thiaphenalenes for bringing sulphur into a quadricovalent state. The preparation of 1,3-dimethylthieno[3,4-c]thiophen (47) has been attempted [22] by the sequence of reactions in Scheme 5. Pyrolysis of the sulphoxide (46)

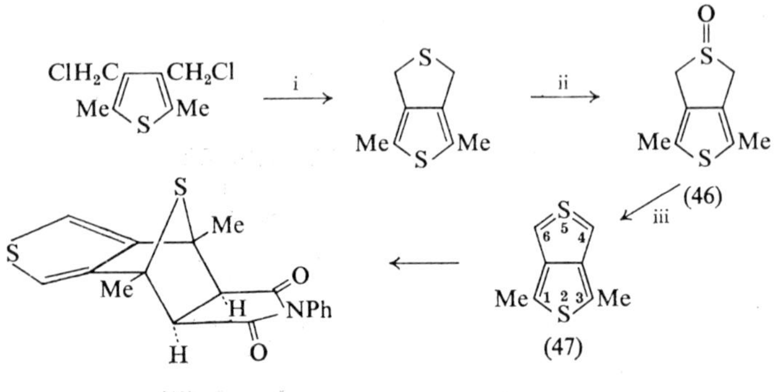

Reagents: i, Na$_2$S–aq. EtOH; ii, NaIO$_4$–MeOH–H$_2$O;
iii, $N$-phenylmaleimide–boiling Ac$_2$O

**Scheme 5**

[21] J. Fabian, Z. Chem., 1969, **9**, 272.
[22] M. P. Cava and N. M. Pollack, J. Amer. Chem. Soc., 1967, **89**, 3639.

over neutral alumina gave a polymer. Compound (47) was not isolated but its transient existence was inferred from the isolation of a mixture of the *exo*-adduct (48) (shown) and the corresponding *endo*-adduct, when the sulphoxide (46) was boiled in acetic anhydride containing N-phenylmaleimide.

A stable thieno[3,4-c]thiophen (49) has been obtained directly by the action of phosphorus pentasulphide on tetrabenzoylethylene in boiling

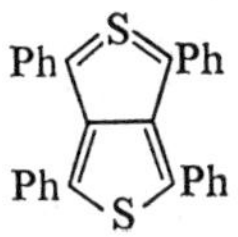

Ph–S–Ph
Ph–S–Ph
(49)

xylene.[23] An alternative, but longer, synthesis, giving better yields of compound (49), was modelled on that described in Scheme 5. Tetraphenylthieno[3,4-c]thiophen (49) forms purple needles, m.p. 245—248 °C, $\lambda_{max}(\varepsilon)$ 553 (13,000), 296 (20,000), and 255 (17,000) nm, whose $^1$H n.m.r. spectrum shows a singlet at $\delta$7·12 suggesting non-coplanarity of the phenyl substituents with the heterocyclic rings. Its mass spectrum shows a molecular ion peak (100%). It reacted with N-phenylmaleimide in an analogous manner to that of the transient compound (47) giving an mixture of *exo*- and *endo*-adducts. Compound (49) is important in being the first stable heterocyclic compound containing sulphur bonded to carbon alone for which only quadricovalent uncharged structures may be written.

## 6 Thieno[3,4-c]-1,2,5-thiadiazoles

A stable derivative (51) of this system has been obtained recently[24] by treatment of the 1,2,5-thiadiazole (50) with phosphorus pentasulphide in

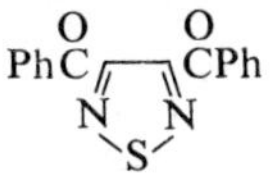

PhC(O)–C(O)Ph
N–N
S
(50)

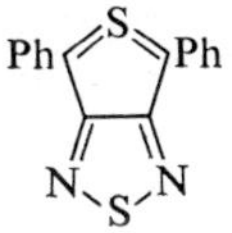

Ph–S–Ph
N–S–N
(51)

dioxan. Forming stable purple crystals, m.p. 146 °C, $\lambda_{max}(\varepsilon)$ 558 (8650), 330 (19,900), 312 (21,600), and 275 (16,700) nm, compound (51) gave a simple mass spectrum with peaks at $m/e$ 294 (M$^+$, 100%) and 121 (C$_7$H$_5$S,$^+$ 30%). It reacted sluggishly with N-phenylmaleimide giving a mixture of *exo*- (52) (shown) and *endo*-adducts, and when irradiated gave a colourless dimer (53).

[23] M. P. Cava and G. E. M. Husbands, *J. Amer. Chem. Soc.*, 1969, **91**, 3952.
[24] J. D. Bower and R. H. Schlessinger, *J. Amer. Chem. Soc.*, 1969, **91**, 6891.

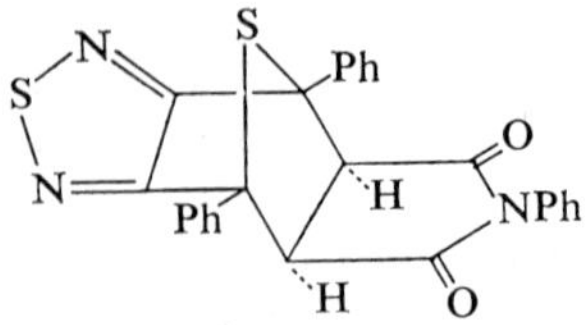

(52)

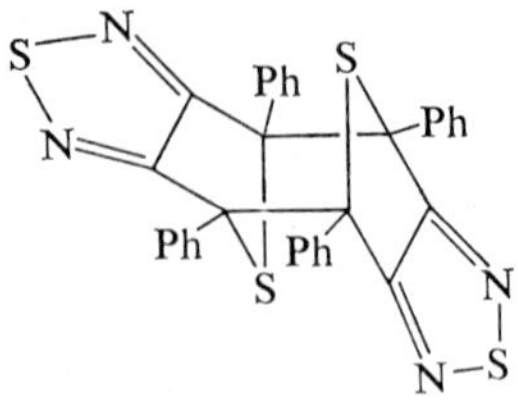

(53)

Attempts were also made to prepare the derivative (55).[24] Boiling solutions of the sulphoxide (54), whose synthesis is described, in acetic anhydride developed a blue colour, $\lambda_{max}$ 645 nm, from which the dimer (56)

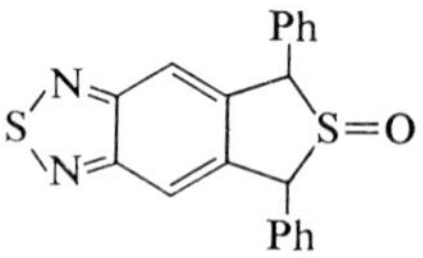

(54)

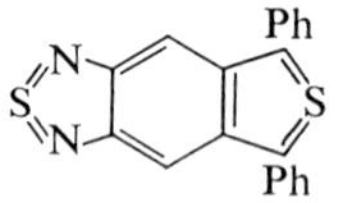

(55)

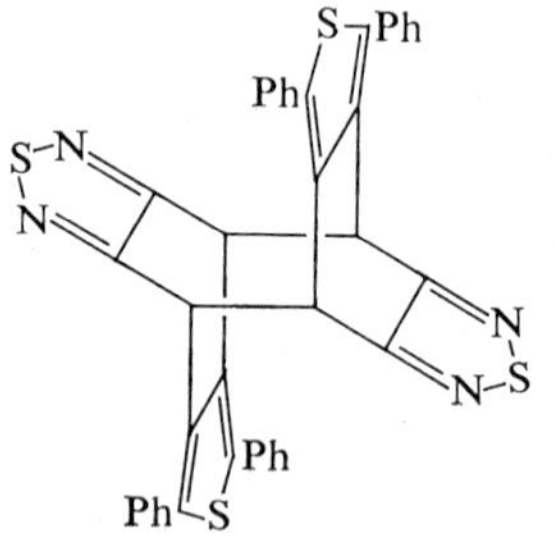

(56)

was isolated, evidently formed from the transient compound (55). When this reaction was carried out in the presence of N-phenylmaleimide, four adducts were isolated corresponding to *exo*- and *endo*-addition to both the thiophen and the benzene rings.

# 8
# 6a-Thiathiophthenes and Related Compounds

## 1 Introduction

The structure and bonding at the sulphur atoms of 6a-thiathiophthene are still unsettled. Formula (1), with the numbering shown, is employed in this review for convenience.

A critical review of the chemistry of 6a-thiathiophthenes has appeared.[1] It serves as a background to the material of this review.

## 2 Structural Studies

Increased interest in the unusual structure and bonding of 6a-thiathiophthenes has resulted in a spate of recent publications on $X$-ray crystal structure determinations of 6a-thiathiophthenes and related compounds. In their pioneering work on the structure of 2,5-dimethyl-6a-thiathiophthene (2), Bezzi, Mammi, and Garbuglio[2] assigned the crystal to the orthorhombic centric space group $P\,nma$ rather than the alternative non-centric $P\,n2_1a$. The former space group requires $C_{2v}$ molecular symmetry for 2,5-dimethyl-6a-thiathiophthene and hence identical S—S distances. These are 2·36 Å, substantially longer than the normal S—S covalent bond length of 2·04—2·08 Å, but well inside the sum of the van der Waals' radii

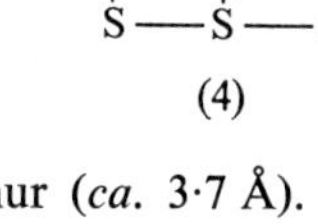

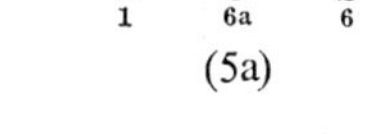

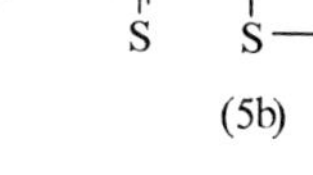

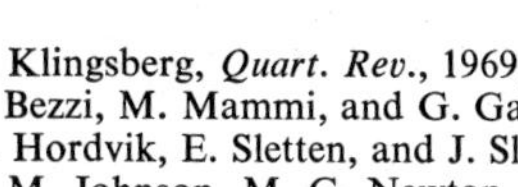

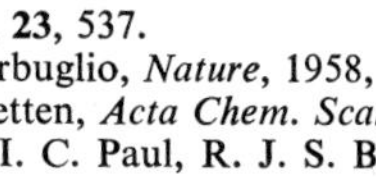

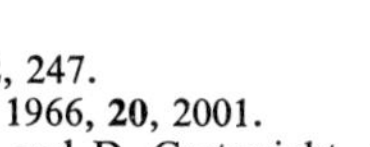

of sulphur (*ca.* 3·7 Å). More recent studies of the structure of the unsymmetrical derivatives (3)[3] and (4)[4] revealed unequal S—S distances in

---

[1] E. Klingsberg, *Quart. Rev.*, 1969, **23**, 537.
[2] S. Bezzi, M. Mammi, and G. Garbuglio, *Nature*, 1958, **182**, 247.
[3] A. Hordvik, E. Sletten, and J. Sletten, *Acta Chem. Scand.*, 1966, **20**, 2001.
[4] S. M. Johnson, M. G. Newton, I. C. Paul, R. J. S. Beer, and D. Cartwright, *Chem. Comm.*, 1967, 1170.

these compounds. It was suggested[4] therefore that 2,5-dimethyl-6a-thiathiophthene might really be a statistically disordered combination of structures (5a) and (5b), with the S—S distances unequal but indicative of some degree of bonding between $S^{6a}$ and $S^6$. Nyburg[5] has now re-examined the structure of 2,5-dimethyl-6a-thiathiophthene with a full three-dimensional refinement of the *P nma* structure in the usual way. He finds no *prima facie* evidence for statistical disordering in this structure, and concludes that if there is disordering, the difference in the two molecular geometries must be too small to detect by *X*-ray structure analysis. Important bond distances and angles are given in Figure 1. Refining the data for the alternative

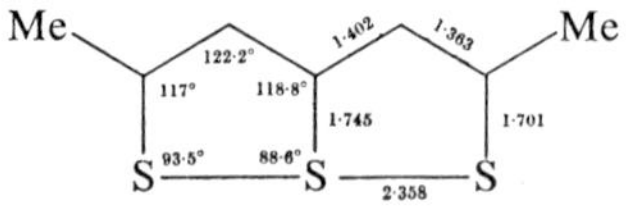

**Figure 1** *Molecular dimensions of 2,5-dimethyl-6a-thiathiophthene*

$P n2_1 a$ space group resulted in excessive asymmetry of the molecule and unacceptable bond lengths.

Full details have appeared[6] of a single-crystal *X*-ray analysis of 3-benzoyl-5-(*p*-bromophenyl)-2-methylthio-6a-thiathiophthene (4), following a preliminary report.[4] There are two crystallographically independent molecules in the crystal unit. Several bond lengths in one molecule differ appreciably from the corresponding bond lengths in the other, especially in the disubstituted ring (Figure 2, A and B). Hordvik[7] has also published full details of the structure of 2,4-diphenyl-6a-thiathiophthene (3), previously reported briefly.[3] The heterocyclic system is virtually planar, and

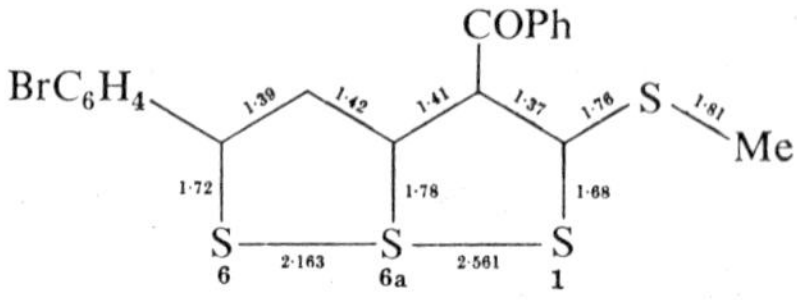

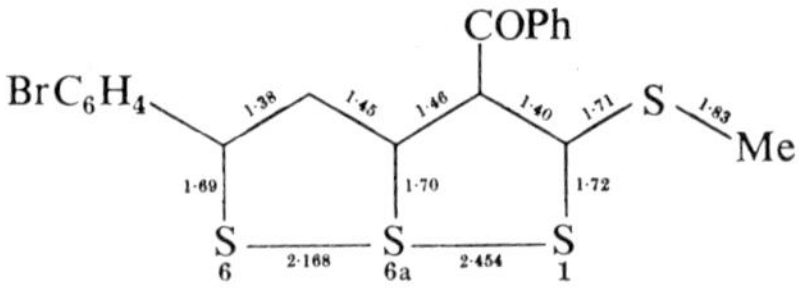

**Figure 2** *Bond lengths of the two crystallographically independent molecules in the crystal unit of 3-benzoyl-5-p-bromophenyl-2-methylthio-6a-thiathiophthene*

[5] F. Leung and S. C. Nyburg, *Chem. Comm.*, 1969, 137.
[6] S. M. Johnson, M. G. Newton, and I. C. Paul, *J. Chem. Soc. (B)*, 1969, 986.
[7] A. Hordvik, E. Sletten, and J. Sletten, *Acta Chem. Scand.*, 1969, **23**, 1852.

comparison of C—C and C—S distances with standard bond lengths indicates stabilisation through $\pi$-bonding (Figure 3). As in compound (4), the S—S distances differ considerably. This difference is rationalised [7] on the basis of the Gleiter–Hoffmann model of 6a-thiathiophthene (see

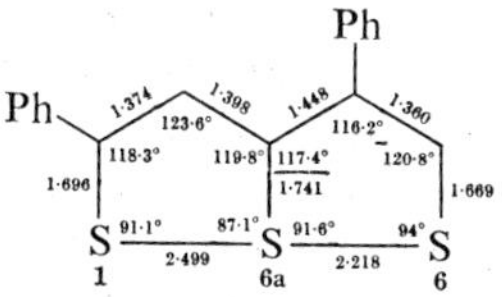

**Figure 3** Bond lengths and angles in 2,4-diphenyl-6a-thiathiophthene

section 3) in which $\sigma$-bonding of the $S^1$–$S^{6a}$–$S^6$ system involves an electron-rich, three-centre bond. It will be recalled that this type of bonding is found in the linear trihalide ions, for which structural investigations and theoretical calculations show that the less electronegative of the terminal halogen atoms forms the stronger bond with the central halogen atom. Since the 2-phenyl group in 2,4-diphenyl-6a-thiathiophthene is closer to $S^1$ than the 4-phenyl group is to $S^6$, $S^1$ becomes more electronegative than $S^6$ and so forms a weaker, and longer, bond with $S^{6a}$. Similar reasoning applied to compound (4) accounts for the $S^1$—$S^{6a}$ bond being longer than the $S^{6a}$—$S^6$ bond. However, it is difficult to rationalise the results of a later report [8] that the $S^1$—$S^{6a}$ and $S^{6a}$—$S^6$ distances are almost identical in 2,4-diphenyl-(3) and 2-methyl-4-phenyl-6a-thiathiophthene (6)

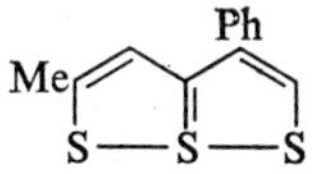

(6)

($S^1$—$S^{6a}$, 2·475; $S^{6a}$—$S^6$, 2·237 Å). The arguments outlined above have also been used to explain the contraction of the S—S bond in the oxygen analogues (7) (2·12 Å)[9] and (8) (2·106 Å). Details of a structure determination of compound (8) have been published.[10] In compounds (7) and (8) the

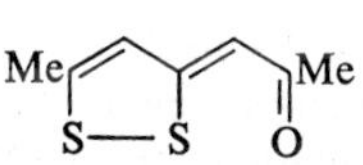

(7) 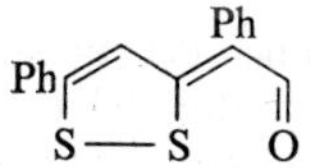                (8)

S—S bond lengths lie close to those of normal disulphides (ca. 2·08 Å), and the S - - - O distances are considerably less than the sum of the van der Waals' radii of sulphur and oxygen.

[8] A. Hordvik and K. Julshamn, *Acta Chem. Scand.*, 1969, **23**, 3610.
[9] M. Mammi, R. Bardi, G. Traverso, and S. Bezzi, *Nature*, 1961, **192**, 1282.
[10] A Hordvik, E. Sletten, and J. Sletten, *Acta Chem. Scand.*, 1969, **23**, 1377.

In view of the problems associated with the interpretation of crystallographic data for 2,5-dimethyl-6a-thiathiophthene, the appearance of structural data for two further symmetrically substituted 6a-thiathiophthenes is welcome. The S—S distances in 2,5-diphenyl-6a-thiathiophthene (9) are almost but not quite identical (2·355 and 2·297 Å).[11] The difference is considerably smaller than the corresponding differences in the unsymmetrical derivatives (3) and (4). It has been suggested that the two S—S distances in any symmetrically substituted 6a-thiathiophthene are exactly equal in an isolated molecule, and the deviation from equality results from the S—S bonds in 6a-thiathiophthenes being unusually sensitive to intermolecular as well as intramolecular effects. This conclusion is consistent with the results of $^1$H n.m.r. spectral studies.[12-14] Symmetrical 6a-thiathiophthenes show magnetic equivalence of ring protons or identical substituents at the pairs of sites C-2, C-5 and C-3, C-4. This demonstrates that these compounds *in solution*, in which intermolecular effects are averaged, possess $C_{2v}$ symmetry and consequently S—S bonds of equal length.

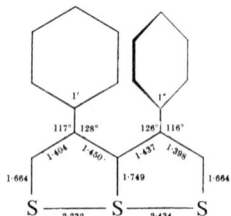

(9)        (10)

Data for the isomeric 3,4-diphenyl-6a-thiathiophthene (10) have also appeared (Figure 4).[15] The S—S distances are considerably more unequal

**Figure 4** *Bond lengths and angles in 3,4-diphenyl-6a-thiathiophthene*

than in the isomer (9), presumably owing to steric clash of the phenyl substituents. Although the 6a-thiathiophthene system is planar, the external bond angles at C-3 and C-4 deviate considerably from 120°. The 'inner' angles (C-1')(C-3)(C-3a) and (C-1")(C-4)(C-3a) widen at the expense of the 'outer' ones, and C-1' and C-1" are displaced from the plane of the 6a-thiathiophthene system in opposite directions. The phenyl substituents

---

[11] A. Hordvik, *Acta Chem. Scand.*, 1968, **22**, 2397.
[12] G. Pfister-Guillouzo and N. Lozac'h, *Bull. Soc. chim. France*, 1964, 3254.
[13] J. G. Dingwall, S. McKenzie, and D. H. Reid, *J. Chem. Soc. (C)*, 1968, 2543.
[14] D. H. Reid and J. D. Symon, *Chem. Comm.*, 1969, 1314.
[15] P. L. Johnson and I. C. Paul, *Chem. Comm.*, 1969, 1014.

## 6a-Thiathiophthenes and Related Compounds

make angles of 70° and 74° with the plane of the 6a-thiathiophthene system.

Compound (11) had previously been synthesised [16] with a view to determining whether a system of four or more sulphur atoms arranged linearly would interact, with formation of S—S bonds. A preliminary communication [17] on the structure of compound (11), shown in Figure 5, indicates that three of the sulphur atoms are bonded in a 6a-thiathiophthene unit (rings A

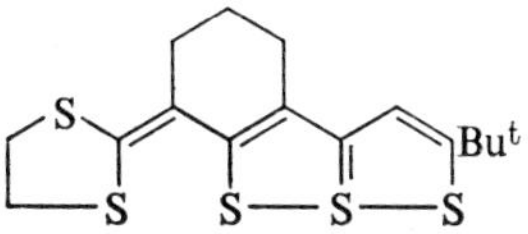

(11)

and B), and the fourth sulphur atom has a negligible influence on their bonding. The internal angles at the carbon atoms in 'ring' C are all greater than 120°, indicating repulsion between the sulphur atoms in that cavity.

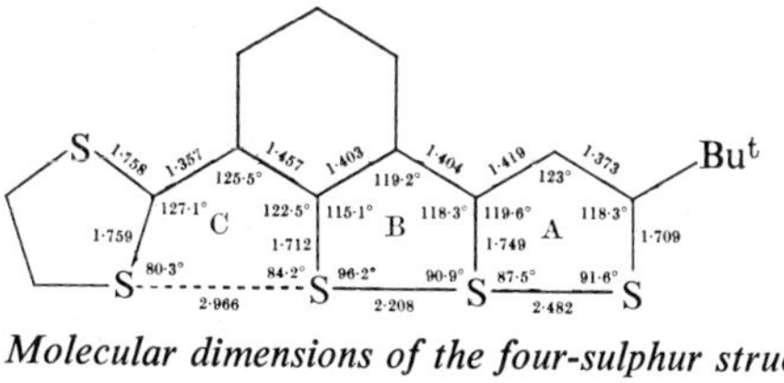

**Figure 5** *Molecular dimensions of the four-sulphur structure* (11)

A related problem is whether 2-thioacylmethylene-2H-1,3-dithioles (12) are best represented by structure (12) or the 'isothiathiophthene' structure

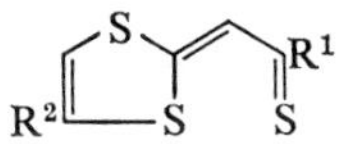

(12) 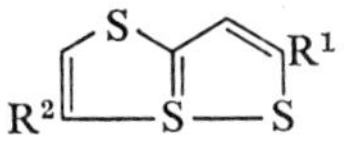                (13)

(13). Recent X-ray crystallographic data for the diphenyl derivative (12; $R^1 = R^2 = Ph$) supports the monocyclic formulation (12).[18] This compound is essentially planar, and the S—S distance of 2·91 Å is very close to the long S—S distance in compound (11).

Aza-derivatives of 3-acylmethylene-3H-1,2-dithioles have been prepared (section 4), and the structures of two members (14) and (15) have been determined.[19] The S—S and S - - - O distances in compound (14) are 2·14 and 2·10 Å, respectively, and the nitroso- rather than the nitro-oxygen is

---

[16] N. Lozac'h and M. Stavaux, *Bull. Soc. chim. France*, 1968, 4273, and unpublished work.
[17] J. Sletten, *Chem. Comm.*, 1969, 688.
[18] R. J. S. Beer, D. Frew, P. L. Johnson, and I. C. Paul, *Chem. Comm.*, 1970, 154.
[19] K. I. G. Reid and I. C. Paul, *Chem. Comm.*, 1970, 329.

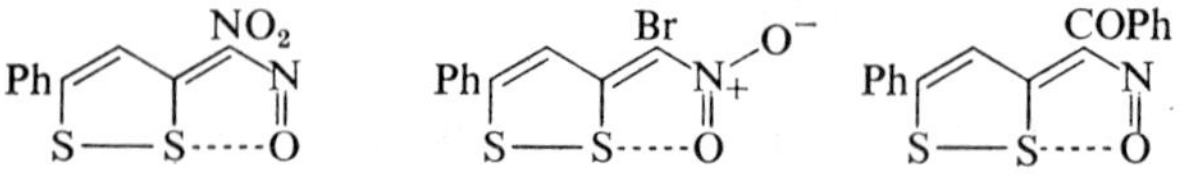

close to the sulphur atom. These results recall earlier findings for compound (16),[20] for which it was shown that the nitroso rather than the carbonyl group is adjacent to the sulphur atom, and for which S—S and S---O distances of 2·178 and 2·034 Å were found, respectively. The S—S bond length in compound (15), like that in compound (14), is in the region of normal covalent S—S distances, but the S---O distance is considerably longer than that found in compounds (14) and (16) (Figure 6). The authors

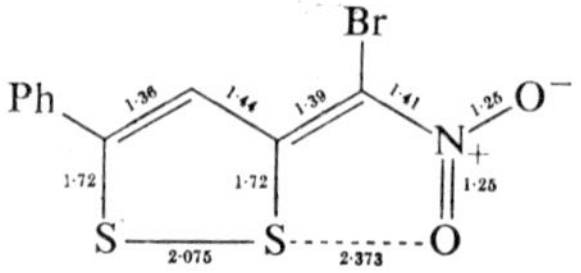

**Figure 6** Bond lengths in 3-bromonitromethylene-5-phenyl-3H-1,2-dithiole

conclude that covalent bonding between sulphur and oxygen in compound (15) must be very weak or non-existent. Comparison of S---O distances in compounds (14)—(16) indicates an order of preference for S---O interaction of nitro < nitroso > carbonyl.

## 3 Bonding

Gleiter and Hoffmann [21] have made a notable contribution to theories of bonding in 6a-thiathiophthene. The essential feature of their theory is to regard the three sulphur atoms as a linear system of atoms which uses three orbitals occupied by four electrons for $\sigma$-bonding (Figure 7), with super-

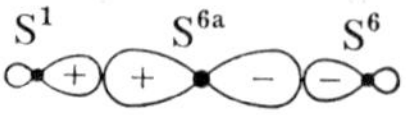

**Figure 7** Electron-rich (four-electron) three-centre bond involving p-orbitals linking the sulphur atoms in 6a-thiathiophthene

imposed $\pi$-bonding. The concept of electron-rich three-centre bonds is familiar in the chemistry of the interhalogens and, in organic chemistry, to describe the transition state for $S_N2$ displacement at a saturated carbon atom. The stabilisation attending the superimposed $\pi$-bonding of the sulphur atoms in 6a-thiathiophthene is not expected to be large, since the equilibrium distance for a three-centre bond involving Second Row elements is reached at a stage where $p_\pi$–$p_\pi$ overlap is still small. The potential energy of the $S^1$–$S^{6a}$–$S^6$ system as a function of the displacement of

[20] R. J. S. Beer, D. Cartwright, R. J. Gait, R. A. W. Johnstone, and S. D. Ward, *Chem. Comm.*, 1968, 688.
[21] R. Gleiter and R. Hoffmann, *Tetrahedron*, 1968, 5899.

$S^{6a}$ from an equidistant location between $S^1$ and $S^6$ was calculated (a) not including d-orbitals, (b) with inclusion of d-orbitals. The former situation favours an unsymmetrical structure, the latter a *nearly* symmetrical one with a very flat minimum. As the σ-bonds in the linear sulphur sequence are fractional, it would not be surprising if the S—S bonds were weaker than normal disulphide bonds and more sensitive to inter- and intra-molecular effects.

Johnstone and Ward[22] have carried out SCF MO calculations on 6a-thiathiophthene, using a 10π-electron model (Figure 8) similar to that previously employed by Maeda.[23] $S^{6a}$ is σ-bonded to $S^1$, $S^6$, and C-3a by $p^2d$ hybrid orbitals $h_1$, $h_2$, and $h_3$. The remaining $p_z$ orbital at $S^{6a}$ serves as

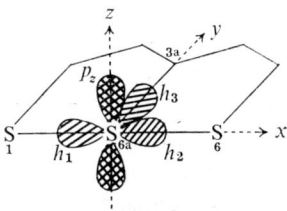

**Figure 8** *Bonding orbitals at the central sulphur atom in* 6a-*thiathiophthene*

the π-orbital. π-Bonding is then provided by one 2p-electron from each carbon atom, one 3p-electron from $S^{6a}$, and two 3p-electrons from each outer sulphur atom. There is satisfactory agreement between predicted and experimental transition energies and ionisation potentials, and it is concluded that π-bonding exists between the three sulphur atoms.

Extended Hückel calculations have been carried out[24] on the 3-acyl-methylene-3H-1,2-dithioles (7) and (17) in order to probe for S---O

(17)

bonding interactions. Compound (7) is known to have a short S---O distance (2·41 Å).[9] Calculated S---O overlaps, with and without charge iteration, and with and without inclusion of d-orbitals in both cases, were close to zero, indicating negligibly weak covalent bonding between sulphur and oxygen.

## 4 Formation and Syntheses

A preparatively useful synthesis of 6a-thiathiophthene (1) involves the ring-opening of 4H-thiapyran-4-thione (18) by sodium sulphide in dipolar

---

[22] R. A. W. Johnstone and S. D. Ward, *Theor. Chim. Acta*, 1969, **14**, 420.
[23] K. Maeda, *Bull. Chem. Soc. Japan*, 1960, **33**, 1466.
[24] J. A. Kapecki and J. E. Baldwin, *J. Amer. Chem. Soc.*, 1969, **91**, 1120.

aprotic solvents, and oxidation of the resulting anion (19) by potassium ferricyanide (Scheme 1).[25] Similarly, ring-opening of 4$H$-thiapyran-4-thiones with sodium hydroxide leads to 3-acylmethylene-3$H$-1,2-dithioles (Scheme 2).

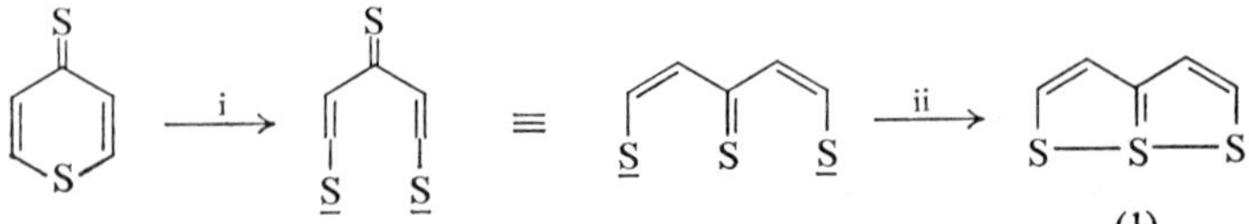

Reagents: i, Na$_2$S–DMSO–H$_2$O; ii, K$_3$Fe(CN)$_6$

**Scheme 1**

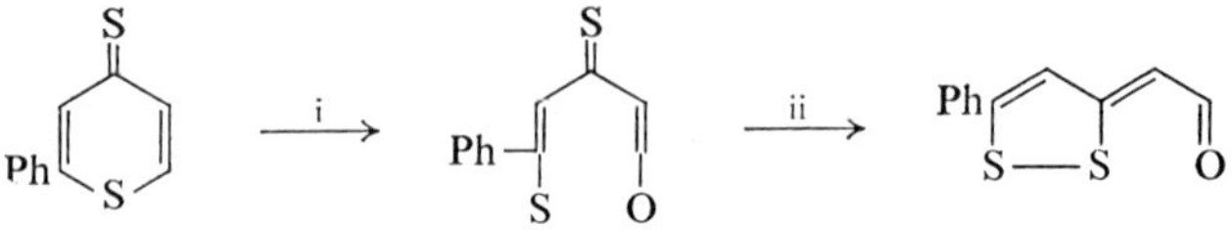

Reagents: i, NaOH–DMF–H$_2$O; ii, K$_3$Fe(CN)$_6$

**Scheme 2**

The salts (21) result from hydride abstraction–dehydrogenation of the 3,4-bridged 6a-thiathiophthenes (20) with trityl fluoroborate.[26] $^1$H n.m.r. studies show that in all four salts (21) the R groups are magnetically equivalent, indicating that the cations have real (structure 21) or time-averaged (structure 22) $C_{2v}$ symmetry.

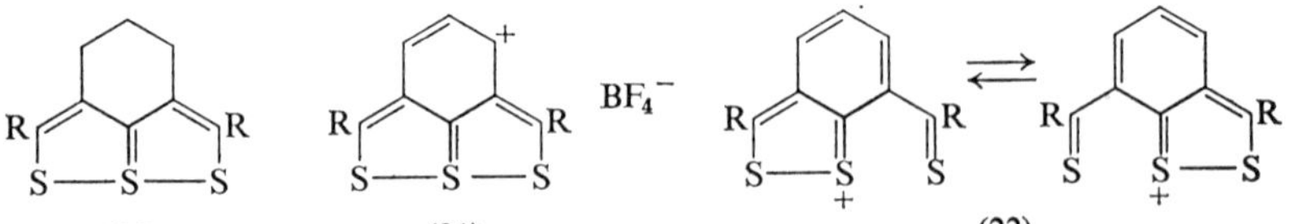

(20)—(22): R = Ph, $p$-MeC$_6$H$_4$, $p$-MeOC$_6$H$_4$, or MeS

The course of the reaction of 3-aryl-1,2-dithiolium salts with aroyldithioacetate esters is dependent on the basicity of the reaction medium.[27] Weakly basic or neutral media promote the formation of 6a-thiathiophthenes (Scheme 3) according to path (a), in which a hydride ion is abstracted from C-3 of the intermediate (23), probably by unreacted 3-aryl-1,2-dithiolium cation, and a proton is then lost from C-1′. Strongly basic media favour exclusive formation of 2$H$-thiapyran-2-thiones via path (b), through initial deprotonation at C-1′ and rearrangement of the resulting intermediate (24).

[25] J. G. Dingwall, D. H. Reid, and J. D. Symon, *Chem. Comm.*, 1969, 466.
[26] E. I. G. Brown, D. Leaver, and T. J. Rawlings, *Chem. Comm.*, 1969, 83.
[27] F. Clesse and H. Quiniou, *Compt. rend.*, 1969, **268**, C, 637.

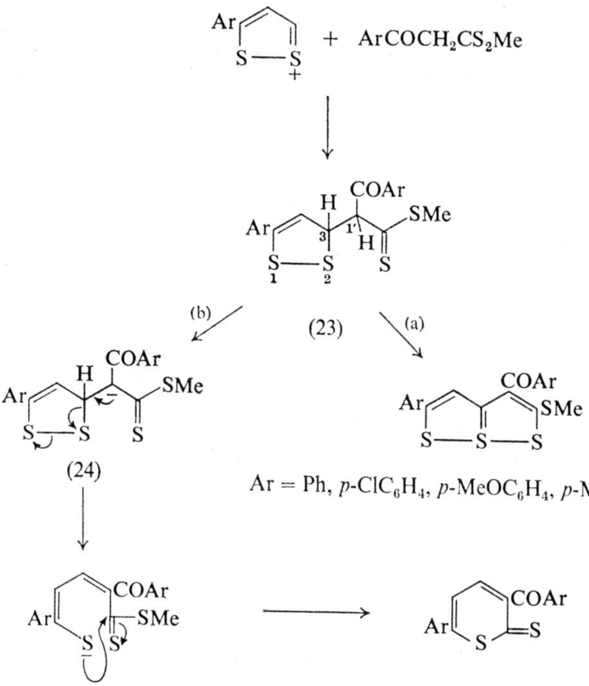

Ar = Ph, p-ClC$_6$H$_4$, p-MeOC$_6$H$_4$, p-MeC$_6$H$_4$

**Scheme 3**

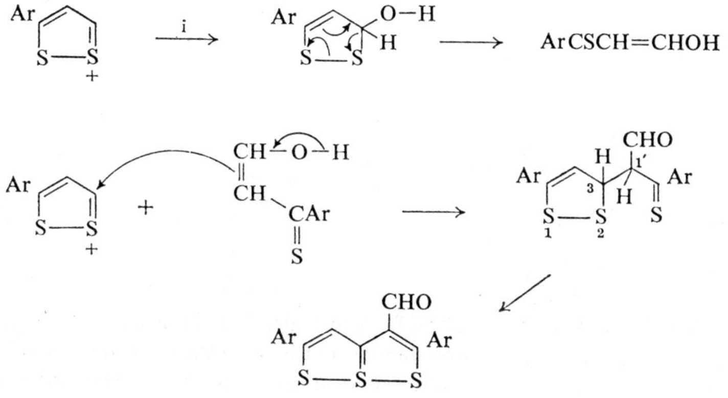

ArCSCH=CHOH

(25) Ar = Ph
(26) Ar = p-MeC$_6$H$_4$

Reagent: i, Et$_3$N–EtOH–H$_2$O

**Scheme 4**

3-Formyl-6a-thiathiophthenes (25) and (26) result from the breakdown of 3-aryl-1,2-dithiolium salts by base according, it is suggested,[28] to Scheme 4. 3-Acylmethylene-3H-1,2-dithioles (27) are obtained by the condensation of 5-aryl-3-methylthio-1,2-dithiolium salts with the sodium enolates of 3-aryl-3-oxopropionaldehydes (Scheme 5).[29, 30]

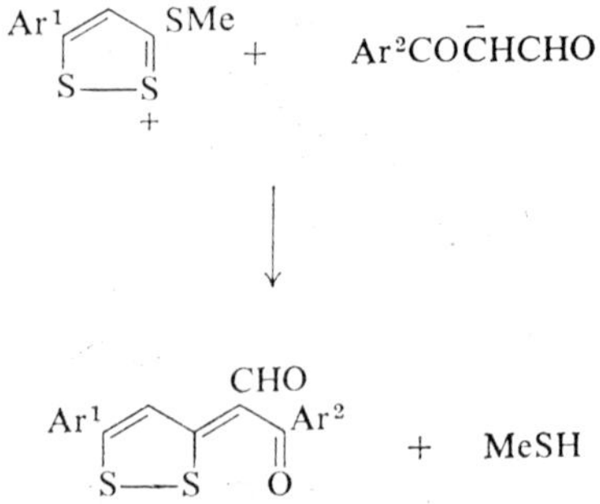

Scheme 5

New types of compounds structurally analogous to 6a-thiathiophthenes or 3-acylmethylene-3H-1,2-dithioles have been prepared.[31] Nitrosation of 2-dimethylamino-5-phenyl-6a-thiathiophthene (28) and 2,5-bismethylthio-6a-thiathiophthene (29) gives, respectively, compounds (30) and (31). These compounds have absorption spectra similar to that of the previously reported [20] nitroso-compound (32), from which it is inferred that all three compounds have the structures shown, that is, with the oxygen atom of the nitroso-group adjacent to the sulphur atom. The same communication describes the synthesis of the 3-nitromethylene-3H-1,2-dithioles (33) and (34) (Scheme 6) and the nitrosation and bromination of compound (33) to

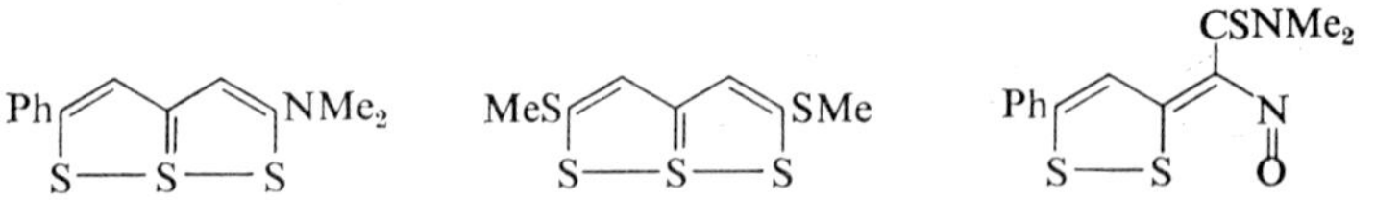

[28] J. Bignebat and H. Quiniou, Compt. rend., 1968, **267**, C, 180.
[29] J. Bignebat and H. Quiniou, Compt. rend., 1969, **269**, C, 1129.
[30] J. Bignebat and H. Quiniou, Compt. rend., 1970, **270**, C, 83.
[31] R. J. S. Beer and R. J. Gait, Chem. Comm., 1970, 328.

give the derivatives (14) and (15). The three nitro-compounds (15), (33), and (34) evidently have similar structures, but different from that of the nitrosonitro-derivative (14). X-Ray crystallographic studies of compounds (14) and (15) are described in section 2. It is concluded that the nitrosogroup takes precedence over the CO·Ph, CS·SMe, CS·NMe$_2$, and NO$_2$ groups in competition for the apparently advantageous position adjacent to the S—S link.

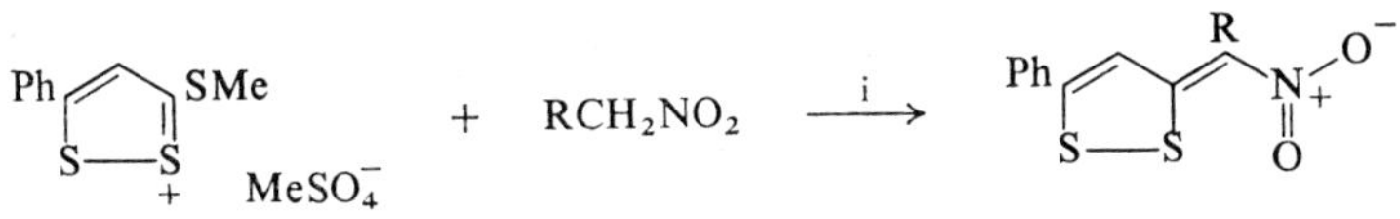

Ph⌒SMe  +  RCH$_2$NO$_2$  →$^i$  Ph⌒[structure]
S—S⁺        MeSO$_4^-$

(33) R = H
(34) R = Me

Reagent: i, RCH$_2$NO$_2$–CH$_3$CO$_2$H–pyridine

**Scheme 6**

Attempts to ascertain whether a linear arrangement of more than three sulphur atoms in suitable heterocyclic structures might be a source of unusual stability[32] have resulted in the synthesis of two compounds containing a sequence of five sulphur atoms. Compound (35), prepared by the reactions shown in Scheme 7, has a symmetrical structure, as evidenced by the simplicity of its $^1$H n.m.r. spectrum.[33] The $C_{2v}$ symmetry could arise from a delocalised structure (35a) ↔ (35b), etc., in which all sulphur atoms interact in a bonding manner, or from a symmetrical localised structure (36). The authors favour the first possibility. An aza-analogue (38) of this system has been obtained[34] by the reactions illustrated in Scheme 8. The electronic spectra of the compound (38) and its precursor (37) show the same relationship as is typical for 6a-thiathiophthenes and their precursors (the 3-acylmethylene-3H-1,2-dithioles); namely, that replacement of oxygen by sulphur causes a bathochromic shift of the long wavelength absorption band in the visible region and weakens it, and introduces a second and much stronger absorption in the u.v. region. No further information is yet available which would establish structure (38) or a geometrically isomeric structure for this compound.

An interesting problem is whether other heterocyclic systems, having properties similar to those of 6a-thiathiophthene, would result by replacement of one or more sulphur atoms in 6a-thiathiophthene by other elements in Groups VI and V. A derivative (39) of the isothiazolo[5,1-e]isothiazole system has been synthesised[14] by the sequence in scheme 9. The $^1$H n.m.r. spectrum is simple, showing only three sharp singlets, and remains unchanged in pattern down to −70 °C. These data are consistent with the

[32] E. Klingsberg, *J. Heterocyclic Chem.*, 1966, **3**, 243.
[33] M. Stavaux and N. Lozac'h, *Bull. Soc. chim. France*, 1969, 4184.
[34] E. Klingsberg, *Chem. and Ind.*, 1968, 1813.

$C_{2v}$-symmetric bicyclic formulation (39), but a rapidly equilibrating system of identical monocyclic structures (40a) and (40b) would also satisfy the $^1$H n.m.r. data.

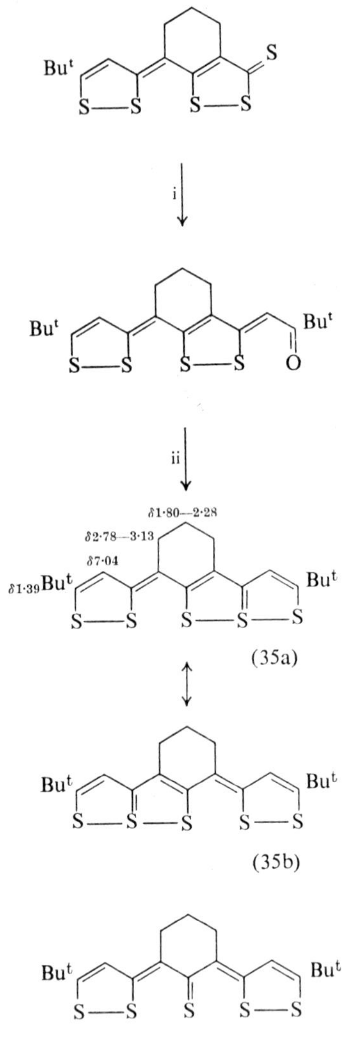

Reagents: i, Bu$^t$COCHN$_2$; ii, P$_2$S$_5$

**Scheme 7**

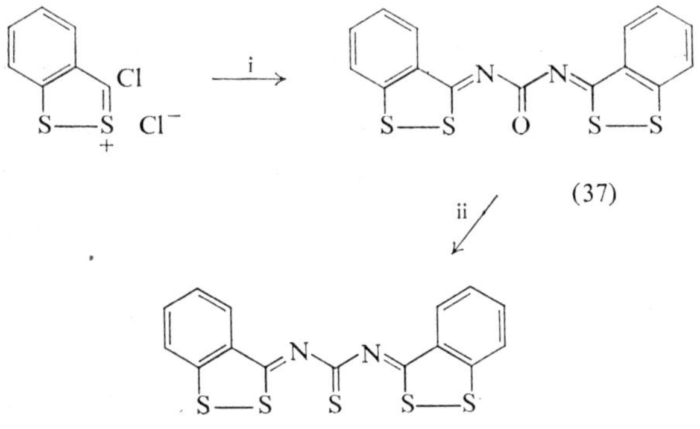

(37)

(38)

Reagents: i, Urea–*o*-chlorotoluene; ii, $P_2S_5$–*o*-chlorotoluene

**Scheme 8**

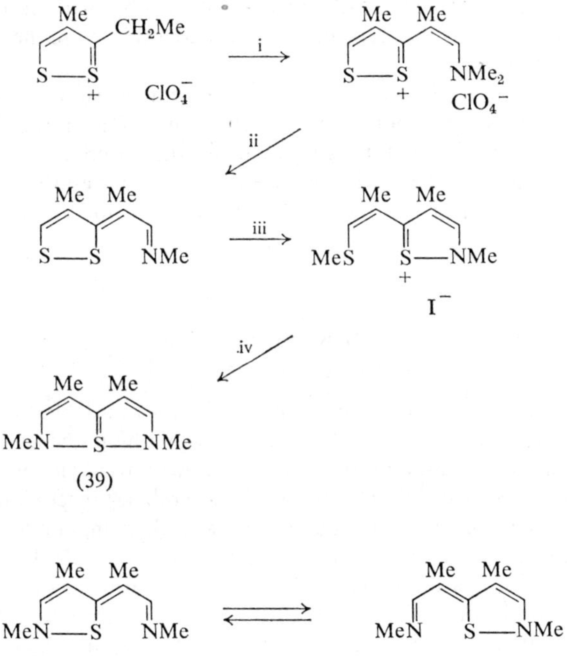

(39)

(40a) (40b)

Reagents: i, $Me_2NCHS$–$Ac_2O$; ii, aq. $MeNH_2$; iii, MeI–MeCN; iv, aq. $MeNH_2$

**Scheme 9**

## 5 Reactions

Little work has been reported on the reactivity of 6a-thiathiophthenes. Formylation of 2-phenyl-6a-thiathiophthene with $NN$-dimethylthioformamide and phosphoryl chloride gave [35] the 4-formyl derivative (41). The use of DMF in place of $NN$-dimethylthioformamide gave a poor yield of the aldehyde (41), but has proved effective in the formylation of 2,5-diaryl-6a-thiathiophthenes in $o$-dichlorobenzene solution.[28,29] The aldehydes

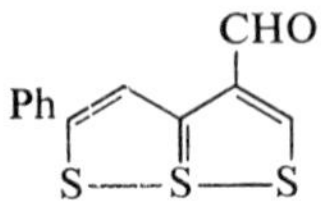

(41)

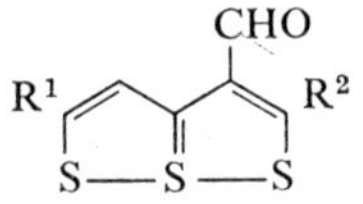

(42) $R^1 = R^2 = Ph$
(43) $R^1 = R^2 = p\text{-MeC}_6H_4$
(44) $R^1 = Ph, R^2 = p\text{-MeOC}_6H_4$

(42)—(44) were obtained. The i.r. carbonyl stretching absorption in these compounds is normal (ca. 1655 cm$^{-1}$, KBr). The heterocyclic ring proton is strongly deshielded by the formyl group, and resonates at ca. $\delta 10\cdot 1$. Substituted 3-acylmethylene-3$H$-1,2-dithioles are formylated [30] at C-1', giving formylketones (27) which are also obtained by synthesis (Scheme 5), and by oxidation of the corresponding formyl-6a-thiathiophthenes (42) and (43) with peracetic acid.

Reactions involving attack at, or replacement of, sulphur atom in 6a-thiathiophthenes are reported.[36] The first case of a 6a-thiathiophthene reacting with a carbonyl reagent involves the reaction of 2-methyl-5-phenyl-6a-thiathiophthene with hydrazine. It was not established whether the hydrazone (45) or (46) was obtained. $S$-Alkylation of unsymmetrical

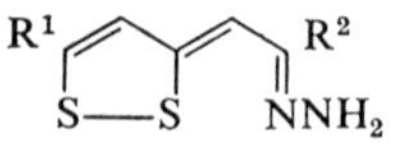

(45) $R^1 = Me, R^2 = Ph$
(46) $R^1 = Ph, R^2 = Me$

6a-thiathiophthenes with methyl iodide had previously been reported,[37] and the $S$-ethylation of two symmetrical 6a-thiathiophthenes (2) and (9) as well as a series of 3-cyano-substituted ones with triethyloxonium fluoroborate is now described.[36] The products (47) and (48) react with aniline to give the phenylimines (49), and condense with compounds containing reactive methylene groups. For example, the salts (47; R = Me) and (47; R = Ph) react with malononitrile to give the products (50).

On the basis of comparative u.v. and $^1$H n.m.r. spectral evidence Behringer favours a dithiole structure (49) for the products from the

---

[35] J. G. Dingwall, D. H. Reid, and K. Wade, *J. Chem. Soc.* (C), 1969, 913.
[36] H. Behringer and J. Falkenberg, *Chem. Ber.*, 1969, **102**, 1580.
[37] E. Klingsberg, *J. Org. Chem.*, 1968, **33**, 2915.

reaction of the salts (47) and (48) with aniline, rather than the isothiazole (51) or bicyclic structure (52). Similar conclusions had been reached by Klingsberg.[37]

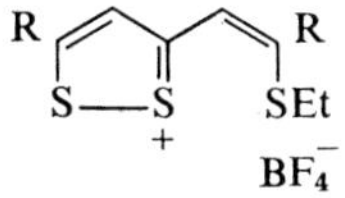

(47)

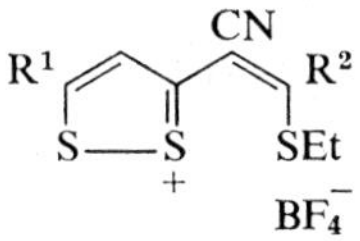

(48)

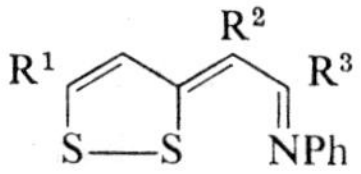

(49)

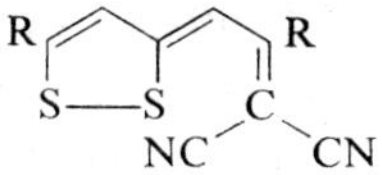

(50)

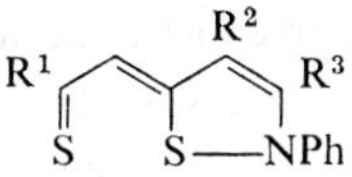

(51)

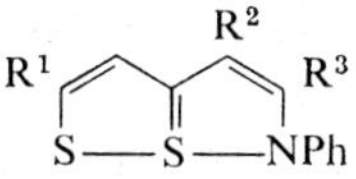

(52)

# 9
# 1,2- and 1,3-Dithioles

## 1 1,2-Dithiole-3-thiones and 1,2-Dithiole-3-ones

The well-known synthesis of 1,2-dithiole-3-thiones involving treatment of β-ketoesters with phosphorus pentasulphide has been extended to compounds of type (1), where $R^2$ is an electron-attracting group (CN, $CO_2Et$, Cl). The corresponding 'trithiones' (2) are obtained in good yield.[1] Esters of 3-hydroxythiophen-2-carboxylic acids give thieno-1,2-dithiole-3-thiones

$R^1 \cdot CO \cdot CH(R^2) \cdot CO_2Et$
(1)

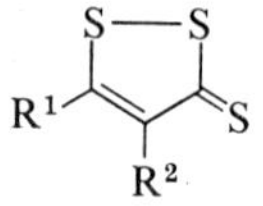

(2)

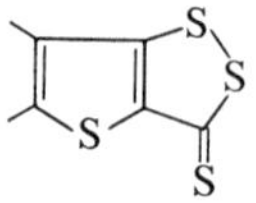

(3)

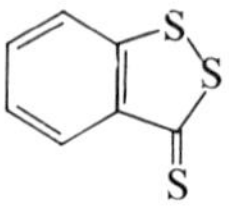

(4)

(3) when heated with phosphorus pentasulphide, but only in poor yield.[2]

A potentially useful one-step synthesis of benzo-1,2-dithiole-3-thione (4) has been described. This compound and the isomeric 1,3-dithiole-2-thione are produced when benzyne is generated in the gas phase (from phthalic anhydride) in the presence of carbon disulphide.[3]

The action of aqueous hydrochloric acid on the dithiolate (5) leads to an interesting dimeric product, formulated as (6).[4] Treatment of the dimer with aniline yields 5-amino-4-carbamoyl-1,2-dithiole-3-thione (7), previously obtained by the prolonged action of methanolic acetic acid on (5). The n.m.r. spectrum of the 4-thioacyl-1,2-dithiole-3-thione (8; $R^1 = R^2 = p\text{-Me}\cdot C_6H_4$) shows that the two aryl groups are not rendered equivalent by 'one-bond, no-bond resonance'. Attempts to prepare the individual compounds represented by (8; $R^1 = Ph$, $R^2 = p\text{-Me}\cdot C_6H_4$) and (8; $R^1 = p\text{-Me}\cdot C_6H_4$, $R^2 = Ph$) lead to identical mixtures of the isomers, which are therefore believed to be interconvertible. However, in cations of

[1] C. Trebaul and J. Teste, *Bull. Soc. chim. France*, 1969, 2456.
[2] J. Brelivet, P. Appriou, and J. Teste, *Compt. rend.*, 1969, **268**, C, 2231.
[3] E. K. Fields and S. Meyerson, *Tetrahedron Letters*, 1970, 629.
[4] L. Henriksen, *Chem. Comm.*, 1969, 1408.

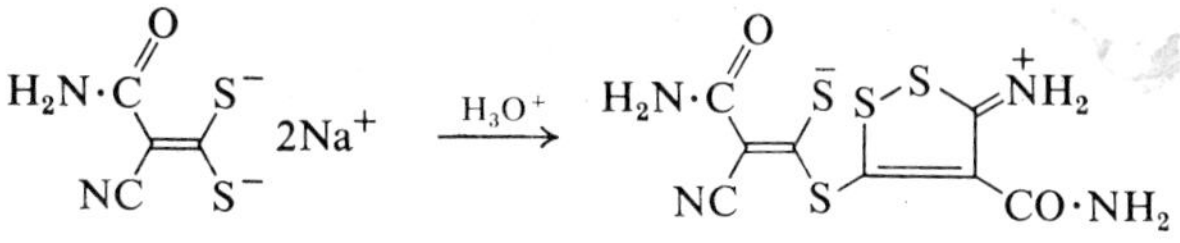

# 1,2- and 1,3-Dithioles

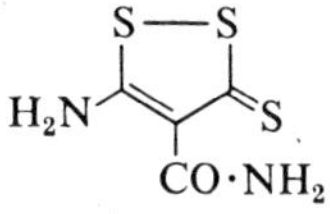

[Structures (5), (6), (7) shown]

type (9; R = p-Me·C$_6$H$_4$, p-MeO·C$_6$H$_4$, or MeS), the n.m.r. spectra suggest that the aryl groups do become equivalent.[5]

Further examples have been reported of cycloaddition reactions between 1,2-dithiole-3-thiones and acetylene derivatives, leading to 1,3-dithiol-2-ylidene-thioketones. 4,5-Diphenyl-1,2-dithiole-3-thione, with methyl

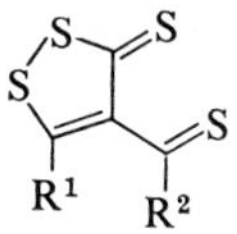

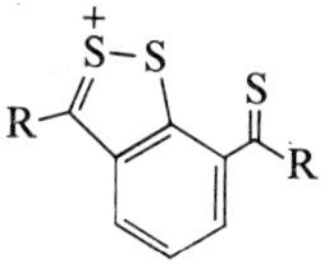

[Structures (8), (9) shown]

acetylenedicarboxylate, yields the 1:1 adduct (10) which can undergo further addition to give the 1:2 adduct (11). With ethyl propiolate, and with ethyl phenylpropiolate, the 1:1 adducts appear to be mixtures of the two possible isomers. 5-Phenyl-3-phenylimino-1,2-dithiole (12) adds methyl acetylenedicarboxylate but the product is the 1:2 adduct (13); with ethyl propiolate and ethyl phenylpropiolate, 1:1 adducts are obtained.[6]

Base-catalysed condensation of carbonyl sulphide with cyclohexeno-1,2-dithiole-3-thione (14) and methylation of the resulting anions with methyl

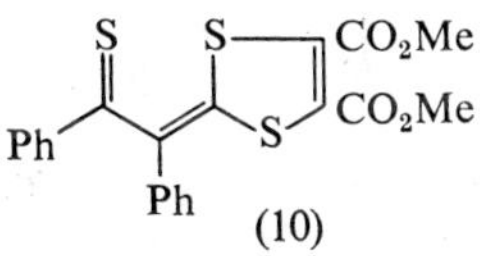

[Structures (10), (11), (12), (13) shown]

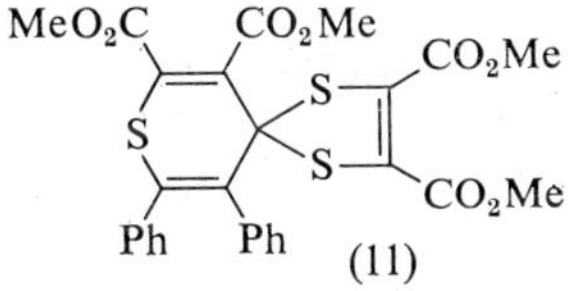

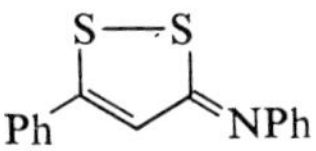

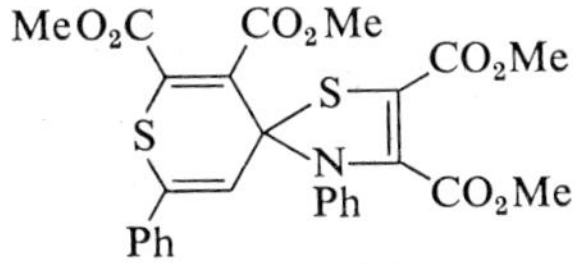

[5] E. I. G. Brown, D. Leaver, and T. J. Rawlins, *Chem. Comm.*, 1969, 83.
[6] J. M. Buchshriber, D. M. McKinnon, and M. Ahmed, *Canad. J. Chem.*, 1969, **47**, 2039.

iodide have been shown to yield two products (15) and (16), the product composition depending on the time interval before the methylation step.[7]

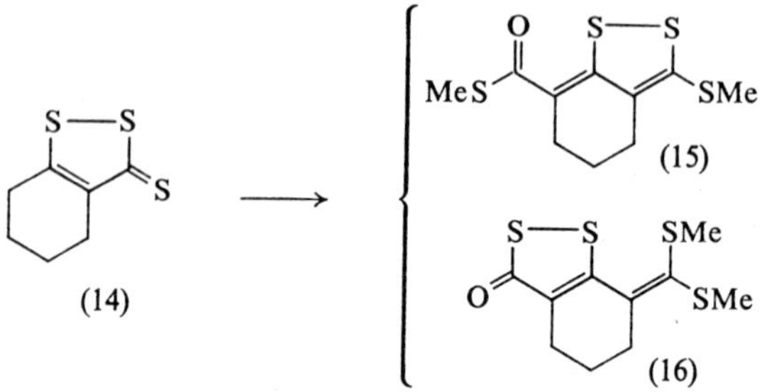

Sulphuration of the dithiolone derivative (16) gives the corresponding thione, an isomer of the product (17) previously obtained from (14) by condensation with carbon disulphide, with subsequent methylation and sulphuration.

The conversion of 4-chloro-5-diethylamino-1,2-dithiole-3-one to the 1,2-dithiin (18), with methanolic sodium methoxide, has been reported,[8] and the action of phenyl magnesium bromide on 4-aryl-5-dialkylamino-1,2-dithiole-3-ones (19) has been shown to bring about fission of the disulphide link, yielding thioamides of type (20), after treatment of the intermediate with water.[9] Benzo-1,2-dithiole-3-one similarly gives the ring-opened product (21).

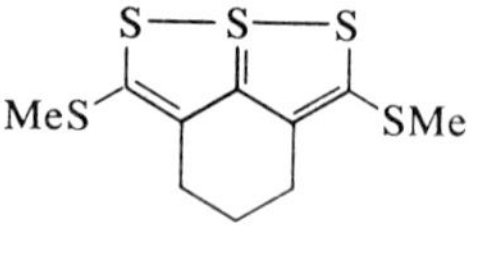

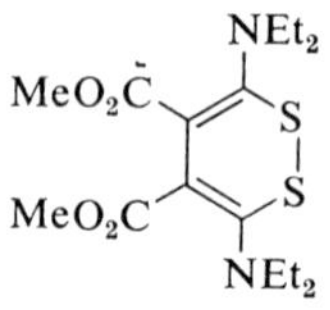

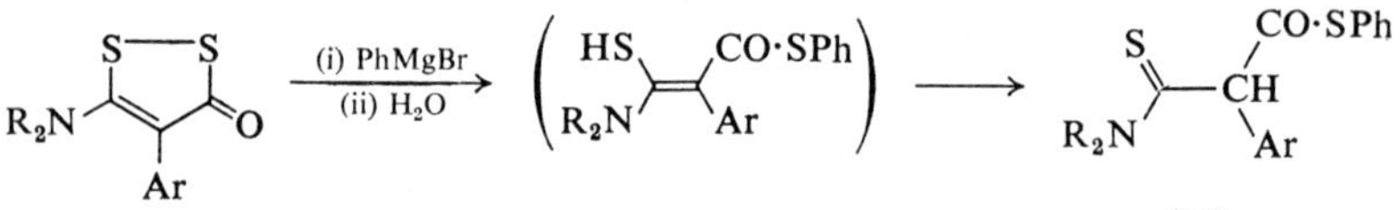

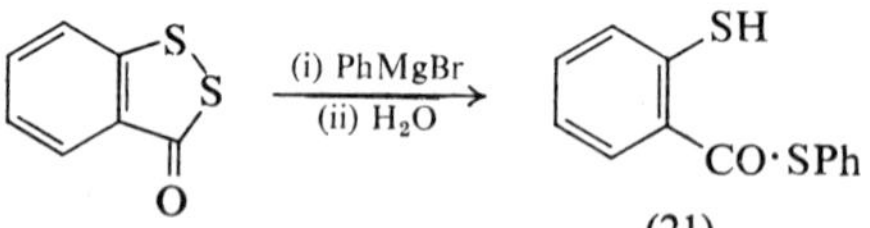

[7] J. L. Burgot and J. Vialle, *Bull. Soc. chim. France*, 1969, 3333.
[8] F. Boberg, H. Niemann, and K. Kirchhoff, *Annalen*, 1969, **728**, 32.
[9] F. Boberg and R. Schardt, *Annalen*, 1969, **728**, 44.

# 1,2- and 1,3-Dithioles

Details have appeared of an $X$-ray crystal structure study of the 1,2-dithiolylidene-aldehyde (22). The S—S···O system is nearly linear with S—S and S···O distances similar to those found previously in the dithiolylidene-ketone (23). In both compounds there is evidence of bond alternation in the conjugated system of carbon–carbon bonds.[10]

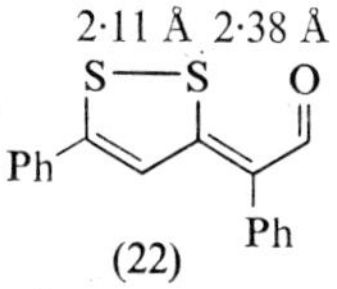

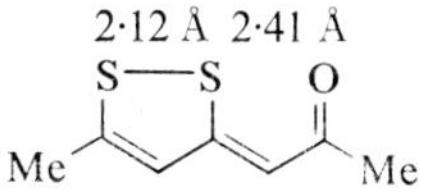

## 2 1,2-Dithiolium Salts

3,5-Diaryl-1,2-dithiolium tri-iodides (25) (and hence the corresponding perchlorates) are conveniently prepared by the addition of iodine to ethanolic solutions of $\beta$-diketones (24; $R^1 = R^2 = Ar$) saturated with hydrogen sulphide and hydrogen chloride.[11]

Acetylacetone (24; $R^1 = R^2 = Me$) and hydrogen sulphide have been found to react with ferrous or ferric ions in ethanol acidified with hydrochloric acid to give violet crystals of 3,5-dimethyl-1,2-dithiolium tetrachloroferrate (26).[12] Analogous highly coloured salts have been prepared from preformed dithiolium chlorides and metal salts. Since neither the dithiolium cations nor the complex anions (*e.g.* $FeCl_4^{2-}$) absorb strongly in the visible region, the intense colour of the solid salts is attributed to charge transfer in the crystals. An $X$-ray crystallographic study[13] shows that, in the tetrachloroferrate (26), there are rather close contacts between chlorine and sulphur atoms (distances 3·28—3·38 Å, significantly shorter than the van der Waals separation of 3·65 Å). The dithiolium cations in the salt are planar and the dimensions do not differ significantly from those in simple iodides.

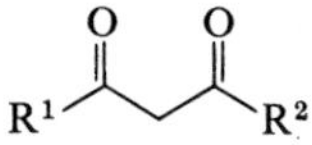

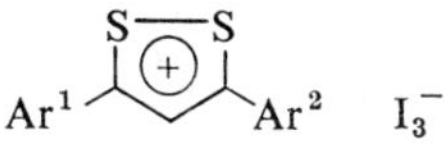

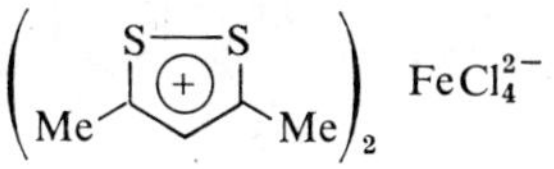

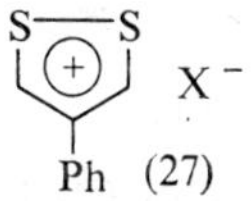

---

[10] A. Hordvik, E. Sletten, and J. Sletten, *Acta Chem. Scand.*, 1969, **23**, 1377.
[11] J. P. Guémas and H. Quiniou, *Compt. rend.*, 1969, **268**, C, 1805.
[12] G. A. Heath, R. L. Martin, and I. M. Stewart, *Chem. Comm.*, 1969, 54.
[13] H. C. Freeman, G. H. W. Milburn, C. E. Nockolds, P. Hemmerich, and K. H. Knauer, *Chem. Comm.*, 1969, 55.

*X*-Ray crystal structure studies have also been reported on 4-phenyl-1,2-dithiolium bromide (27; X = Br)[14] and thiocyanate (27; X = NCS).[15] The bromide, like the iodide studied previously, shows close contacts between halogen and sulphur atoms.

A series of 3-ethoxy-1,2-dithiolium salts of type (29) has been prepared by *O*-ethylation of 1,2-dithiole-3-ones (28) with triethyloxonium fluoro-

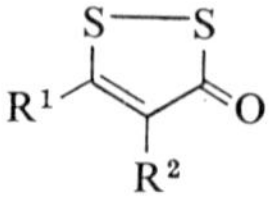

(28) 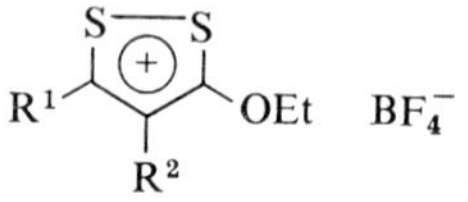  (29)

borate.[16] The salts readily revert to their precursors in warm ethanol and are not obtainable by the action of ethanol on 3-chloro-1,2-dithiolium salts. The electronic absorption spectrum of 1,2-dithiole-3-one (28; $R^1 = R^2 = H$) in concentrated sulphuric acid corresponds closely with that of the 3-ethoxy-1,2-dithiolium cation in the same solvent, indicating that protonation occurs on the carbonyl oxygen atom. Reactions of the 3-ethoxy-1,2-dithiolium salts with nucleophiles are apparently complex.

Extending the well-known reaction of 3-methylthio-1,2-dithiolium salts with active methyl(ene) groups, 3-methylthio-5-phenyl-1,2-dithiolium methosulphate (30; R = MeS, Ar = Ph) has been condensed with nitromethane and with nitroethane to give 1,2-dithiol-3-ylidenenitroalkanes (3-nitromethylene-1,2-dithioles) (31; R = H or Me), required for comparison with a series of structurally similar nitroso-compounds (32; R = H,

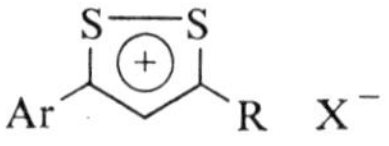

(30) 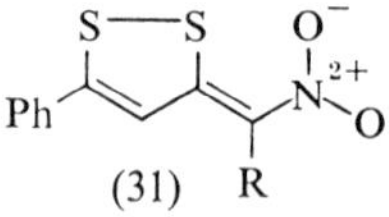  (31)

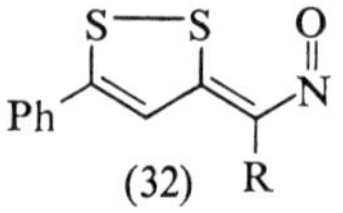

(32) 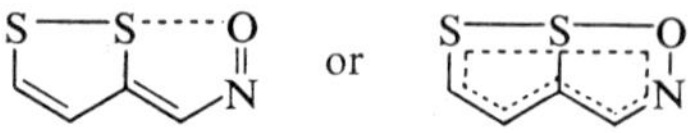  (33)

$CS_2Me$, $CO \cdot Ph$, or $CS \cdot NMe_2$) in which the presence of a bicyclic system (33) is suspected. Nitrosation of (31; R = H) gives (32; R = $NO_2$) and it is clear that the nitroso-group takes precedence over other groups (*e.g.* $CO \cdot Ph$, $CS_2Me$, $NO_2$) in competition for the position *cis* to the dithiole linkage.[17]

*X*-Ray crystallographic studies have shown [18] that the oxygen atom of the nitroso-group in (32; R = $NO_2$) is in close 'contact' with the dithiole

---

[14] A. Hordvik and R. M. Baxter, *Acta Chem. Scand.*, 1969, **23**, 1082.
[15] A. Hordvik and H. M. Kjøge, *Acta Chem. Scand.*, 1969, **23**, 1367.
[16] J. Faust and J. Fabian, *Z. Naturforsch.*, 1969, **24B**, 577.
[17] R. J. S. Beer and R. J. Gait, *Chem. Comm.*, 1970, 328.
[18] K. I. Reid and I. C. Paul, *Chem. Comm.*, 1970, 329.

# 1,2- and 1,3-Dithioles

sulphur atom (S···O distance, 2·10 Å), whereas in the nitro-compound (31; R = Br), obtained [17] by bromination of (31; R = H), the oxygen–sulphur distance is considerably greater (2·37 Å). The dimensions found for compound (31; R = Br) are generally similar to those reported for 1,2-dithiolylidene-ketones.

Further examples of the use of dithiolium salts in synthesis are provided by the preparation of a series of derivatives (34; $n$ = 2, 4, or 5) of cyclic ketones from either 3-aryl-1,2-dithiolium perchlorates (30; R = H, X = $ClO_4$) or 5-aryl-3-methylthio-1,2-dithiolium iodides (30; R = SMe, X = I),[19] and by the condensation of salts of the latter type with aroylacetaldehydes to give formyl-1,2-dithiolylidene-ketones (35), which can also be

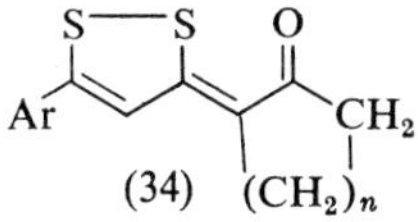

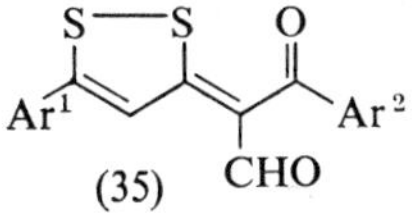

obtained from 3-formyl-6a-thiathiophthenes by reaction with mercuric acetate.[20]

The products obtained by condensing dithiolium salts of type (30; R = H) with methyl aroyldithioacetates depend on the reaction conditions. In neutral or slightly basic media, 3-aroyl-5-aryl-2-methylthio-6a-thiathiophthenes (36) are formed, but under more basic conditions the products are 3-aroyl-6-aryl-2$H$-thiopyran-2-thiones (37). Formation of the thiopyran derivatives presumably involves the intermediates and electron shifts

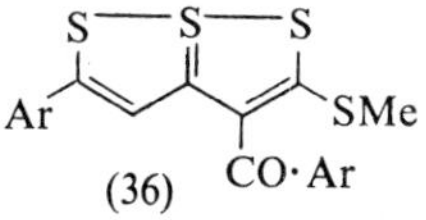

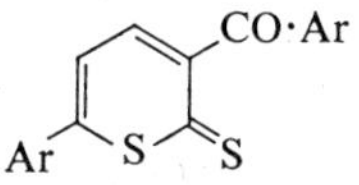

illustrated in structures (38) and (39). This interesting reaction is not confined to 3-aryl-1,2-dithiolium salts, but also occurs with 4-aryl-, 3,5-diaryl-, 3,4-diaryl-, and other substituted 1,2-dithiolium salts.[21]

The structures previously assigned to the products obtained by reaction of 3-aryl-1,2-dithiolium salts with primary aromatic amines have been confirmed by an alternative synthesis of one representative. Spectral data

[19] O. Coulibaly and Y. Mollier, *Bull. Soc. chim. France*, 1969, 3208.
[20] J. Bignebat and H. Quiniou, *Compt. rend.*, 1969, **269**, *C*, 1129; 1970, **270**, *C*, 83.
[21] F. Clesse and H. Quiniou, *Compt. rend.*, 1969, **268**, *C*, 637; 1969, **269**, *C*, 1059.

suggest that these products should be formulated as *cis*-enaminethiones (40).[22]

The chemistry of 3-halogeno-1,2-dithiolium salts (41) has been extended by the observation that these salts react with compounds containing 'positive' halogen, for example $NN$-dichlorosulphonamides (42) and $NN$-dichlorourethanes (43), to give derivatives of 3-imino-1,2-dithioles (44; $R^3 = SO_2R$, $CO_2R$). The resonance-stabilised cation (41; $R^1$ = MeNPh, $R^2$ = Cl) does not react with $NN$-dichlorobenzenesulphonamide.[23]

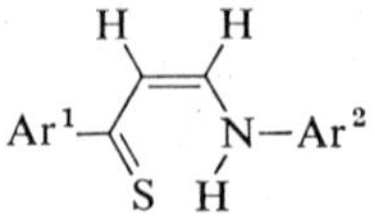

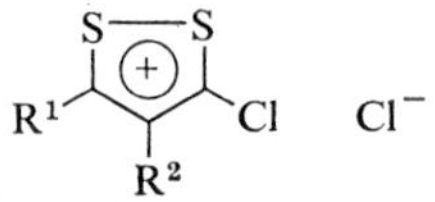

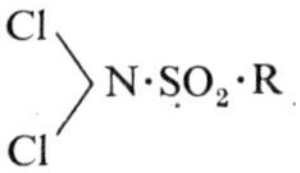

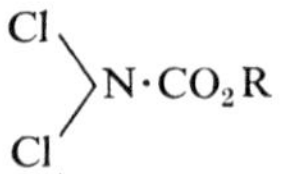

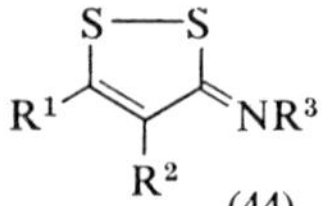

1,2-Dithiolium salts of type (46; R = SEt or $Me_2N$) feature in several reports. Thus, 6a-thiathiophthenes (45) are ethylated by triethyloxonium fluoroborate to give salts (46; R = SEt, X = $BF_4$), which react with aniline to give anils (47; R = Ph). The anils are hydrolysed under acidic conditions to 1,2-dithiol-3-ylideneketones (48).[24] Vilsmeier salts (46; R = $Me_2N$), obtained from 3-methyl(ene)-1,2-dithiolium salts by condensation with dimethylthioformamide,[25] react with hydrosulphide anion yielding 6a-thiathiophthenes and with methylamine to give $N$-methylimines (47;

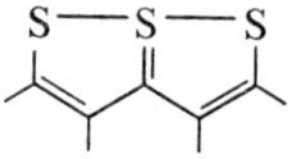

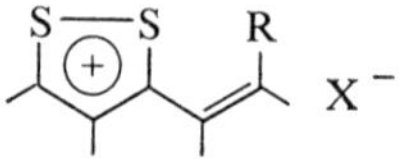

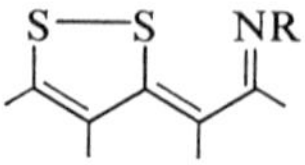

R = Me).[26] This latter sequence has been used in a synthesis of the apparently symmetrical system (49).

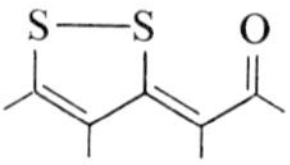

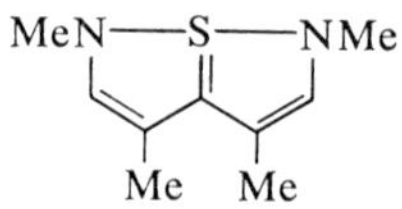

[22] J. Bignebat, H. Quiniou, and N. Lozac'h, *Bull. Soc. chim. France*, 1969, 127.
[23] F. Boberg and R. Wiedermann, *Annalen*, 1969, **728**, 36.
[24] H. Behringer and J. Falkenberg, *Chem. Ber.*, 1969, **102**, 1580.
[25] J. G. Dingwall, D. H. Reid, and K. Wade, *J. Chem. Soc.* (*C*), 1969, 913.
[26] D. H. Reid and J. D. Symon, *Chem. Comm.*, 1969, 1314.

# 1,2- and 1,3-Dithioles

## 3 Derivatives of 1,3-Dithiole

In an interesting study of new types of heterofulvalenes, 2-ethylthio-4-phenyl-1,3-dithiolium perchlorate (50; X = S) and its selenium analogue (50; X = Se) have been shown to combine with 2,4-diphenylpyrrole in acetonitrile at room temperature forming coloured crystalline charge-transfer complexes (51; X = S or Se). In hot acetonitrile, the complexes are slowly converted into condensation products, isolated as the free bases (52; X = S or Se), after treatment with triethylamine. Other examples reported include compounds (53) and (54), containing the imidazole nucleus.[27]

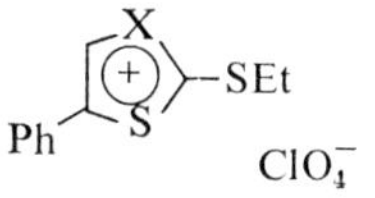

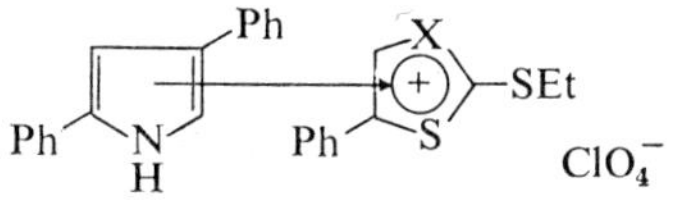

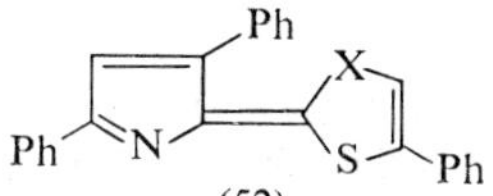

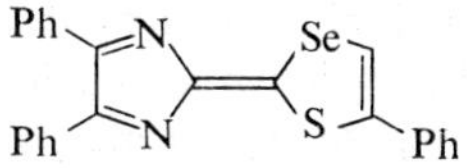

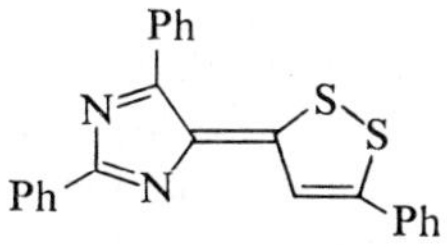

4-Aryl-2-dialkylamino-1,3-dithiolium salts (56; R = $NR^1_2$, X = $BF_4$), conveniently prepared by the action of phosphorus pentasulphide and hydrofluoroboric acid on the readily available dithiocarbamates (55), are reduced with sodium borohydride to 2-dialkylamino-1,3-dithioles (57; R = $NR^1_2$) which immediately yield 4-aryl-1,3-dithiolium hydrogen sulphates or perchlorates (56; R = H) on treatment with the corresponding acid in ethanolic solution.[28]

A detailed study has been made of the reactions of 4-phenyl-1,3-dithiolium hydrogen sulphate (56; R = H, X = $HSO_4$) with a wide range of nucleophiles.[29] The products include 2-dialkylamino-1,3-dithioles (57; R = $NR^1_2$), 2-hydroxy-4-phenyl-1,3-dithiole (57; R = OH), 2-methoxy-4-phenyl-1,3-dithiole (57; R = OMe), and also 4-phenyl-1,3-dithiole (57; R = H), obtained by reaction with lithium aluminium hydride. The reaction between the dithiolium salt and triethylamine yields the bis-1,3-dithiolylidene compound (58), presumably *via* a carbenoid intermediate. 4,5-Diphenyl-

---

[27] R. Weiss and R. Gompper, *Tetrahedron Letters*, 1970, 481.
[28] A. Takamizawa and K. Hirai, *Chem. and Pharm. Bull. (Japan)*, 1969, **17**, 1924.
[29] A. Takamizawa and K. Hirai, *Chem. and Pharm. Bull. (Japan)*, 1969, **17**, 1931.

R$_2$N·CS·S·CH$_2$·CO·Ar
(55)

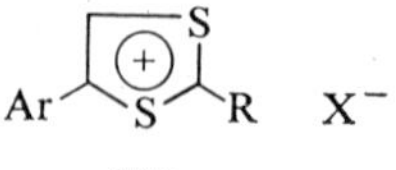

(56)

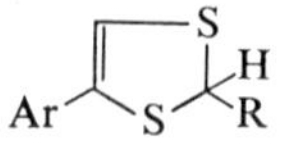

(57)

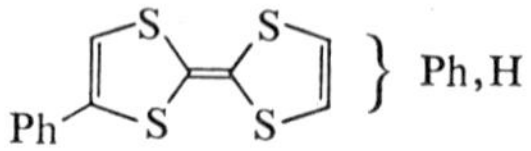

(58) } Ph, H

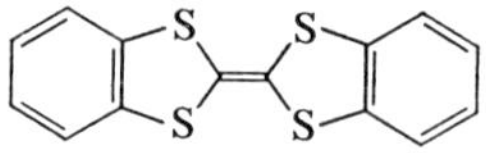

(59)

1,3-dithiolium perchlorate reacts similarly with triethylamine.[30] The bisbenzo-1,3-dithiole derivative (59) has been isolated as a minor product from the benzyne–carbon disulphide vapour-phase reaction[3] mentioned earlier in this chapter.

The product previously thought to be mesionic anhydro-4-hydroxy-2-phenyl-1,3-dithiolium hydroxide (60) has been shown to be incorrectly formulated. The authentic compound (60), prepared by the action of acetic anhydride and triethylamine on the dithioester (61), adds methyl acetylenedicarboxylate with elimination of carbonyl sulphide, giving a thiophen derivative (62).[31]

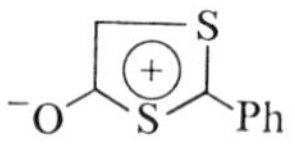

(60)    Ph·CS·S·CH$_2$·CO$_2$H
        (61)

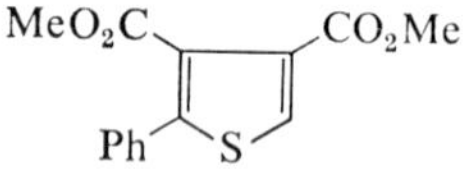

(62)

p-Benzoquinone reacts with the dithiolate (63) to produce the benzo-1,3-dithiole derivative (64), chlorination of which in tetrahydrofuran gives the oxidised product (65). Further reaction of the chlorinated quinone (65) with dithiolate (63) then yields the tricyclic compound (66), which is also obtainable directly from chloranil.[32] Nitrobenzo-1,3-dithioles similar to (64) have also been prepared.[33]

Although 'isothiathiophthenes' show some similarities to 6a-thiathiophthenes, an X-ray crystallographic study of the p-bromophenyl compound (67) suggests that members of the iso-series are probably best regarded as 1,3-dithiol-2-ylidene-thioketones and not as bicyclic systems. The distance between the adjacent thione and dithiole sulphur atoms in compound (67)

[30] K. M. Pazdro, *Roczniki Chem.*, 1969, **43**, 1089.
[31] K. T. Potts and U. P. Singh, *Chem. Comm.*, 1969, 569.
[32] K. Klemm and B. Geiger, *Annalen*, 1969, **726**, 103.
[33] M. Yokoyama, *J. Org. Chem.*, 1970, **35**, 283.

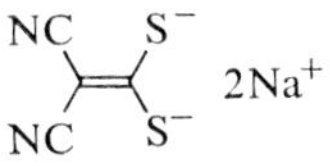

is 2·91 Å.[34] The thione (68; X = S) reacts with triethyloxonium fluoroborate to give a salt which can be converted, by the action of aniline, to the anil (68; X = NPh) and hence, by acid hydrolysis, to the ketone (68; X = O).[24]

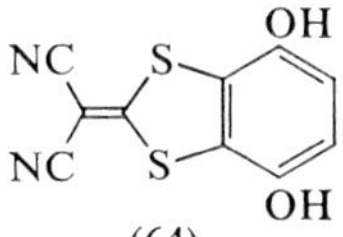

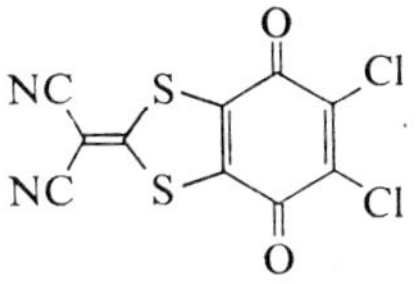

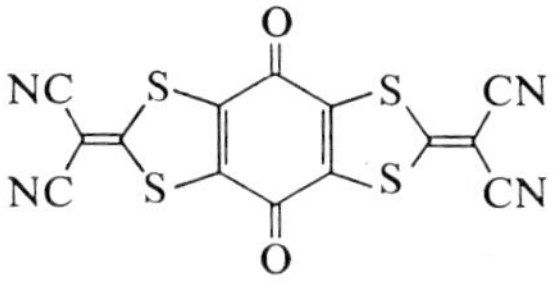

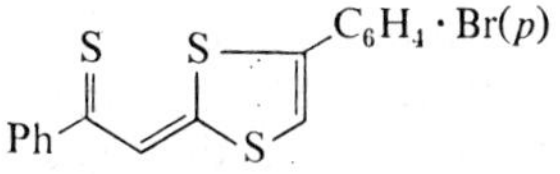

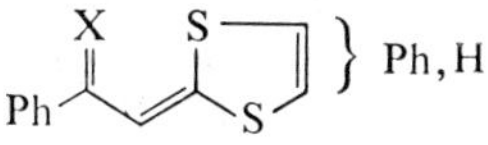

[34] R. J. S. Beer, D. Frew, P. L. Johnson, and I. C. Paul, *Chem. Comm.*, 1970, 154.

# 10
# Thiopyrans

## 1 2H-Thiopyrans and Related Compounds

Acetylenic sulphides of type (1) yield 2H-thiopyrans (3) when heated in polar solvents. The cyclisation proceeds *via* the related allenes (2), which have been isolated in some cases (*e.g.* when $R^1$ = SEt).[1] Applied to propargyl vinyl sulphide (1; $R^1 = R^2 = R^3$ = H), this method gives 2H-thio-

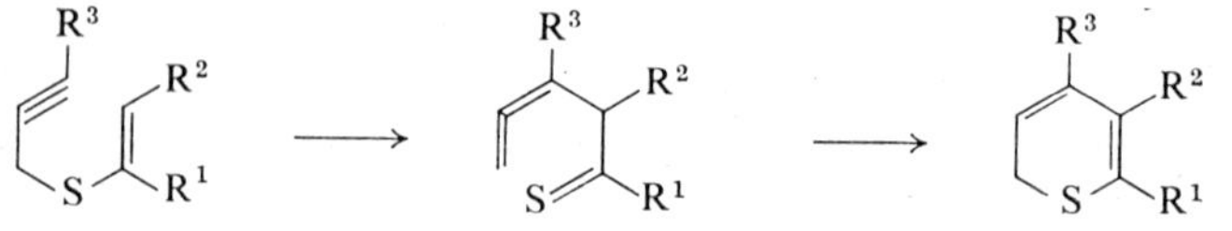

pyran (3; $R^1 = R^2 = R^3$ = H) free from 4H-thiopyran.[2] The two isomeric thieno-2H-thiopyrans, (4) and (5), are similarly obtained from propargyl thienyl sulphides.[3]

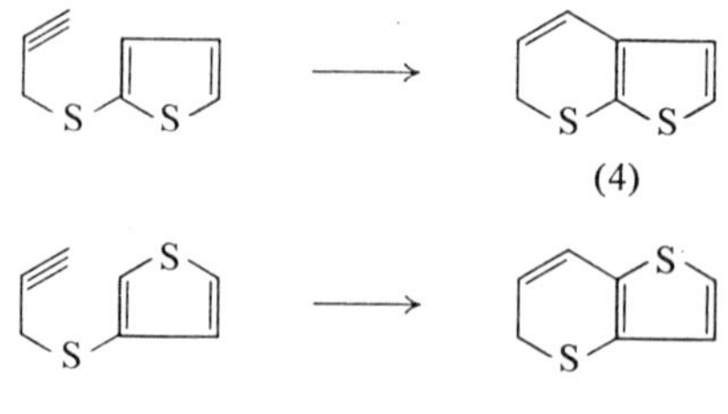

The formation of benzothiopyrylium salts (8) and thiochromans (9) by treatment of γ-oxosulphides (6) with 70% perchloric acid probably involves benzo-2H-thiopyrans (7) as intermediates.[4] It is suggested that intermolecular hydride transfer occurs between the thiochromene (7) and its protonated derivative (10). When the reaction is carried out in the presence of triphenylmethyl chloride, the yield of the thiopyrylium perchlorate is

[1] P. J. W. Schuijl, H. J. T. Bos, and L. Brandsma, *Rec. Trav. chim.*, 1969, **88**, 597.
[2] L. Brandsma and P. J. W. Schuijl, *Rec. Trav. chim.*, 1969, **88**, 30.
[3] L. Brandsma and H. J. T. Bos, *Rec. Trav. chim.*, 1969, **88**, 732.
[4] B. D. Tilak and G. T. Panse, *Indian J. Chem.*, 1969, **7**, 191.

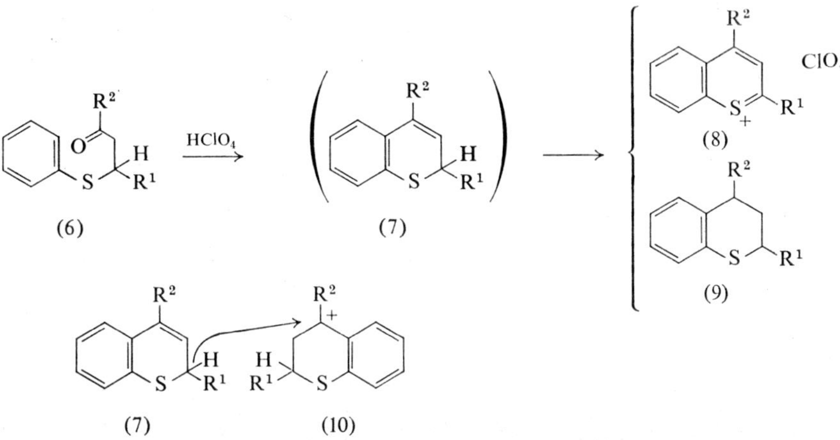

increased. The method has been used to prepare naphthothiopyrylium salts.

3,4-Dimethyl-$\Delta^3$-thiochromene (11) has been shown to undergo the disproportionation reaction shown, with stereoselective hydride transfer, when treated with perchloric acid. The products are the benzothiopyrylium salt (12) and the thiochroman (13; 85% cis-isomer).[5] The same selectivity is observed in the cyclisation of the γ-oxosulphide (14), which gives the thiopyrylium salt (15) and the thiochroman (16; 75% cis-isomer). Reduction of the salt with sodium borohydride yields the thiochromene (17), which is dehydrogenated, when heated with diphenyl disulphide, to the dibenzo-2H-thiopyran (18). Treatment of (18) with triphenylmethyl chloride in acetic acid containing perchloric acid gives the dibenzothiopyrylium perchlorate (19). Other polycyclic thiopyrylium salts have been synthesised by similar routes.[6]

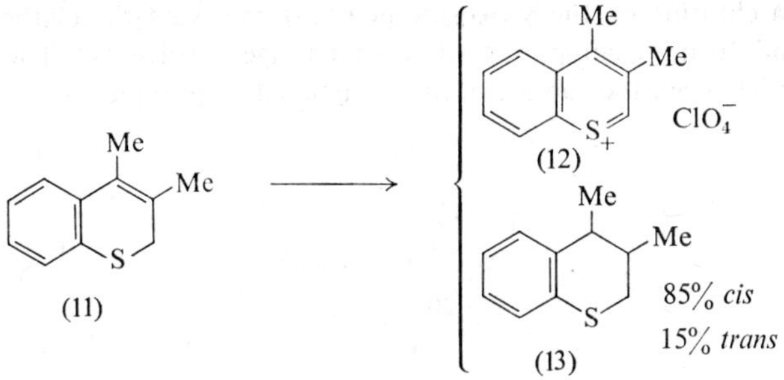

4-Phenylbenzothiopyrylium perchlorate (20) is rapidly decomposed by aqueous sodium carbonate giving 4-phenylthiocoumarin (21) and 4-phenyl-

---

[5] B. D. Tilak, R. B. Mitra, and Z. Muljiani, *Tetrahedron*, 1969, 25, 1939.
[6] S. L. Jindal and B. D. Tilak, *Indian J. Chem.*, 1969, 7, 637.

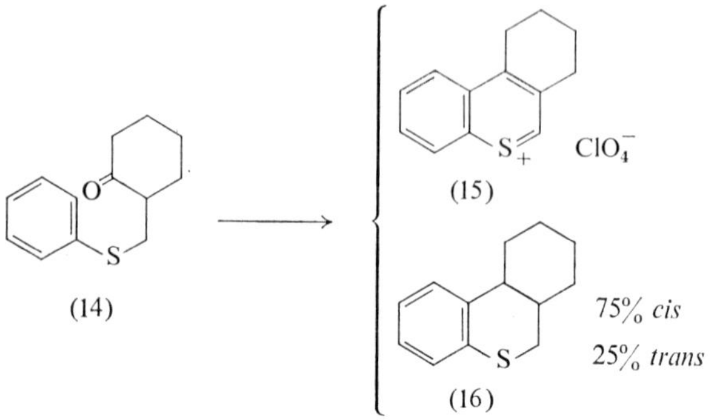

Δ³-thiochromene (22). Hydride transfer again appears to be involved, the suggested mechanism being that shown.⁷

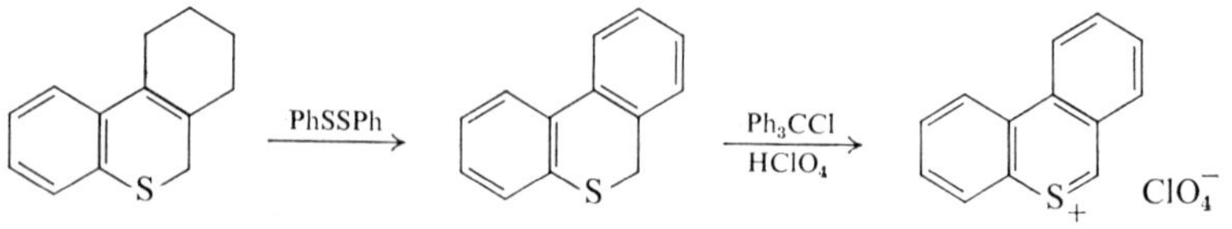

The thiochroman (24), prepared by the action of phenol and anhydrous hydrogen chloride on the γ-oxosulphide (23), is a versatile clathrate-host compound, forming complexes with a wide range of solvents.⁸ The ethanol clathrate has been the subject of an X-ray crystallographic study.⁹

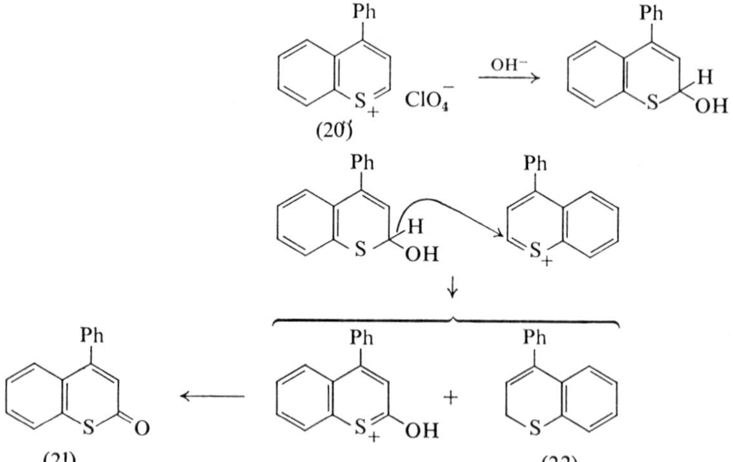

[7] B. D. Tilak and G. T. Panse, *Indian J. Chem.*, 1969, **7**, 315.
[8] D. D. MacNicol, *Chem. Comm.*, 1969, 836.
[9] D. D. MacNicol, H. H. Mills, and F. B. Wilson, *Chem. Comm.*, 1969, 1332.

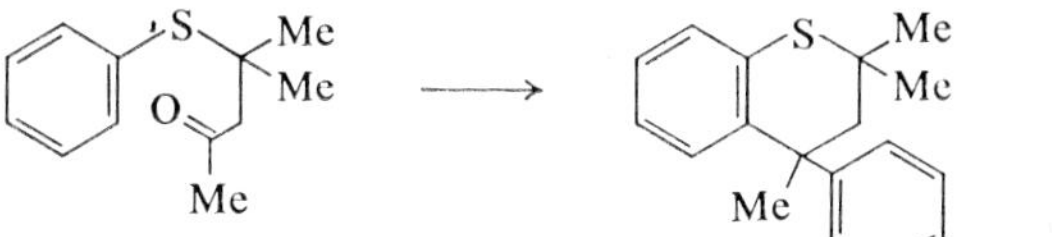

(23) → (24)

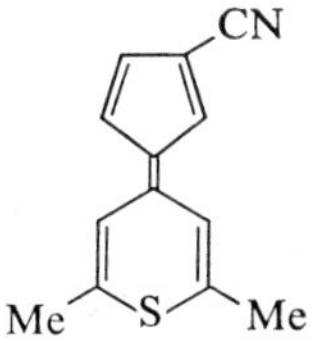

(25)

The cyclopentadienylidene-4H-thiopyran (25) has been shown to have an almost planar structure by X-ray crystallography.[10] Compounds of this general type are discussed in a review of earlier work on hetero-analogues of sesquifulvalenes.[11]

## 2 4H-Thiopyran-4-ones and 4H-Thiopyran-4-thiones

4H-Thiopyran-4-ones (26) undergo base-catalysed hydrogen–deuterium exchange at the 2- and 6-positions. The exchange is rapid at room temperature and much faster than that observed with structurally similar pyridones.[12]

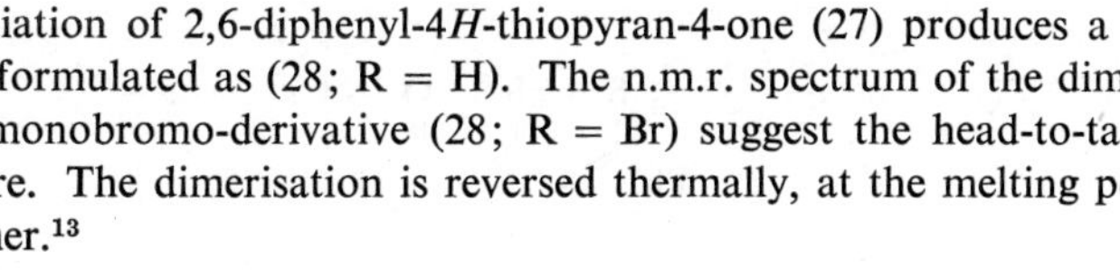

(26)   (27)   (28)

Irradiation of 2,6-diphenyl-4H-thiopyran-4-one (27) produces a yellow dimer, formulated as (28; R = H). The n.m.r. spectrum of the dimer and of its monobromo-derivative (28; R = Br) suggest the head-to-tail, *anti* structure. The dimerisation is reversed thermally, at the melting point of the dimer.[13]

2,3-Cycloalkenothiochromones (31) have been prepared by the action of enamines derived from cyclic ketones on the mixed anhydride (30), formed

---
[10] H. Burzlaff and I. Kawada, *Chem. Ber.*, 1969, **102**, 3632.
[11] G. Seitz, *Angew. Chem. Internat. Edn.*, 1969, **8**, 478.
[12] P. Beak and E. M. Monroe, *J. Org. Chem.*, 1969, **34**, 589.
[13] N. Sugiyama, Y. Sato, H. Kataoka, C. Kashima, and K. Yamada, *Bull. Chem. Soc. Japan*, 1969, **42**, 3005.

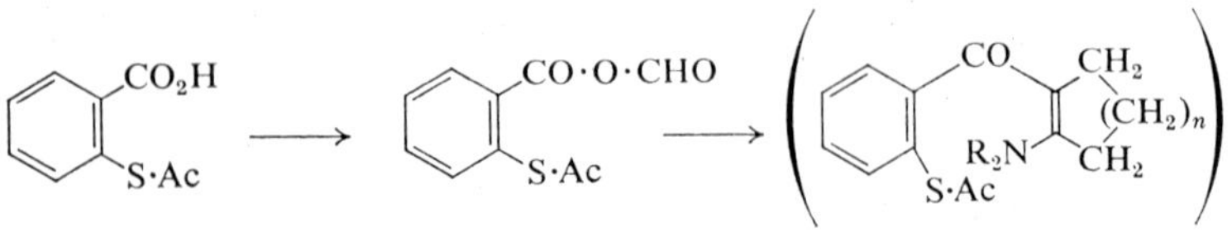

*in situ* by treatment of the acid (29) with ethyl chloroformate and triethylamine.[14] The sulphone (32) is cyclised to the 1,1-dioxothiopyrone derivative (33) when heated (to 270 °C) with concentrated sulphuric or polyphosphoric acid.[15]

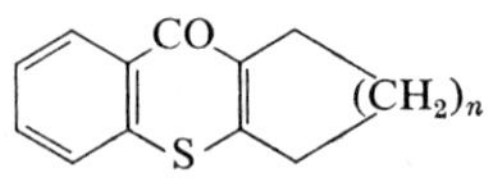

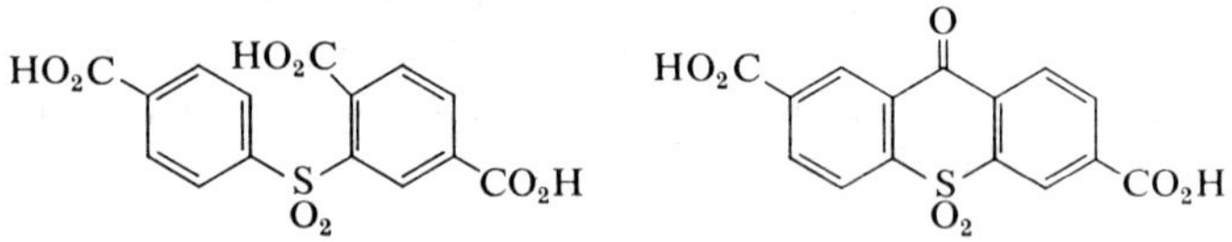

4*H*-Thiopyran-4-thiones (34) are resistant to the action of sodium hydroxide, sodium sulphide, or sodium hydrogen sulphide in aqueous ethanol, but in DMF ring-opening reactions occur. Thus, when the parent thione (34; R = H) is treated with sodium sulphide, a deep red solution presumably containing the dianion (35) is obtained, and addition of potassium ferricyanide causes oxidation to 6a-thiathiophthene (36), which hitherto has been difficult of access. Ferricyanide oxidation of the

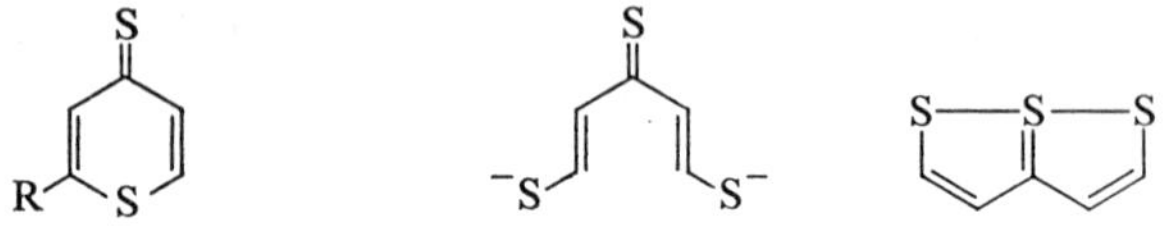

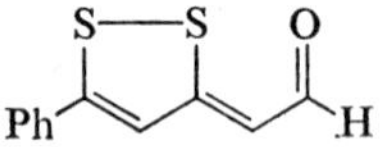

[14] G. V. Boyd, D. Hewson, and R. A. Newberry, *J. Chem. Soc.* (*C*), 1969, 935.
[15] O. F. Bennett and P. Gauvin, *J. Org. Chem.*, 1969, **34**, 4165.

intermediate obtained by the action of alkali on 2-phenyl-4$H$-thiopyran-4-thione (34; R = Ph) gives the dithiolylideneacetaldehyde (37).[16]

## 3 2$H$-Thiopyran-2-ones, 2$H$-Thiopyran-2-thiones, and Related Compounds

4-Hydroxy-6-methyl-2$H$-thiopyran-2-ones of type (38; $R^1$ = Me, $R^2$ = $CO_2Et$, $R^3$ = Ph or $PhCH_2$) undergo condensation reactions with aromatic aldehydes, leading to styryl derivatives (39). Compounds of type (38;

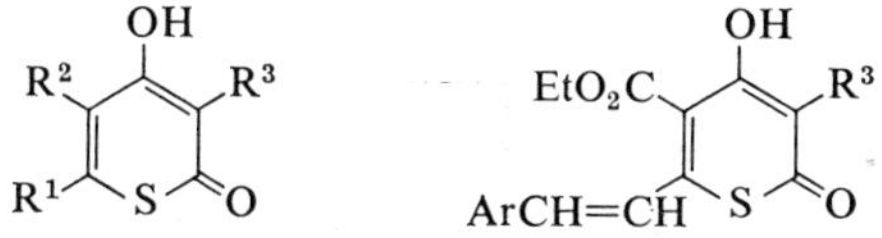

$R^1$ = Me or Ph, $R^2$ = H, $R^3$ = Ph or $PhCH_2$) react with ammonia to give the related pyridones.[17]

4,5,6-Triphenyl-2$H$-pyran-2-thione (40) is converted by the action of aqueous dimethylamine to a mixture of the thioamide (41) and 4,5,6-triphenyl-2$H$-thiopyran-2-one (42). The mechanism suggested involves the tautomeric intermediate (43).[18]

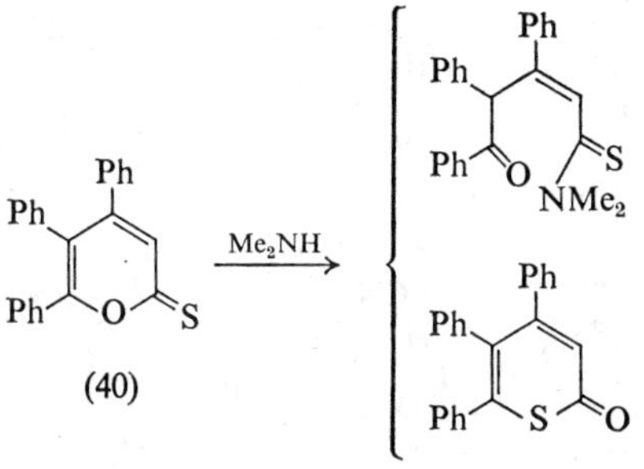

A series of merocyanine dyestuffs has been prepared from 4-hydroxy-thiocoumarin (44; X = S)[19] and a detailed study has been made of the mass spectra of 4-hydroxyselenocoumarin (44; X = Se) and the dicoumarol analogue (45). The fragmentations indicate a strong tendency to form the

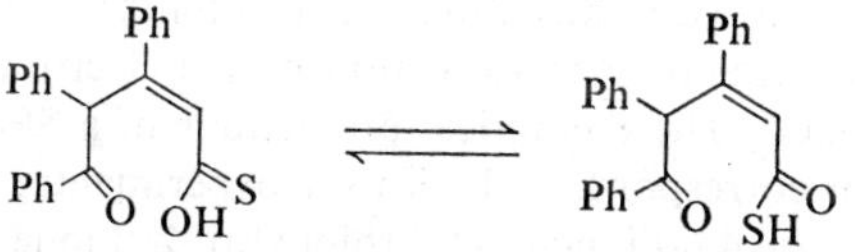

---

[16] J. G. Dingwall, D. H. Reid, and J. D. Symon, *Chem. Comm.*, 1969, 466.
[17] F. K. Splinter and H. Arold, *J. prakt. Chem.*, 1969, 311, 869.
[18] I. E. El-Kholy, F. K. Rafla, and M. M. Mishrikey, *J. Chem. Soc.* (C), 1969, 1950.
[19] P. C. Rath and K. Rajagopal, *Indian J. Chem.*, 1969, 7, 1273.

benzoselenophen ring (46). Compound (45) gave no evidence of anticoagulant activity when tested in rats.[20]

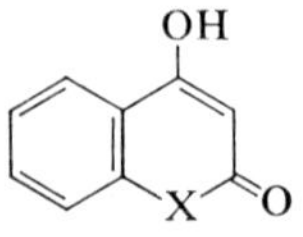

(44)

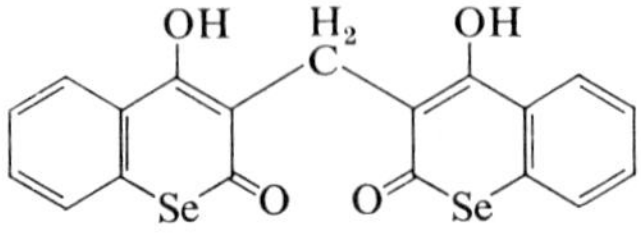

(45)

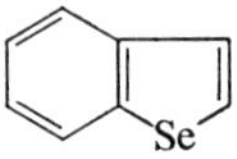

(46)

The reaction of dithio-acids (48; $R^2$ = H, Me, or Ar) with β-chlorovinyl-aldehydes or -ketones (47; $R^1$ = H, Me, or Ar) in the presence of bases provides a general route to 2*H*-thiopyran-2-thiones (49), with a variety of substitution patterns. 3-Cyano-4,6-diphenyl-2*H*-thiopyran-2-one (52; X = O), obtained by the action of the acid chloride (51) on the sodium derivative (50) of benzoylthioacetophenone, yields the corresponding thione (52;

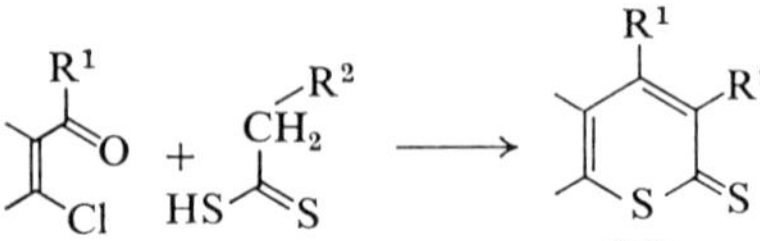

X = S) on treatment with phosphorus pentasulphide.[21] These studies have contributed to an extensive investigation of the electronic spectra of 2*H*-thiopyran-2-thiones. The theoretical predictions of a SCF–PPP molecular orbital treatment correspond well with the experimental data.[22]

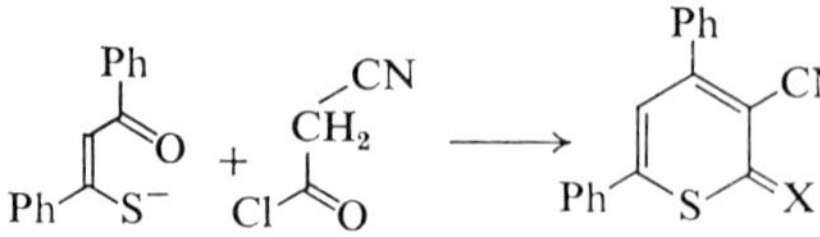

The chemistry of 4,6-diphenyl-2*H*-thiopyran-2-thione (53; X = S) and some of its derivatives has been studied in detail.[23] The thione readily forms thiopyrylium salts (54; R = MeS) with methylating agents; treatment with acetic anhydride yields the 2-(acetylthio)-thiopyrylium cation, isolated as the perchlorate (54; R = SAc, X = $ClO_4$), which is very easily

---

[20] N. P. Buu-Hoï, M. Mangane, O. Périn-Roussel, M. Renson, A. Ruwet, and M. Maréchal, *J. Heterocyclic Chem.*, 1969, **6**, 825.
[21] S. Scheithauer and R. Mayer, *Z. Chem.*, 1969, **9**, 59.
[22] J. Fabian and G. Laban, *Tetrahedron*, 1969, **25**, 1441.
[23] J. Faust, G. Speier, and R. Mayer, *J. prakt. Chem.*, 1969, **311**, 61.

hydrolysed, reverting to the starting material. 2-Halogenothiopyrylium salts (54; R = X = Cl) and (54; R = X = Br) are obtained by the action of oxalyl chloride and oxalyl bromide respectively on (53; X = S), and with phosphorus pentachloride both the thione (53; X = S) and its oxygen analogue (53; X = O) (obtained from the thione by treatment with mercuric acetate) yield the chloride (54; R = X = Cl).

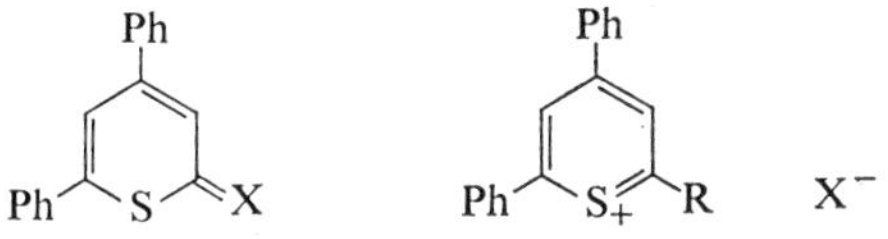

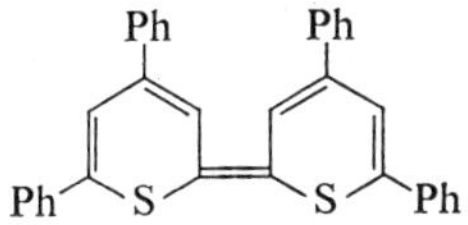

Other reactions of 4,6-diphenyl-2$H$-thiopyran-2-thione include those with primary amines, leading to 2-imino-compounds (53; X = NR), with hydroxylamine and hydrazine, giving the oxime and hydrazone, and the reaction brought about by copper bronze in hot inert solvents, which yields the bis-thiopyranylidene derivative (55). In contrast to the behaviour of the thione, 4,6-diphenyl-2$H$-thiopyran-2-one (53; X = O) yields $N$-alkylpyridones with primary aliphatic amines. The salts (54; R = X = Cl or Br) react with sulphur in pyridine to give the thione (53; X = S).

3-Acyl-2$H$-thiopyran-2-thiones (58) have been isolated from condensations of 1,2-dithiolium salts (56) with methyl acyldithioacetates (57) in basic media [24] (see Chapter 9, p. 341), and 4,6-disubstituted-3-acyl-2$H$-thiopyran-2-thiones (60) are obtained from the reaction of $\beta$-ketodithioacids with acylacetylenes of type (59).[25]

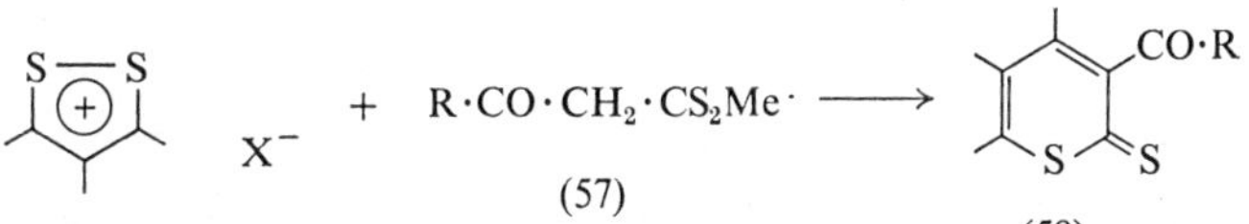

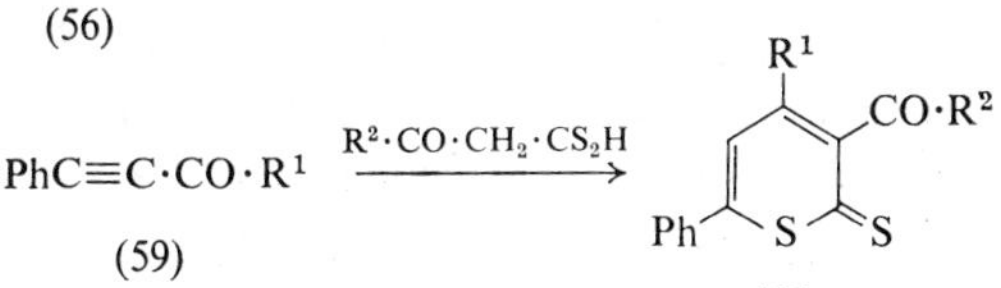

[24] F. Clesse and H. Quiniou, *Compt. rend.*, 1969, **268**, C, 637; 1969, **269**, C, 1059.
[25] J. P. Pradère, A. Guenec, G. Duguay, J. P. Guemas, and H. Quiniou, *Compt. rend.*, 1969, **269**, C, 929.

## 4 Complex Thiopyran and Selenopyran Derivatives

Variously substituted thioxanthenes and selenoxanthenes have been prepared for pharmacological evaluation,[26] and the two naphthoselenopyrans (61) and (62) have been described.[27]

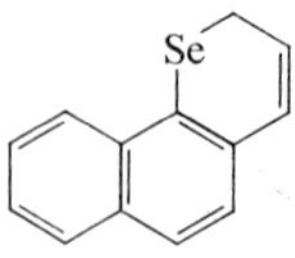

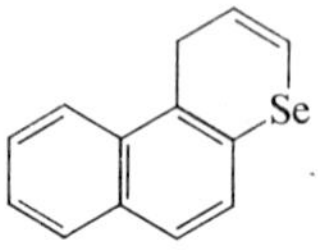

Several papers discuss thienothiopyran derivatives. The alcohol (63; R = H or Me), obtained by sodium borohydride reduction of the corresponding ketone, is converted to the blue-green thienothiopyrylium perchlorate (64; R = H or Me) by the action of triphenylmethyl perchlorate in acetic acid. Similar sequences of reactions lead to compound (65),[28] and to the four intensely-coloured benzothienothiopyrylium salts (66), (67), (68), and (69).[29] Hückel M.O. calculations on this series of compounds give results which are seriously at variance with the observed spectral data.

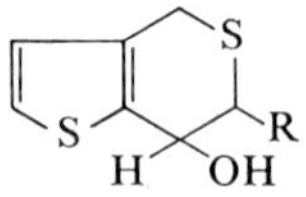

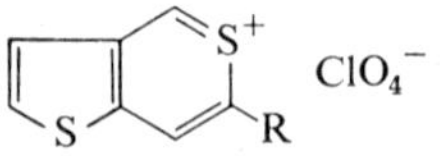

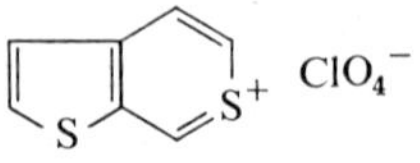

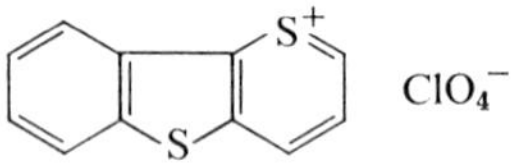

The thienothiochromone (70) has been converted, by conventional routes, to the thiopyrylium salts (71; R = H) and (71; R = Ph).[30]

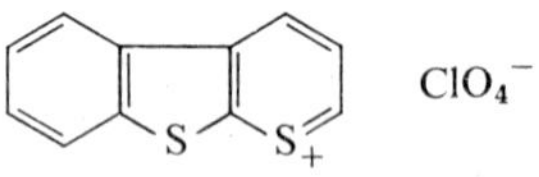

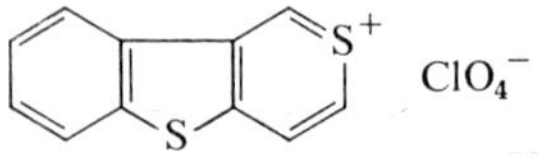

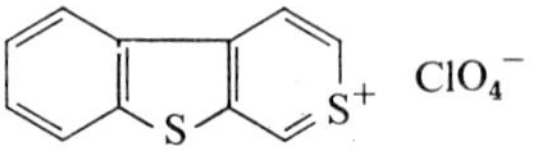

Numerous thiopyran derivatives with fused nitrogen-containing rings have been reported. The unstable pyrido-2H-thiopyran (74) is obtained

---

[26] J. F. Muren and B. M. Bloom, *J. Medicin. Chem.*, 1970, **13**, 14; J. F. Muren, *J. Medicin. Chem.*, 1970, **13**, 140; K. Sindelar, E. Svatek, J. Metysova, J. Metys, and M. Protiva, *Coll. Czech. Chem. Comm.*, 1969, **34**, 3792.
[27] N. Bellinger and P. Cagniant, *Compt. rend.*, 1969, **268**, C, 1385.
[28] T. E. Young and C. R. Hamel, *J. Org. Chem.*, 1970, **35**, 821.
[29] T. E. Young and C. R. Hamel, *J. Org. Chem.*, 1970, **35**, 816.
[30] B. D. Tilak and S. L. Jindal, *Indian J. Chem.*, 1969, **7**, 948.

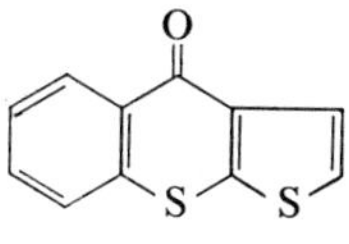

from the β-keto-ester (72) by lithium aluminium hydride reduction to the diol (73), and subsequent cyclisation and dehydration with hydrobromic acid.[31] Application of the Fischer indole synthesis to thiochromanones of type (75; R = Me, MeO, or Cl) gives polycyclic indolo-2H-thiopyrans (76), which are oxidised by the action of picric acid in boiling ethanol to compounds of type (77).[32]

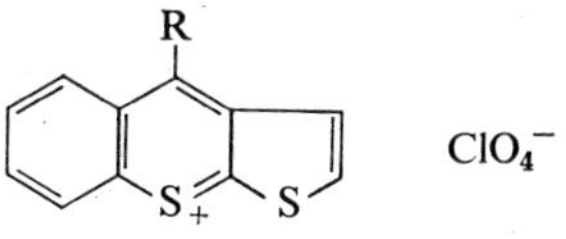

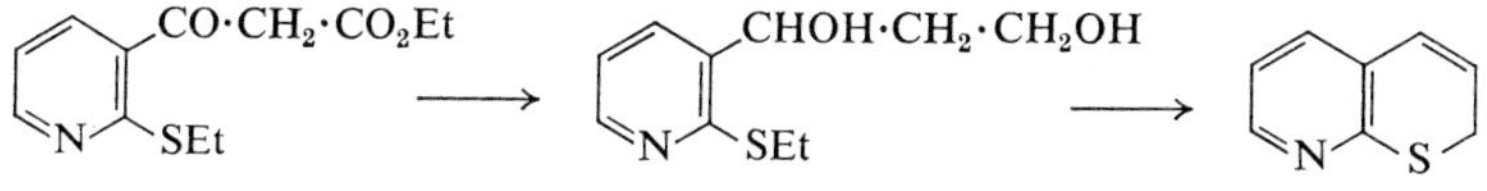

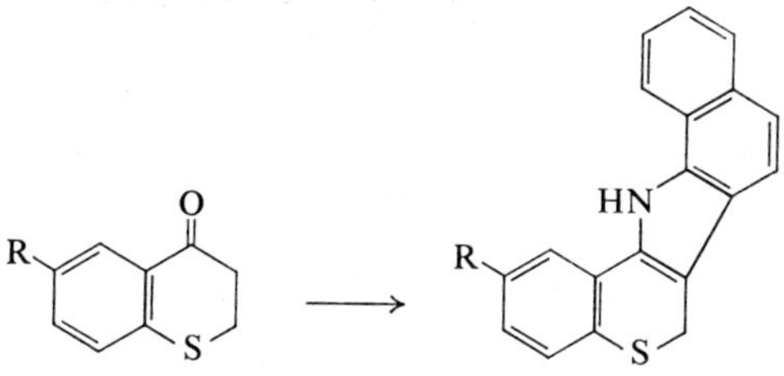

Pyrimidinecarboxylic acids of general structure (78) have been cyclised, via the corresponding acid chlorides, to pyrimidothiochromones (79),[33] and the 2H-thiopyran derivative (80) has been prepared by the action of cyanoguanidine on thiochroman-3-one.[34] 3-Phenyl-2H-thiopyran 1,1-

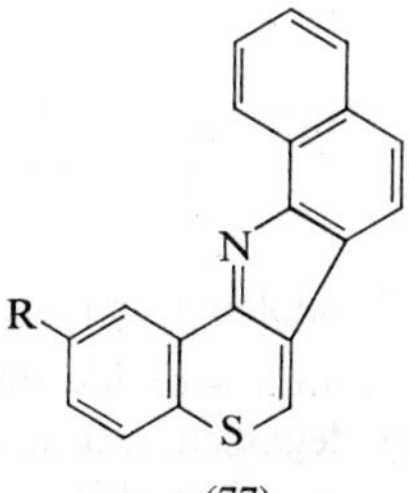

[31] H. Sliwa, Bull. Soc. chim. France, 1970, 642.
[32] N. P. Buu-Hoï, P. Jacquignon, A. Croisy, A. Loiseau, F. Périn, A. Ricci, and A. Martani, J. Chem. Soc. (C), 1969, 1422.
[33] S. C. Bennur and V. V. Badiger, Monatsh., 1969, 100, 2136.
[34] A. Rosowsky, P. C. Huang, and E. J. Modest, J. Heterocyclic Chem., 1970, 7, 197.

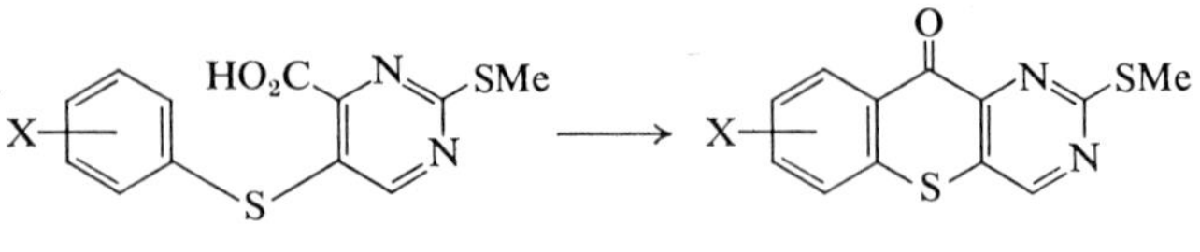

(78)  (79) 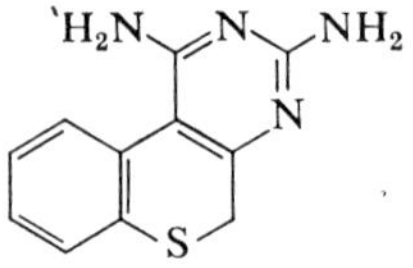  (80)

dioxide (81) adds diazomethane to give the isopyrazoline (82), which isomerises in hot methanol to the pyrazoline (83). This compound is converted by the action of bromine to the pyrazole (84).[35] The diazomethane addition reaction may involve a preliminary isomerisation of compound (81) to 5-phenyl-2H-thiopyran 1,1-dioxide.

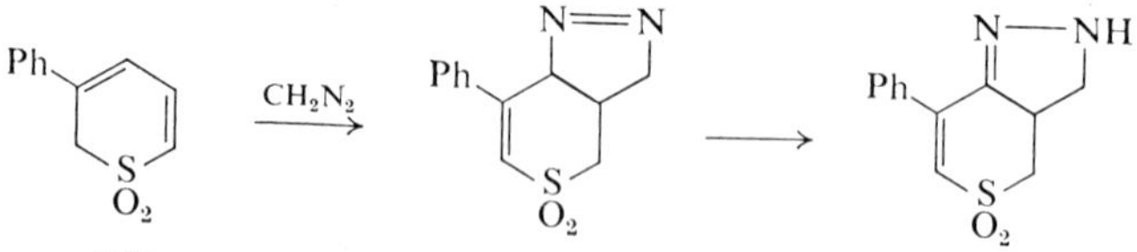

(81)  (82)  (83)

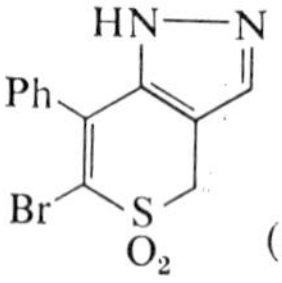

(84)

## 5 'Quadrivalent' Sulphur Compounds

The action of methyl magnesium bromide on the sulphoxide (85), in the absence of light and using degassed solutions, gives a relatively stable, deep-blue solution of the acenaphthothiopyran derivative (86; R = H), the n.m.r. spectrum of which indicates marked deshielding of all the protons relative to those in the precursor.[36] Attempts to isolate the product yielded only yellow polymeric material. The tetraphenyl compound (86; R = Ph) has been obtained in crystalline form, and in high yield, by treatment of 1,4,5,8-tetrabenzoylnaphthalene with phosphorus pentasulphide. Compounds (87; X = Y = O), (87; X = O, Y = S), and (87;

---

[35] S. Bradamante, S. Maiorana, A. Mangia, and G. Pagani, *Tetrahedron Letters*, 1969, 2971.
[36] I. S. Ponticello and R. H. Schlessinger, *Chem. Comm.*, 1969, 1214.

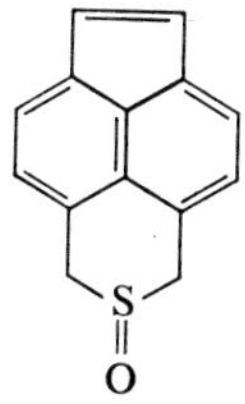

(85) 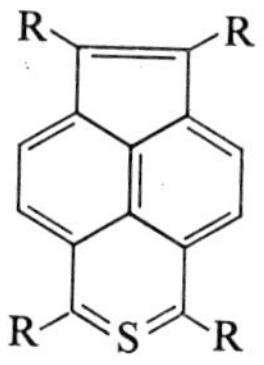 (86) 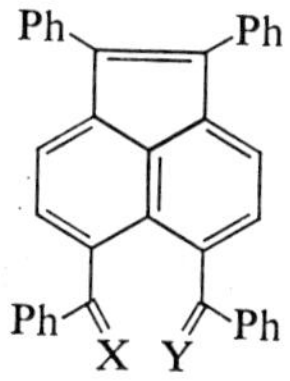 (87)

X = Y = S) are probably intermediates in the formation of (86; R = Ph).[37] $\pi$-Electron densities and bond orders have been calculated for compound (86; R = H) and for the simpler ring system (88), by a partly empirical

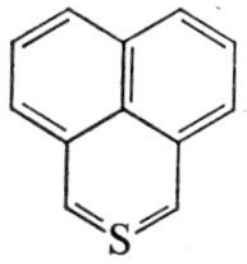

(88)

SCF–MO method, without explicitly including the $d$-orbital participation suggested by structures written with quadrivalent sulphur.[38]

[37] J. M. Hofmann and R. H. Schlessinger, *J. Amer. Chem. Soc.*, 1969, **91**, 3953.
[38] J. Fabian, *Z. Chem.*, 1969, **9**, 272.

# 11
# Thiepines and Dithiins

## 1 Thiepines

Dewar [1] describes the application of his version of the SCF–MO $\pi$-electron approximation to bivalent sulphur compounds including thiepine (1),

(1)  (2)  (3)

benzo[*d*]thiepine (2), and thieno[*c,d*]thiepine (3), for which heats of atomisation, bond lengths, ionisation potentials, charge densities, and resonance energies are calculated. Thiepine is predicted to be antiaromatic, with a negative resonance energy ($-1.45$ kcal mol$^{-1}$) minimised by bond alternation. The calculated bond lengths are close to those predicted for polyene single and double bonds. The same is true for benzo[*d*]thiepine (2) for which a resonance energy of 19.9 kcal mol$^{-1}$ is calculated. This is negative relative to the predicted resonance energy of benzene (22.6 kcal mol$^{-1}$). Bond lengths and charge densities in thiepine and benzo[*d*]thiepine are given in Figures 1 and 2. Neither of the compounds (1) and (2) has yet

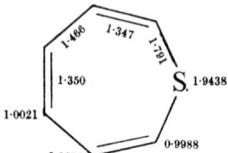

**Figure 1** *Calculated bond lengths (inner numbers) and charge densities (outer numbers) in thiepine*

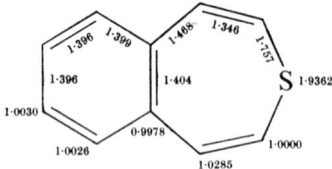

**Figure 2** *Calculated bond lengths (inner numbers) and charge densities (outer numbers) in benzo*[d]*thiepine*

[1] M. J. S. Dewar and N. Trinajstic, *J. Amer. Chem. Soc.*, 1970, **92**, 1453.

been synthesised, although the properties of the oxygen and nitrogen analogues (oxepin and azepin) indicate that neither of them is aromatic.

Low temperature ($-150$ °C) $^1$H n.m.r. (250 MHz) studies [2] of 3-methyl-6-isopropylthiepine 1,1-dioxide (4), using the diastereotopic methyl groups in the isopropyl function, reveal an inversion barrier of the non-planar ring of 6·4 kcal mol$^{-1}$, which is very similar to that of 1,3,5-cycloheptatriene.

Synthetic approaches to 2,7-dihydrothiepine 1,1-dioxides have been described. Sulphur monoxide, generated by thermal decomposition of thiiran 1-oxide in toluene at 110 °C, reacts with the *trans*-triene (5) to form 2,7-dihydro-4,5-diphenylthiepine 1-oxide (6) and with cycloheptatriene to form ditropyl.[3] Oxidation of the sulphoxide (6) with *m*-chloroperbenzoic acid gave the dioxide (7), and reduction with lithium aluminium hydride gave the dihydrothiepine (8). The sulphone (7) was also obtained by addition of sulphur dioxide to the *cis*-triene (9). The reactions involving sulphur monoxide are rationalised in terms of sulphur monoxide reacting at least partly in the $^3\Sigma^-$ state. Ring-expansion of a tetrahydro-4*H*-thiapyran-4-one forms the basis of a synthesis [4] of the 2,7-dihydrothiepine 1,1-dioxides (10) and (11), illustrated in Scheme 1. Thermal decomposition of the dioxides (10) and (11) occurs stereospecifically, giving the *trans,cis,cis*- (12) and *trans,cis,trans*- (13) octatrienes, together with sulphur dioxide. The stereochemical results require that the decomposition be an antarafacial process, indicating that stereoelectronic factors are dominant, and that symmetry considerations may be used as a probe for bonding characteristics in sulphones.

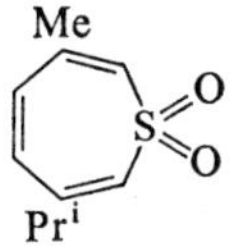

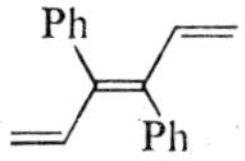

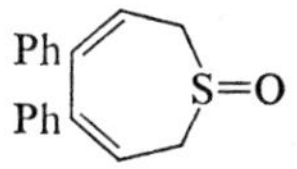

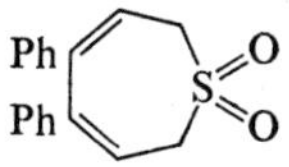

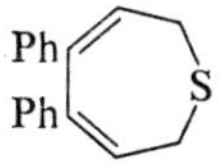

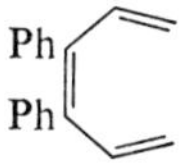

Further work [5] on the benzo[*b*]thiepine system indicates that it does not possess marked $\pi$-electron stabilisation. The derivative (16) was prepared

---

[2] F. A. L. Anet, C. H. Bradley, M. A. Brown, W. L. Mock, and J. H. McCausland, *J. Amer. Chem. Soc.*, 1969, **91**, 7782.
[3] R. M. Dodson and J. P. Neilson, *Chem. Comm.*, 1969, 1159.
[4] W. L. Mock, *J. Amer. Chem. Soc.*, 1969, **91**, 5682.
[5] W. E. Parham and D. G. Weetman, *J. Org. Chem.*, 1969, **34**, 56.

Reagents: i, $Ac_2O$–$HClO_4$–$CHCl_3$; ii, $PhHgCBrCl_2$; iii, $Bu^n_3SnH$; iv, $LiAlH_4$–ether; v, $Et_3N$–benzene; vi, $NaBH_4$–MeOH; vii, aq. $H_2SO_4$

**Scheme 1**

by addition of dichlorocarbene to the thiapyran (14) and ring-expansion of the resulting adduct (15) with methoxide in DMSO. The methoxy-group in the benzothiepine (16) occupies position 3 as a result of an elimination–addition mechanism involving the cyclopropane (15). Compound (16) is unstable and could not be isolated, its presence being demonstrated by the formation of the naphthalene (17) and sulphur in increased amount when the crude product was heated, and by $^1H$ n.m.r. studies which showed the presence and disappearance of the vinyl proton signal. In addition to being thermally unstable, with a tendency to extrude sulphur, the thiepine (16) is readily hydrolysed to the ketones (18) and (19), neither of which enolises to hydroxythiepines.

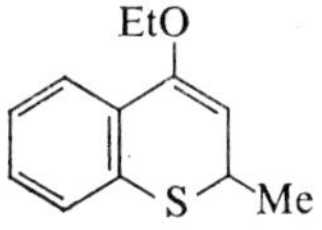

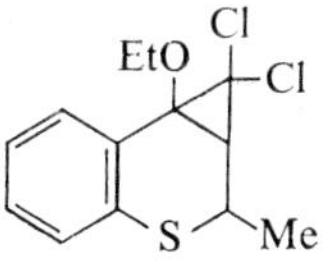

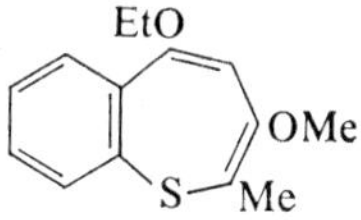

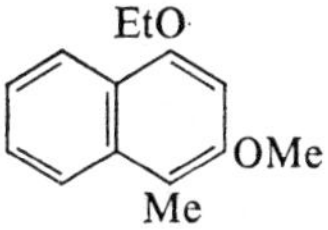

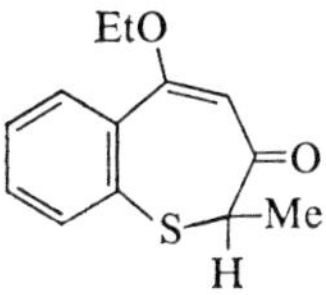

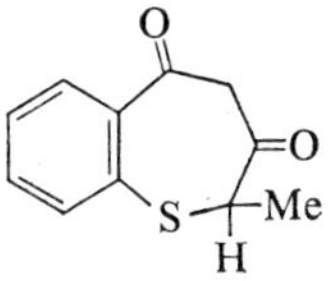

Ready ring-contraction of benzo[b]thiepines has also been reported by other workers.[6] In boiling xylene the benzo[b]thiepine (20) eliminates sulphur to form the naphthalene (22) via the postulated episulphide (21). The reaction is not acid- or base-catalysed, proceeds well in non-polar solvents, and the enol acetate groups are unattacked. When compound (20) is boiled in acetic anhydride containing sodium acetate, the acetylthio-derivative (24) is obtained, evidently via the episulphide (23) as shown. The S-dioxide (25) of the benzo[b]thiepine (20) behaves in a different fashion with sodium acetate in acetic anhydride, giving compound (28) via the intermediates (26) and (27) as shown.

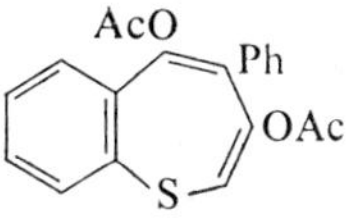

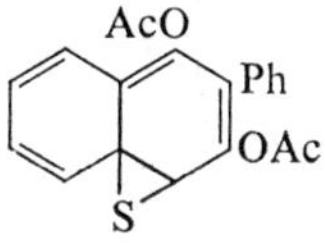

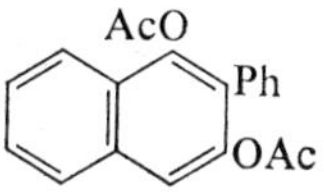

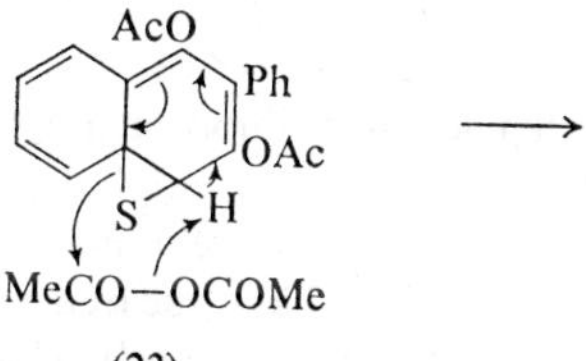

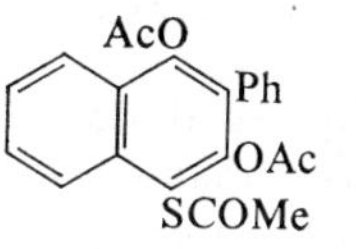

Thieno[c,d]thiepine (3) is predicted[1] to have a small positive resonance energy (3·5 kcal mol$^{-1}$), less than that of thiophen, and the bond lengths (Figure 3) in the thiophen ring also give some indication of delocalisation.

[6] H. Hofmann, H. Westernacher, and H. Haberstroh, *Chem. Ber.*, 1969, **102**, 2595.

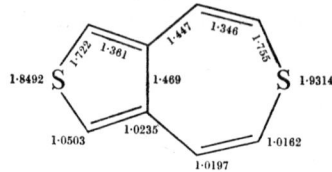

(25) → (26)

(27) → (28)

Thus the 2,3-bond (1·361 Å) is significantly longer than the normal polyene double bond (1·345 Å), and the C—S distance (1·722 Å) is shorter than that in benzo[*d*]thiepine (2). The $\pi$-electron densities (Figure 3) in the thiophen

**Figure 3** *Calculated bond lengths (inner numbers) and charge densities (outer numbers) in thieno*[c,d]*thiepine*

ring are close to those in thiophen itself and imply a greater degree of charge transfer than would be expected in a normal sulphide. Thieno[*c,d*]-thiepine (3) has been synthesised recently [7] and a preliminary report on its crystal structure and that of its 6,6-dioxide (29) has appeared.[8] The thiepine (3) is planar, and the only bond length (central bond) which has been reported (1·46 Å) is in good agreement with the predicted value. The dioxide (29) is non-planar, the seven-membered ring existing in a boat conformation recalling thiepine 1,1-dioxide. Schlessinger argues that the thiepine (3) is probably aromatic because, unlike its dioxide (29), it is planar. However, Dewar [1] brings forward counter arguments, and concludes that thieno[*c,d*]thiepine (3) should resemble benzo[*d*]thiepine (2) in having an aromatic ring attached to a non-aromatic or weakly aromatic thiepine moiety.

[7] R. H. Schlessinger and G. S. Ponticello, *J. Amer. Chem. Soc.*, 1967, **89**, 7138.
[8] T. D. Sakore, R. H. Schlessinger, and H. M. Sobell, *J. Amer. Chem. Soc.*, 1969, **91**, 3995.

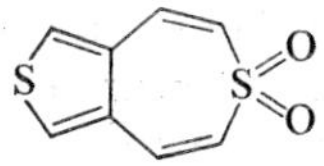

(29) 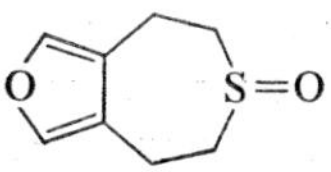  (30)

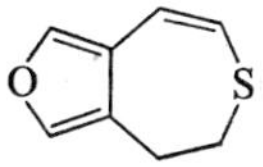

(31) 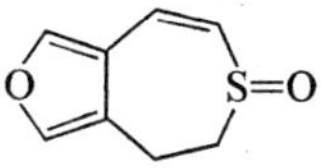  (32) 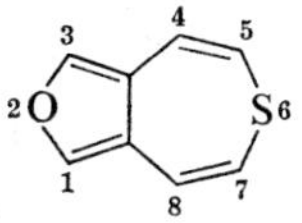  (33)

A synthesis and some properties bearing on the aromatic character of furano[c,d]thiepine (33) are reported.[9] The sulphoxide (30), synthesised from furan-2,5-diacetic acid by standard methods, was converted by boiling acetic anhydride into 4,5-dihydrofurano[c,d]thiepine (31), whose sulphoxide (32) was likewise transformed into furano[c,d]thiepine (33). This compound, although sufficiently stable to be isolated, was readily oxidised to a sulphoxide and thence to a sulphone. It also reacted more readily than thieno[c,d]thiepine (3) with N-phenylmaleimide at 25 °C, losing sulphur to give a mixture of exo- (34) and endo-adducts. The sulphoxide of furano[c,d]thiepine behaved similarly with N-phenylmaleimide, losing sulphur monoxide to give the same adducts. However,

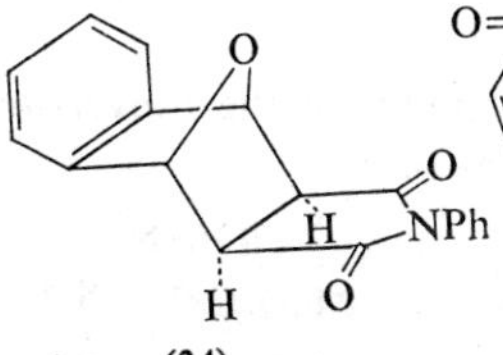

(34) 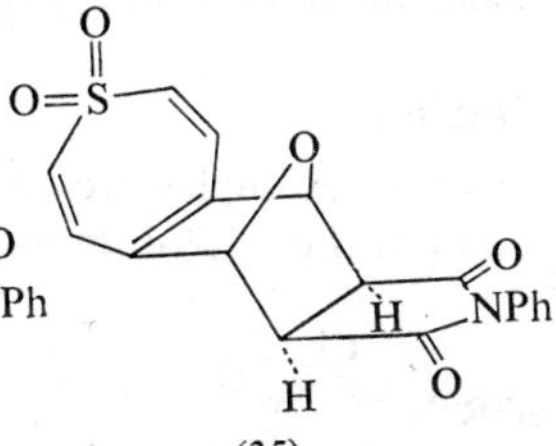  (35)

the sulphone reacted at 25 °C with N-phenylmaleimide with retention of the $SO_2$ bridge, giving the exo- (35) and endo-adducts. Heating the adducts above 100 °C resulted in loss of sulphur dioxide and formation of the same adducts as were obtained from furano[c,d]thiepine and its sulphoxide.

Comparison of the energy barriers to conformational inversion of the seven-membered ring in 10-(dimethylhydroxymethyl)-2,3:6,7-dibenzothiepine (36), in oxygen and nitrogen analogues, and in tropone derivatives, using $^1$H n.m.r. spectroscopy, shows that the change in delocalisation energy associated with the planar seven-membered ring is apparently due to increased Ar—S—Ar (Ar—C=C—Ar) conjugation, and is not a special property of the cyclic conjugated system.[10]

[9] R. H. Schlessinger and G. S. Ponticello, *Tetrahedron Letters*, 1969, 4361.
[10] M. Nógradi, W. D. Ollis, and I. O. Sutherland, *Chem. Comm.*, 1970, 158.

In the course of attempts to obtain the sulphone corresponding to the sulphide (37), oxidation led to 3-methyl-5H-4,1-benzo-oxathiepin-5-one 1,1-dioxide (38), the first example of this ring system.[11]

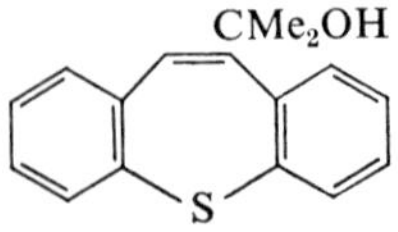

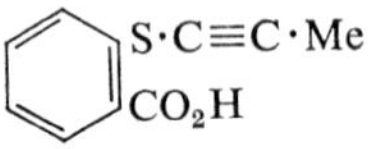

A sequence of reactions leading to 8-trifluoromethyl-10-(4-methylpiperazino)-10,11-dihydrodibenzo[b,f]thiepine (39) is described.[12] The same authors describe the preparation of derivatives of 10,11-dihydrodibenzo[b,f]selenepine and 6,11-dihydrodibenzo[b,e]selenepine.[13]

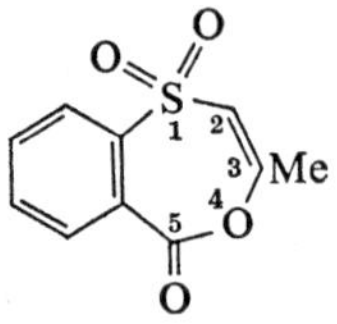

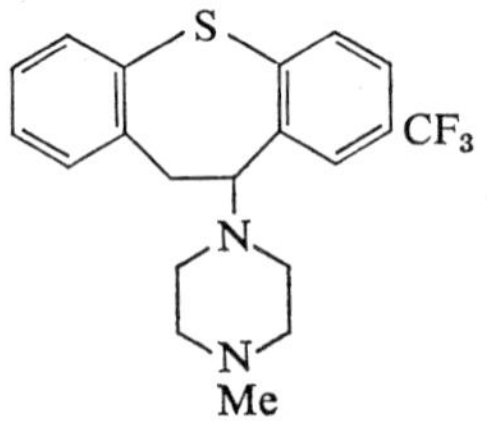

## 2 Dithiins

**1,2-Dithiins.**—Further examples are cited [14] of the conversion of 4-chloro-1,2-dithiol-3-ones (40) into 1,2-dithiins (41).

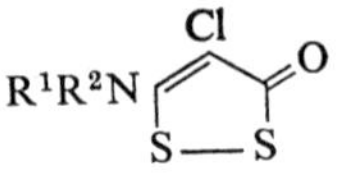

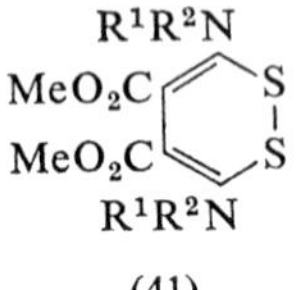

---

[11] R. H. Rynbrandt, *J. Heterocyclic Chem.*, 1970, **7**, 191.
[12] K. Pelz, I. Jirkovsky, J. Metyšová, and M. Protiva, *Coll. Czech. Chem. Comm.*, 1969, **34**, 3936.
[13] K. Šindelář, J. Metyšová, and M. Protiva, *Coll. Czech. Chem. Comm.*, 1969, **34**, 3792; K. Šindelář, J. Metyšová, E. Svatek, and M. Protiva, *Coll. Czech. Chem. Comm.*, 1969, **34**, 2122.
[14] F. Boberg, H. Niemann, and K. Kirchoff, *Annalen*, 1969, **728**, 32.

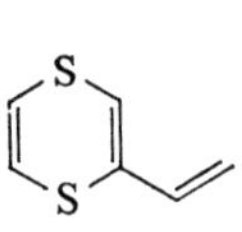

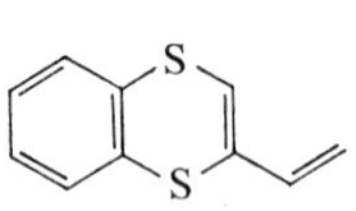

**1,4-Dithiins.**—The disodium salt of *cis*-dimercaptoethylene reacts with buta-1,3-diyne to give vinyl-1,4-dithiin (42) rather than 1,4-dithiocin.[15] 2-Vinyl-1,4-benzodithiin (43) is formed similarly from benzene-1,2-dithiol.

In an important paper [16] Grigg rationalises a number of known reactions in organosulphur chemistry, involving the elimination of sulphur and sulphur dioxide from heterocyclic structures, in terms of a thermal electrocyclic process followed by cheletropic elimination of sulphur or sulphur dioxide. The thermal $(4n+2)$ disrotatory electrocyclic processes for 1,4-dithiins, thiepines, and thiepine 1,1-dioxides, and the subsequent cheletropic eliminations are illustrated in Scheme 2.

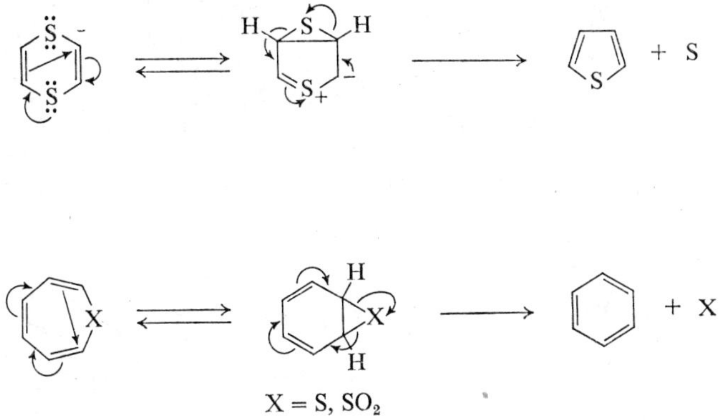

**Scheme 2**

An elegant synthesis of 5,6-dihydro-1,4-dithiins is reported.[17] Reaction of the ethylenethioketal (44) of an α-acetylaminoketone with phosphorus pentoxide gives an intermediate (45) of the type produced in the Vilsmeier reaction. This fragments with loss of acetonitrile to form the 5,6-dihydro-1,4-dithiin (46) (Scheme 3). For example, compound (44; R = 3,4-dimethoxyphenyl) gives the derivative (46; R = 3,4-dimethoxyphenyl) and, interestingly, not the Bischler–Napieralski cyclisation product (47). In an extension of the method, use of the homologous thioketal afforded the seven-membered ring analogue (48). Cyanogen di-*N*-oxide reacts with ethylenedithiol to give the 2,3-dihydro-1,4-dithiin (49) and with toluene-3,4-dithiol to give the benzo-1,4-dithiin derivative (50).[18]

Some reactions of 5,6-dihydro-1,4-dithiin-2,3-dicarboxylic acid anhydride (51) with amines are reported.[19] Primary amines and monosubstituted

[15] W. Schroth, F. Billig, and A. Zschunko, *Z. Chem.*, 1969, **9**, 184.
[16] R. Grigg, R. Hayes, and J. L. Jackson, *Chem. Comm.*, 1969, 1167.
[17] J. L. Massingill, M. G. Reinecke, and J. E. Hodgkins, *J. Org. Chem.*, 1970, **35**, 823.
[18] N. E. Alexandrou and D. N. Nicolaides, *J. Chem. Soc.* (*C*), 1969, 2319.
[19] H. R. Schweizer, *Helv. Chim. Acta*, 1969, **52**, 2221, 2236.

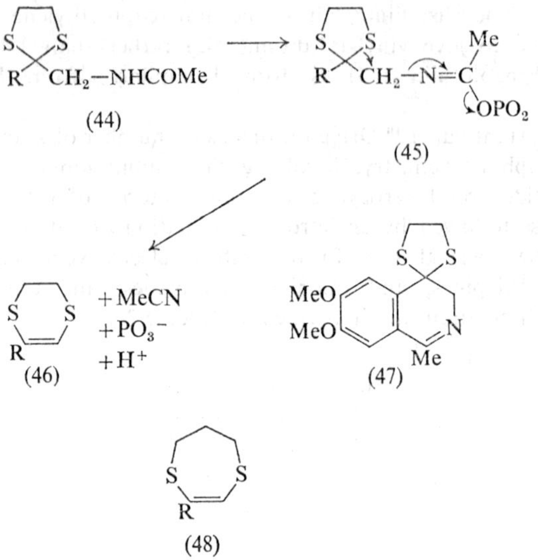

Scheme 3

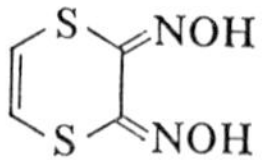

hydrazines afford imides of general structure (52), while o-phenylenediamine gives the novel heterocyclic system (53).

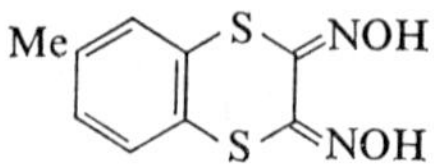

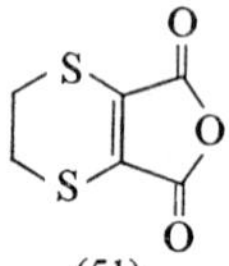

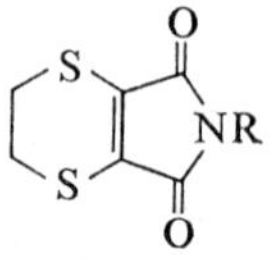

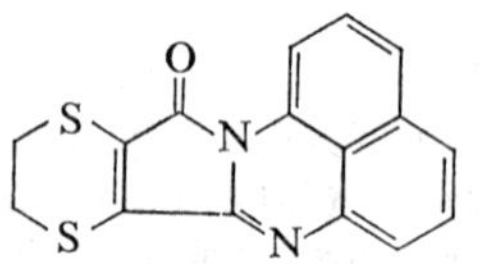

**Benzo- and Dibenzo-1,4-dithiins.**—2,3-Dichloro-1,4-benzoquinones react with the salt of cis-1,2-dimercapto-1,2-dicyanoethylene to give 5,8-dihydroxybenzo-1,4-dithiins, for example (54), oxidised to the corresponding quinones (55).[20] The quinones are ring-opened by secondary amines affording salts of the type (56). The same author describes [21] the reaction of 2,3-dichloro-1,4-benzoquinones with sodium sulphide, which leads to very stable quinones such as (57).

---

[20] K. Fickentscher, *Chem. Ber.*, 1969, **102**, 2378.
[21] K. Fickentscher, *Chem. Ber.*, 1969, **102**, 1739.

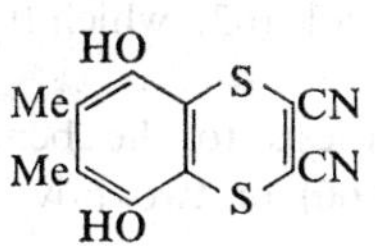

(54)

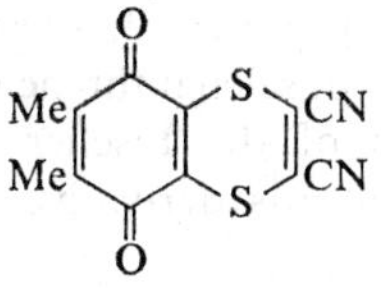

(55)

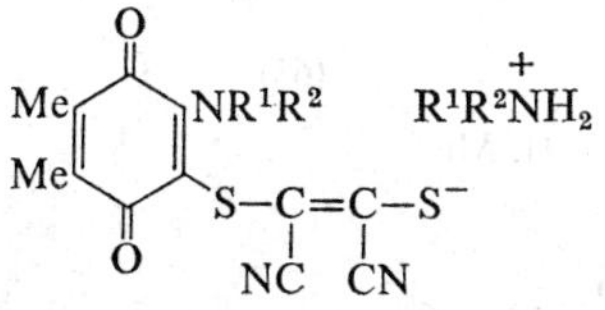

(56)

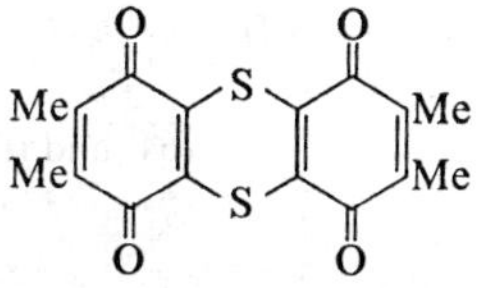

(57)

Kinetic studies of the hydrolysis of thianthrenium perchlorate (58) point to a disproportionation equilibrium followed by attack of the resulting dication (59) by water, giving thianthrene and its sulphoxide (60) in equal amounts (Scheme 4).[22, 23]

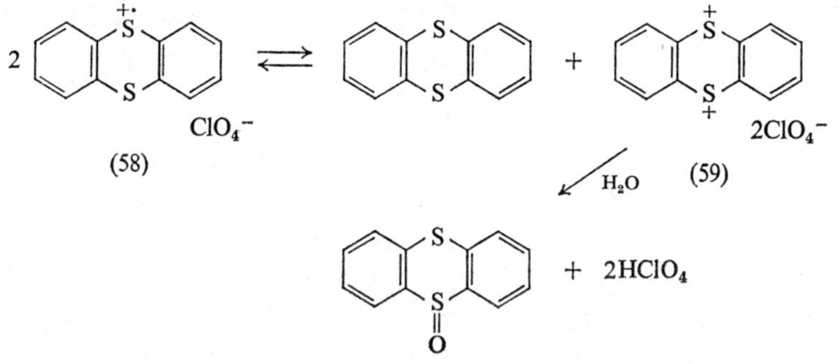

(58) (59) (60)

**Scheme 4**

Selenanthrene (63) has been isolated in low yield from the reaction of selenium dioxide with 1-aminobenzotriazole.[24] It is suggested that the zwitterion (61) resulting from the oxidation reacts with selenium produced

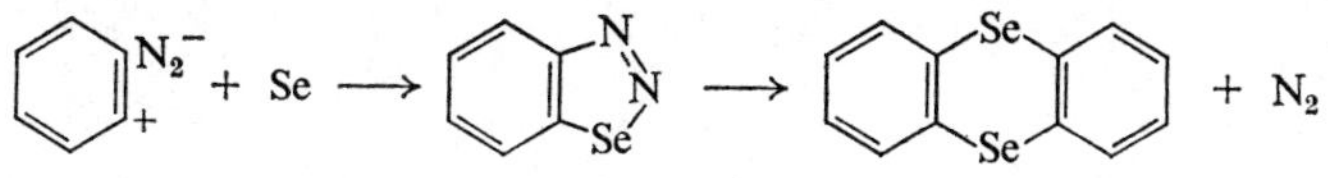

(61) (62) (63)

[22] H. J. Shine and Y. Murata, *J. Amer. Chem. Soc.*, 1969, **91**, 1872.
[23] Y. Murata and H. J. Shine, *J. Org. Chem.*, 1969, **34**, 3368.
[24] C. D. Campbell and C. W. Rees, *J. Chem. Soc.* (*C*), 1969, 752.

concomitantly to form 1,2,3-benzoselenadiazole (62), which then eliminates nitrogen to form selenanthrene (63).

Application of the Fischer indole synthesis to the benzo-1,4-dithiin ketone (64) affords derivatives (65) and (66) of two novel heterocyclic systems.[25]

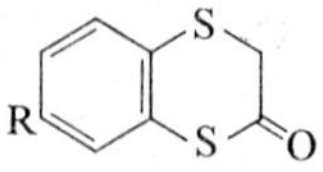

(64)

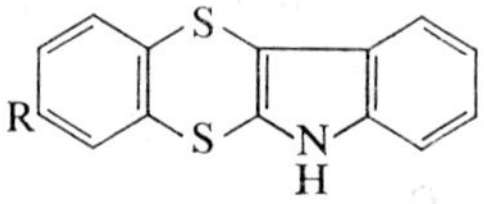

(64) and (65): R = H, Me

(65)

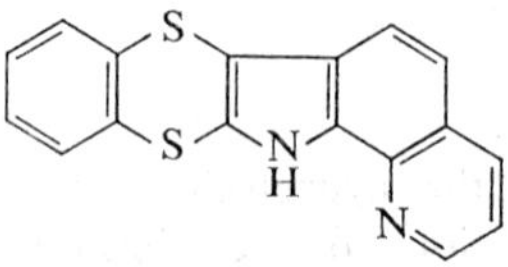

(66)

[25] N. P. Buu-Hoï, P. Jacquignon, L. Ledésert, A. Ricci, and D. Balucani, *J. Chem. Soc. (C)*, 1969, 2196.

# 12
# Isothiazoles

## 1 Introduction

Polycyclic ring systems incorporating the isothiazole structure have been investigated extensively for many years,[1] but contributions to the chemistry of the monocyclic parent isothiazole are of comparatively recent date. Interest in this heterocyclic structure has greatly increased during the past few years, and has occasioned the appearance of two reviews [2,3] of this subject, in 1963 and 1965. Subsequent developments have been briefly referred to in a paper by McGregor et al.[4] The period under review has witnessed further useful progress, both in the provision of new syntheses, and in the further study of the properties of this class of heterocyclic compounds.

## 2 Syntheses

A new general synthesis of isothiazoles[4] employs the corresponding isoxazoles as starting materials. Successive ring-cleavage of the isoxazole (1) by Raney nickel, and treatment of the resulting intermediate enaminoketone (2) with phosphorus pentasulphide and an oxidising agent (e.g.

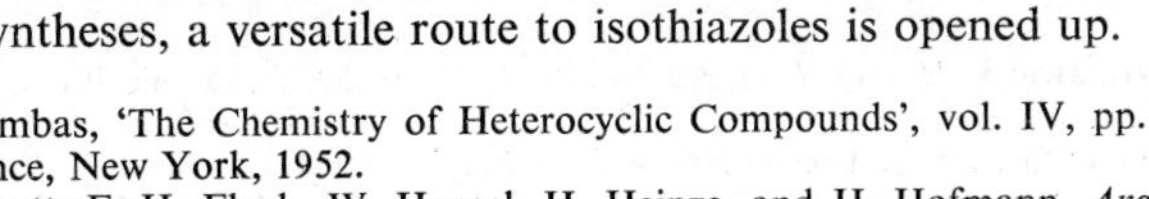

$$
\underset{(1)}{\overset{R^1\quad R^2}{\underset{N\diagdown_O\diagup R^3}{\|\quad\|}}} \longrightarrow \underset{(2)}{\overset{R^1\quad R^2}{\underset{H_2N\quad O\diagup R^3}{\|\quad\|}}} \longrightarrow \underset{(3)}{\overset{R^1\quad R^2}{\underset{N\diagdown_S\diagup R^3}{\|\quad\|}}}
$$

chloranil) affords the corresponding isothiazole (3) in satisfactory yield. Since variously substituted isoxazoles[5] are readily available by numerous reliable syntheses, a versatile route to isothiazoles is opened up.

---

[1] L. L. Bambas, 'The Chemistry of Heterocyclic Compounds', vol. IV, pp. 225—378, Interscience, New York, 1952.
[2] F. Hubenett, F. H. Flock, W. Hansel, H. Heinze, and H. Hofmann, *Angew. Chem. Internat. Edn.*, 1963, **2**, 714.
[3] R. Slack and K. R. H. Woolridge, 'Advances in Heterocyclic Chemistry', ed. A. R. Katritzky, vol. 4, p. 107, Academic Press, New York, 1965.
[4] D. N. McGregor, U. Corbin, J. E. Swigor, and L. C. Cheney, *Tetrahedron*, 1969, **25**, 389.
[5] A. Quilico, 'The Chemistry of Heterocyclic Compounds', ed. A. Weissberger, vol. 7, p. 1, Wiley, New York, 1962; N. K. Kochetkov and S. D. Sokolov, 'Advances in Heterocyclic Chemistry', ed. A. R. Katritzky, vol. 2, p. 365, Academic Press, New York, 1963.

An elegant synthesis of the isothiazole system, due to Woodward and Gougoutas,[6,7] consists of the condensation of a β-crotonic ester (4) and

$$\text{Me}-\underset{\underset{\text{NH}_2}{|}}{\text{C}}=\text{CH}\cdot\text{CO}\cdot\text{R} + \text{CSCl}_2 \xrightarrow{-2\text{HCl}} \underset{\text{N}\diagdown\text{S}}{\text{Me}\diagup\!\!\diagdown\text{COR}}$$

(4) (5)

thiophosgene, affording 3-methylisothiazole-4-carboxylic acid esters (5). The use of 4-iminopentan-2-one in this procedure results in 4-acetyl-3-methylisothiazole (5; R = Me) in one step,[8] and provides a significant improvement over the previous multi-stage synthesis [9] of this compound.

Goerdeler's general synthesis [10] of 5-aminoisothiazoles by the oxidative cyclisation of enamine–isothiocyanate adducts has been successfully extended to the production of 5-sulphonylaminoisothiazoles [11] using arylsulphonyl isothiocyanates, which have recently become available.[12] They generally react smoothly with primary and secondary enamines (6, 9) of the β-aminocrotonic ester type to yield yellow adducts (7, 10), which are readily dehydrogenated by bromine or iodine to sulphonylaminoisothiazoles (8) or sulphonyliminoisothiazolines (11), respectively. Imines which are capable of tautomerising undergo the same sequence of reactions (12 → 13 → 14). The carbethoxy-compound (8; R = Me; A = CO$_2$Et) is readily hydrolysed by alkalis, and decarboxylated to (15), and the corresponding isothiazoline (16) is accessible in an analogous manner.

### 3 Properties

A systematic concise survey of the physical and chemical properties of isothiazoles is given in Slack and Woolridge's review.[3] Electrophilic substitution in the 4-position, and ring cleavage at the N—S bond appear to be among the more dominant properties of this ring system, and are in broad agreement with the electron distribution pattern. These generalisations have been confirmed and others established as more information has become available.

**Alkylation.**—The 5-position of 4-acetyl-3-methylisothiazole, in common with that of analogous isothiazoles,[13] is alkylated by successive treatment

---

[6] R. B. Woodward, *Harvey Lectures 1963—4*, Series 59, Academic Press, New York, 1965, p. 31.
[7] J. Z. Gougoutas, Ph.D. Thesis, Harvard University, 1964.
[8] S. Rajappa, A. S. Akerkar, and V. S. Iyer, *Indian J. Chem.*, 1969, 7, 103.
[9] D. Buttimore, D. H. Jones, R. Slack, and K. R. H. Woolridge, *J. Chem. Soc.*, 1963, 2032.
[10] J. Goerdeler and H. Pohland, *Chem. Ber.*, 1961, 94, 2950, and subsequent papers.
[11] J. Goerdeler and U. Krone, *Chem. Ber.*, 1969, 102, 2273.
[12] K. Dickoré and E. Kühle, *Angew. Chem.*, 1965, 77, 429 (*Internat. Edn.*, 1965, 4, 430); R. Gompper and W. Haegele, *Chem. Ber.*, 1966, 99, 2885; K. Hartke, *Arch. Pharm.*, 1966, 299, 174.
[13] M. P. L. Caton, D. H. Jones, R. Slack, and K. R. H. Woolridge, *J. Chem. Soc.*, 1959, 3061.

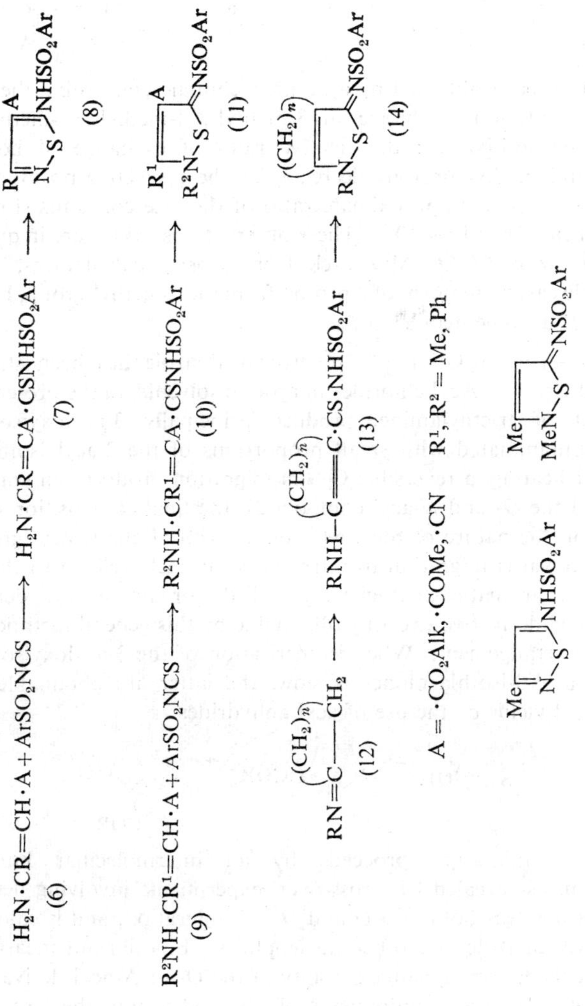

$H_2N \cdot CR=CH \cdot A + ArSO_2NCS \longrightarrow H_2N \cdot CR=CA \cdot CSNHSO_2Ar \longrightarrow$

(6)                      (7)

structure (8): isothiazole ring with R at 5-position, A at 4-position, NHSO$_2$Ar at 3-position

$R^2NH \cdot CR^1=CH \cdot A + ArSO_2NCS \longrightarrow R^2NH \cdot CR^1=CA \cdot CSNHSO_2Ar \longrightarrow$

(9)                      (10)

structure (11): $R^1$ at 4-position, A at 5-position, $R^2N$—, =NSO$_2$Ar on ring

$$RN=C \overset{(CH_2)_n}{\underset{\phantom{X}}{\text{———}}} CH_2 \longrightarrow RNH-C \overset{(CH_2)_n}{=\!=\!=} C \cdot CS \cdot NHSO_2Ar \longrightarrow$$

(12)                      (13)

structure (14): fused ring with (CH$_2$)$_n$, RN–, =NSO$_2$Ar

$A = \cdot CO_2Alk, \cdot COMe, \cdot CN \quad R^1, R^2 = Me, Ph$

structure (15): Me-substituted isothiazole, NHSO$_2$Ar

structure (16): Me, MeN-substituted, =NSO$_2$Ar

with butyl-lithium and alkyl halide, the acetyl keto-group being protected by ketalisation:[8]

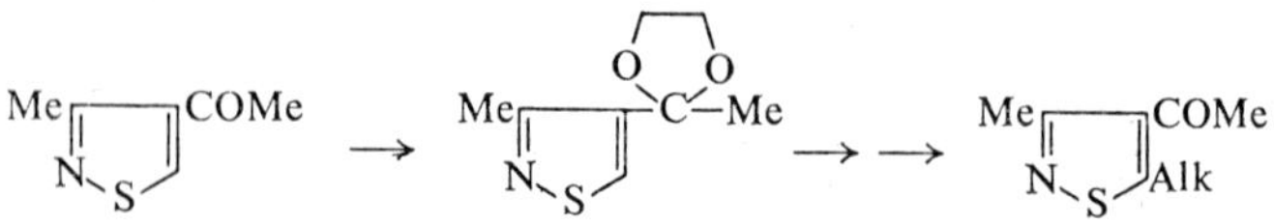

Data have been obtained by a n.m.r. technique concerning the rates of hydrogen–deuterium exchange in 3-methyl-, 4-methyl-, 5-methyl-, and 3,5-dimethyl-isothiazoles: the relative rates of exchange of both ring-protons and methyl protons decrease for the respective positions in the order 5->3->4-, the approximate ratio of the rate constants (for methyl protons) being $100:1:<10^{-4}$. These observations, which are in qualitative agreement with LCAO–MO $\pi$-electron density calculations, and the remarkable ease of removal of a proton from the 5-methyl group, have been discussed [14] in some detail.

**Acylation.**—The acylation of 3-hydroxyisothiazole has been studied by Chan and Crow.[15] Acyl chlorides in aprotic solvents, in the presence of an equivalent of triethylamine, produce principally 3-acyloxyisothiazole, usually contaminated with small proportions of the 2-acyl isomers. On storage or heating, a reversible $O \rightarrow N$ migration produces an equilibrium mixture of the $O$- and $N$-acyl compounds, the final composition of which depends on the nature of the acyl group. With aliphatic acyl groups the tendency towards migration to nitrogen is inversely related to the size of the group; aromatic analogues give little or no $N$-acyl derivatives. $O$-Sulphonyl derivatives, readily obtainable by this general method, do not undergo rearrangement. When isomerisation of the 3-acyloxyisothiazoles to the $N$-acyl-3-isothiazolones is slow, the latter are obtainable rapidly and in good yields by the use of acyl anhydrides.

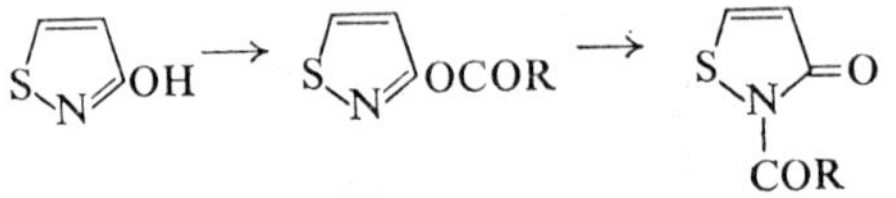

The acyl migration proceeds by an intermolecular bimolecular mechanism, as revealed by 'cross-over' experiments involving deuteriated species; it involves both $O \rightarrow O$ and $O \rightarrow N$ migration, and is catalysed by 3-hydroxyisothiazole and other nucleophiles. Equilibrium measurements show that the thermodynamic stability of the $O$- or $N$-acyl derivatives is a function of the steric requirements of the acyl group, the larger groups being more stable on the oxygen. This is probably due to out-of-plane twisting in the $N$-acyl-3-isothiazolones, which prevents N—CO overlap in all except formyl derivatives, which exist entirely as $N$-formyl isomers at equilibrium.

[14] J. A. White and R. C. Anderson, *J. Heterocyclic Chem.*, 1969, **6**, 199.
[15] A. W. K. Chan and W. D. Crow, *Austral. J. Chem.*, 1968, **21**, 2967.

## Isothiazoles

**Ring-cleavage.**—The S—N bond in isothiazoles is an ambiphilic centre: both electrophilic attack at nitrogen or nucleophilic attack at sulphur are possible.

Nucleophiles reversibly attack the S—N bond of 3-hydroxyisothiazole;[16] the mechanism that operates in the case of the cyanide ion has been established by kinetic studies.[17, 18] A consideration of the transition state (18) suggests that the direction of the reaction (to 17 or 19) depends on the

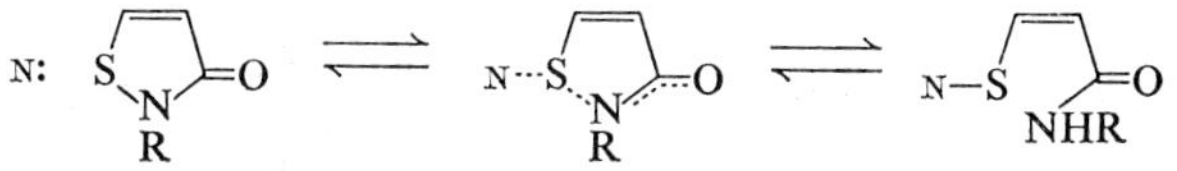

(17)　　　(18)　　　(19)

relative effectiveness of the nucleophile N and the amide-nitrogen as leaving groups and nucleophiles. Nucleophilic attack by cyanide ion is greatly promoted[18] by protonation of the substrate, because the S—N bond of the resulting isothiazolium cation is weakened.

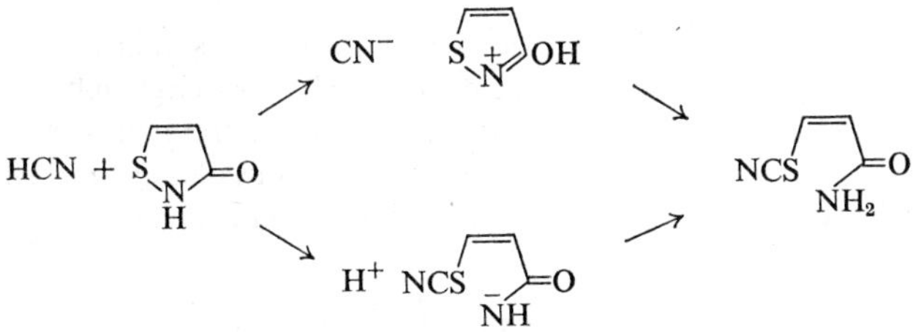

The S—N bond of N-acyl-3-isothiazolones[15] is similarly cleaved in this reaction: the product is simultaneously deacylated in the aqueous medium employed, giving cis-3-thiocyanoacrylamide (21). The corresponding reaction of 3-acyloxyisothiazoles (20) (examined in the case of the trimethylacetyl compound by u.v. spectroscopy) occurs slowly, with formation of the same compound (21), converted in the subsequent isolation procedure into 3-hydroxyisothiazole (22).

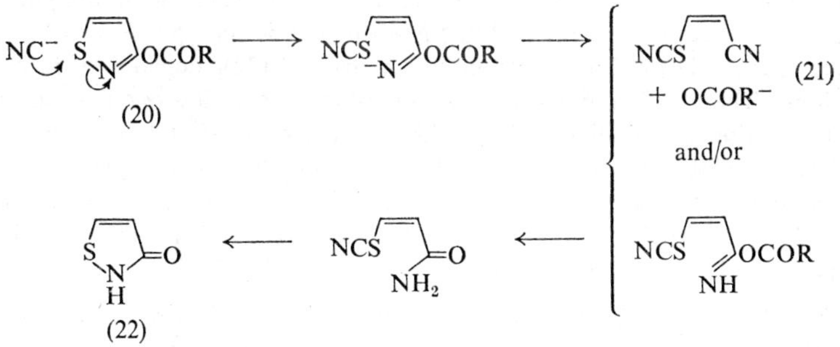

[16] W. D. Crow and N. J. Leonard, *J. Org. Chem.*, 1965, **30**, 2660.
[17] W. D. Crow and I. Gosney, *Austral. J. Chem.*, 1966, **19**, 1693.
[18] W. D. Crow and I. Gosney, *Austral. J. Chem.*, 1967, **20**, 2729.

Nucleophilic attack by carbanions brings about ring-cleavage of $N$-ethyl-3-isothiazolones (23) in an analogous manner. Thus, diethyl malonate in sodium ethoxide produces, *via* the sodium salt (24), the *cis*-alkylmercaptoacrylamide (25); other carboxylic acids behave similarly.[19]

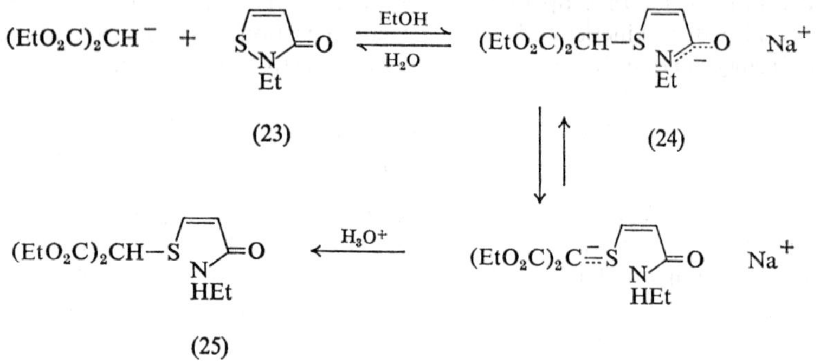

$$(EtO_2C)_2CH^- + \underset{Et}{\underset{(23)}{S-N=O}} \underset{H_2O}{\overset{EtOH}{\rightleftharpoons}} (EtO_2C)_2CH-S-\underset{Et}{N}=O \quad Na^+ \quad (24)$$

$$(EtO_2C)_2CH-S-\underset{HEt}{N}=O \quad \overset{H_3O^+}{\longleftarrow} \quad (EtO_2C)_2C=S-\underset{HEt}{N}=O \quad Na^+$$

(25)

Reaction of $N$-ethyl-3-isothiazolone with resonance-stabilised carbanions derived from primary carbon acids (E—CH$_3$, E = electrophile, *e.g.* in acetone) yields mono- (26), 1,1-di- (27), and more highly substituted products (*e.g.* 28), involving one, two, or more molecules of the isothiazole. The mechanism of these and further analogous reactions was discussed in detail.[19]

MeCOCH$_2$·SCH=CHCONHEt  (26)

MeCOCH(SCH=CHCONHEt)$_2$  (27)

(EtNHCO·CH=CHS)$_2$CH·CO·CH(SCH=CHCONHEt)$_2$  (28)

**Ring-cleavage with Ring-expansion.**—A ring-expansion of 5-aroylamido-isothiazoles, by the nucleophilic fission of their N—S bond, to pyrimidine derivatives has been observed by Machoń.[20] 4-Carbethoxy-3-methyl-5-benzamidoisothiazole (29) is converted, on treatment with anhydrous ethanolic hydrazine, into 2-phenyl-4-methyl-6-mercapto-5-carbethoxy-pyrimidine (31) or the corresponding hydrazide (32), depending on conditions. The N—S ring-fission of (29) may involve an intermediate species (30). Action of alkali on (29) yields the corresponding free carboxylic acid (33), while action of acid produces the 6-hydroxy-analogue (34).

In contrast, isothiazoles (29) containing an electron-withdrawing substituent in their 5-aroyl group (*e.g. o-* or *p-*Cl) did not undergo ring-expansion, being merely hydrolysed by acids to (35).[20]

**Other Reactions.**—A number of interconversions of isothiazoles by conventional reactions that have been reported include the conversion of

---

[19] W. D. Crow and I. Gosney, *Austral. J. Chem.*, 1969, **22**, 764.
[20] Z. Machoń, *Roczniki Chem.*, 1968, **42**, 2091.

(35) (29)

(30)

(31) (32)

(34) (33)

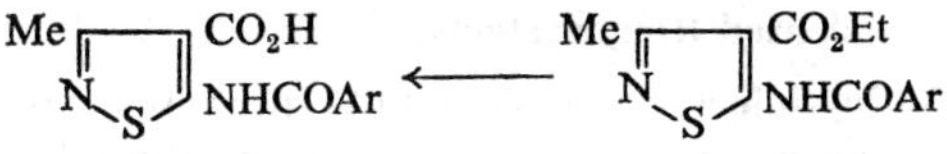

3-methylisothiazole-4-carboxylic ester (36) into the corresponding 4-cyanomethyl compound (37)[8] and 3-methylisothiazole-4-carboxylic acid (38) into the 4-amine (39) by the Curtius reaction.[8]

(37) (36) (38)

(39)

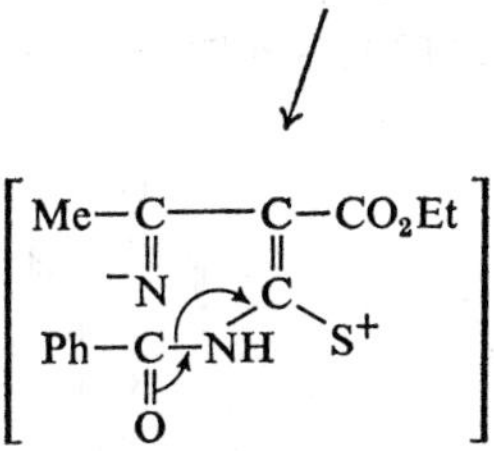

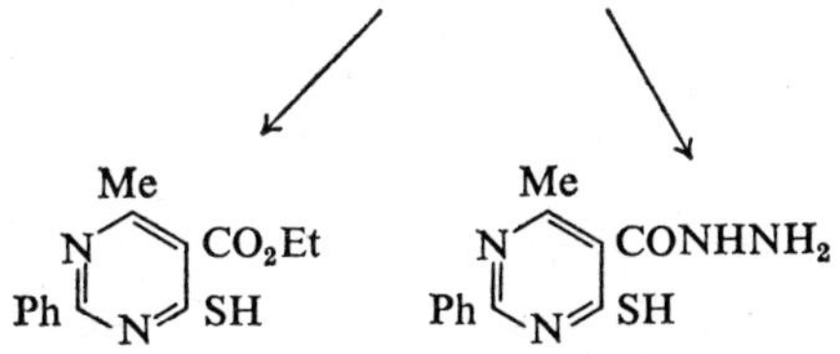

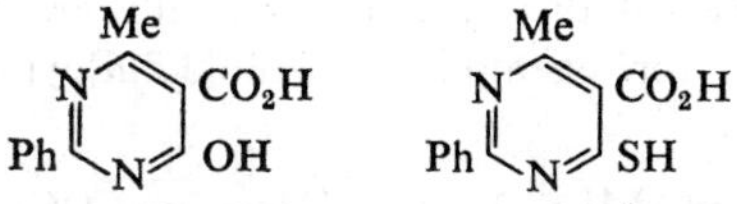

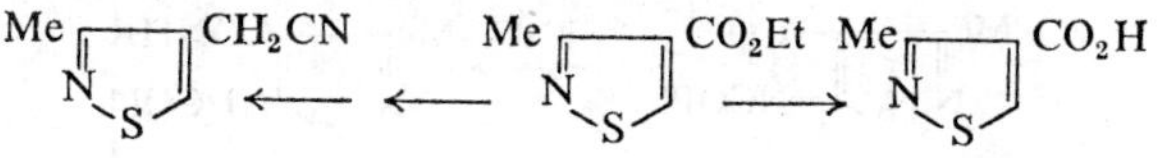

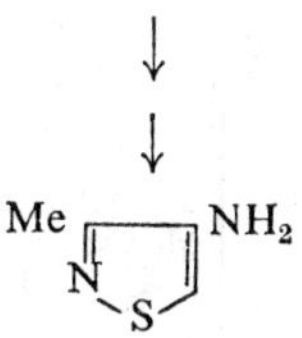

## 4 Condensed Ring Systems

**Isothiazolo[2,3-a]pyridines.**—An attempt to synthesise the isothiazolo[2,3-a]-pyridinium structure, starting from 3-cyanoisothiazole (40), has been reported.[21] It involved the interaction of this nitrile (40) with 3-ethoxypropylmagnesium bromide, followed by cyclisation of the resulting ketone (41) to the tetrahydro-derivative (42) of the desired compound (43).

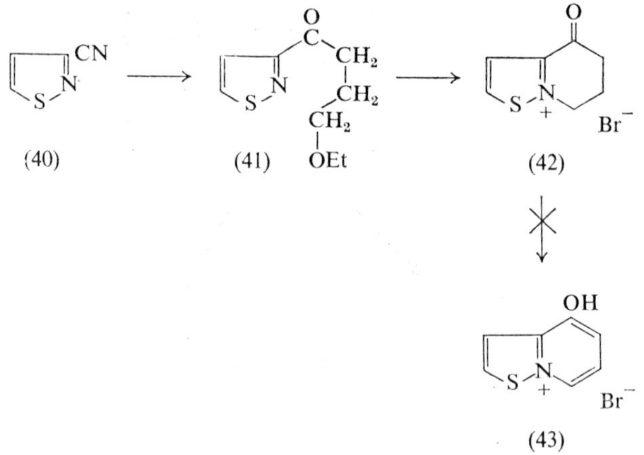

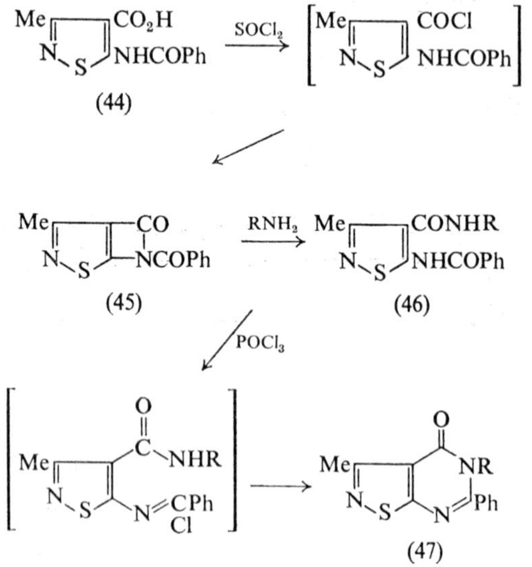

Although cyclic ketones of this type are generally aromatised by acetic anhydride (providing, for example, thiazolo[3,2-a]pyridinium salts by this method [22]), the present intermediate (42) was decomposed by this reagent, probably due to fission of the isothiazole ring.[21] (See also thiazolo[3,4-a]pyridinium salts, this Report, p. 434.)

**Isothiazolo[5,4-d]pyrimidines.**—5-Benzoylamino-4-carbonamido-3-methylisothiazoles (46), which are accessible from the carboxylic acids (44) *via* the lactams (45), are cyclisable to 5-substituted 3-methyl-4-keto-6- phenyl-4,5-dihydroisothazoilo[5,4-d]pyrimidines (47).[23]

[21] D. G. Jones and G. Jones, *J. Chem. Soc. (C)*, 1969, 707.
[22] D. G. Jones and G. Jones, *J. Chem. Soc. (C)*, 1967, 515.
[23] Z. Machoń, *Dissertationes Pharm.*, 1969, **21**, 325 (*Index Chemicus*, 1969, **35**, 123,852).

# 13
# Thiazoles and Related Compounds

## 1 Introduction

Although the chemistry of thiazoles is one of the most intensively investigated areas in the heterocyclic field, it is possibly less well served with monographs and reviews than are other groups of heterocyclic compounds of comparable importance: the formidable extent of the accumulated knowledge may partly be responsible for this situation. Sprague and Land's [1] excellent yet concise survey issued in 1957, comprising some 1200 references, is still the fullest and most authoritative account of the chemistry of thiazole and benzothiazole. Syntheses of thiazoles have been carefully reviewed by Wiley, England, and Behr,[2] and thiazolidinones have been dealt with in great detail by Brown.[3] A number of books on penicillin, the monumental volume edited by Clarke, Johnson, and Robinson [4] foremost among them, contain much information of the reduced thiazole system.[4,5] Useful correlations of spectral and other physical properties are given in Katritzky's well-known monographs.[6] Anyone concerned with the chemistry of thiazole is bound to look forward to the appearance of the relevant volumes in the series of monographs edited by A. Weissberger.

## 2 Thiazoles

The current literature shows clearly that interest in thiazole continues unabated. The great bulk of the contributions is primarily concerned with the synthesis of the thiazole system, and less so with its reactions, but an increasing awareness of the importance of the correlation between structure, physical properties, and reactivity is discernible. Metzger,[7] for example,

---

[1] J. M. Sprague and A. H. Land, 'Heterocyclic Compounds', ed. R. C. Elderfield, Wiley, New York, 1957, vol. 5, pp. 484—722.
[2] R. H. Wiley, D. C. England, and L. C. Behr, 'Organic Reactions', Wiley, New York, 1951, vol. 6, pp. 367—409.
[3] F. Brown, *Chem. Rev.*, 1961, **61**, 463.
[4] H. T. Clarke, J. R. Johnson, and Sir Robert Robinson, 'The Chemistry of Penicillin', Princeton University Press, Princeton, N.J., and Oxford, 1949.
[5] F. P. Doyle and J. H. C. Nayler, 'Penicillins and Related Structures', in 'Advances in Drug Research', ed. N. J. Harper and A. B. Simmonds, Academic Press, New York, 1964, vol. 1 pp. 2—63.
[6] 'Physical Methods in Heterocyclic Chemistry', ed. A. R. Katritzky, Academic Press, New York, 1963.
[7] J. Metzger, *Z. Chem.*, 1969, **9**, 99.

## Thiazoles and Related Compounds

has reviewed the work the aim of which is the elucidation of the detailed geometry and electron densities of the thiazole structure, based on HMO and PPP calculations, and its consequences on the reactivities of the individual positions in the molecule.

**Syntheses.**—In the following account, the synthetic routes are classified according to the nature of the 'skeleton' of the components which join to form the ring system, thus following essentially Sprague and Land's [1] scheme.

*Hantzsch's Synthesis:*(*Type A*, C—C + S—C—N). The exceptional versatility and usefulness of Hantzsch's synthesis, first described in 1887,[8] in the production of thiazoles, has been paid tribute previously,[1,2] and its importance is clearly reflected by its continued wide use, accounting in one form or another for a large portion of current synthetic work.

(i) *Mechanism.* The intermediates in Hantzsch's synthesis from thioamides and α-halogenocarbonyl compounds, generally assumed to be open-chain α-thioketones, have recently been recognised, in certain cases as 4-hydroxythiazolinium salts.[9-11] To prevent the acid-catalysed elimination of the elements of water from these intermediates, the use of reactants incorporating a basic centre or the presence of a base are necessary.

In further work along these lines,[12] the reaction of *N*-methyl-(*p*-dimethylamino)thiobenzamide with halogenoketones or halogenoaldehydes gave stable 4-hydroxythiazolinium salts (1), which were dehydrated by methanolic hydrogen chloride to the thiazolium salts (2). The results of n.m.r. spectroscopy revealed that the intermediates (1) existed as both possible diastereoisomers, which are in dynamic equilibrium *via* the acyclic α-thioketone. The observations thus suggest that hydroxythiazolines, in equilibrium with the corresponding α-thioketones, are concerned in the mechanism of the Hantzsch synthesis. Similar conclusions have also been briefly reported by Metzger *et al.*[13] and are implicit in Baganz and Rüger's [14] production of 4-formylthiazoles by this synthesis (see below).

$R^2 \cdot CHX \cdot COR^1$
+
$MeNHCS \cdot C_6H_4 \cdot NMe_2$

→ (1) → (2)

[8] A. Hantzsch *et al.*, *Ber.*, 1887, **20**, 3118, 3336 and subsequent papers.
[9] K. M. Muraveva and M. N. Shchukina, *Zhur. obshchei. Khim.*, 1960, **30**, 2327, 2334, 2340, 2344.
[10] A. Takamizawa, K. Hirai, T. Ishiba, and Y. Matsumoto, *Chem. and Pharm. Bull. (Japan)*, 1967, **15**, 731.
[11] B. M. Regan, F. T. Galysh, and R. N. Morris, *J. Medicin. Chem.*, 1967, **10**, 649.
[12] R. S. Egan, J. Tadanier, D. L. Garmaise, and A. P. Gaunce, *J. Org. Chem.*, 1968, **33**, 4422.
[13] A Babadjamian, M. Chanon, and J. Metzger, *Compt. rend.*, 1968, **267**, C, 918.
[14] H. Baganz and J. Rüger, *Chem. Ber.*, 1968, **101**, 3872.

(ii) *Applications.* The interaction of *sym*-dichloroacetone (4) with aryl thioamides [15] or aryl thioureas [16] yields substituted thiazoles, either directly or on cyclisation of the primary condensation products. Thiosemicarbazones (3) [17] similarly yield 2-arylidenehydrazino-4-chloromethylthiazoles (5), which are successively acetylated to (6) and brominated to (7). These derivatives are convertible into the free acetylhydrazino-compounds (8, 9) by their interaction with dinitrophenylhydrazine in acid media. The halogen atom of the 4-(chloromethyl)group in both (8) and (9) is very readily replaceable: thus, the corresponding 4-methyl- and 4-hydroxymethyl-thiazoles are accessible by reduction or hydrolysis, and are employed in reaction sequences similar to the ones outlined in the reaction scheme.[17]

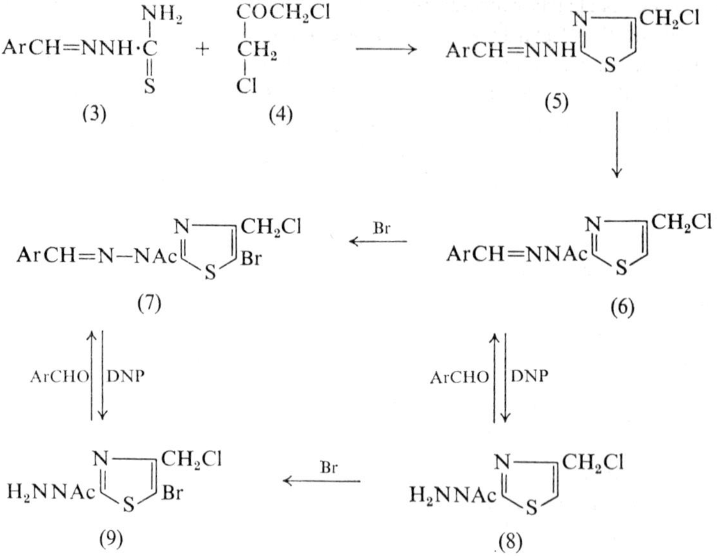

The use of the readily available [18] 1,1,3-tribromoacetone (10) in this general synthesis yields 2-substituted-4-dibromomethylthiazoles (12), which

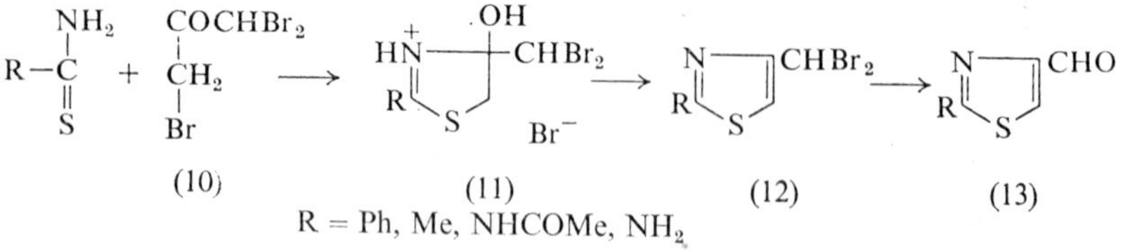

R = Ph, Me, NHCOMe, $NH_2$

[15] A. Silberg, I. Simiti, and H. Mantsch, *Chem. Ber.*, 1961, **94**, 2887.
[16] I. Simiti, M. Farkas, and S. Silberg, *Chem. Ber.*, 1962, **95**, 2672.
[17] I. Simiti and M. M. Coman, *Bull. Soc. chim. France*, 1969, 3276.
[18] C. Rappe, *Arkiv Kemi*, 1963, **21**, 503.

are a convenient source of the corresponding aldehydes (13).[14] The isolation of the intermediate thiazolinium salts (11) is noteworthy (see mechanism on p. 379).

The potential usefulness of thiazolecarboxylic acids and their derivatives as intermediates in the production of thiazolo[4,5-d]pyridazines has prompted a detailed study of their synthesis.[19] The condensation of a thioamide and diethyl α-chloro-β-ketosuccinate (14) in boiling ethanol affords the desired compounds (15), which are suitable precursors of 4,7-dioxo-4,5,6,7-tetrahydrothiazolo[4,5-d]pyridazines (16). The diesters (15) are hydrolysed, under appropriate conditions, to the free acids (17), the monoesters (21), or the monopotassium salts (19), which yield, in their turn, the cyclic anhydrides (20) under the influence of thionyl chloride. Pyrolysis of (17, R = α-$C_4H_3S$, *i.e.* thienyl) results in its decarboxylation to (18). A detailed study and interpretation of the i.r. spectra of these heterocyclic acids was provided.[19]

The mono- (21) and di-esters (15) yield the expected mono- and di-amides and -hydrazides by the conventional methods; the hydrazinolysis of monoesters (*e.g.* 21, R = α-$C_4H_3S$, thienyl) by methylhydrazine yields a single product, which is formulated as (22) (and not as 24) since it fails to

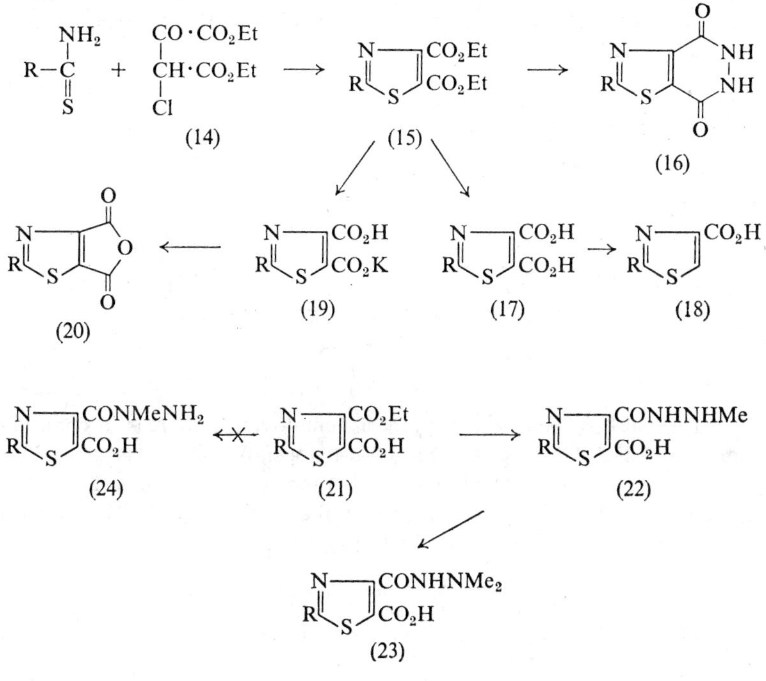

[19] M. Robba and Y. LeGuen, *Bull. Soc. chim. France*, 1969, 1762.

yield ketonic derivatives; the product is convertible, on further methylation, into the dimethylhydrazino-compound (23).[20]

The use of bromopyruvic ester (25) in this synthesis similarly affords thiazole-4-carboxylic acid esters (26), which are convertible into 4-hydrazides, azides, and other functional derivatives in the usual manner. Analogously, 1-bromo-3-oximinobutan-2-one yields the 4-oximinoethyl compound (27), while chloromalonic dialdehyde gives rise to the 5-formylthiazole (28).[21]

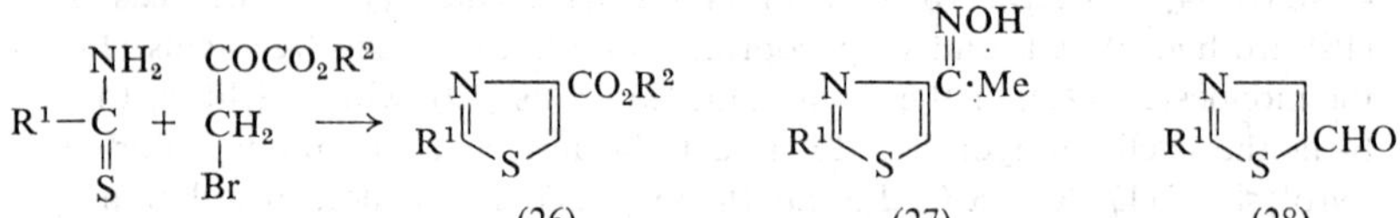

$$R^1-\underset{S}{\overset{NH_2}{\underset{\|}{C}}} + \underset{Br}{\overset{COCO_2R^2}{\underset{|}{CH_2}}} \longrightarrow R^1\underset{S}{\overset{N}{\rceil}}CO_2R^2 \quad R^1\underset{S}{\overset{N}{\rceil}}\overset{NOH}{\underset{\|}{C}\cdot Me} \quad R^1\underset{S}{\overset{N}{\rceil}}CHO$$

(25)      (26)      (27)      (28)

3-Iminothiobutyramide (29), containing four nucleophilic centres, any two of which might react with two electrophilic sites in phenacyl bromide, undergoes the Hantzsch reaction preferentially, yielding the enamine (30) in dry dioxan, or (4-phenylthiazol-2-yl)acetone (31) in isopropanol. Other enamines are obtainable from the ketone (31) by standard methods. Reduction of the enamine (30) hydrobromide by sodium borohydride gives a dimeric product, which was formulated as (32) on the basis of spectroscopic evidence.[22]

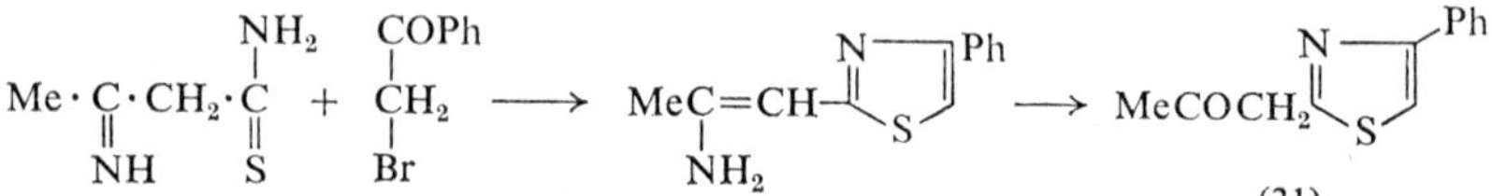

$$Me\cdot\underset{NH}{\overset{NH_2}{\underset{\|}{C}}}\cdot CH_2\cdot\underset{S}{\overset{\|}{C}} + \underset{Br}{\overset{COPh}{\underset{|}{CH_2}}} \longrightarrow Me\underset{NH_2}{\overset{}{\underset{|}{C}=CH}}\underset{S}{\overset{N}{\rceil}}Ph \longrightarrow MeCOCH_2\underset{S}{\overset{N}{\rceil}}Ph$$

(29)      (30)      (31)

$$Ph\underset{S}{\overset{N}{\rceil}}CH=CH-NH-\underset{Me}{\overset{}{\underset{|}{CH}}}-CH_2\underset{S}{\overset{N}{\rceil}}Ph$$

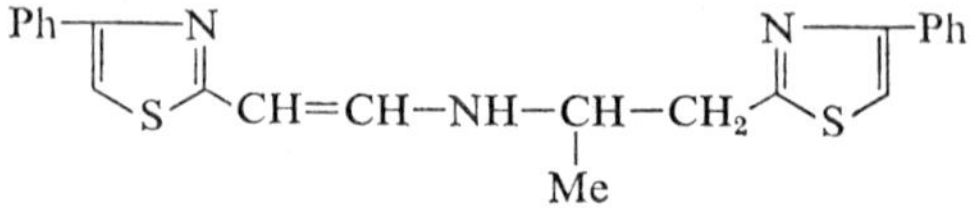

(32)

A variation[23] of Hantzsch's synthesis, employing thioureas in conjunction with α-diazoketones in place of α-halogenoketones, has proved to be generally applicable, providing a wide range of 4-substituted 2-alkyl(or aryl)aminothiazoles[24] and 2-arylimino-3,4-diarylthiazolines.[25] The reaction involving thiourea may occur in the following stages:

[20] M. Robba and Y. LeGuen, *Bull. Soc. chim. France*, 1969, 2152.
[21] A. Benkö and A. Levente, *Annalen*, 1968, **717**, 148.
[22] T. R. Govindachari, S. Rajappa, and K. Nagarajan, *Helv. Chim. Acta*, 1968, **51**, 2102.
[23] C. King and F. M. Miller, *J. Amer. Chem. Soc.*, 1949, **71**, 367.
[24] W. Hampel and I. Müller, *J. prakt. Chem.*, 1968, **38**, 320; W. Hampel, *Z. Chem.*, 1969, **9**, 61.
[25] W. Hampel and I. Müller, *J. prakt. Chem.*, 1969, **311**, 684.

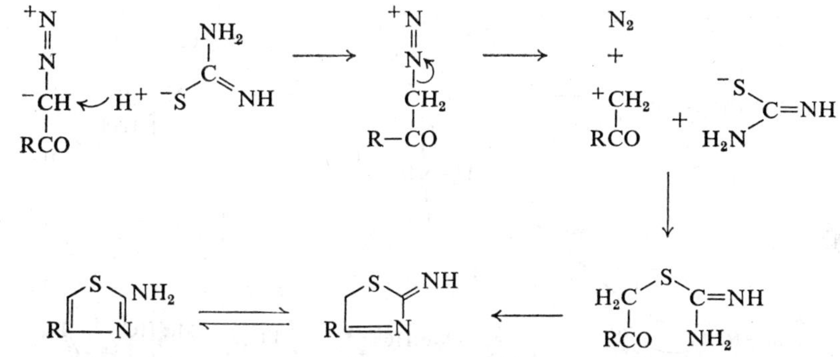

(iii) *Labelled Thiazoles.* 1-(5-Nitrothiazol-2-yl)-2-imidazolidinone (33), an antiparasitic drug, incorporating labelled carbon in either hetero-ring (33a or 33b), has been synthesised for biological and metabolic studies.[26] One of the reaction sequences employed, starting with [$^{14}$C]thiourea, is outlined:

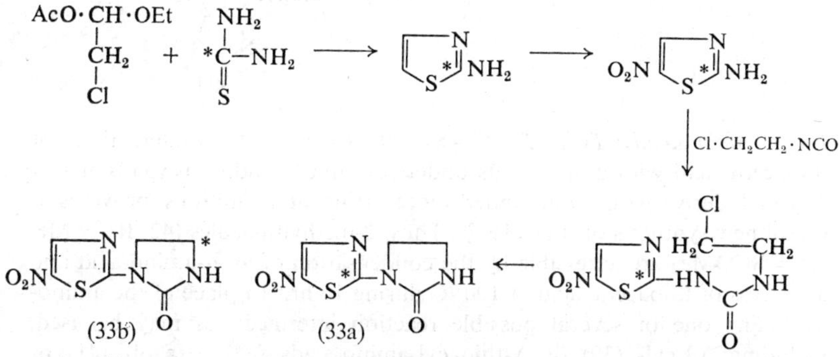

(33b) 2·5 mCi/mmole

(33a) 0·33 mCi/mmole

*From 1-Thiocyanato-2-isothiocyanatoethylenes:*(Type A, C—C + S—C—N). In the course of wider studies of the interaction of thiocyanogen and phosphorylenes, the use of β-oxoalkylenephosphoranes (Ph$_3$P=CR$^1$COCHR$^2$R$^3$) was found to afford novel 1-thiocyanato-2-isothiocyanatoethylenes, *e.g.* (35), which are suitable precursors of thiazoles.[27] Thus, aniline (or dimethylamine) attacks 2-methyl-3-isothiocyanato-4-thiocyanatohept-3-ene (35) at its isothiocyanato-group, yielding 2-anilino-4-propyl-5-isopropylthiazole (36). Methanolic alkali, reacting initially at the SCN function, produces 2-thiolo-4-propyl-5-isopropylthiazole (37).

[26] J. W. Faigle and H. Keberle, *J. Labelled Compounds*, 1969, **5**, 173.
[27] E. Zbiral and H. Hengstberger, *Annalen*, 1969, **721**, 121.

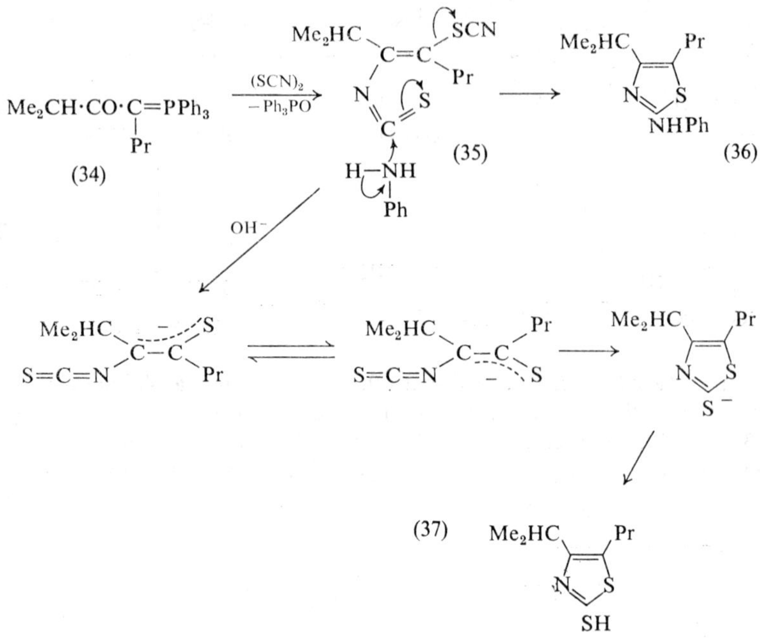

*From Amino-acids:*(Type B, C—S + N—C—C). The interaction of thioacetic acid with amino-acids under restrained conditions yields merely *N*-acetyl derivatives,[28] but under more stringent conditions provides a useful new synthesis of thiazoles.[29] Thus, 2-methylthiazoles (42, $R^1$ = Me, $R^3$ = SCOMe) are accessible by the condensation of an α-amino-acid and an excess of thioacetic acid at 100 °C during 16 hr. In place of the amino-acid, any one of several possible reaction intermediates may be used, including *N*-acyl- (39) or *N*-thioacyl-amino-acids (38), oxazol- (41) or thiazol-5(4*H*)-ones (40), or 5-acetoxythiazoles (42; $R^3$ = OCOMe). The reaction scheme clearly shows the versatility of this group of reactions. Thiobenzoylproline (43; R = H) yields the mesionic 2-phenylthiazole-5-thione (45), which is also obtainable from the corresponding thiazolone (44) and carbon disulphide.[29]

2-Phenylthiazolones are intermediates in a new stepwise degradation of polypeptides in which their *N*-thiobenzoyl derivatives (46), prepared by the action of thiobenzoylthioglycollic acid, are employed.[30] Under the influence of trifluoroacetic acid, their terminal thiobenzoylamino-acid unit is detached and ring-closed to the 2-phenylthiazol-5(4*H*)-one trifluoro-

---

[28] M. W. Farlow, *J. Biol. Chem.*, 1948, **176**, 71; A. Stoll and E. Seebeck, *Helv. Chim. Acta*, 1948, **31**, 189.
[29] G. C. Barrett, A. R. Khokhar, and J. R. Chapman, *Chem. Comm.*, 1969, 818; compare also H. Behringer and K. Kuchinka, *Annalen*, 1961, **650**, 179.
[30] G. C. Barrett and A. R. Khokhar, *J. Chromatog.*, 1969, **39**, 47.

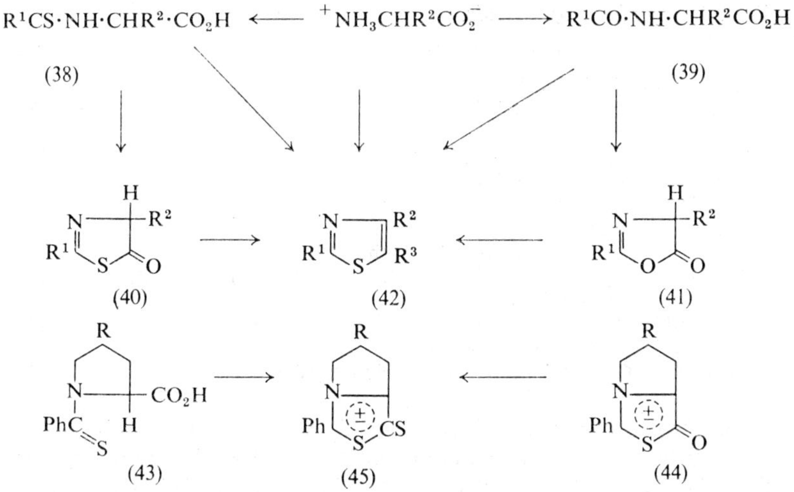

$R^1CS \cdot NH \cdot CHR^2 \cdot CO_2H$ ⟵ $^+NH_3CHR^2CO_2^-$ ⟶ $R^1CO \cdot NH \cdot CHR^2CO_2H$

(38)                                                                       (39)

(40)                 (42)                 (41)

(43)                 (45)                 (44)

acetate (47); this is identified mass-spectrometrically,[31] or chromatographically, either directly or after being cleaved to the *N*-thiobenzoylamino-acid anilide (48) by brief treatment with aniline in boiling toluene. Proline and hydroxyproline yield the mesionic thiazones (44; R = H or OH) in this process; they are not cleaved by aniline, but show a distinctive behaviour during thin-layer chromatography.[30]

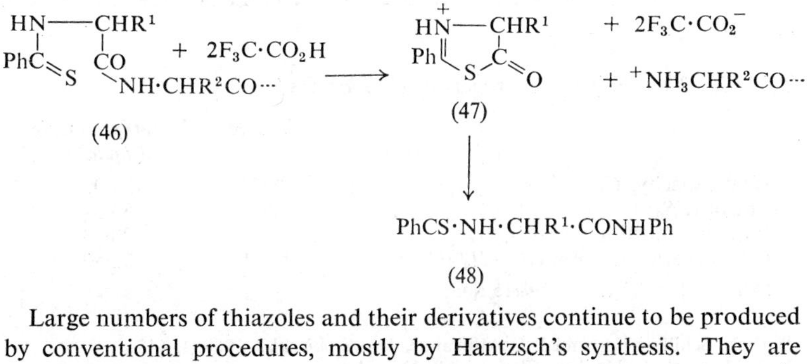

(46)                                   (47)

$PhCS \cdot NH \cdot CHR^1 \cdot CONHPh$

(48)

Large numbers of thiazoles and their derivatives continue to be produced by conventional procedures, mostly by Hantzsch's synthesis. They are listed briefly in Table 1.

**Physical Properties.**—The spectral properties of the majority of new thiazoles currently synthesised are normally determined as a matter of course and are occasionally instrumental in the assignment of structures. The u.v., i.r., and n.m.r. spectra of eight 2,4-disubstituted thiazoles have been recorded and correlated with the structural features of the homologues concerned.[32]

[31] G. C. Barrett and J. R. Chapman, *Chem. Comm.*, 1968, 335.
[32] A. Babadjamian and J. Metzger, *Bull. Soc. chim. France*, 1968, 4878.

**Table 1** *Synthesis of thiazoles*

| Type of thiazole | Ref. |
|---|---|
| Lower 2,4-dialkylthiazoles | 32 |
| 4-Vinyl- and 4-isopropenyl-thiazoles | 33 |
| Thiazol-5-ylbenzimidazoles | 34 |
| 2-Substituted 4-chloromethylthiazoles | 35 |
| Substituted 2-aminothiazoles | 36 |
| 2-Aminothiazoles, sulphonyl derivatives | 37 |
| 2-Amino-4-(fluorophenyl)thiazoles | 38 |
| 2-Amino(or hydroxy)-4-(4-alkylselenophenyl)thiazoles | 39 |
| 4,5-Dialkyl-2-($p$-dimethylaminophenyl)-3-methylthiazolium iodides | 40 |
| Thiazolylthiocarbanilides | 41 |
| 5-Nitrothiazole-2-carboxaldehydes | 42 |
| 5-Nitrothiazole-2-carboxylic acids | 42 |
| 2-(4-Chlorophenyl)thiazol-4-ylacetic acid ('Myalex') | 43 |

The heat capacity and thermodynamic functions of thiazole and its 2-methyl homologue have been determined [44] by adiabatic calorimetry from 5 to 340 K, with the results collected in Table 2. Thiazole melts at 239·53 K with an entropy increment of 9·57 cal mol⁻¹ K⁻¹. The agreement of the third-law ideal gas entropy at 298·15 K (66·46 cal mol⁻¹ K⁻¹) with the spectroscopic value (66·41 cal mol⁻¹ K⁻¹) indicates that the crystal is an ordered form at 0 K. In the case of 2-methylthiazole, the phase equilibria, and consequently the thermal behaviour, are complicated by the existence of both stable and metastable forms, melting at 248·42 and 246·50 K, with entropies of melting, $\Delta S_m$ 11·70 and 11·00 cal mol⁻¹ K⁻¹, respectively.[44]

**Table 2** *Thermodynamic properties of thiazoles* [44]

| | Thiazole | 2-Methylthiazole (liquid) |
|---|---|---|
| Heat capacity, $C_s$ | 28·92 | 36·01 |
| Entropy, $S_0$ | 40·62 | 50·64 |
| Enthalpy function $(H^0 - H_0^0)/T$ | | 28·13 |
| Gibbs energy function $(G^0 - H_0^0)/T$ | −18·01 | −22·50 |
| (All in cal mol⁻¹ K⁻¹ at 298·15 K) | | |

[33] C. L. Schilling jun. and J. E. Mulvaney, *Macromolecules*, 1968, **1**, 445.
[34] J. M. Singh, *J. Medicin. Chem.*, 1969, **12**, 553.
[35] A. Silberg and Z. Frenkel, *Rev. Roumaine Chim.*, 1968, **13**, 1251.
[36] F. Gagiu, G. Czavassy, and E. Bebesel, *J. prakt. Chem.*, 1969, **311**, 168.
[37] R. D. Desai, G. S. Saharia, and H. S. Sodhi, *J. Indian Chem. Soc.*, 1969, **46**, 115.
[38] S. Giri and A. Mahmood, *J. Indian Chem. Soc.*, 1969, **46**, 441.
[39] E. Hannig and H. Ziebandt, *Pharmazie*, 1968, **23**, 552.
[40] D. L. Garmaise, C. H. Chambers, and R. C. McCrae, *J. Medicin. Chem.*, 1968, **11**, 1205.
[41] B. S. Kulkarni, B. S. Fernandez, M. R. Patel, R. A. Bellare, and C. V. Deliwala, *J. Pharm. Sci.*, 1969, **58**, 852.
[42] D. W. Henry, *J. Medicin. Chem.*, 1969, **12**, 303.
[43] W. Hepworth, B. B. Newbould, D. S. Platt, and G. J. Stacey, *Nature*, 1969, **221**, 582.
[44] P. Goursot and E. F. Westrum jun., *J. Chem. and Eng. Data*, 1968, **13**, 468, 471.

**Chemical Properties.**—The more prominent chemical reactions of the heteroaromatic thiazole system are electrophilic substitution in the thiazole nucleus (provided it is suitably activated) and replacement reactions involving substituents, including those of reactive 2-alkyl groups. Current work has been concerned with each of these aspects; for reasons of expediency, a few reactions have already been mentioned in the section dealing with syntheses. The growing interest in the function of thiazoles, benzothiazoles, and their derivatives as ligands in the formation of metal complexes is noteworthy. Attention is again drawn to continued efforts to correlate observed reactivities and calculated electron densities in thiazole, and to the masterly discussion of this problem by Metzger.[7]

*Radical Reactions.* The decomposition of thiazol-2-yldiazonium salts in a variety of media at 0 °C in presence of alkali generates thiazol-2-yl radicals. Their strongly electrophilic character is shown by their power to substitute at positions of high electron density of the media in which they are formed (*e.g.* toluene, anisole, nitrobenzene, pyridine). In agreement with expectation, the reactivity of the media participating in the substitution process is increased by electron-releasing groups (*e.g.* Me, MeO) and reduced by electron-attracting groups (*e.g.* $NO_2$, Cl).[45]

Vinyl- and isopropenyl-thiazoles undergo radical polymerisation, and copolymerisation with styrene. 2-Vinylthiazole is much more reactive towards its own radical or towards a styryl radical than is the 4-vinyl monomer; an explanation for this observation has been sought in terms of the higher resonance stabilisation of the former isomer.[33]

*Nitration and Halogenation.* It is well known that thiazole is not readily substituted electrophilically unless the molecule is suitably activated, *e.g.* by electron-releasing 2-substituents ($NH_2$, MeCONH, HO). The nitration of 2-methylthiazole by nitronium tetrafluoroborate has been studied by Asato[46] and that of alkylthiazoles by nitric and sulphuric acids by Metzger *et al.*,[47] whose observations are summarised in Table 3. Amongst several series of alkylated thiazoles, 2,4-dialkylthiazoles are seen to be the most reactive, being nitrated (70%) in the 5-position, *i.e.* the centre of greatest electron density. The position of the nitro-group in the individual products was deduced from the nature of their u.v. spectra (5-substitution: $\lambda_1$ 215 nm, $\lambda_2$ 285 nm; $\varepsilon_1 < \varepsilon_2$; 4-substitution: $\lambda_1$ 218 nm, $\lambda_2$ 280 nm; $\varepsilon_1 > \varepsilon_2$)

Nitrothiazoles have also been prepared by the conversion of a 2-amino- into a 2-nitro-group by established techniques. Thus, 2-amino-5-formyl-thiazole is diazotised in fluoroboric acid, and the resulting diazonium salt is treated with excess of nitrite in presence of copper powder to afford 50% crude yields of 2-nitro-5-formylthiazole.[48]

---

[45] G. Vernin, B. Barré, H. Dou, and J. Metzger, *Compt. rend.*, 1969, **268**, C, 2025.
[46] G. Asato, *J. Org. Chem.*, 1968, **33**, 2544.
[47] H. J. Dou, G. Vernin, and J. Metzger, *J. Heterocyclic Chem.*, 1969, **6**, 575.
[48] G. Asato, G. Berkelhammer, and E. L. Moon, *J. Medicin. Chem.*, 1969, **12**, 374.

**Table 3** *Nitration of alkylthiazoles*[47]

| Alkylthiazole | Reactivity | Yield | Product |
|---|---|---|---|
| 2-Methylthiazole | | 12% | 5-Nitro(77%); 4-nitro(23%) |
| 2-n-Propylthiazole | | 14% | 5-Nitro(71%); 4-nitro(29%) |
| 4-Alkylthiazoles | 0·066 | | 5-Nitro |
| 5-Alkylthiazoles | 0·04 | | 4-Nitro |
| 2,4-Dialkylthiazoles | 1 | 70% | 5-Nitro |
| 2,5-Dialkylthiazoles | 0·5 | | 4-Nitro |

For purposes of approximate comparison, the 'reactivity' of 2,4-dimethylthiazole is taken as standard (1).

2-Phenyl-4-dibromomethylthiazole (49) is brominated in its 5-position in presence of catalytic quantities of dibenzoyl peroxide, presumably by a free-radical mechanism. The product (50) is hydrolysed nearly quantitatively by aqueous ethanol to the 4-formyl compound (51). While nitration of (49) occurs preferentially in the aryl substituent, resulting (with simultaneous hydrolysis) in (52), that of 2-acetamido-4-dibromomethylthiazole (53) yields the expected 5-nitro-derivative (54) in 65% yield.[14]

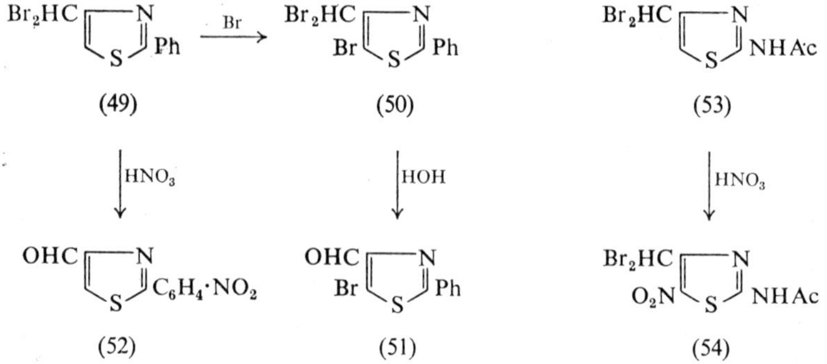

*Other Substitution Reactions.* Typical replacement reactions of heteroaromatic systems are exemplified in the following multi-stage synthesis of 2-isopropenylthiazole from the 2-amine:[33]

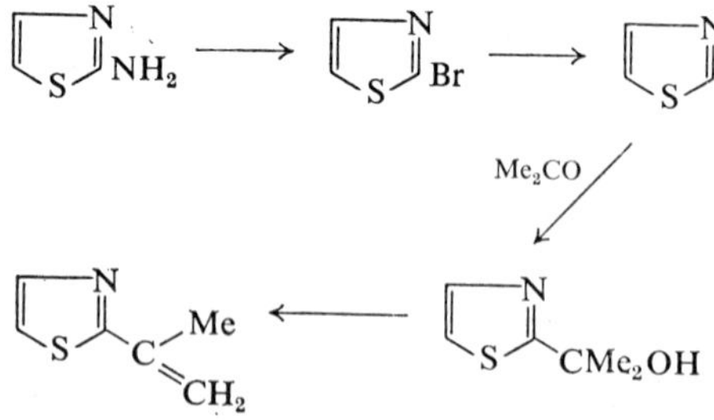

The ready aminolysis of 2-bromothiazoles has been utilised in the production of a large and varied series of 2-alkyl(and aryl)amino-5-

nitrothiazoles (55) for biological evaluation.[49] In this nucleophilic replacement, the 2-halogeno-5-nitrothiazoles (56) may, in a parallel reaction, undergo ring cleavage to products that have been formulated [50] as (1-nitro-2-aminovinyl)thiocyanates (57) on the basis of spectroscopic evidence.

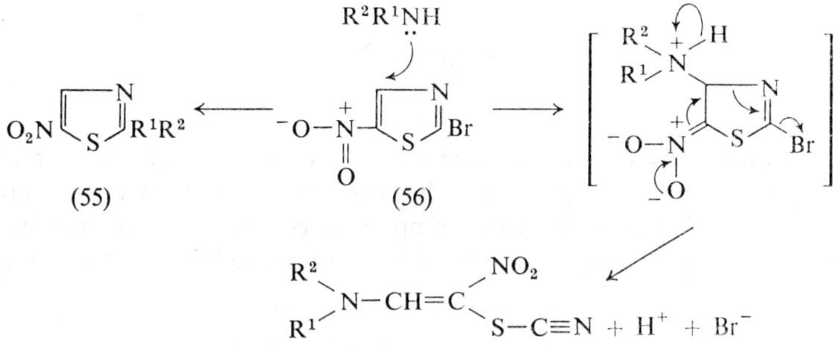

$$O_2N\underset{(55)}{\underset{S}{\overset{N}{\bigsqcup}}}R^1R^2 \longleftarrow\ ^-O-\overset{+}{\underset{\underset{O}{\parallel}}{N}}\underset{(56)}{\underset{S}{\overset{N}{\bigsqcup}}}Br \longrightarrow \left[\underset{R^1\diagdown\overset{+}{N}}{\overset{R^2\diagup H}{\phantom{N}}}\ \ \ ^-O-\overset{+}{\underset{\underset{O}{\mid}}{N}}\underset{S}{\overset{N}{\bigsqcup}}Br\right]$$

$$\underset{R^1}{\overset{R^2}{\diagdown}}N-CH=C\underset{S-C\equiv N}{\overset{NO_2}{\diagup}} + H^+ + Br^-$$
(57)

The ring fission may indeed become the dominant process, being favoured by the use of sterically hindered, strongly basic secondary aliphatic amines and highly polar solvents, such as dimethyl sulphoxide.[50]

The facile isomerisation in alkaline media of substituted 2,5-diaminothiazoles to 2-thiolo-5-aminoimidazoles has, by analogy with the behaviour of 2,5-diamino-1,3,4-thiadiazoles (see p. 450), been interpreted by the following concerted mechanism:[51]

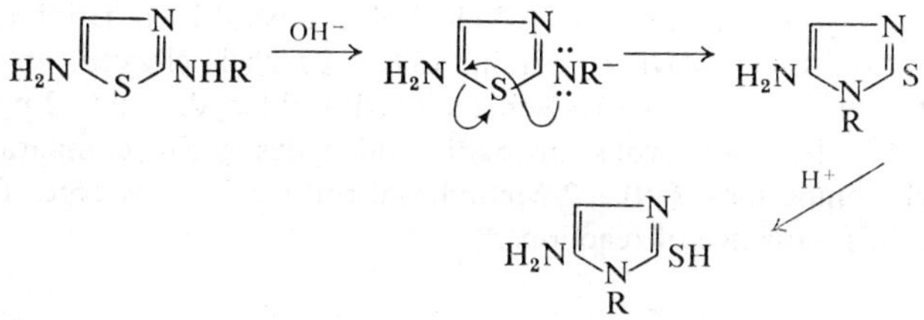

$$H_2N\underset{S}{\overset{N}{\bigsqcup}}NHR \xrightarrow{OH^-} H_2N\underset{S}{\overset{N}{\bigsqcup}}NR^- \longrightarrow H_2N\underset{N}{\overset{N}{\bigsqcup}}S^-$$
$$\underset{R}{\phantom{H_2N}} \xrightarrow{H^+} H_2N\underset{\underset{R}{N}}{\overset{N}{\bigsqcup}}SH$$

The readily accessible halogenoalkylthiazoles are useful intermediates in the production of a variety of substituted alkyl-derivatives,[1] and are capable of alkylating active methylene compounds such as malonic esters.[1] Thus, in a synthesis of thiazolyl-amino-acids,[52] 2-diethylamino-4-chloromethylthiazole is converted into the appropriate substituted alanine by its condensation with sodioacetamidomalonic ester and subsequent acid hydrolysis:[53]

[49] L. M. Werbel, E. F. Elslager, A. A. Phillips, D. F. Worth, P. J. Islip, and M. C. Neville, *J. Medicin. Chem.*, 1969, **12**, 521, and numerous references given therein.
[50] A. O. Ilvespää, *Helv. Chim. Acta*, 1968, **51**, 1723.
[51] Y. R. Rao, *Indian J. Chem.*, 1969, **7**, 836.
[52] A. Silberg, M. Ruse, H. Mantsch, and S. Csontos, *Chem. Ber.*, 1964, **97**, 1767.
[53] A. Silberg, M. Ruse, E. Hamburg, and I. Cirdan, *Chem. Ber.*, 1969, **102**, 423.

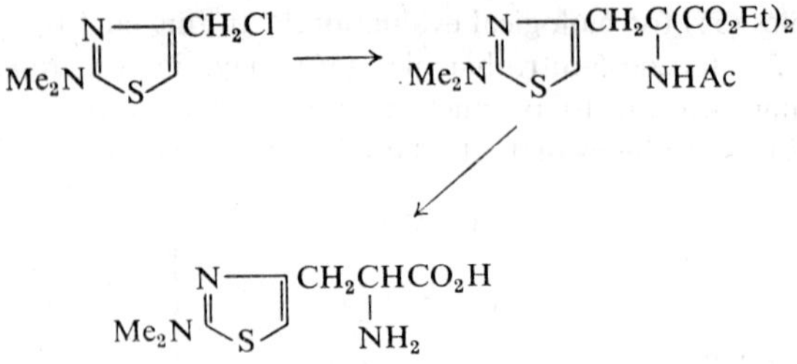

$$\underset{Me_2N}{\overset{N}{\rceil}}\underset{S}{\overset{}{\rfloor}}CH_2Cl \longrightarrow \underset{Me_2N}{\overset{N}{\rceil}}\underset{S}{\overset{}{\rfloor}}\underset{NHAc}{\overset{CH_2C(CO_2Et)_2}{|}}$$

$$\underset{Me_2N}{\overset{N}{\rceil}}\underset{S}{\overset{}{\rfloor}}\underset{NH_2}{\overset{CH_2CHCO_2H}{|}}$$

Substituted 4-chloromethylthiazoles are convertible, by means of the Delépine–Sommelet reaction, into the corresponding 4-formyl compounds, which react in their turn with compounds containing active methylene groups to yield azlactones, acrylic acids, and pyrazolones.[35] The *cis*- and

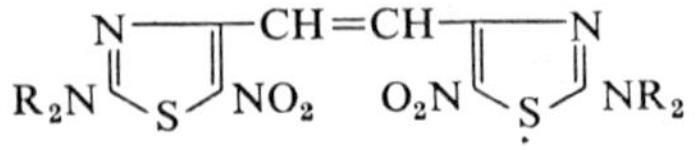

$$\underset{R_2N}{\overset{N}{\rceil}}\underset{S}{\overset{}{\rfloor}}\underset{NO_2}{\overset{CH=CH}{}}\underset{O_2N}{\overset{N}{\rceil}}\underset{S}{\overset{}{\rfloor}}NR_2$$

(58)

*trans*-isomers of (58) are formed from 4-chloromethyl-2-diethylamino-5-nitrothiazole under the influence of sodium ethoxide.[53]

*Alkylthiazoles.* 2-Methyl groups are highly reactive in thiazoles, and still more so in thiazolium salts, but methyl groups are inert in the 4- or 5-positions.[1] Thus, the condensation of 2-methyl-5-nitrothiazole and 2-pyridinecarboxaldehyde in propyl alcohol, catalysed by piperidine, yields 2-[2-(5-nitro-2-thiazolyl)vinyl]pyridine (59),[46] which is cleaved, on ozonolysis at $-40\,°C$, to 5-nitro-2-thiazolecarboxaldehyde and 2-pyridinecarboxaldehyde. Numerous aromatic aldehydes yield comparable β-arylvinyl compounds (59). 2-Methyl-4-nitrothiazole, however, fails to undergo this sequence of reactions.[48]

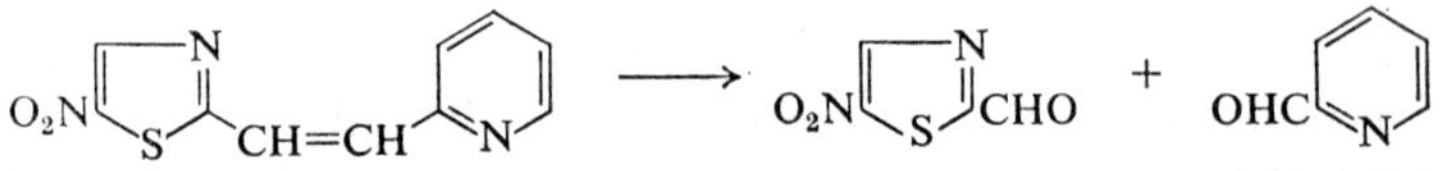

$$O_2N\underset{S}{\overset{N}{\rceil}}CH=CH\underset{N}{\bigcirc} \longrightarrow O_2N\underset{S}{\overset{N}{\rceil}}CHO + OHC\underset{N}{\bigcirc}$$

(59)

$$Ar\underset{S}{\overset{\overset{+}{N}Me\ I^-}{\rceil}}CH=CHR \quad Ar\underset{S}{\overset{\overset{+}{N}Me\ I^-}{\rceil}}CH=NR \quad Ar\underset{S}{\overset{\overset{+}{N}Me\ I^-}{\rceil}}N=NR$$

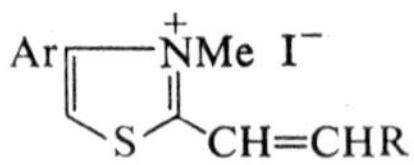

(60) 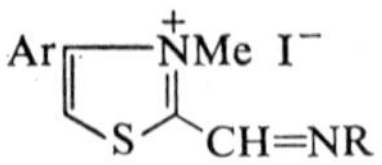 (61) 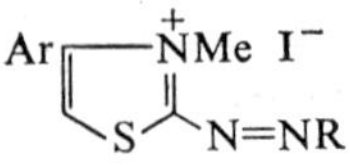 (62)

With *p*-dimethylaminobenzaldehyde, quaternised 2-methyl-4-arylthiazoles yield 2-(*p*-dimethylaminostyryl)-3-methyl-4-arylthiazolium salts (60). Their aza- (61) and diaza-analogues (62) are accessible from the same starting materials. The replacement of the CH=CH by the CH=N and N=N system in the chromophores of these dyes results in a progressive bathochromic displacement (by 35—70 nm) of their light-absorption maxima.[54] The quaternisation of 2-ethylaminophenylazothiazoles occurs preferentially at their hetero-nitrogen, as revealed by a study of the u.v. spectra of the products.[55]

*Complexes with Metals.* 4-(2-Pyridyl)thiazole (A) and its 2-isomer (B) form complexes with a variety of metallic ions. The acid dissociation constants of the protonated species $AH^+$ (p$K$, 4·10) and $BH^+$ (p$K$, 2·17) have been measured, and the formation constants for the species $MA^{2+}$, $MA_2^{2+}$, and $MA_3^{2+}$, as well as $MB^{2+}$, $MB_2^{2+}$, and $MB_3^{2+}$ (M = Cu, Ni, Co, Fe, and Zn) calculated from data obtained by potentiometric, spectrophotometric, and partition techniques. The bonding in these metal complexes is considered to involve the two nitrogen atoms of the pyridylthiazoles, giving the familiar chelating α-di-imide structure

$$-N=C-C=N-.$$[56]

2,4-Bis(2-pyridyl)thiazole forms mono- and bis-complexes with bivalent cobalt, nickel, and copper.[57] Bis[2,4-bis(2-pyridyl)thiazole]cobalt(II), an example of this series, is a paramagnetic compound.[58] Mono- and bis-iron(II) complexes are similarly available:[58] the physical properties of the mono-complexes suggest that they are five-co-ordinate.

*Biochemical Aspects.* A new amino-acid, incorporating the thiazole nucleus, has been isolated from the fungus *Xerocomus subtomentosus* by chromatographic techniques, and was identified as 1-amino-4-(4-carboxythiazol-2-yl)butyric acid (63) by its spectral and chemical properties. Its

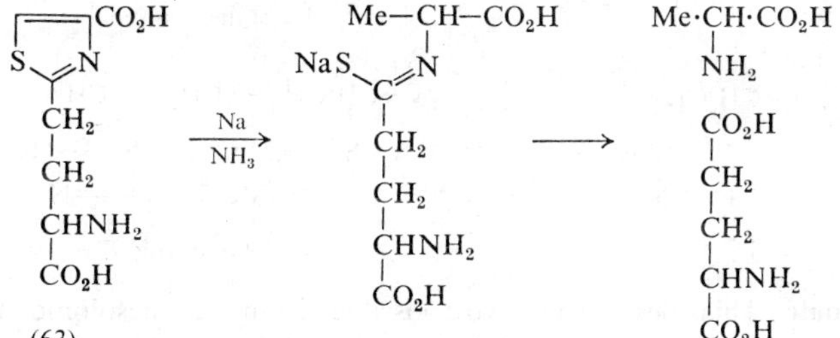

degradation, by sodium in liquid ammonia, followed by hydrolysis, yields alanine and glutamic acid; a synthesis of (63) from the latter was also briefly described in outline.[59]

[54] P. L. Nayak, P. B. Tripathy, and M. K. Rout, *Indian J. Chem.*, 1968, **6**, 572.
[55] L. Pentimalli and L. Greci, *Gazzetta*, 1968, **98**, 1369.
[56] W. J. Eilbeck, F. Holmes, T. W. Thomas, and G. Williams, *J. Chem. Soc. (A)*, 1968, 2348.
[57] H. A. Goodwin, *Austral. J. Chem.*, 1964, **17**, 1366.
[58] H. A. Goodwin and R. N. Sylva, *Austral. J. Chem.*, 1968, **21**, 2881.
[59] J. Jadot, J. Casimir, and R. Warin, *Bull. Soc. chim. belges*, 1969, **78**, 299.

2-(p-Chlorophenyl)thiazol-4-ylacetic acid ('Myalex'; 64), a compound possessing anti-inflammatory properties, is metabolised (rat) to the phenols (65) and (66) (in the ratio 1:2). The minor metabolite (65) arises by the 'NIH-shift'; as previously observed with p-halogenated anilines, acetanilides, and aromatic amino-acids, this comprises microsomal hydroxylation,

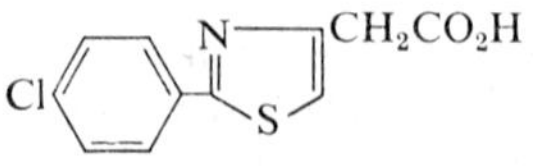

(64) 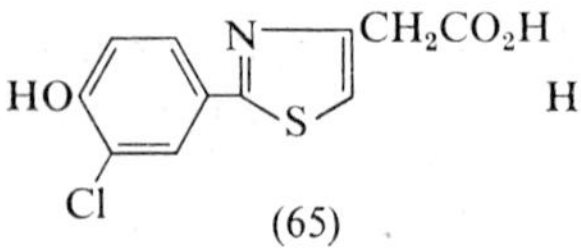   (65) 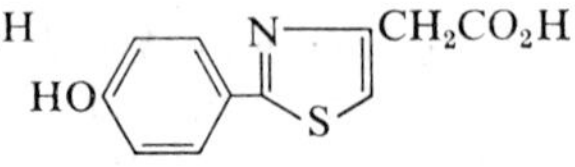   (66)

and the migration of the halogen function. The same reaction (64 → 65) occurs *in vitro*, though in lower yield, under the influence of pertrifluoroacetic acid.[60]

The well-known function of thiamine pyrophosphate as a coenzyme for enzymic reactions, and as catalyst for thiamine-dependent non-enzymic reactions[61] has prompted the examination of the catalytic activity of polymers incorporating thiazolium units. Quaternisation of 4-vinylthiazole with methyl iodide resulted in the direct formation of a quaternised polymer. Partially quaternised polymers of 4-vinylthiazole and 5-vinyl-4-methylthiazole were obtained by treating the respective poly(vinylthiazoles) with methyl iodide. The latter polymers showed some catalytic activity for the acyloin condensation of furfural to furoin.[62]

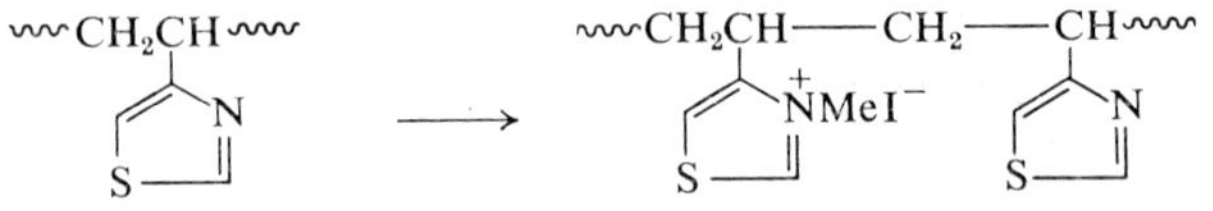

42% Quaternised

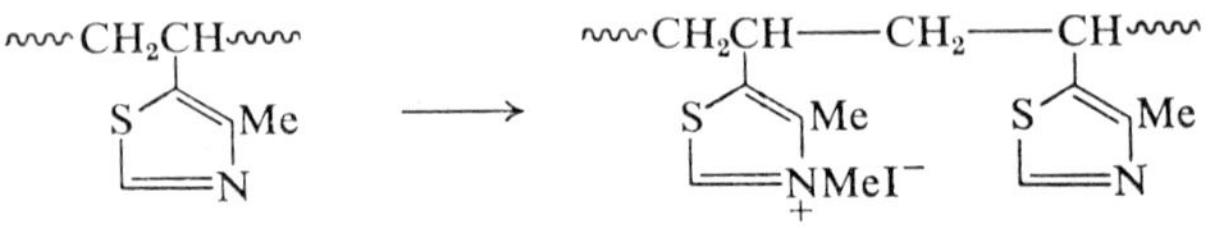

78% Quaternised

**Mesoionic Thiazoles.**—The two possible isomeric mesoionic thiazole structures are the anhydro-2,3-disubstituted-5(or 4)-hydroxythiazolium hydroxides (67 and 68).

The latter compounds (68) are accessible[63] by the cyclisation of acids of type (69) (obtained from thiobenzanilides and bromoacetic acid) under very carefully controlled conditions: the reaction is performed in the minimum

[60] D. M. Foulkes, *Nature*, 1969, **221**, 582.
[61] R. Breslow, *J. Amer. Chem. Soc.*, 1957, **79**, 1762; *ibid.*, 1958, **80**, 3719; R. Breslow and E. McNelis, *ibid.*, 1959, **81**, 3080.
[62] C. L. Schilling jun. and J. E. Mulvaney, *Macromolecules*, 1968, **1**, 452.
[63] K. T. Potts, U. P. Singh, and E. Houghton, *Chem. Comm.*, 1969, 1128.

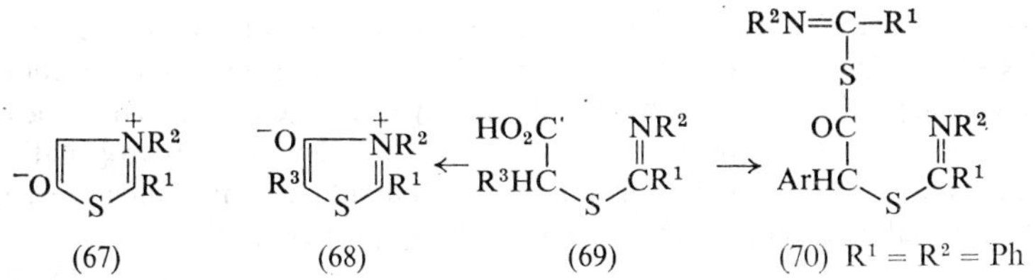

## Thiazoles and Related Compounds

$$R^2N=C-R^1$$
$$|$$
$$S$$
$$|$$

[structures (67), (68), (69), (70) shown]

(67) $\phantom{xx}$ (68) $\phantom{xx}$ (69) $\phantom{xxxxx}$ (70) $R^1 = R^2 = Ph$

volume of acetic anhydride–triethylamine (1 : 3) at room temperature, and the product induced to crystallise as rapidly as possible. In the absence of these critical precautions, the product is a linear anhydrosulphide (70), erroneously formulated as (68) in a previous investigation.[64] Both (68) and (70) yield the *same* anhydro-5-acetyl-2,3-diphenyl-4-hydroxythiazolium hydroxide (68; $R^1 = R^2 = Ph$; $R^3 = COMe$) on treatment with hot acetic anhydride.[63]

Cycloaddition reactions of mesoionic ring systems are, with few exceptions, accountable in terms of 1,3-dipolar intermediates, but 1,4-additions of dipolarophiles have been observed (for literature, see reference 65). The foregoing mesoionic thiazoles (71a, b) undergo a very ready 1,4-addition with dimethyl acetylenedicarboxylate in refluxing benzene, the nature of the product depending on the degree of substitution of the starting material. The hypothetical primary cycloadduct (72; R = H, from 71a) yields, with extrusion of sulphur, dimethyl 1,6-diphenyl-2-pyridone-4,5-dicarboxylate

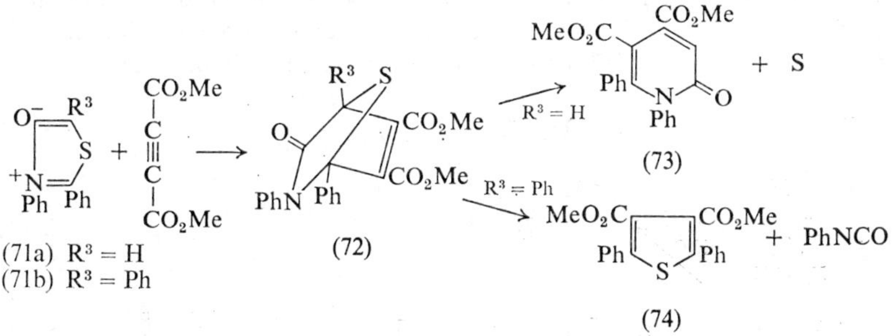

[structures (71a) $R^3 = H$, (71b) $R^3 = Ph$, (72), (73), (74) shown]

(73), but that from (71b) (*i.e.* 72; $R^3 = Ph$) eliminates phenyl isocyanate, forming dimethyl 2,5-diphenylthiophen-3,4-dicarboxylate (74).[65]

### 3 2-Thiazolines

Of the two possible series of dihydrothiazoles, the 2-thiazolines ($\Delta^2$-thiazolines, A) have been investigated more frequently and extensively than the $\Delta^4$-isomers (B). Although this tendency continues to be reflected in the preponderance of current work on 2-thiazolines, significant contributions are being made to the chemistry of 4-thiazolines.

[64] M. Ohta, H. Chosho, C. Chin, and K. Ichimura, *J. Chem. Soc. Japan*, 1964, **85**, 440.
[65] K. T. Potts, E. Houghton, and U. P. Singh, *Chem. Comm.*, 1969, 1129.

2-Thiazolines incorporating appropriate 2-substituents (C, e.g. X = S, O, RN) are tautomers of thiazolidines (D), but are considered, on account of their properties, more appropriately as dihydrothiazoles (C). This type of tautomerism may occur theoretically at all three levels of hydrogenation of the ring system, and the formulation of individual compounds must in each case take account of their predominating properties.[1]

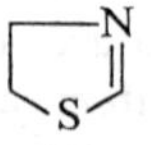

**Syntheses.**—*From* N-(β-*Halogenoalkyl*)*imidohalides*:[66] (*Type C*, C—C—N—C + S). A general synthesis of 2-azolines takes advantage of the reactivity of N-(β-halogenoalkyl)imidohalides, which are convertible into 2-oxazolines, 2-thiazolines, or 2-imidazolines on further reaction with water, hydrogen sulphide, or amines, respectively. This versatile synthesis, providing a wide variety of substituted five-membered heterocyclic compounds, has the added advantage of being performed as a one-stage process.[66]

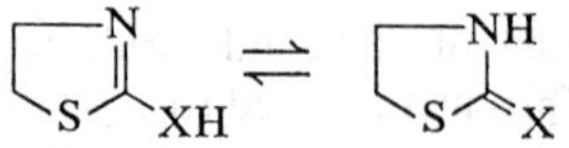

N-(β-Halogenoalkyl)iminohalides (76) are accessible from olefins and halogens in the presence of nitriles; their imino-halogen is the more mobile, being displaced hydrolytically at room temperature with formation of N-(β-halogenoalkyl)amides (77). Under more vigorous conditions, nucleophilic substitution (*e.g.* by aqueous alkaline sulphide solutions) of both halogen atoms of (76) results in 2-thiazolines (79) in one stage.[66] This

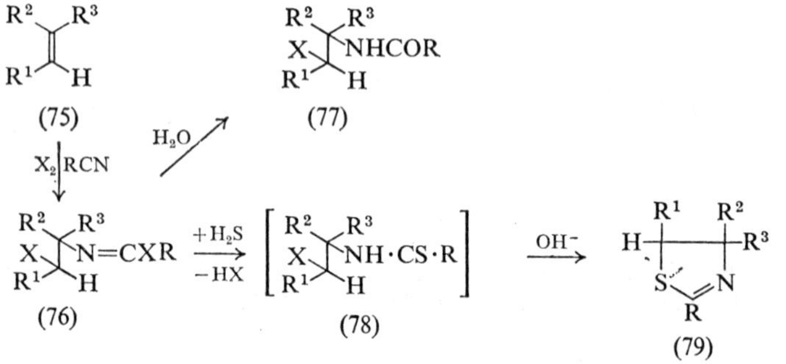

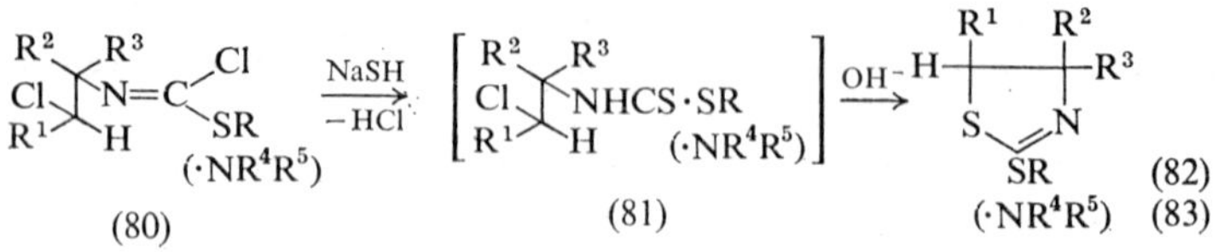

[66] J. Beger, D. Schöde, and J. Vogel, *J. prakt. Chem.*, 1969, **311**, 408.

three-component halogenation of olefins is known [67] to proceed stereo-specifically: accordingly, *cis*- or *trans*-but-2-ene (75; $R^1$, $R^2$ = Me, $R^3$ = H) and acetonitrile yield the pure (g.l.c.) *cis*- or *trans*-2,4,5-trimethyl-2-thiazoline stereoisomer (79; R, $R^1$, $R^2$ = Me; $R^3$ = H), respectively.[66] On replacing the nitrile by thiocyanate esters or dialkylcyanamides, the sequence of reactions (80 → 81 → 82, 83) gives rise to 2-alkylthio- (82) or 2-dialkylamino-2-thiazolines (83).[68] Although yields are generally only moderate, the synthesis is of considerable interest because of its wide applicability and simplicity of operation.

*Condensation of α-Thiolo-acids and Cyanogen:*[69] (*Type D;* C—C—S+C—N). The reaction of α-thiolocarboxylic acids (84) and cyanogen affords good yields of 2-[(S-carboxyalkyl)thioimidoyl]-2-thiazolin-4-ones (86), presumably by cyclodehydration of the primary adducts (85). Treatment with acetic anhydride cyclises the remaining

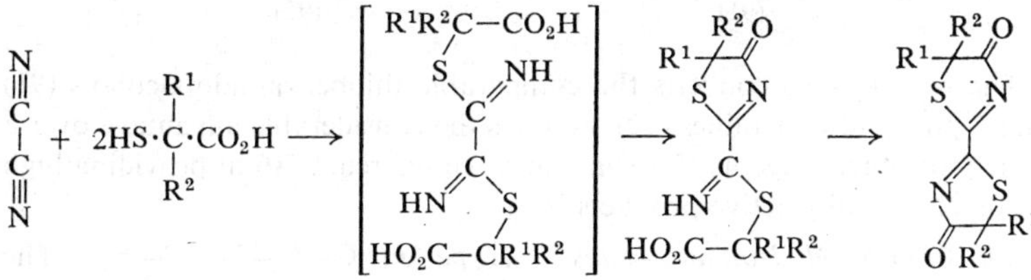

linear moiety of the product (86), giving the symmetrical dicyclic substituted 4,4′-diketo-2-bisthiazolinyl (87), the assigned structure of which agrees with its physical properties. The results of an X-ray crystallographic study of the dienol diacetate of 87 ($R^1$ = H, $R^2$ = Me) provided accurate information on the dimensions of this molecule, and excluded possible alternative six-membered structures (*e.g.* 88) for the dicyclic (87) and therefore also the monocyclic product (86).[69]

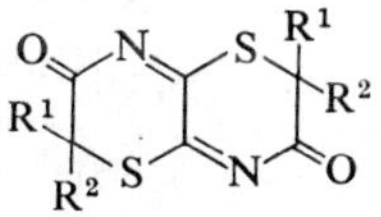

In the following two syntheses, a preformed chain containing carbon, nitrogen, and sulphur atoms in the correct sequence is cyclised directly to the heterocyclic system.

---

[67] T. L. Cairns, P. J. Graham, P. L. Barrick, and R. S. Schreiber, *J. Org. Chem.*, 1952, **17**, 751; J. Beger, K. Günther, and J. Vogel, *J. prakt. Chem.*, 1969, **311**, 15.
[68] J. Beger, *J. prakt. Chem.*, 1969, **311**, 549.
[69] S. C. Mutha and R. Ketcham, *J. Org. Chem.*, 1969, **34**, 2053.

*From* N-*Thiobenzoyl-α-amino-acids:*[70] (*Type F, C—C—N—C—S*). Trifluoroacetic acid is an excellent reagent for cyclising N-thiobenzoyl-α-amino-acids (89) to 2-phenyl-2-thiazol-5-ones (90). The reaction proceeds particularly rapidly when N-thiobenzoyl-α-amino-acid amides (89, X = NH$_2$) or peptides (89, X = NH—peptide) are employed.

<pre>
    HN——CHR                          N——CHR
     |    |              →           ‖    |
   PhC    COX                      Ph C   CO
      S                                S
      
      (89)                             (90)

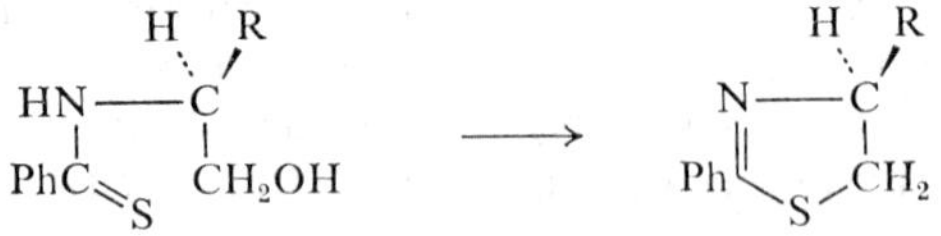

         H   R                          H   R
         ·. /                           ·. /
    HN——C                           N——C
     |   |               →          ‖   |
   PhC   CH$_2$OH                 Ph C   CH$_2$
      S                                S
      
      (91)                             (92)
</pre>

The reagent also converts the comparable thiobenzamido-alcohols (91) into 2-phenyl-2-thiazolines (92), and has the considerable advantage over a variety of other reagents (for summary, see reference 70) in providing high yields of optically active products.[70]

*From Chloroacyl Isothiocyanates:*[71] (*Type F, C—C—N—C—S*). The interaction of chloroacetyl isothiocyanate (93) with aniline (and arylamines generally) yields chloroacetanilides by the usual nucleophilic *substitution*. However, the expected N-chloroacetyl-N'-arylthioureas (94) which are formed in the parallel nucleophilic *addition* are cyclised, with loss of

<pre>
   ClCH$_2$CO·NCS  +  H$_2$NAr   ——→   ClCH$_2$CO·NHCNHAr
                                                      ‖
        (93)                              (94)        S
</pre>

<pre>
         O
         ‖
          ⟩—N
              ⟩—NHAr
         S
         
        (95)
</pre>

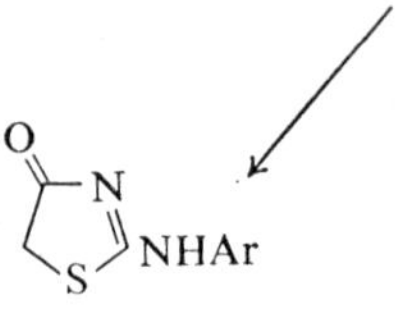

hydrogen chloride, to 2-arylamino-2-thiazolin-4-ones (95) (*e.g.* 'N-phenylpseudothiohydantoin') in 10—25% yield. α-Chloropropionyl (and α-bromoisovaleroyl) isothiocyanate similarly give the corresponding 5-methyl(or isopropyl)-2-thiazolines, but α-bromoisobutyryl isothiocyanate produces stable substituted thioureas that generally show no tendency to cyclisation.[71]

---

[70] G. C. Barrett and A. R. Khokhar, *J. Chem. Soc.* (*C*), 1969, 1117.
[71] R. Pohloudek-Fabini and E. Schröpl, *Pharmazie*, 1968, **23**, 561.

## Thiazoles and Related Compounds

**From Aziridines.**[72-74] The acid-catalysed rearrangement of 1-thioacylaziridines to 2-substituted 2-thiazolines is a well-known reaction,[72] as is the corresponding route to oxazolines. The isomerisations probably involve nucleophilic substitution reactions at a saturated carbon atom; several mechanisms that have been discussed have remained largely unsupported by experimental studies (for a summary, see reference 73).

Recently, the acid-catalysed isomerisation of cis-1-(N-phenylthiocarbamoyl)-2,3-dimethylaziridine (96) to the 2-thiazoline derivatives (97) (as well as that of the oxygen analogues) has been carefully studied and the mechanism of the cleavage of the aziridine ring examined stereochemically. The reaction was performed in a variety of solvents, in the presence of an excess of one of the following acids: toluene-p-sulphonic, 2,4,6-trinitrobenzenesulphonic, and picric acids, 2,6(and 2,4)-dinitrophenol, and trifluoroacetic acid. In the sulphonic-acid-induced isomerisation, complete retention of configuration (97a) is observed in all media, and is accounted for by a double inversion mechanism ($S_N2$). Under the influence of picric acid in weakly nucleophilic solvents, such as benzene or nitromethane, almost complete racemisation occurs, suggesting the operation of an $S_N1$ mechanism involving carbonium ions.[73]

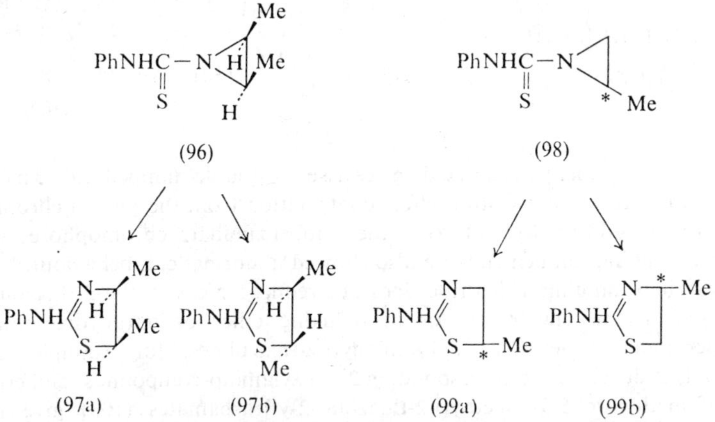

Further relevant observations have become available from a study[74] of the acid-catalysed isomerisation of (S)-1-(N-phenylthiocarbamoyl)-2-methylaziridine (98), which gives 1:1-mixtures of 2-anilino-5-methyl-2-thiazoline (99a) and its 4-methyl isomer (99b). The mode of the ring-opening of the aziridine appears to be almost independent of the acid and solvent employed. The results of experiments using optically active (98)

---

[72] See, for example, A. S. Deutsch, and P. E. Fanta, *J. Org. Chem.*, 1967, **32**, 3069.
[73] T. Nishiguchi, H. Tochio, A. Nabeya, and Y. Iwakura, *J. Amer. Chem. Soc.*, 1969, **91**, 5835.
[74] T. Nishiguchi, H. Tochio, A. Nabeya, and Y. Iwakura, *J. Amer. Chem. Soc.*, 1969, **91**, 5841.

suggest that the formation of (99a) proceeds by an $S_N2$ or $S_N1$ mechanism, depending on the acid and solvent used. The detailed arguments and discussion of the extensive data given in the papers [73,74] should be consulted for further information.

**Properties.**—The i.r. spectra of 2-thiazolines generally include characteristic bands at *ca.* 1560, 1200, and 1000 cm$^{-1}$, the first of which is assigned to a C=N frequency.[14,75]

Thiobenzimidates (101) derived from *N*-thiobenzoyl-L-α-amino-acids (100) show negative circular dichroism associated with their characteristic high-intensity absorption maximum near 270 nm. Since the thiobenzimidate chromophore in a 4-substituted 2-phenyl-2-thiazoline (102) adopts a relatively fixed relationship with the asymmetric centre, the c.d. behaviour of the two series of compounds (101, 102) may be usefully compared.[76]

4-Methyl-2-phenyl-2-thiazoline (102, $R^1$ = H; $R^2$ = Me), prepared from (+)-alaninol [70] and therefore configurationally related to L-alanine, shows negative c.d., with features similar to those of the negative c.d. curves of the *S*-methylthiobenzimidate (101; $R^1$ = $R^2$ = Me) derived from L-alanine. The 4-phenyl analogue (102, $R^1$ = Ph, $R^2$ = H), configurationally related

PhCS·NHCHR$^1$CO$_2$H   ⟶

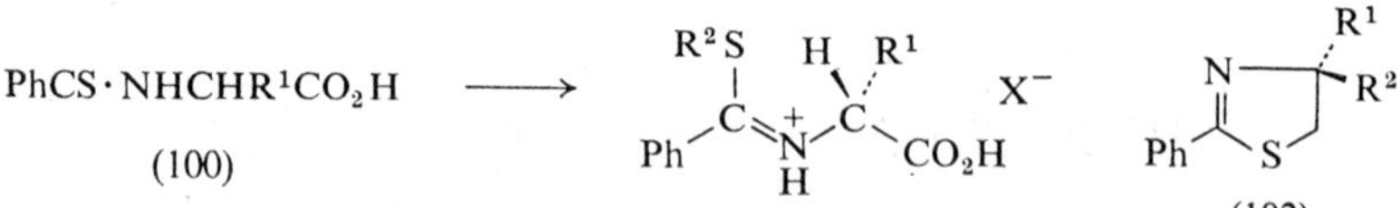

(100)

(101)

(102)

to D-α-aminophenylacetic acid, gives rise to double-humped c.d. curves, probably reflecting a Cotton effect contribution from the phenyl chromophore in addition to that from the thiobenzimidate chromophore; the corresponding thiobenzimidate also showed abnormal c.d. behaviour.[76]

Lithium aluminium hydride does not reduce or cleave the 2-thiazoline ring at room temperature or in refluxing ether or tetrahydrofuran.[77] 2-Benzamido-5-methyl(or 5,5-dimethyl)-2-thiazolines, for example, are merely reduced to the corresponding 2-benzylamino-compounds, and ethyl *N*-[5-methyl(or 5,5-dimethyl)-2-thiazolin-2-yl]carbamates (103) give the corresponding 2-methylamino-2-thiazolines (104). The structures of the products were confirmed by established synthetic routes.[77]

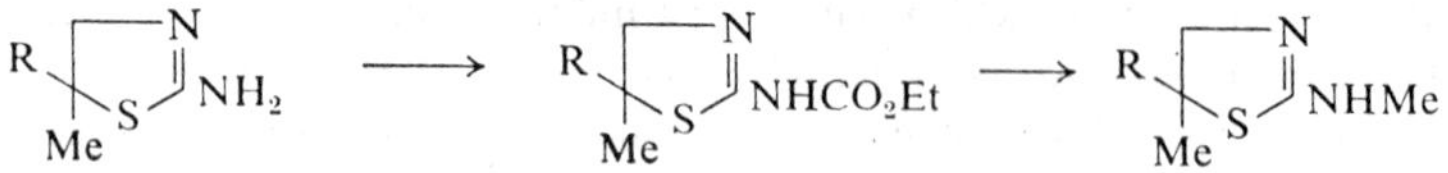

(103)   (104)

[75] W. Otting and F. Drawert, *Chem. Ber.*, 1955, **88**, 1469.
[76] G. C. Barrett and A. R. Khokhar, *J. Chem. Soc.* (C), 1969, 1120.
[77] D. S. Petrova and H. P. Penner, *Croat. Chem. Acta*, 1968, **40**, 189.

The condensation of dimethyl acetylenedicarboxylate with thiazole and its alkyl homologues yields a range of fused hetero-dicyclic systems,[78] and 2-thiazolines similarly yield diadducts.[79, 80] Depending on the nature of the starting materials, the products have been formulated as (105) or (106), the former in analogy with the 9a-$H$-quinolizines accessible from pyridines by the same reaction.

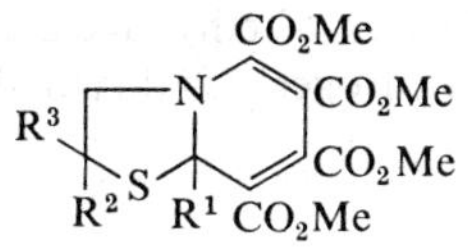

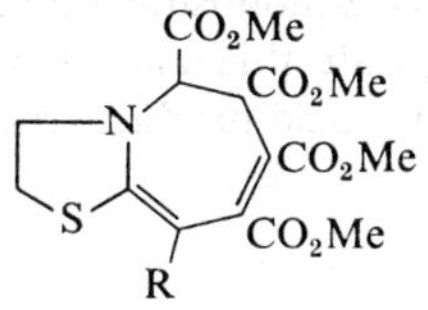

**Metal Complexes.**—Based on previous work [81] concerned with copper polysulphide derivatives of thiourea (of type 107), Peyronel et al.[82] have prepared substances, by the interaction of ethanolamine, sulphur, carbon disulphide, and cupric salts, which they regard as co-ordination complexes of copper polysulphide and 2-thiolo-2-thiazoline (108, 109). Similar products, formulated as (110) and (111), were obtainable directly from 2-thiolo-2-thiazoline, as well as 2-thiolobenzthiazole. The substances were not fully described, nor were more detailed structural proposals given.

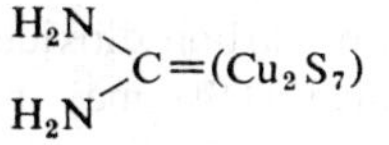

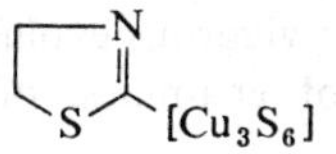

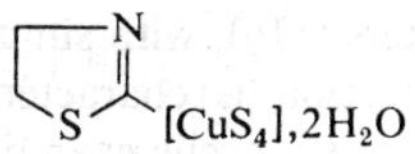

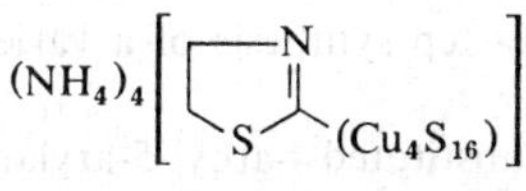

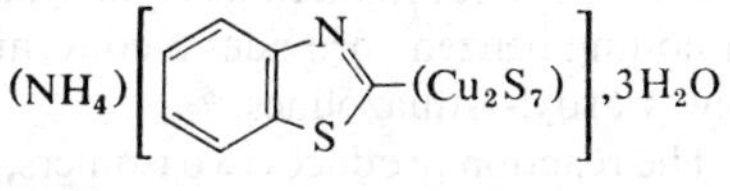

### 4 4-Thiazolines

**Syntheses.**—*Condensation of α-Thioloketones and Schiff's Bases: (Type D, C—C—S+N—C)*. The general synthesis [83] of 4-thiazolines by the interaction of α-thioloketones and Schiff's bases (or their components) has been

---

[78] D. H. Reid, F. S. Skelton, and W. Bonthrone, *Tetrahedron Letters*, 1964, 1797; R. M. Acheson, M. W. Foxton, and G. R. Miller, *J. Chem. Soc.*, 1965, 3200.
[79] J. Roggero and C. Divorne, *Compt. rend.*, 1969, **268**, *C*, 870.
[80] C. Divorne and J. Roggero, *Compt. rend.*, 1969, **269**, *C*, 615.
[81] G. Peyronel and D. de Filippo, *Gazzetta*, 1958, **88**, 271, and subsequent papers; G. Peyronel, D. de Filippo, C. Preti, and G. Marcotrigiano, *Ricerca Sci.*, 1964, **34**, Ser. A, 449.
[82] G. Peyronel, D. de Filippo, and C. Preti, *Gazzetta*, 1969, **99**, 476.
[83] K. Rühlmann, M. Thiel, and F. Asinger, *Annalen*, 1959, **622**, 94

extended to include the use of amino-acids.[84] Thus, 2-thiolopentan-3-one (112), used as model compound, reacts with ethylglycine and the appropriate oxo-compound, or the Schiff's base (113) derived therefrom, with loss of water, to afford the appropriate 4-thiazolin-3-ylacetic acid esters (114). On thermolysis, 2-monosubstituted examples (114; $R^1$ = H) are aromatised to the appropriate thiazoles (115), with loss of ethyl acetate. Reduction with 86% formic acid produces the fully hydrogenated thiazolidinyl-3-acetic

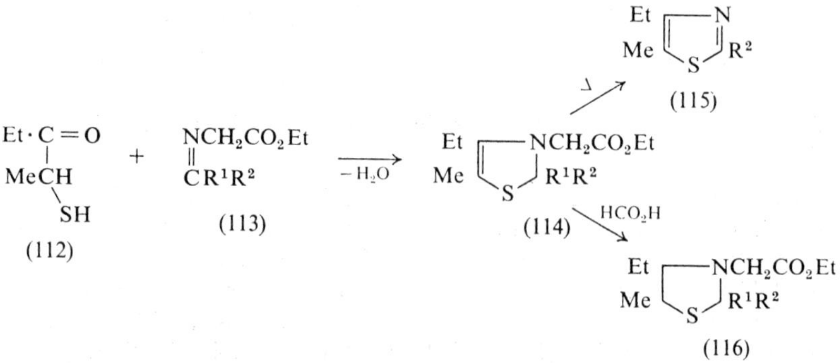

$$\begin{array}{c} \text{Et-C=O} \\ | \\ \text{MeCH} \\ \setminus \\ \text{SH} \\ (112) \end{array} + \begin{array}{c} \text{NCH}_2\text{CO}_2\text{Et} \\ \| \\ \text{CR}^1\text{R}^2 \\ (113) \end{array} \xrightarrow{-H_2O} \begin{array}{c} \text{Et} \\ \text{Me} \end{array}\!\!\left[\begin{array}{c} \text{N} \\ \text{S} \end{array}\right]\!\!R^2 \quad (115)$$

$$\begin{array}{c} \text{Et} \\ \text{Me} \end{array}\!\!\left[\begin{array}{c} \text{NCH}_2\text{CO}_2\text{Et} \\ \text{S} \end{array}\right]\!\!R^1R^2 \xrightarrow{HCO_2H} \begin{array}{c} \text{Et} \\ \text{Me} \end{array}\!\!\left[\begin{array}{c} \text{NCH}_2\text{CO}_2\text{Et} \\ \text{S} \end{array}\right]\!\!R^1R^2$$
(114) \hspace{3cm} (116)

acid esters (116), with simultaneous vigorous evolution of carbon dioxide. This reaction is characteristic[85] of enamines, of which (114) may be regarded as a cyclic example.

*Condensation of 3-Aroylaziridines and Aryl Isothiocyanates:*(Type C—N—C+S—C). The interaction of 3-aroylaziridines (117) and aryl isothiocyanates in boiling benzene provides a convenient one-step synthesis of a variety of new 4-aroyl-4-thiazolines.[86]

The reaction produces two isomers, *viz.* disubstituted 4-aroyl-5-arylamino-(118) and 2-arylimino-4-aroyl-4-thiazolines (119). The former usually predominate, but the proportion of isomers formed is controlled, in some measure, by the nature of the N-substituent of the aziridine. Like numerous comparable additions of aziridines, the first reaction (117 → 118) is probably initiated by the thermal cleavage of the aziridine to an azomethine ylide; this undergoes a concerted (2+3) cycloaddition with the C=S moiety of the isothiocyanate, yielding the final product (118) by tautomerisation of the intermediate. The parallel reaction (117 → 119) is interpreted in terms of the initial nucleophilic attack of the isothiocyanate group at the 2-position of the aziridine ring; the resulting thiazolidine would be dehydrogenated to the observed product (119) either during the reaction or in the subsequent working-up. The thiazolidine was indeed isolable in one case (120); however, its dehydrogenation did not yield the thiazoline (119):

[84] F. Asinger, H. Offermanns, and C. Dudeck, *Monatsh.*, 1968, **99**, 1428.
[85] N. J. Leonard and F. P. Hauck jun., *J. Amer. Chem. Soc.*, 1957, **79**, 5279.
[86] J. W. Lown, G. Dallas, and T. W. Maloney, *Canad. J. Chem.*, 1969, **47**, 3557.

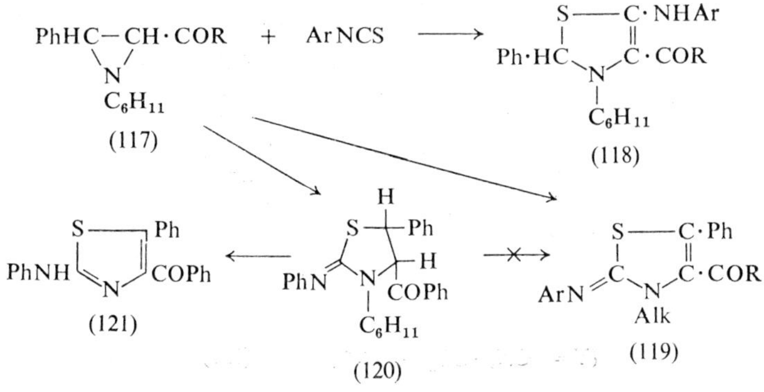

## Thiazoles and Related Compounds

[Scheme showing structures (117), (118), (119), (120), (121):

PhHC—CH·COR + ArNCS → (118) S—C·NHAr / Ph·HC—C·COR, N-C$_6$H$_{11}$
\N/
C$_6$H$_{11}$
(117)

(121) PhNH-thiazole with Ph, COPh ← (120) PhN-thiazoline with Ph, H, COPh, C$_6$H$_{11}$

(119) ArN=C(S)—C(Ph)=C(COR)—N-Alk ─×→ ]

tetrachloro-1,2-benzoquinone, for example, caused more extensive decomposition to the 2-arylaminothiazole (121).[86]

*Direct Cyclisation:*[87] (*Type* C—N—C—C—S). Thiamine disulphide (122), on being heated in triethylamine or an aromatic amine, yields the 4-thiazoline-2-thione (123) as principal product in high yields. Thiochrome (124) and other products are formed in small quantities. The reaction

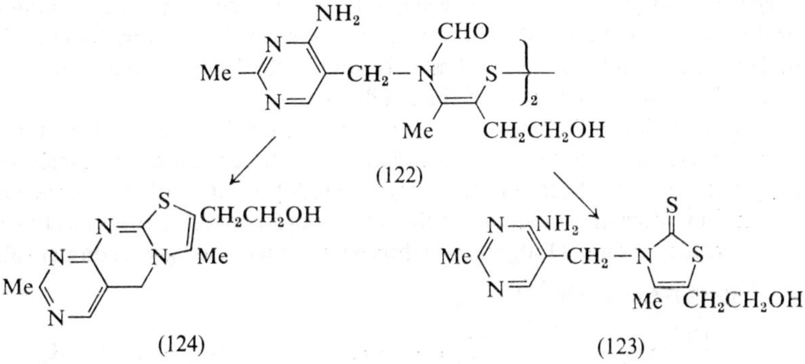

[Structures (122), (123), (124) shown: (122) thiamine disulphide with pyrimidine-CH$_2$-N(CHO)-C(Me)=C(CH$_2$CH$_2$OH)-S- dimer; (124) thiochrome fused ring system with CH$_2$CH$_2$OH and Me; (123) Me-pyrimidine-NH$_2$-CH$_2$-N(thiazoline-2-thione)-Me, CH$_2$CH$_2$OH]

appears to be initiated by the nucleophilic scission of the disulphide link by the amine, and proceeds by liberation, from one of the resulting scission-fragments, of sulphur, which is utilised in a thiolation step associated with the final cyclisation to (123).[87]

**Chemical Reactions.**—The base-catalysed conjugate addition of 4-thiazoline-2-thiones (125) to acrylonitrile or methyl acrylate occurs at the —CS—NH— moiety of the heterocycle, by a kinetically controlled *S*-addition, or by a thermodynamically controlled *N*-addition (affording 126 or 127, respectively).[88] This work has now been supplemented by an examination of the

---

[87] R. G. Cooks and P. Sykes, *J. Chem. Soc.* (C), 1968, 2871.
[88] W. J. Humphlett, *J. Heterocyclic Chem.*, 1968, **5**, 387.

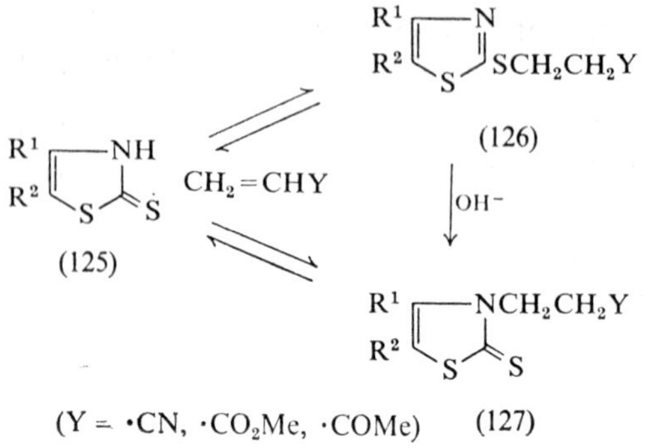

(125) R¹R²C(S)-S-C(NH)... ⇌ CH₂=CHY ⇌ (126) R¹R²C with N, SCH₂CH₂Y

|OH⁻
↓

(127) R¹R²C with NCH₂CH₂Y, S, S

(Y = •CN, •CO₂Me, •COMe)

behaviour of methyl vinyl ketones, and an attempt made to correlate the numerous observations with the relative powers of the individual 4-thiazoline-2-thiones as vinyl acceptors, the role of the base catalyst, and the occurrence of steric hindrance.[89]

## 5 Thiazolidines

**Syntheses.**—*Condensations Involving Thioamido- or Thiocyanato-compounds:* (*Type A*, C—C + S—C—N). The versatile mode of condensing thioamido- or thiocyanato-compounds with two-carbon systems, which forms the basis of Hantzsch's and related syntheses, is applicable, by a suitable choice of reactants, to the production of thiazolidines.

One variation of this route employs α-chloroketones and metallic thiocyanates as starting materials. Thus, the thiocyanation of α-chloropropionanilide (128) (or α-bromobutyrylanilide) occurs with simultaneous cyclisation, affording moderate yields of 4-oxo-2-imino-5-methyl(or ethyl)-3-phenylthiazolidine (130).[90] α-Halogenated acylureas give comparable

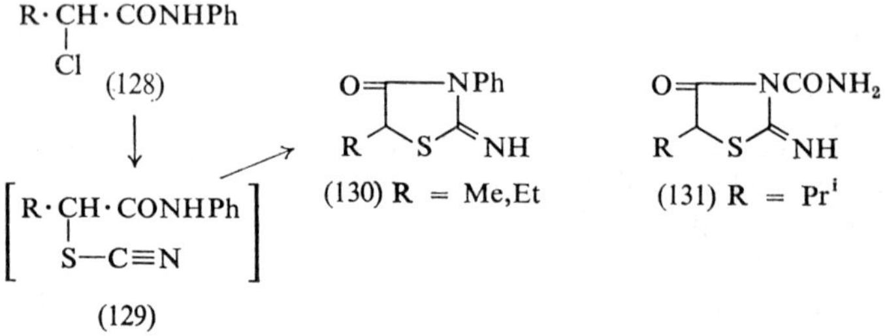

R·CH·CONHPh
|
Cl     (128)

↓

[R·CH·CONHPh
|
S—C≡N]
(129)

(130) R = Me, Et — O=C, NPh, R, S, NH

(131) R = Prⁱ — O=C, NCONH₂, R, S, NH

results: α-bromoisovaleroylurea, for example, yields 4-oxo-2-imino-5-isopropylthiazolidine (21%), formed by loss of the *N*-amido group from the primary cyclisation product (131).[91]

[89] W. J. Humphlett, *J. Heterocyclic Chem.*, 1969, **6**, 397.
[90] E. Schröpl and R. Pohloudek-Fabini, *Pharmazie*, 1968, **23**, 597; compare also G. Frerichs and H. Beckurts, *Arch. Pharm.*, 1900, **238**, 615.
[91] R. Pohloudek-Fabini and E. Schröpl, *Pharmazie*, 1969, **24**, 96.

Oxalyl chloride is a useful source of the two-carbon component of the heterocyclic ring, affording substituted thiazolidine-4,5-diones. Its reaction with ammonium dithiocarbamate (or $N$-monosubstituted derivatives thereof) under nitrogen at 0 °C produces (3-substituted) 2-thioxothiazolidine-4,5-diones (132) in varying yields.[92] The use of thioamides or $NN$-disubstituted thioureas[93] or thioamides incorporating an acidic function similarly provides the appropriately substituted thiazolidine-4,5-diones, e.g. (133).[94]

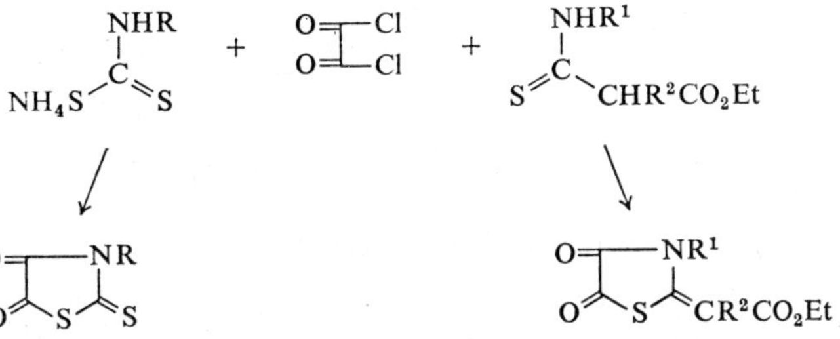

*From Ethanolamines:*(Type B, S—C+N—C—C). $N$-Thiocarbamoyl derivatives of 2-aminoethyl and 3-aminopropyl alcohols readily undergo ring-closure to N—S heterocyclic compounds.[95] The use of hydrazinoethanol in this reaction provides a route to thiazolidines, but yields are low.[96]

Thus, addition of phenyl isothiocyanate occurs at the substituted nitrogen of both hydrazinoethanol (134a) and hydrazinoethyl hydrogen sulphate (134b); the resulting thiocarbamoyl derivatives (135a, b) are cyclisable to 3-amino-2-phenyliminothiazolidine (137) (yield, 26—30%). Diaddition of phenyl isothiocyanate to (134a) provides the $NN'$-bis(thiocarbamoyl) derivative (136a), which is ring-closed, by concentrated hydrochloric acid, to (137) (18%) and (139) (13%). The analogous diadduct (135b), prepared *in situ*, is cyclised by alkali to (138) (38%). The structures of the individual products were mostly inferred from spectral data.

In contrast, the reaction of (134b) with *o*-methoxycarbonylphenyl isothiocyanate yields, after hydrolysis, a 1,3,4-thiadiazine derivative; its formation presupposes addition of the isothiocyanate to the unsubstituted nitrogen of the hydrazino-group of (134b) in the initial stage. Its preferred formation is ascribed to steric crowding in the transition state of the alternative adduct.[96]

[92] V. Hahnkamm, G. Kiel, and G. Gattow, *Naturwiss.*, 1968, **55**, 80, 650.
[93] J. Goerdeler and H. Schenk, *Chem. Ber.*, 1965, **98**, 2954; J. Goerdeler and K. Jonas, *Chem. Ber.*, 1966, **99**, 3572.
[94] G. Barnikow and G. Saeling, *Z. Chem.*, 1969, **9**, 145.
[95] E. Cherbuliez, O. Espejo, B. Willhalm, and J. Rabinowitz, *Helv. Chim. Acta*, 1968, **51**, 241, and preceding papers.
[96] J. Rabinowitz, S. Chang, J. M. Hayes, and F. Woeller, *J. Org. Chem.*, 1969, **34**, 372.

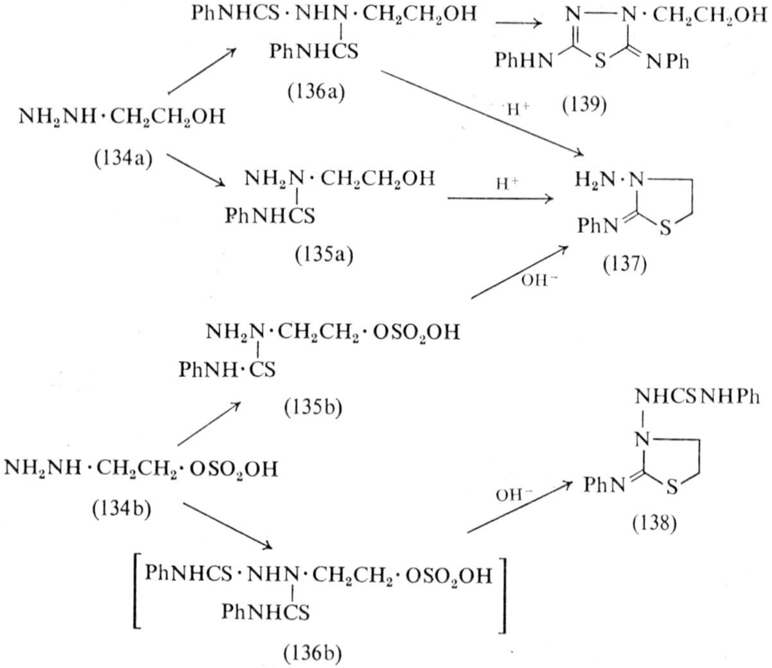

$$
\begin{array}{c}
\text{PhNHCS·NHN·CH}_2\text{CH}_2\text{OH} \\
\text{|} \\
\text{PhNHCS} \\
(136a)
\end{array}
\longrightarrow
\begin{array}{c}
\text{N}\!=\!\!=\!\text{N·CH}_2\text{CH}_2\text{OH} \\
\text{PhHN}\diagup\text{S}\diagdown\text{NPh} \\
(139)
\end{array}
$$

NH$_2$NH·CH$_2$CH$_2$OH (134a)

$$
\begin{array}{c}
\text{NH}_2\text{N·CH}_2\text{CH}_2\text{OH} \\
\text{|} \\
\text{PhNHCS} \\
(135a)
\end{array}
\xrightarrow{H^+}
\begin{array}{c}
\text{H}_2\text{N·N}\!\!-\!\! \\
\text{PhN}\diagup\text{S} \\
(137)
\end{array}
$$

$$
\begin{array}{c}
\text{NH}_2\text{N·CH}_2\text{CH}_2\text{·OSO}_2\text{OH} \\
\text{|} \\
\text{PhNH·CS} \\
(135b)
\end{array}
$$

NH$_2$NH·CH$_2$CH$_2$·OSO$_2$OH (134b)

$$
\begin{array}{c}
\text{NHCSNHPh} \\
\text{|} \\
\text{N} \\
\text{PhN}\diagup\text{S} \\
(138)
\end{array}
$$

$$
\left[\begin{array}{c}
\text{PhNHCS·NHN·CH}_2\text{CH}_2\text{·OSO}_2\text{OH} \\
\text{|} \\
\text{PhNHCS}
\end{array}\right]
$$
(136b)

The interaction of 1-substituted 2-amino-alcohols (140) with carbon disulphide in sodium hydroxide yields exclusively 2-oxazolidinethiones (142). In contrast, the isomeric 2-substituted 2-amino-alcohols (141) afford

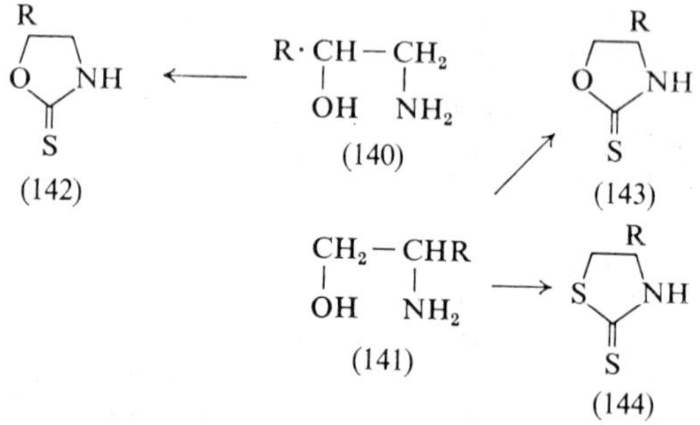

$$
\begin{array}{c}
R \\
\diagup\diagdown \\
O\quad NH \\
\diagdown\!\!=\!\!\diagup \\
S \\
(142)
\end{array}
\longleftarrow
\begin{array}{c}
R·CH\!-\!CH_2 \\
|\quad\quad| \\
OH\quad NH_2 \\
(140)
\end{array}
$$

$$
\begin{array}{c}
R \\
\diagup\diagdown \\
O\quad NH \\
\diagdown\!\!=\!\!\diagup \\
S \\
(143)
\end{array}
$$

$$
\begin{array}{c}
CH_2\!-\!CHR \\
|\quad\quad| \\
OH\quad NH_2 \\
(141)
\end{array}
\longrightarrow
\begin{array}{c}
R \\
\diagup\diagdown \\
S\quad NH \\
\diagdown\!\!=\!\!\diagup \\
S \\
(144)
\end{array}
$$

mixtures of 4-substituted oxazolidine-2-thiones (143), and thiazolidine-2-thiones (144), which are readily separable by chromatography.[97]

*From a Sulphur Dioxide–Keten Adduct:*[98] (*Type D,* C—C—S+C—N). The existence at low temperatures of an adduct of keten and sulphur

---
[97] D. Lednicer and D. E. Emmert, *J. Medicin. Chem.*, 1968, **11**, 1258.
[98] A. de Souza Gomez and M. M. Joullié, *J. Heterocyclic Chem.*, 1969, **6**, 729; *Chem. Comm.*, 1967, 935.

dioxide is revealed by n.m.r. studies, and by 1,3-dipolar additions which it undergoes *in situ*. Thus, the passage of keten into a solution of benzylideneaniline in anhydrous sulphur dioxide at $-70\,°C$ affords 2,3-diphenylthiazolidin-4-one 1,1-dioxide (146) in one stage; analogous products are

$$PhN=CHPh$$
$$+$$

$$\underset{(145a)}{O=C-SO_2 \atop \diagdown / \atop CH_2} \rightleftharpoons \underset{(145b)}{O=\overset{+}{C} \;\; SO_2^- \atop \diagdown \;\; / \atop CH_2} \longrightarrow \underset{(146)}{PhN-Ph \atop O=\diagdown \;\; SO_2}$$

similarly obtainable in 50—60% yield. Their structure is confirmed by an alternative synthesis, and by the reductive removal of the 1,1-oxygen atoms by lithium aluminium hydride. The keten–sulphur dioxide adduct may be represented as an equilibrium mixture of the cyclic 2-oxothiiran 1,1-dioxide (145a) and the linear 1,3-dipolar ion (145b), but more information is needed before the nature of the product can be fully established.[98]

*From Ethyl Isothiuronium Compounds:*[99] (*Type* C—S—C—C—N, *or Type A*). In the course of the development of a synthesis of cystine, α-acetaminoacrylic acid (147) was condensed with thiourea in acetic anhydride to yield the azlactone (148). This intermediate (for which alternative methods of preparation are available) is readily convertible, either directly or *via* the acyclic isothiuronium salt (149), by boiling 20% hydrochloric acid, into 2-iminothiazolidine-4-carboxylic acid (152), possibly by the stages (150 → 151) shown.[99] The reaction resembles the well-established rearrangement of *S*-(β-aminoethyl)isothiourea to 2-iminothiazolidine.[100]

*Intramolecular Radical Addition:* (*Type* C—N—C—C—S). Intramolecular radical addition processes involving ethylenic thiols yield sulphur-containing heterocycles.[101] Applied to *N*-allylaminoethanethiol (153)[102] and its homologues,[103] the reaction is a useful route to thiazolidines.

Thus, u.v. irradiation of an *N*-allylaminoethanethiol (153) in boiling cyclohexane under nitrogen during 5—8 hr affords 60—70% yields of the thiazolidine (157). The reaction is probably initiated by the abstraction of an allylic hydrogen atom, yielding (154), followed by double bond migration, and final cyclisation of the intermediate enamine (155), or imine (156, when R = H). Small quantities of the *N*-allylthiazolidine (158) are formed as by-product (10—22%) when the parent *N*-allylaminoethanethiol (153, R = H) is the starting material.[102]

[99] P. Rambacher, *Chem. Ber.*, 1968, **101**, 3433.
[100] D. G. Doherty, R. Shapira, and W. T. Burnett jun., *J. Amer. Chem. Soc.*, 1957, **79**, 5667.
[101] J. M. Surzur, M. P. Crozet, and C. Dupuy, *Compt. rend.*, 1967, **264**, C, 610.
[102] J. M. Surzur and M. P. Crozet, *Compt. rend.*, 1969, **268**, C, 2109.
[103] R. J. Wineman, M. H. Gollis, J. C. James, and A. M. Pomponi, *J. Org. Chem.*, 1962, **27**, 4222.

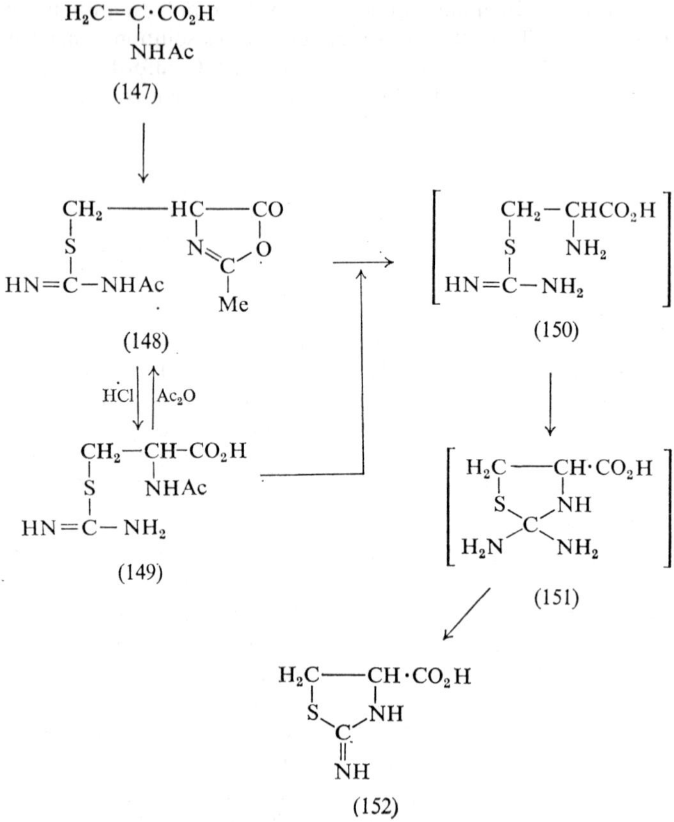

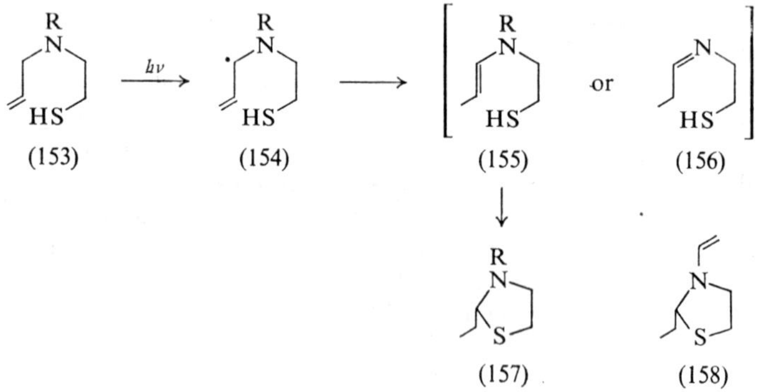

The photolysis of saturated aminoethanethiols also affords thiazolidines, though by different mechanisms. *NN*-Dibutylaminoethanethiol (159), a tertiary amine, slowly produces 2-propyl-3-butylthiazolidine (160) in low yields, possibly as follows:

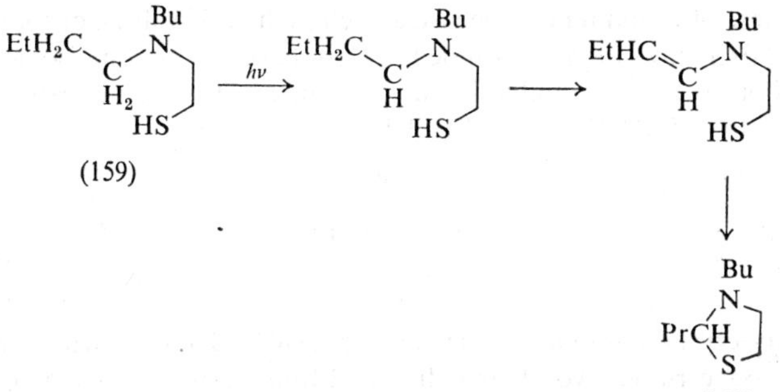

# Thiazoles and Related Compounds

(159) → (160)

In contrast, the ready conversion of the secondary amines (161) into (162) in good yield has been interpreted by the following mechanism:[102]

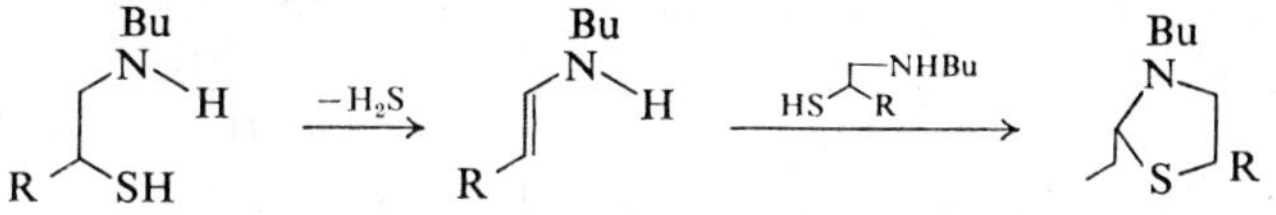

(161) → (162)

The preparation of numerous reduced thiazoles by established syntheses continues to be reported; these classes of compounds are listed in Table 4.

**Table 4** *Synthesis of reduced thiazoles*

| Type of compound | Ref. |
|---|---|
| 2-Thiazolines | |
|   2-Alkyl-2-thiazolines (Alk = $C_2$—$C_{11}$) | 104 |
|   2-Benzylamino-2-thiazolines, etc. | 105 |
| 4-Thiazolines | |
|   4-Methyl-3-aryl-2-arylimino-4-thiazolines | 106 |
|   2-Imino-3-aryl-4-amino-5-carbethoxy-4-thiazolines | 107 |
|   3-Substituted 5-carbethoxymethyl-4-methyl-4-thiazoline-2-thiones | 108 |
| Thiazolidines | |
|   4-(and 5)-Substituted thiazolidines | 109 |
|   2-Mono- and 2,2-di-substituted thiazolidines | 110 |
|   *m*-Bis(2-iminothiazolidin-4-on-2-yl)benzene, etc. | 111 |
|   5-Phenylazo(and amino)-2-arylimino-3-arylthiazolidin-4-ones | 112 |
|   5-Chloroacetamido(and 5-diethylaminoacetamido)-2-arylimino-3-arylthiazolidin-4-ones | 112 |

[104] K. Thewald and G. Renckhoff, *Fette, Seifen, Anstrichm.*, 1968, **70**, 648.
[105] D. S. Petrova and H. P. Penner, *Croat. Chem. Acta*, 1968, **40**, 189.
[106] J. Bartoszewski, *Roczniki Chem.*, 1969, **43**, 319.
[107] A. Singh, H. Singh, A. S. Uppal, and K. S. Narang, *Indian J. Chem.*, 1969, **7**, 884.
[108] G. Westphal, *J. prakt. Chem.*, 1969, **311**, 527.
[109] J. L. Larice, T. Torrese, and M. Roggero, *Compt. rend.*, 1969, **269**, C, 650.
[110] B. J. Sweetman, M. Bellas, and L. Field, *J. Medicin. Chem.*, 1969, **12**, 888.
[111] H. S. Chaudhary and H. K. Pujari, *Indian J. Chem.*, 1968, **6**, 488.
[112] P. N. Bhargava and M. R. Chaurasia, *J. Pharm. Sci.*, 1969, **58**, 896.

**Properties.**—I.r. and n.m.r. spectra, as well as the $\Delta G^{\ddagger}$ values, of compounds (163)—(165), have been determined in a study of the thermal isomerisation rates about the C=C double bonds of conjugated keten mercaptoaminals of type $MeS(Me_2N)C=C(CO_2Et)_2$.[113]

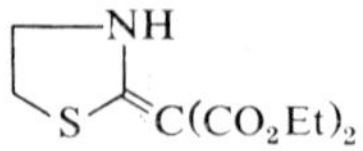

(163) 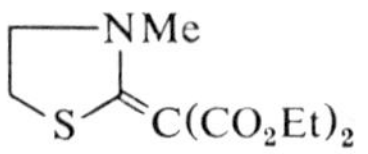   (164) 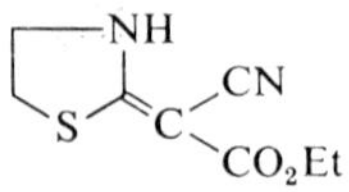   (165)

X-Ray crystallographic studies have provided detailed structural information concerning two thiazolidines. 2-Imino-5-phenylthiazolidin-4-one possesses the zwitterionic structure (166), as shown by the location of two hydrogen atoms on the exocyclic nitrogen atom, and the short C—N bond-lengths in the five-membered ring.[114] 2-Aminothiazolidin-4-one-5-acetic

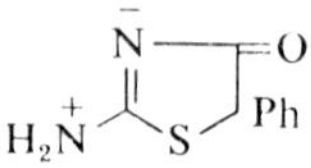

(166) 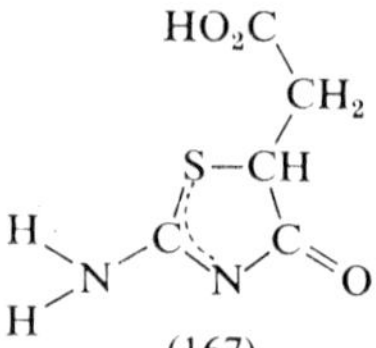   (167)

acid has a predominantly amino- (167) rather than imino-structure. Its bond-lengths and angles indicate the partial double-bond character of the S—C(2), C(2)—N(3), and C(2)—N(amino) bonds.[115]

While the alkylation of thiazolidine-2-thione (168) with alkyl ω-bromo-alkanoates in dimethylformamide and potassium carbonate as acid acceptor normally yields ω-(2-thiazolin-2-ylthio)alkanoates (169), methyl 3-bromopropionate reacts at the ring nitrogen giving methyl 2-thioxo-3-thiazolidinepropionate (171). The S-alkyl compounds (169) are cleaved hydrolytically to S-2-aminoalkyl-S'-ω-carboxyalkyl dithiocarbonate hydrochlorides (170), but the N-alkyl analogue resists ring-opening, giving merely the free acid (172).[116] The reported[117] synthesis of (172) by the cyanoethylation of (168) with acrylonitrile suggests that the anomalous alkylation with methyl 3-bromopropionate is the result of the intermediate formation of methyl acrylate.[116]

When rhodanine (2-thionothiazolidin-4-one, 173) is alkylated by dialkylphosphorochloridates [$(RO)_2POCl$; R = Me, Et, Bu$^n$] in inert solvents in presence of triethylamine, both N- and S-alkylation occurs, the ratio of the isomers formed depending on conditions.[118]

[113] Y. Shvo and I. Belsky, *Tetrahedron*, 1969, **25**, 4649.
[114] L. A. Plastas and J. M. Stewart, *Chem. Comm.*, 1969, 811.
[115] V. Amirthalingam and K. V. Muralidharan, *Chem. Comm.*, 1969, 986.
[116] T. P. Johnston, C. R. Stringfellow jun., and J. R. Piper, *J. Medicin. Chem.*, 1968, **11**, 1257.
[117] R. J. Gaul, W. J. Fremuth, and M. N. O'Connor, *J. Org. Chem.*, 1961, **26**, 5106.
[118] A. I. Ginak, K. A. Vyunov, and E. G. Sochilin, *Zhur. obshchei Khim.*, 1969, **39**, 1180 (Eng. trans., p. 1150).

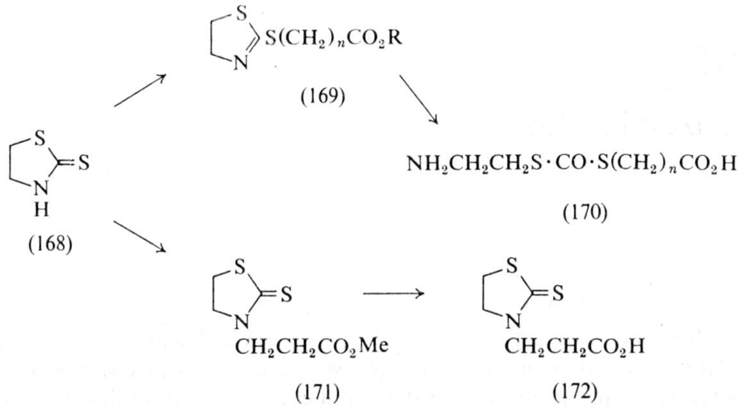

(168) → (169) S(CH₂)ₙCO₂R

(168) structure: thiazolidine-2-thione with NH

(169) → (170) NH₂CH₂CH₂S·CO·S(CH₂)ₙCO₂H

(168) → (171) CH₂CH₂CO₂Me → (172) CH₂CH₂CO₂H

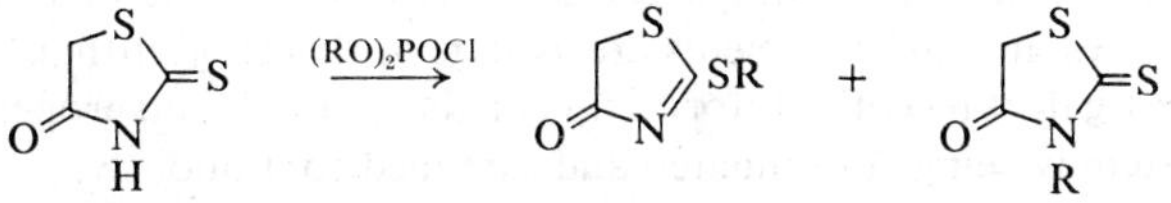

(173) + (RO)₂POCl → SR derivative + R-substituted derivative

*Biological Activity.* Extensive data concerning the antimicrobial action of more than 200 substituted thiazolines and thiazolidines, chiefly of the rhodanine type (173) have been tabulated, and their significance discussed.[119] Certain 5-diethylaminoacetamido-2-arylimino-3-arylthiazolidin-4-ones possess local anaesthetic properties, as shown by tests using the sciatic plexus of the frog.[112]

[119] T. Zsolnai, *Arzneim.-Forsch.*, 1969, **19**, 558.

# 14
# Benzothiazoles

## 1 Introduction

The chemistry of benzothiazole, the beginnings of which may be traced back to Hofmann's work in 1879, has been very thoroughly investigated over the years, but problems of interest are still encountered from time to time. Benzothiazole-spiranes, metal complexes, and the possible role of benzothiazole derivatives in animal pigments are aspects of this subject that have concerned investigators recently. Information on the synthesis and properties of this ring system is being consolidated and extended continuously.

## 2 Synthesis

Syntheses of benzothiazoles and benzothiazolines are dealt with side by side, because of the similarity of the available routes. Moreover, the dihydro compounds are often precursors of benzothiazoles, being readily converted into the latter by dehydrogenation or another suitable elimination. Syntheses are again classified according to the nature of the fragments used in constructing the desired ring pattern.

**Synthesis from o-Aminothiophenols.**—(*Type A*, $S-C_6H_4-N+C$). A useful extension [1] of the well-known benzothiazole synthesis involving the insertion of a carbon atom between the *N*- and *S*-function of an *o*-aminothiophenol employs mixed anhydrides (2) derived from carboxylic acids and half-esters. These reagents, which have recently attracted increasing attention in peptide syntheses,[2] are prepared *in situ* from the appropriate carboxylic acid and isobutyl chloroformate in tetrahydrofuran–*N*-methylmorpholine at −20 °C under nitrogen. They react rapidly with *o*-methylaminothiophenol (1) at low temperatures to yield the acylated intermediates (3) which are instantly cyclised to benzothiazolium salts (4) on addition of perchloric acid. Phenylpropiolic acid produces (6) (*trans*-isomer), which arises by the addition of hydrogen chloride to the strongly polarised triple bond of the primary product (5).

Ferrocenecarboxylic acid gives the brownish-violet explosive perchlorate (7), apparently the first example of a metal π-complex of a heterocyclic

---

[1] H. Quast and E. Schmitt, *Chem. Ber.*, 1969, **102**, 568.
[2] G. W. Anderson, J. E. Zimmerman, and F. M. Callahan, *J. Amer. Chem. Soc.*, 1967, **89**, 5012.

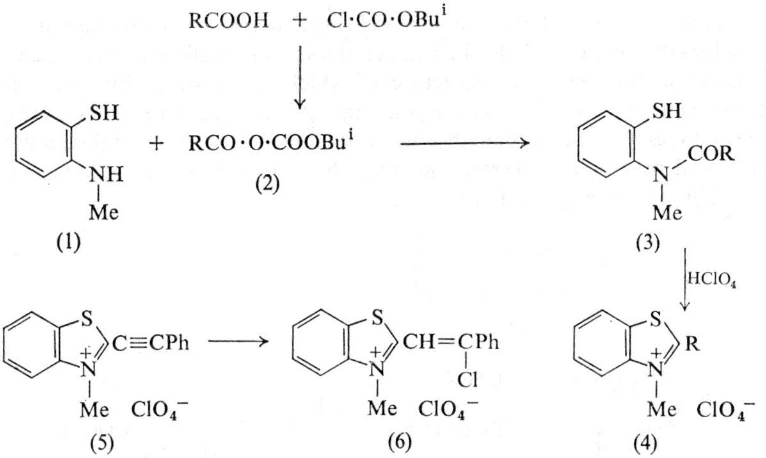

analogue of sesquifulvalene; it is comparable, in its spectral properties, with the $\pi$-complex (8).

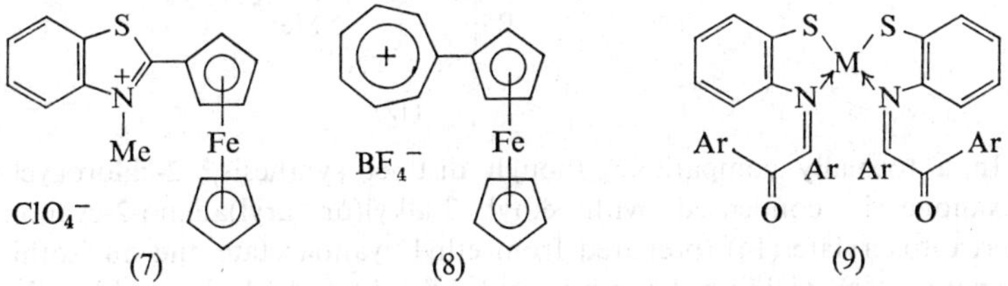

2-Aminothiophenol reacts with equimolar proportions of aromatic 1,2-diketones (*e.g.* α-furil, benzil, *etc.*) in boiling methanol to yield 2-aryl-2-aroylbenzothiazolines. These are converted in boiling glacial acetic acid, with loss of aromatic aldehyde, into 2-arylbenzothiazoles.[3] The elucidation of this sequence of reactions corrects a previous erroneous interpretation.[4]

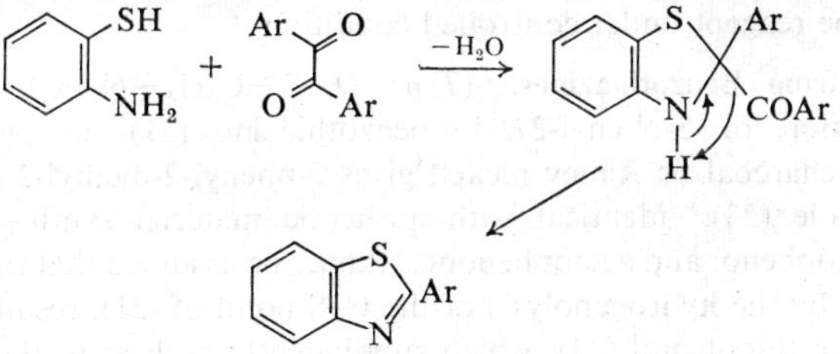

With cadmium acetate, 2-aryl-2-aroylbenzothiazolines yield red complexes, the spectroscopic properties of which favour structures of type (9).

**Jacobson–Hugershoff Synthesis.**—(*Type B*, $C_6H_5-N-C-S$). In a variation of the Jacobson–Hugershoff benzothiazole synthesis, the thioanilides of

[3] E. Bayer and E. Breitmaier, *Chem. Ber.*, 1969, **102**, 728.
[4] H. Jadamus, Q. Fernando, and H. Freiser, *J. Amer. Chem. Soc.*, 1964, **86**, 3057.

α-cyanomalonic ester (10) are used as starting materials.[5] These compounds, which exist preferentially in their thioenol form, are cyclised by bromine to benzothiazolines (11), the structure of which is confirmed by their decarboxylation to the corresponding 2-methylbenzothiazoles (12). Disulphides (13) are obtainable without difficulty from the thioanilides (10), and may be reconverted into them by reduction, or into the benzothiazolines (11) by oxidation.[5]

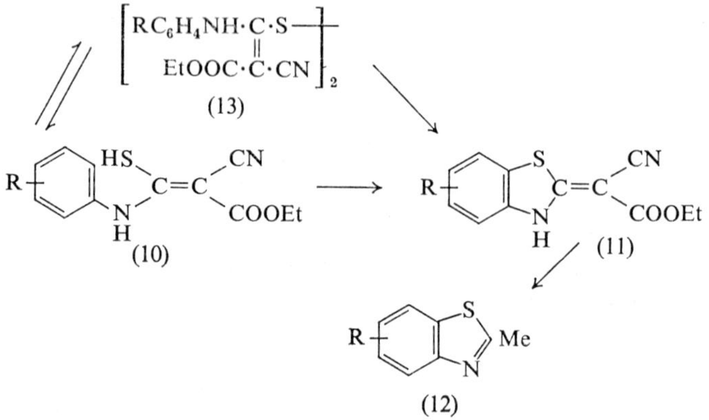

$$\left[\begin{array}{l} RC_6H_4NH\cdot C\cdot S- \\ EtOOC\cdot C\cdot CN \end{array}\right]_2$$
(13)

(10) → (11) → (12)

In a formally comparable, though distinct synthesis,[6] 2-chlorocyclohexanone is condensed with ethyl 3-alkyl(or aryl)amino-2-cyano-3-mercaptoacrylate (14) (prepared from ethyl cyanoacetate and an isothiocyanate ester), yielding the substituted 4,5,6,7-tetrahydrobenzothiazolines (17); they are formulated, on the basis of their u.v. and i.r. spectra, as *cis*-isomers (ester-amine groups). Under sufficiently mild conditions, the intermediate hydroxybenzothiazolidines (16) may be isolated. Concentrated sulphuric acid hydrolyses the nitrile (17) to the amide (18), which is in turn convertible into the substituted acetamide (19) or into the ester (20) by the same reagent under controlled conditions.[6]

**Synthesis from Benzothiazines.**—(*Type D*, S—$C_6H_4$—N—C). Catalytic hydrogenation of 3-phenyl-2*H*-1,4-benzothiazine (21), in presence of palladium charcoal or Raney nickel, gives 2-phenyl-2-methyl-2,3-dihydrobenzothiazole (23),[7] identical with authentic material synthesised from *o*-aminothiophenol and acetophenone. It may be assumed that the reaction is initiated by the hydrogenolysis of the C–S bond of (21), resulting in the intermediate thiophenol (22), which subsequently cyclises to the observed product (23). This sequence is consistent with the mechanism postulated for a hydrogenolytic thiazine-thiazoline rearrangement.[8]

[5] H. Kunzek and G. Barnikow, *Chem. Ber.*, 1969, **102**, 351.
[6] R. Laliberté, H. Warwick, and G. Médawar, *Canad. J. Chem.*, 1968, **46**, 3643.
[7] M. Wilhelm and P. Schmidt, *J. Heterocyclic Chem.*, 1969, **6**, 635.
[8] D. Sica, C. Santacroce, and R. A. Nicolaus, *Gazzetta*, 1968, **98**, 488.

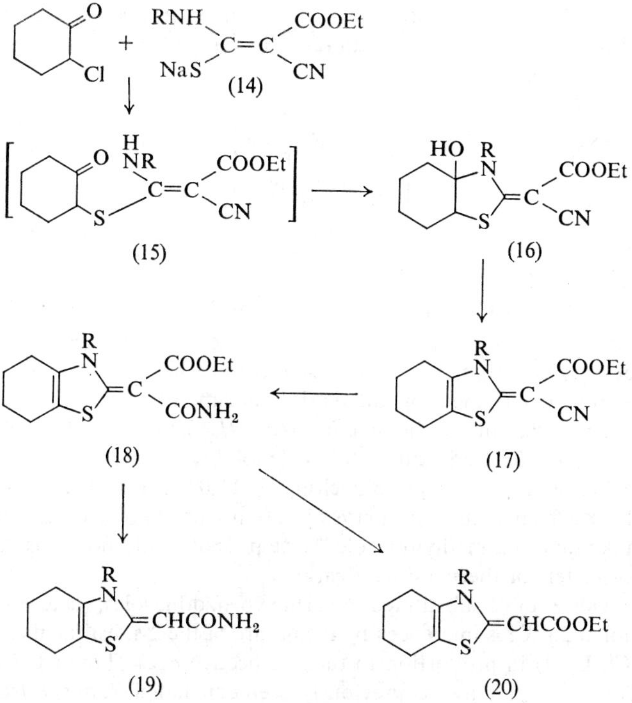

# Benzothiazoles

[Scheme showing structures (14)–(20) for benzothiazole synthesis]

**Table 1** Synthesis of benzothiazoles and benzothiazolines

| Type of compound | Ref. |
|---|---|
| 2-Aminobenzothiazoles | 9 |
| 2-Aminobenzothiazoles, sulphonyl-derivatives | 10 |
| 2-Thioureidobenzothiazoles | 11 |
| Halogeno-2-aminobenzothiazoles | 12 |
| 2-Aryl-3-methylbenzothiazolium perchlorates | 13 |
| 2-Substituted 3-alkylbenzothiazolium salts | 14 |
| Cyanine dyes incorporating benzothiazole moieties | 15 |
| $N$-Methyl-2-acylidenebenzothiazolines | 16 |
| $N$-Substituted benzothiazoline-2-thiones (Mannich bases) | 17 |

[9] L. Pentimalli, *Gazzetta*, 1969, **99**, 362.
[10] R. D. Desai, G. S. Saharia, and H. S. Sodhi, *J. Indian Chem. Soc.*, 1969, **46**, 115.
[11] M. U. Malik, P. K. Srivastava, and S. C. Mehra, *J. Chem. and Eng. Data*, 1969, **14**, 110.
[12] F. H. Jackson and A. T. Peters, *J. Chem. Soc. (C)*, 1969, 268.
[13] Z. N. Timofeeva, M. V. Petrova, M. Z. Girshovich, and A. V. Eltsov, *Zhur. obshchei Khim.*, 1969, **39**, 54 (Eng. trans., p. 46).
[14] D. L. Garmaise, G. Y. Paris, J. Komlossy, C. H. Chambers, and R. C. McCrae, *J. Medicin. Chem.*, 1969, **12**, 30.
[15] J. C. Banerji and S. N. Sanyal, *Indian J. Chem.*, 1968, **6**, 346.
[16] G. Ciurdaru and V. I. Denes, *Rev. Roumaine Chim.*, 1969, **14**, 1063.
[17] R. S. Varma and W. L. Nobles, *J. Pharm. Sci.*, 1969, **58**, 497.

Large numbers of benzothiazoles and their derivatives continue to be synthesised by conventional procedures. The types of compounds produced are listed in Table 1.

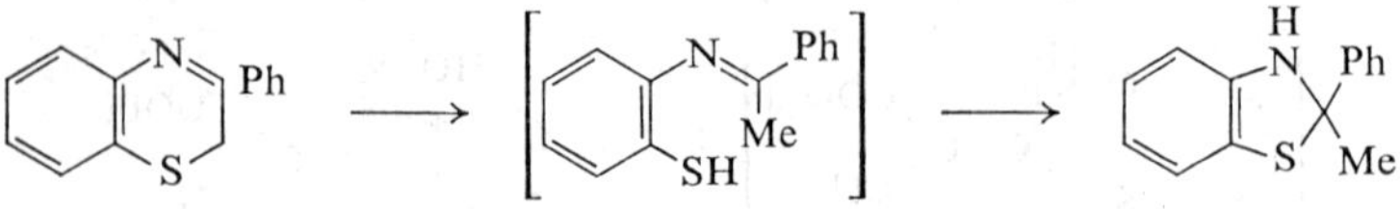

(21) → (22) → (23)

### 3 Physical Properties of Benzothiazoles

The thermodynamic properties of benzothiazole have been studied by Goursot and Westrum.[18] The low temperature heat capacity of benzothiazole was determined by adiabatic calorimetry. The heat capacity ($C$), entropy ($S^0$), and Gibbs function ($G^c - H_0^0/T$) of the liquid phase at 298·15 K are 45·30, 50·15, and −20·58 cal mol$^{-1}$ K$^{-1}$. The stable form melts at 275·65 K with an entropy of melting of 11·09 cal mol$^{-1}$ K$^{-1}$. A metastable form which occurred on crystallisation from the melt was detected, but, unlike that of 2-methylthiazole [19] (see p. 386), could not be isolated in the calorimeter for thermal investigation.

The acidity of 2-substituted 6-carboxybenzothiazoles, determined in methanol at 25 °C, is influenced by the nature of the 2-substituents (H, Me, OMe, Cl, Br, I) in proportion to their respective $\sigma_{para}$ Hammett values.[20] The effects are small, as has previously been concluded from spectroscopic measurements. The relative differences in the observed p$K$ values are comparable with those of the isomeric 6-substituted 2-carboxybenzothiazoles.[21]

### 4 Chemical Properties of Benzothiazoles

**Reduction and Oxidation.**—The polarographic reduction of 2-aryl-3-methylbenzothiazolium perchlorates (24) yields the corresponding benzothiazolines (25). The effect of substituents in the 2-aryl group on the reduction potential may be correlated satisfactorily with their Hammett $\sigma$-values.[13]

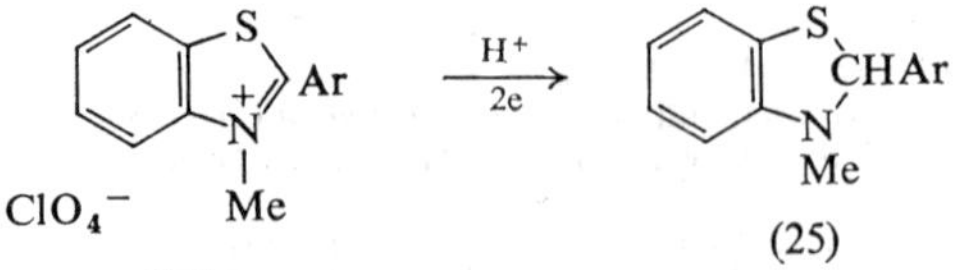

(24) → (25)

---

[18] P. Goursot and E. F. Westrum jun., *J. Chem. and Eng. Data,* 1969, **14**, 1.
[19] P. Goursot and E. F. Westrum jun., *J. Chem. and Eng. Data,* 1968, **13**, 468.
[20] M. Bosco, L. Forlani, and P. E. Todesco, *Ann. Chim. (Italy),* 1968, **58**, 1148.
[21] P. E. Todesco and P. Vivarelli, *Boll. Sci. Fac. Chim. Ind. Bologna,* 1964, **22**, 1.

The most convenient route to benzimidazole N-oxides, viz. the reductive cyclisation of N-substituted N-acetyl-o-nitranilide with sodium borohydride in presence of a reductive catalyst, is not applicable to their sulphur analogues (26 —×→ 27), which give merely o-nitrophenol.[22] However, 2-methylbenzothiazole (28, R = H) affords the N-oxide, albeit in low yields, on treatment with suitable oxidising agents, preferably monopermaleic or trifluoroperacetic acid. The product is separated chromatographically from substantial quantities of tars and orthanilic acid and its N-acetyl derivative arising by oxidative ring-cleavage of the thiazole ring of (28). N-Oxide formation from 6-substituted 2-methylbenzothiazoles is favoured by electron-releasing substituents (28, R = Me, OMe) and inhibited by electron-attracting substituents (28, R = Cl, $NO_2$). Certain 2-functions (SH, SMe, OMe, COOR) prevent the desired oxidation entirely. The assigned N- (rather than the S-) oxide structure of the products agrees with their u.v. and i.r. spectra, and is confirmed by a comparison of their calculated and observed dipole moments.[22]

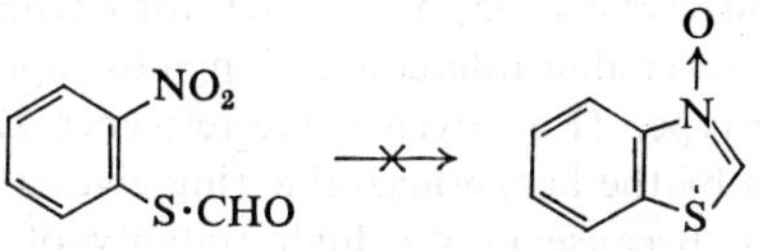

## 2-Methylenebenzothiazoles.[23]

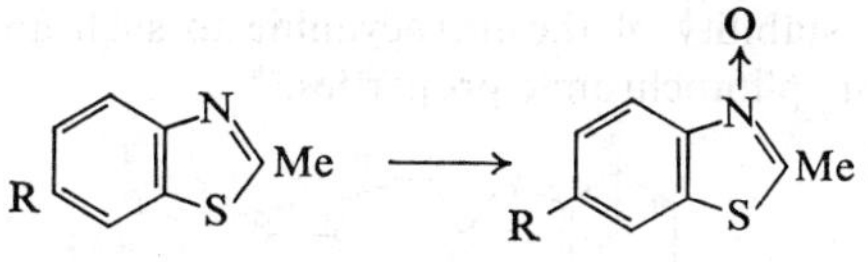

The synthesis of a 2-methylene base (31) from 2,3-dimethylbenzothiazolium salts (30), claimed by König and Meier,[24] yields in fact [25, 26] the dimeric product (32). The true monomeric 2-methylene bases (31, R = Me, Et) have now been obtained,[23] by being generated in the absence of quaternary salts, thus preventing the dimerisation step from taking place. This is achieved experimentally by treating the benzothiazolium salt with a large excess of a strong base (e.g. tetramethylguanidine) in a solvent in which the quaternary salt is only sparingly soluble. The monomers are distillable in a vacuum, and are stable at low temperatures for several weeks, but solidify rapidly to the dimers (32) on being exposed to air at room temperature. Examples of analogous methylene bases derived from a naphtho[2,1]thiazoline and from the

---

[22] S. Takahashi and H. Kano, *Chem. Pharm. Bull. Japan*, 1969, **17**, 1598.
[23] J. R. Owen, *Tetrahedron Letters*, 1969, 2709.
[24] W. König and W. Meier, *J. prakt. Chem.*, 1925, **109**, 334.
[25] H. Larivé and R. Dennilauler, *Chimia*, 1961, **15**, 115.
[26] J. Metzger, H. Larivé, E. Vincent, and R. Dennilauler, *J. Chem. Phys.*, 1963, **60**, 944.

related selenazoline are also on record (33, 34). The n.m.r. spectra of the compounds agree with the assigned structures; in particular, the terminal methylene protons give rise to an AB pattern at δ 3·9—4·2.

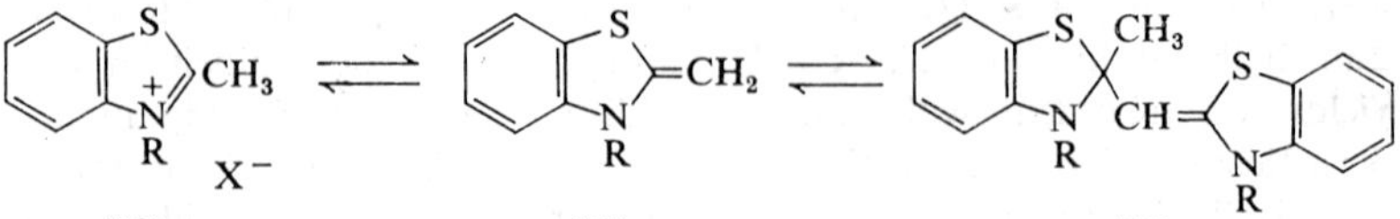

(30)  (31)  (32)

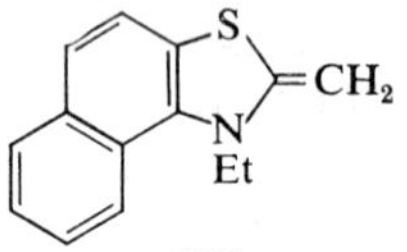

(33) 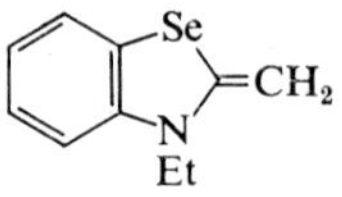  (34)

**Photochromic Benzothiazole Spiranes.**—Spiropyran photochromes containing different heterocyclic nuclei joined to the benzopyran moiety are well known. Indolinospiropyrans (35, Y = $CH_2$), for example, exhibit photochromism by their reversible interconversion into 'open-form' structures of the merocyanine type. (For reviews, see reference 27.) Substitution of the indoline nucleus by the benzothiazoline ring gives a non-photochromic system (35, Y = S), because of the high stability of the corresponding merocyanine derivative.[28] However, introduction of a methyl group at C-3′ reduces the stability of the merocyanine to such an extent that the system recovers its photochromic properties.[29]

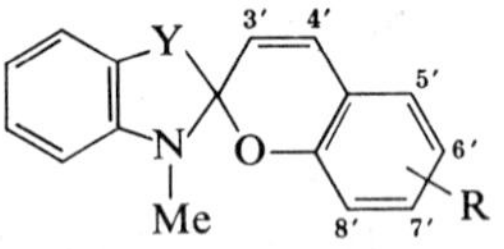

(35)

In continuation of their extensive work [29] on this subject, Metzger et al.[30] have prepared a series of 3-substituted benzothiazole-spiropyrans (38) by the condensation of quaternary 2-alkylbenzothiazolium salts (36) and 2-hydroxy-3-methoxy-5-nitrobenzaldehyde (37). U.v. light cleaves these spiranes to merocyanines (39), which are reconverted thermally, or by visible light, into their precursors (38) and hence decolorised. Further examples of (39) have been prepared by the same reaction by Smets et al.,[31]

[27] R. Dessauer and J. Paris, 'Advances in Photochemistry', Interscience, New York, 1963, Vol. 1, pp. 275—321; W. Luck and H. Sand, *Angew. Chem. Internat. Edn.*, 1964, 3, 570; R. Exelby and R. Grinter, *Chem. Rev.*, 1965, 65, 247; E. Fischer, *Fortschr. Chem. Forsch.*, 1967, 7, 605.
[28] O. Chaude, *Cahiers Phys.*, 1954, 50, 17, and subsequent papers.
[29] R. Guglielmetti and J. Metzger, *Bull. Soc. chim. France*, 1967, 2824; R. Guglielmetti, E. J. Vincent, and J. Metzger, *ibid.*, 4195.
[30] R. Guglielmetti and J. Metzger, *Bull. Soc. chim. France*, 1969, 3329.
[31] P. H. Vandewyer, J. Hoefnagels, and G. Smets, *Tetrahedron*, 1969, 25, 3251.

and the decolorisation kinetics of the open-ring merocyanines studied by spectrophotometric methods. The nature and position of the substituents in the benzopyran and benzothiazoline moieties influence the u.v. absorption properties and the first-order decolorisation rate-constants of the individual compounds. Activation energies and entropies were evaluated. Most of the observations may be interpreted in terms of the electron-attracting and electron-releasing properties of the substituents, but steric effects become significant in certain examples.[31]

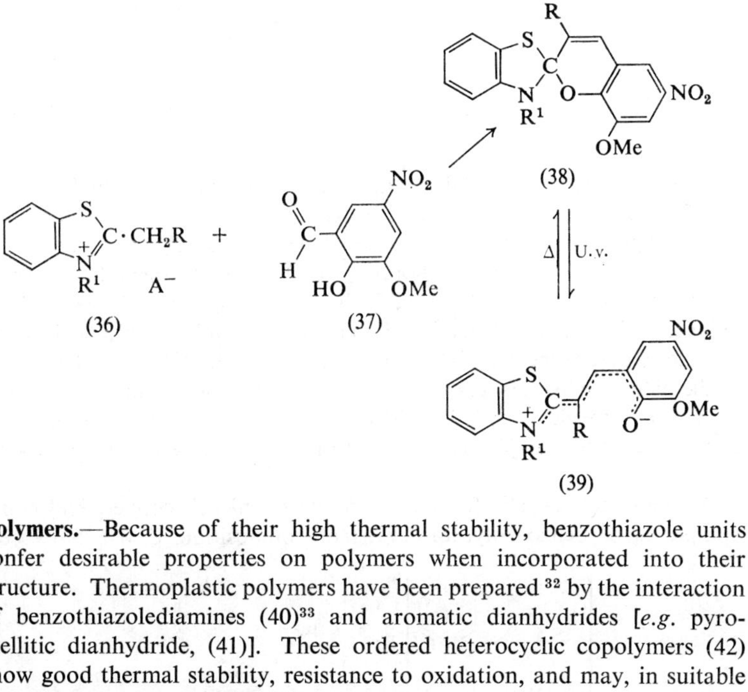

**Polymers.**—Because of their high thermal stability, benzothiazole units confer desirable properties on polymers when incorporated into their structure. Thermoplastic polymers have been prepared [32] by the interaction of benzothiazolediamines (40)[33] and aromatic dianhydrides [*e.g.* pyromellitic dianhydride, (41)]. These ordered heterocyclic copolymers (42) show good thermal stability, resistance to oxidation, and may, in suitable cases, be spun into excellent fibres.

**Metal Complexes.**—Benzothiazole (L) forms well-characterised tetrahedral complexes $ML_2X_2$ with $ZnX_2$ (X = Cl, or Br) and $CoX_2$ (X = Cl, Br, I, or NCS), and tetrahedral or octahedral complexes with $NiX_2$ (X = Cl, Br, or I). In these compounds, benzothiazole appears to co-ordinate through the nitrogen hetero-atom, except when acting as a bridging ligand, when both heteroatoms are involved.[34, 35] Iron complexes are of type $ML_2X_2$

---

[32] J. Preston, W. F. DeWinter, and W. B. Black, *J. Polymer Sci.*, Part A–1, *Polymer Chem*, 1969, **7**, 283.
[33] J. Preston, W. F. DeWinter, and W. L. Hofferbert jun., *J. Heterocyclic Chem.*, 1968, **5**, 269.
[34] E. J. Duff, M. N. Hughes, and K. J. Rutt, *J. Chem. Soc. (A)*, 1968, 2354.
[35] N. N. Y. Chan, M. Goodgame, and M. J. Weeks, *J. Chem. Soc. (A)*, 1968, 2499.

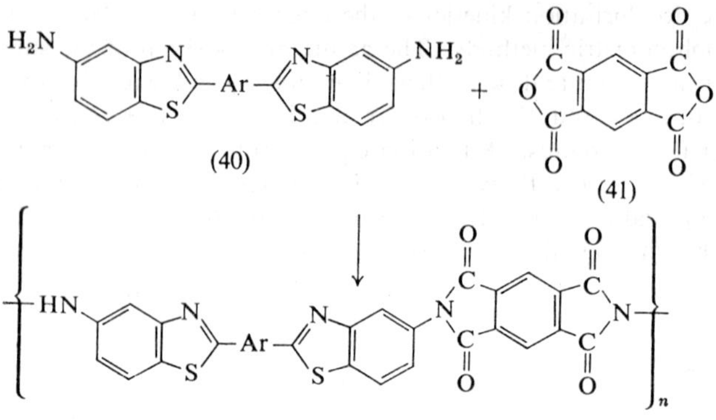

[Structure (40): H₂N-benzothiazole-Ar-benzothiazole-NH₂]

[Structure (41): benzene-dianhydride]

[Structure (42): polymer with repeating benzothiazole-Ar-benzothiazole-bisimide units]

(tetrahedral) or $ML_4X_2$ (octahedral).[35] A series of analogous complexes from 2-amino-, 2-chloro-, and 2-methyl-benzothiazoles have been studied from the same point of view,[36, 37] and others have been briefly described.[38]

Bidentate ligands ($L_2$), incorporating sulphur and nitrogen donor atoms, react with dinitrosylcobalt bromide $[Co(NO)_2Br]_2$, with cleavage of the halogen bridge and formation of diamagnetic ionic cobalt–nitric oxide complexes of the type $[Co(NO)_2(L_2)]^+Br^-$. Complexes (43) derived from more acidic bidentate ligands (e.g. 2-mercaptobenzothiazole) continue to interact with water, to yield, with loss of hydrogen bromide, the non-ionic diamagnetic compound $[Co^+(NO)_2(L_2)^-]$ (44), containing a covalent S—Co bond.[39]

Amongst heterocyclic formazan complexes, nickel, copper, and cobalt complexes of benzoazolyl-formazans of composition $L_2M^{2+}$ have been

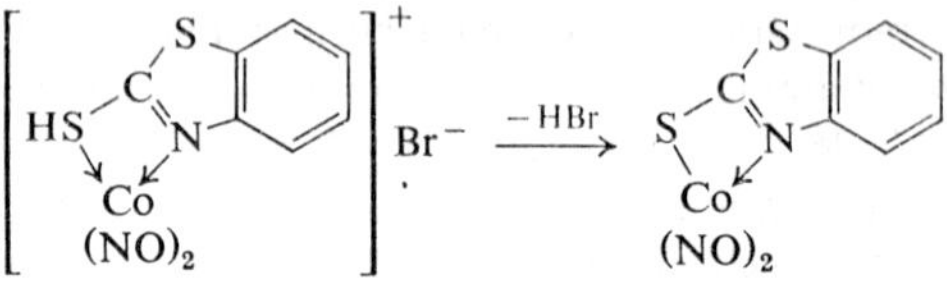

[Structures (43) and (44): cobalt nitrosyl complexes with mercaptobenzothiazole ligand]

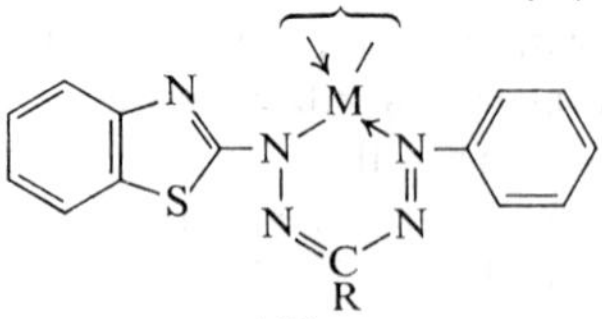

[Structure (45): benzothiazolyl-formazan metal complex]

[36] E. J. Duff, M. N. Hughes, and K. J. Rutt, *J. Chem. Soc.* (*A*), 1969, 2101, 2126.
[37] M. J. M. Campbell, D. W. Card, and R. Grzeskowiak, *Inorg. Nuclear Chem. Letters*, 1969, **5**, 39.
[38] W. U. Malik, P. K. Srivastava, and S. C. Mehra, *J. Indian Chem. Soc.*, 1969, **46**, 486.
[39] W. Hieber and K. Kaiser, *Z. anorg. Chem.*, 1968, **362**, 169.

prepared without difficulty. A detailed study of their visible and i.r. spectra, as well as magnetic properties, has provided a basis for a discussion of their structure; in general terms, the complexes may be represented as (45).[40]

**Biochemical Aspects: Phaeomelanins.**—Gallophaeomelanin-1, one of the reddish-brown pigments of the feathers of the chicken, has been recognised as a macromolecule incorporating β-(4-hydroxybenzothiazol-6-yl)alanine units (48). Its oxidation by alkaline permanganate followed by methylation yields 4,5-di(methoxycarbonyl)- (46) and 2,4,5-tri(methoxycarbonyl)- (47) and related thiazoles. The sum of the indirect evidence suggests that it is the 2-position of the ring system (48) that is concerned in the formation of the polymeric pigment.[41]

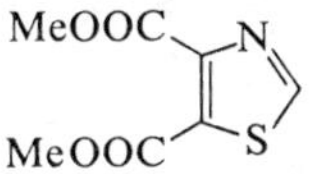

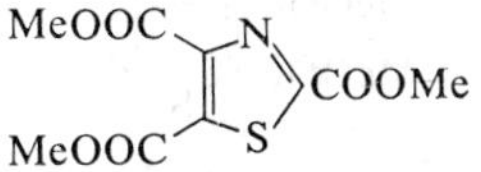

A more detailed examination[42] of the oxidative degradation of this pigment resulted in the identification of four additional products (49—52). The presence of structural units such as (53) (and analogues not shown) in the macromolecular pigment would account for the appearance of all the observed degradation products. A possible simple biosynthesis of these structural units (53) from (48) (R = CHO) by a Mannich reaction was suggested.[42]

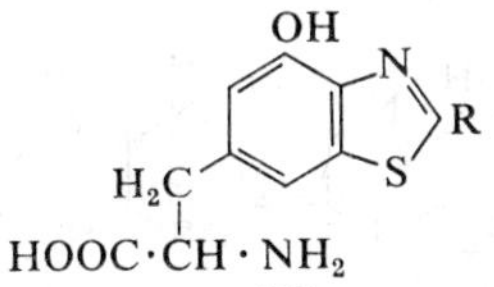

In a study[43] which sought to establish structures (54) (R = COOH) for a yellow-orange pigment of the feathers of New Hampshire chickens, and (54) (R = H) for that of a compound extractable from red human hair (a 'trichosiderin'), a number of model compounds incorporating the benzothiazole structure were prepared. They included condensation products of 2-formylbenzothiazole with benzothiazines and thiazines (55, 56) obtained as shown, which served as patterns that incorporate trichosiderin-like chromophores. Their spectral properties were correlated with those of the natural products.

[40] G. N. Lipunova, E. I. Krylov, N. P. Bednyagina, and V. A. Sharov, *Zhur. obshchei Khim.*, 1969, **39**, 1293 (Eng. trans., p. 1264).
[41] E. Fattorusso, L. Minale, G. Cimino, S. de Stefano, and R. A. Nicolaus, *Gazzetta*, 1969, **99**, 29.
[42] L. Minale, E. Fattorusso, G. Cimino, S. de Stefano, and R. A. Nicolaus, *Gazzetta*, 1969, **99**, 431.
[43] R. A. Nicolaus, G. Prota, C. Santacroce, G. Scherillo, and D. Sica, *Gazzetta*, 1969, **99**, 323.

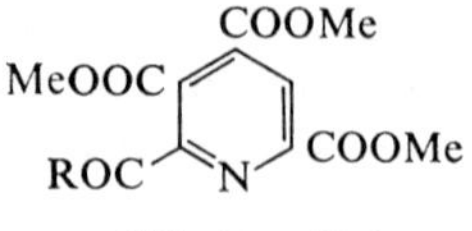

(49) R = OMe
(50) R = NH$_2$

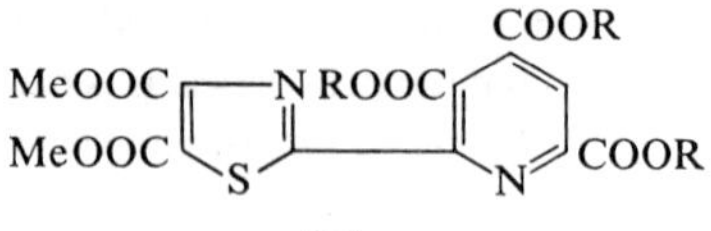

(51)

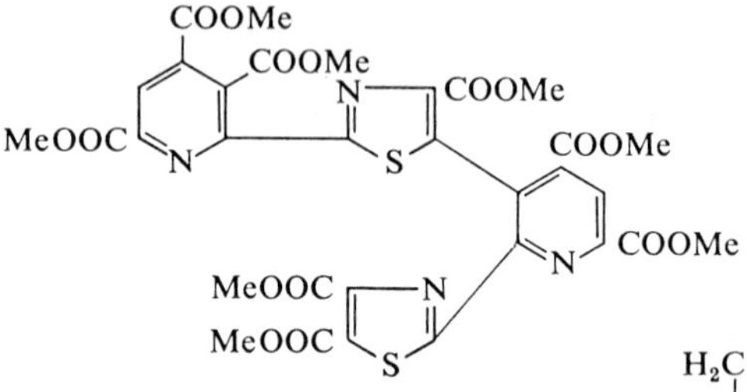

(52)

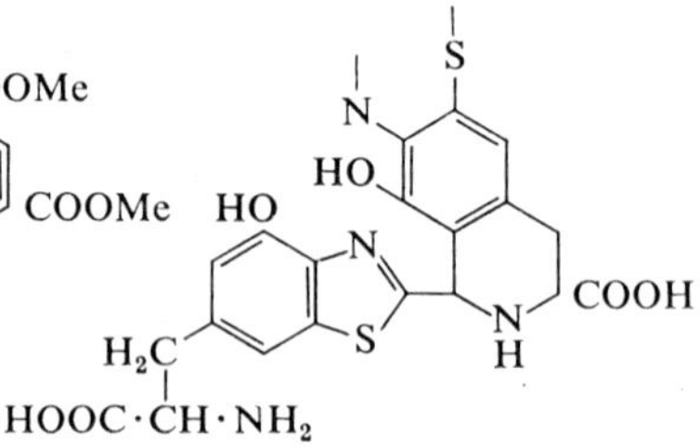

(53)

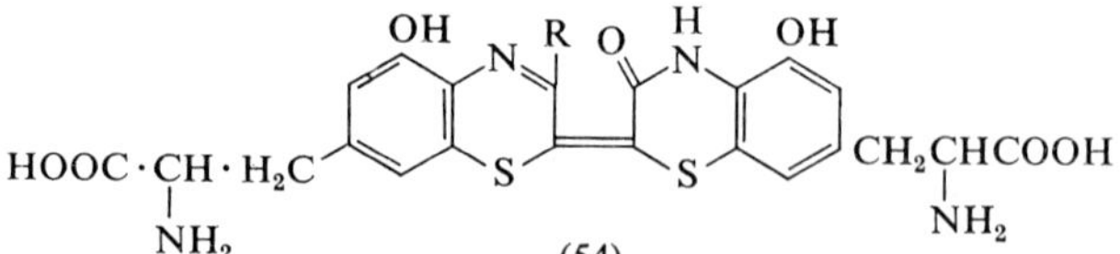

(54)

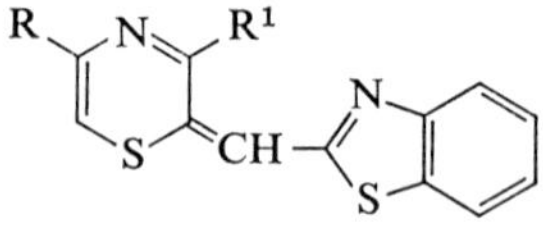

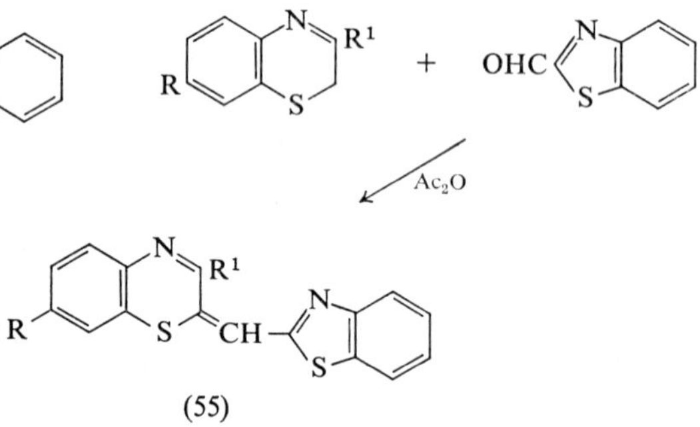

(55)

(56)

## 5 Chemical Properties of Benzothiazolines

**Alkylation of Benzothiazoline-2-thiones.**—Benzothiazoline-2-thione (57) normally undergoes *S*-alkylation, exceptions being Mannich type reactions [44] and vinyl additions,[45,46] which yield *N*-substituted derivatives.

[44] K. N. Ayad, E. B. McCall, A. J. Neale, and L. M. Jackman, *J. Chem. Soc.*, 1962, 2070.
[45] H. Irai, S. Shima, and N. Murato, *Kogyo Kagaku Zasshi*, 1959, **62**, 82.
[46] I. Y. Postovskii and N. N. Vereshchagina, *Khim. geterotsikl. Soedinenii Akad. Nauk Latv. S.S.R.*, 1965, 621 (*Chem. Abs.*, 1966, **64**, 5070).

Ethylene oxide in dioxan effects S-hydroxyethylation (forming 58), but causes N-hydroxyethylation in acetic acid, producing, with simultaneous thiono-oxo replacement, 3-(2-hydroxyethyl)-benzothiazolin-2-one (59). The same product also arises from S-alkylthiobenzothiazoles (60), and from the S-hydroxyethylthio-analogue (58), suggesting that the latter is an intermediate in this reaction.[47]

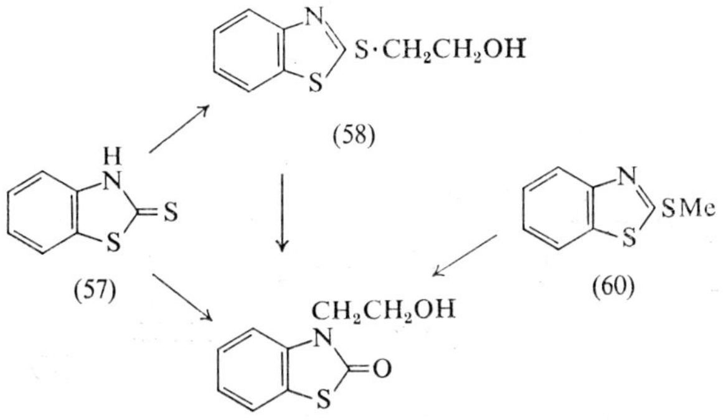

**Thionation of Benzothiazolin-2-ones.**—3-Substituted benzothiazolin-2-ones are usually convertible into the corresponding 2-thiones by the conventional procedure, using phosphorus pentasulphide in boiling pyridine. This reaction is useful in the production of N-substituted benzothiazoline-2-thiones, which unlike their 2-oxo-analogues are normally not accessible by N-alkylation, reaction occurring preferentially at the more strongly nucleophilic sulphur atom (see preceding section).

Thus, in the course of work aiming at the synthesis of benzothiazoline-2-thiones bearing a reactive N-(3)-two-carbon substituent, the readily available 3-(2-acetoxyethyl)benzothiazolin-2-one (61) was successfully thionated, and hydrolysed to the desired 3-(2-hydroxyethyl)benzothiazoline-2-thione (63). The course of the thionation may be monitored spectroscopically: the strong carbonyl absorption in the 6 μ region of the i.r. spectrum disappears as the reaction goes to completion. U.v. absorption spectra are particularly useful in confirming the site of substitution in the tautomeric benzothiazoline–thione system: N-substituted thiones give rise to an absorption maximum at 325 nm, which is absent in S-alkylated isomers and 2-oxo-analogues.[48]

**Miscellaneous Reactions.**—The action of amines at ca. 100 °C on benzothiazolin-2-one cleaves the thiazoline ring and yields, with simultaneous oxidation of (64), the appropriate disulphides (65).[49]

[47] P. Sohar, G. H. Denny jun., and R. D. Babson, *J. Heterocyclic Chem.*, 1968, **5**, 769.
[48] P. Sohar, G. H. Denny jun., and R. D. Babson, *J. Heterocyclic Chem.*, 1969, **6**, 163.
[49] D. Simov and A. Antonova, *Doklady Bolgar. Akad. Nauk*, 1968, **21**, 881.

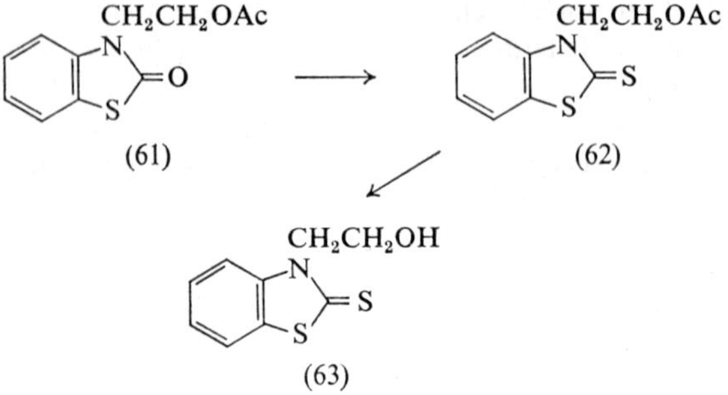

[Structures (61), (62), (63), (64), (65) shown]

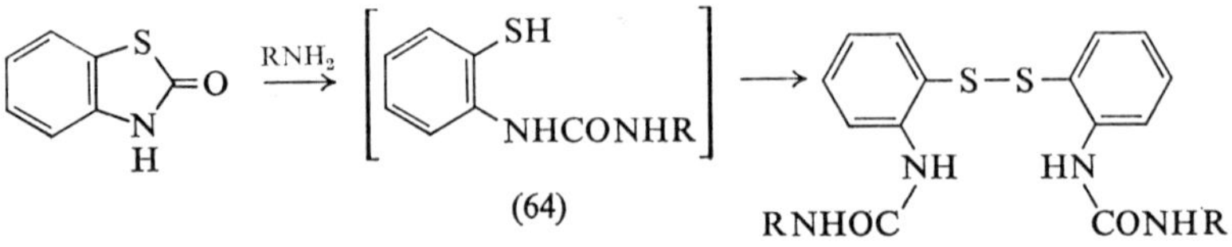

In a wider study [50] of the reaction between isocyanates derived from phosphoric acids with methylene bases, ketens, and related compounds, the following addition, affording the *N*-phosphorylated amide (66) was carried out, but not described in detail:

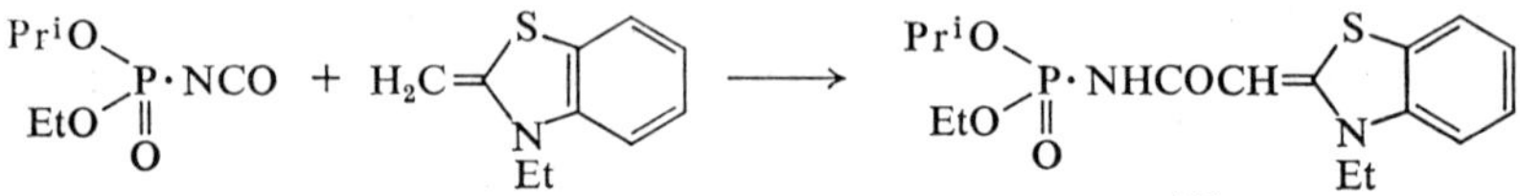

[Reaction scheme leading to structure (66)]

The self-condensation of 2,2-dimethylbenzothiazoline-5-carboxamide has been examined under various conditions, but gave only intractable polymers of variable molecular weight.[51]

**Uses.**—3-Methylbenzothiazolin-2-one hydrazone (HMBT) is known to yield blue solutions with phenols [52] and carbonyl compounds [52,53] in oxidising media, and has been used [54] in the colorimetric estimation of phenols, probably by the production of compounds such as (67).

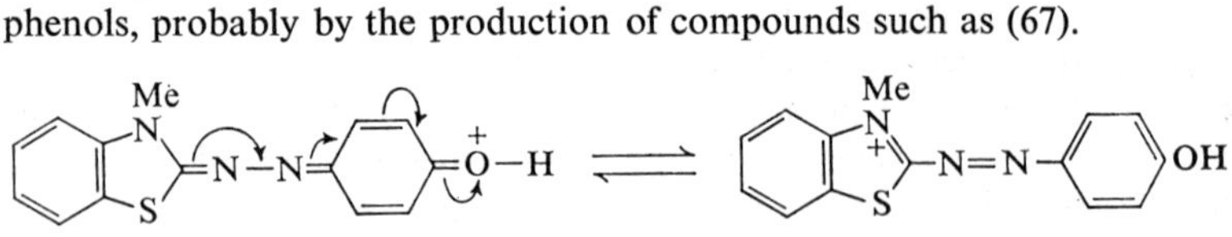

[Structure (67) equilibrium shown]

[50] Z. M. Ivanova, S. K. Mikhailik, V. A. Shokol, and G. I. Derkach, *Zhur. obshchei Khim.*, 1969, **39**, 1504.
[51] P. M. Hergenrother and H. H. Levine, *J. Polymer. Sci.*, 1968, A1, **6**, 2939.
[52] S. Hünig and K. G. Fritsch, *Annalen*, 1957, **609**, 172.
[53] M. Pays, *Ann. pharm. franç.*, 1966, **24**, 763; E. Sawicki, T. R. Hauser, T. W. Stanley, and E. Walter, *Analyt. Chem.*, 1961, **33**, 93.
[54] M. Pays and R. Bourdon, *Ann. pharm. franç.*, 1968, **26**, 681.

4-Chloro-2-oxobenzothiazolin-3-yl acetic acid (Benazolin) is used as an ingredient of herbicides. Measurements employing the $^{14}$C-carboxyl labelled compound showed that it was readily removed from soils by water, being lost from crops within 4—8 weeks after application, and that no detectable residues of it were contained in field-grown cereals.[55]

[55] D. K. Lewis, *J. Sci. Food Agric.*, 1969, **20**, 185.

# 15
## Condensed Ring Systems Incorporating Thiazole

### 1 Introduction

Fused ring systems incorporating thiazole are well known and numerous. The first known thiazoles were benzothiazole derivatives,[1] and the long list[2] of multi-ring structures based on thiazole is continually augmented by reports of new examples.

A general synthesis of particularly wide scope and usefulness in this field merits special mention at the outset. It involves the condensation of a cyclic thioamide with an α-halogenoketone or α-halogenoaldehyde or other suitable two-carbon system[3,4] to yield thiazolo-azinium salts, and is therefore an extension of Hantzsch's synthesis. Since any available cyclic thioamide may reasonably form the 'foundation' for the added thiazole ring (which may itself bear various substituents depending on the choice of the α-halogenoketo-compound), the route is clearly highly versatile, and other synthesis. With 1,2-dibromoethane, cyclic thioamides yield the expected thiazoline heterocycles.

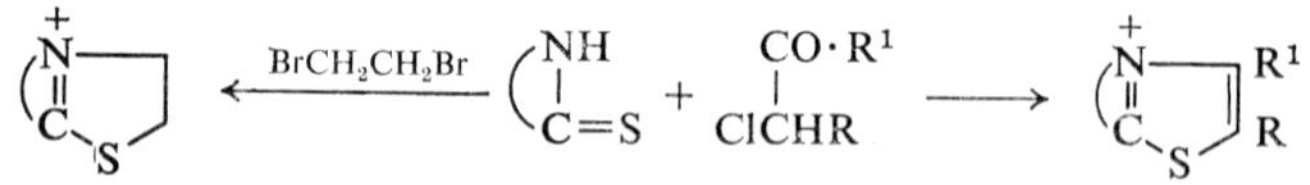

The number of possible polyheterocyclic patterns appears to be almost limitless, and provides a tempting field for mere synthetic exercises. A number of significant and systematic investigations, however, are on record; amongst them, Undheim's elegant and purposeful work on thiazolo[3,2-a]pyridines, initially stimulated by the isolation of a compound of this structure from natural sources, might justifiably be singled out for special mention.

In the present account, the polyheterocyclic structures are arranged in order of ascending complexity and are classified in accordance with the system of the Ring Index.[2] Although this method is not unexceptionable for the present purpose, separating from one another ring systems that might be dealt with advantageously as a group, it has the merit of presenting

---
[1] A. W. Hofmann, *Ber.*, 1879, **12**, 1126, 2359.
[2] A. M. Patterson, L. T. Capell, and D. F. Walker, 'The Ring Index', American Chemical Society, 2nd ed., Washington, 1960.
[3] C. K. Bradsher and D. F. Lohr, *Chem. and Ind.*, 1964, 1801.
[4] F. S. Babichev and V. N. Bubnovskaya, *Ukrain. khim. Zhur.*, 1964, **30**, 848.

# Condensed Ring Systems Incorporating Thiazole

the information most systematically for reference and retrieval, and is adopted for this reason. Slight deviations from this scheme have on occasion been expedient, *e.g.* when a range of closely related ring systems are accessible by a single reaction. The chemistry of benzothiazole forms the subject of Chapter 14.

## 2 Structures Comprising Two Five-membered Rings (5,5)

**Thiazolo[3,2-*d*]tetrazoles.**—[$CN_4$–$C_3NS$]. The condensation of mercaptotetrazoles (1) and α-halogenoketones (*e.g.* ω-bromoacetophenone) yields, by the extension of Hantzsch's synthesis, 5-acylmethylmercaptotetrazoles (2), which are cyclised by concentrated sulphuric acid to the thiazolo-[3,2-*d*]tetrazolylium salts (3); the perchlorates decompose explosively at their m.p.s, but are stable at room temperature.[5]

**Imidazo[2,1-*b*]thiazoles.**—[$C_3NS$–$C_3N_2$]. 3-Substituted and 2,3-disubstituted 5,6-dihydro-4*H*-imidazo[2,1-*b*]thiazoles (5) are formed in fair yields by the Hantzsch synthesis, using 2-mercaptoimidazoline (4) and α-halogenoketones (or ketones and halogens) as starting materials.[6]

In their efforts to develop an improved route to 6-phenyl-2,3,5,6-tetrahydroimidazo[2,1-*b*]thiazole (9), Turner *et al.*[7] have taken advantage of the high reactivity of the readily available [8] 1-arylsulphonylaziridines. 2-Phenyl-1-(toluene-*p*-sulphonyl)aziridine (6), for example, which is generated *in situ* from styrene and *N*,*N*-dichloro-*p*-toluenesulphonamide in toluene, reacts with ethanolamine in dioxan to yield chiefly the normal adduct (7) which is convertible, by way of the chloride and thiocyanate, into the substituted 2-iminothiazolidine (8). The latter is also obtainable from the aziridine and 2-aminothiazoline directly. The final ring closure to the desired end product (9) occurs under the influence of concentrated sulphuric acid at room temperature. The cleavage of the sulphonamide at the C–N bond, rather than its usual N–S fission, has been observed in strong acid, when either a secondary or tertiary arylcarbonium ion is generated in this reaction.[9]

**Thiazolo(or Oxazolo)[2,3-*b*]thiazole.**—[$C_3NS$(or O)–$C_3NS$]. The use of a keten–sulphur dioxide adduct in the synthesis of thiazolidines is described elsewhere in this Report (see p. 405). Applied to 2-phenylthiazoline and

---

[5] R. Neidlein and J. Tauber, *Tetrahedron Letters*, 1968, 6287.
[6] V. K. Chadha and H. K. Pujari, *Canad. J. Chem.*, 1969, **47**, 2843.
[7] T. Bailey, T. P. Seden, and R. W. Turner, *J. Heterocyclic Chem.*, 1969, **6**, 751.
[8] T. P. Seden and R. W. Turner, *J. Chem. Soc. (C)*, 1968, 876.
[9] S. Searles and S. Nukina, *Chem. Rev.*, 1959, **59**, 1077.

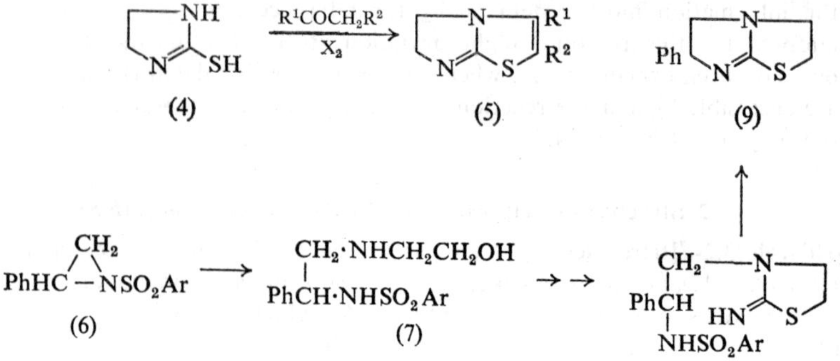

2-(p-nitrophenyl)oxazoline (10), this route [10] provides excellent yields of the rather labile dihydro-7a-aryl-7aH-thiazolo(or oxazolo)[2,3-b]thiazol-3(2H)-one 1,1-dioxide (11).

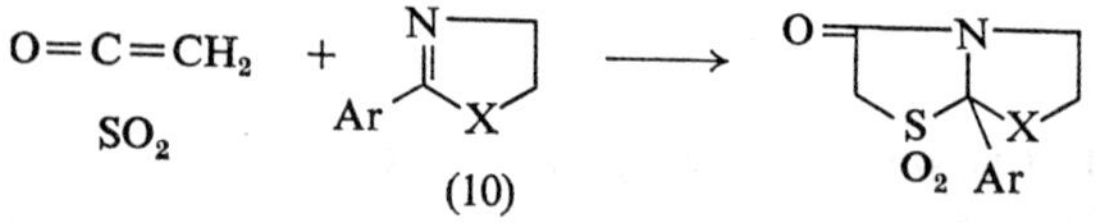

## Thiazolidino[4,5-d]thiazoles.—[C₃NS–C₃NS]

Thiazolidino[4,5-d]thiazoles.—[$C_3NS$–$C_3NS$]. A series of 2-(*N*-substituted)-amino-5-ketothiazolidino[4,5-*d*]thiazoles (13) has been prepared [11] in 30—45% yield by the condensation of 2,4-diketothiazolidine (12) and a thiourea in the presence of an equivalent of iodine. The method is yet another extension of the Hantzsch synthesis of thiazoles from thiourea, iodine, and acetoacetic ester,[12] the $-CH_2CO-$ function of which participates in the formation of the ring structures.

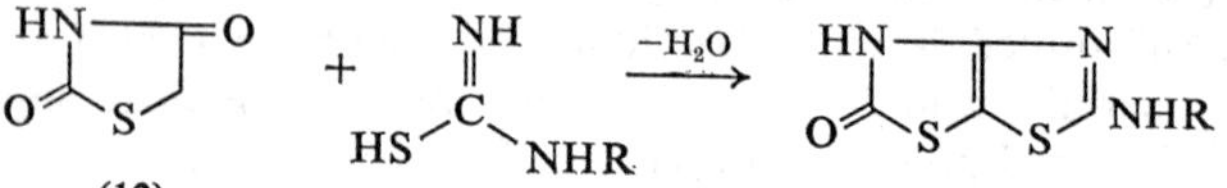

### 3 Structures Comprising One Five-membered and One Six-membered Ring (5,6)

Thiazolo[3,2-*a*]-*s*-triazines.—[$C_3NS$–$C_3N_3$]. 2-Amino-2-thiazoline, which yields the expected thiourea (14) on treatment with phenyl isothiocyanate,[13] reacts with an excess of this reagent to yield a product (83%), that has

---

[10] A. de Souza Gomez and M. M. Joullié, *J. Heterocyclic Chem.*, 1969, **6**, 729; *Chem. Comm.*, 1967, 935.
[11] J. M. Singh, *Bull. Chem. Soc. Japan*, 1968, **41**, 2183.
[12] R. M. Dodson and W. King, *J. Amer. Chem. Soc.*, 1945, **67**, 2242.
[13] D. L. Klayman, J. J. Maul, and G. W. A. Milne, *J. Heterocyclic Chem.*, 1968, **5**, 517.

been formulated as 2-phenylimino-3-phenyl-4-thioxothiazolo[3,2-a]tetrahydro-s-triazine (17).[14] The postulated mechanism of the reaction, involving the probable formation of the bis(phenylthiocarbamoyl) derivative (15), is shown below. On acid hydrolysis, the thiazoline ring of the condensed system (17) is cleaved, giving successively the substituted triazine (18) and cyanuric acid (19), depending on conditions. The mass spectrum of (17), as well as its X-ray crystallographic analysis,[15] is in agreement with its assigned structure.

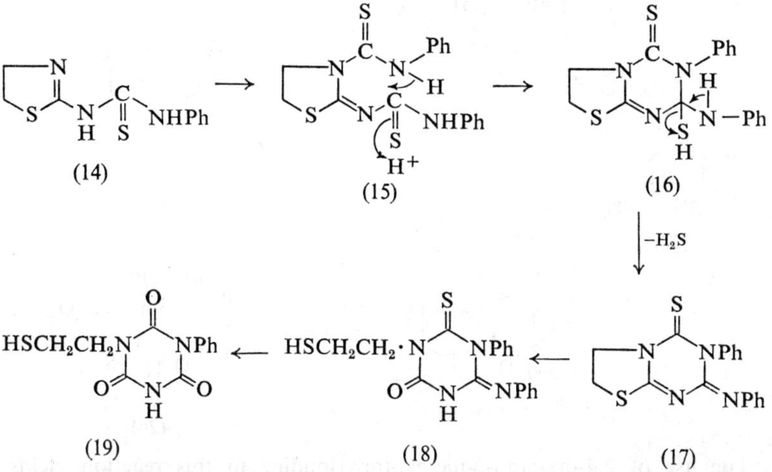

A selenium analogue of (17) (i.e. 2-phenylimino-3-phenyl-4-thioxoselenazolo[3,2-a]tetrahydro-s-triazine) was obtained by refluxing N-phenyl-N'-(2-selenazolin-2-yl)thiourea with excess of phenyl isothiocyanate in benzene.[14]

**Thiazolo[3,2-a(and c)]pyrimidines.**—[$C_3NS$–$C_4N_2$]. The ready availability of mercaptopyrimidines makes a variety of thiazolo-pyrimidines easily accessible. 3-Substituted and 2,3-disubstituted 4,5,6,7-tetrahydrothiazolo-[3,2-a]pyrimidines (21), for example, are produced in fair yields by the interaction of 2-mercapto-3,4,5,6-tetrahydropyrimidine (20) and α-halogenoketones (or ketones and halogen).[6]

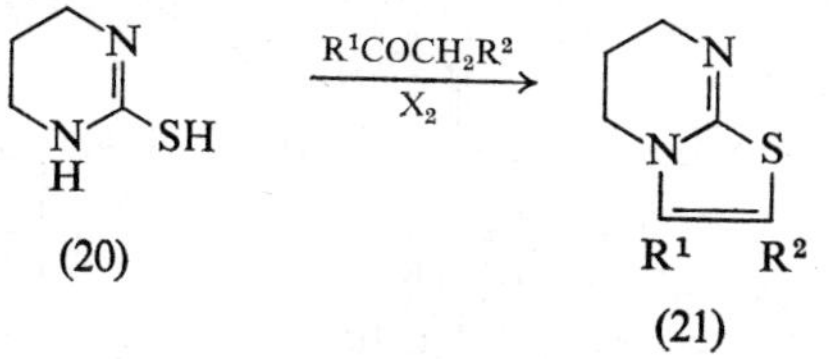

---

[14] D. L. Klayman and G. W. A. Milne, *Tetrahedron*, 1969, **25**, 191.
[15] J. Karle, J. Flippen, and I. L. Karle, *Z. Krist.*, 1967, **125**, 115.

The same condensation involving 4,4,5,6-substituted 2-thionotetrahydropyrimidines (22) yields the quaternary 7H-thiazolo[3,2-a]pyrimidinium salts (23) which are convertible into the free bases (24). At room temperature, the alternative cyclisation yields 7-hydroxythiazolo[3,2-a]-pyrimidinium salts (25); the corresponding bases were not isolated, since alkalis cleaved their pyrimidine ring, with formation of 2-imino-3-(γ-oxoalkyl)-Δ4-thiazolines (26).[16]

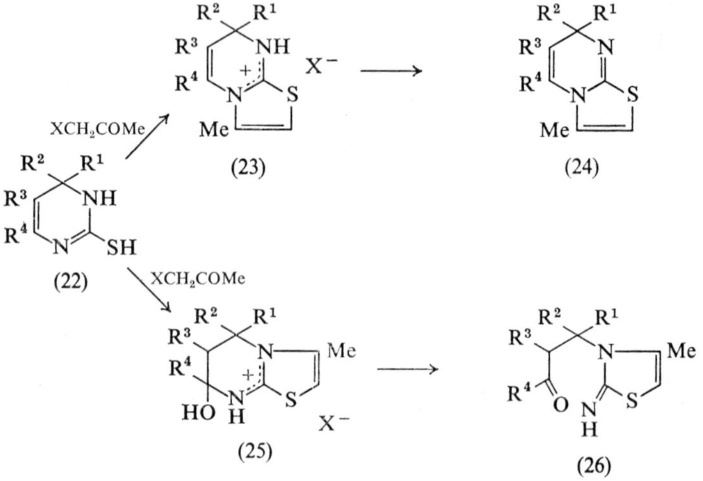

The use of 2,4-diamino-6-mercaptopyrimidine in this reaction yields intermediate pyrimidyl sulphide derivatives (27) that are cyclisable to the substituted thieno[2,3-d]pyrimidine (28) thermally and, in one example, to the 3-substituted 5,7-diaminothiazolo[3,2-c]pyrimidine (29) in concentrated sulphuric acid at 100 °C. The product is isolated as its quaternary salt; it is cleaved, by dilute alkali, in its pyrimidine ring, affording a mixture of thiazol-2-yl acetamide (30) and the corresponding amidine.[17]

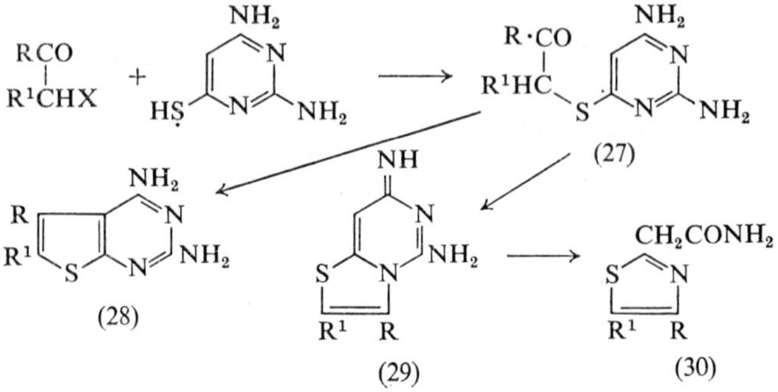

[16] G. Jaenecke, H. J. Mallon, and H. Raepke, Z. Chem., 1968, 8, 462.
[17] B. Roth, J. Medicin. Chem., 1969, 12, 227.

### Condensed Ring Systems Incorporating Thiazole

**Thiazolo[4,5-*d*]pyrimidines.**—[$C_3NS$–$C_4N_2$]. The action of alkalis on 2,4-diamino-6-hydroxypyrimidine-5-isothiuronium bromide (31) yields the corresponding disulphide (32),[18] and not the thiazolo[4,5-*d*]pyrimidine, as claimed by Horiuchi.[19] The true 2,5-diamino-7-hydroxythiazolo[4,5-*d*]-pyrimidine (33) is obtainable by thiocyanation of 2,4-diamino-6-hydroxypyrimidine, and cyclisation of the resulting 5-thiocyanato-compound by successive treatment with acetic anhydride and deacetylation.[18]

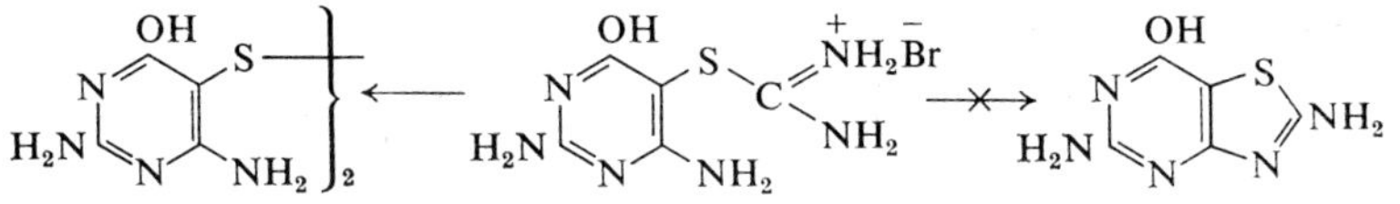

(32)  (31)  (33)

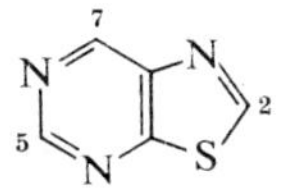

(34) 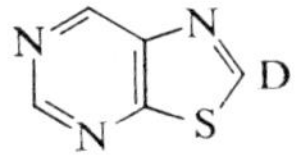  (34a) 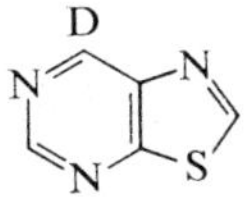  (34b)

The n.m.r. spectrum of thiazolo[5,4-*d*]pyrimidine (34) includes peaks at 551, 552, and 568 c/s.[20, 21] Those of the 2- (34a) and 7-deuteriated (34b) products, specially synthesised for these studies,[20] lack the first and last peak respectively. It is therefore concluded, contrary to previous assignments,[21] that the correct order of chemical shifts, going down field, is H-2, H-5, and H-7.[20]

**Thiazolo[4,5-*d*]pyridazine.**—[$C_3NS$–$C_4N_2$]. One example of a reduced thiazolo[4,5-*d*]pyridazine has been prepared from 1-(*p*-nitrophenyl)-2-methyl-5-bromopyridazine-3,6-dione and thiourea as follows:[22]

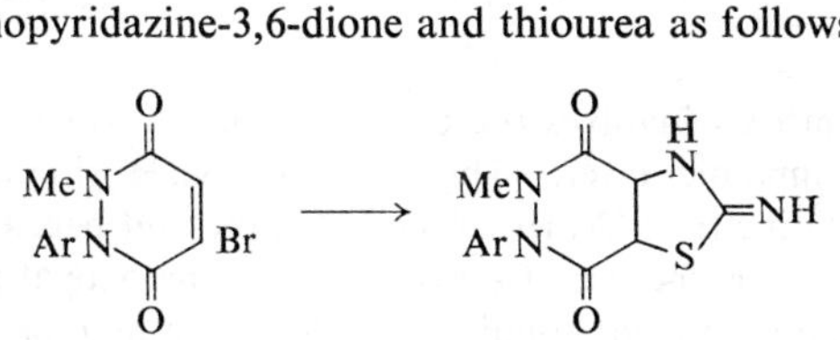

**Thiazolo[3,2-*a*]pyridines.**—[$C_3NS$–$C_5N$]. A variety of thiazolo[3,2-*a*]-pyridines has been produced by at least three distinct syntheses, starting with either the preformed thiazole or pyridine ring.

The interaction of mercaptoamines and keto acids provides a general route to condensed thiazole systems. An application of this approach is

---

[18] J. A. Baker and P. V. Chatfield, *J. Chem. Soc.* (*C*), 1969, 603.
[19] M. Horiuchi, *Chem. Pharm. Bull. Japan*, 1959, **7**, 393.
[20] M. Benedek-Vamos and R. Promel, *Tetrahedron Letters*, 1969, 1011.
[21] E. Suzuki, S. Sugiura, T. Naito, and S. Inoue, *Chem. Pharm. Bull. Japan*, 1968, **16**, 750.
[22] S. Baloniak, *Roczniki Chem.*, 1968, **42**, 1867.

the condensation of 2-mercaptoethylamine (35) (cysteamine) and 4-benzoylbutyric acid (36) to 8a-phenylhexahydro-5H-thiazolo[3,2-a]pyridin-5-one (37) in 67% yield.[23] The use of 2-benzoylbenzoic acid in conjunction with 2-mercaptoethylamine or 2-aminothiophenol in this synthesis provides the thiazolo[2,3-a]isoindol-5(9bH)-one (38a) (68%), or the isoindolo[1,2-b]benzothiazol-11-one (38b) (70%). The experimental procedure is simple and convenient, the reactants being refluxed in toluene or xylene, until the excess amine and water, collected in a Dean–Stark tube, is completely removed (5—24 h).[23]

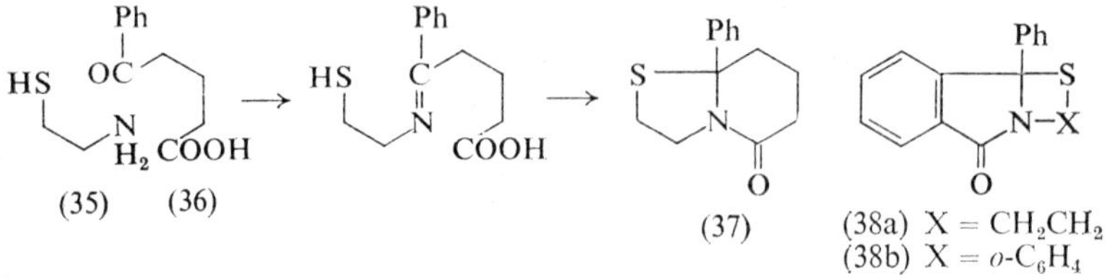

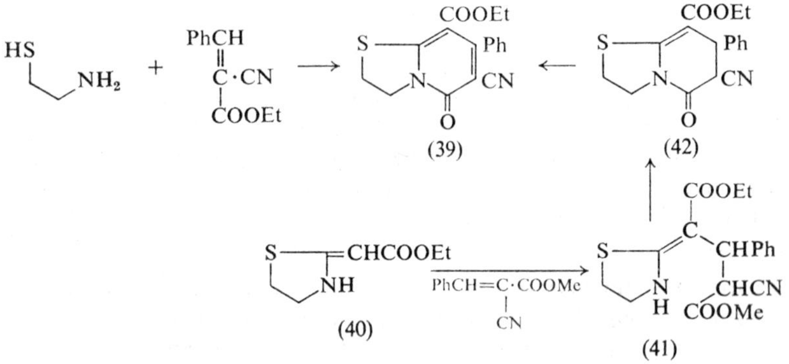

A related synthesis involves the condensation of 2-mercaptoethylamine with a cyanocinnamic ester. Thus, the known[24] simple synthesis of 2-phenylthiazolidine from 2-mercaptoethylamine and benzaldehyde[24] may also be performed by the use of ethyl α-cyanocinnamate at room temperature. Under more stringent conditions, however, ammonia is evolved and 5-oxo-7-phenyl-8-ethoxycarbonyl-6-cyano-2,3-dihydro-5H-thiazolo[3,2-a]pyridine (39) (45%) is formed.[25] Its structure is supported by its alternative three-stage synthesis, by the addition of methyl α-cyanocinnamate to ethyl 2-thiazolinyl-2-acetate (40), followed by cyclisation of the adduct (41) by sodium ethylate to (42). The final dehydrogenation stage (42 → 39) under the influence of hydrogen peroxide was only inferred from changes in the u.v. spectra of solutions treated with this reagent.[25]

[23] P. Aeberli and W. J. Houlihan, *J. Org. Chem.*, 1969, **34**, 165.
[24] R. Kuhn and F. Drawert, *Annalen*, 1954, **590**, 55.
[25] G. Dietz, W. Fiedler, and G. Faust, *Chem. Ber.*, 1969, **102**, 522.

Thiazolo[3,2-a]pyridinium salts (44) are produced by the condensation of N-substituted 2-methylthiazolium salts (43) and 1,2-diketones. By a suitable choice of the latter (e.g. 9,10-phenanthraquinone, acenaphthaquinone), polycyclic products are formed.[26] The structure of the salts (44) agrees with their spectroscopic properties, and with their degradation, by Raney nickel to 3,4-disubstituted pyridines (45).[27]

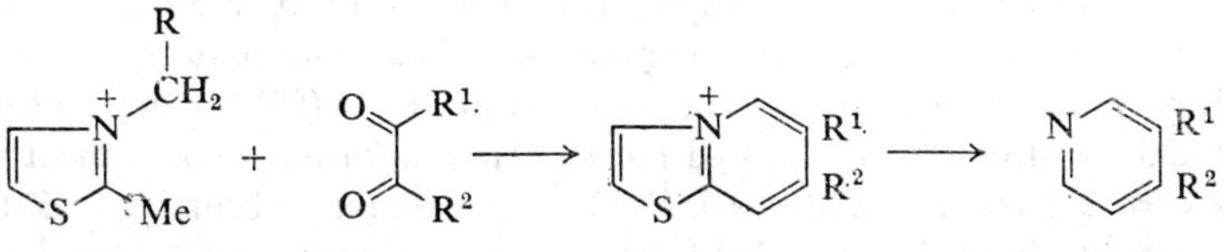

(43) R = COPh, CN, or COOEt    (44)    (45)

The isolation from tissue of a natural product having a thiazolo[3,2-a]pyridine structure has stimulated extensive and significant work by Norwegian investigators in this field. Specifically, acid hydrolysis of bovine liver had given a fluorescent compound [28] of molecular formula $C_9H_9NO_3S$, which was characterised, by spectroscopic and degradative evidence,[29] as L(−)-8-hydroxy-5-methyldihydrothiazolo[3,2-a]pyridinium-3-carboxylate (46). Its desulphuration by Raney nickel gave α-(3-hydroxy-6-methylpyridinium)propionic acid (47) and 2-methyl-5-pyridinol (48).[29]

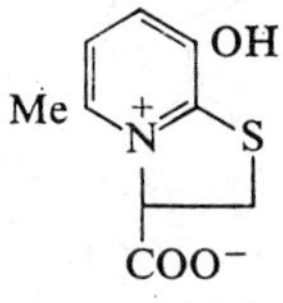

(46) 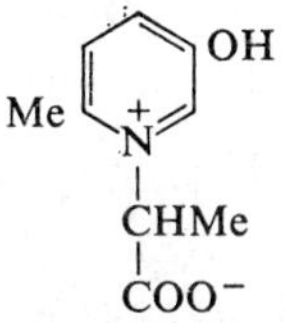    (47) 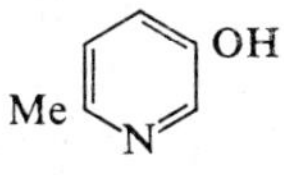    (48) 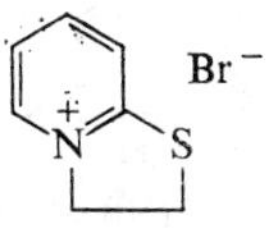    (49)

This formulation is confirmed by an unequivocal synthesis of the natural product (46) and of its analogues.[30] Thus, the established synthesis of the parent dihydrothiazolo[3,2-a]pyridinium salt (49) [31] from pyridine-2-thione and 1,2-dibromoethane, on being extended to 3-hydroxy-6-methylpyridine-2-thione (50), affords 5-methyldihydrothiazolo[3,2-a]pyridinium-8-oxide (52), identical with the decarboxylated compound from bovine liver hydrolysate. The cyclisation is initiated at the sulphur atom of (50), its strongest nucleophilic centre, and is completed by nucleophilic attack by the nuclear nitrogen on the carbon atom bearing the second bromine atom;

[26] O. Westphal and A. Joos, Angew. Chem., 1969, 81, 84 (Internat. Edn., 1969, 8, 74).
[27] O. Westphal, G. Feix, and A. Joos, Angew. Chem., 1969, 81, 85 (Internat. Edn., 1969, 8, 74).
[28] P. Laland, J. O. Alvsaker, F. Haugli, J. Dedichen, S. Laland, and N. Thorsdalen, Nature, 1966, 210, 917.
[29] K. Undheim and V. Nordal, Acta Chem. Scand., 1969, 23, 371.
[30] K. Undheim, V. Nordal, and K. Tjonneland, Acta Chem. Scand., 1969, 23, 1704.
[31] R. W. Balsiger, J. A. Montgomery, and T. P. Johnston, J. Heterocyclic Chem., 1965, 2, 97.

15*

the reactive intermediate is isolable in favourable cases (*e.g.* 53). By condensing the thiolactam (50) with methyl 2,3-dibromopropionate in strong alkali, followed by hydrolysis, the racemate of the natural product (46) is directly accessible; in weak bases, the 2-carboxy-isomer is obtained instead. The n.m.r. spectra of all the intermediates and products were recorded and discussed.[30]

The scope of the synthesis is further extended [32] by the use of α-bromo-α,β-unsaturated carbonyl compounds. Thus, α-bromoacrylic acid (54) (and its functional derivatives) give the products (56) directly in excellent yield. α-Bromomaleic anhydride similarly affords *trans*-2,3-dicarboxy-8-hydroxy-5-methyldihydrothiazolo[3,2-*a*]pyridinium bromide (57), the configuration being revealed by its n.m.r. spectrum.

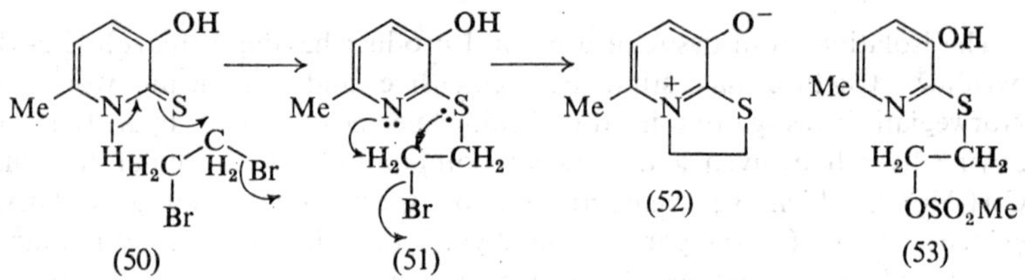

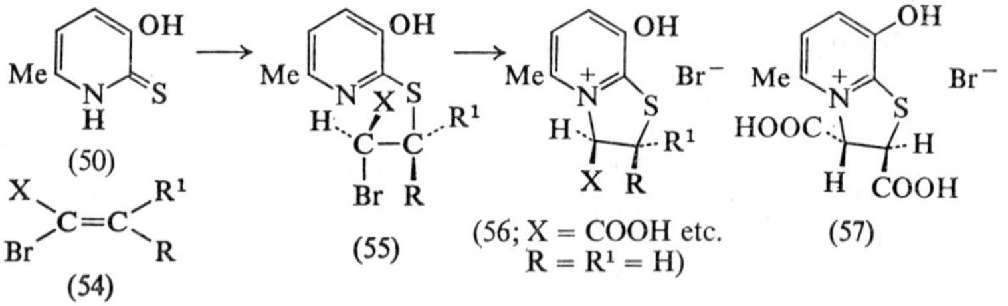

The action of peracids on L(−)-8-hydroxy-5-methyldihydrothiazolo-[3,2-*a*]pyridinium-3-carboxylate (46) and a number of its derivatives yields sulphoxides [29, 33] stereoselectively. According to the chemical evidence and the n.m.r. spectra, the *cis*-isomers (58) are the chief reaction products. The mechanism of the reaction is thought to involve the preliminary oxidation of the 3-carboxy to an acyl peroxide group: since sulphoxide formation involves a nucleophilic attack by sulphur on peroxide oxygen by a concerted electronic displacement,[34] the peroxy-group so formed would be correctly spaced for intramolecular *cis* oxidation of the sulphur. Decarboxylation of (58) under mild conditions removes the original asymmetric centre, giving a strongly dextrorotatory sulphoxide enantiomer

[32] K. Undheim and L. Borka, *Acta Chem. Scand.*, 1969, **23**, 1715.
[33] K. Undheim and V. Nordal, *Acta Chem. Scand.*, 1969, **23**, 1966.
[34] D. Barnard, L. Bateman, and J. I. Cuneen, 'Organic Sulphur Compounds', ed. N. Kharasch, Pergamon, 1961, vol. I, p. 230.

(46) (58) (59)

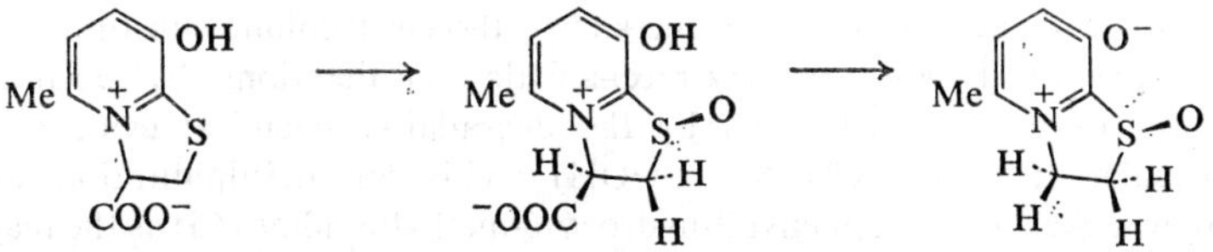

which, on the basis of synthetic and mechanistic considerations, is likely to have the (*S*)-configuration (59).

Electrophilic substitutions undergone by thiazolo[3,2-*a*]pyridines were also studied.[35] 5-Methyldihydrothiazolo[3,2-*a*]pyridinium-8-oxide (and its 3-carboxy-derivative) are halogenated and nitrated in their 7-position. The 7-halogeno-substituent thus introduced is in turn displaceable by nucleophilic reagents without disrupting the thiazoline ring, yielding, for example, the thione (62) with potassium hydrogen sulphide in dimethylformamide.

(46) (60) (61) $Y = Br, I, NO_2$ (62) (63) (64) (65)

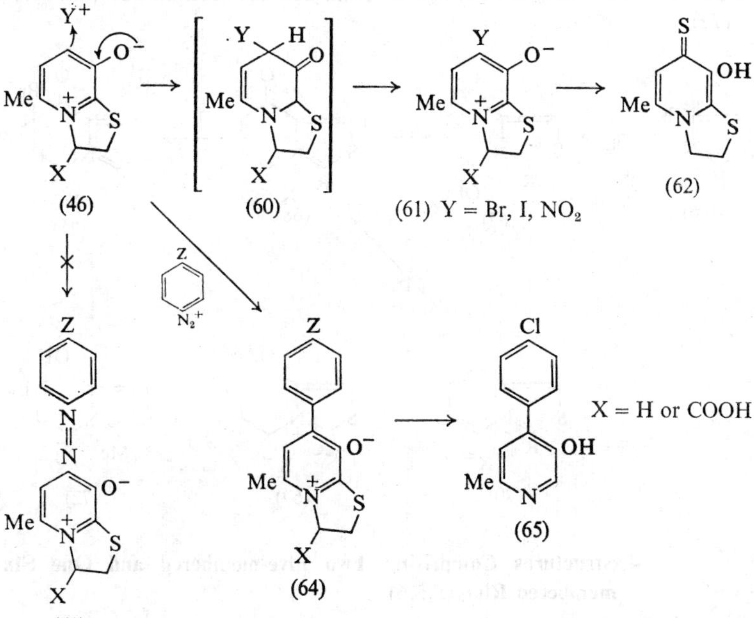

Interaction of (46) with the fairly weakly electrophilic diazonium salts yields (64) by arylation, and not the expected azo-derivatives (63). The reaction proceeds presumably by the initial electrophilic attack of the reagent on the phenolate oxygen of (46), the intermediate then rearranging *via* a six-membered transition state, with expulsion of nitrogen.[35]

[35] K. Undheim and V. Nordal, *Acta Chem. Scand.*, 1969, **23**, 1975.

The correctness of assigning the site of the electrophilic substitution to the 7-position is confirmed by mechanistic considerations, by the n.m.r. spectra of the products, and by the degradative chemical evidence. In particular (64) (Z = Cl; X = COOH) yields, on desulphuration with Raney nickel, 4-chlorophenyl-3-hydroxy-6-methylpyridine (65) as the major product.[35]

**Thiazolo[3,4-a]pyridines.**—[$C_3NS-C_5N$]. A successful synthesis of thiazolo-[3,2-a]pyridinium salts[36] has been extended to the [3,4-a]-isomers with interesting results.[37] The tetrahydro-derivatives (68) of the desired products (70) were synthesised from 4-cyano-2-methylthiazole (66) as outlined; on being treated with acetic anhydride, however, instead of being aromatised to (70) as expected, they gave the enol-acetates (69). This observation supports the view, previously advanced, that enol-acetates are concerned as intermediates in this and related aromatisations. An example of the required fully aromatic bicyclic system was in fact obtainable from (68) upon successive bromination and thermal dehydrobromination (68) → (71) → (72).

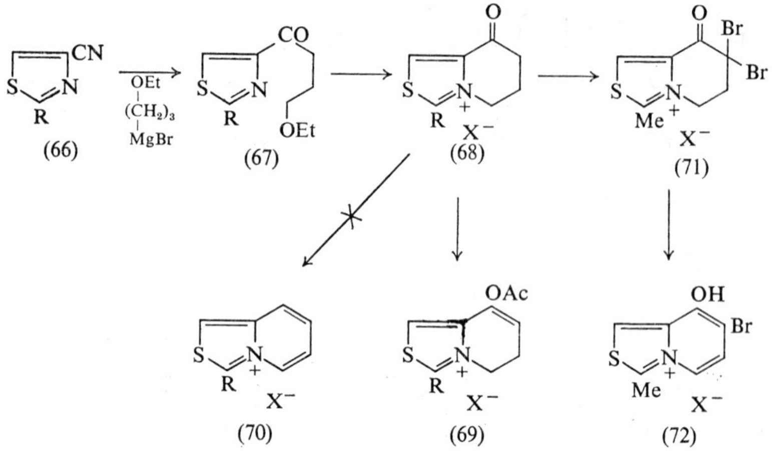

## 4 Structures Comprising Two Five-membered and One Six-membered Ring (5,5,6)

**Tetrazolo[5,1-b]benzothiazoles.**—[$CN_4-C_3NS-C_6$]. Tetrazolobenzothiazoles (73), and their thiazole analogue (76) react smoothly with triphenylphosphine in benzene at 50—60 °C to yield 2-(triphenylphosphoranylidene)-amino derivatives (75) of benzothiazoles.[38] The reaction probably occurs

---

[36] D. G. Jones and G. Jones, *J. Chem. Soc.* (C), 1967, 515.
[37] D. G. Jones and G. Jones, *J. Chem. Soc.* (C), 1969, 707.
[38] I. N. Zhmurova, A. P. Martynyuk, and A. V. Kirsanov, *Zhur. obshchei Khim.*, 1969, 39, 1223 (Eng. trans., p. 1193).

# Condensed Ring Systems Incorporating Thiazole

by the usual [39] oxidative imination of the tertiary phosphines by intermediate benzothiazole-azides, arising from the starting materials by tautomerisation. The products are readily quaternised; of their possible alternative structures (77a or b), the latter are established by the acidic hydrolysis of a representative salt to triphenylphosphine oxide and 2-imino-3-methylbenzothiazoline.[38]

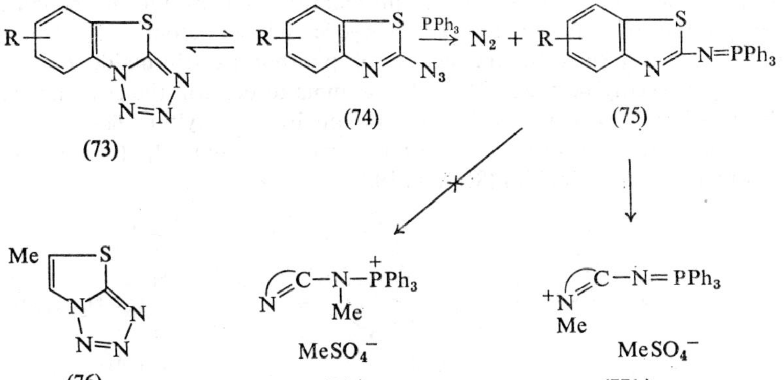

**Bisthiazolo[3,2-a ; 5',4'-e]pyrimidines.**—[$C_3NS$–$C_3NS$–$C_4N_2$]. The interaction of o-aminocarbethoxy compounds with α-thiocyanoacetophenones (79) is known to yield thiazolopyrimidones.[40] The use of the substituted 4-amino-5-carbethoxy-4-thiazolines (78) in this reaction [41] yields, in a first stage, partly cyclised condensation products (80). These are further cyclisable, by means of polyphosphoric acid at 120 °C, to products that

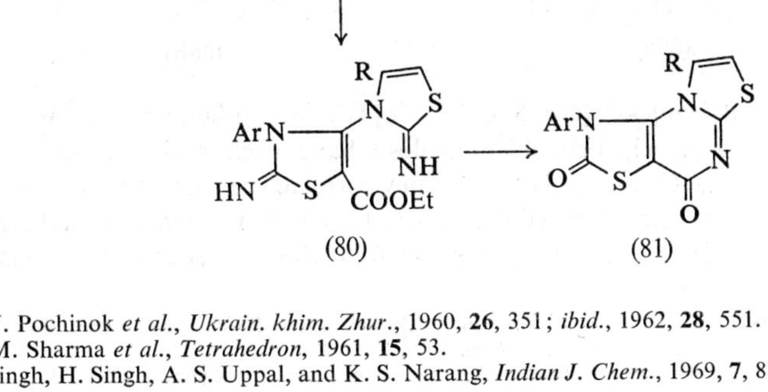

---

[39] V. Y. Pochinok et al., *Ukrain. khim. Zhur.*, 1960, **26**, 351; ibid., 1962, **28**, 551.
[40] G. M. Sharma et al., *Tetrahedron*, 1961, **15**, 53.
[41] A. Singh, H. Singh, A. S. Uppal, and K. S. Narang, *Indian J. Chem.*, 1969, **7**, 884.

have been formulated as 9$H$-2,3-dihydro-3,5-diarylbisthiazolo[3,2-$a$; 5′,4′-$e$]-pyrimidine-2,9-diones (81).[41]

**Thiazolo[2,3-$b$]benzothiazoles.**—[$C_3NS$–$C_3NS$–$C_6$]. The action of thionyl chloride on 3-(2-hydroxyethyl)benzothiazolin-2-one simply yields the chloroethyl derivative. When the corresponding 2-thione is so treated, however, the replacement is followed by immediate cyclisation to 2,3-dihydrothiazolo[2,3-$b$]benzothiazolium chloride (83), by the nucleophilic substitution of halogen by the exocyclic sulphur atom. The same product also results by the action of phosphorus tribromide on the $S$-(2-hydroxyethyl)-isomer (84).[42] In attempts to confirm the structure of the fused ring system (83), its degradation by an alkylthio fission[43] by alkali carbonate or sulphide gave the benzothiazolium salts (85a and b) and thence the disulphides (86a and b).[42]

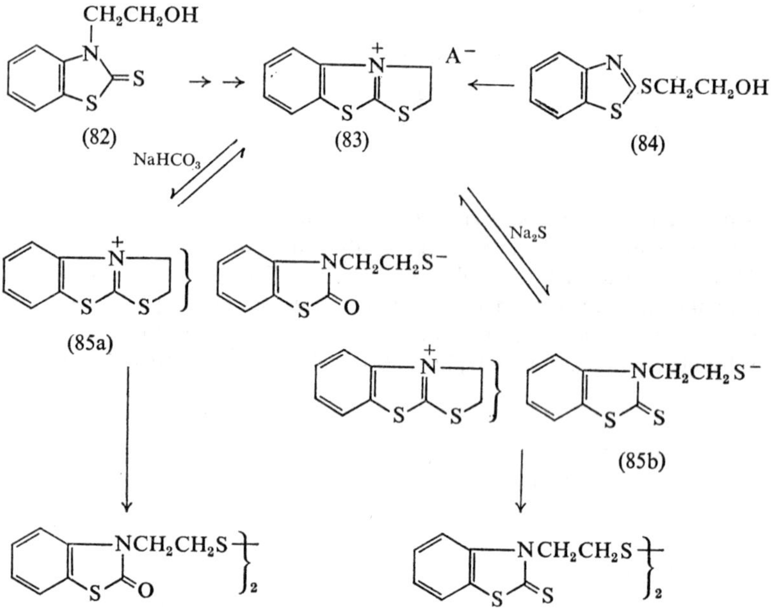

**Benzobisthiazoles.**—[$C_3NS$–$C_3NS$–$C_6$]. The synthetic routes to the five possible isomeric benzobisthiazoles[2] have been reviewed and critically discussed by Grandolini et al.[44] They corrected certain erroneous structural assignments and prepared examples of the only isomers that had remained unknown [e.g. (87) and (88)] by the following sequence of reactions:

[42] P. Sohar, G. H. Denny jun., and R. D. Babson, *J. Heterocyclic Chem.*, 1969, **6**, 163.
[43] Compare W. A. Sexton, *J. Chem. Soc.*, 1939, 473.
[44] G. Grandolini, A. Martani, and A. Fravolini, *Ann. Chim. (Italy)*, 1968, **58**, 1248.

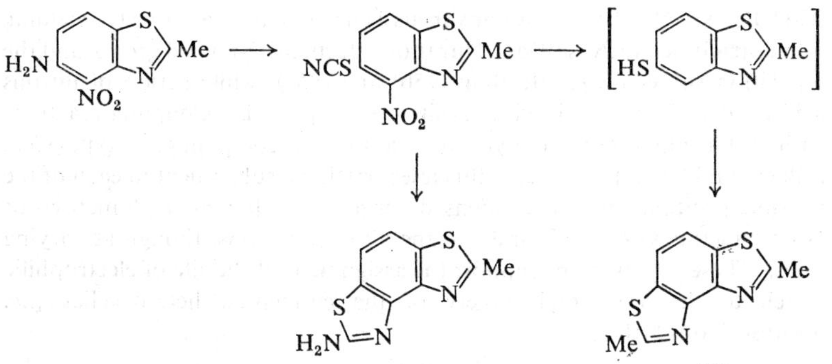

U.v., i.r., and n.m.r. spectra of all five isomeric benzobisthiazoles, specially prepared for this purpose, were recorded and briefly discussed.

**Thiazolo[3,2-a]benzimidazoles.**—[$C_3NS-C_3N_2-C_6$]. The thiazolo[3,2-a]-benzimidazole system has been synthesised from the appropriate cyclic thioamide by the general route. Thus, 2-mercaptobenzimidazole (89) reacts with chloroacetaldehyde (dimethylacetal) to yield 3-hydroxy-thiazolidino[3,2-a]benzimidazole (90), which is oxidised by chromium trioxide to 3-oxothiazolidino[3,2-a]benzimidazole (91), or dehydrated by polyphosphoric acid to thiazolo[3,2-a]benzimidazole (92). By a suitable choice of α-halogenoketo-compounds, or by the use of ethylene dichloride, the scope of the method is greatly extended, and affords related compounds based on this ring system.[45, 46]

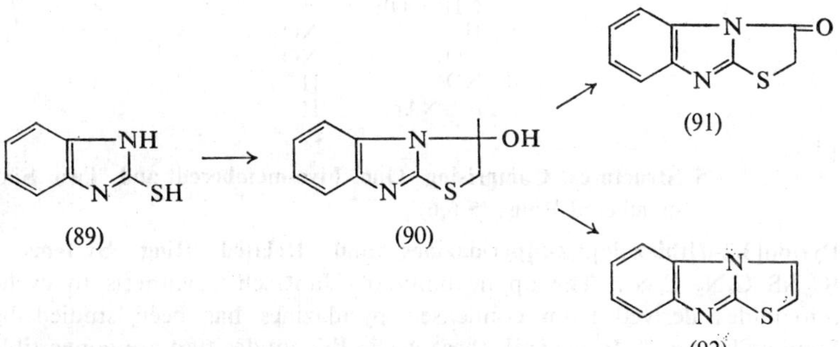

**Imidazolo[2,1-b]benzothiazoles.**—[$C_3NS-C_3N_2-C_6$]. Substituted 2-phenyl-imidazolo[2,1-b]benzothiazoles (94), obtained from 2-aminobenzothiazoles (93) and phenacyl bromide,[47] undergo addition and electrophilic substitution at their 3-position.[48] Thus, diethyl azodicarboxylate

[45] A. E. Alper and A. Taurins, *Canad. J. Chem.*, 1967, **45**, 2903.
[46] H. Ogura, T. Itoh, and Y. Shimada, *Chem. Pharm. Bull. Japan*, 1968, **16**, 2167.
[47] L. Pentimalli and A. M. Guerra, *Gazzetta*, 1967, **97**, 1286.
[48] L. Pentimalli, *Gazzetta*, 1969, **99**, 362.

(EtOOC·N=N·COOEt) in anhydrous benzene reacts additively, yielding 3-hydrazino-derivatives (95a). Nitration affects the 2-phenyl group, and the 3-position successively, affording (95b) and (95c), while action of nitrous acid in glacial acetic acid gives 3-nitroso-compounds. Coupling reactions with aryldiazonium salts in pyridine result in azo-compounds of type (95e). 2-Phenylimidazolo[2,1-b]benzothiazoles bearing a substituent in each of the possible positions in their condensed benzene ring [i.e. 6-, 7-, 8-methyl- or 5- or 7-methoxy- (94)] all undergo the above reactions, though at varying rates. These observations support the assignment of the site of electrophilic attack to the imidazolyl moiety of the condensed heterocyclics [i.e. position 3 of (94)].[48]

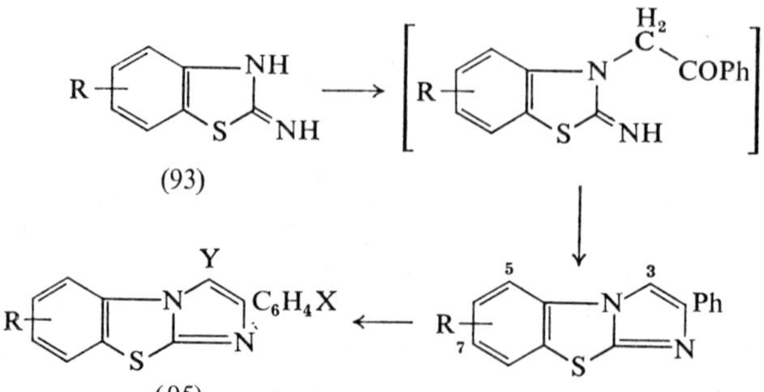

(93) → [intermediate with COPh] → (94), (95)

|     | Y         | X      |
|-----|-----------|--------|
(95) a | ·NCOOEt   | H      |
|     | NHCOOEt   |        |
| b   | H         | $NO_2$ |
| c   | $NO_2$    | $NO_2$ |
| d   | NO        | H      |
| e   | N=NAr     | H      |

## 5 Structures Comprising One Five-membered and Two Six-membered Rings (5,6,6)

**Pyrido[3,2-d]thiazolo[3,2-b]pyridazines and Related Ring Systems.—** [$C_3NS–C_4N_2–C_5N$]. The applicability of Hantzsch's synthesis to cyclic thioamides derived from condensed pyridazines has been studied by Stanovnik et al.[49] In general, these cyclic thioamides (96) are convertible into the corresponding ketosulphides (97) by the action of α-halogenoketones in presence of sodium alkoxide, but yield the intermediate hydroxythiazolines (98) in the absence of the base. The latter are reconverted into the ketosulphides (97) when crystallised from ethanolic dimethylformamide, but both (97) and (98) readily yield the final thiazolium heterocycles (99) on treatment with concentrated sulphuric acid at room temperature. The demonstrated existence of the intermediates (98),

[49] B. Stanovnik, M. Tišler, and A. Vrbanič, J. Org. Chem., 1969, **34**, 996.

previously only inferred, provides support for the concerted mechanism of this reaction that had been suggested.[50]

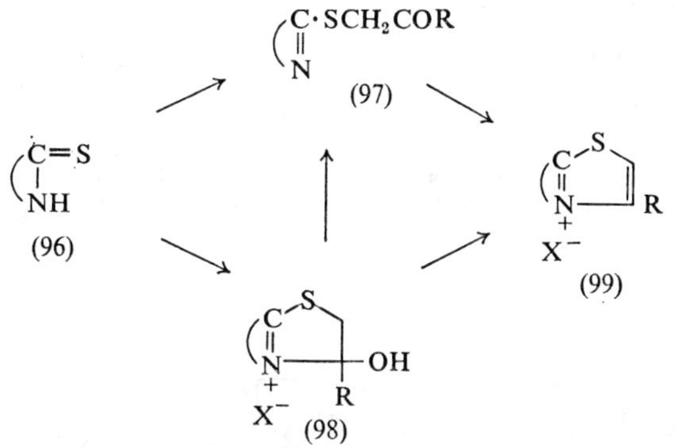

Applied to pyrido[2,3-*d*]pyridazine-5(6*H*)-thione (104), to the isomeric 8(7*H*)-thione, to *s*-triazolo[4,3-*b*]pyridazine-6(5*H*)-thione, and to 6-chloropyridazine-3(2*H*)-thione, the reaction affords the four ring systems (100)—(103). Its course is illustrated for the first example, resulting in a 3-phenylpyrido[3,2-*d*]thiazolo[3,2-*b*]pyridazin-4-ium salt (100).

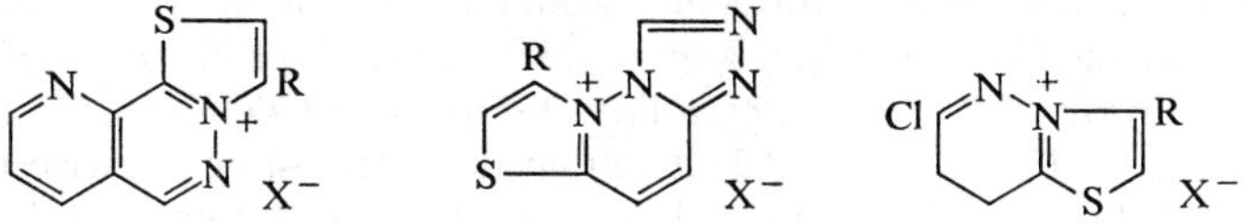

**Thiazolo[3,2-*a*]quinazoline.**—[C₃NS–C₄N₂–C₆]. In attempts to prepare (107) by the condensation of the substituted chloroacetamido-4-thiazolines (106) and thiourea, the reaction gave merely the deacetylated amine (109) and pseudohydantoin (110), presumably by the hydrolytic cleavage of the intermediate (108) and cyclodehydration of the severed thioureido side chain. In contrast, *N*-(α-chloroacetyl)anthranilic acid (111) gives rise to an intermediate (112) which undergoes preferred *intra*molecular cyclisation to the thiazolo[3,2-*a*]quinazoline-3,9-dione (113).[51]

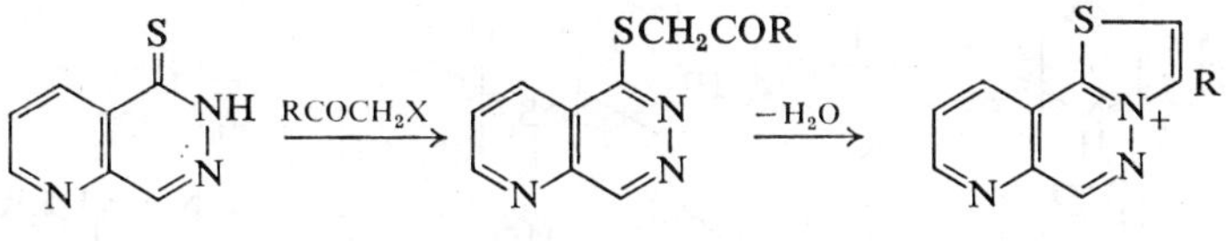

[50] C. K. Bradsher and D. F. Lohr, *J. Heterocyclic Chem.*, 1966, **3**, 27.
[51] A. Singh, A. S. Uppal, and K. S. Narang, *Indian J. Chem.*, 1969, **7**, 881.

(107) (109) (110) (106) (108)

(111) (112) (113)

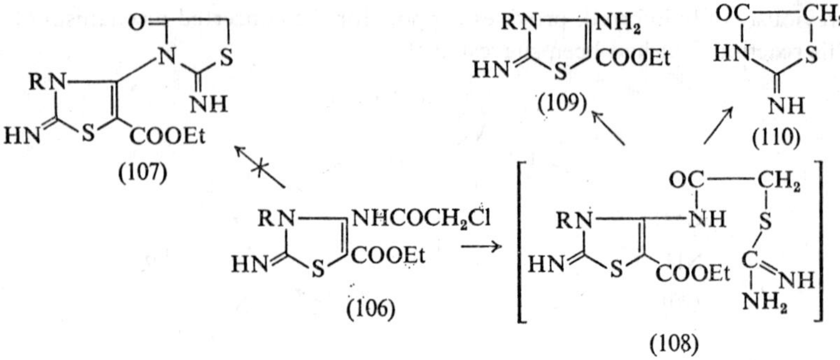

**Piperidino[5,4-*c*]thiazoline-spiro-piperidines.**—[$C_3NS$–$C_5N$–$C_6$]. In an extension of their recent thiazoline synthesis [52] from ketones, sulphur, and ammonia, Asinger *et al.*[53] have prepared 5,5,7,7-tetramethylpiperidino[5,4-*c*]thiazoline-2-spiro-4′-(2′2′6′6′-tetramethyl)-piperidine (115) from 2,2,6,6-tetramethylpiperid-4-one (triacetone-amine) (114) by this method. The product (115) is converted hydrolytically into (116) and thence, by desulphuration, into the starting

(114) (115) (116) (117)

(118)

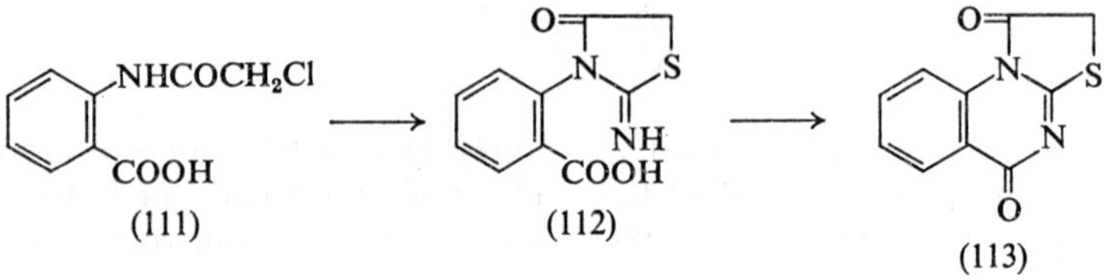

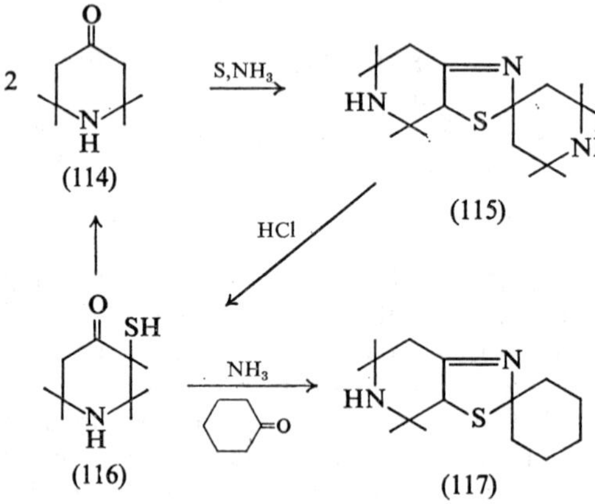

[52] F. Asinger and H. Offermanns, *Angew. Chem.*, 1967, **79**, 953 (*Internat. Edn.*, 1967, **6**, 907).
[53] F. Asinger, A. Saus, and E. Michel, *Monatsh.*, 1968, **99**, 1436.

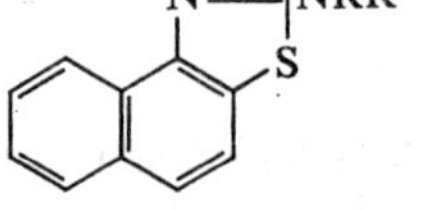

material (114). Like α-mercaptoketones in general,[54] the cyclic mercaptoketone (116), which is isolable as the hydrochloride, yields 2-substituted 3-thiazolines with aldehydes or ketones in the presence of ammonia: a number of examples synthesised by this reaction includes the 2,2-pentamethylene compound (117). The alternative synthesis of the spirocompound (115) by this route supports its formulation.[53]

**Naphtho[1,2-d]thiazoles.**—[$C_3NS-C_6-C_6$]. The quaternisation of 2-dimethylamino-, 2-morpholino-, and 2-anilino-naphtho[1,2-d]thiazoles (118)[55] occurs preferentially at the heterocyclic nitrogen; the absorption spectra of the quaternary salts resemble those of the protonated bases.[56]

### 6 Structures Comprising Four Rings

**Thiazolo[3,2-a]perimidines.**—[$C_3NS-C_4N_2-C_6-C_6$]. A synthesis of thiazolo-[3,2-a]perimidin-3(2H)-one (121) has been described: the condensation of 2-mercaptoperimidine (119) with chloroacetic acid yields perimidin-2-thiolacetic acid (120) which is cyclised to (121) by acetic anhydride.[57]

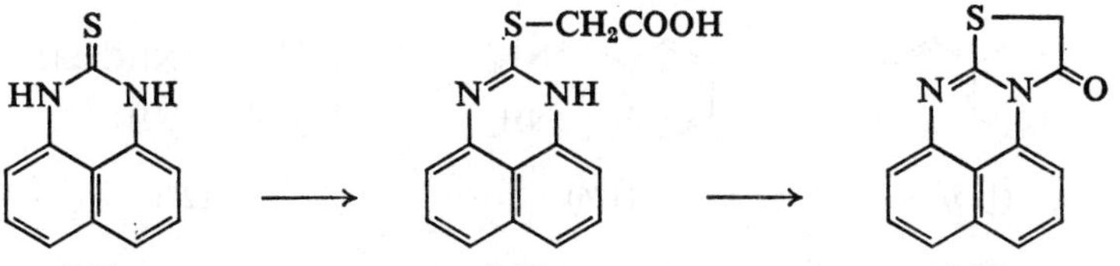

(119)    (120)    (121)

**Benzothiazolo[2,3-b]quinazolines.**—[$C_3NS-C_4N_2-C_6-C_6$]. The synthesis of benzothiazoloquinazolines has been performed by the route (122) → (124). Isatoic anhydride (122) is condensed with o-aminothiophenols in benzene to the substituted benzanilides (123), which are cyclised on prolonged

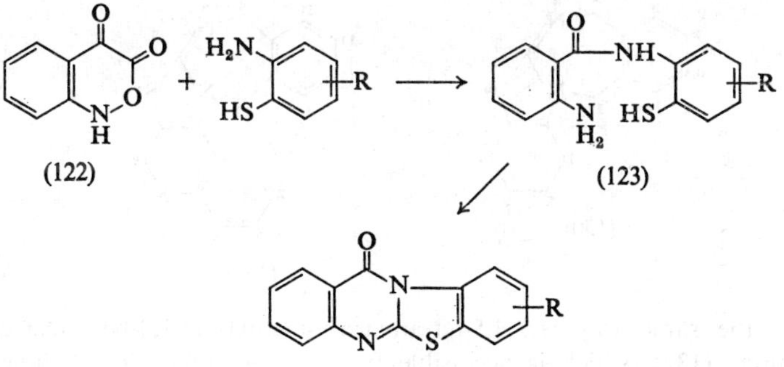

(122)    (123)

(124)

---

[54] F. Asinger and M. Thiel, *Angew. Chem.*, 1958, **70**, 667.
[55] L. Pentimalli and V. Passalocqua, *Boll. sci. fac. chim. ind. Bologna*, 1967, **25**, 125.
[56] L. Pentimalli and L. Greci, *Gazzetta*, 1968, **98**, 1369.
[57] H. S. Chaudhary and H. K. Pujari, *Indian J. Chem.*, 1969, **7**, 767.

boiling with ethyl chloroformate in the same solvent to 12*H*-3-substituted benzothiazolo[2,3-*b*]quinazolin-12-ones (124) in 75—85% yield.[58]

**Thiazolo[3,2-*b*][2,4]- and Thiazolo[2,3-*b*][1,3]-benzodiazepines.**—1,2,4,5-Tetrahydro-3*H*-2,4-benzodiazepine-3-thione (128) is obtainable without difficulty in excellent yield from phthalazine (125) by successive reductive ring cleavage to (126), addition of carbon disulphide [yielding (127)], and thermal cyclisation. This is a versatile starting material for synthesising numerous thiazolo[3,2-*b*][2,4]benzodiazepines.[59] Its condensation with 1,2-dibromoethane, for example, produces 2,3,5,10-tetrahydrothiazolo-[3,2-*b*][2,4]benzodiazepine (129), while other derivatives of this ring system are obtainable, as expected, by the use of chloropropan-2-one, ethyl 2-chloroacetoacetate, ethyl chloroacetate, and ethyl 2-bromohexanoate. Two more extended fused ring systems arise by the interaction of the intermediate (128) and 2-chlorocyclohexanone and 3-bromothiochroman-4-one, resulting in (130) and (131), respectively. Brief data were recorded concerning the i.r. and n.m.r. spectra of the compounds, but they were otherwise not further investigated.[59]

(125)

(126)

(127)

(128)

(129)

(130)

(131)

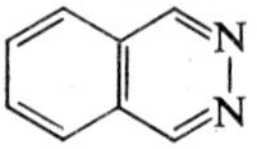

In the same way, 1,3,4,5-tetrahydro-5-methyl-2*H*-1,3-benzodiazepine-2-thione (132), which is accessible from β-methylphenylethylamine and carbon disulphide, provides access to a variety of thiazolo[2,3-*b*][1,3]-

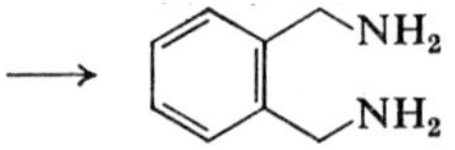

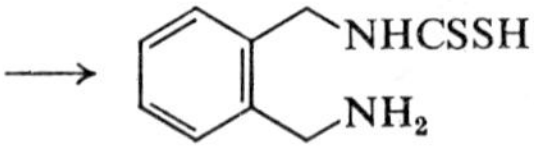

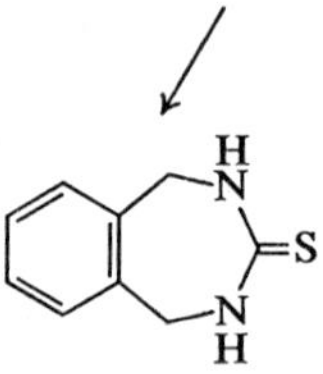

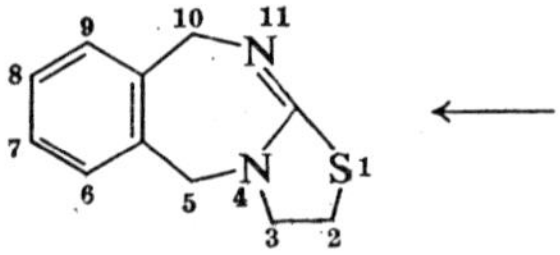

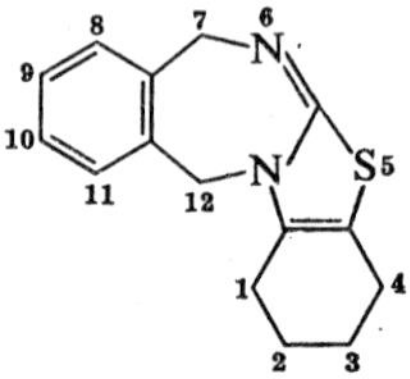

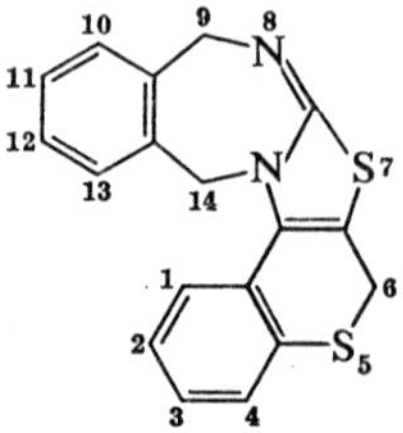

[58] A. N. Kaushal, A. P. Taneja, and K. S. Narang, *Indian J. Chem.*, 1969, **7**, 444.
[59] E. F. Elslager, D. F. Worth, N. F. Haley, and S. C. Perricone, *J. Heterocyclic Chem.*, 1968, **5**, 609.

benzodiazepines. Its interaction with ethyl chloroacetate, ethyl 2-bromohexanoate, ethyl 2-chloroacetoacetate, 2-bromo-2′-methoxyacetophenone, and 2-bromoacetophenone gives 5,6-dihydro-6-methylthiazolo[2,3-*b*][1,3]-benzodiazepin-3(2*H*)-one (133), and the appropriate analogues (134)—(137), respectively. The use of 2-chlorocyclo-pentanone and -hexanone similarly yields the novel ring systems (138) and (139). The structural assignments for the products (133)—(137) are supported by the spectroscopic (n.m.r.) evidence. Hopes that the compounds might prove to be useful anthelmintics were not realised.[60]

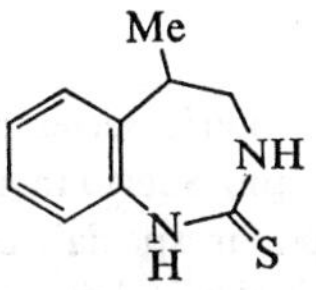

(132)

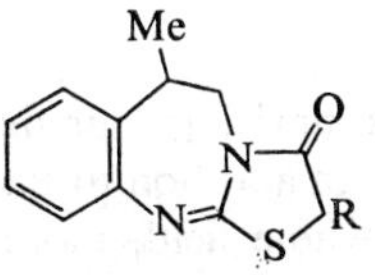

(133) R = H
(134) R = (CH$_2$)$_3$Me

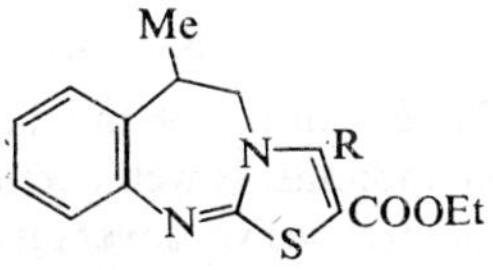

(135) R = Me
(136) R = *o*-MeOC$_6$H$_4$
(137) R = Ph

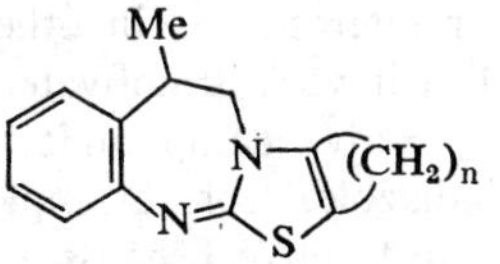

(138) n = 3
(139) n = 4

[60] E. F. Elslager, D. F. Worth, and S. C. Perricone, *J. Heterocyclic Chem.*, 1969, **6**, 491.

# 16
# Thiadiazoles

## 1 Introduction

The chemistry of thiadiazoles, comprising four distinct isomeric series of compounds, is well documented. In addition to Bambas'[1] and Sherman's[2] comprehensive coverage of the whole field, more detailed or specialised reviews have dealt with 1,2,4-,[3] 1,2,5-,[4] and 1,3,4-thiadiazoles.[5] Jensen[6] has summarised the work of his group on heterocyclic compounds, some of it yet unpublished, which included investigations of thiazoles and 1,3,4-thiadiazoles. Studies of mesionic thiadiazoles may be singled out as being currently of particular interest. As in other areas of heterocyclic chemistry, much of the published work is motivated, at least in part, by the hope of finding biologically active compounds: 2-amino-5-(1-methyl-5-nitro-2-imidazolyl)-1,3,4-thiadiazole, for example, has recently been claimed[7] to be one of the most active broad-spectrum antibacterial and antiparasitic compounds known.

## 2 1,2,3-Thiadiazoles

Pechmann's synthesis[8] of 5-anilino-1,2,3-thiadiazole from phenyl isothiocyanate and diazomethane, though not applicable to methyl isothiocyanate,[9] has been extended successfully to imidoyl isothiocyanates (2), obtainable from the corresponding halides (1) by double decomposition. These undergo 1,2-cycloaddition with equimolar proportions of diazomethane, yielding 5-guanidino-1,2,3-thiadiazoles (3) in satisfactory yields.[10]

---

[1] L. L. Bambas, 'The Chemistry of Heterocyclic Compounds', vol. IV, Interscience Publishers Inc., New York, 1952.
[2] W. A. Sherman, 'Heterocyclic Compounds', ed. R. C. Elderfield, Wiley, New York, 1961, vol. 7, p. 541.
[3] F. Kurzer, 'Advances in Heterocyclic Chemistry', ed. A. R. Katritzky, Academic Press, New York, 1965, vol. 5, p. 119.
[4] L. M. Weinstock and P. I. Pollack, 'Advances in Heterocyclic Chemistry', ed. A. R. Katritzky, Academic Press, New York, 1968, vol. 9, p. 107.
[5] J. Sandström, 'Advances in Heterocyclic Chemistry', ed. A. R. Katritzky, Academic Press, New York, 1968, vol. 9, p. 165.
[6] K. A. Jensen, *Z. Chem.*, 1969, **9**, 121.
[7] G. Berkelhammer and G. Asato, *Science*, 1968, **162**, 1146.
[8] H. v. Pechmann and A. Nold, *Ber.*, 1896, **29**, 2588.
[9] J. C. Sheehan and P. T. Izzo, *J. Amer. Chem. Soc.*, 1949, **71**, 4059.
[10] J. Goerdeler and D. Weber, *Tetrahedron Letters*, 1964, 799; *Chem. Ber.*, 1968, **101**, 3475.

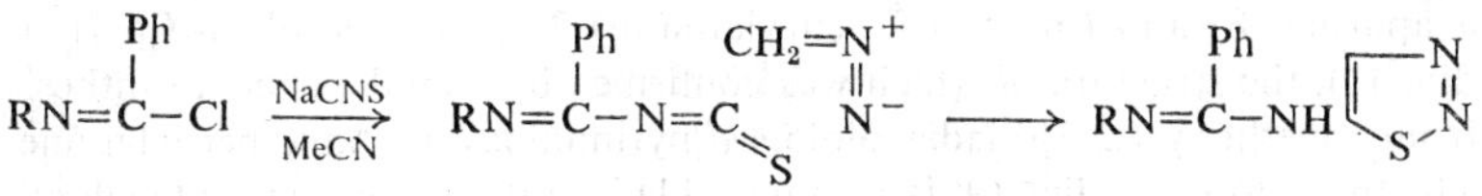

## Thiadiazoles

$$\underset{(1)}{\text{RN}=\text{C}-\text{Cl}} \xrightarrow[\text{MeCN}]{\text{NaCNS}} \underset{(2)}{\text{RN}=\text{C}-\text{N}=\text{C}\underset{S}{\overset{\text{CH}_2=\text{N}^+}{\overset{\|}{\text{N}^-}}}} \longrightarrow \underset{(3)}{\text{RN}=\text{C}-\text{NH}\underset{S}{\overset{N}{\underset{\|}{\text{N}}}}}$$

Polymers of type (4), incorporating 1,2,3-thiadiazoles, are obtainable by the condensation of bisphenol–epichlorohydrin copolymers and a 1,2,3-thiadiazole-4-carboxylic acid under restrained conditions.[11] These macromolecules are rendered insoluble by photochemically induced cross-linking and are therefore useful in the printing and the electronic industries. The cross-linking process is ascribed to the action of thioketocarbene radicals of type (5) which are known to arise from 4-phenyl-1,2,3-thiadiazole under the influence of light energy.[12]

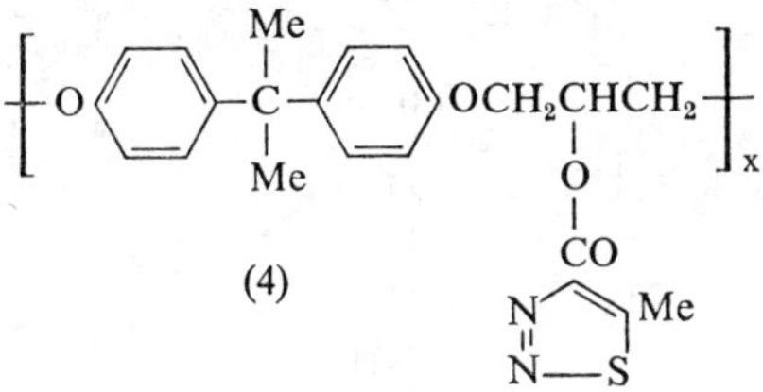

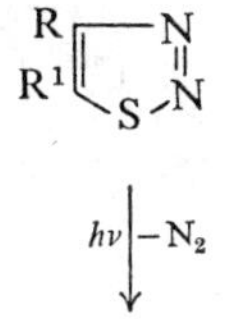

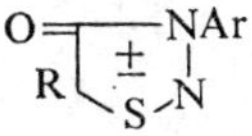

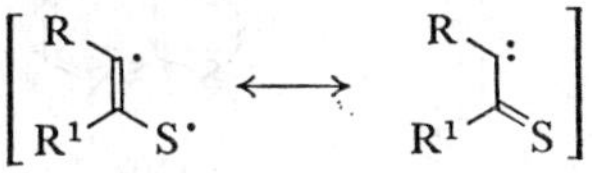

The mesionic anhydro-3-aryl-4-hydroxy-1,2,3-thiadiazolium hydroxides (6), prepared by known methods, are monoamine oxidase inhibitors both *in vitro* and *in vivo*,[13] and share this property with the comparable mesionic *N*-arylsydnones.[14] Neither of these mesionic heterocyclics is related to any known structural type exerting this activity. Their action is visualised as depending on the interaction of the heteroaromatic ring with the π-electron-bonding surfaces of the enzyme, and that of the anionic charge (acting through the *exo*-oxygen) with an electrophilic enzyme group. The necessary spatial requirements were illustrated and briefly discussed.[13]

**1,2,3-Thiadiazolo[5,4-d]pyrimidines.**—In the course of synthetic studies concerning 8-azapurines, Albert [15] discovered a reversible rearrangement of 6-mercapto-8-azapurines to thiadiazolopyrimidines. Specifically, 9-benzyl-6-hydroxy-8-azapurine (7), on being boiled with phosphorus pentasulphide in pyridine, gave an equilibrium mixture of 9-benzyl-6-mercapto-8-

---

[11] G. A. Delzenne and U. Laridon, *Ind. chim. belge*, 1969, **34**, 395.
[12] W. Kirmse and L. Horner, *Annalen*, 1958, **614**, 4.
[13] E. H. Wiseman and D. P. Cameron, *J. Medicin. Chem.*, 1969, **12**, 586.
[14] D. P. Cameron and E. H. Wiseman, *J. Medicin. Chem.*, 1968, **11**, 820.
[15] A. Albert, *J. Chem. Soc. (C)*, 1969, 152; see also A. Albert and K. Tratt, *Angew. Chem. Internat. Edn.*, 1966, **5**, 587.

azapurine (8) and (mainly) 7-benzylamino[1,2,3]thiadiazolo[5,4-*d*]pyrimidine (9), the structure of which was confirmed by its independent synthesis from 7-methoxy[1,2,3]thiadiazolo[5,4-*d*]pyrimidine (10)[16] and benzylamine. The thiazolopyrimidine (9) is reconvertible into the azapurine (8) by alkali, while the reverse reaction is promoted by heating in a solvent at 80 °C; the rearrangements proceed by ring-opening of the five-membered rings of (8) and (9) as indicated. The same reactions occur with the parent 6-hydroxy-8-azapurine [(7) → (8) ↔ (9); R = H], and the 9-methyl homologue [(7) → (8) ↔ (9); R = Me].

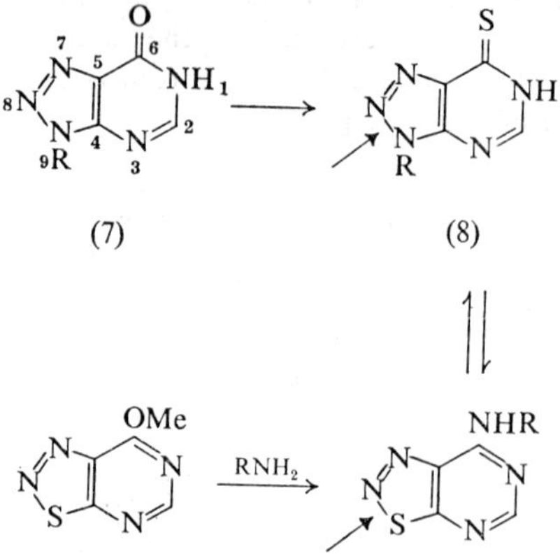

## 3 1,2,4-Thiadiazoles

Several interesting aspects of the chemistry of 1,2,4-thiadiazoles have been the subject of recent investigations.

The ring sulphur of heterocyclic systems may bear one or two exocyclic oxygen atoms; *S*-dioxides of this type are accessible by two broad approaches, *viz.* by cyclisation of precursors already incorporating the oxidised sulphur function, as in the classical synthesis of saccharin,[1, 17] or by the oxidation of the pre-formed sulphur-heteroaromatic ring,[1, 18] usually by hydrogen peroxide or potassium permanganate. The former approach, which has already provided $\Delta^3$-1,2,3-thiadiazoline-1,1-dioxides,[19] has now yielded $\Delta^2$-1,2,4-thiadiazoline-1,1-dioxides.[20, 21]

Thus, *N*-(iodomethylsulphonyl)benzamidine (11), produced from benzamidine under Schotten–Baumann conditions, is cyclised by alkali, with loss of hydrogen iodide, to 3-phenyl-$\Delta^2$-1,2,4-thiadiazoline-1,1-dioxide (12).

---

[16] E. C. Taylor and E. E. Garcia, *J. Org. Chem.*, 1964, **29**, 2121.
[17] I. Remsen and C. Fahlberg, *Ber.*, 1879, **12**, 469; *Amer. Chem. J.*, 1879, **1**, 426.
[18] E. W. McClelland *et al.*, *J. Chem. Soc.*, 1923, 170; *ibid.*, 1926, 921; *ibid.*, 1939, 760.
[19] P. Mazak and J. Suszko, *Roczniki Chem.*, 1929, **9**, 431.
[20] A. Lawson and R. B. Tinkler, *J. Chem. Soc.* (*C*), 1969, 652.
[21] A. Lawson and R. B. Tinkler, *J. Chem. Soc.* (*C*), 1970, 1429

# Thiadiazoles

The same ring closure occurs under the influence of acyl halides in alkaline media, and results, with simultaneous acylation, in 4-acyl-3-phenyl-$\Delta^2$-1,2,4-thiadiazoline 1,1-dioxides (13). The assigned ring structures agree with the observed spectroscopic properties and are supported by the alternative synthesis of one example of this series (13; Ac = $MeSO_2$) from iodomethanesulphonamide and the substituted methanesulphinimidoyl chloride (14) in pyridine. Evidence is available for the correctness of the $\Delta^2$- (12) rather than the isomeric $\Delta^3$-structure (12a) of the compounds concerned.[20]

$N^1$-Substituted $N^2$-(iodomethanesulphonyl)benzamidines similarly afford the 3,4-disubstituted homologues (15), while iodomethanesulphonyl-guanidines (17) (prepared *in situ* from iodomethanesulphonamide (16) and

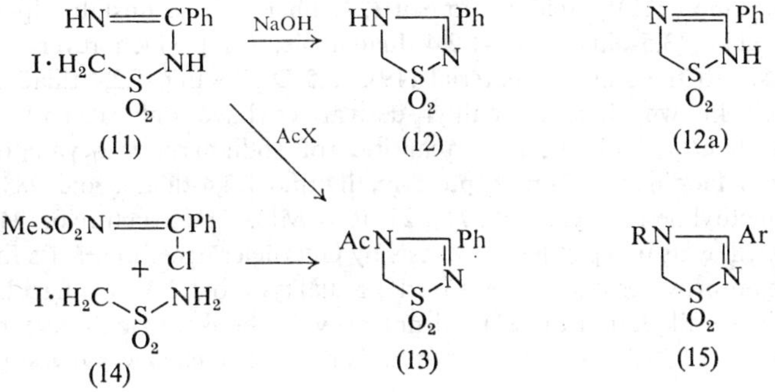

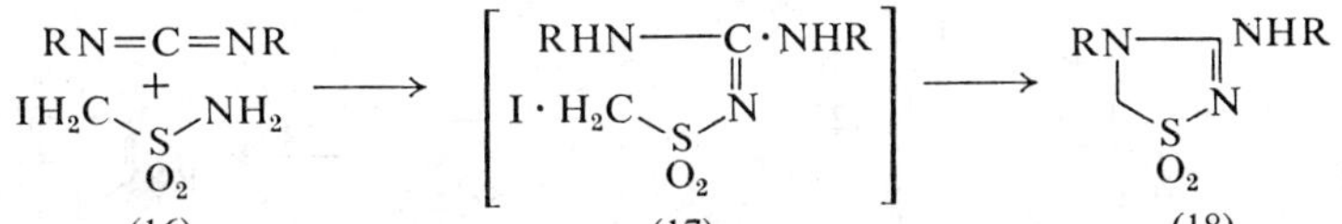

diarylcarbodi-imides) cyclise spontaneously to 4-substituted 3-arylamino-$\Delta^2$-1,2,4-thiadiazoline 1,1 dioxides (18).[21]

The u.v. spectra of all the 1,2,4 thiadiazoline 1,1-dioxides contain a strong band at *ca.* 235 m$\mu$ (log $\varepsilon$, *ca.* 4) attributed to the $-SO_2-N=\overset{|}{C}-Ph$ grouping. A study of the mass spectra of three representative compounds provided data for a suggested fragmentation pattern of this ring system, the decomposition of which appears to be initiated by the expulsion of sulphur dioxide.[21]

Grube and Suhr [22] have compared the rates of nucleophilic substitution of halogen by a range of primary and secondary amines in a series of ten chloro-substituted heterocycles. 5-Chloro-3-phenyl-1,2,4-thiadiazole proved one of the most reactive species. Replacement of a CH=CH group by sulphur in the hetero-rings raises the reaction rate: thus, the ratios of the velocity constants for the following pairs of compounds:

[22] H. Grube and H. Suhr, *Chem. Ber.*, 1969, **102**, 1570.

chlorothiazole-chloropyridine, chlorobenzothiazole-chloroquinoline, and chlorothiadiazole-chloropyrimidine are 20, 350, and 400 respectively. Since the inductive effect of sulphur is small, the activation is probably due to a stabilisation of the transition state. Repeated on a preparative scale, the experiments afforded a range of substituted 5-amino-3-phenyl-1,2,4-thiadiazoles in excellent yields.[22]

The vexing problem of the relationship between 'perthiocyanic' and 'isoperthiocyanic' acids has been settled in recent years.[3, 23] The interaction of hydrochloric acid and ammonium thiocyanate gives 3-amino-5-thiono-1,2,4-dithiazole ('isoperthiocyanic acid') (19), which isomerises under the influence of alkali (with transient liberation of sulphur) to the perthiocyanate anion (22). Acidification of (22) liberates the unstable 'perthiocyanic acid' [3,5-dimercapto-1,2,4-thiadiazole, (21)] which reverts spontaneously to the starting material (19). 3,5-Dialkylthio-1,2,4-thiadiazoles are well known, but monoalkyl derivatives have only recently been investigated.[24] Equimolecular quantities of sodium perthiocyanate (22) and methyl iodide give 3-mercapto-5-methylthio-1,2,4-thiadiazole (28%; 24) and dimethyl perthiocyanate (27%; 23, R = Me). The structure of (24) was confirmed by its independent synthesis by Goerdeler's method.[23] Its further alkylation provides a series of mixed 3,5-dialkylthio-1,2,4- thiadiazoles.

Since the alkylation of (22) did not provide the isomeric 3-mercapto-5-methylthio-1,2,4-thiadiazole, its synthesis by other means was investigated.

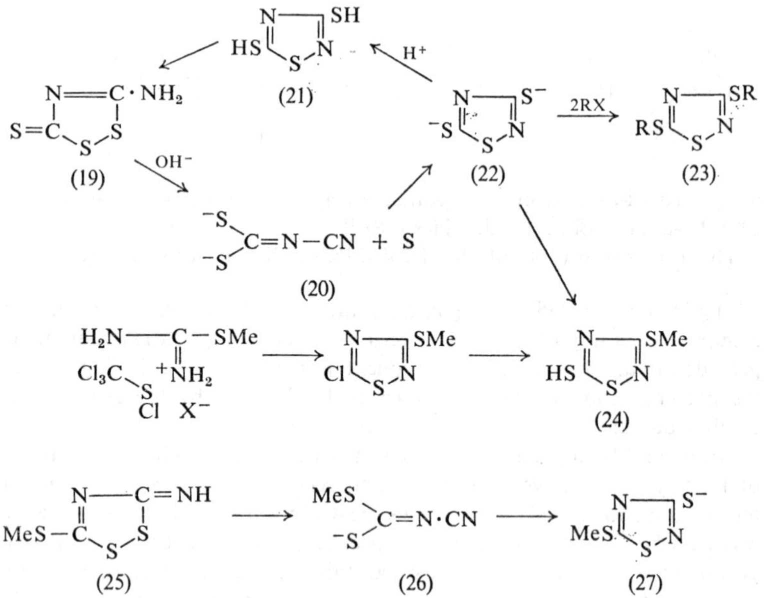

[23] J. Goerdeler and G. Sperling, *Chem. Ber.*, 1957, **90**, 892.
[24] R. Seltzer, *J. Org. Chem.*, 1969, **34**, 2562.

The reaction of 3-imino-5-methylthio-1,2,4-dithiazole (25) [25] with hydroxide ions gave, by a reaction analogous to that of isoperthiocyanic acid (19), the salts of S-methyl cyanodithioimidocarbonate (84%, 26) and 3-mercapto-5-methylthio-1,2,4-thiadiazole (7%, 27); however, these were not isolated, being identified only by their further methylation *in situ*, and gas chromatography.

## 4 1,3,4-Thiadiazoles

**Synthesis.**—The use of well-established syntheses of 1,3,4-thiadiazoles [1, 2, 5] continues to provide examples of this series of compounds. These have recently included 2-benzylamino-5-aryl-1,3,4-thiadiazoles,[26] as well as 2-trifluoroacetamido-[27] and 2-halogenoacetamido-analogues,[28] 2-amino-5-(9-acridyl)-1,3,4-thiadiazole and its homologues,[29] and 2-β-(benzimidazol-2-yl)ethylamino-5-mercapto-1,3,4-thiadiazole.[30]

Imidoyl isothiocyanates (2), useful in the production of 1,2,4-thiadiazoles (see foregoing section), are also suitable for building up the 1,3,4-thiadiazole structure. Thus, they undergo a 1,2-cycloaddition with N-phenylbenzhydrazide, affording good yields of the substituted Δ²-1,3,4-thiadiazolines (28).[10]

$$\begin{array}{c}
\text{PhN} \overset{-}{\longrightarrow} \overset{+}{\text{N}} \\
\text{PhC}-\text{N}\!=\!\text{C}\diagdown_{\text{S}} \overset{|||}{\text{CPh}} \\
\underset{\text{NR}}{\|} \\
(2)
\end{array}
\longrightarrow
\begin{array}{c}
\text{PhN}\!-\!\text{N} \\
\text{PhC}\cdot\text{N}\diagdown_{\text{S}}\diagup\text{Ph} \\
\underset{\text{NR}}{\|} \\
(28)
\end{array}$$

The interaction of *p*-nitrophenyl isothiocyanate and amidrazones [31] (29) in ether at room temperature yields adducts (30), which cyclise quantitatively in warm ethanol to 1,3,4-thiadiazoles or 1,3,4-thiadiazolines, depending on the degree of substitution of the starting materials (29).[32] The use of benzoyl isothiocyanate similarly yields the corresponding 2-benzamido-derivatives (32). Phenyl isothiocyanate and benzamidrazone (29, R¹, R² = Ph) yield 2-anilino-5-phenyl-1,3,4-thiadiazole (55%, 31; R = Ph, R¹ = H) and 3-mercapto-4,5-diphenyl-1,2,4-triazole (40%, 33, R = Ph, R¹ = H) side by side.[32]

Bis(trifluoromethyl)diazomethane reacts readily with perfluorothiocarbonyl compounds [such as hexafluorothioacetone or bis(trifluoromethyl)thioketen] to yield Δ³-1,3,4-thiadiazolines [*e.g.* (34), (34a)].[33] The

[25] R. E. Allen, R. S. Shelton, and M. G. Van Campen jun., *J. Amer. Chem. Soc.*, 1954, **76**, 1158.
[26] T. R. Vakula, V. R. Rao, and V. R. Srinivasan, *Indian J. Chem.*, 1969, **7**, 577.
[27] F. Gagiu, C. Daicoviciu, U. Binder, and Z. Györfi, *Pharmazie*, 1969, **24**, 98.
[28] F. Gagiu, T. Suciu, O. Henegaru, C. Daicoviciu, and U. Binder, *Ann. pharm. franç.*, 1969, **27**, 73.
[29] I. S. Ioffe and A. V. Tomchin, *Zhur. obshchei Khim.*, 1969, **39**, 1151 (Eng. trans., p. 1120).
[30] V. S. Misra and N. S. Agarwal, *J. prakt. Chem.*, 1969, **311**, 697.
[31] T. Bany, *Roczniki Chem.*, 1968, **42**, 252.
[32] G. Barnikow and W. Abraham, *Z. Chem.*, 1969, **9**, 183.
[33] W. J. Middleton, *J. Org. Chem.*, 1969, **34**, 3201.

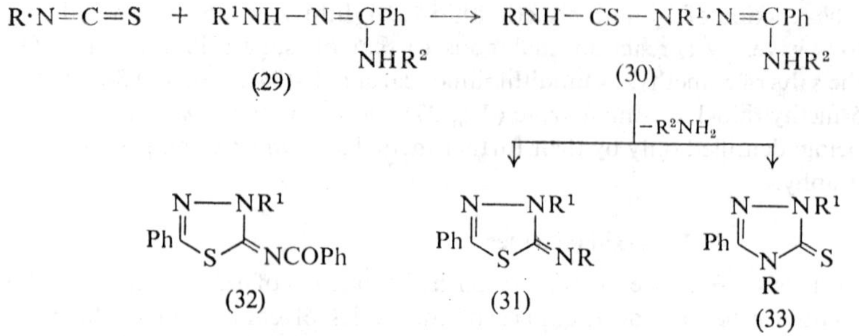

$$R \cdot N=C=S + R^1NH-N=CPh \longrightarrow RNH-CS-NR^1 \cdot N=CPh$$
$$\underset{(29)}{\overset{|}{NHR^2}} \qquad\qquad \underset{(30)}{\overset{|}{NHR^2}}$$

$$\Bigg\downarrow -R^2NH_2$$

(32) Ph, N—NR$^1$, S, NCOPh

(31) Ph, N—NR$^1$, S, NR

(33) Ph, N—NR$^1$, N(R), S

products, being stabilised by the fluorine, may be isolated and characterised; on pyrolysis, they readily undergo ring contraction, with loss of nitrogen, to the episulphides (35), (35a). These results confirm the role of labile

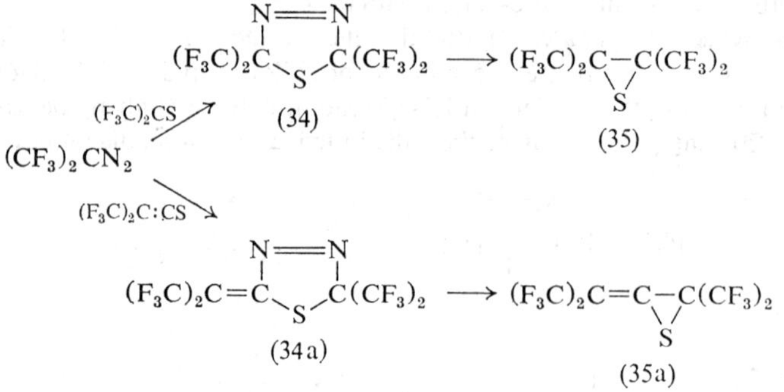

$$(CF_3)_2CN_2 \begin{array}{c} \nearrow (F_3C)_2CS \\ \searrow (F_3C)_2C\!:\!CS \end{array}$$

$$(F_3C)_2C\underset{S}{\overset{N=\!\!=\!\!N}{\underset{|}{\diagup}}}C(CF_3)_2 \longrightarrow (F_3C)_2C\underset{S}{\overset{\diagdown\diagup}{-}}C(CF_3)_2$$
(34) (35)

$$(F_3C)_2C=C\underset{S}{\overset{N=\!\!=\!\!N}{\underset{|}{\diagup}}}C(CF_3)_2 \longrightarrow (F_3C)_2C=C\underset{S}{\overset{\diagdown\diagup}{-}}C(CF_3)_2$$
(34a) (35a)

thiadiazolines as intermediates in the synthesis of episulphides by this route, as has previously been postulated.[34]

The well-known isomerisation of substituted 2,5-diamino-1,3,4-thiadiazoles (36) to 4-substituted 3-mercapto-5-alkyl(or aryl)amino-1,2,4-triazoles (37) in alkaline media appears to be accounted for by the following concerted mechanism, initiated by proton abstraction by the base from the reactant, at its more 'acidic' substituent:

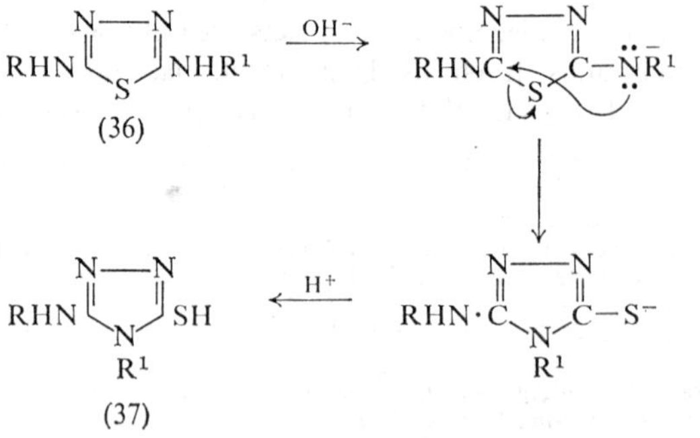

(36) RHN, N—N, S, NHR$^1$ $\xrightarrow{OH^-}$ RHNC, N—N, S, C—NR$^1$

$\Bigg\downarrow$

(37) RHN, N—N, N(R$^1$), SH $\xleftarrow{H^+}$ RHN·C, N—N, N(R$^1$), C—S$^-$

[34] H. Staudinger and J. Siegwart, *Helv. Chim. Acta*, 1920, **3**, 833.

Accordingly, the isomerisation takes the following direction: (i) When one amino-group bears an aryl residue (36, $R^1$ = Ar), while the other is either alkyl or unsubstituted (36, R = H or Alk), it is the arylamino-group which exchanges its position with the sulphur. (ii) When both R and $R^1$ are aromatic, both possible 1,2,4-triazoles are formed: electron-releasing p-substituents in the aryl residue promote, while electron-withdrawing p-substituents suppress the participation of the arylamino-group in the exchange, as is reflected in the yields of the representative isomers obtained.[35]

**Analytical.**—With platinum(IV) salts, 2,5-dimercapto-1,3,4-thiadiazole forms the yellow complexes $Pt(C_2HN_2S_3)_2$, which appear suitable for use in the photometric determination of platinum.[36]

**Mesoionic 1,3,4-Thiadiazoles.**—Mesoionic 1,3,4-thiadiazole-2-thiones were first prepared in 1895 by Busch,[37] but were formulated correctly as (40) only many years later.[38] Their preparation and study have attracted increasing attention in recent years.[39-42]

A novel synthesis affording a wide range of mesoionic 1,3,4-thiadiazole-2-thiones (40) in high yields involves the interaction of 1-substituted thiohydrazides (38) with carbon disulphide, or less favourably, with thiophosgene.[43] (See also references 41, 42, 44.) The same compounds result from the interaction of N-substituted acylhydrazines (39) and carbon disulphide, but this reaction has not been examined in many cases so far.[43]

The mesoionic 1,3,4-thiadiazole-2-thiones (40) thus produced show the usual strong absorption [39, 41, 42] in the 1345 $cm^{-1}$ region of their i.r. spectra. They are readily cleaved by aqueous alkalis at room temperature yielding 1-substituted acylhydrazines, isolable as the p-nitrobenzylidene derivatives (41). Their thermal isomerisation at 220 °C affords good yields of 5-substituted 2-alkylthio-1,3,4-thiadiazoles (42).

Mesoionic 1,3,4-thiadiazol-2-ones (43) are similarly readily obtainable[45] (see also reference 44) in good yields from (38) by the action of phosgene, or of methyl chloroformate–triethylamine.

Degradative ring-opening of the mesionic compound (43a) by aniline at 120 °C proceeds by nucleophilic attack at C–2, resulting in ring-opening to the thiobenzoylsemicarbazide (44), which is in turn cleaved by the excess of

[35] Y. R. Rao, *Indian J. Chem.*, 1969, **7**, 836.
[36] S. Gregorowicz and Z. Klima, *Coll. Czech. Chem. Comm.*, 1968, **33**, 3880.
[37] M. Busch, *Ber.*, 1895, **28**, 2635.
[38] A. Schönberg, *J. Chem. Soc.*, 1938, 824; K. A. Jensen and A. Friediger, *Kgl. Danske Videnskab. Selskab. Mat. Fys. Medd.*, 1943, **20**, 1.
[39] For summaries, see W. Baker and W. D. Ollis, *Quart. Rev.*, 1957, **11**, 15; L. B. Kier and E. B. Roche, *J. Pharm. Sci.*, 1967, **56**, 149.
[40] M. Ohta, H. Kato, and T. Kaneko, *Bull. Chem. Soc. Japan*, 1967, **40**, 579.
[41] L. B. Kier and M. K. Scott, *J. Heterocyclic Chem.*, 1968, **5**, 277.
[42] K. T. Potts and C. Sapino, *Chem. Comm.*, 1968, 672.
[43] R. Grashey, M. Baumann, and W. D. Lubos, *Tetrahedron Letters*, 1968, 5881.
[44] A. Y. Lazaris, *Zhur. org. Khim.*, 1968, **4**, 1849.
[45] R. Grashey, M. Baumann, and W. D. Lubos, *Tetrahedron Letters*, 1968, 5877.

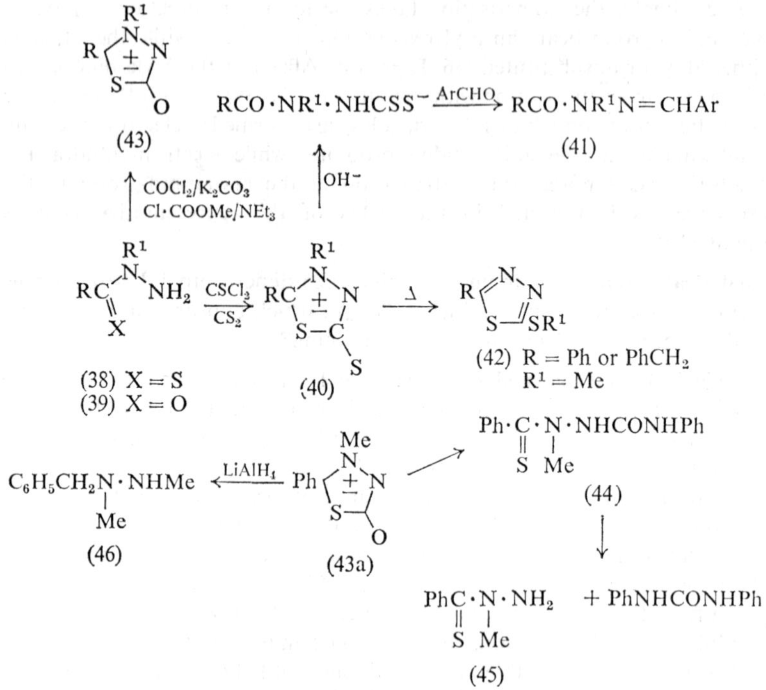

aniline to N-methyl-N-thiobenzoylhydrazine (45) (75%) and carbanilide (82%). The degradation by alkali at room temperature proceeds analogously, while reduction by lithium aluminium hydride eliminates the sulphur function, resulting in N-benzyl-NN'-dimethylhydrazine (46).[45]

**1,3,4-Thiadiazolo[2,3-a]isoquinolines.**—The production of substituted tetrahydro-1,3,4-thiadiazolo[2,3-a]isoquinolines (47) via 3,4-dihydroisoquinoline-N-imines has been briefly announced.[46] Dehydrogenation of (47) (R = H) by iodine in triethylamine yields the mesionic compound (48)[47] which is degraded by alkali in the usual way [43, 48] to 2-amino-3,4-dihydroisoquinolone (49), isolated as its p-nitrobenzylidene derivative (85%).

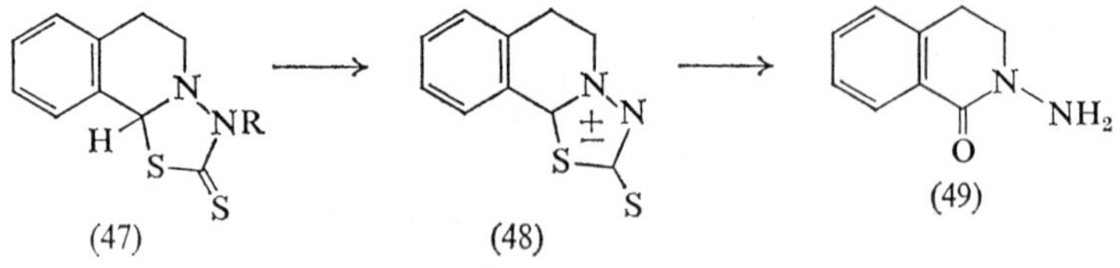

[46] R. Grashey and M. Baumann, Tetrahedron Letters, 1969, 2173.
[47] R. Grashey and M. Baumann, Tetrahedron Letters, 1969, 2177.
[48] R. M. Moriarty, J. M. Kliegemann, and R. B. Desai, Chem. Comm., 1967, 1255.

## 5 1,2,5-Thiadiazoles

A new synthesis of the relatively little known 1,2,5-thiadiazole ring system has been made possible by the recent availability [49] of $NN'$-diarylsulphurdiimides (50).

$NN'$-Di($p$-ethoxycarbonylphenyl)sulphurdi-imide, a particularly reactive member of this series, rapidly undergoes 1,3-cycloaddition with diphenylketen in benzene to yield 3,3-diphenyl-2,5-di($p$-ethoxycarbonylphenyl)-1,2,5-thiadiazolidone (51) [50] (see also reference 51). The product is

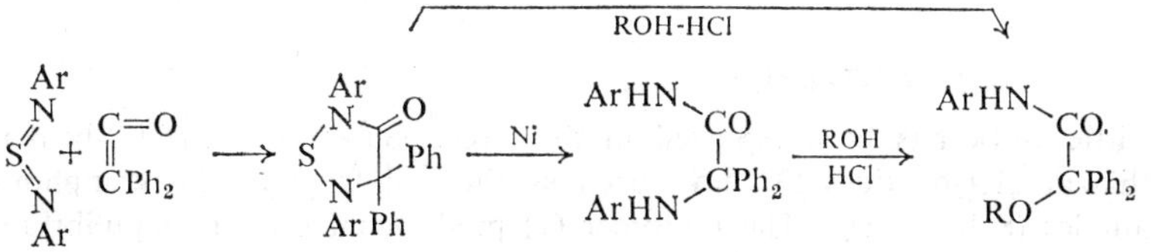

$$\underset{(50)}{\overset{Ar}{\underset{Ar}{\overset{N}{\underset{N}{\overset{\diagup}{\diagdown}}}S}}} + \overset{C=O}{\underset{CPh_2}{\|}} \longrightarrow \underset{(51)}{\overset{Ar}{\underset{ArPh}{\overset{N}{\underset{N}{\diagup}}S\diagdown\overset{O}{\underset{Ph}{\diagdown}}}}} \overset{Ni}{\longrightarrow} \underset{(52)}{\overset{ArHN}{\underset{ArHN}{\diagdown\overset{CO}{\underset{CPh_2}{|}}}}} \overset{ROH}{\underset{HCl}{\longrightarrow}} \underset{(53)}{\overset{ArHN}{\underset{RO}{\diagdown\overset{CO.}{\underset{CPh_2}{|}}}}}$$

R = Et   Ar = $p$-EtOOC·C$_6$H$_4$·

desulphurised by Raney nickel, to $NN'$-bis($p$-ethoxycarbonylphenyl)-2-amino-2,2-diphenylacetamide (52), and solvolysed by ethanolic hydrochloric acid to $N$-$p$-ethoxycarbonylphenyl-2,2-diphenyl-2-ethoxyacetamide, (53).[50] Thermolysis [51] gives rise to aryl isocyanate, benzophenone anil, and sulphur.

**2,1,3-Benzothiadiazole.**—Alkyl thiocyanates react with dialkyl hydrogen or trialkyl phosphites, to yield $OO,S$-trialkyl thiophosphates by a rearrangement of the Michaelis–Arbuzov type.[52, 53] Applied to 4-nitro-5-thiocyanato-2,1,3-benzothiadiazole (54), the reaction with diethyl hydrogen phosphite proceeds in the expected manner,[54] affording the thiophosphate (55).

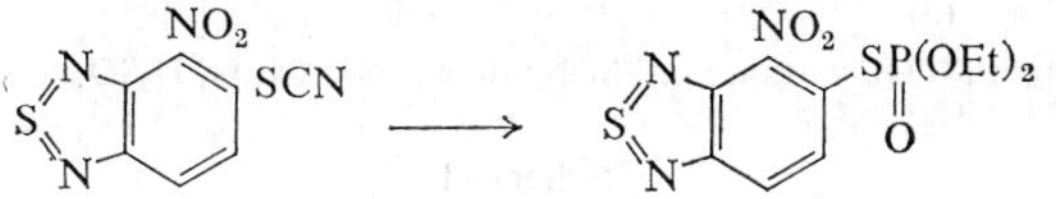

[49] H. H. Hörhold and K. D. Flossmann, *Z. Chem.*, 1967, **7**, 345; H. H. Hörhold, K. D. Flossmann, and J. Beck, German patent (DDR) 64,705.
[50] H. H. Hörhold and H. Eibisch, *Tetrahedron*, 1969, **25**, 4277.
[51] T. Minami, O. Aoki, H. Miki, Y. Oshiro, and T. Agawa, *Tetrahedron Letters*, 1969, 447.
[52] J. Michalski and J. Wieczorkowski, *Roczniki Chem.*, 1959, **33**, 105; W. A. Sheppard, *J. Org. Chem.*, 1961, **26**, 1460.
[53] G. Hilgetag and H. Teichman, *Chem. Ber.*, 1963, **96**, 1465.
[54] I. G. Vitenberg and V. G. Pesin, *Zhur. obshchei Khim.*, 1969, **39**, 1270 (Eng. trans., p. 1241).

# 17
Thiazines and Thiazepines

## 1 1,2-Thiazines

Little work has been reported in this area. Substituted 2H-1,2-benzothiazine 1,1-dioxides (3) have been synthesised from o-toluenesulphonamides (Scheme 1).[1] The tautomer (1) predominates in the equilibrium

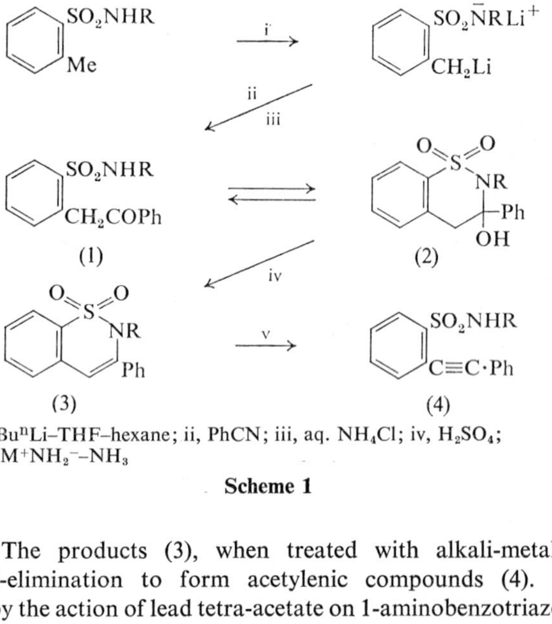

Reagents: i, Bu$^n$Li–THF–hexane; ii, PhCN; iii, aq. NH$_4$Cl; iv, H$_2$SO$_4$; v, M$^+$NH$_2^-$–NH$_3$

**Scheme 1**

(1) ⇌ (2). The products (3), when treated with alkali-metal amides, undergo β-elimination to form acetylenic compounds (4). Benzyne, generated by the action of lead tetra-acetate on 1-aminobenzotriazole, reacts with thionylaniline *in situ* to form 6H-dibenzo[c,e][1,2]thiazine 5,5-dioxide (6) (Scheme 2).[2] The primary product (5) is evidently oxidised further under the reaction conditions. Intervention of singlet state sulphonyl nitrenes is suggested to account for the products from the thermal decomposition of aryl sulphonyl azides. Biphenyl-2-sulphonyl azide (7) decomposes in dodecane at 150 °C to give the dibenzo[c,e][1,2]thiazine 5,5-dioxide (9) as

---

[1] H. Watanabe, C.-L. Mao, I. T. Barnish, and C. R. Hauser, *J. Org. Chem.*, 1969, **34**, 919.
[2] C. D. Campbell and C. W. Rees, *J. Chem. Soc. (C)*, 1969, 748.

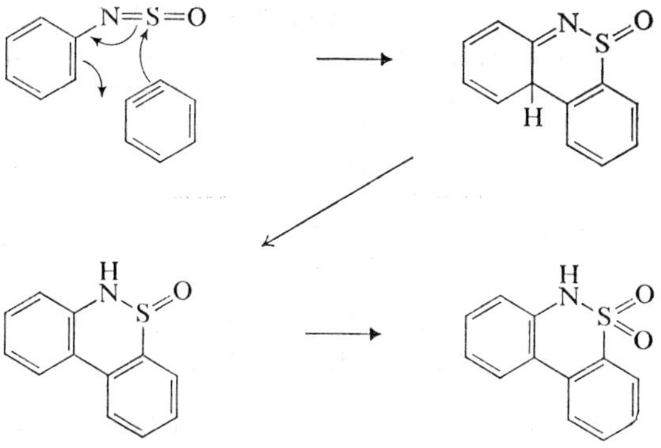

[Scheme 2 diagram with structures (5) and (6)]

**Scheme 2**

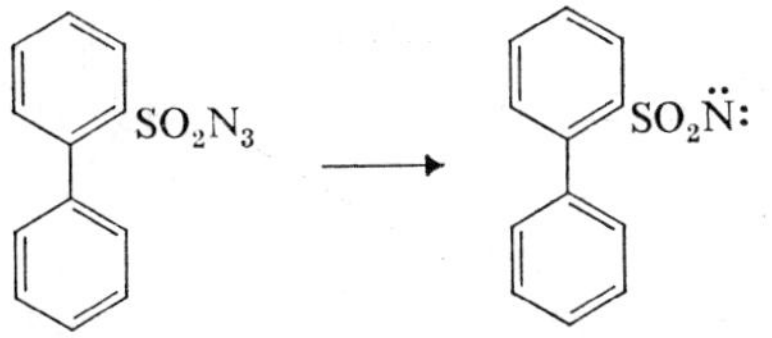

[Diagram with structures (7), (8), (9), (10)]

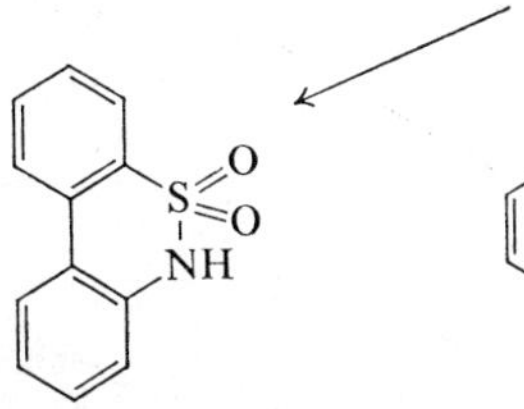

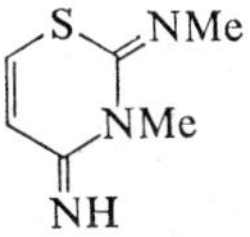

the sole reaction product, probably *via* the nitrene (8) which undergoes intramolecular C–H insertion by nitrene nitrogen.[3]

## 2 1,3-Thiazines

**Simple 1,3-Thiazines.**—The use of acetylenic compounds in the synthesis of 1,3-thiazines is reported by two groups.[4,5] Cyanoacetylene reacts with $NN'$-dimethylthiourea to give the derivative (10), thereby extending the range of acetylenes from which 1,3-thiazines can be obtained.[4] In a more detailed account Cain and Warrener[5] showed that propiolic acid reacts

---

[3] R. A. Abramovitch, C. I. Azogu, and I. T. McMaster, *J. Amer. Chem. Soc.*, 1969, **91**, 1219.
[4] T. Sasaki, K. Kanematsu, and K. Shoji, *Tetrahedron Letters*, 1969, 2371.
[5] E. N. Cain and R. N. Warrener, *Austral. J. Chem.*, 1970, **23**, 51.

with dithiocarbamic acid and its *N*-methyl and *N*-ethyl derivatives to form intermediate acrylic acids (11), shown to have *cis* stereochemistry. Cyclisation of these acids with acetic anhydride–sulphuric acid, and acid hydrolysis

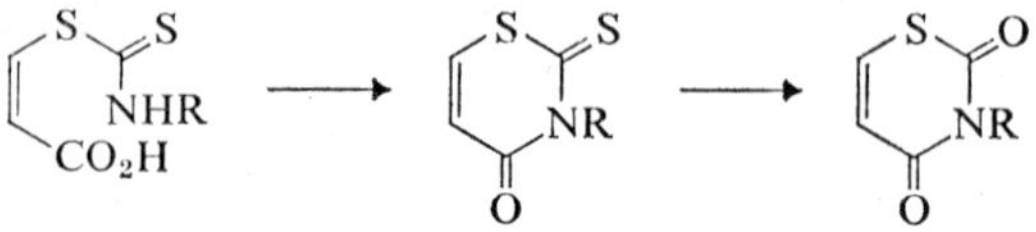

(11)–(13); R = H, Me, Et

(6M-HCl) of the resulting thiones (12), gives sulphur analogues of uracil (1-thiauracils) (13). A sulphur analogue of thymine, 1-thiathymine (14), was obtained by the sequence of reactions in Scheme 3.

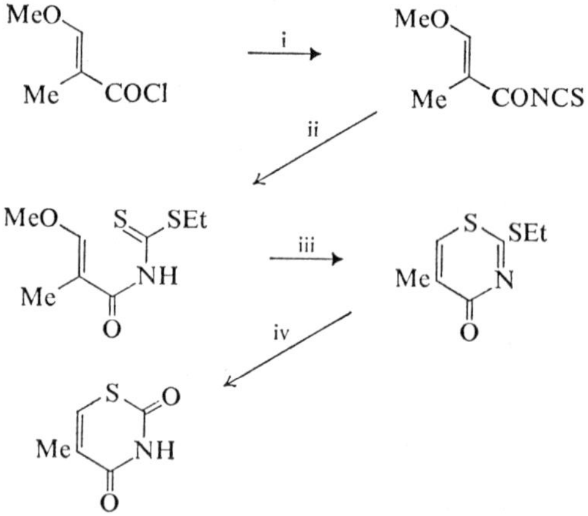

Reagents: i, KSCN–MeCN; ii, EtSH–Et$_3$N–ether; iii, conc. H$_2$SO$_4$; iv, conc. HCl

Scheme 3

The occurrence of the 1,3-thiazine structure in the cephalosporins has led to the development of novel syntheses of dihydro-1,3-thiazines related to cephalosporins. Condensation of the amine (15) with benzaldehyde, and acid-catalysed cyclisation and detritylation of the resulting imine (16), gave the thiazine (17) (Scheme 4).[6] Use of the aldehyde (18) in place of benzaldehyde in a similar sequence of reactions led to compound (19), a known intermediate for the synthesis of deacetylcephalosporin lactones. In boiling acetic anhydride, penicillin *S*-oxides (20) undergo ring-enlargement to 1,3-thiazines in an intramolecular oxidation–reduction involving

[6] J. E. Dolfini, J. Schwartz, and F. Weisenborn, *J. Org. Chem.*, 1969, **34**, 1582.

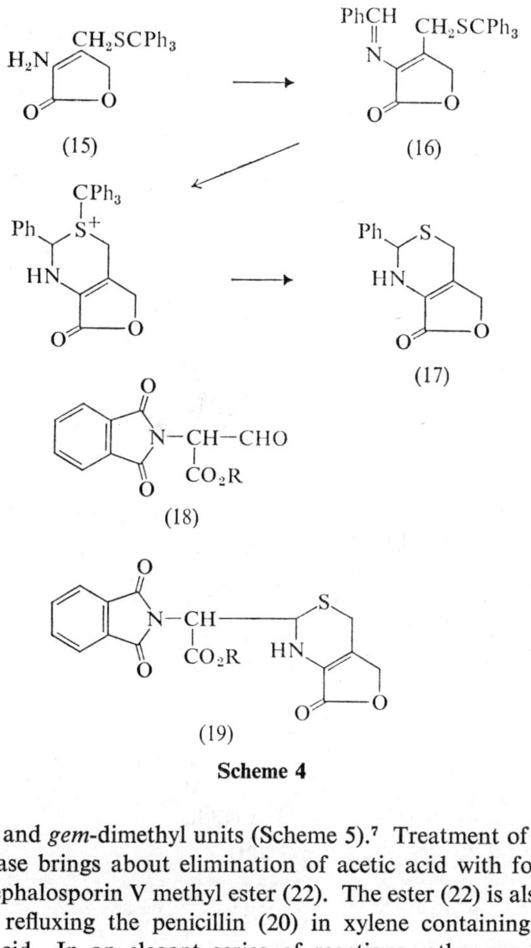

**Scheme 4**

the *S*-oxide and *gem*-dimethyl units (Scheme 5).[7] Treatment of the acetate (21) with base brings about elimination of acetic acid with formation of deacetoxycephalosporin V methyl ester (22). The ester (22) is also obtained directly by refluxing the penicillin (20) in xylene containing *p*-toluenesulphonic acid. In an elegant series of reactions, other workers[8] have converted the ester (22) into cephalosporin V.

A novel rearrangement of a penicillanic acid derivative into a 1,4-thiazepine and thence into a 1,4-thiazine is reported.[9] The sequence of reactions in Scheme 6 is suggested to account for these results. The thiazepine (24) is transformed into the thiazine (25) less rapidly than it is formed from the penicillanic acid (23), thus allowing its isolation (see also section 5).

---

[7] R. B. Morin, B. G. Jackson, R. A. Mueller, E. R. Lavagnino, W. B. Scanlon, and S. L. Andrews, *J. Amer. Chem. Soc.*, 1969, **91**, 1401.
[8] J. A. Webber, E. M. Van Heyningen, and R. T. Vasileff, *J. Amer. Chem. Soc.*, 1969, **91**, 5674.
[9] J. R. Jackson and R. J. Stoodley, *Chem. Comm.*, 1970, 14.

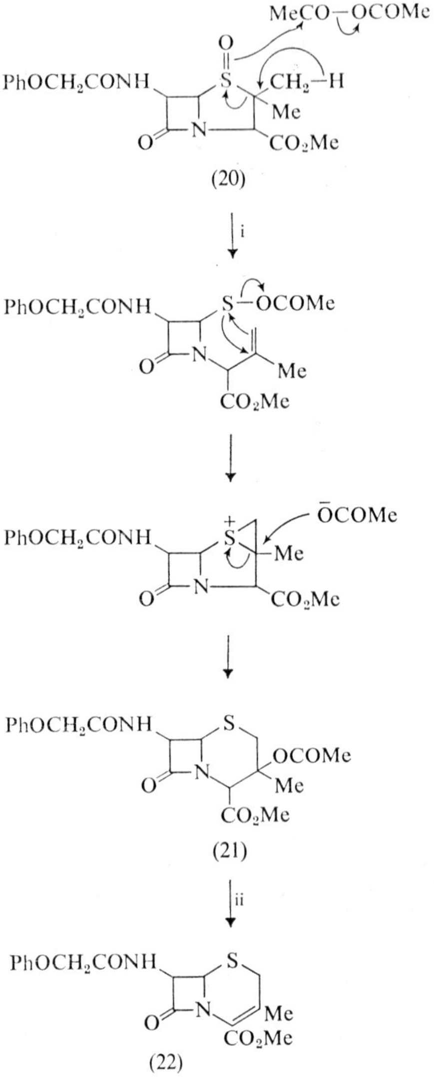

Scheme 5 (structures 20, 21, 22)

Reagents: i, Ac$_2$O; ii, Base.

**Scheme 5**

**Benzo-1,3-thiazines.**—2-Methyl-4,5-benzo-6$H$-1,3-thiazin-6-one (26) is formed quantitatively[10] from anthranilic acid in thioacetic acid at 100 °C. The reaction of 4,5-benzo-1,3-thiazin-6-thiones with secondary amines has

[10] G. C. Barrett, J. R. Chapman, and A. R. Khokhar, *Chem. Comm.*, 1969, 818.

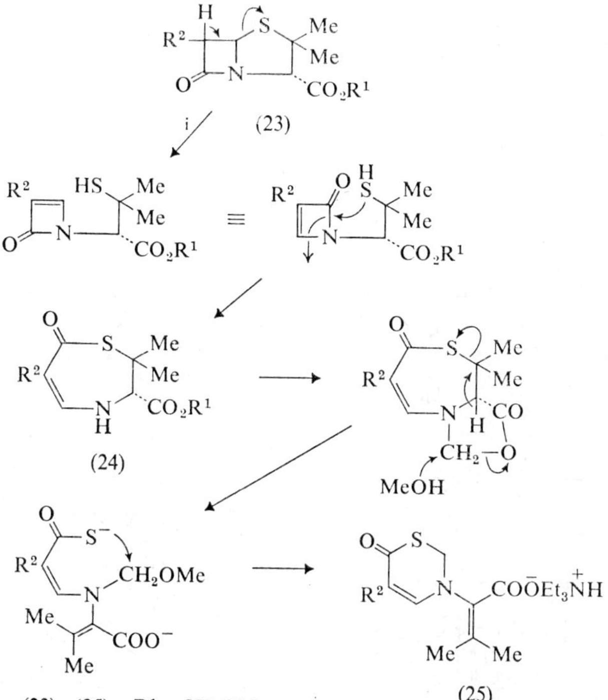

## Scheme 6

(23)–(25); $R^1 = CH_2OMe$
$R^2 = p\text{-}NO_2\cdot C_6H_4\cdot CH=N$

Reagent: i, $Et_3N$–$CHCl_3$

**Scheme 6**

been studied.[11] The course of the reaction is solvent dependent. Typically, the thione (27) is ring-opened by dimethylamine in benzene to give the thioamide (28), whereas in ethanol the ring sulphur atom is replaced by the Me–N unit to give the quinazoline thione (29). The remarkable incorporation of the primary amino-group into the product was confirmed using methylamine in place of dimethylamine, when the same product (29) was obtained. 2-Chloro-5,6-benzo-4$H$-1,3-thiazin-4-thiones (30) are obtained by the action of hydrogen chloride on 2-thiocyanatobenzoyl chlorides, or by the action of phosphorus pentachloride on the corresponding acids.[12] The selenium analogue (31) is obtained analogously. These compounds are readily hydrolysed, and the 2-chloro substituent undergoes ready nucleophilic displacement.

---

[11] C. Denis-Garez, L. Legrand, and N. Lozac'h, *Bull. Soc. chim. France*, 1969, 3727.
[12] G. Simchen and J. Wenzelburger, *Chem. Ber.*, 1970, **103**, 413.

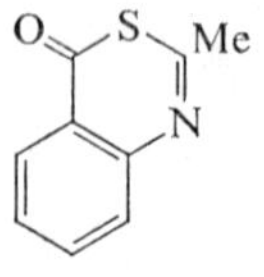

(26) 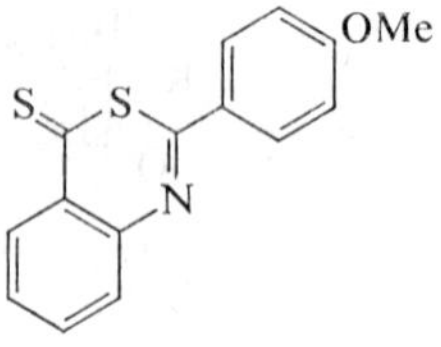 (27) 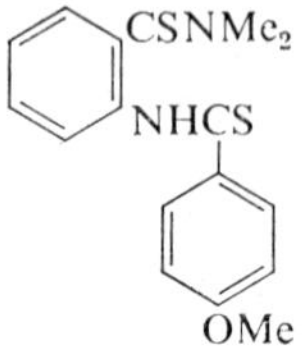 (28) 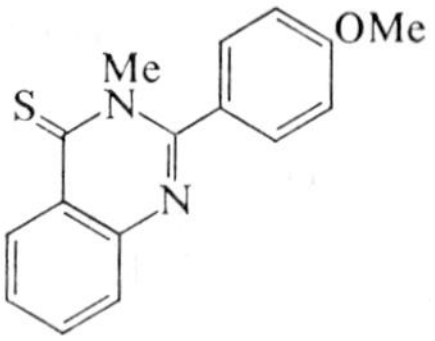 (29) 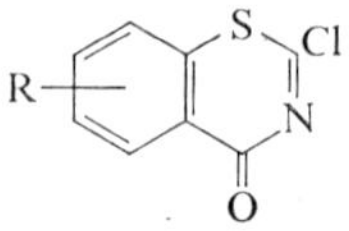 (30) R = H, 7-chloro, 7-methoxy, 6-chloro, 6-methoxy, 7-nitro, 6-nitro. (31) 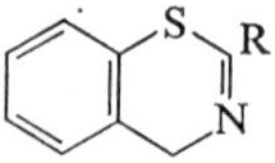 (32) R = Ph (33) R = Me (34) R = Et (35) 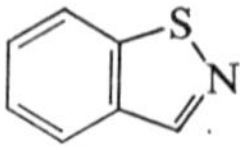 (36)

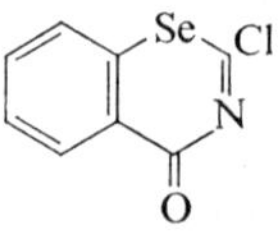

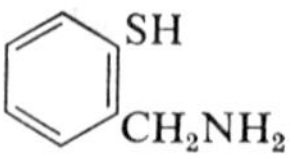

The chemical behaviour of 2-alkyl- and 2-aryl-5,6-benzo-1,3-thiazines has been studied.[13] The heterocyclic ring is fairly resistant to ring-opening. The 2-phenyl derivative (32) is unaffected by concentrated acids and boiling alkali, although boiling dilute acid leads to benzoisothiazole (36) by the sequence (32) → (35) → (36). The 2-methyl derivative (33) is ring-opened by concentrated hydrochloric acid giving o-aminomethylbenzenethiol (35), isolated as its dibenzoyl derivative. Depending on the reagent, oxidation may occur at sulphur giving a sulphoxide, or at C-4 giving a 5,6-benzo-1,3-thiazin-4-one. Treatment of the 2-ethyl derivative (34) with silver nitrate, followed by methyl iodide, effects methylation at C-4.

[13] D. Bourgoin-Legay and R. Boudet, *Bull. Soc. chim. France*, 1969, 2524.

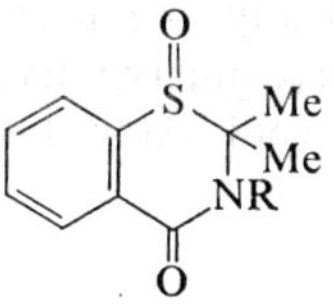

(37) R = H
(38) R = Me

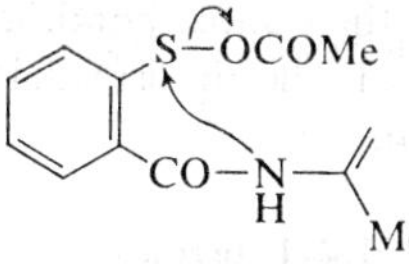

(39)

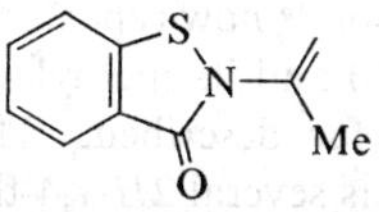

(40)

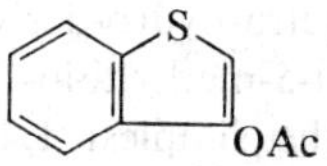

(41)

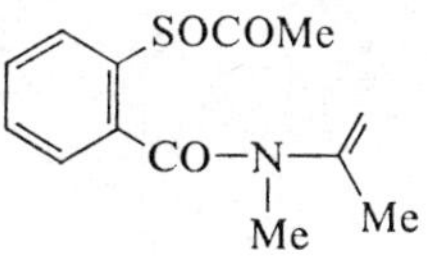

(42)

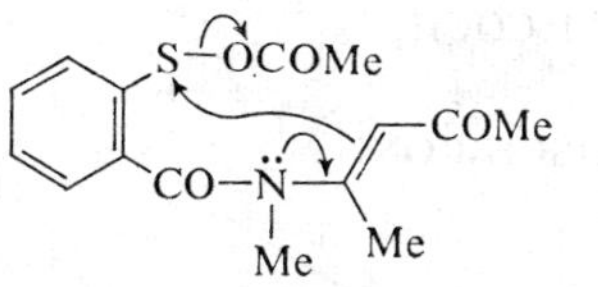

(43)

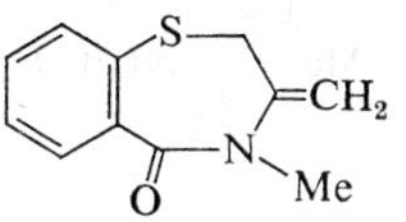

(44)

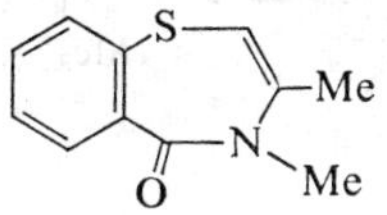

(45)

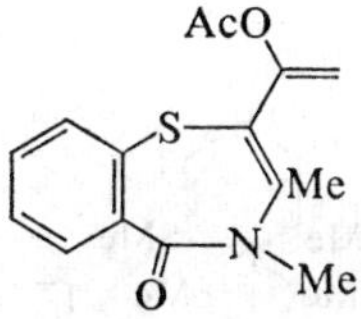

(46)

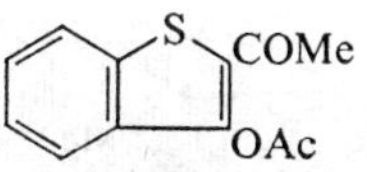

(47)

Following their work on the enlargement of the thiazolidine ring in penicillins,[7] Morin and co-workers have studied the behaviour of the 5,6-benzothiazine sulphoxides (37) and (38) with boiling acetic anhydride.[14] Compound (37) rearranges to a mixture of the isothiazolone (40) and the benzothiophen (41) via, it is suggested, the intermediate (39) (cf. Scheme 5). The N-methyl derivative (38) behaves differently. The tertiary nitrogen atom in the related intermediate (42) is unable to react effectively with the sulphenic acid moiety generated in the ring-opening process. Instead, attack on the double bond leads to the four products (44)—(47). Compound (44) is converted into compound (45) under the conditions of

[14] R. B. Morin and D. O. Spry, *Chem. Comm.*, 1970, 335.

the reaction. The intermediate (42) is sufficiently long-lived to undergo acetylation of the double bond, leading to the secondary intermediate (43). Ring-closure of the intermediate (43) and enol acetylation gives the thiazepine (46).

## 3 1,4-Thiazines

**Monocyclic 1,4-Thiazines.**—Although a substantial volume of work has been published on ring homologues of 1,4-thiazine, little is known of the chemistry of the parent heterocycle, and most published work is concerned with its di- or tetra-hydro-derivatives. Johnson [15] now reports the synthesis of 3,4-dimethyl-5-methylthio-1,4-thiazine (48) and its methylation product (49), two of the simplest 1,4-thiazines so far described. The reactions leading to compounds (48) and (49), as well as several 2$H$-1,4-thiazines, are shown in Scheme 7. A $^1$H n.m.r. study [16] of the sulphone (50) and its anion

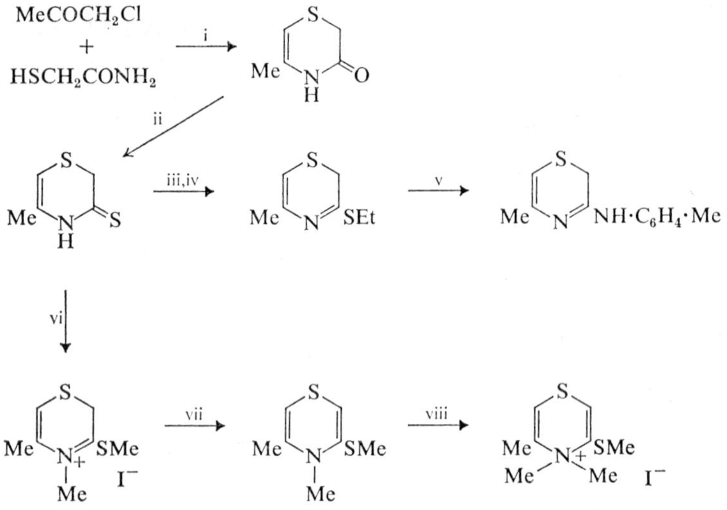

Reagents: i, Et$_3$N; ii, P$_2$S$_5$; iii, Et$_3$O$^+$BF$_4^-$; iv, K$_2$CO$_3$; v, $p$-MeC$_6$H$_4$NH$_2$
vi, MeI–acetone; vii, KOBu$^t$–ether; viii, MeI

**Scheme 7**

(51) in DMSO-$d_6$ shows greater deshielding of the *ortho* protons in the anion (51) compared with those in the thiazine (50), in spite of the negative charge in the former. This is taken to be due to a greater ring-current in the anion than in the thiazine. Compound (50) is remarkably stable, being unchanged by boiling triethylamine after 2 weeks.

[15] C. R. Johnson and C. B. Thanawalla, *J. Heterocyclic Chem.*, 1969, **6**, 247.
[16] I. Satayt, *J. Org. Chem.*, 1969, **34**, 250.

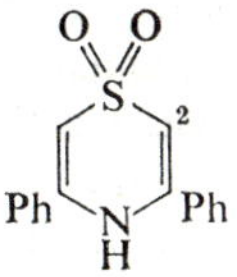

(50) 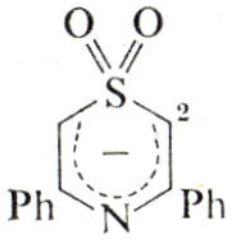  (51)

2-Methyl-3-phenacylthiazolium bromide with aqueous sodium hydrogen carbonate does not give 6-phenylpyrrolo[2,1-*b*]thiazole (52).[17] Re-examination of the reaction[18] has revealed the formation of the 2,3-dihydro-1,4-thiazine (55), evidently *via* the carbinol (53) and its ring-opened form (54), which recyclises to the thiazine (55).

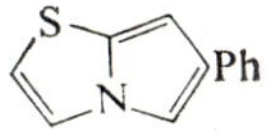

(52) 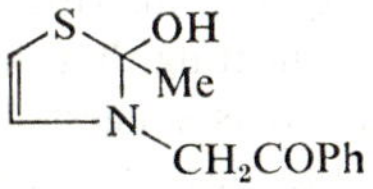  (53)

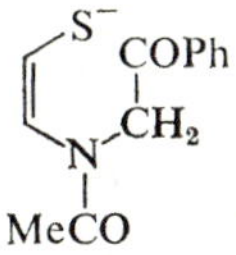

(54) 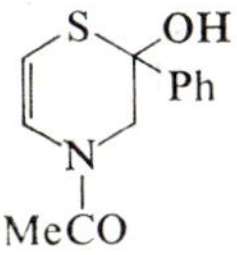  (55)

$^1$H N.m.r. studies of conformational equilibria of 3-substituted 6-methoxycarbonyl-2,3-dihydro-1,4-thiazines (56) have been carried out.[19] A clear preference emerges for the axial conformation (56b) when the

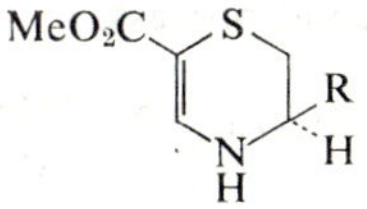

(56)

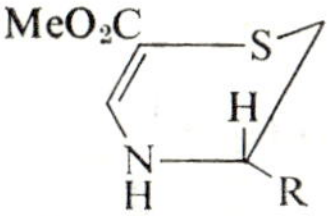

(56a) 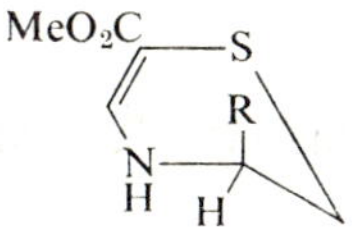  (56b)

[17] B. B. Molloy, D. H. Reid, and F. S. Skelton, *J. Chem. Soc.*, 1965, 65.
[18] D. J. Adam and M. Wharmby, *Tetrahedron Letters*, 1969, 3063.
[19] A. R. Dunn and R. J. Stoodley, *Tetrahedron Letters*, 1969, 2979.

is transformed by concentrated hydrobromic acid into the complex thiazinium salt (65).[23]

**Phenothiazines (dibenzo-1,4-thiazines) and Related Compounds.**—M.O. calculations have been performed on phenothiazine, phenoxathiin, and phenoxazine, using the SCF–PPP approximation.[24] It is well known that the nitrogen atom in phenothiazine plays a key role in determining the reactivity towards electrophilies, position 3 being about as or slightly more reactive than position 1. The reactivity indices (Figure 1) agree with the

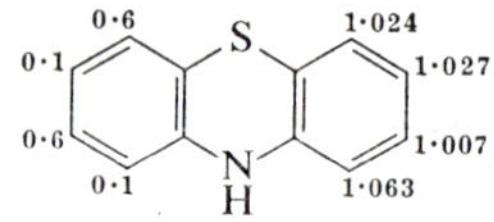

**Figure 1** $\pi$-Electronic charge (right-hand side) and localisation energies in electron-volts (left-hand side) for electrophilic substitution of phenothiazine

facts. The calculations indicate that the two carbocyclic rings are quite independent, and the chemistry of phenothiazine resembles that of the corresponding o-disubstituted benzene derivatives.

Thionyl chloride (one molecular equivalent) reacts with triphenylamine in boiling benzene to give 10-phenylphenothiazine 5-oxide, and with diphenylamine to give a mixture of phenothiazine, phenothiazine 5-oxide,

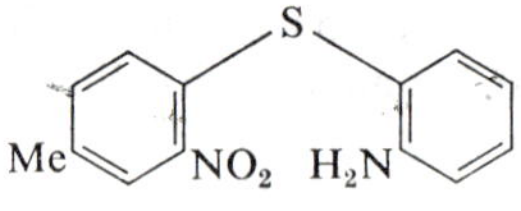

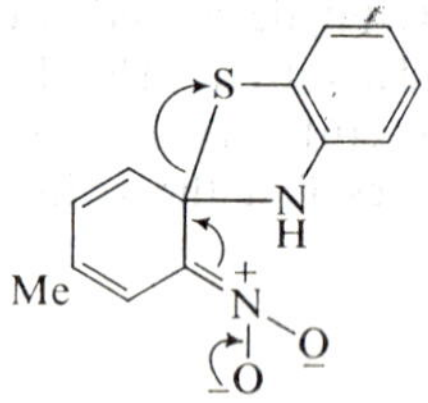

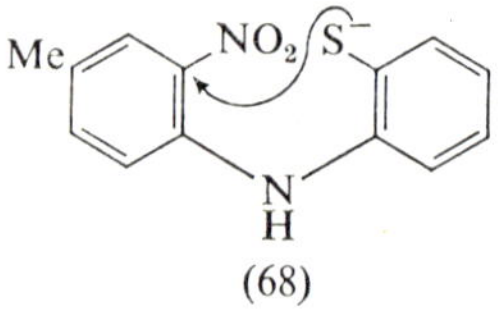

and 1,3,7,9-tetrachlorophenothiazine.[25] Two rearrangements involving diaryl sulphides, leading to phenothiazines, are reported. Heating 2,2′-aminonitrodiphenyl sulphide (66) at 190 °C, alone or in NN-dimethylacetamide, gives [26] a mixture of the phenothiazine (69) and the dibenzo-

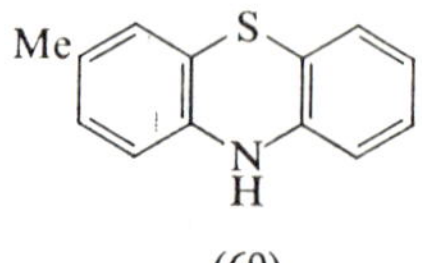

---

[23] J. Fröhlich, U. Habermalz, and F. Kröhnke, *Tetrahedron Letters*, 1970, 271.
[24] H. H. Mantsch and J. Dehler, *Canad. J. Chem.*, 1969, **47**, 3173.
[25] B. C. Smith and M. E. Sobeir, *Chem. and Ind.*, 1969, 621.
[26] F. A. Davis and R. B. Wetzel, *Tetrahedron Letters*, 1969, 4483.

thiophen (70). The reaction is envisaged as proceeding *via* the intermediates (67) and (68) in turn, in a thermal variation of the Smiles rearrangement, and is completed by intramolecular displacement of nitrite ion by sulphide. Formation of 2-methyl-4-nitrodibenzothiophen (70) is considered to result from cyclisation of the intermediate (71), formed by

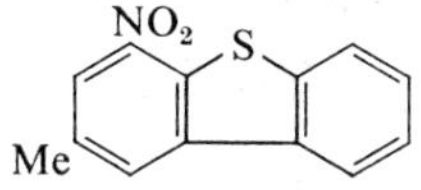

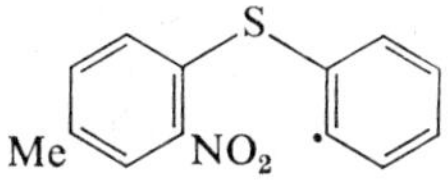

homolytic cleavage of the C–NH$_2$ bond or by loss of nitrogen from the diazonium ion (72), itself produced by the action of expelled nitrite on the amine (66). The sulphide (73) undergoes the Smiles rearrangement in the presence of hydroxide ion, and leads to 1-trifluoromethylphenothiazine (74; R = CF$_3$).[27] The normally very stable trifluoromethyl group undergoes in this compound a new reaction in methanolic potassium hydroxide, giving the orthoester [74; R = C(OMe)$_3$].

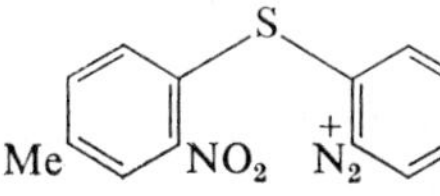

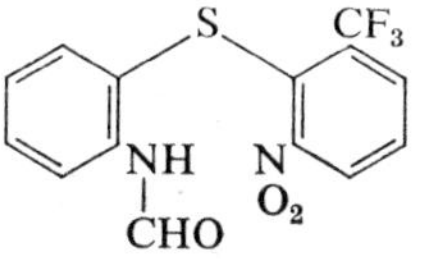

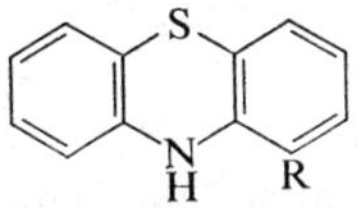

Formation of the phenothiazine cation-radical results from the action of iodine in DMSO on phenothiazine.[28] Subsequent further reaction of the cation-radical produces mainly the dimer (75), together with small amounts of the ketone (76) and 3-iodophenothiazine. Combined u.v., e.s.r., and kinetic studies of the bromination of phenothiazine show that the pheno-

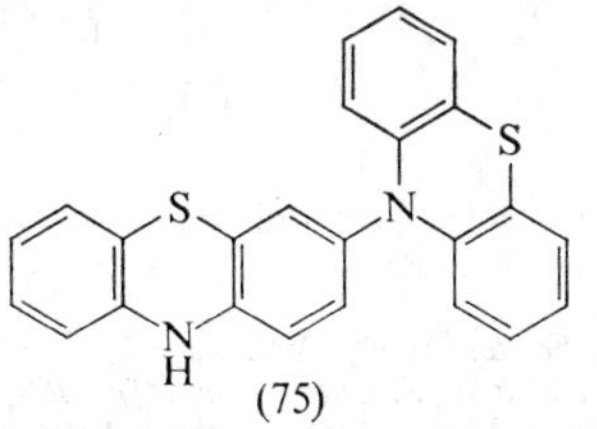

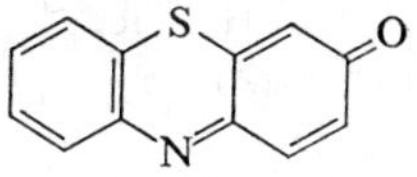

[27] A. J. Saggiomo, M. Asai, and P. M. Schwartz, *J. Heterocyclic Chem.*, 1969, **6**, 631.
[28] Y. Tsujino, *Tetrahedron Letters*, 1969, 763.

thiazine cation-radical is first formed in a rapid irreversible step and is subsequently brominated.[29]

12-Methylbenzo[a]phenothiazine is demethylated and desulphurised to benzo[a]carbazole by lithium in THF.[30] Benzo[a]phenothiazine reacts thermally or photochemically with heterocyclic azides to give azomethine dyes.[31] For example, 2-azidobenzothiazole yields compound (77). Triplet

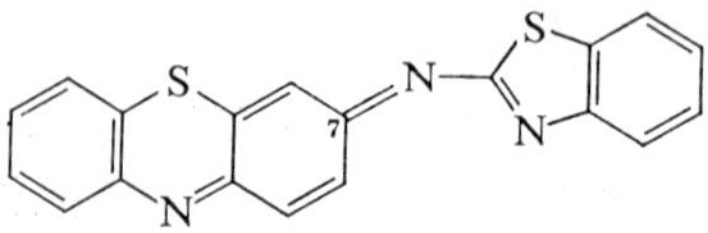

(77)

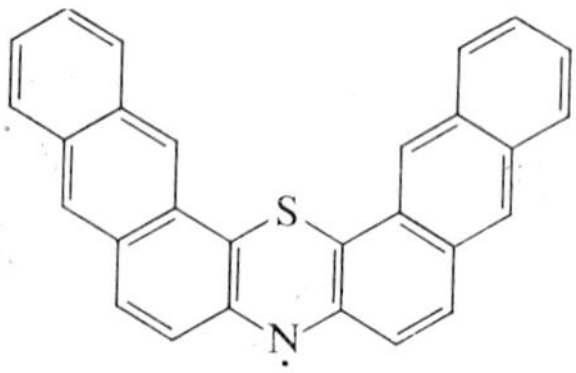

(78)

nitrenes are invoked, which attack the benzophenothiazine at position 7. The green radical (78), prepared by oxidation of the corresponding phenothiazine derivative with lead dioxide in boiling xylene, has been characterised by e.s.r. and i.r. spectroscopy and by M.W. determination.[32] A synthesis of 2H-pyrano[6,5-b]phenothiazines (79), members of a novel

(79)

heterocyclic system, is described.[33] The pyrimidobenzothiazines (80) in

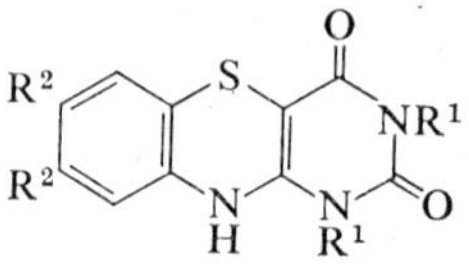

(80) 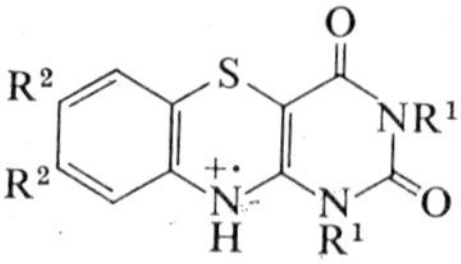  (81)

(80) and (81); $R^1 = R^2 = Me$; and
$R^1 = Me, R^2 = H$

[29] P. Alcais and M. Rau, Bull. Soc. chim. France, 1969, 3390.
[30] T. G. Jackson, S. R. Morris, and B. W. Martin, J. Chem. Soc. (C), 1969, 1728.
[31] J. A. Van Allen, G. A. Reynolds, and D. P. Maier, J. Org. Chem., 1969, 34, 1691.
[32] M. Zander and W. H. Franke, Tetrahedron Letters, 1969, 5107.
[33] N. P. Buu-Hoi, B. Lobert, and G. Saint-Ruf, J. Chem. Soc. (C), 1969, 2137.

concentrated sulphuric acid generate the corresponding cation-radicals (81). Formation of the neutral radical (82) is also described. These radicals were characterised by their e.s.r. spectra.[34] Synthesis of the compound (83) the first reported dithieno[2,3-*b* : 2′,3′-*e*]-1,4-thiazine, and the thieno-[3,2-*b*]benzo-1,4-thiazine (84) are reported.[35]

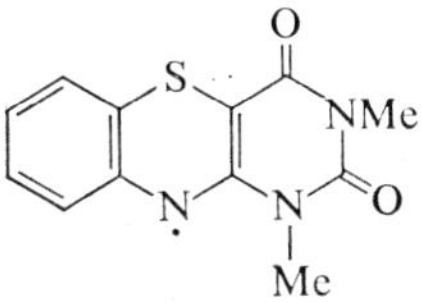

(82)

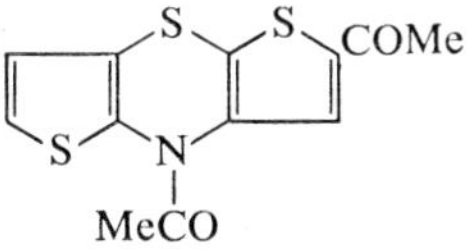

(83)

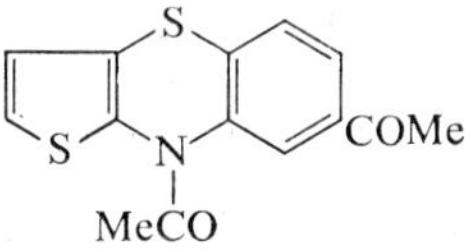

(84)

## 4 Phosphorus Analogues of Thiazines

Synthetic studies of dibenzo[*b*,*e*]thiaphosphorins have been described. Simultaneous reduction and cyclisation to compound (86) occur when the sulphone (85) is treated with phosphorus trichloride in the presence of

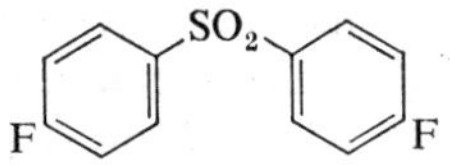

(85)

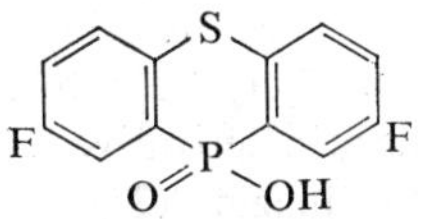

(86)

aluminium chloride.[36] 2,7-Disubstituted phenothiaphosphinic acids (87) are formed when 4,4′-disubstituted diaryl sulphides are treated with phosphorus trichloride in the presence of aluminium chloride.[37] The reaction of 4,4′-dimethyldiphenyl sulphide with methylphosphonous dichloride gave the intermediate (88; R = Me), which was not isolated but was oxidised with peracetic acid to the stable sulphone (89; R = Me). The more stable phenyl derivative (88; R = Ph) was isolated before being oxidised to the sulphone (89; R = Ph). Reaction of the ethyl ester of the phosphinic acid (87; $R^1 = R^2 = Me$) with $Me_2N \cdot (CH_2)_3MgCl$ in THF gave the phosphine oxide (90) which was reduced with trichlorosilane to the tertiary phosphine (91).

---

[34] H. Fenner, *Tetrahedron Letters*, 1970, 617.
[35] C. J. Gool and J. S. Faber, *Rec. Trav. chim.*, 1970, **89**, 68.
[36] I. Granoth and A. Kalir, *J. Chem. Soc.* (*C*), 1969, 2424.
[37] I. Granoth, A. Kalir, Z. Pelah, and E. D. Bergmann, *Tetrahedron*, 1969, **25**, 3919.

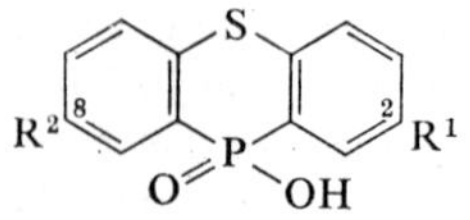

(87) — dibenzo system with P(=O)OH, substituents R² (position 8) and R¹ (position 2)

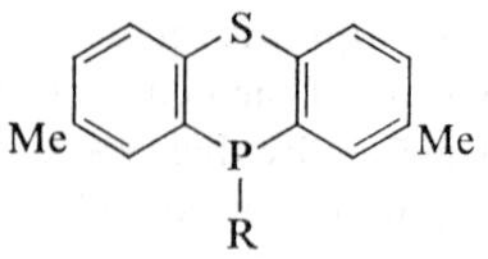

(88) — dibenzo-S,P system with Me, Me and P–R

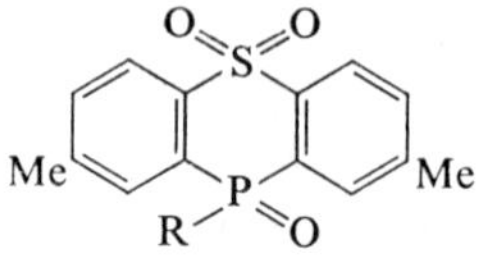

(89) — with SO₂ bridge, P(R)=O, Me, Me

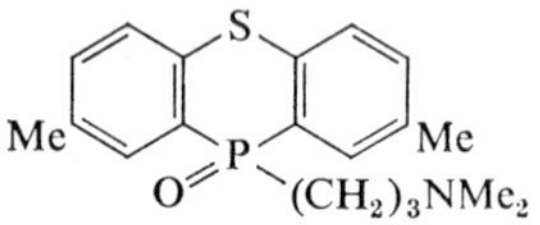

(90) — Me, Me, P(=O)(CH₂)₃NMe₂

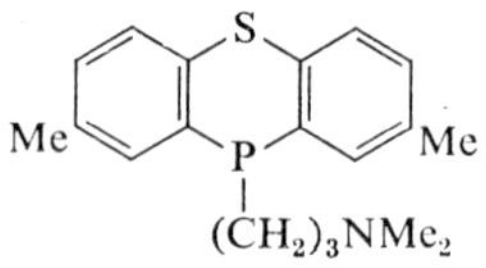

(91) — Me, Me, P–(CH₂)₃NMe₂

## 5  1,4-Thiazepines

Few, but nevertheless interesting results have been reported in this area. The isolation [9] of the thiazepine (24; $R^1 = CH_2OMe$, $R^2 = p\text{-}NO_2 \cdot C_6H_4 \cdot CH{=}N$) as an intermediate in the rearrangement of the penicillanic acid derivative (23; $R^1 = CH_2OMe$, $R^2 = p\text{-}NO_2C_6H_4 \cdot CH{=}N$) into the 1,4-thiazine (25), has been described in section 2. In an almost identical study,[38] the conversion of the derivative (23; $R^1 = Me$, $R^2 =$ phthalimido) into the thiazepine (24; $R^1 = Me$, $R^2 =$ phthalimido) is described. In this reaction thiazepine formation appears to be quantitative. Formation [14] of the benzo-1,4-thiazepines (44)—(47) in the rearrangement of the benzo-1,3-thiazine (38) in boiling acetic anhydride has also been described in section 2.

Studies [39] of the equilibrium in aqueous solution between the pseudo-equatorial and the pseudo-axial forms (these differ in the conformational arrangement of the sulphoxide group) of the dibenzo- 1,4-thiazepine (92) show that the former is 0·75 kcal mol⁻¹ more stable than the latter, with calculated free energies of activation of 24·6 and 23·9 kcal mol⁻¹.

It had previously been shown that thermal decomposition of aryl 2-azidophenyl sulphides [40] and phosphite deoxygenation of 2-nitrophenyl

---

[38] O. K. J. Kovacs, B. Ekström, and B. Sjöberg, *Tetrahedron Letters*, 1969, 1863.
[39] W. Michaelis and R. Gauch, *Helv. Chim. Acta*, 1969, **52**, 2486.
[40] J. I. G. Cadogan, S. Kulik, and M. J. Todd, *Chem. Comm.*, 1968, 736.

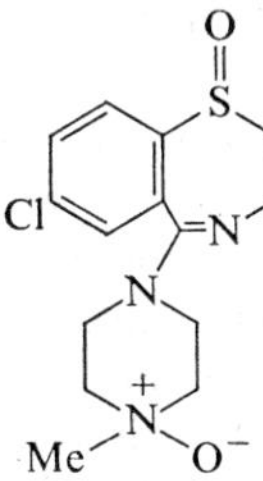

(92)

aryl sulphides [41] give rearranged phenothiazines (95), probably *via* intermediates of the type (94) derived from a nitrene (93). In an extension [42] of this work, thermal decomposition of 2-azidophenyl-2,6-dimethylphenyl sulphide (96; R = $N_3$), in which the essential *ortho* positions are blocked,

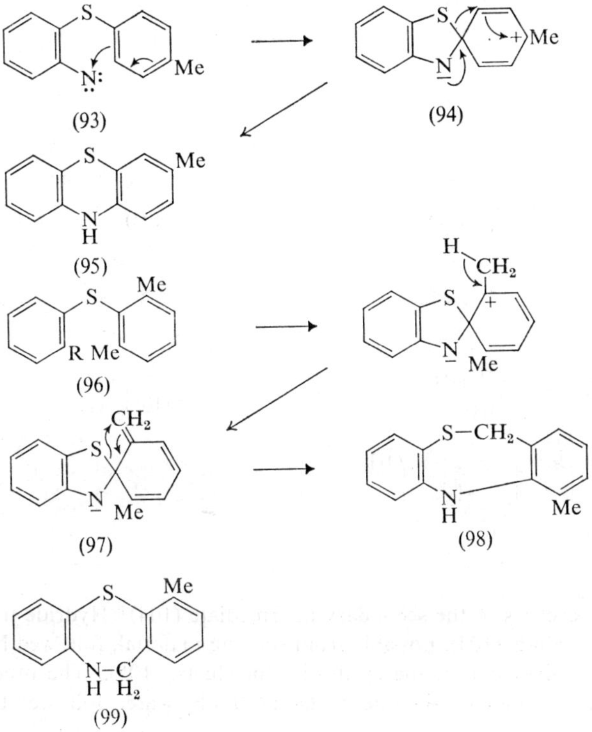

(93)  (94)

(95)

(96)

(97)

(98)

(99)

resulted in formation of the 5,11-dihydrodibenzo[b,e]-1,4-thiazepine ring system (98). It is suggested that the nitrene (96; R = $\ddot{N}$:), rather than insert into a methyl C–H bond to give the [b,f]thiazepine (99) (which was

[41] J. I. G. Cadogan, *Quart. Rev.*, 1968, **22**, 222.
[42] J. I. G. Cadogan and S. Kulik, *Chem. Comm.*, 1970, 233.

not detected), attacks the adjacent benzene ring at the nucleophilic 1-position to give the spiro-intermediate (97) which undergoes a sigmatropic rearrangement to give the observed product (98).

A kinetic study of the acid-catalysed rearrangement of the thiadiazepine (100) is reported.[43] Concurrent disproportionation and ring-contraction occurs. The rate of decomposition in acid solution is proportional to the product of $h_0$ and the concentration of the thiadiazepine. It is proposed that monoprotonation of compound (100) is followed by a rate-determining cleavage to give the primary intermediate (101). This either disproportionates to a mixture of the diamine (102) and the dehydrogenation product

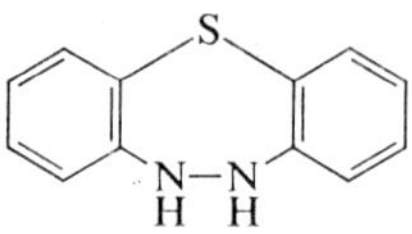

(100)

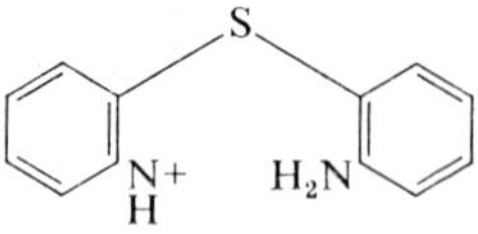

(101)

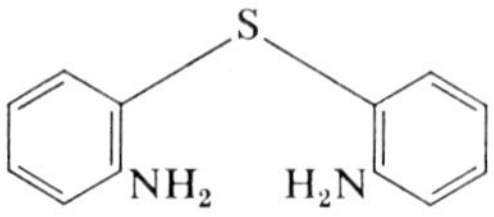

(102)

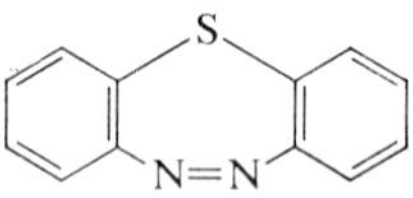

(103)

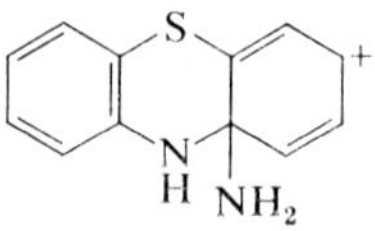

(104)

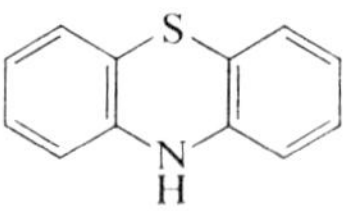

(105)

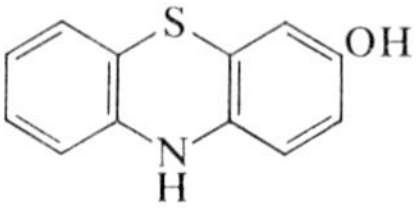

(106)

(103), or cyclises to the secondary intermediate (104). Hydride-transfer to the intermediate (104), possibly from starting material, followed by loss of ammonia, gives one of the cyclisation products, (105). The other, (106), results from attack on the intermediate (104) by water, followed by loss of ammonia.

[43] L. D. Hartung and H. J. Shine, *J. Org. Chem.*, 1969, **34**, 1013.

# 18
# Thiadiazines

A review of recent advances in the chemistry of 1,3,4-thiadiazines has appeared.[1]

Recent work on thiadiazines has been mainly synthetic. 2H-Benzo-[e][1,2,3]thiadiazine 1,1-dioxide (1), the parent compound of a heterocyclic system hitherto known only in the form of 4-hydrazino-derivatives, has been synthesised.[2] The hydrazone of o-formylbenzene sulphonic acid reacted with phosphorus pentachloride to give compound (1) in high yield. The 2H-isomer (1) is more stable than the 4H-isomer (2), not being isomerised by acid or base. Chlorination of the benzothiadiazine (1) results in the

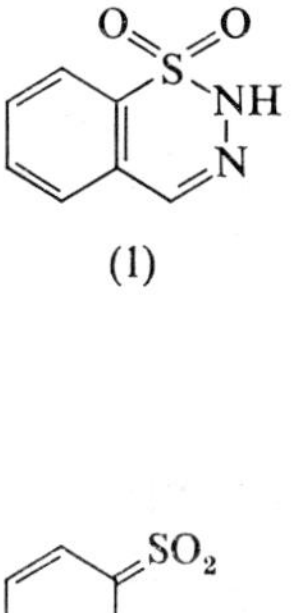

(1)

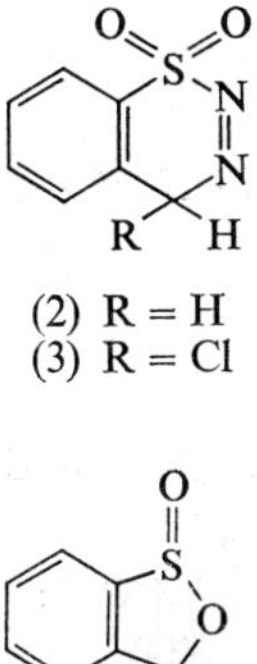

(2) R = H
(3) R = Cl

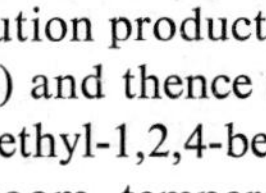

(4)

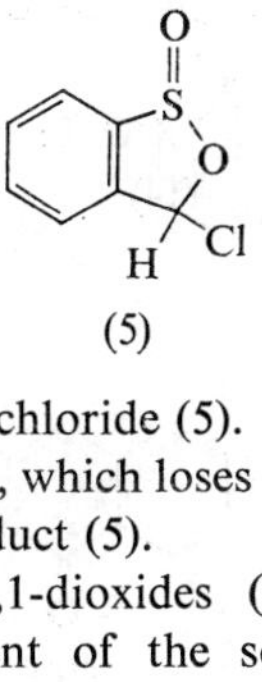

(5)

elimination of nitrogen and formation of the chloride (5). It is postulated that the substitution product (3) is first formed, which loses nitrogen to give the sulphene (4) and thence the observed product (5).

3-Dichloromethyl-1,2,4-benzothiadiazine 1,1-dioxides (7) are formed smoothly at room temperature by treatment of the sodium salts of o-aminobenzenesulphonamides with methyl dichloroacetate in methanol.[3] The intermediate sulphonamides (6), which may be isolated by working at low temperatures, cyclise spontaneously at room temperature. The use

[1] H. Beyer, Z. Chem., 1969, 9, 361.
[2] J. F. King, A. Hawson, D. M. Deaken, and J. Komery, Chem. Comm., 1969, 33.
[3] A. D. B. Sloan, Chem. and Ind., 1969, 1305.

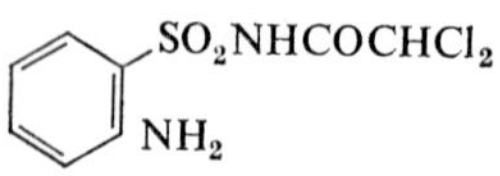

(6) C6H4(SO2NHCOCHCl2)(NH2)

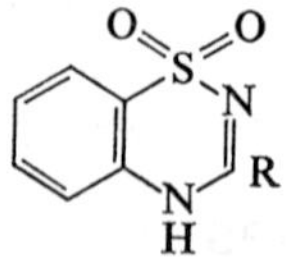

(7) R = CHCl2
(8) R = CCl3

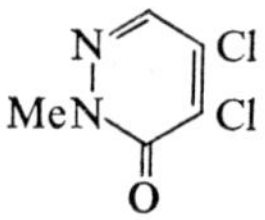

(9) 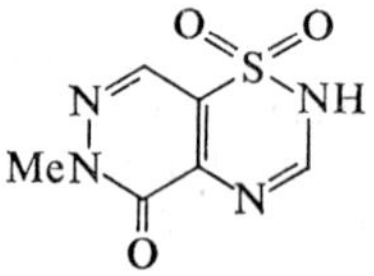
(10) 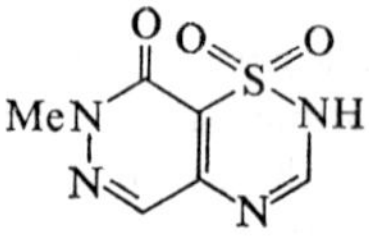
(11)

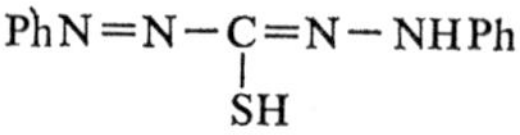

(12) PhN=N–C(SH)=N–NHPh

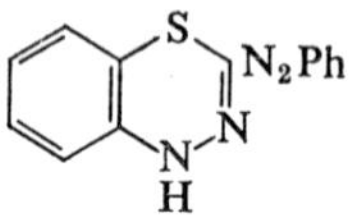

(13)

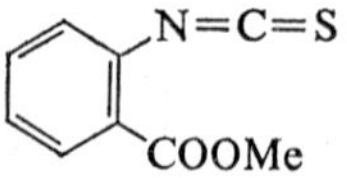

(14) C6H4(N=C=S)(COOMe)

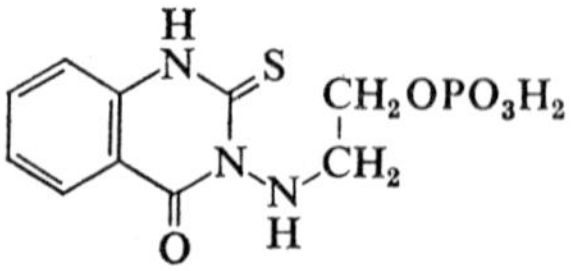

(15)

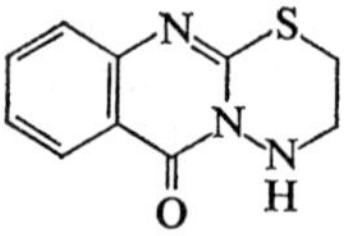

(16)

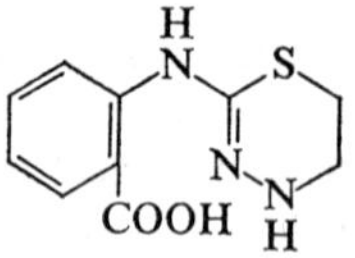

(17)

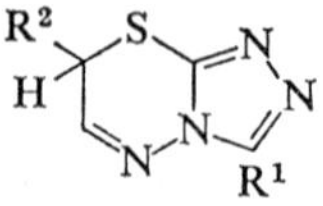

(18)

of methyl trichloroacetate leads to the 3-trichloromethyl derivatives (8), also in high yield.[4]

Syntheses of the pyridazino[4,5-e][1,2,4]thiadiazine 1,1-dioxides (10) and (11) by different routes from the same starting material (9) are described.[5]

Dithizone (12) undergoes ready oxidative cyclisation in boiling acetic acid to give the 1,3,4-thiadiazine (13), whose structure was established by X-ray crystallographic analysis.[6]

Sodium hydrazinoethyl phosphate reacts with the isothiocyanate (14) to give the quinazolinethione (15), which is cyclised by 0·5M hydrochloric acid to the dihydro-1,3,4-thiadiazine (17) *via* the intermediate tricyclic thiadiazine (16).[7] Triazolo[3,4-b][1,3,4]thiadiazines (18) are formed by the reaction of 3-mercapto-4-amino-1,2,4-triazoles with α-bromoketones.[8]

[4] A. J. Foubister and A. D. B. Sloan, *Chem. and Ind.*, 1970, 27.
[5] R. F. Meyer, *J. Heterocyclic Chem.*, 1969, **6**, 407.
[6] W. S. McDonald, H. M. N. H. Irving, G. Raper, and D. C. Rupainwar, *Chem. Comm.*, 1969, 392.
[7] J. Rabinowitz, S. Chang, and N. Capurro, *Helv. Chim. Acta*, 1969, **52**, 250.
[8] G. Westphal and P. Henklein, *Z. Chem.*, 1969, **9**, 111.

# Author Index

Aamisepp, M., 150
Abbott, D. J., 73
Abdel-Mageb, I. M., 226
Abdulvaleeva, F. A., 73
Abe, Y., 104
Abraham, D. J., 293
Abraham, W., 217, 449
Abramovitch, R. A., 56, 99, 203, 300, 455
Abu-Samra, A., 50
Acampora, M., 285
Acheson, R. M., 399
Adam, D. J., 463
Adamek, J. P., 100
Adams, G. E., 85, 105
Adolph, H., 293
Advani, B. G., 206
Aeberli, P., 430
Agami, C., 284
Agarwal, N. S., 220, 449
Agarwala, U., 244
Agawa, T., 306, 453
Agbalyan, S. G., 95
Ager, E., 56, 98
Ahlbrecht, H., 254
Ahmed, M., 337
Ahmed, R., 186
Ahlquist, D., 235
Aimar, N., 126, 136
Ajayi, S. O., 247
Akaboshi, S., 203, 245
Akano, M., 220
Akerkar, A. S., 370
Aki, O., 211
Akiba, K., 210
Aksnes, G., 257
Alabaster, R. J., 87
Albert, A., 207, 445
Albitskaya, V. M., 54
Albrecht, H. P., 83
Albrecht, R., 307
Albright, J. D., 275
Alcais, P., 468
Alden, C. K., 55
Aldrich, J. E., 85, 105
Alexandrou, N. E., 144, 365
Ali, Y., 83
Alkin, C. W., 246
Allen, E. A., 174
Allen, L. C., 2
Allen, R. E., 449
Allenmark, S., 65, 74, 75, 78
Almqvist, S.-O., 127, 150
Alper, A. E., 437
Alper, H., 60, 102, 152
Altman, T., 135

Altona, C., 134
Alves, A., 65
Alvsaker, J. O., 431
Amaral, M. J. S. A., 78
Amberger, E., 224
Amel, R. T., 254, 257
Amirthalingam, V., 408
Ammon, H. L., 138
Amosova, S. V., 56
Amrik Singh., 215
Andersen, K. K., 73, 157, 305
Anderson, G. W., 410
Anderson, P. J., 57
Anderson, R. C., 372
Ando, W., 66, 67, 250, 260
Andreades, S., 148, 224
Andreev, L. N., 183
Andrews, S. L., 153, 457
Andrieu, C., 188, 242
Anet, F. A. L., 359
Angier, R. B., 104, 177, 206
Angst, C., 103
Anisuzzman, A. K. M., 144
Ansell, G. B., 239
Anteunis, M., 106, 167
Anthoni, U., 156, 215, 216, 217, 220, 226, 229, 230, 237, 239, 244
Antonakis, K., 82
Antonova, A., 421
Aoki, O., 453
Appel, R., 300, 303, 304
Appriou, P., 336
Arai, I., 217
Araki Y., 226
Arbusov, B. A., 114, 162, 213
Archer, R. A., 155
Aren, A. K., 215
Arens, J. F., 61, 63, 65, 70
Argrett, L., 50
Aristov, L. I., 232
Ariyan, Z. S., 49
Armitage, D. A., 89, 185
Armstrong, D. R., 3
Arnold, K., 235
Aroia, S. S., 152
Arold, H., 351
Aroney, M. J., 106
Arshinova, R. P., 162
Arya, V. P., 209
Asaba, M., 220
Asai, M., 467
Asato, G., 387, 444
Ash, A. A., 82

Asinger, F., 179, 202, 399, 400, 440, 441
Askani, R., 101
Asovskaya, V. A., 55
Aspila, K. I., 230
Asquith, R. S., 105
Atavin, A. S., 56, 69
Atkins, G. M., 306
Atkins, R. C., 61
Auzzi, G., 223
Ayad, K. N., 420
Aytkhodzhaeva, M. Zh., 232
Azerbaev, I. N., 232
Azman, A., 242
Azogu, C. I., 300, 455
Azovskaya, V. A., 64

Babadjamian, A., 379, 385
Babichev, F. S., 202, 424
Babson, R. D., 233, 421, 436
Babushkina, T. A., 87
Babuyeva, T. M., 206
Badiger, V. V., 355
Baer, Y., 31
Baganz, H., 379
Bahr, G., 138, 235
Bailey, T., 425
Bain, P. J. S., 60, 191
Baisheva, A. U., 92
Baker, A. D., 31
Baker, J. A., 429
Baker, R., 79
Baker, W., 451
Baldwin, J. E., 61, 141, 197, 262, 264, 327
Baldwin, M. J., 88
Ball, D. H., 96
Ballard, R. E., 241
Baloniak, S., 429
Balsiger, R. W., 431
Balucani, D., 368
Bambas, L. L., 369, 444
Banerji, J. C., 413
Bank, K. C. 28, 167, 186
Banks, R. E., 56, 88
Bany, T., 449
Baranov, S. N., 176, 207
Baranovskaya, V. F., 99
Barba, N. A., 217
Barbee, R. B., 102, 141, 164
Barbieri, G., 72
Bardi, R., 323
Barefoot, R. D., 238
Bargigia, G. 145, 231

Barlin, G. B., 65
Barnard, D., 432
Barnes, A. N. M., 246
Barnes, W. T., 84
Barnett, R. E., 56
Barnikow, G., 204, 206, 217, 234, 403, 412, 449
Barnish, I. T., 454
Barraclough, G., 189
Barré, B., 387
Barrett, G. C., 52, 78, 206, 236, 242, 384, 385, 396, 398, 458
Barrick, P. L., 395
Barron, R., 197
Barry, W. J., 87
Barsegov, R. G., 59
Barton, D. H. R., 154
Barton, T. J., 87, 101
Bartoszewski, J., 215, 407
Bassery, L., 69
Bassett, R. N., 153
Basu, R., 41
Bateman, L., 432
Bates, R. D., 263
Battisti, A., 158
Bauer, L., 120
Baumenova, M., 219, 451, 452
Baumgarten, R. J., 100
Baxter, R. M., 340
Bayer, E., 411
Beak, P., 207, 349
Beare, S. D., 75
Bebesel, E., 386
Beck, A. K., 71
Beck, G., 85
Beck, J., 102, 453
Becke, F., 90, 203
Becker, H., 287
Becker, H. D., 81
Becker, R. F., 201, 205, 246
Beckhurts, H., 402
Bednyagina, N. P., 419
Bedos, B., 221
Beer, R. J. S., 321, 325, 326, 330, 340, 345
Beeson, J., 284
Beger, J., 206, 394, 395
Begland, R. W., 72, 124, 179, 201
Behn, N. S., 270
Behr, L. C., 378
Behringer, H., 334, 342, 384
Beilan, H. S., 299
Belaits, I. L., 242
Belaventsev, M. A., 133
Bellamy, L. J., 182
Bellare, R. A., 386
Bellas, M., 150, 407
Bellinger, N., 354
Belsky, I., 152, 220, 408
Bender, M. L., 95
Benedek-Vamos, N., 429
Benkö, A., 382
Bennett, O. F., 350
Bennett, P., 283
Bennett, R. H., 137

Bennur, S. C., 355
Benschop, H. P., 90
Bentley, H. R., 301
Bentz, F., 99
Berchtold, G. A., 54, 64, 126
Bergein, M. T., 236
Bergeron, R., 100
Bergmann, E. D., 469
Bergmann, F., 203
Bergmark, T., 31
Berkelhammer, G., 387, 444
Bernat, J., 177, 242
Bernhard, S. A., 54
Berse, C., 113
Beskov, S. D., 244
Besprozvannaya, M. M., 205
Bestmann, H. J., 84, 202
Betteridge, D., 31
Betts, M. J., 56
Beveridge, D. L., 2, 35
Beverly, G. M., 59
Bewley, T. A., 105
Beyer, H., 473
Beyl, V., 98
Bezmenova, T. E., 137
Bezzi, S., 321, 323
Bhandhari, K. S., 152
Bhargava, H. D., 209
Bhargava, P. N., 407
Bhatnagar, G. M., 105
Bhattacharya, S. N., 72
Bibler, J. P., 93
Bichiashvili, A. D., 59
Bielefeld, M. J., 9
Biffin, M. E. C., 80
Bignebat, J., 236, 238, 330, 341, 342
Billig, F., 54, 365
Billig, F. A., 49, 182
Binder, U., 449
Binenfeld, Z. J., 102
Binsch, G., 169
Birkett, D. J., 57
Bite, D. V., 215
Bixon, R., 96
Black, D. St. C., 179
Black, W. B., 417
Blackburn, G. M., 262, 263
Blanc-Guenee, J., 157, 287
Blandamer, M. J., 121
Bleisch, S., 138, 235
Bloch, J. C., 248
Block, E., 63, 64, 134, 159
Bloom, B. M., 354
Bloom, R. K., 100
Bly, R. K., 281
Bly, R. S., 270, 281
Boberg, F., 203, 338, 342, 364
Bobrov, A. V., 78
Boccu, E., 89
Bochmann, G., 72
Bode, K.-D., 183, 184
Boekelheide, U., 169
Boelens, H., 53
Boer, F. P., 2
Boerma, J. A., 177

Bogdanov, V. S., 236
Bogdanowicz, M. J., 225
Bogdonowicz-Szwed, K., 202
Bogemann, M., 183, 184
Boggio, R. J., 100
Bognar, R., 206
Bogolyubov, G. M., 102
Bohlmann, F., 79
Bohman, O., 65, 74, 75
Bohme, H., 260
Boiko, Y. A., 60
Boldyrev, B. G., 82, 89, 106
Bolluyt, S. C., 219
Bonaccorsi, R., 2
Bonet, G. B., 120
Bonthrone, W., 399
Bonnett, R., 83, 120
Boots, S. G., 280
Bordwell, F. G., 118
Borisevich, A. N., 205, 206
Borka, L., 432
Bos, H. J. T., 61, 65, 208, 346
Bosco, M., 414
Bossa, M., 41, 242
Boter, H. L., 90
Boudet, R., 460
Bourdon, R., 422
Bourgeois, P., 93
Bourgoin-Legay, D., 460
Boustany, K. S., 55
Bower, J. D., 179, 319
Bowie, J. H., 144, 245
Box, H. C., 215
Boyd, G. V., 350
Bracha, P., 87
Bradamante, S., 356
Bradley, C. H., 359
Bradley, D. C., 244
Bradsher, C. K., 424, 439
Brady, D. G., 201, 295
Braibanti, A., 240
Branch, R. F., 69
Brandsma, L., 53, 63, 65, 66, 70, 186, 208, 221, 223, 346
Braude, G. L., 300
Bravo, P., 54, 285, 286
Breitfeld, B., 214
Breitmaier, E., 411
Brelivet, J., 336
Breslow, D. S., 99
Breslow, R., 392
Breuer, E., 282
Brewar, A. D., 207
Briault, J., 69
Brice, M. D., 148
Bridgeford, D., 225
Brigardo, J., 244
Brimacombe, J. S., 82
Brinkoff, H. C., 239
Brisson, H., 284
Brook, A. G., 169
Brouwer, D. M., 94
Brown, B. R., 56
Brown, B. T., 238
Brown, C., 88, 90, 120
Brown, D. J., 80

# Author Index

Brown, E. I. G., 328, 337
Brown, F., 378
Brown, H. C., 55
Brown, K., 206
Brown, K. S., 96
Brown, M. A., 359
Brown, M. T., 59
Brown, P. O., 140
Brown, P. R., 190
Brown, R. D., 33
Brown, R. K., 88
Brown, W. V., 65
Browning, J., 190
Brownlee, P. J. E., 58
Brundle, C. R., 15
Bubnovskaya, V. N., 424
Bucher, D., 292
Buchholt, H. C., 102
Buchler, G., 300
Buchshriber, J. M., 337
Budde, W. L., 215
Budeshinski, Z., 209
Budzinski, E. E., 215
Bulat, A. D., 72, 78
Bunnett, J. F., 56
Buntin, G. A., 209
Burakevich, J. V., 205
Burden, M. G., 276
Burgess, E. M., 306
Burgot, J. L., 338
Burke, N., 225
Burnett, W. T., jun., 405
Burns, G., 34
Burzlaff, H., 349
Busch, D. H., 179
Busch, M., 451
Bushweller, C. H., 172, 173
Butler, E., 171
Butler, P. E., 62
Buttimore, D., 370
Buu-Hoï, N. P., 352, 355, 368, 468
Buys, H. R., 134, 163, 174
Bycroft, B. W., 67, 208
Bystritskii, G. I., 242

Cadogan, J. I. G., 68, 470, 471
Cagniant, P. 354
Cahill, P. J., 158
Cain, E. N., 455
Cairns, T. L., 395
Calas, R., 93
Caldwell, S. M., 153
Callahan, F. M., 410
Callear, A. B., 185
Calo, V., 88
Cambi, L., 145, 231
Cameron, D. P., 445
Campaigne, E., 172, 182, 183
Campbell, B. J., 105
Campbell, C. D., 102, 367, 454
Campbell, M. J., 247
Campbell, M. J. M., 418
Campbell, R. W., 201, 291, 295
Campbell, T. W., 184
Cao, N. H., 72

Capell, L. T., 424
Caplin, G. A., 283
Cappozzi, G., 122
Capuano, L., 210
Capurro, N., 475
Card, D. W., 418
Carleton, P. S., 81
Carlson, R. G., 270
Carlsson, L. O., 215
Carmack, M., 155
Carr, D., 301
Carroll, N. M., 289
Cartwright, D., 321, 326
Casanova, J., 254, 265
Casaux, M., 175
Caserio, M. C., 62, 255, 265
Casimir, J., 391
Castro, C. E., 62
Casy, A. F., 69
Cato, V., 122
Caton, M. P. L., 370
Cava, M. P., 47, 314, 318, 319
Cavalchi, B., 64
Ceccon, A., 85, 105, 189
Cecil, R., 57
Cerfontain, H., 49, 94
Chabrier, P., 183
Chadha, V. K., 152, 425
Chadpa, M. S., 175
Chakrabarti, C. L., 230
Chalkley, G. R., 79
Chambers, C. H., 386, 413
Chan, A. W. K., 197, 372
Chan, N. N. Y., 417
Chande, M. S., 229
Chandra, G., 80
Chandramohan, M. R., 152
Chang, P. L. F., 122, 200
Chang, S., 217, 403, 475
Chanon, M., 379
Chapman, J. R., 52, 206, 384, 385, 458
Chary, C. A. I., 243
Chatfield, P. V., 429
Chaude, O., 416
Chaudhary, H. S., 152, 407, 441
Chaudhuri, A., 87
Chaurasia, M. R., 407
Chauvin, J., 202
Chaykovsky, M., 279
Cheeseman, G. W. H., 94
Chen, C. T., 82
Chen, J.-Y., 197
Cheney, S. C., 191, 369
Cheng, J. C., 154
Cheong, K.-K., 242
Cherbuliez, E., 403
Cheung, K. K., 139
Chevrechenko, T. M., 209
Chiang, Y. H., 91, 146
Chiesi, A., 234
Chin, C., 393
Chinone, A., 206
Chio, H., 106
Chmutova, G. A., 63
Chollar, B. H., 158

Chong Hoe Pak., 241
Chopard, P. O., 268
Chosho, H., 393
Choudhury, C. B., 41
Chow, L. C., 62, 265
Christ, B. J., 65, 172
Christ, D. R., 120
Christensen, A. T., 252
Christensen, B. W., 76, 302, 305
Christiaens, L., 84
Chuchani, G., 56
Ciganek, E., 111, 145, 191
Cignitti, M., 41
Cimino, G., 75, 419
Cinquini, M., 72, 77, 79
Cirdan, I., 389
Ciuffarin, E., 50, 89
Ciurdaru, G., 413
Claeson, G., 135
Claisse, J. A., 55
Clark, D. T., 3, 9, 33, 36, 41
Clark, M. J., 89, 185
Clark, R. N., 232
Clarke, H. T., 378
Clarkson, R. B., 87
Clauss, K., 101
Clayton, J. P., 153
Clement, W. H., 82
Clementi, E., 2, 15
Clesse, F., 188, 220, 236, 238, 328, 341, 353
Clezy, P. S., 189
Clifford, A. F., 299
Closson, W. D., 98
Coates, R. M., 146
Coffen, D. L., 28, 71, 141, 143, 167, 186, 249
Cogliano, J. A., 300
Cohen, E., 135
Cohen, S. G., 63
Cohn, W. E., 86, 204
Colaitis, D., 244
Colin, J. E., 172
Collins, C. J., 86
Collis, M. H., 405
Colombo, L., 145, 231
Colonna, S., 72, 77, 79
Coman, M. M., 380
Comer, F., 154
Conde-Caprace, G., 172
Connor, J., 185
Connor, R., 49
Cook, A. F., 251
Cook, D. B., 37
Cooks, R. G., 245, 401
Cooper, R. D. G., 68, 153, 154
Cope, A. C., 59
Coppinger, G. M., 184
Corbin, U., 191, 369
Corey, E. J., 61, 63, 64, 65, 159, 250, 269, 279, 287, 288
Correa, W. S., 103
Corson, F. P., 85
Cory, M., 278
Cosani, A., 95
Cotton, F. A., 100

Coulibaly, O., 341
Coulson, C. A., 3
Couture, A., 186
Cox, J. M., 85
Cox, M. E., 58
Coxon, J. M., 148, 149
Crabbé, P., 85
Cragg, R. H., 156
Cram, D. J., 81, 305
Cramer, F., 242
Crampton, M. R., 56
Crandall, R. K., 225
Crewther, W. G., 105
Cristophersen, C., 216, 220
Crochet, K. L., 127, 187
Croisy, A., 355
Crooks, S. C., 118
Crow, W. D., 197, 372, 373, 374
Crowell, T. I., 83
Crozet, M. P., 126, 136, 405
Cruickshank, D. W. J., 1, 33
Csizmadia, I. G., 2, 3
Csontos, S., 389
Cundall, R. B., 85, 105
Cundy, C. S., 190
Cuneen, J. I., 432
Curci, R., 71
Curtis, R. F., 139
Cusachs, L. C., 33
Czaja, R. F., 52
Czavassy, G., 386

Dagga, F. A., 179, 202
Dahm, J., 220
Daicoviciu, C., 449
Dahl, B. M., 229
Dahl, O., 217, 230
Dallas, G., 84, 400
Dalziei, J. A. W., 170
Damanski, A. F., 102
Damerau, W., 79
Danby, G. J., 15
Daneke, J., 52, 211
D'Angeli, F., 209
Dangieri, T. J., 82
Danilova, T. A., 136
Danstead, E., 148
Danyushevskii, Ya. L., 236
Darling, T. R., 148
Darwish, D., 253
Das, B. C., 215
Daunis, J., 219
Davenport, D. A., 80
Davenport, R. W., 305
Davidson, J. S., 215
Davies, D. I., 55
Davis, F. A., 466
Davis, M., 198
Davoli, V., 72
Day, J., 305
Deaken, D. M., 201, 473
Dean, F. M., 188
De Angelis, G. C., 273
De Boer, A., 52
De Christopher, P. J., 100
Dedichen, J., 431
De Filippo, D., 92, 399

Degener, E., 85
De Graw, J. I., 278
De Groot, A. E., 177
Deguay, G., 222
Dehler, J., 241, 466
De La Camp, U., 73
de la Mater, G., 183
Delavarenne, S. Y., 62
Deliwala, C. V., 386
Delzenne, G. A., 445
De Marco, P. V., 154, 155, 244
De Marinis, R. M., 54
de Mayo, P., 133, 199, 292
Dembech, P., 92
Denes, V. I., 413
Denis-Garez, C., 223, 459
Dennilauler, R., 415
Denny, G. H., jun., 233, 421, 436
De Prett, R., 71
Derkach, G. I., 234, 422
Desai, R. B., 452
Desai, R. D., 386, 413
de Savignac, A., 210
Deshpande, P. H., 206, 218
de Souza Gomez, A., 404, 426
Dessauer, R., 416
Desseyn, H. O., 244
de Stefano, S., 419
Deutsch, A. S., 397
Devitt, F. H., 268
Devlin, J. F., 226
Dewar, M. J. S., 2, 41, 358
De Winter, W. F., 417
de Wolf, N., 174
Dhami, K. S., 152
Diaz, A., 96
Di Bello, C., 209
Dickinson, P. H., 215
Dickoré, K., 370
Dickson, D. R., 185
Diebert, C., 115
Diefenbach, H., 82, 137, 251
Diekmann, J., 250, 296
Dietz, G., 430
Dillon, P. J., 169
Dines, M. B., 88, 91, 149, 156
Dingwall, J. G., 53, 185, 324, 328, 334, 342, 351
Distler, H., 148
Dittli, C., 207
Dittmer, D. C., 84, 115, 122, 133, 186, 195, 200
Dively, W. R., 209
Divorne, C., 151, 399
Dmitriev, B. A., 82
Doane, W. M., 86, 143, 227
Dobosh, P. A., 2, 35
Doddrell, D., 244
Dodson, R. M., 158, 359, 426
Doering, W. von E., 82
Doherty, D. G., 86, 204, 405
Dolfini, J. E., 456

Dombrovskii, A. V., 202
Dominiano, P., 240
Donaldson. J. D., 172
Donnelly, J. A., 283, 284
Donohue, J., 240
Dou, H. J. M., 206, 387
Doucet, J., 81
Douglas, A. W., 2
Douglass, I. B., 87
Doyle, F. P., 378
Drawert, F., 398, 430
Drefahl, G., 269
Dremetskaya, L. D., 244
Dresdner, R. D., 50
Dronov, V. I., 92
Drott, H. R., 103
Drozd, V. N., 79
Dryhurst, G., 91
Dubenko, R. G., 202, 205
Dubini Paglia, E., 145, 231
DuBose, C. M., 270
Dudeck, C., 400
Duff, E. J., 417, 418
Duguay, G., 353
Dumaitre, B., 86, 98
du Manoir, J., 146, 289
Dumas, J., 246
Dunbar, J. E., 89, 229
Dunina, V. V., 229
Dunn, A. R., 463, 464, 465
Dunn, M. F., 54
Dupuy, C., 126, 136, 405
Durand, H., 236
Durgaprasad, G., 243
Durig, J. R., 128
Durst, A., 81
Durst, T., 146, 287, 288, 289, 290, 291, 295
Duus, F., 138, 188, 235, 245
Dworak, G., 150
Dyatkin, B. L., 209
Dyer, E., 207, 216
Dykman, E., 87
Dyson, N. H., 283
Dzantiev, B. G., 54

Eaborn, C., 72
Eagles, P. A. M., 57
Earley, J. V., 203
Eberhardt, M. K., 56
Ebersberg, J., 162
Ebner, W., 210
Eccleston, G., 175
Eckroth, D. R., 123, 201
Edelson, S. S., 148
Edgerley, P. S., 246
Edmondson, R. C., 93
Edward, J. T., 58, 183, 237, 242
Edwards, F. B., 105
Edwards, J. A., 283
Edwards, J. D., 71
Edwards, J. O., 140, 190
Effenberger, F., 101, 308
Egan, R. S., 379
Egginton, C. D., 99
Ehlers, A., 207
Eibisch, H., 102, 155, 453
Eilbeck, W. J., 391

## Author Index

Eilingsfeld, H., 202
Eisele, B., 57
Ekström, B., 152, 470
Eland, J. H. D., 15, 31
El-Faghi, M. S., 97, 201
Elguero, J., 207
Eliel, E. L., 134, 164
El-Kholy, I. E.-S., 162, 351
Elliott, D. L., 95
Elofson, R. M., 245
Eloy, F., 295
Elser, W., 69
Elslager, E. F., 389, 443, 444
El'tsov, A. V., 203, 205, 233, 413
Emmert, D. E., 404
Enemark, J., 176
Engelhardt, J. E., 59
England, D. C., 378
Erastov, O. A., 162, 170
Esayan, G. T., 95
Espejo, O. 403
Etienne, A., 96, 98
Etlis, V. S., 209
Euchner, H., 251
Evans, E., 96
Exelby, R., 416
Exner, O., 92
Eyring, H., 242

Faber, J. S., 469
Fabian, J., 41, 182, 183, 240, 241, 318, 340, 352, 357
Fackler, J. P., jun., 226
Fahlberg, C., 446
Fahlman, A., 31, 33
Faigle, J. W., 383
Fairweather, R., 85
Faisst, W., 166
Falkenberg, J., 334, 342
Fanta, P. E., 397
Farissey, W. J., 81
Farkas, I., 206
Farkas, M., 380
Farlow, M. W., 384
Fattorusso, E., 419
Faulstich, H., 89
Faust, G., 430
Faust, J., 340, 352
Fava, A., 50, 85, 188
Fava Gasparri, G., 240
Favstritsky, N. A., 107
Fayter, R. G., 85
Fedin, I. M., 209
Feit, P. W., 111
Feix, G., 431
Feld, D., 263
Felhaber, H. W., 303
Fenner, H., 469
Fenselau, A. H., 275
Fenselau, C., 144
Ferguson, G., 139
Fernandez, B. S., 386
Fernando, Q., 240, 411
Ferrier, R. J., 76
Feuer, H., 289
Fickentscher, K., 366

Fiedler, W., 430
Field, L., 56, 102, 141, 150, 164, 407
Fieldhouse, J. W., 201, 295
Fields, E. K., 336
Fife, T. U., 147
Filira, F., 209
Filleux-Blanchard, M. L., 236, 238
Finegold, H., 148
Fink, S. C., 97, 201
Finkelstein, N. P., 227
Finley, K. T., 225
Firestone, R. A., 63
Fischer, E., 416
Fischer, J. G., 299
Fischer, K., 294
Fish, R. H., 62, 255, 265
Fitton, A. O., 228
Fitts, D. D., 9
Fleming, I., 60
Fleur, A. H. M., 172
Flippen, J. L., 240, 427
Fliszar, S., 256
Flock, F. H., 369
Flossmann, K. D., 453
Flygare, W. H., 109
Fogleman, W. W., 33
Folli, U., 75
Follweiler, D. H., 311
Folting, K., 240
Fontana, A., 89, 210
Ford, P. W., 80
Forkey, D. M., 239
Forlani, L., 414
Forrester, A. R., 50
Forshult, S., 79, 138
Forward, G. C., 51, 145
Foubister, A. J., 475
Foulkes, D. M., 392
Fourquet, J. L., 228
Foxton, M. W., 399
Franck, R. W., 101
Frank, J. A. K., 189
Franke, W. H., 468
Franzen, G. R., 169
Fravolini, A., 436
Freedman, R. B., 95
Freeman, H. C., 339
Freeman, J. P., 123, 199, 200
Freidlina, R. K., 56, 211
Freiser, H., 411
Fremuth, W. J., 408
Frenkel, Z., 386
Frerichs, G., 402
Frese, E., 108, 110, 187, 227, 245
Freund, H. G., 215
Frew, D., 325, 345
Freymanis, Ya. A., 236
Friebolin, H., 166
Fried, J. H., 283
Friediger, A., 451
Friedman, A., 215
Fritsch, K. G., 422
Fritz, K., 241
Fröhlich, J., 466
Fromm, E., 51
Frydman, N., 96

Fryer, R. I., 203
Fu, Y.-C., 242
Fuganti, C., 76
Fujino, Y., 108, 141
Fujisawa, T., 89
Fujita, E., 224
Fujita, S., 82
Fujita, T., 224
Fukuyama, M., 113
Fuller, G. C., 57
Furukawa, M., 108, 141
Furukawa, N., 76, 301
Fusco, R., 301

Gabe, E. J., 240
Gabrio, T., 234
Gadallah, F. F., 245
Gadallah, L. A., 245
Gagiu, F., 386, 449
Gagnaire, D., 81
Gait, R. J., 326, 330, 340
Galama, P., 92
Galante, J. J., 100
Galiazzo, G., 72, 93
Galoyan, G. A., 95
Galstukhova, N. B., 209
Galysh, F. T., 379
Gambaryan, N. P., 201
Ganter, C., 159
Garbuglio, G., 321
Garcia, E. E., 446
Garmaise, D. L., 379, 386, 413
Garrett, P. E., 143, 167, 186
Gašpert, B., 216
Gassman, P. G., 287
Gatica, H. S. E., 152, 215
Gattow, G., 126, 152, 225, 403
Gauch, R., 470
Gaudiano, G., 54, 285, 286
Gaul, R. J., 408
Gaunce, A. P., 379
Gauthier, R., 113
Gautier, J. A., 157
Gauvin, P., 350
Gazda, E. S., 246
Geens, A., 106, 167
Geiger, B., 223, 344
Gelan, J., 167
Gelius, U., 3, 8, 31, 33
Gelting, N. C., 230
Georgadis, M. P., 172
George, J. W., 86
George, T. A., 80
George, T. J., 208
Georgiadis, K., 226
Gerhard, S., 137
Gerlach, D. H., 189
Germain, G., 139
Gertsema, A., 301
Geske, D. H., 71
Gewald, K., 183
Ghate, S. P., 209
Ghersetti, S., 198
Ghiringhelli, D., 76
Ghosh, R., 85

Gibbs, G. J., 219
Gibian, M. J., 95, 102, 139
Gibson, R. E., 138
Gierer, J., 94
Gilbert, E. E., 50, 207
Gilman, R. E., 225
Gilmore, W. F., 232
Ginak, A. I., 408
Ginsberg, A. P., 197
Ginsburg, D., 135
Giovini, R., 72
Giri, S., 386
Girshovich, M. Z., 233, 413
Gisler, H. J., 144
Gitlitz, M. H., 244
Giudicelli, J. E., 218
Givens, E. N., 86
Gleason, J. G., 61, 126, 145
Gleiter, R., 33, 326
Glemser, O., 299
Glidewell, C., 229
Gliniecki, F., 228
Gloede, J., 69
Gloss, R. A., 300
Glue, S., 88
Gmiro, V. E., 105
Gnad, J., 90
Goddard, D. R., 247
Goel, R. G., 299
Goerdeler, J., 84, 209, 370, 403, 444, 448
Goethals, E. J., 131, 141
Goggin, C. B., 97
Goldfarb, Y. L., 64, 83
Goldfart, Ya. A., 236
Goldman, I. M., 198
Goldman, L., 275
Golini, J., 173
Golinkin, H. S., 121
Gollnick, K., 71, 109
Gombler, W., 87
Gomes, A. de S., 120, 151
Gompper, R., 69, 251, 343, 370
Gomwalk, U. D., 214
Goodchild, J., 188
Goodgame, M., 417
Goodwin, H. A., 391
Gool, C. J., 469
Goralski, C. T., 271
Gorbalenko, V. I., 215
Gorbenko, E. F., 205
Gordon, W., 51
Gorokhovskaya, V. I., 246
Gorushkina, G. I., 83
Gosney, I., 197, 373, 374
Gosselck, J., 137, 254, 368, 272
Gouesnard, J. P., 136
Gougoutas, J. Z., 370
Goursot, P., 386, 414
Govindachari, T. R., 382
Graham, D. M., 54
Graham, P. J., 395
Grambal, F., 237
Gramstad, T., 241
Grand, A. F., 241
Grandberg, I. I., 207
Grandolini, G., 436
Granoth, I. 60, 469

Grant, D. W., 103
Grashey, R., 219, 451, 452
Grassetti, D. R., 58, 204
Grayshan, R., 169
Graziani, M., 93
Greci, L., 391, 441
Green, D., 101
Green, M., 190, 196
Green, M. J., 76
Gregorowicz, S., 451
Greidamus, J. W., 188
Greth, W. E., 207
Griffin, G. W., 149
Griffith, R. J., 225
Grigg, R., 365
Grindley, T. B., 180
Grin'stein, V. Y., 209
Grinter, R., 416
Grobov, L. N., 209
Groen, S. H., 261
Groeneveld, W. L., 172
Grohe, K., 205
Gronebaum, J., 304
Gronow, B., 58
Gross, H., 69
Grossert, J. S., 201
Grossoni, G., 234
Grotens, A. M., 239
Grube, H., 447
Gruber, R., 140, 186
Grunwell, J. R., 103
Grzeskowiak, R., 418
Guaraldi, G., 89, 107
Guémas, J. P., 222, 339, 353
Guenec, A., 222, 353
Günther, W., 200, 395
Guerra, A. M., 437
Guglielmetti, R., 416
Gunther, D., 101, 150
Gupta, P. C., 183
Gupta, P. L., 50
Gupta, S. K., 82
Gusarov, A. V., 56
Guseva, F. F., 114
Gustafsson, R., 88, 120
Gutfreund, H., 57
Guttenplan, J., 63
Guy, R. G., 83, 120
Gvozdeva, E. A., 64
Györfi, Z., 449


Haake, M., 277, 301
Haas, F., 51
Haas, W., 184
Haase, L., 69
Habermalz, U., 466
Haberstroh, H., 361
Hackler, R. E., 61
Hackley, B. E., 82
Hadjimihalakis, P. M., 144
Hadzi, D., 242
Haegele, W., 370
Haffer, G., 79
Haffner, H. E., 100
Hagen, H., 203
Hagitani, A., 217
Hahnkamm, V., 152, 403

Haines, A. H., 242
Haley, N. F., 443
Hall, R. J., 50
Hallam, H. E., 243
Haller, R., 162
Halls, D. J., 184
Halpern, B., 85
Ham, G. E., 62
Hamada, A., 98
Hamblin, D. C., 175
Hambly, A. N., 65, 172
Hamburg, E., 389
Hamel, C. R., 354
Hamer, J., 182
Hamid, A. M., 297
Hammond, W. B., 148
Hampel, M., 200
Hampel, W., 215, 219, 382
Hamrin, K., 31, 33
Handford, B. O., 58
Hankovsky, O. H., 216
Hannig, E., 386
Hansel, W., 369
Hansen, E., 197
Hansen, F., 240
Hanssgen, D., 303
Hantzsch, A., 379
Hardy, F. E., 71, 113
Hardy, P. M., 78
Hardy, W. R., 95
Harell, D., 58
Harjit Singh, 209
Harmon, R. E., 82
Harnish, D. P., 183, 184
Harpp, D. N., 61, 126, 145
Harrington, D. E., 225
Harris, B. R., 95
Harris, R. E., jun., 207
Harris, R. L. N., 52, 211
Hartke, K., 210, 370
Hartmann, A. G., 287
Hartmann, H., 41, 221, 258
Hartshorn, M. P., 148, 149
Hartung, L. D., 472
Harvey, A. B., 128
Haszeldine, R. N., 56, 88
Hatch, M. J., 268, 269
Hauck, F. P., jun., 400
Haugli, F. B., 431
Haugwitz, R. D., 210
Hauser, C. R., 289, 454
Hauser, R. W., 199
Hauser, T. R., 422
Haustein, W., 92
Havinga, E., 134
Havlin, R., 62
Hawley, D. M., 139
Hawson, A., 201, 473
Hay, J. M., 50
Hayashi, S., 108, 141
Hayashi, Y., 71, 261, 274
Hayasi, Y., 259, 268
Hayasi, T., 273
Hayatsu, H., 94, 207
Hayes, J. M., 403
Hayes, R., 365
Hazell, R. G., 240
Hazen, J. R., 96, 162
Heap, J. M., 51, 136
Heasley, L., 107

## Author Index

Heath, G. A., 339
Heckler, R. E., 262, 264
Heckmann, K. S., 56
Hedblom, M. O., 142
Heden, P. F., 31
Hedgley, E. J., 59, 223
Hedman, J., 31
Hegarty, B., 93
Heidberg, J., 221
Heine, H. G., 103, 228
Heinze, H., 369
Heitzer, H., 205
Hejsek, M., 237
Helferich, B., 176
Hell, P. M., 89, 91, 205
Hellstrom, N., 150
Helmer, F., 56
Helmholtz, L., 35
Helmkamp, G. K., 33, 95, 121
Hemmerich, P., 339
Hendrickson, J. B., 100
Henegaru, O., 449
Hengstberger, H., 383
Hengstenberg, W., 59
Henian, R. S., 133
Henion, J. D., 68, 225
Henklein, P., 475
Henriksen, L., 139, 223, 336
Henry, D. W., 386
Henson, P. D., 76
Hepworth, W., 386
Herbert, R. B., 95
Herbrandson, H. F., 178
Hergenrother, P. M., 422
Herman, M. A., 244
Hermann, H. J. A., 138
Hershenson, F. H., 120
Herzog, K., 239
Hess, H. J., 273
Hetzel, F. H., 231
Heukelbach, E., 300
Hewson, D., 350
Hideg, K., 216
Hieber, W., 418
Higgins, W., 76, 93
Highman, F. E., 232
Hilgetag, G., 100, 453
Hill, A. W., 188
Hiller, S. A., 215
Hillier, I. H., 3, 8
Hino, T., 203, 245
Hirai, K., 84, 132, 150, 200, 229, 232, 343, 379
Hirao, I., 218
Hiraoka, H., 185
Hiraoka, T., 279
Hirst, L., 105
Hiskey, R. G., 58, 59, 103
Hitch, M. J., 170, 172
Hiyama, T., 82
Ho, A. W. W., 146
Ho, K. C., 56
Hochrainer, A., 248, 251, 256
Hodgkins, J. E., 365
Höbold, W., 200
Hoefnagels, J., 416
Hönigschmid-Grossich, R., 224

Hoerhold, H. H., 102, 155, 453
Hoff, S., 63
Hofferbert, W. L., jun., 417
Hoffmann, J. M., jun., 191
Hoffmann, J. M., 149, 221, 316
Hoffmann, R., 2, 33, 326
Hofmann, A. W., 424
Hofmann, H., 361, 369
Hofmann, J. F., 357
Hoganson, E. D., 97, 201
Hogenboom, B. E., 201
Hogg, D. R., 86, 88, 120
Hojo, M., 97, 143
Holcomb, W. D., 99
Holland, R. J., 255
Holliman, F. G., 95
Hollis, P. C., 37
Holm, A., 226
Holm, R. H., 189
Holmes, F., 391
Holorik, S., 60
Holt, B., 278, 280, 312
Holtschmidt, H., 85
Holzbacher, Z., 243
Homer, G. D., 73
Honda, K., 307
Honwad, V. K., 62
Honzl, J., 88
Hoogenboom, B. E., 97
Hooper, T. R., 50
Hope, D. B., 91
Hope, J., 73, 240
Hoppen, H. D., 304
Horak, V., 258
Hordvik, A., 321, 323, 324, 339, 340
Hori, M., 309
Horii, T., 104
Horiuchi, M., 429
Horner, L., 91, 445
Hornyak, G., 216
Hortmann, A. G., 312
Horton, D., 69
Hosoi, K., 61
Hottschmidt, H., 205
Houghton, E., 392, 393
Houlihan, W. J., 430
House, H. O., 280
Houser, R. W., 123, 159
Howard, J., 278, 312
Howe, R. K., 219
Hoyer, G. A., 56
Hruby, V. J., 259
Huang, P. C., 355
Hubble, W., 101
Hubenett, F., 369
Hudson, R. F., 256
Hübner, H., 200
Huegli, F., 103
Hühnerfüss, H., 238
Hünig, S., 422
Huennekens, F. M., 58
Hughes, M. N., 417, 418
Huisman, H. O., 280
Humphlett, W. J., 233, 401, 402
Hung-Yin Lin, G., 240

Hunter, J. M., 310
Hurd, R. N., 183
Hurlock, R. J., 228
Hurwic, J., 246
Husbands, G. E. M., 47, 319
Hutchins, R. O., 164
Hutchinson, B. J., 157
Huylebroeck, J., 131, 141
Huzinaga, S., 2

Iarossi, D., 75
Ichi, T., 97, 143
Ichibori, K., 66, 250
Ichikawa, T., 65
Ichimura, K., 393
Iddon, B., 56, 98
Ide, J., 279, 280, 285, 287, 312
Ignat'eva, S. N., 162, 170
Ilvespää, A. O., 389
Imamoto, T., 220
Inamoto, N., 210
Inch, T. D., 134
Indelicato, J. M., 95
Ingles, D. L., 94
Inglis, A. S., 58
Inoue, S., 429
Ioffe, I. S., 219, 237, 449
Ioffe, N. T., 89
Ionescu, M., 127
Irai, H., 420
Irgolic, K., 184
Irving, H. M. N. H., 475
Ishii, Y., 84, 148, 224
Ishiba, T., 132, 150, 379
Ishibe, N., 144
Ishidate, M., 82
Ishihara, S., 111
Ishikawa, F., 150
Islip, P. J., 389
Ismaili, I., 222
Isono, K., 94
Isono, M., 92
Itani, H., 111
Ito, Y., 295
Itoh, K., 84
Itoh, T., 437
Ivanov, I. A., 210
Ivanov, C., 59
Ivanov, O. C., 59
Ivanova, T. A., 207
Ivanova, Z. M., 422
Iwakura, H., 111
Iwakura, Y., 397
Iyer, V. S., 370
Izzo, P. T., 444

Jackman, L. M., 420
Jackson, B. G., 153, 457
Jackson, F. H., 241, 413
Jackson, J. L., 365
Jackson, J. R., 152, 457
Jackson, T. G., 468
Jacob, W. A., 244
Jacobsen, N. W., 215
Jacot-Guillarmod, A., 55
Jacquier, R., 207, 219
Jacquignon, P., 355, 368
Jadamus, H., 411

Jadot, J., 391
Jaenecke, G., 428
Jaggard, J. F., 228
Jagt, J. C., 85
Jahnke, U., 52, 211
Jakobsen, P., 239
James, B. R., 60
James, J. C., 405
James, R., 126, 136
Jamieson, W. D., 68
Janczewski, M., 75
Janiga, E. R., 277
Jankowski, K., 113
Janku, J., 163
Jansing, J., 101
Janssen, M., 182
Janzen, E. G., 287
Jao, L. K., 147
Jarchow, O. H., 240
Jautelat, M., 250, 269
Jefferson, A., 73
Jencks, W. P., 56, 218
Jenkins, C. S. P., 242
Jensen, H., 101
Jensen, K. A., 183, 226, 444, 451
Jensen, L. H., 240
Jernow, J. L., 71
Ji, S., 98
Jindal, S. L., 347, 354
Jirkovsky, I., 364
Jogdeo, P. S., 175
Johansson, G., 31, 33
Johnson, A. Wm., 248, 254, 255, 257, 259
Johnson, B. J., 89
Johnson, C. D., 275
Johnson, C. R., 71, 128, 131, 134, 137, 157, 195, 207, 277, 284, 301, 305, 462
Johnson, D. A., 277
Johnson, J. R., 378
Johnson, L. N., 57
Johnson, N. P., 174
Johnson, P. L., 324, 325, 345
Johnson, P. Y., 64
Johnson, S. M., 321, 322
Johnsson, H., 78
Johnston, T. P., 408, 431
Johnstone, R. A. W., 41, 326, 327
Jolles-Bergeret, B., 92
Jonas, K., 403
Jonassen, H. B., 33
Jones, C. M., 243
Jones, D. G., 376, 434
Jones, D. H., 370
Jones, D. M., 139
Jones, D. N., 76, 77, 93
Jones, F. B., 90
Jones, F. N., 61, 148, 224
Jones, G., 376, 434
Jones, J. B., 169
Jones, J. H., 85
Jones, N. D., 154
Jones, V. K., 280
Jones, W. C., 59, 103
Jonsson, H. G., 135

Joos, A., 431
Jori, G., 72, 93
Joris, S. J., 230
Joullié, M. M., 120, 151, 404, 426
Joynson, M. A., 57
Julia, M., 284
Julia, S., 284
Julshamm, K., 323
Junek, H., 202
Just, G., 200
Just, H., 100

Kabuss, S., 166
Kahn, A. A., 301
Kai, K., 94
Kaiser, C., 284
Kaiser, E. M., 289
Kaiser, E. T., 201
Kaiser, K., 418
Kakese, T., 214
Kalabin, G. A., 69
Kalbag, S. M., 209
Kalinkin, M. I., 87, 89
Kalinowski, H. O., 211
Kalinowski, M. K., 245
Kalir, A., 60, 469
Kalman, A., 300
Kaluzin, G. N., 210
Kammel, G., 302
Kanaoka, Y., 207
Kandror, I. I., 56
Kaneko, T., 451
Kanematsu, K., 212, 455
Kano, H., 415
Kapecki, J. A., 141, 197, 327
Kapovits, I., 299, 300
Kapp, M., 219
Kappe, T., 206, 215
Kapps, M., 262, 263
Karabanov, Yu. V., 209
Karelina, L., 50
Karger, M. H., 95
Karle, I. L., 240, 427
Karle, J., 427
Karlsson, S. E., 31
Karmann, H.-G., 195
Karo, A. M., 2
Karo, W., 49
Karsa, D. R., 56
Kashihara, M., 218
Kashima, C., 349
Kasperek, G. J., 107
Kataev, E. G., 61, 63, 212
Kataeva, L. M., 61
Kataoka, H., 349
Katayama, H., 224
Katekar, G. F., 238
Kato, H., 33, 125, 216, 451
Kato, K., 214
Kato, T., 65
Katritzky, A. R., 157, 243
Kaufman, J. A., 110, 193
Kaushal, A. N., 209, 442
Kawada, I., 349
Kawai, K., 244
Kawakami, J. H., 55
Kawakatsu, Y., 301
Kawakishi, S., 85

Kawalek, B., 243
Kawamoto, K., 91
Kawamura, S., 104
Kawaoka, H., 232
Kay, I. T., 84, 88
Kayser, M., 144
Keane, D. D., 284
Kearney, P. C., 62
Keberle, H., 383
Keiser, J. E., 134, 137, 195, 301
Kelly, D. P., 61, 262, 264
Kelly, J. F., 101
Kelly, P., 102
Kemula, W., 245
Kendall, R. V., 233
Kenney, G. W. J., 90
Keogh, H. J., 243
Kerfanto, M., 69
Kessler, H., 90, 106, 211
Ketcham, R., 204, 395
Keung, E. C. H., 60
Khachaturova, G. T., 240
Kharasch, N., 49, 182, 183
Khasanova, M. N., 52, 211
Khazemova, L. A., 54
Khlystunova, E. V., 234
Khodair, A. I., 49, 183
Khokhar, A. R., 52, 206, 236, 384, 396, 398, 458
Khromova, Z. I., 156
Khromov-Borisov, N. V., 105
Kice, J. L., 107
Kiefer, G., 308
Kiel, G., 403
Kier, L. B., 451
Kil'desheva, O. V., 111, 225
Kimling, H., 177
King, C., 382
King, J. F., 98, 133, 199, 200, 201, 290, 291, 292, 295, 473
King, R. B., 61
King, R. D., 281
King, W., 426
Kingsbury, W. D., 131, 137
Kingston, D. G. I., 68
Kinoshita, T., 155, 213
Kinstle, T. H., 79, 198
Kirchhoff, K., 338, 364
Kirsanov, A. V., 434
Kirsanova, N. A., 234
Kirienko, S. S., 209
Kirmse, W., 262, 263, 445
Kise, M., 76
Kishida, Y., 63, 279, 285, 287, 312
Kiss, J., 97
Kitamura, Y., 218
Kitching, W., 93
Kiyoshima, Y., 53, 150, 226
Kjaer, A., 76, 302, 305
Kjellin, G., 229, 237
Kjøge, H. M., 340
Klaboe, P., 172
Klasek, A., 60
Klauke, E., 205

# Author Index

Klayman, D. L., 215, 233, 426, 427
Kleeman, M., 292
Klein, W., 177
Klemm, K., 223, 344
Kless, M., 56
Kliegemann, J. M., 452
Klima, Z., 451
Klimova, L. I., 246
Klingebiel, U. I., 62
Klingler, T. C., 289
Klingsberg, E., 41, 321, 331, 334
Klinowski, C. W., 230
Klivenyi, F., 228
Kloosterziel, H., 92
Klopman, G., 2
Klose, G., 235
Knauer, K. H., 339
Knaus, E. E., 203
Knaus, G. N., 99
Knipe, A. C., 65
Knöfel, W., 108, 110, 187, 227
Knoll, F., 304
Knunyants, I. L., 62, 89, 111, 133, 183, 209, 225
Knyazhanskii, M. I., 236
Kobayashi, M., 307
Kobayashi, Y., 148, 224
Kober, E. H., 83
Kobori, T., 89
Koch, H. P., 26
Kochetkov, N. K., 82
Kodachenko, G. F., 99
Kodama, Y., 244
Koeberg-Telder, A., 94
König, H., 313
Koenig, P. F., 240
König, W., 415
Kohout, L., 258
Kohtz, J., 71
Koizumi, T., 242
Kojima, A., 98
Kolb, R., 128, 210
Kolenbrander, H. M., 139
Kolesnik, Y. A., 50
Kolesnikova, S. A., 89
Kolesova, M. B., 203
Kollenz, G., 215
Kolmakova, L. E., 89
Komaritsa, I. D., 207
Komeno, T., 111
Komery, J., 201, 473
Komlossy, J., 413
Komori, O., 53, 150, 226
Kondo, K., 113, 116, 251, 256, 259, 264, 268
Konig, H., 278
Konishi, H., 33
Konizer, G. B., 270, 281
Konovalova, L. K., 212
Kopecky, J. J., 216
Kopylova, B. V., 52, 211
Korcek, S., 60
Korostova, S. E., 56
Korsloot, J. G., 210
Korte, F., 139
Korytnyk, W., 210
Kosbahn, W., 88

Koslev, N. S., 210
Kosower, E. M., 103
Kosower, N. S., 103
Kost, A. A., 82
Koto, H., 121
Koutek, B., 94
Kovacs, O. K. J., 152, 470
Koval, I. V., 99
Kovalev, L. S., 156
Kozuka, S., 97
Krakow, M. H., 138
Kramolowsky, R., 230
Krapf, H., 195
Krause, M., 91
Krause, W., 260
Krebs, A., 177
Krebs, B., 240, 299
Kremlev, M. M., 99
Kresze, G., 88, 306, 307
Kreutzberger, A., 209
Krishnaswami, N., 209
Kristean, P., 177, 242
Kriul'kev, V. I., 236
Krivis, A. F., 246
Kröhnke, F., 466
Krohn, J., 204
Krollpfeiffer, F., 258
Krone, U., 84, 370
Krow, G. R., 101
Krueger, J. H., 107, 213
Kruglyak, Yu. L., 156
Krylov, E. I., 419
Kryszczynska, H., 245
Ku, A. T., 222
Kubersky, H. P., 235
Kuchar, M., 237
Kuchinka, K., 384
Kucsman, A., 299, 300
Kudrya, T., 206
Kühle, E., 87, 146, 183, 370
Kuhlmann, G. E., 84, 115, 186
Kuhn, R., 430
Kukolja, S., 68, 153
Kukota, S., 206
Kuleshova, N. D., 183
Kulik, S., 68, 470, 471
Kulkarni, B. S., 386
Kumamoto, T., 61
Kunieda, N. 75, 76
Kunieda, T., 285
Kunin, B. S., 197
Kunz, D., 138, 235
Kunze, U., 221
Kunzek, H., 206, 412
Kupin, B. S., 53, 60
Kurtz, A. N., 52, 145
Kurusu, T., 97
Kurzer, F., 183, 218, 220, 444
Kushi, Y., 240
Kuszmann, J., 111
Kwart, H., 86, 208, 301
Kwee, S., 206

Laba, V. I., 72
Laban, G., 41, 352
Lablache-Combier, A., 186
Lack, R. E., 75, 180
Lagercrantz, C., 79, 138

Laird, R. M., 97
Laird, T., 264
Laland, P., 431
Laland, S., 431
Laliberté, R., 412
Lamazouère, A.-M., 221
Lambert, J. P. F., 50
Lambie, A. J., 99
Lamon, R. W., 206, 219
Land, A. H., 378
Landa, S., 163
Landini, D., 64, 75
Landon, W., 67, 208
Landor, S. R., 283
Landquist, J. K., 219
Lang, L., 216
Langermann, N. R., 207
Lanigan, D., 83, 120
Lanigan, T., 95
Lappert, M. F., 80
Large, G. B., 107
Larice, J. L., 407
Laridon, U., 445
Larivé, H., 415
La Rochelle, R., 61, 263
Larsen, C., 156, 215, 216, 217, 220, 229, 230, 237, 239, 244
Larsen, E., 204
Larsen, O., 230
Lassmann, G., 79
Last, W. D., 91
Latif, N., 115
Latscha, H. P., 177
Lattes, A., 124, 210
Lattrell, R., 101
Lauerer, D., 156, 237
Laufer, R. J., 102
Lauffer, M. A., 95
Laughlin, R. G., 300, 303
Lautenschlaeger, F. K., 102, 109, 111, 120, 136
Lavagnino, E. R., 153, 457
Lawesson, S. O., 138, 188, 235, 245
Lawson, A., 446
Lazaris, A. Y., 451
Leaver, D., 328, 337
Leaver, I. H., 59
Le Bel, N. A., 52
Le Berre, A., 96, 98
Leddicotte, G. W., 50
Ledésert, L., 368
Lednicer, D., 404
Lee, H. K., 147
Lee, J. B., 245
Lee, J. T., jun., 207
Lee, L. T. C., 274, 286
Lee, T. W. S., 98, 200, 292
Le Fèvre, R. J. W., 106
Le Floc'h, Y., 69
Legrand, L., 221, 223, 459
Le Guen, Y., 206, 381, 382
Lehmann, H. G., 282
Leibetseder, J., 50
Leibovskaya, G. A., 156
Leistner, S., 237
Leitz, H. F., 169
Lemire, A. I., 239

Lempert, K., 207, 216, 243, 244
Leon, N. H., 59, 223
Leonard, N. J., 120, 373, 400
Leppin, E., 109
Leroy, C., 69
Le Roy Salerni, O., 215
Lesko, J., 60
Leung, F., 322
Levander, O., 50
Levente, A., 382
Levin, G. O., 88
Levin, J.-O., 120
Levin, V. S., 87
Levine, H. H., 422
Levison, K. A., 33
Levy, G. C., 115
Lewis, A. J., 149
Lewis, D. K., 423
Li, C. H., 105
Liener, I. E., 78
Lilga, K. T., 215
Lillya, C. P., 257
Lindberg, B. J., 31, 33
Lindberg, S. E., 97, 201
Lindell, T. J., 97, 201
Lindgram, I., 31
Lindner, A., 221
Lindner, E., 195
Lindner, P., 109
Lindsay, L. F., 182
Linke, K.-H.,220, 234, 243
Linkova, M. G., 62, 111, 183, 225
Linn, C. J., 97, 201
Linn, W. J., 111, 145, 191
Lippsmeier, B., 94, 95
Lipscomb, W. N., 2
Lipunova, G. N., 419
Litvinov, V. P., 64
Liu, J. K., 237, 242
Liu, L. K., 96, 127, 200
Liu, T. Y., 58
Liveris, M., 56
Lloyd, D., 250, 257, 308
Lobert, B., 468
Lodam, B. D., 247
Loder, J. W., 141
Logan, T. J., 302
Logiudice, F., 240
Lohaus, G., 101
Lohr, A. D., 209
Lohr, D. F., 424, 439
Lohs, K., 79
Loiseau, A., 355
Lonatin, V. E., 205
Lopez Castro, A., 240
Louis-Ferdinand, R. T., 57
Love, A. M., 205
Love, G. M., 123, 201
Lovejoy, D. J., 88
Lovett, W. E., 63
Lovins, R. E., 84
Lowe, D. A., 278
Lowe, P. A., 280, 312
Lown, J. W., 84, 400
Lozac'h, N., 182, 188, 221, 222, 236, 324, 325, 331, 342, 459

Lozinskii, M., 206
Lozinskii, M. O., 50, 209
Lubos, W. D., 451
Luck, W., 416
Lufoff, J. S., 146
Lukman, B., 242
Luloff, J. S., 91
Lumma, W. C., 126
Lunazzi, L., 198
Lund, M., 206
Lustig, E., 197
Lustig, M., 299
Lutsenko, L. N., 82
Lwowski, W., 186
Lyon, G. D., 100
Lyubimova, N. B., 244

McAuley, A., 214
Maccagnani, G., 198
McCall, E. B., 420
McCalla, D. R., 103
McCants, D., 301
McCasland, G. E., 144
McCausland, J. H., 359
McClean, I. A., 179
McClelland, E. W., 446
McConnell, W. B., 68
McCormick, J. E., 135
McCrae, R. C., 386, 413
McDermott, E. E., 301
McDonald, W. S., 475
McElhinney, R. S., 135
McFarland, J. W., 101
McGregor, D. N., 191, 369
MacGregor, W. S., 81
Machoń, Z., 374, 376
McIntosh, C. L., 133, 199
Mackensen, G., 242
McKenzie, S., 184, 324
McKinnon, D. M., 337
McMaster, I. T., 300, 455
McMurdy, C. W., jun., 33
McMurray, C. H., 57
McNelis, E., 392
MacNicol, D. D., 348
McWeeny, R., 37
Madding, G. D., 279
Maeda, K., 327
Märkl, G., 311
Magerlein, B. J., 68
Magnusson, B., 173
Mahapatra, N., 215
Mahmood, A., 386
Maier, D. P., 468
Maier, L., 202
Maiorana, S., 294, 356
Mairanovskii, S. G., 72
Majerski, Z., 46
Majes, J. M., 217
Makarina-Kibak, L. Ya., 216
Makisumi, Y., 66, 67, 207
Makoveer, P. S., 138
Maksimets, V. P., 237
Maksimova, L. I., 203
Makucha, M. P., 216
Malekin, S. I., 156
Malewski, G., 100
Malik, W. U., 413, 418
Malisheva, E. N., 236

Mallams, A. K., 82
Mallon, H. J., 428
Malmovskii, M. S., 215
Maloney, T. W., 84, 400
Malpass, J. R., 101
Malte, A., 62
Malte, A. M., 74, 78, 160
Mamdapur, V. R., 175
Mammi, M., 321, 323
Mangane, M., 352
Mangia, A., 356
Mangini, A., 198
Manhas, M. S., 206
Manmade, A., 149
Manme, R., 31
Mannafov, T. G., 61
Mannervik, B., 106
Manotti Lanfredi, A. M., 240
Mantione, R., 65
Mantsch, H., 380, 389
Mantsch, H. H., 466
Mantz, I. B., 130
Manya, P., 80
Manyo, N. P., 219
Mao, C.-L., 454
Marathe, K. G., 284
Marcotrigiano, G., 399
Marcus, E., 200, 201
Marcus, F., 124
Marcus, N. L., 150
Maréchal, M., 352
Marino, J. P., 71, 276
Markov, P., 41
Markovac, A., 82
Markovits-Korms, R., 216
Maroni, P., 175
Marsden, J. C., 58
Marshall, D. L., 78
Marshall, J. A., 169
Martani, A., 355, 436
Martensson, O., 109
Martin, B. W., 468
Martin, D., 228
Martin, D. J., 228
Martin, G. J., 136, 236, 238
Martin, M., 69
Martin, P. K., 212
Martin, R. B., 153
Martin, R. L., 59, 189, 339
Martynov, I. V., 156
Martynyuk, A. P., 434
Marziano, N., 75
Mashkina, A. V., 138
Mason, J., 103
Mason, S. F., 182
Massingill, J. L., 365
Mastbrook, D., 197
Mataga, N., 35
Mathew, M., 240
Matsakura, H., 141
Matsuda, I., 84
Matsui, M., 82
Matsukura, H., 108
Matsumoto, H., 306
Matsumoto, M., 125, 216
Matsumoto, S., 84, 112, 150
Matsumoto, Y., 379
Matsumura, Y., 125, 216

# Author Index

Matsuura, T., 203
Matsuyama, H., 67
Matthews, W. S., 78, 160
Matubara, S., 268
Matuszak, C. A., 98
Mauger, R., 246
Maul, J. J., 426
Mautner, H. G., 138
Mayer, K. H., 156, 237
Mayer, R., 95, 138, 182, 183, 222, 235, 239, 241, 245, 352
Mayers, D. F., 201
Mayman, T., 135
Maynard, J. R., 60
Mazak, A., 446
Mazarguil, H., 124
Mazharuddin, M., 207, 237
Maziarczyk, H., 75
Mazur, Y., 95, 96
Meaney, D. C., 284
Médawar, G., 412
Meehan, G. V., 97, 161
Mehlhorn, A., 41
Mehlhorn, L., 240
Mehra, S. C., 413, 418
Mehren, R., 54
Mehrishi, J. N., 58
Mehta, S. R., 202
Meier, W., 415
Melloni, G., 122
Mel'nikova, N. V., 243
Melumad, D., 282
Melvas, B. W., 245
Menczel, G., 216
Menin, J., 218
Menyhart, M., 206
Meppelder, F. H., 90
Merkel, W., 83, 207
Merritt, L. L., 240
Merritt, M. V., 71
Mertes, M. P., 144, 147
Messing, A. W., 73
Metallidis, A., 202
Metys, J., 354
Metyšová, J., 354, 364
Metzger, J., 41, 206, 215, 378, 379, 385, 387, 415, 416
Metzner, W., 103, 228
Meyer, R. F., 475
Meyers, C. Y., 74, 78, 160
Meyerson, S., 336
Michaelis, W., 470
Michalski, J., 453
Michel, E., 440
Middleton, W. J., 110, 193, 449
Midgley, J. M., 168
Migalina, Yu. V., 215
Migita, T., 66, 67, 250, 260
Mikhailik, S. K., 422
Mikhailova, V. N., 72, 78
Mikhaleva, A. I., 69
Mikhel'son, M. G., 205
Miki, H., 306, 453
Mikol, A. J., 50
Mikol, G. J., 287
Mikolajczyk, M., 75

Mikulla, W. D., 126
Milburn, G. H. W., 339
Mileshina, L. A., 240
Milewich, L., 144
Millar, A. W., 78
Millard, B. J., 168
Miller, D. J., 33
Miller, E. F., 257
Miller, F. M., 382
Miller, G. R., 399
Miller, J., 56, 80
Miller, L., 197
Mills, H. H., 348
Milne, G. W. A., 215, 426, 427
Minale, L., 419
Minami, T., 306, 453
Minkin, J. A., 240
Minkin, V. I., 236, 246
Minnier, C. E., 207, 216
Miocque, M., 157, 287
Miotti, U., 85, 105, 189
Mirek, J., 243
Mirza, J., 207
Misawa, S., 245
Mishrikey, M. M., 351
Mishriky, N., 115
Miskidzh'yan, S. P., 209
Miskow, M. H., 174
Mislow, K., 60, 75, 76, 129, 256
Misra, V. S., 220, 449
Mitamura, S., 112
Mitchell, D. J., 240
Mitchell, K. A. R., 3, 249
Mitchell, R. H., 169
Mitra, R. B., 347
Mitsunobu, O., 214
Miyake, A., 179
Miyasaki, K., 216
Miyoshi, H., 91
Mizukami, F., 244
Mock, W. L., 359
Modena, G., 50, 71, 75, 88, 122
Modest, E. J., 355
Moffatt, J. G., 54, 83, 136, 251, 275, 276
Moffitt, W. E., 26
Mohl, H. R., 200, 291
Mojé, S., 62
Mollier, Y., 188, 202, 242, 341
Mollin, J., 237
Molloy, B. B., 463
Momicchioli, F., 92
Mondeshka, D., 55
Mondon, A., 207
Monroe, E. M., 349
Montanari, F., 50, 64, 72, 77, 79
Montesano, R. M., 93
Montgomery, J. A., 431
Montijn, P. P., 65
Moon, E. L., 387
Moore, D. W., 239
Moore, H. W., 83
Moran, T., 301
Moravek, J., 216
Moretti, I., 72

Morgan, C. D., 91
Morgenstern, J., 95, 182, 183
Mori, K., 53, 226
Mori, M., 53, 150, 226
Moriarty, R. M., 144, 203, 452
Moriconi, E. J., 101
Morimoto, H., 121
Morin, R. B., 68, 153, 457, 461
Morita, H., 114
Morita, Z., 251
Morris, R. N., 379
Morris, S. R., 468
Morrisett, J. D., 103
Morrissey, A. C., 128
Morton, W. D., 88
Mosalenko, V. A., 207
Moser, J. F., 159
Moskowitz, H., 157, 287
Moss, D. B., 80
Moule, D. C., 243
Mruthyunjaya, H. C., 214
Müller, I., 215, 382
Mueller, R. A., 153, 457
Mueller, W. H., 50, 62, 87, 88, 91, 115, 120, 149, 156
Müller-Hagen, J., 200
Mukaiyama, T., 61
Mukherjee, R., 203
Mulder, R. J., 291
Muljiani, Z., 347
Muller, A., 299
Muller, H., 282
Muller, P., 201
Mulvaney, J. E., 386, 392
Mundy, D., 76, 77
Muntz, R. L., 75
Murabayashi, A., 66, 67, 207
Muralidharan, K. V., 408
Muramatsu, I., 217
Murata, Y., 367
Murato, N., 420
Muraveva, K. M., 379
Murayama, K., 86
Murdock, K. C., 104, 177
Muren, J. F., 354
Murnietze, D. Y., 236
Murray, J. F., 58, 204
Musher, J. I., 3
Musina, A. A., 63
Mutha, S. C., 204, 395
Myhre, R. V., 96
Myron, M., 226

Nabeya, A., 397
Nagabhushan, T. L., 216
Nagai, T., 72, 125, 290
Nagano, M., 142, 224
Nagao, Y., 224
Nagarajan, K., 209, 382
Nahlovska, Z., 136
Nahlovsky, B., 136
Naik, A. R., 295
Naik, S. R., 144
Nair, M. D., 202, 209
Naito, T., 150, 429

Najer, H., 218
Nakabayashi, T., 104
Nakagawa, M., 203, 245
Nakagawa, Y., 84, 95
Nakai, T., 142, 143, 232
Nakaido, S., 67, 250
Nakamura, K., 251
Nakamura, S., 61
Nakamura, Y., 53, 226
Nakayama, K., 66
Namiki, M., 85
Nanbata, T., 94
Nanjo, M., 227
Nanobashvili, E. M., 59
Narang, A. S., 209
Narang, K. S., 209, 215, 407, 435, 439, 442
Narang, N. S., 151
Nardelli, M., 234, 240
Nash, B. W., 84, 230
Naumann, K., 60
Nayak, P. L., 391
Nayak, U. G., 51, 135
Nayler, J. H. C., 378
Neale, A. J., 60, 191, 420
Neale, R. S., 150
Neelakantan, P., 216
Neeter, R., 280
Negishi, A., 113, 116
Neidle, S., 302
Neidlein, R., 300, 425
Neilson, J. P., 359
Nelander, B., 141
Nelson, D. J., 97, 201
Nemes, E. N., 206
Nesynov, E. P., 205
Neuray, D., 202
Neville, M. C., 389
Newberry, R. A., 84, 206, 230, 350
Newbould, B. B., 386
Newburg, N. R., 99
Newlands, M. J., 93
Newman, H., 206
Newman, M. S., 231
Newton, M. G., 321, 322
Ng, F. T. T., 60
Nguyen Kong-Zinh, 217
Nguyen Tkhy Nguyet Fyong, 202
Nicholson, D. G., 172
Nicolaides, D. N., 365
Nicolaus, R. A., 412, 419
Niederprüm, H., 98
Nielsen, P. H., 156, 215, 216, 217, 220, 229, 230, 237, 239, 241
Niem, T. P. F., 98
Niemann, H., 338, 364
Nigam, S. N., 68
Nikonova, L. A., 79
Nikonova, L. Z., 213
Nischk, G.-E., 99
Nise, G., 106
Nishiguchi, T., 397
Nishimoto, K., 35
Nishino, H., 242
Nishio, M., 74
Nisimura, T., 290
Nivopozhkin, L. E., 236

Nobles, W. L., 413
Nockolds, C. E., 339
Nógradi, M., 363
Nold, A., 444
Nomura, R., 107
Norcross, B. E., 59
Nordal, V., 207, 431, 432, 433
Nordberg, R., 31, 33
Nordling, C., 31, 33
Norell, J. R., 200, 295
Norton, R. V., 87
Noth, H., 232
Noyori, R., 65, 288
Nozaki, H., 71, 82, 112, 251, 259, 264, 268, 273, 288, 290
Nudelman, A., 81
Nudelman, N. S., 56
Nuhn, P., 237, 245
Nukina, S., 425
Nuretdinova, O. N., 111, 114, 213
Nurmukhametov, R. N., 240, 242
Nyburg, S. C., 322
Nyitrai, J., 207, 216

Oae, S., 65, 75, 76, 97, 107, 114, 228, 249, 301
Oates, N., 197
Obata, N., 98
O'Brien, J. P., 169
Ochrymowycz, L. A., 69, 127
O'Connor, M. N., 408
Oda, R., 261, 274, 295
Offermanns, H., 179, 202, 400, 440
Ogata, Y., 92
Ogino, K., 97
Ogura, H., 437
Ohnishi, Y., 63, 125, 171, 193, 195, 242
Ohno, A., 63, 125, 171, 193, 195, 242
Ohno, K., 35
Ohshiro, Y., 306
Ohta, M., 206, 393, 451
Oiry, J., 234
Ojima, L., 210
Okada, M., 82
Okahara, M., 99, 299
Okano, M., 295
Okano, S., 100
Okawara, M., 142, 143, 232
Okawara, R., 195
Olah, G. A., 222
Olekhnovich, L. P., 236
Oliver, W. R., 79, 198
Olivier, J. L., 237
Ollis, W. D., 262, 263, 283, 363, 451
Olofson, R. A., 71, 233, 276
Olsen, B., 204
Olson, J. O., 97, 201
Olsson, K., 127
O'Neil, J. W., 173

Opitz, G., 183, 200, 290, 291, 292, 294
Oppolzer, W., 250
Orlov, A. M., 111, 225
Osborn, R. B. L., 196
Oshiro, Y., 453
Ostermann, G., 260
Ostroumov, Yu. A., 236
O'Sullivan, W. I., 289
Otting, W., 398
Ottmann, G., 83
Overend, W. G., 281
Owatari, H., 65
Owen, J. R., 415
Owen, L. N., 51, 136
Owen, T. C., 59, 140
Owsley, D. C., 33, 95, 121
Oxenius, R., 207

Paar, J. E., 289
Pace, J., 301
Pacifici, J. G., 115
Packer, J. E., 59
Padovan, M., 85, 189
Padwa, A., 140, 158, 186
Padzera, J., 206
Paetzold, R., 72
Pagani, G., 356
Pak, K. A., 79
Pak, P. D., 210
Pal, B. C., 86, 204
Palenik, G. J., 240
Palit, S. R., 214
Pankow, B., 52, 211
Panse, G. T., 346
Panyushkin, V. T., 236
Paolini, L., 41
Papa, I., 85
Papini, C., 223
Paquer, D., 110, 188, 193
Paquette, L. A., 50, 72, 97, 101, 123, 124, 138, 159, 161, 179, 199, 200, 201, 294
Para, M., 75
Paranjpe, M. G., 218
Parasaran, T., 309
Parfitt, L. T., 55
Parham. W. E., 261, 359
Pariaud, J. C., 246
Parikh, J. R., 82
Paris, G. Y., 413
Paris, J., 416
Pariser, R., 45
Parker, W. L., 80
Parkhomenko, P. I., 137
Parks, T. E., 85
Parr, R. G., 45
Parrish, F. W., 96
Passalocqua, V., 441
Passerini, R., 75
Pasto, D. J., 146, 275
Patel, C. C., 243
Patel, M. R., 386
Patsch, M., 202
Patterson, A. L., 240
Patterson, A. M., 424
Patterson, J. I. H., 225
Patton, W., 85

# Author Index

Paul, B., 210
Paul, D. B., 80
Paul, I. C., 189, 321, 322, 324, 325, 340
Paulsen, H., 238
Pavlova, E. V., 229
Pavlova, L. V., 67
Payne, G. B., 266, 268, 271, 272
Pays, M., 422
Pazdro, K. M., 228, 344
Pearson, P. S., 87
Pedersen, C. T., 147
Pedersen, E. B., 138, 188
Peel, J. B., 33
Peggion, E., 95
Pelah, Z., 60, 469
Pel'kis, P. S., 152, 202, 205, 206, 209
Pelz, K., 364
Penlynev, V. M., 202
Penner, H. P., 398, 407
Pentimalli, L., 215, 391, 413, 437, 441
Perelman, D., 81
Peretyazhko, M. Z., 152
Perez Rodriguez, M., 240
Perham, R. N., 57
Périn, F., 355
Périn-Roussel, O., 352
Perkins, A. W., 51, 135
Perkins, P. G., 33
Perkone, A. Y., 215
Perricone, S. C., 443, 444
Pesin, V. G., 453
Peters, A. T., 241, 413
Peters, G. C., 60
Petersen, J. B., 135
Peterson, P. E., 95
Pethrick, R. A., 175
Petke, J. D., 2
Petrov, A. A., 53, 60, 102, 156
Petrov, K. A., 183
Petrov, M. L., 53, 197
Petrova, D. S., 398, 407
Petrova, M. V., 413
Petrova, R. G., 56
Petrovskii, P. V., 56
Petterson, D., 107
Pews, R. G., 85
Peyronel, G., 399
Pfister-Guillouzo, G., 324
Pfitzner, K. E., 276
Pffeiderer, W., 207
Pfohl, S., 84, 202
Phan Chi Dao., 222
Philbin, E. M., 284
Philips, J. C., 138
Phillips, A. A., 389
Phillips, G. M., 310
Phillips, W. G., 275
Pickering, T. L., 126
Pickles, R., 84, 230
Pickworth Glusker, J., 240
Piers, K., 133, 199
Pilipenko, A. T., 243
Piper, J. R., 408
Pirkle, W. H., 75
Plastas, L. A., 408
Plat, M., 157
Platenburg, D. H. J. M., 90
Platt, D. S., 386
Plevachuk, N. Y., 207
Plimmer, J. R., 62
Pochinok, V. Y., 435
Podgorski, M., 75
Podszun, W., 101
Pohland, H., 370
Pohloud ek-Fabini, R., 84, 396, 402
Pollack, N. M., 47, 314, 318
Pollack, P. I., 444
Polezzo, S., 2
Polievktov, M. K., 72
Polk, M., 309, 311
Polovina, L. N., 215
Polovinkina, N. I., 138
Pomerantz, I., 197
Pomponi, A. M., 405
Ponsold, K., 269
Ponticello, G. S., 362, 363
Ponticello, I. S., 314, 315, 316, 356
Popilin, O. N., 237
Pople, J. A., 2, 35
Popp, F. D., 230
Portnyagina, V. A., 144
Postavskii, I. Ya., 246, 420
Potts, K. T., 206, 224, 344, 392, 393, 451
Pradère, J. P., 222, 353
Praefcke, K., 71, 108, 110, 187, 195, 227
Pramanick, D., 214
Pranskiene, T., 62
Preston, J., 417
Preti, C., 399
Prevost, C., 284
Price, C. C., 249, 309, 311
Price, E., 238
Price, N. C., 57
Prilezhaeva, E. N., 55, 64, 72, 216
Prinsen, A. J., 94
Pritzkow, W., 200
Procter, A. R., 51
Proinov, I., 218
Proksch, G., 219
Promel, R., 429
Prota, G., 419
Protiva, M., 354, 364
Pryor, S., 182
Pujari, H. K., 152, 407, 425, 441
Punja, N., 283
Puskas, J., 243
Pye, E. L., 238
Pyler, R. E., 51, 162

Quast, H., 410
Quilico, A., 369
Quiniou, H., 188, 220, 222, 236, 328, 330, 339, 341, 342, 353
Quintana, J., 259

Raban, M., 90
Rabinowitz, H. N., 189
Rabinowitz, J., 215, 217, 403, 475
Rachinskii, F. Y., 67
Rachlin, A. I., 169
Rachlin, S., 176
Rackow, S., 228
Radda, G. K., 57, 95
Radeglia, R., 138, 235, 245
Rademaker, P. D., 280
Radford, D. V., 106
Radhakrishnamurti, D. S., 269
Radha Vakula, T., 219, 233
Raepke, H., 428
Rafla, F. K., 162, 351
Ragulin, L. I., 133
Rahman, A. U., 152
Rahman, R., 68, 103, 178
Rahman, U. R., 215
Raimondi, D. L., 2
Rajagopal, K., 351
Rajappa, S., 206, 370, 382
Ramachandra Rao, Y., 215
Rambacher, P., 405
Ramiah, K. V., 243
Ramsay, A. A., 144
Ramsay, G. C., 59
Ramsey, K. C., 144
Randall, F. J., 255
Randau, G., 213
Ranga Rao, V., 219
Rankin, D. W. H., 229
Rao, G. V., 173
Rao, P. B., 244
Rao, V. R., 449
Rao, V. V., 206
Rao, V. R., 176, 389, 451
Raper, G., 475
Rapoport, H., 212
Rappe, C., 88, 120, 380
Rashi, M., 203
Rasteikiene, L., 62
Rath, P. C., 351
Ratts, K. W., 250, 256, 258, 261
Rau, M., 468
Rauk, A., 2, 3
Rautenstrauch, V., 73
Rave, T. W., 302
Rawlings, T. J., 60, 191, 328, 337
Rayner, D., 305
Razumova, N. A., 156
Reedjik, J., 172
Rees, C. W., 102, 367, 454
Regan, B. M., 379
Reich, I. L., 96
Reid, D. H., 53, 184, 185, 324, 328, 334, 342, 351, 399, 463
Reid, E. E., 49, 182, 183, 184
Reid, K. I. G., 325, 340
Reinecke, M. G., 365
Remers, W. A., 219
Remijnse, J. D., 234
Remirov, A. B., 170

Rempel, G. L., 60
Remsen, I., 446
Renard, S. H., 183
Renckhoff, G., 407
Renfrow, W. B., 99
Rennerfeldt, L., 97, 201
Renson, M., 84, 239, 352
Repella, D. A., 314
Replogle, L. L., 60
Reshef, N., 135
Rettig, M. F., 33, 121
Reubke, K. J., 237
Reynaud, P., 222
Reynolds, G. A., 468
Rheinbolt, H., 184
Rhodes, J. E., 80
Ricci, A., 355, 368
Richards, F. F., 84
Richards, K. E., 148, 149
Richardson, A. C., 83, 96
Richer, J.-C., 81
Richmond, G. D., 287
Rickett, F. E., 56
Ried, W., 83, 207
Rieth, K., 292
Rigau, J. J., 71, 157, 301, 305
Rigby, A., 228
Ringsdorf, H., 82, 251
Riordan, J. F., 58
Rippberger, H., 242
Rist, C. E., 86, 143, 227
Ritchie, R. K., 241, 244
Robba, M., 206, 381, 382
Robert, A., 81
Robertson, R. E., 98, 121
Robins, R. K., 242
Robinson, C. H., 144
Robinson, (Sir) R., 378
Robinson, W. T., 148
Robson, B., 280
Robson, P., 71, 113
Robson, R., 249
Roche, E. B., 451
Rodmar, B., 173
Rodmar, S., 173
Roebke, H., 169
Røe, J., 207
Rogers, D., 302
Rogers, J. H., 89, 229
Rogers, S. J., 57
Roggero, J., 152, 399
Roggero, M., 407
Rohr, W., 183
Rolle, W., 200
Romano, U., 75
Romanovski, B. V., 136
Rombauer, R. B., 105
Romers, C., 134
Roos, B., 2, 3
Rosen, M., 294
Rosen, M. H., 120, 201
Rosen, W. A., 179
Rosevear, D. T., 174
Rosmus, P., 239
Rosnati, V., 80
Rosowsky, A., 355
Ross, B., 304
Ross, S. D., 170, 172
Rossall, B., 98

Rossetto, O., 85, 188
Rossi, S., 294
Roth, B., 428
Rothe, L., 206
Rout, M. K., 391
Rowley, P. J., 55
Rowsell, D. G., 126, 136
Roy, J., 51
Roy, P. D., 94
Rozhkov, I. N., 89
Rüger, J., 379
Rühlmann, K., 399
Ruess, K. P., 213
Ruff, F., 299, 300
Ruiz, E. B., 261
Rukhadze, E. J., 229
Rumpf, P., 80
Rundel, W., 52, 106
Rupainwar, D. C., 475
Rusakov, E. A., 219
Ruse, M., 389
Russell, C. R., 86, 143, 227
Russell, G. A., 50, 69, 80, 127, 287, 288
Russell, G. B., 141
Rutolo, D. A., 254, 265
Rutt, K. J., 417, 418
Ruwet, A., 239, 352
Rydon, H. N., 78
Rynbrandt, R. H., 364

Sabol, M. A., 73, 305
Sabourin, E. T., 80, 287
Sadokage, T., 228
Saeling, G., 204, 206, 403
Saenger, W., 240
Safe, S., 103, 104, 178
Saggiomo, A. J., 467
Saharia, G. S., 386, 413
Saika, D., 299
Saikachi, H., 61, 84
Saint-Ruf, G., 468
Saito, I., 203
Saito, M., 82
Saito, S., 117
Sakai, K., 75
Sakai, S., 84, 148, 224
Sakai, Y., 2
Sakic, A., 102
Sakore, T. D., 362
Salamon, G., 210
Saldabol, N. O., 215
Salmane, R., 50
Salmon, A. G., 57
Salmond, W. G., 3, 309
Salomone, R., 84, 101
Salvadori, G., 256
Salvatori, T., 54
Sammes, P. G., 154
Sand, H., 416
Sandler, S. R., 49
Sandström, J., 182, 215, 229, 237, 238, 241, 444
Sannicolo, F., 80
Sans, A., 179
Santacroce, C., 412, 419
Santry, D. P., 2
Sanyal, S. N., 413
Sapino, C., 451

Saraswathi, N., 246
Sardessai, M. S., 152
Sarlo, E., 95
Sarma, P. K., 87
Sartori, P., 94, 95
Sasaki, T., 98, 150, 212, 455
Sasatani, T., 67, 207
Sasha, B. S., 143
Sastri, V. S., 230
Sataty, I., 462
Sathyanarayana, D. N., 243
Sato, E., 207
Sato, T., 239
Sato, Y., 349
Satoh, Y., 217
Sauer, J., 195
Saunders, V. R., 3, 8
Saus, A., 202, 440
Sausins, A. E., 209
Saville, B., 58
Sawacki, Y., 92
Sawicki, E., 422
Sayed, M. I., 82
Sayer, J. M., 218
Sayigh, A. A. R., 102
Scanlon, J. F., 245
Scanlon, W. B., 153, 457
Schäfer, W., 183
Schardt, R., 203, 338
Scharf, H. D., 139
Scharp, J., 243
Schaumann, E., 238
Scheele, R. B., 95
Scheinmann, F., 73
Scheit, K. H., 240
Scheithauer, S., 138, 222, 235, 239, 241, 245, 352
Scheller, G., 106
Schempp, H., 293
Schenk, H., 254, 403
Scherillo, G., 419
Schick, H., 269
Schilling, C. L., jun., 386, 392
Schindlbauer, H., 96
Schinski, W. L., 130
Schipper, E., 91, 146
Schlatmann, J. L. M. A., 210
Schlessinger, R. H., 72, 113, 149, 179, 191, 198, 221, 296, 314, 315, 316, 319, 356, 357, 362, 363
Schleyer, P. von R., 96
Schmeisser, M., 94, 95
Schmid, H. G., 166
Schmidbaur, H., 278, 302
Schmidt, G., 272
Schmidt, M., 228
Schmidt, P., 412, 465
Schmidt, R. R., 192
Schmitt, E., 410
Schöberl, A., 183
Schöde, D., 206, 394
Schoen, J., 202
Schönberg, A., 71, 99, 108, 110, 182, 183, 187, 195, 227, 245, 451

## Author Index

Schoeps, J., 163
Schollhorn, R., 303
Schollkopf, U., 260
Schossig, G., 260
Schowen, R. L., 147
Schrauzer, G. N., 189
Schreiber, R. S., 395
Schrepfer, J., 210
Schroeck, C. W., 284
Schroeder, D. C., 183
Schröders, H. H., 209
Schröpl, E., 396, 402
Schroll, G., 245
Schroth, W., 54, 365
Schuijl, P. J. W., 53, 70, 186, 208, 221, 223, 346
Schuijl-Laros, D., 66, 208, 223
Schulenberg, S., 98
Schultz, A. G., 72, 113, 198, 296, 314
Schulz, G., 216
Schulz, U., 103
Schumann, H., 228
Schumann, D., 245
Schut, J., 210
Schwalm, W. J., 188
Schwartz, I. L., 51
Schwartz, J., 456
Schwartz, N. V., 102, 109
Schwartz, P. M., 467
Schweizer, H. R., 365
Schweizer, P., 232
Schwenker, G., 128
Schwenker, R., 210
Schwille, D., 192
Scoffone, E., 72, 93
Scorrano, G., 75, 88
Scott, F. L., 97
Scott, M. K., 451
Scott, R. B., jun., 183
Scrocco, E., 2
Searle, R. J. G., 268
Searles, S., 425
Seden, T. P., 301, 425
Seebach, D., 71, 169
Seebeck, E., 384
Seel, F., 87
Seetharaman, P. A., 206
Segal, G. A., 2
Seidel, P. R., 207
Seidel, W. C., 226
Seip, H. M., 136
Seitz, G., 126, 251, 349
Sekera, A., 80
Selchau, M., 84
Seltzer, R., 448
Semin, G. K., 87
Sengupta, T. K., 214
Senning, A., 102, 307
Sensse, K., 242
Serratosa, F., 259
Servis, R. E., 94
Seshadri, S., 152
Sexton, W. A., 436
Seyferth, D., 110, 193
Sgarabotto, P. S., 240
Shah, S. R., 152
Shaikh, A., 206
Shakshooki, S., 225

Shamshurin, A. A., 232
Shankaranarayana, M. L., 234
Shapiro, R., 94, 405
Sharma, G. M., 435
Sharov, V. A., 419
Sharp, J. C., 71, 134, 137, 195
Shasha, B. S., 86, 227
Shatenshtein, A. I., 64
Shaw, M. R., 56
Shchukina, M. N., 209, 379
Sheehan, J. C., 444
Sheetz, D. P., 62
Shefter, E., 158
Shekhtman, R. I., 55
Shelemina, N. V., 136
Sheppard, W. A., 296, 453
Sherman, W. A., 444
Shevchuk, M. I., 202
Shibuya, K., 94
Shields, T. C., 52, 145
Shigorin, D. N., 240, 242
Shima, S., 420
Shimada, Y., 437
Shimakawa, Y., 101
Shimizu, K., 121
Shimonin. L. I., 216
Shine, H. J., 367, 472
Shine, R. J., 233
Shirota, T., 72
Shirota, Y., 125, 290
Shishkov, A. V., 54
Shlyk, Y. N., 102
Shoji, K., 150, 455
Shokol, V. A., 422
Shostakovskii, S. M., 78
Shrab, A., 209
Shulman, J. I., 61
Shu'lman, V. M., 234
Shur, A. M., 217
Shvo, Y., 152, 220, 408
Sica, D., 412, 419
Siddall, T. H., 238
Sidebottom, H. W., 88
Sieber, A., 58
Siegbahn, K., 31, 33
Siegbahn, P., 2, 3
Siegl, W. O., 128
Siegwart, J., 450
Siekmann, L., 304
Silberg, A., 380, 386, 389
Silhan, W., 256
Silhanek, J., 87, 113, 198
Silverthorn, W. E., 197
Simanek, V., 60
Simchen, G., 85, 459
Sime, J. G., 139
Simiti, I., 380
Simiti, J. M., 218
Simonetta, M., 2
Simov, D., 421
Simpson, R. E., 50
Sindelar, K., 354, 364
Sineokov, A., 209
Singer, E., 99
Singer, M. I. C., 250, 257 308

Singer, R. J., 91
Singh, A., 407, 435, 439
Singh, B., 244
Singh, H., 152, 176, 407, 435
Singh, I., 152
Singh, J. M., 386, 426
Singh, S., 176
Singh, U. P., 206, 224, 344, 392, 393
Sinnott, M. L., 96
Sirakawa, K., 211
Siskin, M., 311
Sivaramakrishnan, R., 269
Sjoberg, K., 287
Sjöberg, B., 152, 470
Skancke, P. N., 41
Škaric, V., 216
Skelton, F. S., 399, 463
Skold, C. N., 72, 198
Skupin, D., 234
Slack, R., 369, 370
Slesarchuk, L. P., 106
Sletten, E., 240, 321, 322, 323, 339
Sletten, J., 216, 240, 321, 322, 323, 325, 339
Sliwa, H., 355
Sliwinski, W. F., 96
Sloan, A. D. B., 57, 109, 473, 475
Sloan, M. F., 99
Smale, G., 36
Smallcombe, S. H., 255
Smentowski, F., 287
Smets, G., 416
Smith, B. C, 466
Smith, C., 264
Smith, C. J., 168
Smith, D. J. H., 133, 199
Smith, H. W., 212
Smith, J. H., 86
Smith, R. L., 103, 149
Smolanka, I. V., 215, 219
Smolders, W., 131, 141
Smushkevich, Y. I., 206
Smutny, E. J., 206
Smythe, G. A., 189
Snodin, D. J., 79
Snyder, J. P., 61
Sobeir, M. E., 466
Sobell, H. M., 362
Sochilin, E. G., 408
Sodhi, H. S., 386, 413
Sodtke, U., 195
Sohar, P., 233, 244, 421, 436
Soji, K.. 212
Sokolovsky, M., 58
Sokol'ski, G. A., 133
Soldan, F., 101, 150
Solomko, Z. F., 215
Soltys, J. F., 54
Sommer, L. H., 73
Son, P., 293
Songstad, J., 257
Sorm, M., 88
Soroka, T. N., 209
Sotiropoulos, J., 221

Soundararajan, S., 246
Sparrow, J. T., 58
Speakman, P. R. H., 71, 113, 249, 280
Speckamp, W. N., 280
Spedding, H., 241, 244
Speier, G., 352
Spence, M. J., 97
Spence, R. A., 210
Sperling, G., 448
Spillane, W. J., 97
Spillett, M. J., 79
Spinnler, M. A., 33
Splinter, F. K., 351
Sprague, J. M., 378
Sprecher, M., 96
Spry, D. O., 154, 461
Spurlock, L. A., 85
Spurlock, S. N., 121
Srivastava, P. K., 214, 413, 418
Srinivasan, V. R., 219, 233, 449
Staab, H. A., 183
Stabilini, M. P., 2
Stacey, G. J., 386
Stang, P. J., 95
Staninets, V. I., 215
Stanley, T. W., 422
Stanovnik, B., 207, 438
Stapf, W., 221
Staudinger, H., 450
Stavaux, M., 325, 331
Steger, E., 239
Steggerda, J. J., 173, 231, 239
Steiger, W., 206, 215
Stein, W., 54
Stepanyants, G. U., 55, 64
Stephen, J. F., 124, 200, 201
Stephen, I. W., 155
Sterk, H., 150
Sterlin, S. R., 209
Sternbach, L. H., 203
Stetter, H., 91, 163
Steudel, R., 50, 106
Stevens, C. L., 82
Stevens, G., 79
Stevens, T. S., 260
Stewart, D. G., 249
Stewart, I. M., 189, 339
Stewart, J. M., 408
Stewart, W. E., 238
Stirling, C. J. M., 65, 73, 78
Stoddart, J. F., 180
Stokely, P. F., 100
Stoll, A., 384
Stone, F. G. A., 190, 196
Stoodley, R. J., 152, 457, 463, 464, 465
Stork, G., 139
Stotter, P. L., 139
Stoyanovich, F. M., 83
Strasorier, L., 95
Strating, J., 117, 134, 198, 243, 291, 296, 297, 298
Strehlke, P., 216
Strehlow, W., 198

Stringfellow, C. R., jun., 408
Strof, J., 237
Strukov, O. G., 156
Stuart, A., 207
Stumbreviciute, Z., 62
Su, T. M., 96
Subramaniam, C. R., 243
Succardi, D., 206
Suciu, T., 449
Sugamori, S. E., 98
Sugiura, S., 429
Sugiyama, N., 349
Suhr, H., 447
Sukiasyan, A. N., 64
Suld, G., 309
Summerville, R., 95
Suprunchuk, T., 209, 243
Surzur, J. M., 126, 136, 405
Suschitzky, H., 56
Suszko, J., 446
Sutherland, I. O., 264, 283, 363
Sutter, D. H., 109
Sutton, G. J., 244
Sutton, L. E., 246, 310
Suvorov, N. N., 206
Suydam, F., 207
Suzuki, E., 59, 429
Suzuki, S., 94
Svatek, E., 354, 364
Swaelens, G., 106, 167
Swain, C. G., 259
Swallow, W. H., 149
Swan, J. M., 210
Sweetman, B. J., 56, 150, 407
Swern, D., 99, 299
Swigor, J. E., 191, 369
Sy, M., 234
Syamal, A., 244
Sykes, P., 401
Sylva, R. N., 391
Symon, J. D., 53, 324, 328, 342, 351
Szabó, A. E., 228
Szabo, I. F., 206
Szarek, W. A., 180
Szilagyi, L., 206
Shelton, R. S., 449

Tada, H., 195
Tadanier, J., 379
Taddei, F., 74, 75
Tagaki, W., 97, 107
Taguchi, T., 53, 150, 226
Takahashi, S., 415
Takahashi, Y., 237, 239
Takahishi, Y., 136
Takai, K., 84
Takaku, M., 71, 112, 251, 259, 264, 268, 273
Takamizawa, A., 84, 112, 132, 150, 229, 232, 343, 379
Takashina, N., 133
Takeda, T., 111
Takeshima, T., 220
Takeuchi, H., 72

Takizawa, T., 98
Tam, S. W., 245
Tamaru, Y., 97, 143
Tamres, M., 241
Tamura, C., 155, 213
Tanaka, M., 94
Taneja, A. P., 442
Tang, R., 129
Tarasoff, L., 75, 180
Tarbell, D. S., 63, 96, 162, 183, 184
Tauber, J., 425
Taurins, A., 437
Taylor, A., 68, 103, 104, 178
Taylor, E. C., 279, 446
Taylor, M. R., 240
Taylor, P. J., 84, 152
Tedder, J. M., 88
Teichman, H., 453
Teitel, S., 169
Tempel, E., 294
Tenconi, F., 301
Terada, A., 63
Terent'ov, A. P., 229
Ternay, A. L., 74
ter Wiel, J., 79
Teste, J., 336
Thaler, W. A., 171
Thanawalla, C. B., 207, 462
Thelman, J. P., 274
Thewald, K., 407
Thiel, M., 399, 441
Thijs, L., 117, 134, 198, 296, 297, 298
Thoennes, D. J., 101
Thomas, A. M., 103
Thomas, P. H., 235
Thomas, T. W., 391
Thompson, J. W., 299
Thompson, R. C., 78
Thomson, J. C., 239
Thomson, R. H., 50
Thomson, T., 260
Thorn, G. D., 197
Thornton, E. R., 259
Thorsdalen, N., 431
Thyagarajan, B. S., 49, 183
Thyagarajan, G., 207, 216, 237
Ticozzi, C., 285, 286
Tien, H. J., 82
Tien, J. M., 82
Tilak, B. D., 346, 347, 348, 354
Timofeeva, Z. N., 413
Ting, J. S., 82
Tinkler, R. B., 446
Tiripicchio, A., 240
Tiripicchio Camellini, M., 240
Tišler, M., 207, 438
Tjønneland, K., 207, 431
Tobler, E., 54, 113
Tochio, H., 397
Todd, P., 58
Todesco, P. E., 414
Tokizawa, M., 207
Tokura, N., 72, 125, 200, 290

# Author Index

Toldy, L., 210
Tomachewski, G., 210, 214
Tomalia, D. A., 62
Tomasi, J., 2
Tomchin, A. B., 219, 237
Tomchin, A. V., 449
Tomita, K., 142, 224
Tomlinson, R. L., 253
Tomoeda, M., 179
Tompa, A. S., 238
Tonatello, U., 188
Tonellato, U., 85, 105, 122, 189
Toniolo, C., 210
Torre, G., 72, 75
Torrese, T., 407
Torsell, K., 275
Tosato, M. L., 41
Tóth, G., 210
Tóth, I., 210
Tozune, S., 250
Tratt, K., 445
Traverso, G., 323
Traynelis, V. J., 81
Trebaul, C., 336
Treibs, A., 221
Treindl, L., 98
Trinajstic, N., 41, 46, 358
Trinderup, P., 204
Tripathy, P. B., 391
Trippett, S., 297
Trofimov, B. A., 56, 69
Trofimova, T. A., 106
Troitskaya, V. S., 207
Tronich, W., 110, 193, 278
Trost, B. M., 61, 114, 130, 254, 263, 270, 284
Truce, W. E., 96, 127, 200, 201, 279, 289, 291, 293, 295
Tsoy, L. A., 232
Tsuchihashi, G., 89, 116, 125, 171, 193, 195, 242
Tsuda, T., 211
Tsujihara, K., 301
Tsujikawa, T., 211
Tsujino, Y., 467
Tsuneoka, K., 203, 245
Tsurugi, J., 104
Tucker, B., 102
Tufariello, J. J., 274, 286
Tuleen, D., 137
Tull, R., 201
Tuman, R. W., 82
Tunemoto, D., 251, 268
Turbak, A., 225
Turkevich, N. M., 175, 207
Turley, R., 243
Turner, D. W., 15
Turner, J. A., 182
Turner, R. W., 301, 425
Turro, N. J., 148

Ueda, S., 274
Ueda, T., 242
Ueno, K., 150
Ueno, Y., 142, 143, 232
Uhlemann, E., 235
Ulrich, H., 102

Umani-Ronchi, A., 285, 286
Umbach, U., 54
Undheim, K., 207, 431, 432, 433
Uneyama, K., 228
Unukovich, M. S., 54
Upham, R. A., 59
Uppal. A. S., 215, 407, 435, 439
Urna, V., 99
Usenko, Yu. N., 202
Ustynyuk, Y. A., 79
Uteg, K.-H., 231
Uziel, M., 86, 204

Vachek, J., 237
Vagaonescu, M., 127
Vakula, T. R., 449
Vanag, E. V., 209
Van Allen, J. A., 468
Van Bekkum, H., 234
Van Campen, M. G., jun., 449
Van den Bosch, G., 61
Van der Gen, A., 53
Vanderhoek, J. Y., 64
Vandewyer, P. H., 416
Van Doorn, J. A., 94
Van Drunen, J. A. A., 92
Van Heyningen, E. M., 457
van Leusen, A. M., 85, 279, 291
Van Overstraeten, A., 295
Vang Thang, K., 237
Varand, V. L., 234
Vargha, L., 111
Varma, R. S., 413
Vasileff, R. T., 457
Vasilev, N. P., 69
Vasilevskii, V. L., 226
Vasudeva Murthy, A. R., 214
Vedenskii, V. M., 207, 216
Veillard, A., 2
Veldhuis, B., 207
Vengrinovich, L. M., 175, 207
Venier, C. G., 107
Venkateswarlu, A., 209
Verbeek, M., 139
Verdun, D. L., 199, 292
Vereshchagina, N. N., 420
Verheyden, J. P. H., 54, 136
Vernin, G., 206, 387
Verny, M., 61
Veronese, A. C., 209
Veronese, F. M., 89
Vesely, V., 60
Vessiere, R., 61
Vialle, J., 110, 188, 193, 338
Viallefont, P., 219
Vid, L. V., 106
Vidoni, M. E., 234
Viehe, H. G., 62
Vincent, E. J., 415, 416
Vinkler, E., 228
Vinokurov, V. G., 207

Vipond, P. W., 86
Vitenberg, I. G., 453
Vivaldi, R., 206
Vivarelli, P., 92, 414
Vladzimirskaya, E. V., 175
Vladzimirskaya, Y. V., 207
Vlasova, L. A., 246
Vogel, J., 206, 394, 395
Volka, K., 243
Volpp. S. P., 205
Vollrath, R., 144
von Chamier, C., 211
von Pechmann, H., 444
Von Schriltz, D. M., 305
Vorbrüggen, H., 216
Voss, P., 98
Vrbanic, A., 207, 438
Vrencur, D. J., 295
Vyunov, K. A., 408

Waclawek, W., 246
Wade, K., 334
Wade, K. O., 185
Wadsworth, E. M., 83
Wälti, M., 91
Wagner, A., 183
Wagner, G., 206, 237, 245
Wahren, M., 200
Walinsky, S. W., 71
Walker, D. F., 424
Walker, H. G., 184
Walker, L. A., 240
Wallace, T. J., 183, 290
Wallenfals, K., 57, 59, 92
Wallmark, I., 138
Walsh, J. A., 80
Walter, E., 422
Walter, R., 51
Walter, W., 89, 91, 183, 184, 201, 204, 205, 213, 235, 237, 238, 239, 246
Walther, H., 100
Walton, D. R. M., 72
Walton, J. C., 88
Walz, G., 292
Wanatabe, K., 251
Wander, J. D., 69
Wanzlick, H.-W., 52, 211
Warburton, W. K., 84, 230
Ward, S. D., 41, 326, 327
Wardell, J. L., 103
Warin, R., 391
Warrener, R. N., 455
Warwick, R., 412
Wassermann, H. H., 198
Wasson, F. I., 153
Watanabe, H., 454
Watanabe. T., 274
Wataya, Y., 94
Waterfield, M. D., 84
Watson, K. J., 204
Webb, N. C., 300
Webber, J. A., 457
Weber, D., 444
Weber, H., 195
Weber, U., 59
Webster, B. C., 1, 33
Weeks, M. J., 417
Weetman, D. G., 359

Wegler, R., 183
Weichmann, R. L., 87
Weiner, S. A., 288
Weininger, S. J., 110, 193
Weinstock, H., 201
Weinstock, J., 284
Weinstock, L. M., 444
Weintraub, P. M., 232
Weisenborn, F., 456
Weiss, H., 239
Weiss, M. J., 219
Weiss, R., 343
Wejroch, M., 246
Welcher, M., 94
Wellington, K. A., 97, 201
Wells, D. V., 187
Wells, J., 281
Welsh, D. V., 127
Wen, R. Y., 155
Wenkert, E., 244
Wenzelburger, J., 85, 459
Wepster, B. M., 234
Werbel, L. M., 389
Weringa, W. D., 191
Werme, L. O., 31
Wertheim, B., 103
Wessely, F., 251
Westernacher, H., 361
Westphal, G., 231, 407, 475
Westphal, O., 431
Westrum, E. F., jun., 386, 414
Wetzel, R. B., 466
Whalley, W. B., 168
Wharmby, M., 463
Whistler, R. H., 162
Whistler, R. L., 51, 135
White, A. M., 222
White, A. W., 198
White, J. A., 372
White, J. G., 101
White, P. Y., 144
White, R. F. M., 175
Whitehead, J. K., 301
Whitehouse, R. D., 76, 77
Whiting, D. A., 51
Whiting, J., 58
Whiting, M. C., 79, 145
Whitman, E. S., 195
Whitmore, W. G., 252
Whitten, J. L., 2
Wiechert, R., 282
Wiecko, J., 59
Wieczorkowski, J., 453
Wiedermann, R., 342
Wiekenkamp, R. H., 51
Wieland, T., 58
Wiersema, A. K., 183
Wijers, H. E., 53
Wikholm, R. J., 83
Wilbraham, A. C., 140
Wildschut, G. A., 65, 70
Wiles, D. M., 209, 243
Wiley, R. H., 378
Wilhelm, M., 412, 465
Wilhelms, R. E., 251
Wilkinson, M., 183, 218, 220

Wilkinson, W., 174
Willems, J. F., 183
Willemse, J., 173, 231
Willett, J. E., 78
Willhalm, B., 403
Williams, D. T., 76
Williams, G., 391
Williams, H., 264
Williams, H. B., 127, 187
Williams, J. L., 259
Williams, J. R., 148
Williams, N. R., 281
Williams, R. E., 82
Willson, R. L., 85, 105
Wilson, D. A., 82
Wilson, F. B., 348
Winchester, R. V., 59
Wineman, R. J., 405
Winkelman, D. V., 102, 139
Winkelmann, E., 209
Winkler, A., 268
Winstein, S., 93, 96
Winter, H., 200
Winterfeldt, E., 266
Wintner, C., 59
Wise, L. D., 97, 161
Wiseman, E. H., 445
Witkop, B., 285
Wittig, G., 37
Wittman, H., 150
Witwicki, J., 58
Wobig, D., 84, 209
Woeller, F., 217, 403
Wojcicki, A., 93
Wojtkowski, P., 274, 275
Wolfe, S., 3, 153
Wolff, M. E., 135
Wolfsberg, M., 35
Wong, J. L., 212
Wong, K. W., 56
Wood, G., 174
Wood, H. C. S., 207
Wood, M. S., 162
Wood, R. H., 178
Woodward, R. B., 80, 370
Woolridge, K. R. H., 369, 370
Worth, D. F., 389, 443, 444
Wortmann, J., 126, 225
Wragg, R. T., 51, 65, 113, 141, 179
Wright, S. H. B., 210
Wu, B. K., 289
Wucherpfennig, W., 129, 306
Wudl, F., 140
Wunsh, K. H., 207
Wynberg, H., 177, 243
Wyn-Jones, E., 175
Wynne, K. J., 86, 87

Yabe, K., 136, 237, 239
Yagihara, T., 250, 260
Yakovleva, Y. Y., 209
Yamada, K., 349
Yamamoto, Y., 288, 290
Yamasaki, T., 227
Yamashita, M., 218

Yamauchi, K., 81
Yan, S. J., 82
Yang, C. S., 58
Yang, S.-F., 65
Yano, M., 207
Yano, Y., 65
Yao, A. N., 250, 261
Yao, L. Y., 101
Yao, N., 258
Yarbrough, K. N., 127, 187
Yarkova, E. G., 63, 212
Yarovskii, D. F., 209
Yasuda, K., 195
Yellin, W., 300
Yokoyama, M., 76, 127, 220, 223, 344
Yonezawa, T., 33, 121, 125, 216
Yoshida, H., 226
Yoshida, Z., 91, 97, 143
Yoshikawa, Y., 107
Yoshimine, M., 269
Yoshioka, T., 86, 150
Young, G. T., 58
Young, I. M., 56
Young, T. E., 354
Young, W. G., 93
Yuldasheva, L. K., 162
Yura, Y., 280

Zaborsky, O. R., 201
Zabrodskaya, I. S., 202
Zahradnik, R., 1, 41
Zaitseva, G. I., 54
Zakrewski, K., 58
Zanati, G., 135
Zander, M., 468
Zanke, D., 210, 214
Zauer, K., 244
Zbiral, E., 383
Zbirovsky, M., 87, 113, 198
Zecchi, G., 80
Zefirov, N. S., 73
Zefirova, L. I., 79
Zefman, Y. V., 201
Zeid, I., 176
Zeigler, E., 202, 206, 215
Zenarosa, C. V., 82
Zhdanov, Yu. A., 236
Zhitar, B. E., 176, 207
Zhmurova, I. N., 434
Zhuarlkova, L. G., 209
Zhukova, E. N., 237
Ziebandt, H., 386
Ziegler, E., 150, 215
Zika, R. G., 87
Ziman, S., 114
Zimenovskii, B. S., 207
Zimmerman, J. E., 410
Zinner, G., 144
Zon, G., 60
Zschunke, A., 54, 245, 365
Zsolani, T., 409
Zumach, G., 87, 146
Zwanenburg, B., 79, 117, 134, 198, 296, 297, 298
Zwkovic, T., 41

QD
305
S3
O 68
v. 1

AUG 27 1973